PLANAR MICROWAVE ENGINEERING

Modern wireless communications hardware is underpinned by RF and microwave design techniques. This insightful book contains a wealth of circuit layouts, design tips, and measurement techniques for building and testing practical gigahertz systems. The book covers everything you need to know to design, build, and test a high-frequency circuit. Microstrip components are discussed, including tricks for extracting good performance from cheap materials. Connectors and cables are also described, as are discrete passive components, antennas, low-noise amplifiers, oscillators, and frequency synthesizers. Practical measurement techniques are presented in detail, including the use of network analyzers, sampling oscilloscopes, spectrum analyzers, and noise figure meters. Throughout the focus is practical, and many worked examples and design projects are included. A CD-ROM included with the book contains a variety of design and analysis programs. The book is packed with indispensable information for students taking courses on RF or microwave circuits and for practicing engineers.

Thomas H. Lee received his Sc.D. from the Massachusetts Institute of Technology and is an Associate Professor of Electrical Engineering at Stanford University. He has been a Distinguished Lecturer of both the IEEE Solid-State Circuits Society and the IEEE Microwave Theory and Techniques Society. He is the winner of four "best paper" awards at international conferences as well as a Packard Foundation Fellowship winner. Professor Lee has written more than a hundred technical papers and authored the acclaimed text *The Design of CMOS Radio-Frequency Integrated Circuits,* now in its second edition. He holds thirty-five U.S. patents and has co-founded several companies, including Matrix Semiconductor.

PLANAR MICROWAVE ENGINEERING

A Practical Guide to Theory, Measurement, and Circuits

THOMAS H. LEE
Stanford University

CAMBRIDGE
UNIVERSITY PRESS

PUBLISHED BY THE PRESS SYNDICATE OF THE UNIVERSITY OF CAMBRIDGE
The Pitt Building, Trumpington Street, Cambridge, United Kingdom

CAMBRIDGE UNIVERSITY PRESS
The Edinburgh Building, Cambridge CB2 2RU, UK
40 West 20th Street, New York, NY 10011-4211, USA
477 Williamstown Road, Port Melbourne, VIC 3207, Australia
Ruiz de Alarcón 13, 28014 Madrid, Spain
Dock House, The Waterfront, Cape Town 8001, South Africa

http://www.cambridge.org

© Cambridge University Press 2004

This book is in copyright. Subject to statutory exception and
to the provisions of relevant collective licensing agreements,
no reproduction of any part may take place without
the written permission of Cambridge University Press.

First published 2004

Printed in the United States of America

Typeface Times 10.75/13.5 and Futura   *System* AMS-TEX   [FH]

A catalog record for this book is available from the British Library.

Library of Congress Cataloging in Publication data
Lee, Thomas H., 1959–
Planar microwave engineering : a practical guide to theory, measurement, and circuits /
Thomas Lee.
p.   cm.
Includes bibliographical references and index.
ISBN 0-521-83526-7
1. Microwave circuits. 2. Microwave receivers. 3. Microwave devices.   I. Title.

TK7876.L424   2004
621.381'32 – dc22   2004050811

ISBN 0 521 83526 7 hardback

for Angelina

CONTENTS

Preface	*page* xiii

1 A MICROHISTORY OF MICROWAVE TECHNOLOGY — 1
1. Introduction — 1
2. Birth of the Vacuum Tube — 11
3. Armstrong and the Regenerative Amplifier/Detector/Oscillator — 14
4. The Wizard War — 18
5. Some Closing Comments — 27
6. Appendix A: Characteristics of Other Wireless Systems — 27
7. Appendix B: Who Really Invented Radio? — 29

2 INTRODUCTION TO RF AND MICROWAVE CIRCUITS — 37
1. Definitions — 37
2. Conventional Frequency Bands — 38
3. Lumped versus Distributed Circuits — 41
4. Link between Lumped and Distributed Regimes — 44
5. Driving-Point Impedance of Iterated Structures — 44
6. Transmission Lines in More Detail — 46
7. Behavior of Finite-Length Transmission Lines — 51
8. Summary of Transmission Line Equations — 53
9. Artificial Lines — 54
10. Summary — 58

3 THE SMITH CHART AND S-PARAMETERS — 60
1. Introduction — 60
2. The Smith Chart — 60
3. S-Parameters — 66
4. Appendix A: A Short Note on Units — 69
5. Appendix B: Why 50 (or 75) Ω? — 71

4 IMPEDANCE MATCHING — 74
1. Introduction — 74
2. The Maximum Power Transfer Theorem — 75
3. Matching Methods — 77

5 CONNECTORS, CABLES, AND WAVEGUIDES — 108
1. Introduction — 108
2. Connectors — 108
3. Coaxial Cables — 115
4. Waveguides — 118
5. Summary — 120
6. Appendix: Properties of Coaxial Cable — 121

6 PASSIVE COMPONENTS — 123
1. Introduction — 123
2. Interconnect at Radio Frequencies: Skin Effect — 123
3. Resistors — 129
4. Capacitors — 133
5. Inductors — 138
6. Magnetically Coupled Conductors — 147
7. Summary — 157

7 MICROSTRIP, STRIPLINE, AND PLANAR PASSIVE ELEMENTS — 158
1. Introduction — 158
2. General Characteristics of PC Boards — 158
3. Transmission Lines on PC Board — 162
4. Passives Made from Transmission Line Segments — 178
5. Resonators — 181
6. Combiners, Splitters, and Couplers — 183
7. Summary — 230
8. Appendix A: Random Useful Inductance Formulas — 230
9. Appendix B: Derivation of Fringing Correction — 233
10. Appendix C: Dielectric Constants of Other Materials — 237

8 IMPEDANCE MEASUREMENT — 238
1. Introduction — 238
2. The Time-Domain Reflectometer — 238
3. The Slotted Line — 246
4. The Vector Network Analyzer — 254
5. Summary of Calibration Methods — 264
6. Other VNA Measurement Capabilities — 265
7. References — 265
8. Appendix A: Other Impedance Measurement Devices — 265
9. Appendix B: Projects — 268

9 MICROWAVE DIODES — 275
1. Introduction — 275
2. Junction Diodes — 276
3. Schottky Diodes — 279
4. Varactors — 281
5. Tunnel Diodes — 284
6. PIN Diodes — 287
7. Noise Diodes — 289
8. Snap Diodes — 290
9. Gunn Diodes — 293
10. MIM Diodes — 295
11. IMPATT Diodes — 295
12. Summary — 297
13. Appendix: Homegrown "Penny" Diodes and Crystal Radios — 297

10 MIXERS — 305
1. Introduction — 305
2. Mixer Fundamentals — 306
3. Nonlinearity, Time Variation, and Mixing — 312
4. Multiplier-Based Mixers — 317

11 TRANSISTORS — 341
1. History and Overview — 341
2. Modeling — 351
3. Small-Signal Models for Bipolar Transistors — 352
4. FET Models — 361
5. Summary — 368

12 AMPLIFIERS — 369
1. Introduction — 369
2. Microwave Biasing 101 — 370
3. Bandwidth Extension Techniques — 381
4. The Shunt-Series Amplifier — 395
5. Tuned Amplifiers — 413
6. Neutralization and Unilateralization — 417
7. Strange Impedance Behaviors and Stability — 420
8. Appendix: Derivation of Bridged T-Coil Transfer Function — 427

13 LNA DESIGN — 440
1. Introduction — 440
2. Classical Two-Port Noise Theory — 440
3. Derivation of a Bipolar Noise Model — 445
4. The Narrowband LNA — 451
5. A Few Practical Details — 455

	6. Linearity and Large-Signal Performance	457
	7. Spurious-Free Dynamic Range	462
	8. Cascaded Systems	464
	9. Summary	467
	10. Appendix A: Bipolar Noise Figure Equations	468
	11. Appendix B: FET Noise Parameters	468
14	**NOISE FIGURE MEASUREMENT**	472
	1. Introduction	472
	2. Basic Definitions and Noise Measurement Theory	472
	3. Noise Temperature	477
	4. Friis's Formula for the Noise Figure of Cascaded Systems	479
	5. Noise Measure	480
	6. Typical Noise Figure Instrumentation	481
	7. Error Sources	487
	8. Special Considerations for Mixers	491
	9. References	492
	10. Appendix: Two Cheesy Eyeball Methods	492
15	**OSCILLATORS**	494
	1. Introduction	494
	2. The Problem with Purely Linear Oscillators	494
	3. Describing Functions	495
	4. Resonators	515
	5. A Catalog of Tuned Oscillators	519
	6. Negative Resistance Oscillators	524
	7. Summary	528
16	**SYNTHESIZERS**	529
	1. Introduction	529
	2. A Short History of PLLs	529
	3. Linearized PLL Model	532
	4. PLL Rejection of Noise on Input	536
	5. Phase Detectors	537
	6. Sequential Phase Detectors	542
	7. Loop Filters and Charge Pumps	544
	8. Frequency Synthesis	551
	9. A Design Example	561
	10. Summary	564
	11. Appendix: Inexpensive PLL Design Lab Tutorial	565
17	**OSCILLATOR PHASE NOISE**	574
	1. Introduction	574
	2. General Considerations	576

3. Detailed Considerations: Phase Noise ... 579
   4. The Roles of Linearity and Time Variation in Phase Noise ... 582
   5. Circuit Examples – *LC* Oscillators ... 592
   6. Amplitude Response ... 597
   7. Summary ... 599
   8. Appendix: Notes on Simulation ... 600

18 **MEASUREMENT OF PHASE NOISE** ... 601
   1. Introduction ... 601
   2. Definitions and Basic Measurement Methods ... 601
   3. Measurement Techniques ... 604
   4. Error Sources ... 611
   5. References ... 612

19 **SAMPLING OSCILLOSCOPES, SPECTRUM ANALYZERS, AND PROBES** ... 613
   1. Introduction ... 613
   2. Oscilloscopes ... 614
   3. Spectrum Analyzers ... 625
   4. References ... 629

20 **RF POWER AMPLIFIERS** ... 630
   1. Introduction ... 630
   2. Classical Power Amplifier Topologies ... 631
   3. Modulation of Power Amplifiers ... 650
   4. Additional Design Considerations ... 679
   5. Summary ... 687

21 **ANTENNAS** ... 688
   1. Introduction ... 688
   2. Poynting's Theorem, Energy, and Wires ... 690
   3. The Nature of Radiation ... 691
   4. Antenna Characteristics ... 695
   5. The Dipole Antenna ... 697
   6. The Microstrip Patch Antenna ... 707
   7. Miscellaneous Planar Antennas ... 720
   8. Summary ... 721

22 **LUMPED FILTERS** ... 723
   1. Introduction ... 723
   2. Background – A Quick History ... 723
   3. Filters from Transmission Lines ... 726
   4. Filter Classifications and Specifications ... 738
   5. Common Filter Approximations ... 740

	6. Appendix A: Network Synthesis	766
	7. Appendix B: Elliptic Integrals, Functions, and Filters	774
	8. Appendix C: Design Tables for Common Low-pass Filters	781

23 MICROSTRIP FILTERS 784

1. Background 784
2. Distributed Filters from Lumped Prototypes 784
3. Coupled Resonator Bandpass Filters 803
4. Practical Considerations 841
5. Summary 843
6. Appendix: Lumped Equivalents of Distributed Resonators 844

Index 847

PREFACE

First, it was called *wireless,* then *radio.* After decades in eclipse *wireless* has become fashionable once again. Whatever one chooses to call it, the field of RF design is changing so rapidly that textbook authors, let alone engineers, are hard pressed to keep up. A significant challenge for newcomers in particular is to absorb an exponentially growing amount of new information while also acquiring a mastery of those foundational aspects of the art that have not changed for generations. Compounding the challenge is that many books on microwave engineering focus heavily on electromagnetic field theory and never discuss actual physical examples, while others are of a cookbook nature with almost no theory at all. Worse still, much of the lore on this topic is just that: an oral tradition (not always correct), passed down through the generations. The rest is scattered throughout numerous applications notes, product catalogs, hobbyist magazines, and instruction manuals – many of which are hard to find, and not all of which agree with each other. Hobbyists are almost always unhappy with the theoretical bent of academic textbooks ("too many equations, and in the end, they still don't tell you how to *make* anything"), while students and practicing engineers are often unhappy with the recipe-based approaches of hobby magazines ("they don't give you the theory to show how to *change* the design into what I actually need"). This book is a response to the students, hobbyists, and practicing engineers who have complained about the lack of a modern reference that balances theory and practice.

This book is a much-expanded version of notes used in the teaching of EE414, a one-term advanced graduate laboratory course on gigahertz transceiver design at Stanford University. Even so, it is intended to satisfy a much broader audience than simply those seeking a Ph.D. In EE414, students spend approximately nine weeks designing, building, and testing every building block of a 1-GHz transceiver using microstrip construction techniques. These building blocks include various antennas and microstrip filters as well as a low-noise amplifier, mixer, PLL-based frequency synthesizer and FM modulator/demodulator, and power amplifier. The "final exam" is a successful demonstration of two-way communications with these FM transceivers. I am deeply grateful to the students of the first class in particular, who graciously and enthusiastically served as guinea pigs as the course material and notes were being developed in real time (at best). The present form of the text owes a great deal to their suggestions.

It is true that you will find here a certain number of theoretical discussions of measurement techniques and microwave design, replete with complex transfer functions and transforms galore. That's necessary because an important function of a Ph.D. program is the mental torment of graduate students. But hobbyists who just want to "get to the good stuff" right away are free to ignore all of the equations and to focus instead on the many practical rules of thumb or on the numerous ways to build and characterize microwave circuits with inexpensive components and equipment. Every effort is made to provide verbal and physical explanations of what the equations *mean*. The weekend experimenter may enjoy in particular the projects included in many of the chapters, ranging from homemade diodes to a sub-$10 microwave impedance measurement system. The practicing engineer will find much useful information about how to extract the most reliable data from leading-edge instrumentation, with a special focus on understanding calibration methods (and their limitations) in order to avoid making subtle, but surprisingly common, errors. Younger engineers may also enjoy finding answers to many of their questions about everyday RF items (Does *BNC* really stand for "baby N connector"?[1] Why is everything 50 Ω? Who was Smith, and why did he invent a chart?). Readers are invited to pick and choose topics to suit their tastes; this book is a smorgasbord, as is clear from the following brief descriptions of the chapters.

Chapter 1 provides a short history of RF and microwave circuits. It is impossible to provide anything remotely approaching a comprehensive overview of this vast topic in one chapter, and we don't even try. Our hope is that it provides some entertainment while establishing a context for the rest of the book.

Chapter 2 introduces some definitions and basic concepts. We try to devise a less arbitrary definition of *microwave* than the simple frequency-based one offered in many other places. We also try very hard to avoid actually solving Maxwell's equations anywhere, preferring instead to appeal to physical intuition. Again, this book is about design and testing, not pure analysis, so any analysis we perform is dedicated to those aims.

The Smith chart and S-parameters are staples of classical microwave design, so Chapter 3 provides a brief introduction to them. The complex appearance of the chart is off-putting to many who would otherwise be interested in RF and microwave design. We therefore offer a brief history of the Smith chart to explain why it's useful, even today. In the process, we hope to make the Smith chart a little less intimidating.

At high frequencies, power gain is hard to come by, so impedance matching is a standard task of every microwave engineer. Chapter 4 presents a number of impedance matching methods, along with a brief explanation of the Bode–Fano limit, which helps let us know when we should quit trying (or when we shouldn't even try).

---

[1] No.

After having witnessed eager students inadvertently destroy expensive fixturing, it became clear that a chapter on the care and feeding of connectors was necessary. Chapter 5 surveys a number of popular connectors, their historical origins, their domain of application, and the proper ways to care for them. Cables and their characteristics are discussed as well.

Chapter 6 examines the characteristics of lumped passive elements at microwave frequencies. Simple circuit models (appropriate for design, but not necessarily accurate enough for analysis) are presented to alert the student to issues that may prevent circuits from functioning as desired. To no small degree, an important lesson in RF and microwave design is that there are always irreducible parasitics. Rather than conceding defeat, one must exploit them as circuit elements. Retaining simple mental models of parasitics allows you to devise clever ways of accomplishing this goal.

In Chapter 7 we introduce the most common way of building microwave circuits (either in discrete or integrated form): microstrip. Although the chapter focuses on this one particular method of implementing planar circuits, we also spend a little time discussing coplanar waveguide and coplanar strips, as well as stripline. After introducing basic concepts and some simple (but reasonably accurate) design equations, we examine a large number of passive components that may be realized in microstrip form.

Once you've designed and built some circuits, you'll need to characterize them. One of the most basic measurements you will make is that of impedance. Chapter 8 presents several methods for making impedance measurements, ranging from time-domain reflectometry to vector network analysis. We spend considerable time describing various calibration techniques, for maladroitness here often causes engineers to obtain $1 answers from a $100,000 instrument. For those who don't want to choose between buying a home and buying a network analyzer, we present a simple "slotted line" measurement device that can be fashioned for about $10 yet functions up to at least several gigahertz. The 40-dB cost reduction is not accompanied by a 40-dB utility reduction, fortunately. For example, once calibrated, the device can even be used to determine frequency within 1–2%.

Chapter 9 is devoted to microwave diodes. Engineers accustomed to lower-frequency design are often surprised at the wide variety of functions that diodes can perform. Especially surprising to many newcomers is that some diodes are capable of amplification and even of oscillation.

Chapter 10 builds on that foundation to describe numerous mixers, heart of the modern superheterodyne transceiver. Lumped and distributed implementations are presented, as are active and passive circuits. Depending on the available technology and the design constraints, any of these may be appropriate in a given situation.

Active circuits are more interesting, of course (at least to the author), and so Chapter 11 presents a survey of transistors. The device physicists have been working overtime for decades to give us JFETs, MOSFETs, MESFETs, HEMTs, VMOS, UMOS, LDMOS, HBTs,..., and the list just keeps growing. We attempt to provide a somewhat unified treatment of these transistors, and focus on just two types (MOSFETs and bipolars) as representative of a much wider class.

Chapter 12 considers how to squeeze the most out of whatever transistor technology you are given. We spend a few pages describing how to bias transistors because, at microwave frequencies, many of the lower-frequency techniques have serious implementation consequences (parasitics, again). We then describe methods for extending bandwidth by factors of 2–3 with modest increases in circuit complexity. And for the first time anywhere (to the author's knowledge), we present a detailed derivation of the transfer function and optimum conditions for the bridged T-coil bandwidth boost network.

In Chapter 13, we shift our goals from "give me all the bandwidth you can" to "give me the lowest-noise amplification possible." We discuss noise models and then present the theory of noise matching. We discover that the conditions that maximize power transfer almost never coincide with those that minimize noise factor, and so a compromise strategy is necessary. Again, although we focus the discussion on just one or two types of transistors, the general concepts presented apply to all amplifiers.

Once you've built what you believe is a low-noise amplifier, you have to prove it. Chapter 14 describes the principles underlying noise figure measurement, along with descriptions of how to get the wrong answer (it's very easy). Depending on your objectives (making your LNA look good, or your competitor's look bad), you can either commit or avoid those errors.

Chapter 15 describes how to produce controlled instability to build oscillators. The old joke among frustrated microwave engineers is that "amplifiers oscillate, and oscillators amplify." We hope that the simplified presentation here allows you to design oscillators that really do oscillate, and even on frequency.

Virtually every modern transceiver has a frequency synthesizer somewhere. Chapter 16 describes phase-locked loop synthesizers, along with an extended discussion of spur-producing design defects (and their mitigation). Although the theoretical discussion can get very complex, the design examples should help the impatient hobbyist put together a working design without having to understand every equation.

Chapter 17 analyzes the important subject of phase noise. It's not sufficient for an oscillator to oscillate. The putative scarcity of spectrum obligates all transmitters to follow a "good neighbor" policy and not transmit much energy outside of its assigned band. All oscillators are imperfect in this regard, so Chapter 17 identifies where phase noise comes from and how you can reduce it.

Chapter 18 describes phase noise measurement methods. As with noise figure, there are many subtle (and not so subtle) ways to bungle a phase noise measurement, and we try to steer you clear of those.

Chapter 19 describes spectrum analyzers, oscilloscopes, and probes. Too often, engineers place all of their faith in the instrument, forgetting that connecting their circuit to these devices is up to them. And if done with insufficient care, the quality of the measurement can degrade rapidly. This chapter highlights the more common of these errors – and ways to avoid them. Also, since high-frequency probes are so expensive, we offer a way to build one for a few dollars. To supplement the probe, we also offer a couple of fast pulse-generator circuits with which you may test probe and oscilloscope combinations.

Chapter 20 presents numerous ways to implement power amplifiers at RF and microwave frequencies. At one time, it was sufficient to design for a particular gain and output power. Unfortunately, demands have increased steadily as the telecommunications revolution has unfolded, and now one must achieve high efficiency, low cost, high robustness to varying load conditions, and high linearity. We attempt to survey numerous methods for achieving all of these goals.

The aim of Chapter 21 is showing how to get power into and out of the air. Antennas are mysterious, and we hope to take at least some of the mystery out of antenna design with the material in this chapter. We focus on microstrip patch antennas, but we precede that discussion with an examination of classic nonplanar antennas (e.g., the dipole) to identify important concepts.

Finally, Chapters 22 and 23 focus on the design of passive filters. The presentation is divided roughly into lumped design in Chapter 22 and microstrip distributed filters in Chapter 23. Throughout, we make a concerted effort to focus on practical details, such as the effect of component tolerance, or finite Q. We hope that the numerous design examples and simulation results will illuminate the design procedures and allow you to converge rapidly on an acceptable, repeatable design.

Again, these chapters are ordered in a quasirandom way. You are in no way obligated to read them linearly in the sequence presented. Skip around as you like. It's *your* book.

This text has been informed by the many wonderful mentors, colleagues, and students who have generously shared their knowledge and viewpoints over the years. Stanford Professor Malcolm McWhorter (now Emeritus) oversaw the development of a delightfully unorthodox BNC microstrip mounting arrangement for EE414's prerequisite, an introductory microwave laboratory course called EE344. As discussed later in this book, this mounting method is ideal for students and hobbyists because it allows for the rapid and inexpensive prototyping of circuits in the low-gigahertz frequency range. Howard Swain and Dieter Scherer, both formerly of Hewlett-Packard and both virtuoso designers of widely used microwave instruments, helped to create EE344 and have continued to help teach it. The present mentor of the course, Professor Donald Cox, has graciously communicated lessons learned from teaching EE344 over the years, and the contents of this book have been adjusted as a direct result. I am also greatly indebted to Professor David Rutledge of Caltech, not only for generously allowing the inclusion of *Puff* in the CD-ROM collection of software accompanying this text but also for the great wealth of knowledge he has imparted to me, both in person and from his many publications. I've also benefitted enormously from the fact that simple proximity to David Leeson automatically increases your knowledge of microwave systems by several dB. Having so knowledgeable a faculty colleague has been a godsend.

I have also been the beneficiary of the hard work of several dedicated graduate students who built and tested most of the projects described in this book. Stanford Ph.D. candidates Sergei Krupenin, Arjang Hassibi, Talal Al-Attar, Moon-Jung Kim, and Michael Mulligan merit special mention in particular for their efforts. Rob Chavez of Agilent, while nominally a student in EE414, worked long hours to help other students and the teaching assistant. Future EE414 students, as well as readers of this book, owe him thanks for the insights and suggestions that have shaped this material.

Funding is the fuel that keeps the academic engine purring, and here I have been most fortunate. Generous equipment grants from Hewlett-Packard and Agilent Technologies have given generations of students the privilege of hands-on work with leading-edge gear that not many schools can afford. Support from the William G. Hoover Faculty Scholar Chair and The David and Lucile Packard Foundation has given me tremendous freedom, making it possible to develop new courses, pursue some crazy research ideas, and write textbooks.

Finally, I am most deeply grateful to my loving wife, Angelina, for her patient support during the writing of this book, and for otherwise living up to her name in every way. Without her forbearance, it would have been impossible to complete two book manuscripts in one long year.

CHAPTER ONE

A MICROHISTORY OF MICROWAVE TECHNOLOGY

1.1 INTRODUCTION

Many histories of microwave technology begin with James Clerk Maxwell and his equations, and for excellent reasons. In 1873, Maxwell published *A Treatise on Electricity and Magnetism,* the culmination of his decade-long effort to unify the two phenomena. By arbitrarily adding an extra term (the "displacement current") to the set of equations that described all previously known electromagnetic behavior, he went beyond the known and predicted the existence of electromagnetic waves that travel at the speed of light. In turn, this prediction inevitably led to the insight that light itself must be an electromagnetic phenomenon. Electrical engineering students, perhaps benumbed by divergence, gradient, and curl, often fail to appreciate just how revolutionary this insight was.[1] Maxwell did not introduce the displacement current to resolve any outstanding conundrums. In particular, he was not motivated by a need to fix a conspicuously incomplete continuity equation for current (contrary to the standard story presented in many textbooks). Instead he was apparently inspired more by an aesthetic sense that nature simply should provide for the existence of electromagnetic waves. In any event the word *genius,* though much overused today, certainly applies to Maxwell, particularly given that it shares origins with *genie*. What he accomplished was magical and arguably ranks as the most important intellectual achievement of the 19th century.[2]

Maxwell – genius and genie – died in 1879, much too young at age 48. That year, Hermann von Helmholtz sponsored a prize for the first experimental confirmation of Maxwell's predictions. In a remarkable series of investigations carried out between

---

[1] Things could be worse. In his treatise of 1873, Maxwell expressed his equations in terms of *quaternions*. Oliver Heaviside and Josiah Willard Gibbs would later reject quaternions in favor of the language of vector calculus to frame Maxwell's equations in the form familiar to most modern engineers.

[2] The late Nobel physicist Richard Feynman often said that future historians would still marvel at Maxwell's work, long after another event of that time – the American Civil War – had faded into merely parochial significance.

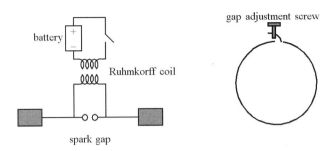

FIGURE 1.1. Spark transmitter and receiver of Hertz

1886 and 1888 at the Technische Hochschule in Karlsruhe, Helmholtz's former pupil, Heinrich Hertz, verified that Maxwell was indeed correct. Another contestant in the race, Oliver Lodge (then a physics professor at University College in Liverpool), published his own confirmation one month after Hertz, having interrupted his work in order to take a vacation. Perhaps but for that vacation we would today be referring to *lodgian waves* with frequencies measured in *megalodges*. Given that *Hertz* is German for *heart* and that the human heart beats about once per second, it is perhaps all for the best that Lodge didn't win the race.

How did Hertz manage to generate and detect electromagnetic waves with equipment available in the 1880s? Experimental challenges certainly extend well beyond the mere generation of some sort of signal; a detector is required, too. Plus, to verify wave behavior, you need apparatus that is preferably at least a couple of wavelengths in extent. In turn, that requirement implies another: sufficient lab space to contain apparatus of that size (and preferably sufficient to treat the room as infinitely large, relative to a wavelength, so that unwanted reflections from walls and other surfaces may be neglected). Hertz, then a junior faculty member, merited a modest laboratory whose useful internal dimensions were approximately 12 m by 8 m.[3] Hertz understood that the experimental requirements forced him to seek the generation of signals with wavelengths of the order of a meter. He accomplished the difficult feat of generating such short waves by elaborating on a speculation by the Irish physicist George Francis FitzGerald, who had suggested in 1883 that one might use the known oscillatory spark discharge of Leyden jars (capacitors) to generate electromagnetic waves. Recognizing that the semishielded structure of the jars would prevent efficient radiation, Hertz first modified FitzGerald's idea by "unrolling" the cylindrical conductors in the jars into flat plates. Then he added inductance in the form of straight wire connections to those plates in order to produce the desired resonant frequency of a few hundred megahertz. In the process, he thereby invented the dipole antenna. Finally, he solved the detection problem by using a ring antenna with an integral spark gap. His basic transmitter–receiver setup is shown in Figure 1.1. When the

---

[3] Hugh G. J. Aitken, *Syntony and Spark*, Princeton University Press, Princeton, NJ, 1985.

switch is closed, the battery charges up the primary of the Ruhmkorff coil (an early transformer). When the switch opens, the rapid collapse of the magnetic field induces a high voltage in the secondary, causing a spark discharge. The sudden change in current accompanying the discharge excites the antenna to produce radiation.

Detection relies on the induction of sufficient voltage in the ring resonator to produce a visible spark. A micrometer screw allows fine adjustment, and observation in the dark permits one to increase measurement sensitivity.[4]

With this apparatus (a very longwave version of an optical interferometer), Hertz demonstrated essential wave phenomena such as polarization and reflection.[5] Measurements of wavelength, coupled with analytical calculations of inductance and capacitance, confirmed a propagation velocity sufficiently close to the speed of light that little doubt remained that Maxwell had been right.[6]

We will never know if Hertz would have gone beyond investigations of the pure physics of the phenomena to consider practical uses for wireless technology, for he died of blood poisoning (from an infected tooth) in 1894 at the age of 36. *Brush and floss after every meal, and visit your dentist regularly.*

Maxwell's equations describe electric and magnetic fields engaged in an eternal cycle of creation, destruction, and rebirth. Fittingly, Maxwell's death had inspired von Helmholtz to sponsor the prize which had inspired Hertz. Hertz's death led to the publication of a memorial tribute that, in turn, inspired a young man named Guglielmo Marconi to dedicate himself to developing commercial applications of wireless. Marconi was the neighbor and sometime student of Augusto Righi, the University of Bologna professor who had written that tribute to Hertz. Marconi had been born into a family of considerable means, so he had the time and finances to pursue his dream.[7] By early 1895, he had acquired enough apparatus to begin experiments in and around his family's villa, and he worked diligently to increase transmission distances. Marconi used Hertz's transmitter but, frustrated by the inherent limitations of a spark-gap detector, eventually adopted (then adapted) a peculiar creation that had been developed by Edouard Branly in 1890. As seen in Figure 1.2, the device, dubbed a *coherer* by Lodge, consists of a glass enclosure filled with a loosely packed and perhaps slightly oxidized metallic powder. Branly had accidentally discovered that the resistance of this structure changes dramatically when nearby

---

[4] Hertz is also the discoverer of the photoelectric effect. He noticed that sparks would occur more readily in the presence of ultraviolet light. Einstein would win his Nobel prize for providing the explanation (and not for his theory of relativity, as is frequently assumed).

[5] The relative ease with which the waves were reflected would inspire various researchers to propose crude precursors to radar within a relatively short time.

[6] This is not to say that everyone was immediately convinced; they weren't. Revolutions take time.

[7] Marconi's father was a successful businessman, and his mother was an heiress to the Jameson Irish whiskey fortune. Those family connections would later prove invaluable in gaining access to key members of the British government after Italian officials showed insufficient interest. The British Post Office endorsed Marconi's technology and supported its subsequent development.

FIGURE 1.2. Branly's coherer

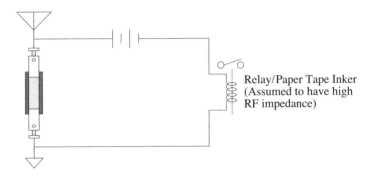

FIGURE 1.3. Typical receiver with coherer

electrical apparatus is in operation. It must be emphasized that the detailed principles that underlie the operation of coherers remain mysterious, but that ignorance doesn't prevent us from describing their electrical behavior.[8]

A coherer's resistance generally has a large value (say, megohms) in its quiescent state and then drops to kilohms or less when triggered by some sort of an EM event. This large resistance change in turn may be used to trigger a solenoid to produce an audible click, as well as to ink a paper tape for a permanent record of the received signal. To prepare the coherer for the next EM pulse, it has to be shaken (or stirred) to restore the "incoherent" high-resistance state. Figure 1.3 shows how a coherer can be used in a receiver. It is evident that the coherer is a digital device and therefore unsuitable for uses other than radiotelegraphy.

The coherer never developed into a good detector, it just got less bad over time. Marconi finally settled on the configuration shown in Figure 1.4. He greatly reduced the spacing between the end plugs, filled the intervening space with a particular mixture of nickel and silver filings of carefully selected size, and partially evacuated the tube prior to sealing the assembly. As an additional refinement in the receiver, a solenoid provided an audible indication in the process of automatically whacking the detector back into its initial state after each received pulse.

Even though many EM events other than the desired signal could trigger a coherer, Marconi used this erratic device with sufficient success to enable increases

---

[8] Lodge named these devices *coherers* because the filings could be seen to stick together under some circumstances. However, the devices continue to function as detectors even without observable physical movement of the filings. It is probable that oxide breakdown is at least part of the explanation, but experimental proof is absent for lack of interest in these devices.

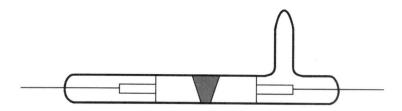

FIGURE 1.4. Marconi's coherer

in communication range to about three kilometers by 1896. As he scaled upward in power, he used progressively larger antennas, which had the unintended side effect of lowering the "carrier" frequencies to below 100 kHz from his initial frequencies of ~100 MHz. This change was most fortuitous, because it allowed reflections from the ionosphere (whose existence was then unknown) to extend transmission distances well beyond the horizon, allowing him to claim successful transatlantic wireless communications by 12 December 1901.[9] Wireless technology consequently ignored the spectrum above 1 MHz for nearly two more decades, thanks to a belief that communication distances were greatest below 100 kHz.

As the radio art developed, the coherer's limitations became increasingly intolerable, spurring the search for improved detectors. Without a body of theory to impose structure, however, this search was haphazard and sometimes took bizarre turns. A human brain from a fresh cadaver was once tried as a coherer, with the experimenter claiming remarkable sensitivity for his apparatus.[10]

That example notwithstanding, most detector research was based on the vague notion that a coherer's operation depends on some mysterious property of imperfect contacts. Following this intuition, a variety of experimenters stumbled, virtually simultaneously, on various types of point-contact crystal detectors. The first patent application for such a device was filed in 1901 by the remarkable Jagadish Chandra Bose for a detector using galena (lead sulfide).[11] See Figures 1.5 and 1.6. This detector exploits a semiconductor's high temperature coefficient of resistance, rather than rectification.[12] As can be seen in the patent drawing, electromagnetic

---

[9] Marconi's claim was controversial then, and it remains so. The experiment itself was not double-blind, as both the sender and the recipient knew ahead of time that the transmission was to consist of the letter *s* (three dots in Morse code). Ever-present atmospheric noise is particularly prominent in the longwave bands he was using at the time. The best modern calculations reveal that the three dots he received had to have been noise, not signal. One need not postulate fraud, however. Unconscious experimenter bias is a well-documented phenomenon and is certainly a possibility here. In any case, Marconi's apparatus evolved enough within another year to enable verifiable transatlantic communication.

[10] A. F. Collins, *Electrical World and Engineer*, v. 39, 1902; he started out with brains of other species and worked his way up to humans.

[11] U.S. Patent #755,840, granted 19 March 1904. The patent renders his name Jagadis Chunder Bose. The transliteration we offer is that used by the academic institution in Calcutta that bears his name.

[12] Many accounts of Bose's work confuse his galena balometer with the point-contact rectifying ("catwhisker" type) detectors developed later by others and thus erroneously credit him with the

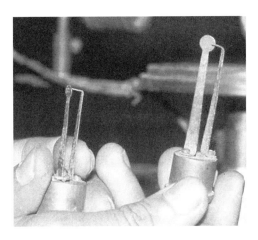

FIGURE 1.5. Actual detector mounts used by Bose (galena not shown) [*courtesy of David Emerson*]

radiation is focused on the point contact, and the resistance change that accompanies the consequent heating registers as a change in current flowing through an external circuit. This type of detector is known as a *bolometer*. In refined form, bolometers remain useful as a means of measuring power, particularly of signals whose frequency is so high that there are no other means of detection. Bose used this detector in experiments extending to approximately 60 GHz, about which he first published papers in 1897.[13] His research into millimeter-wave phenomena was decades ahead of his time.[14] So too was the recognition by Bose's former teacher at Cambridge, Lord Rayleigh, that hollow conductors could convey electromagnetic energy.[15] Waveguide transmission would be forgotten for four decades, but Rayleigh had most of it worked out (including the concept of a cutoff frequency) in 1897.

---

invention of the semiconductor diode. The latter functions by rectification, of course, and thus does not require an external bias. It was Ferdinand Braun who first reported asymmetrical conduction in galena and copper pyrites (among others), back in 1874, in "Ueber die Stromleitung durch Schwefelmetalle" [On Current Flow through Metallic Sulfides], *Poggendorff's Annalen der Physik und Chemie*, v. 153, pp. 556–63. Braun's other important development for wireless was the use of a spark gap in series with the primary of a transformer whose secondary connects to the antenna. He later shared the 1909 Nobel Prize in physics with Marconi for contributions to the radio art.

[13] J. C. Bose, "On the Determination of the Wavelength of Electric Radiation by a Diffraction Grating," *Proc. Roy. Soc.*, v. 60, 1897, pp. 167–78.

[14] For a wonderful account of Bose's work with millimeter waves, see David T. Emerson, "The Work of Jagadis Chandra Bose: 100 Years of MM-Wave Research," *IEEE Trans. Microwave Theory and Tech.*, v. 45, no. 12, 1997, pp. 2267–73.

[15] Most scientists and engineers are familar with Rayleigh's extensive writings on acoustics, which include analyses of ducting (acoustic waveguiding) and resonators. Far fewer are aware that he also worked out the foundations for electromagnetic waveguides at a time when no one could imagine a use for the phenomenon and when no one but Bose could even generate waves of a high enough frequency to propagate through reasonably small waveguides.

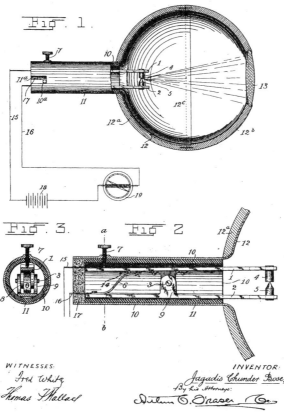

FIGURE 1.6. Bose's bolometer patent (first page)

This patent appears to be the first awarded for a semiconductor detector, although it was not explicitly recognized as such because semiconductors were not yet acknowledged as a separate class of materials (indeed, the word *semiconductor* had not yet been coined). Work along these lines continued, and General Henry Harrison Chase Dunwoody filed the first patent application for a rectifying detector using carborundum (SiC) on 23 March 1906, receiving U.S. Patent #837,616 on 4 December of that year. A later application, filed on 30 August 1906 by Greenleaf Whittier Pickard (an MIT graduate whose great-uncle was the poet John Greenleaf Whittier) for a silicon (!) detector, resulted in U.S. Patent #836,531 just ahead of Dunwoody, on 20 November (see Figure 1.7).

As shown in Figure 1.8, one connection consists of a small wire (whimsically known as a catwhisker) that makes a point contact to the crystal surface. The other connection is a large area contact canonically formed by a low–melting-point alloy

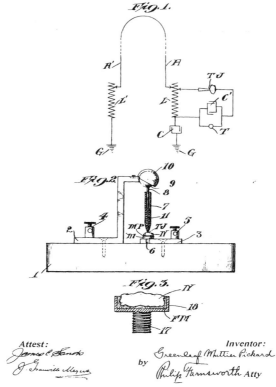

FIGURE 1.7. The first silicon diode patent

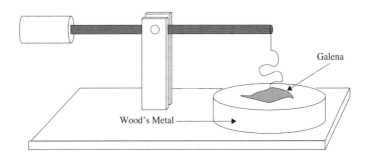

FIGURE 1.8. Typical crystal detector

(usually a mixture of lead, tin, bismuth, and cadmium known as Wood's metal, which has a melting temperature of under 80°C), that surrounds the crystal.[16] One might call a device made this way a point-contact Schottky diode, although measurements

---

[16] That said, such immersion is unnecessary. A good clamp to the body of the crystal usually suffices, and it avoids the use of toxic metals.

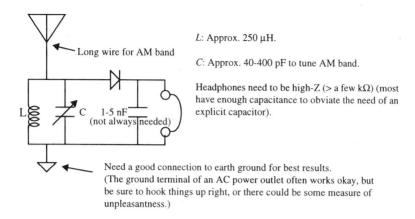

FIGURE 1.9. Simple crystal radio

are not always easily reconciled with such a description. In any event, we can see how the modern symbol for the diode evolved from a depiction of this physical arrangement, with the arrow representing the catwhisker point contact.

Figure 1.9 shows a simple crystal radio made with these devices.[17] An *LC* circuit tunes the desired signal, which the crystal then rectifies, leaving the demodulated audio to drive the headphones. A bias source is not needed with some detectors (such as galena), so it is possible to make a "free-energy" radio.[18] As we'll see, someone who had been enthralled by the magic of crystal radios as a boy would resurrect point-contact diodes to enable the development of radar. Crystal radios remain a focus of intense interest by a corps of dedicated hobbyists attracted by the simple charm of these receivers.

Pickard worked harder than anyone else to develop crystal detectors, eventually evaluating over 30,000 combinations of wires and crystals. In addition to silicon, he studied iron pyrites (fool's gold) and rusty scissors. Galena detectors became quite popular because they are inexpensive and need no bias. Unfortunately, proper adjustment of the catwhisker wire contact is difficult to maintain because anything other than the lightest pressure on galena destroys the rectification. Plus, you have to hunt around the crystal surface for a sensitive spot in the first place. On the other hand, although carborundum detectors need a bias of a couple of volts, they are more

---

[17] Today, *crystal* usually refers to quartz resonators used, for example, as frequency-determining elements in oscillators; these bear no relationship to the crystals used in crystal radios. A galena crystal may be replaced by a commercially made diode (such as the germanium 1N34A), but purists would disapprove of the lack of charm. An ordinary U.S. penny (dated no earlier than 1983), baked in a kitchen oven for 15 minutes at about 250°C to form CuO, exhibits many of the relevant characteristics of the galena (e.g., wholly erratic behavior). Copper-based currencies of other nations may also work (the author has verified that the Korean 10-won coin works particularly well). The reader is encouraged to experiment with coins from around the world and inform the author of the results.

[18] Perhaps we should give a little credit to the human auditory system: the threshold of hearing corresponds to an eardrum displacement of about the diameter of a hydrogen atom!

mechanically stable (a relatively high contact pressure is all right) and found wide use on ships as a consequence.[19]

At about the same time that these crude semiconductors were first coming into use, radio engineers began to struggle with the interference caused by the ultrabroad spectrum of a spark signal. This broadband nature fits well with coherer technology, since the dramatically varying impedance of the latter makes it difficult to realize tuned circuits anyway. However, the unsuitability of spark for multiple access was dramatically demonstrated in 1901, when three separate groups (led by Marconi, Lee de Forest, and Pickard) attempted to provide up-to-the-minute wireless coverage of the America's Cup yacht race. With three groups simultaneously sparking away, no one was able to receive intelligible signals, and race results had to be reported the old way, by semaphore. A thoroughly disgusted de Forest threw his transmitter overboard, and news-starved relay stations on shore resorted to making up much of what they reported.

In response, a number of engineers sought ways of generating continuous sine waves at radio frequencies. One was the highly gifted Danish engineer Valdemar Poulsen[20] (famous for his invention of an early magnetic recording device), who used the negative resistance associated with a glowing DC arc to keep an *LC* circuit in constant oscillation.[21] A freshly minted Stanford graduate, Cyril Elwell, secured the rights to Poulsen's arc transmitter and founded Federal Telegraph in Palo Alto, California. Federal soon scaled up this technology to impressive power levels: an arc transmitter of over 1 *megawatt* was in use shortly after WWI!

Pursuing a different approach, Reginald Fessenden asked Ernst F. W. Alexanderson of GE to produce radio-frequency (RF) sine waves at large power levels with huge alternators (*very* big, very high-speed versions of the thing that recharges your car battery as you drive). This dead-end technology culminated in the construction

---

[19] Carborundum detectors were typically packaged in cartridges and were often adjusted by using the delicate procedure of slamming them against a hard surface.

[20] Some sources persistently render his name incorrectly as "Vladimir," a highly un-Danish name!

[21] Arc technology for industrial illumination was a well-developed art by this time. The need for a sufficiently large series resistance to compensate for the arc's negative resistance (and thereby maintain a steady current) was well known. William Duddell exploited the negative resistance to produce audio (and audible) oscillations. Duddell's "singing arc" was perhaps entertaining but not terribly useful. Efforts to raise the frequency of oscillation beyond the audio range were unsuccessful until Poulsen switched to hydrogen gas and employed a strong magnetic field to sweep out ions on a cycle-by-cycle basis (an idea patented by Elihu Thompson in 1893). Elwell subsequently scaled up the dimensions in a bid for higher power. This strategy sufficed to boost power to 30 kW, but attempts at further increases in power through scaling simply resulted in larger transmitters that still put out 30 kW. In his Ph.D. thesis (Stanford's first in electrical engineering), Leonard Fuller provided the theoretical advances that allowed arc power to break through that barrier and enable 1-MW arc transmitters. In 1931, as chair of UC Berkeley's electrical engineering department – and after the arc had passed into history – Fuller arranged the donation of surplus coil-winding machines and an 80-ton magnet from Federal for the construction of Ernest O. Lawrence's first large cyclotron. Lawrence would win the 1939 Nobel Prize in physics with that device.

of an alternator that put out 200 kW at 100 kHz! It was completed just as WWI ended and was already on its way to obsolescence by the time it became operational.[22]

The superiority of the continuous wave over spark signals was immediately evident, and it stimulated the development of better receiving equipment. Thankfully, the coherer was gradually supplanted by a number of improved devices, including the semiconductor devices described earlier, and was well on its way to extinction by 1910 (although as late as the 1950s there was at least one radio-controlled toy truck that used a coherer).

Enough rectifying detectors were in use by late 1906 to allow shipboard operators on the East Coast of the United States to hear, much to their amazement (even with a pre-announcement by radiotelegraph three days before), the first AM broadcast by Fessenden himself on Christmas Eve. Delighted listeners were treated to a recording of Handel's *Largo* (from *Xerxes*), a fine rendition of *O Holy Night* by Fessenden on the violin (with the inventor accompanying himself while singing the last verse), and his hearty Christmas greetings to all.[23] He used a water-cooled carbon microphone to modulate a 500-W (approximate), 50-kHz (also approximate) carrier generated by a prototype Alexanderson alternator located at Brant Rock, Massachusetts. Those unfortunate enough to use coherers missed out on the historic event. Fessenden repeated his feat a week later, on New Year's Eve, to give more people a chance to get in on the fun.

1.2 BIRTH OF THE VACUUM TUBE

The year 1907 saw the invention, by Lee de Forest, of the first electronic device capable of amplification: the triode vacuum tube. Unfortunately, de Forest didn't understand how his invention actually worked, having stumbled upon it by way of a circuitous (and occasionally unethical) route.

The vacuum tube traces its ancestry to the humble incandescent light bulb of Thomas Edison. Edison's bulbs had a problem with progressive darkening caused by the accumulation of soot (given off by the carbon filaments) on the inner surface. In an attempt to cure the problem, he inserted a metal electrode, hoping somehow to attract the soot to this plate rather than to the glass. Ever the experimentalist, he applied both positive and negative voltages (relative to one of the filament connections) to this plate, and noted in 1883 that a current mysteriously flows when the plate is positive but not when negative. Furthermore, the current that flows depends on filament temperature. He had no theory to explain these observations (remember, the word *electron* wasn't even coined by George Johnstone Stoney until 1891, and the particle itself wasn't unambiguously identified until J. J. Thomson's experiments of

---

[22] Such advanced rotating machinery so stretched the metallurgical state of the art that going much above, say, 200 kHz would be forever out of the question.

[23] "An Unsung Hero: Reginald Fessenden, the Canadian Inventor of Radio Telephony," ⟨http://www.ewh.ieee.org/reg/7/millennium/radio/radio_unsung.html⟩.

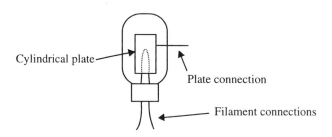

FIGURE 1.10. Fleming valve

1897), but Edison went ahead and patented in 1884 the first electronic (as opposed to electrical) device, one that exploits the dependence of plate current on filament temperature to measure line voltage indirectly.[24] This instrument never made it into production, given its inferiority to a standard voltmeter; Edison just wanted another patent, that's all (that's one way he ended up with 1093 of them).

At about this time, a consultant to the British Edison Company named John Ambrose Fleming happened to attend a conference in Canada. He took this opportunity to visit both his brother in New Jersey and Edison's lab. He was greatly intrigued by the "Edison effect" (much more so than Edison, who was a bit puzzled by Fleming's excitement over so useless a phenomenon), and eventually he published papers on the effect from 1890 to 1896. Although his experiments created an initial stir, the Edison effect quickly lapsed into obscurity after Röntgen's announcement in January 1896 of the discovery of X-rays as well as the discovery of natural radioactivity later that same year.

Several years later, though, Fleming became a consultant to British Marconi and joined in the search for improved detectors. Recalling the Edison effect, he tested some bulbs, found out that they worked satisfactorily as RF rectifiers, and patented the Fleming valve (vacuum tubes are thus still known as valves in the U.K.) in 1905 (see Figure 1.10).[25] The nearly deaf Fleming used a mirror galvanometer to provide a visual indication of the received signal and included this feature as part of his patent.

While not particularly sensitive, the Fleming valve is at least continually responsive and requires no mechanical adjustments. Various Marconi installations used them (largely out of contractual obligations), but the Fleming valve never was popular – contrary to the assertions of some histories – thanks to its high power, poor filament life, high cost, and low sensitivity when compared with well-made crystal detectors.

De Forest, meanwhile, was busy in America setting up shady wireless companies to compete with Marconi. "Soon, we believe, the suckers will begin to bite," he wrote hopefully in his journal in early 1902. And, indeed, his was soon the largest wireless company in the United States after Marconi Wireless. Never one to pass up an opportunity, de Forest proceeded to steal Fleming's diode and even managed to

---

[24] U.S. Patent #307,031, filed 15 November 1883, granted 21 October 1884.
[25] U.S. Patent #803,684, filed 19 April 1905, granted 7 November 1905.

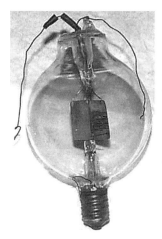

Dual-plate triode used by Armstrong in his regenerative receiver of 1912 (*from the Houck Collection, courtesy Michael Katzdorn*).

FIGURE 1.11. De Forest triode audion

receive a patent for it in 1906 (#836,070, filed 19 May, granted 13 November). He simply replaced Fleming's mirror galvanometer with a headphone and then added a huge forward bias (thus reducing the sensitivity of an already insensitive detector). Conclusive evidence that de Forest had stolen Fleming's work outright came to light when historian Gerald Tyne obtained the business records of H. W. McCandless, the man who made all of de Forest's first vacuum tubes (de Forest called them *audions*).[26] The records clearly show that de Forest had asked McCandless to duplicate some Fleming valves months before he filed his patent. Hence there is no room for a charitable interpretation that de Forest independently invented the vacuum tube diode.

His next achievement was legitimate and important, however. He added a zigzag wire electrode, which he called the grid, between the filament and wing (later known as the plate), and thus the triode was born (see Figure 1.11). This three-element audion was capable of amplification, but de Forest did not realize this fact until years later. In fact, his patent only mentions the triode audion as a detector, not as an amplifier.[27] Motivation for the addition of the grid is thus still curiously unclear. He certainly did not add the grid as the consequence of careful reasoning, as some histories claim. The fact is that he added electrodes all over the place. He even tried "control electrodes" outside of the plate! We must therefore regard his addition of the grid as merely the result of quasirandom but persistent tinkering in his search for a detector to call his own. It would not be inaccurate to say that he stumbled onto the triode, and it is certainly true that others would have to explain its operation to him.[28]

---

[26] Gerald F. J. Tyne, *Saga of the Vacuum Tube*, Howard W. Sams & Co., 1977.

[27] U.S. Patent #879,532, filed 29 January 1907, granted 18 February 1908. Curiously enough, though, his patent for the two-element audio *does* imply amplification.

[28] Aitken, in *The Continuous Wave* (Princeton University Press, Princeton, NJ, 1985) argues that de Forest has been unfairly accused of not understanding his own invention. However, the bulk of the evidence contradicts Aitken's generous view.

From the available evidence, neither de Forest nor anyone else thought much of the audion for a number of years (annual sales remained below 300 units until 1912).[29] At one point, he had to relinquish interest in all of his inventions following a bankruptcy sale of his company's assets. There was just one exception: the lawyers let him keep the patent for the audion, thinking it worthless. Out of work and broke, he went to work for Fuller at Federal.

Faced with few options, de Forest – along with Federal engineers Herbert van Etten and Charles Logwood – worked to develop the audion and discovered its amplifying potential in late 1912, as did others almost simultaneously (including rocket pioneer Robert Goddard).[30] He managed to sell the device to AT&T that year as a telephone repeater amplifier, retaining the rights to wireless in the process, but initially had a tough time because of the erratic behavior of the audion.[31] Reproducibility of device characteristics was rather poor and the tube had a limited dynamic range. It functioned well for small signals but behaved badly upon overload (the residual gas in the tube would ionize, resulting in a blue glow and a frying noise in the output signal). To top things off, the audion filaments (then made of tantalum) had a life of only about 100–200 hours. It would be a while before the vacuum tube could take over the world.

1.3 ARMSTRONG AND THE REGENERATIVE AMPLIFIER/DETECTOR/OSCILLATOR

Thankfully, the audion's fate was not left to de Forest alone. Irving Langmuir of GE Labs worked hard to achieve a more perfect vacuum, thus eliminating the erratic behavior caused by the presence of (easily ionized) residual gases. De Forest had specifically warned against high vacua, partly because he sincerely believed that it would reduce the sensitivity but also because he had to maintain the fiction – to himself and others – that the lineage of his invention had nothing to do with Fleming's diode.[32]

---

[29] Tyne, *Saga of the Vacuum Tube*.
[30] Goddard's U.S. Patent #1,159,209, filed 1 August 1912 and granted 2 November 1915, describes a primitive cousin of an audion oscillator and thus actually predates even Armstrong's documented work.
[31] Although he was officially an employee of Federal at the time, he negotiated the deal with AT&T independently and in violation of the terms of his employment agreement. Federal chose not to pursue any legal action.
[32] Observing that the gas lamp in his laboratory seemed to vary in brightness whenever he used his wireless apparatus, de Forest speculated that flames could be used as detectors. Further investigation revealed that the lamps were responding only to the acoustic noise generated by his spark transmitter. Out of this slender thread, de Forest wove an elaborate tale of how this disappointing experiment with the "flame detector" nonetheless inspired the idea of gases as being responsive to electromagnetic waves and so ultimately led him to invent the audion independently of Fleming.

Whatever his shortcomings as an engineer, de Forest had a flair for language. Attempting to explain the flame detector (U.S. Patent #979,275), he repeatedly speaks of placing the gases in

1.3 THE REGENERATIVE AMPLIFIER/DETECTOR/OSCILLATOR

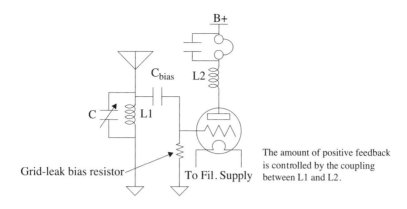

FIGURE 1.12. Armstrong regenerative receiver (see U.S. Patent #1,113,149)

Langmuir's achievement paved the way for a bright engineer to devise useful circuits to exploit the audion's potential. That bright engineer was Edwin Howard Armstrong, who invented the regenerative amplifier/detector[33] in 1912 at the tender age of 21. This circuit (a modern version of which is shown in Figure 1.12) employs positive feedback (via a "tickler coil" that couples some of the output energy back to the input with the right phase) to boost the gain and Q of the system simultaneously. Thus high gain (for good sensitivity) and narrow bandwidth (for good selectivity) can be obtained rather simply from one tube. Additionally, the nonlinearity of the tube may be used to demodulate the signal. Furthermore, overcoupling the output to the input turns the thing into a wonderfully compact RF oscillator.

Armstrong's 1914 paper, "Operating Features of the Audion,"[34] presents the first correct explanation for how the triode works, backed up with ample experimental evidence. A subsequent paper, "Some Recent Developments in the Audion Receiver,"[35] describes the operation of the regenerative amplifier/detector and also shows how overcoupling converts the amplifier into an RF oscillator. The paper is a model of clarity and is quite readable even to modern audiences. The degree to which it enraged de Forest is documented in a remarkable printed exchange immediately following the paper. One may read de Forest's embarrassingly feeble attempts to find fault with Armstrong's work. In his frantic desperation, de Forest blunders badly, demonstrating difficulty with rather fundamental concepts (e.g., he makes statements that are

---

a "condition of intense molecular activity." In his autobiography (*The Father of Radio*), he describes the operation of a coherer-like device (which, he neglects to mention, he had stolen from Fessenden) thus: "Tiny ferryboats they were, each laden with its little electric charge, unloading their etheric cargo at the opposite electrode." Perhaps he hoped that their literary quality would mask the absence of any science in these statements.

[33] His notarized notebook entry is actually dated 31 January 1913, mere months after de Forest's own discovery that the audion could amplify.
[34] *Electrical World,* 12 December 1914.
[35] *Proc. IRE,* v. 3, 1915, pp. 215–47.

equivalent to asserting that the average value of a sine wave is nonzero). He thus ends up revealing that he does not understand how the triode, his own invention (more of a discovery, really), actually works.

The bitter enmity that arose between these two men never waned.

Armstrong went on to develop circuits that continue to dominate communications systems to this day. While a member of the U.S. Army Signal Corps during World War I, Armstrong became involved with the problem of detecting enemy planes from a distance, and he pursued the idea of trying to home in on the signals naturally generated by their ignition systems (spark transmitters again). Unfortunately, little useful radiation was found below about 1 MHz, and it was exceedingly difficult with the tubes available at that time to get much amplification above that frequency. In fact, it was only with extraordinary care that Henry J. Round achieved useful gain at 2 MHz in 1917, so Armstrong had his work cut out for him.

He solved the problem by building upon a system patented by Fessenden, who sought to solve a problem with demodulating CW (continuous wave) signals. In Fessenden's *heterodyne* demodulator, a high-speed alternator acting as a *local oscillator* converts RF signals to an audible frequency, allowing the user to select a tone that cuts through the interference. By making signals from different transmitters easily distinguished by their different pitches, Fessenden's heterodyne system enabled unprecedented clarity in the presence of interference.

Armstrong decided to employ Fessenden's heterodyne principle in a different way. Rather than using it to demodulate CW directly, Armstrong's *superheterodyne* uses the local oscillator to convert an incoming high-frequency RF signal into one at a lower but still superaudible frequency, where high gain and selectivity can be obtained with relative ease. This lower-frequency signal, known as the intermediate frequency (IF), is then demodulated after much filtering and amplification at the IF has been achieved. Such a receiver can easily possess enough sensitivity so that the limiting factor is actually atmospheric noise (which is quite large in the AM broadcast band). Furthermore, it enables a single tuning control, since the IF amplifier works at a fixed frequency.

Armstrong patented the superheterodyne in 1917 (see Figure 1.13). Although the war ended before Armstrong could use the superhet to detect enemy planes, he continued to develop it with the aid of several talented engineers (including his lifelong friend and associate, Harry Houck), finally reducing the number of tubes to five from an original complement of ten (good thing, too: the prototype had a total filament current requirement of 10 A). David Sarnoff of RCA eventually negotiated the purchase of the superhet rights; as a consequence, RCA came to dominate the radio market by 1930.

The demands of the First World War, combined with the growing needs of telephony, drove a rapid development of the vacuum tube and allied electronics. These advances in turn enabled an application for wireless that went far beyond Marconi's original vision of a largely symmetrical point-to-point communications system that

1.3 THE REGENERATIVE AMPLIFIER/DETECTOR/OSCILLATOR

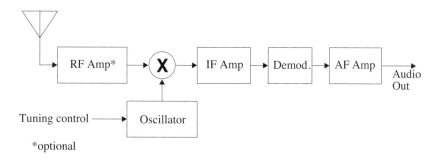

FIGURE 1.13. Superheterodyne receiver block diagram

mimicked the cable-based telegraphy after which it was modeled. Once the technology for radiotelephony was in place, pioneering efforts by visionaries like Fessenden and "Doc" Herrold highlighted the commercial potential for wireless as a point-to-*multipoint* entertainment medium.[36] The lack of any historical precedent for this revolutionary idea forced the appropriation of a word from agriculture to describe it: *broadcasting* (the spreading of seeds). Broadcast radio rose so rapidly in prominence that the promise of wireless seemed limitless. Hundreds of radio start-up companies flooded the market with receivers in the 1920s, at the end of which time the superheterodyne architecture had become important. Stock in the leader, RCA, shot up from about $11 per share in 1924 to a split-adjusted high of $114 as investors poured money into the sector. Alas, the big crash of 1929 precipitated a drop to $3 a share by 1932, as the wireless bubble burst.

With the rapid growth in wireless came increased competition for scarce spectrum, since frequencies commonly in use clustered in the sub–1-MHz band thought to be most useful. A three-way conflict involving radio amateurs ("hams"), government interests, and commercial services was partly resolved by relegating hams to frequencies above 1.5 MHz, a portion of spectrum then deemed relatively unpromising. Left with no options, dedicated hams made the best of their situation. To everyone's surprise, they discovered the enormous value of this "shortwave" spectrum, corresponding to wavelengths of 200 meters and below.[37] By freeing engineers to imagine the value of still-higher frequencies, this achievement did much to stimulate thinking about microwaves during the 1930s.

---

[36] Charles "Doc" Herrold was unique among radio pioneers in his persistent development of radio for entertainment. In 1909 he began regularly scheduled broadcasts of music and news from a succession of transmitters located in and near San Jose, California, continuing until the 1920s when the station was sold and moved to San Francisco (where it became KCBS). See *Broadcasting's Forgotten Father: The Charles Herrold Story*, KTEH Productions, 1994. The transcript of the program may be found at ⟨http://www.kteh.org/productions/docs/doctranscript.txt⟩.

[37] See Clinton B. DeSoto, *Two Hundred Meters and Down*, The American Radio Relay League, 1936. The hams were rewarded for their efforts by having spectrum taken away from them not long after proving its utility.

1.4 THE WIZARD WAR

Although commercial broadcasting drove most wireless technology development after the First World War, a growing awareness that the spectrum above a few megahertz might be useful led to better vacuum tubes and more advanced circuit techniques. Proposals for broadcast television solidified, and development of military communications continued apace. At the same time, AT&T began to investigate the use of wireless technology to supplement their telephone network. The need for additional spectrum became increasingly acute, and the art of high-frequency design evolved quickly beyond 1 MHz, first to 10 MHz and then to 100 MHz by the mid-1930s.

As frequencies increased, engineers were confronted with a host of new difficulties. One of these was the large high-frequency attenuation of cables. Recognizing that the conductor loss in coaxial cables, for example, is due almost entirely to the small diameter of the center conductor, it is natural to wonder if that troublesome center conductor is truly necessary.[38] This line of thinking inspired two groups to explore the possibility of conveying radio waves through hollow pipes. Led respectively by George C. Southworth of Bell Labs and Wilmer L. Barrow of MIT, the two groups worked independently of one another and simultaneously announced their developments in mid-1936.[39] Low-loss waveguide transmission of microwaves would soon prove crucial for an application that neither Southworth nor Barrow envisioned at the time: radar. Southworth's need for a detector of high-frequency signals also led him to return to silicon point-contact (catwhisker) detectors at the suggestion of his colleague, Russell Ohl. This revival of semiconductors would also have a profound effect in the years to come.

A reluctant acceptance of the inevitably of war in Europe encouraged a reconsideration of decades-old proposals for radar.[40] The British were particularly forward-looking and were the first to deploy radar for air defense, in a system called Chain Home, which began operation in 1937.[41] Originally operating at 22 MHz, frequencies increased to 55 MHz as the system expanded in scope and capability, just in time to play a crucial role in the Battle of Britain. By 1941 a 200-MHz system, Chain Home Low, was functional.

The superiority of still higher frequencies for radar was appreciated theoretically, but a lack of suitable detectors and high-power signal sources stymied practical

---

[38] Heaviside had thought about this in the 1890s, for example, but could not see how to get along without a second conductor.

[39] See e.g. G. C. Southworth, "High Frequency Waveguides – General Considerations and Experimental Results," *Bell System Tech. J.*, v. 15, 1936, pp. 284–309. Southworth and Barrow were unaware of each other until about a month before they were scheduled to present at the same conference, and they were also initially unaware that Lord Rayleigh had already laid the theoretical foundation four decades earlier.

[40] An oft-cited example is the patent application for the "Telemobilskop" filed by Christian Hülsmeyer in March of 1904. See U.S. Patent #810,510, issued 16 January 1906. There really are no new ideas.

[41] The British name for radar was RDF (for radio direction finding), but it didn't catch on.

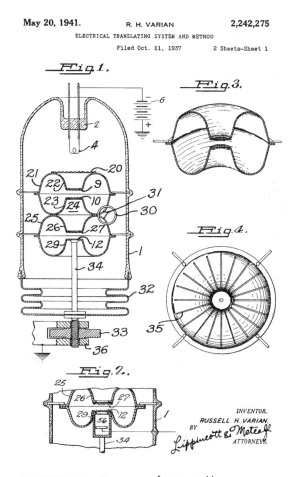

FIGURE 1.14. First page of Varian's klystron patent

development of what came to be called *microwaves*. At that time, the word connoted frequencies of approximately 1 GHz and above. Ordinary vacuum tubes suffer from fundamental scaling limitations that make operation in the microwave bands difficult. The finite velocity of electrons forces the use of ever-smaller electrode spacings as frequencies increase in order to keep carrier transit time small relative to a period (as it must be for proper operation). In turn, small electrode spacings reduce the breakdown voltage, thereby reducing the power-handling capability of the tube. Because power is proportional to the square of voltage, the output power of vacuum tubes tends to diminish quadratically as frequency increases.

In 1937, Russell Varian invented a type of vacuum tube that exploits transit time effects to evade these scaling limits.[42] See Figure 1.14. Developed at Stanford University with his brother Sigurd and physicist William Hansen, the *klystron* first accelerates electrons (supplied by a heated cathode) to a high velocity (e.g., 10% of the

---

[42] U.S. Patent #2,242,275, filed 11 October 1937, granted 20 May 1941.

speed of light). The high-velocity electron beam then passes through the porous parallel grids of a cavity resonator. A signal applied across these grids accelerates or decelerates the electrons entering the cavity, depending on the instantaneous polarity of the grid voltage. Upon exiting, the electrons drift in a low-field region wherein faster electrons catch up with slower ones, leading to periodic bunching (the Greek word for which gives us *klystron*). The conversion of a constant electron density into a pulsatile one leads to a component of beam current at the signal frequency. A second resonator then selects this component (or possibly a harmonic, if desired), and a coupling loop provides an interface to the external world. The klystron suffers less from transit delay effects: partly because the electrons are accelerated first (allowing the use of a larger grid spacing for a given oscillation period) and then subsequently controlled (whereas, in a standard vacuum tube, grid control of electron current occurs over a region where the electrons are slow); and partly because transit delay is essential to the formation of electron bunches in the drift space. As a result, exceptionally high output power is possible at microwave frequencies.

The klystron amplifier can be turned into an oscillator simply by providing for some reflection back to the input. Such reflection can occur by design or from unwanted mismatch in the second resonator. The reflex klystron, independently invented (actually, discovered) by Varian and John R. Pierce of Bell Labs around 1938 or 1939, exploits this sensitivity to reflections by replacing the second resonator with an electrode known as a *repeller*. Reflex klystrons were widely used as local oscillators for radar receivers owing to their compact size and to the relative ease with which they could be tuned (at least over a useful range).

Another device, the *cavity magnetron,* evolved to provide staggering amounts of output power (e.g., 100 kW on a pulse basis) for radar transmitters. The earliest form of magnetron was described by Albert W. Hull of GE in 1921.[43] Hull's magnetron is simply a diode with a cylindrical anode. Electrons emitted by a centrally disposed cathode trace out a curved path on their way to the anode thanks to a magnetic field applied along the axis of the tube. Hull's motivation for inventing this *crossed-field* device (so called because the electric and magnetic fields are aligned along different directions) had nothing whatever to do with the generation of high frequencies. Rather, by using a magnetic field (instead of a conventional grid) to control current, he was simply trying to devise a vacuum tube that would not infringe existing patents.

Recognition of the magnetron's potential for much more than the evasion of patent problems was slow in coming, but by the mid-1930s the search for vacuum tubes capable of higher-frequency operation had led several independent groups to reexamine the magnetron. An example is a 1934 patent application by Bell Labs engineer Arthur L. Samuel.[44] That invention coincides with a renaissance of magnetron-related developments aimed specifically at high-frequency operation. Soon after, the

---

[43] See *Phys. Rev.,* v. 18, 1921, p. 31, and also "The Magnetron," *AIEE J.,* v. 40, 1921, p. 715.
[44] U.S. Patent #2,063,341, filed 8 December 1934, granted 8 December 1936.

brilliant German engineer Hans E. Hollmann invented a series of magnetrons, some versions of which are quite similar to the cavity magnetron later built by Henry A. H. Boot and John T. Randall in 1940.[45]

Boot and Randall worked somewhat outside of the mainstream of radar research at the University of Birmingham, England. Their primary task was to develop improved radar detectors. Naturally, they needed something to detect. However, the lack of suitable signal sources set them casting about for promising ideas. Their initial enthusiasm for the newly developed klystron was dampened by the mechanical engineering complexities of the tube (indeed, the first ones were built by Sigurd Varian, who was a highly gifted machinist). They decided to focus instead on the magnetron (see Figure 1.15) because of its relative structural simplicity. On 21 February 1940, Boot and Randall verified their first microwave transmissions with their prototype magnetron. Within days, they were generating an astonishing 500 W of output power at over 3 GHz, an achievement almost two orders of magnitude beyond the previous state of the art.[46]

The magnetron depends on the same general bunching phenomenon as the klystron. Here, though, the static magnetic field causes electrons to follow a curved trajectory from the central cathode to the anode block. As they move past the resonators, the electrons either accelerate or decelerate – depending on the instantaneous voltage across the resonator gap. Just as in the klystron, bunching occurs, and the resonators pick out the fundamental. A coupling loop in one of the resonators provides the output to an external load.[47]

The performance of Boot and Randall's cavity magnetron enabled advances in radar of such a magnitude that a prototype was brought to the United States under cloak-and-dagger circumstances in the top-secret Tizard mission of August 1940.[48]

---

[45] U.S. Patent #2,123,728, filed 27 November 1936, granted 12 July 1938. This patent is based on an earlier German application, filed in 1935 and described that year in Hollmann's book, *Physik und Technik der Ultrakurzen Wellen, Erster Band* [*Physics and Technology of Ultrashort Waves*, vol. 1]. Hollmann gives priority to one Greinacher, not Hull. This classic reference had much more influence on wartime technological developments in the U.K. and the U.S. than in Germany.

[46] As with other important developments, there is controversy over who invented what, and when. It is a matter of record that patents for the cavity magnetron predate Boot and Randall's work, but this record does not preclude independent invention. Russians can cite the work of Alekseev and Maliarov (first published in a Russian journal in 1940 and then republished in *Proc. IRE*, v. 32, 1944); Germans can point to Hollmann's extensive publications on the device; and so on. The point is certainly irrelevant for the story of wartime radar, for it was the Allies alone who exploited the invention to any significant degree.

[47] This explanation is necessarily truncated and leaves open the question of how things get started. The answer is that noise is sufficient to get things going. Once oscillations begin, the explanation offered makes more sense.

[48] During the war, British magnetrons had six resonant cavities while American ones had eight. One might be tempted to attribute the difference to the "not invented here" syndrome, but that's not the explanation in this case. The British had built just one prototype with eight cavities, and that was the one picked (at random) for the Tizard mission, becoming the progenitor for American magnetrons.

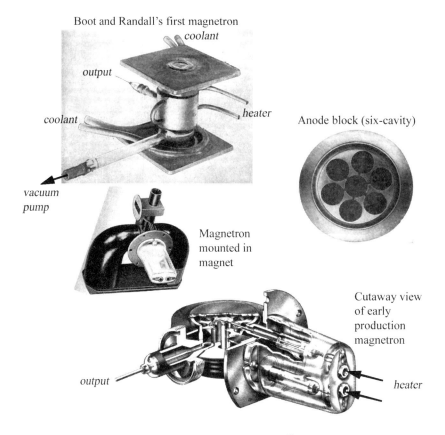

FIGURE 1.15. Magnetrons[49]

The magnetron amazed all who witnessed its operation, and Tizard returned to the United Kingdom with a guarantee of cooperative radar development from the (officially) still-neutral Americans.

The success of Tizard's mission rapidly led to formation of the Radiation Laboratory at the Massachusetts Institute of Technology. By mid-November the "Rad Lab," as it was (and is) known, was an official entity. Eventually, it would actively compete with the Manhattan Project for both funds and personnel. During the course of the war, more money would be spent on developing radar than on developing the atomic bomb.

The magnetron enabled startling advances in both airborne and ground-based radar, with many groups working together (for the most part, anyway). Bell Labs and the Rad Lab were both rivals and comrades, for example. The famous SCR-584 gun-laying radar combined Rad Lab radar with Bell Lab's analog computer to enable both the tracking of a target and the automated aiming and firing of artillery.

---

[49] George B. Collins, *Microwave Magnetrons* (MIT Rad. Lab. Ser., vol. 6), McGraw-Hill, New York, 1948.

By the war's end, magnetrons were producing 100-kW pulses at 10 GHz, and 24-GHz radar was in development.[50] The breathtaking speed with which the state of the art advanced on so many fronts simultaneously is evident upon inspection of the 27-volume set of books written by the Rad Lab staff shortly after the war.[51] Radar technology would soon be adapted for everything from cooking to radioastronomy. The evolution of Pickard's silicon catwhisker diodes into sophisticated centimeter-wave mixer diodes would set the stage for the invention of the transistor and the start of the semiconductor era. In a significant sense, the modern age of electronics could fairly be dated to the Second World War.

The war also saw the rapid development of the radio art well beyond radar. The utility of portable wireless communications was appreciated early on, and in 1940 Motorola delivered the handheld Handie-Talkie AM transceiver to the U.S. Army Signal Corps (which dubbed it the SCR-536).[52] By 1941, commercial two-way mobile FM communications systems had appeared, with its battlefield counterpart following in 1943 (the 15-kg SCR-300 backpack transceiver, the first to be called a Walkie-Talkie).[53]

The value of mobile communications was proven time and again during the war, so it was natural for that appreciation to stimulate the development of postwar mobile wireless. The city of St. Louis, Missouri, became the first to enjoy a commercial mobile radiotelephone service dubbed, appropriately enough, the Mobile Telephone Service (MTS).[54] Operating in the 150-MHz band with six channels spaced 60 kHz apart, the transceivers used FDD, *frequency-division duplexing* (i.e., one frequency each for uplink and downlink), and frequency modulation.[55] Because different frequencies allowed multiple users to communicate simultaneously, this system also represents an early use of frequency-division multiple access (FDMA) in a mobile wireless network.[56]

---

[50] It would fail because of the high atmospheric absorption by water vapor near that frequency. That failure would give birth to the field of microwave spectroscopy, though, so there's a happy ending to the story.

[51] If you count the index then there are 28 volumes. In any case, the entire set of Rad Lab books is now available as a 2–CD-ROM set from Artech House.

[52] The designation *SCR* stands for Signal Corps Radio, by the way. The Handi-Talkie used a complement of five tubes and operated on a single crystal-selectable frequency between 3.5 MHz and 6 MHz. It would soon become an icon, recognizable in countless newsreels and movies about the Second World War. Similarly, the SCR-584 and its rotating antenna would serve as set dressing for many postwar science-fiction "B" movies, scanning the skies for *them*.

[53] The ever-patriotic Armstrong, who had served in the Army Signal Corps during the First World War, offered his FM patents license-free to the U.S. government for the duration of the war.

[54] June 17, 1946, to be precise. See "Telephone Service for St. Louis Vehicles," *Bell Laboratories Record,* July 1946.

[55] Nevertheless, the service offered only half-duplex operation: the user had to push a button to talk and then release it to listen. In addition, all calls were mediated by operators; there was no provision for direct dialing.

[56] To underscore that there really are no new ideas, Bell himself had invented a primitive form of FDMA for his "harmonic telegraph," in which a common telegraph line could be shared by many

In the 1950s, radar continued to develop, and microwave ovens (patented by Raytheon engineer Percy Spencer) made their debut. Powered by magnetrons and operating at 900 MHz, these early ovens were quite a bit larger than the ovens now found in home kitchens.[57]

Planar microwave circuits also debuted in that decade as printed circuit boards became increasingly common. The successful Soviet launching of *Sputnik* in October of 1957 kicked off the space race and gave rise to aerospace applications for microwave technology. The widening availability of ever-better microwave equipment also led to important achievements in radioastronomy, including the first measurements of the cosmic background radiation, starting around 1955. However, the significance of those experiments went unappreciated until much later, so the field was left clear for Arno Penzias and Robert Wilson of Bell Labs to win the Nobel Prize in physics.[58]

As transistors improved throughout the 1950s and early 1960s, they came to displace vacuum tubes with increasing frequency, at ever-increasing frequencies. Perfectly suited for realization in planar form and for aerospace applications, solid-state amplifiers came to dominate low-power microwave technology. At the same time, the never-ending quest for still better performance at higher frequencies led to the development of transistors in gallium arsenide (GaAs). Although the superior mobility of electrons in GaAs had been appreciated in the 1950s, the difficulty of economically producing GaAs of sufficiently high purity and low defect density delayed significant commercialization until the late 1960s. Carver Mead of Caltech succeeded in demonstrating the first Schottky-gate GaAs FET in 1965; the metal–semiconductor FET (MESFET) would eventually dominate cell-phone power amplifiers through the 1990s.

Cellular finally made its debut in limited fashion in early 1969 in the form of payphones aboard a train running between New York City and Washington, D.C. The 450-MHz system, limited as it was to this single route, nonetheless possessed the defining features of cellular: frequency re-use and handoff.[59] A few years later, Motorola filed a patent that is often cited as the first expression of the cellular idea as it is practiced today.[60] By 1975, the Bell System had finally received FCC approval

---

users who were differentiated by frequency. Individually tuned tuning forks assured that only the intended recipient's telegraph would respond.

[57] Contrary to widespread belief, the 2.45-GHz frequency used by most microwave ovens does not correspond to any resonance with the water molecule. Higher frequencies are more strongly absorbed but penetrate less deeply. Lower frequencies penetrate to a greater depth but don't heat as effectively. A broad frequency range exists over which there is a reasonable balance between depth of penetration and speed of heating.

[58] See H. Kragh, *Cosmology and Controversy,* Princeton University Press, Princeton, NJ, 1996.

[59] C. E. Paul, "Telephones Aboard the Metroliner," *Bell Laboratories Record,* March 1969.

[60] Martin Cooper et al., U.S. Patent #3,906,166, filed 17 October 1973, granted 16 September 1975. Bell and Motorola were in a race to realize the cellular concept. Although Bell had been working on the theoretical aspects over a longer period, Motorola was allegedly the first to build an actual system-scale prototype and also the first to complete a cellular call with a *handheld* mobile phone

to offer trial service, but it didn't receive permission to *operate* it until 1977. Trial service finally began in 1978 in Chicago, Illinois, with a transition to full service finally taking place on 12 October 1983. Dubbed AMPS, for Advanced Mobile Phone Service, the analog FM-based system operated in a newly allocated band around 800 MHz (created by reclaiming spectrum previously assigned to upper UHF television channels). Just as MTS and IMTS had, AMPS used frequency-division multiple access (FDMA), in which multiple users may communicate simultaneously by assignment to different frequencies. It also used frequency-division duplexing (FDD), as had IMTS, to enable a user to talk while listening, just as with an ordinary phone. Recall that, in FDD, different frequencies are used for transmitting and receiving.

Certainly other countries had been designing similar systems as well. There are too many to name individually, but it is particularly noteworthy that the 450-MHz Nordic Mobile Telephone System (NMT-450, inaugurated in 1981) was the first multinational cellular system – serving Finland, Sweden, Denmark, and Norway. Aside from the frequency range, its characteristics are very similar to those of AMPS. Within a decade, the first generation of cellular service had become pervasive.[61] This unanticipated rapid growth has happily driven the growth of microwave systems. Previously reserved for military and aerospace applications, the growth in consumer microwave systems continues to force important innovations aimed at cost reductions and mass production.

Looking to the future, one might imagine a sort of Moore's law for spectrum driving carrier frequencies ever upward. However, one must consider that signal attenuation due to absorptive effects in the atmosphere (see Figure 1.16) start to become significant in dry air at tens of gigahertz. Below about 40–50 GHz, atmospheric absorption at sea level is typically below 1 dB/km, but heavy rainfall may exacerbate the loss considerably.[62] There are strong absorption peaks centered at around 22 GHz and 63 GHz (give or take a gigahertz here and there). The lower-frequency absorption peak is due to water, and the higher-frequency one is due to oxygen. The oxygen absorption peak contributes a path loss in excess of 20 dB/km, so it is quite significant. This attenuation, however, may be turned into an attribute if one wishes to permit re-use of spectrum over shorter distances. This property is exploited in various proposals for the deployment of picocells and other short-distance services at 60 GHz.

(on 3 April 1973, according to Cooper, as reported by Dan Gillmor in the 29 March 2003 *San Jose Mercury News*).

[61] This growth surprised almost everyone. In a famous (notorious?) study by McKinsey and Company commissioned by AT&T around 1982, the total U.S. market for cell phones was projected to saturate at 900,000 well-heeled subscribers by 2000. In fact, there were over 100 million U.S. subscribers in 2000, so the prediction was off by over 40 dB. Today, more than a million cell phones are sold worldwide each *day*, and the total number of subscribers exceeds one billion (double the installed base of PCs). Acting on the implications of the McKinsey study, AT&T sold its cellular business unit early on – only to pay $11.5 billion to re-enter the market in 1993–1994 when the magnitude of its error had finally become too large to ignore.

[62] These values are in addition to the Friis path loss.

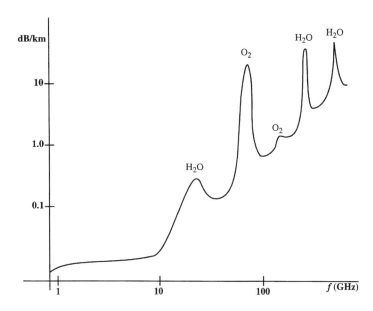

FIGURE 1.16. Approximate atmospheric attenuation vs. frequency at sea level, in dry air[63]

The large amount of spectrum offers high data rates, and the poor propagation is turned to advantage in forcing high frequency re-use ("it's not a bug, it's a feature").

More recently, the FCC has allocated spectrum between the two lowest oxygen absorption peaks for commercial broadband wireless applications. Dubbed "E-band," the spectrum spans 71–76 GHz, 81–86 GHz, and 92–95 GHz and is to be used for high-speed point-to-point wireless LANs and for broadband internet access. Again, the relatively high atmospheric absorption (perhaps 10–20 dB/km in heavy rain) is deemed an attribute to permit maximal spectral re-use while minimizing interference. Another advantage is the ease with which such short-wavelength emissions are formed into tight beams with antennas of compact size.

The development of microwave technology was driven for decades by military, aerospace, and radioastronomical applications. In the last twenty years, however, there has been a decided shift toward consumer microwave products. The days of handcrafted, low-volume, price-is-no-object microwave electronics has largely given way to the disposable cell phone and WLAN card of today. The annual sales volume of products operating in the classic microwave frequency range is staggering. In excess of 1.3 million cell phones are sold daily, and there seems to be no end to the demand for communication devices. Wireless LANs, pagers, satellite television services, GPS, ultrawideband (UWB) links, automobile anticollision radar, and RF ID tags constitute but a small subset of consumer microwave applications. And we must

---

[63] After *Millimeter Wave Propagation: Spectrum Management Implications,* Federal Communications Commission, Bulletin no. 70, July 1997.

not forget nuclear magnetic resonance (an outgrowth of post-WWII microwave research, abetted by the wide availability of postwar surplus radar gear), which begat magnetic resonance imaging. If history is any guide to the future, we can rest assured that we will be surprised by the ways in which microwave technology develops.

1.5 SOME CLOSING COMMENTS

From examining the history of mobile and portable wireless, it might appear that the band from approximately 500 MHz to about 5 GHz is overrepresented. The appearance of favoritism is not an artifact of selective reporting. That segment of spectrum *is* popular, and for excellent reasons.

First, let's consider what factors might constrain operation at low carrier frequencies. One is simply that there's less spectrum there. More significant, however, is that antennas cannot be too small (relative to a wavelength) if they are to operate efficiently. Efficient low-frequency antennas are thus long antennas. For mobile or portable applications, one must choose a frequency high enough that efficient antennas won't be *too* long. At 500 MHz, whose free-space wavelength is 60 cm, a quarter-wave antenna would be about 15 cm long. That value is readily accommodated in a handheld unit.

As frequency increases, we encounter a worsening path loss. One causal factor is the increasing tendency for reflection, refraction, and diffraction, but another can be anticipated from the Friis formula. Increasing the frequency tenfold, to 5 GHz, increases the Friis path loss factor by 20 dB. At these frequencies, interaction with biological tissues is nonnegligible, so simply increasing power by a factor of 100 to compensate is out of the question. Operation at higher frequency is accompanied by an ever-decreasing practical radius of communications.

Thus we see that there is an approximate decade span of frequency, ranging from 500 MHz to 5 GHz, that will *forever* remain the sweet spot for large-area mobile wireless. Unlike Moore's law, then, useful spectrum does not expand exponentially over time. In fact, it is essentially fixed. This truth explains why carriers went on a lunatic tear in the late 1990s, bidding hundreds of billions of dollars for 3G spectrum (only to find the debt so burdensome that many carriers have been forced to make "other arrangements"). No doubt, there will be ongoing efforts to maximize the utility of that finite spectrum – and also to reclaim spectrum from other services (e.g., UHF television) that arguably use the spectrum less efficiently.

1.6 APPENDIX A: CHARACTERISTICS OF OTHER WIRELESS SYSTEMS

It is impossible to list all the services and systems in use, but here we provide a brief sampling of a few others that may be of interest. (A detailed U.S. spectrum allocation chart may be downloaded for free from ⟨http://www.ntia.doc.gov/osmhome/allochrt.pdf⟩.) The first of these is the unlicensed ISM (industrial-scientific-medical) band;

Table 1.1. *ISM band allocations and summary*

Parameter	900 MHz	2.4 GHz	5.8 GHz
Frequency range	902–928 MHz	2400–2483.5 MHz	5725–5850 MHz
Total allocation	26 MHz	83.5 MHz	125 MHz
Maximum power	1 W	1 W	1 W
Maximum EIRP[a]	4 W	4 W (200 W for point-to-point)	200 W

[a] EIRP stands for "effective isotropically radiated power" and equals the product of power radiated and the antenna gain.

Table 1.2. *UNII band allocations and summary*

Parameter	Indoor	Low-power	UNII/ISM
Frequency range	5150–5250 MHz	5250–5350 MHz	5725–5825 MHz
Total allocation	100 MHz	100 MHz	100 MHz
Maximum power	50 mW	250 mW	1 W
Maximum EIRP	200 mW; unit must have integral antenna	1 W	200 W

Table 1.3. *Random sampling of some broadcast systems*

Service/system	Frequency span	Channel spacing
AM radio	535–1605 kHz	10 kHz
TV (ch. 2–4)	54–72 MHz	6 MHz
TV (ch. 5–6)	76–88 MHz	6 MHz
FM radio	88.1–108.1 MHz	200 kHz
TV (ch. 7–13)	174–216 MHz	6 MHz
TV (ch. 14–69)	470–806 MHz	6 MHz

see Table 1.1. Microwave ovens, transponders, RF ID tags, some cordless phones, WLANs, and a host of other applications and services use these bands. Notice that these bands reside within the "sweet spot" for mobile and portable wireless identified earlier.

Another unlicensed band has been allocated recently in the United States. The *unlicensed national information infrastructure* (UNII) band adds 200 MHz to the existing 5-GHz ISM band and also permits rather high EIRPs in one of the bands. See Table 1.2.

Because mobile and cellular systems are not the only uses for wireless, Table 1.3 gives a brief sampling of other (broadcast) wireless systems.

1.7 APPENDIX B: WHO REALLY INVENTED RADIO?

The question is intentionally provocative, and is really more of a Rorschach test than purely a matter of history (if there is such a thing). Frankly, it's an excuse simply to consider the contributions of some radio pioneers, rather than an earnest effort to offer a definite (and definitive) answer to the question.

First, there is the matter of what we mean by the words *radio* or *wireless*. If we apply the most literal and liberal meaning to the latter term, then we would have to include technologies such as signaling with smoke and by semaphore, inventions that considerably predate signaling with wires. You might argue that to broaden the definition that much is "obviously" foolish. But if we then restrict the definition to communication by the radiation of Hertzian waves, then we would have to exclude technologies that treat the atmosphere as simply a conductor. This collection of technologies includes contributions by many inventors who have ardent proponents. We could make even finer distinctions based on such criteria as commercialization, practicality, forms of modulation, and so forth. The lack of agreement as to what the words *radio* and *invention* mean is at the core of the controversy, and we will not presume to settle that matter.

One pioneer who has more than his fair share of enthusiastic supporters is dentist Mahlon Loomis, who patented in 1872 a method for wireless electrical communication.[64] In a configuration reminiscent of Benjamin Franklin's electrical experiments, Loomis proposed a system of kites to hold wires aloft. A sufficiently high voltage applied to these wires would allow electrical signals to be conducted through the atmosphere, where a receiver would detect induced currents using a galvanometer. Allegedly, experiments conducted by Loomis in his home state of West Virginia were successful, but there is no accepted primary evidence to support this claim, and calculations based on modern knowledge cast tremendous doubt in any case.[65] Supporters of Loomis have a more serious difficulty, for William Henry Ward had patented much the same idea (but one using more sophisticated apparatus) precisely three months earlier; see Figure 1.17.[66] Needless to say, reliably conducting enough DC current through the atmosphere to produce a measurable and unambiguous response in a galvanometer is basically hopeless, and neither Loomis nor Ward describes a workable system for wireless telegraphy.

---

[64] U.S. Patent #129,971, granted 30 July 1872. This one-page patent has no drawings of any kind. It may be said that one with "ordinary skill in the art" would not be able to practice the invention based on the information in the patent. Loomis was a certifiable crackpot whose writings on other subjects make for entertaining reading.

[65] Seemingly authoritative reports of successful tests abound (in typical accounts, senators from several states are present for the tests, in which communications between two mountaintops 22 km apart is allegedly demonstrated and then independently verified). However, I have never been able to locate information about these tests other than what Loomis himself provided. Others quoting these same results apparently have had no better success locating a primary source, but continue to repeat them without qualification.

[66] U.S. Patent #126,536, granted 30 April 1872.

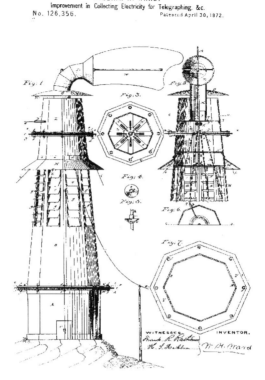

FIGURE 1.17. First page of Ward's patent

Then there's David Edward Hughes, who noticed that – as a result of a loose contact – his homemade telephone would respond to electrical disturbances generated by other apparatus some distance away. After some experimentation and refinement, he presented his findings on 20 February 1880 to a small committee headed by Mr. Spottiswoode, the president of the Royal Society. The demonstration included a portable wireless receiver, the first in history. One Professor Stokes declared that, although interesting, the phenomenon was nothing more than ordinary magnetic induction in action, not a verification of Maxwell's predictions. So strong a judgment from his esteemed colleagues caused Hughes to abandon further work on wireless.[67]

That same year, Alexander Graham Bell invented the *photophone*,[68] a device for optical wireless communication that exploited the recent discovery of selenium's photosensitivity.[69] Limited to daylight and line-of-sight operation, the photophone

---

[67] Hughes was sufficiently discouraged that he did not even publish his findings. The account given here is from Ellison Hawks, *Pioneers of Wireless* (Methuen, London, 1927), who in turn cites a published account given by Hughes in 1899.

[68] A. G. Bell and S. Tainter, U.S. Patent #235,496, granted 14 December 1880.

[69] This property of selenium also stimulated numerous patents for television around this time. Such was the enthusiasm for selenium's potential that *The Wireless & Electrical Cyclopedia* (Catalog no. 20 of the Electro Importing Company, New York, 1918) gushed: "Selenium will solve many

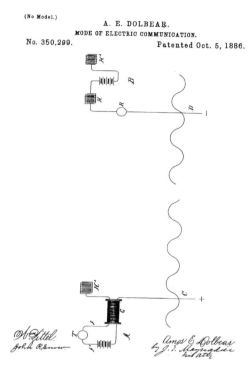

FIGURE 1.18. First page of Dolbear's patent

never saw commercial service, and it remains largely a footnote in the history of wireless. Bell himself thought it important enough that four of his eighteen patents are related to the photophone.

Wireless telegraphy based on atmospheric conduction continued to attract attention, though. Tufts University professor Amos E. Dolbear patented another one of these systems in 1886; see Figure 1.18.[70] This invention is notable chiefly for its explicit acknowledgment that the atmosphere is a shared medium. To guarantee fair access to this resource by multiple users, Dolbear proposed assigning specific time slots to each user. Thus, Dolbear's patent is the first to describe time-division multiple access (TDMA) for wireless communications. Marconi would later purchase the rights to this patent.

We must also not forget Heinrich Hertz. The apparatus he constructed for his researches of 1886–1888 is hardly distinguishable from that used by later wireless pioneers. His focus on the fundamental physics, coupled with his premature passing, is the reason others get the credit for the invention of wireless communication.

Like Hertz, Lodge did not initially focus on applications of wireless technology for communications. For example, his demonstration in 1894 at a meeting of the Royal

---

problems during this century. It is one of the most wonderful substances ever discovered." I suppose that's true, as long as you overlook its toxicity....

[70] U.S. Patent #350,299, granted 5 October 1886.

Institution in London (titled "The Work of Hertz" and which marked the public debut of the coherer in wireless technology) did not involve the transmission or reception of intentional messages.[71] Lodge himself later acknowledged that his initial lack of interest in wireless communications stemmed from two biases. One was that wired communications was an established and reliable technology; it was hard to imagine how wireless technology could ever achieve parity. The other bias was perhaps the result of knowing too much and too little at the same time. Having proven the identity of Hertzian waves and light, Lodge erroneously concluded that wireless would be constrained to line-of-sight communications, limiting the commercial potential of the technology. Lodge was hardly alone in these biases; most "experts" shared his views. Nonetheless, he continued to develop the technology, and he patented the use of tuned antennas and circuits for wireless communication (see Figure 1.19) years before the development of technology for generating continuous waves. He coined the term "syntony" to describe synchronously tuned circuits. As the reader may have noted, the term didn't catch on. Poor Lodge; almost nothing of what he did is remembered today.

Lodge published extensively, and his papers inspired Alexander Popov to undertake similar research in Russia.[72] Popov demonstrated his apparatus to his colleagues of the Russian Physical and Chemical Society on 7 May 1895, a date which is still celebrated in Russia as Radio Day although, like Lodge's a year earlier, his demonstration did not involve actual communication.

According to anecdotal accounts written down thirty years after the fact, Popov then demonstrated wireless telegraphy on 24 March 1896, with the transmission and reception of the message "Heinrich Hertz" achieved over a distance of approximately 250 meters. He followed this up with the first ship-to-shore communication one year later. Continuing refinements in his apparatus enabled the first wireless-assisted naval rescue in 1899–1900.[73]

Unlike Hughes, Dolbear, Hertz, Lodge, and Popov, who were all members of an academic elite, young Guglielmo Marconi was a member of a social elite. He began to work in earnest in December of 1894, shortly after reading Righi's obituary of Hertz, and had acquired enough knowledge and equipment by early 1895 to begin experiments in and around his family's villa (the Griffone). Ever mindful of commercial prospects for his technology, he applied for patents early on, receiving his first (British #12,039) on 2 June 1896.

---

[71] Aitken, *Syntony and Spark*.

[72] Also rendered as *Aleksandr Popoff* (and similar variants) elsewhere.

[73] The range of dates reflects one of the problems with establishing the facts surrounding Popov's contributions. Different sources with apparently equal credibility cite dates ranging from 1899 to 1901 for the rescue of the battleship *General-Admiral Apraksin* in the Gulf of Finland. And it is unfortunate that so significant an achievement as allegedly occurred on 24 March 1896 (still other sources give different dates, ranging over a two-week window) would have gone undocumented for three decades. See Charles Susskind, "Popov and the Beginnings of Radiotelegraphy," *Proc. IRE*, v. 50, October 1962.

1.7 APPENDIX B: WHO REALLY INVENTED RADIO?

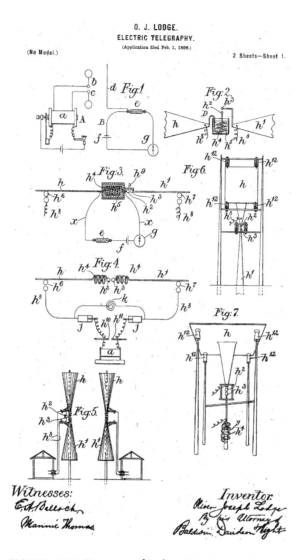

FIGURE 1.19. First page of Lodge, U.S. Patent #609,154 (filed 1 February 1898, granted 10 August 1898)

From the documented evidence, Marconi demonstrated true wireless communications before Popov – although initially to small groups of people without academic or professional affiliations. Neither Marconi nor Popov used apparatus that represented any particular advance beyond what Lodge had a year earlier. The chief difference was the important shift from simply demonstrating that a wireless effect could be transmitted to the conscious choice of using that wireless effect to communicate.

So, does the question of invention reduce to a choice between Marconi and Popov? Or between Marconi and Lodge? Lodge and Popov? What about Tesla?

Tesla?

Nikola Tesla's invention of the synchronous motor made AC power a practicality, and the electrification of the world with it. Tesla subsequently became obsessed with the idea of transmitting industrially significant amounts of power wirelessly. Based on his experience with gases at low pressure, he knew they were readily ionized and thus rendered highly conductive (this behavior is the basis for neon and fluorescent lights). Just as Loomis and Ward had before him, Tesla decided to use the atmosphere as a conductor. Deducing that the upper atmosphere, being necessarily of low pressure, must also be highly conductive, Tesla worked to develop the sources of exceptionally high voltage necessary to produce a conductive path between ground level and the conductive upper atmosphere. Tesla estimated that he would need tens of megavolts or more to achieve his goals.[74] Ordinary step-up transformers for AC could not practically produce these high voltages. The famous Tesla coil (a staple of high school science fairs for a century, now) resulted from his efforts to build practical megavolt sources. Based on his deep understanding of resonant phenomena, the Tesla coil uses the significant voltage boosts that tuned circuits can provide.

Tesla's first patent in this series (see Figure 1.20) is U.S. #645,576, filed 9 September 1897 and granted 20 March 1900. It specifically talks about the conduction of electrical energy through the atmosphere, but not about the transmission of intelligence.[75]

This patent is among several cited in a famous 1943 U.S. Supreme Court decision (*320 US 1*, argued April 9–12 and decided on June 21) that is frequently offered as establishing that Tesla was the inventor of radio. The background for this case is that the Marconi Wireless Telegraph Corporation of America had asserted some of its wireless patents against the United States government shortly after the First World War, seeking damages for infringement. The decision says very clearly that *Marconi's patent for the four-resonator system* is invalid because of prior art. Of three other patents also asserted against the U.S., one was held not to be infringed, another to be invalid, and a third to be both valid and infringed, resulting in a judgment against the U.S. government in the trivial sum of approximately $43,000. The 1943 decision put that narrow matter to rest by citing prior inventions by Lodge, Tesla, and one John Stone Stone in invalidating the four-circuit patent (which had begun life as British patent #7,777). The decision thus certainly declares that Marconi is not the inventor of this *circuit,* but it does not quite say that Marconi didn't invent *radio.* It does note that the four-resonator system enabled the first practical spark-based wireless communications (the four-resonator system is largely irrelevant for continuous

---

[74] Later, he would begin construction of a huge tower on Long Island, New York, for transmitting power wirelessly. Designed by renowned Gilded Age architect Stanford White (whose murder was chronicled in *Ragtime*), the Wardenclyffe tower was to feature an impressive array of ultraviolet lamps, apparently to help create a more conductive path by UV ionization. Owing to lack of funds, it was never completed. Parts were eventually sold for scrap, and the rest of the structure was demolished.

[75] His later patents do discuss transmission of intelligence, but his claims *specifically exclude* the use of Hertzian waves. He was completely obsessed with using the earth as one conductor – and the atmosphere as the other – for the transmission of power.

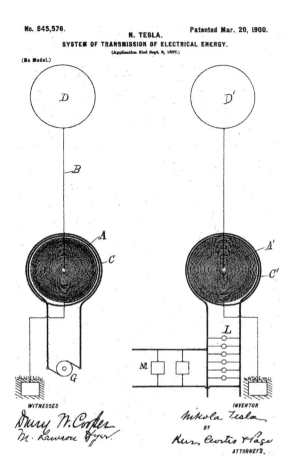

FIGURE 1.20. Tesla's first wireless patent?

wave systems), but the Court does not then make the leap that either Lodge, Tesla, or Stone was therefore the inventor of radio.[76] The oft-cited decision thus actually makes no affirmative statements about inventorship, only negative ones. One might speculate that the Court consciously avoided making broader statements precisely because it recognized the ambiguity of the words *invent* and *radio*.

What we can say for certain is that, of these early pioneers, Marconi was the first to believe in wireless communications as more than an intellectual exercise. He certainly did not innovate much in the circuit domain (and more than occasionally, shall we say, *adapted* the inventions of others), but his vision and determination to make wireless communications a significant business were rewarded, for he quickly made the critically important discovery that wireless is not necessarily limited to line-of-sight communication, proving the experts wrong. Marconi almost single-handedly made wireless an important technology by believing it could be an important business.

---

[76] The reader is invited to verify the author's assertion independently. The entire case is available on-line at ⟨http://www.uscaselaw.com/us/320/1.html⟩.

So, who invented radio? As we've said, it depends on your definition of *invention* and *radio*. If you mean the first to conceive of using some electrical thing to communicate wirelessly, then Ward would be a contender. If you mean the first to build the basic technical apparatus of wireless using waves, then Hertz is as deserving as anyone else (and since light is an electromagnetic wave, we'd have to include Bell and his photophone). If you mean the first to use Hertzian waves to send a message intentionally, either Popov or Marconi is a credible choice (then again, there's Bell, who used the photophone explicitly for communication from the very beginning). If you mean the first to appreciate the value of tuning for wireless, then Lodge and perhaps Tesla are inventors, with Lodge arguably the stronger candidate.

Given the array of deserving choices, it's not surprising that advocacy of one person or another often has nationalistic or other emotional underpinnings, rather than purely technical bases. Situations like this led President John F. Kennedy to observe that "success has many fathers." Wireless certainly has been a huge success, so it's not surprising to encounter so many claimants to fatherhood.

CHAPTER TWO

INTRODUCTION TO RF AND MICROWAVE CIRCUITS

2.1 DEFINITIONS

The title of this chapter should raise a question or two: Precisely what is the definition of *RF*? Of *microwave*? We use these terms in the preceding chapter, but purposely without offering a quantitative definition. Some texts use absolute frequency as a discriminator (e.g., "microwave is anything above 1 GHz"). However, the meaning of those words has changed over time, suggesting that distinctions based on absolute frequency lack fundamental weight. Indeed, in terms of engineering practice and design intuition, it is far more valuable to base a classification on a comparison of the physical dimensions of a circuit element with the wavelengths of signals propagating through it.

When the circuit's physical dimensions are very small compared to the wavelengths of interest, we have the realm of ordinary circuit theory, as we will shortly understand. We will call this the *quasistatic, lumped,* or *low-frequency* realm, regardless of the actual frequency value. The size inequality simplifies Maxwell's equations considerably, allowing one to invoke the familiar concepts of inductances, capacitances, and Kirchhoff's "laws" of current and voltage.

If, on the other hand, the physical dimensions are very large compared to the wavelengths of interest, then we say that the system operates in the classical optical regime – whether or not the signals of interest correspond to visible light. Devices used to manipulate the energy are now structures such as mirrors, polarizers, lenses, and diffraction gratings. Just as in the quasistatic realm, the size inequality enables considerable simplifications in Maxwell's equations.

If the circuit's physical dimensions are roughly comparable to the wavelengths of interest, then we have what we will term the microwave or distributed realm. In this intermediate regime, we generally cannot truncate Maxwell's equations much (if at all), complicating the acquisition of design insight.[1] Here, we might be able

---

[1] Optical fibers have cross-sectional dimensions comparable to a wavelength, and thus their analysis falls in this regime rather than in conventional optics – this is the reason for our distinction between classical optics and this intermediate realm.

Table 2.1. *Radio frequency band designations*

Band	Frequency range	Wavelength range
Extremely low frequency (ELF)	<30 Hz	>10,000 km
Super low frequency (SLF)	30 Hz to 300 Hz	10,000 km to 1000 km
Ultra low frequency (ULF)	300 Hz to 3 kHz	1000 km to 100 km
Very low frequency (VLF)	3 kHz to 30 kHz	100 km to 10 km
Low frequency (LF)	30 kHz to 300 kHz	10 km to 1 km
Medium frequency (MF)	300 kHz to 3 MHz	1 km to 100 m
High frequency (HF)	3 MHz to 30 MHz	100 m to 10 m
Very high frequency (VHF)	30 MHz to 300 MHz	10 m to 1 m
Ultra high frequency (UHF)	300 MHz to 3 GHz	1 m to 10 cm
Super high frequency (SHF)	3 GHz to 30 GHz	10 cm to 1 cm
Extremely high frequency (EHF)	30 GHz to 300 GHz	1 cm to 1 mm
Ludicrously high frequency (LHF)	>300 GHz	<1 mm

to discuss inductances and capacitances at the same time we speak of reflections, for example. It may also be the case that it is inappropriate or even impossible to identify individual inductances and capacitances, because energy may be stored in both electric and magnetic fields that share the same region of space. In the microwave regime we need to accommodate *transmission line* behavior, and radiation of energy (whether wanted or unwanted) potentially becomes significant. The primary focus of this book is on bridging the lumped and distributed realms.

The techniques and intuitions that guide the design of systems are thus best classified according to these normalized wavelength regimes. Consequently there may be considerable overlap in frequency (one can make perfectly respectable Fresnel lenses that operate in the gigahertz frequency range, for example, if you are willing to build relatively large structures), which explains why a classification system based on arbitrary frequency limits is not as useful as we would wish.

As a final comment on this subject, we will generally ignore the optical regime in this textbook. However, the reader should not infer from this neglect that optical techniques are not useful or relevant for what is conventionally called RF circuit design. It is simply that this book has to terminate somewhere.

2.2 CONVENTIONAL FREQUENCY BANDS

In seeming contradiction with the foregoing argument that arbitrary frequency boundaries have a weak physical justification, in Table 2.1 we now list frequency bands and their common (but by no means universal) designations. Not all sources agree on the precise frequency limits of these bands (particularly for the bands below VLF and above EHF), so it's best to supplement these band designations with actual frequency values.

In relating wavelength to frequency, just remember that the product of frequency (in hertz) and wavelength (in meters) is the speed of light (very nearly 3×10^8 m/s).

Table 2.2. *Microwave band designations (IEEE 521-1984)*

Band	Frequency range
L	1.0 GHz to 2.0 GHz
S	2.0 GHz to 4.0 GHz
C	4.0 GHz to 8.0 GHz
X	8 GHz to 12 GHz
Ku	12 GHz to 18 GHz
K	18 GHz to 27 GHz
Ka	27 GHz to 40 GHz
V	40 GHz to 75 GHz
W	75 GHz to 110 GHz

So, a 1-MHz signal has a free-space wavelength of almost exactly 300 m; a 1-GHz signal has a 300-mm wavelength. It's useful to note that the frequency in gigahertz, multiplied by the wavelength in millimeters, is about 300.

An alternative classification system has its origins in radar work during World War II. Based on letters originally chosen at random to confuse the enemy, a lack of standardization has succeeded in confusing just about everyone.[2] The frequency ranges associated with the letters have changed somewhat over time, and they vary from country to country (and even within a country) and also from company to company. For these reasons, the letter-based designations are perhaps best considered obsolete and should be avoided (or, at least, supplemented with actual frequency values, as is the case with the previous designations). Nevertheless, they are still used, so we offer here a table of such bands (Table 2.2) as documented in IEEE 521-1984, the only international standard of which the author is aware. The designations Ku and Ka arose from "under K" and "above K," respectively.

Other systems of letter designations you may encounter are the waveguide bands and those due to organizations and companies such as NASA, Hewlett-Packard, Sperry, Motorola, Narda, Raytheon, and others. These designations are all a bit different, and they may include bands designated by additional letters and omit others. Bear this in mind as you survey the literature.

Ordinary AM radio signals are in the MF band, and FM radio operates in the VHF band – as do the lower television channels (2–13, in three noncontiguous bands that straddle FM radio). Television channels 14–69 use the UHF part of the spectrum, along with all current cellular telephone systems, the Global Positioning System (GPS, at 1.575 GHz), microwave ovens (2.45 GHz) and most cordless telephones (at 900 MHz and 2.4 GHz). Police radar has operated at progressively higher frequencies

---

[2] Wartime secrecy concerns were so great that the various groups in the Allied radar design community didn't even standardize nomenclature among themselves. We are still living with the resulting legacy of confusion, and probably always will.

over time, starting with 10-GHz X-band systems and moving to K- and then Ka-band (and even to laser-based optical systems) in order to stay ahead of improved radar detector technology available to consumers.

The seemingly high occupation rate of the UHF band in the foregoing list is not an artifact of selective reporting. In sharp contrast to the world of microprocessors, where Moore's law regularly delivers exponentially greater computing resources, the UHF frequency band is a "sweet spot" for kilometer-scale terrestrial wireless and shall remain so forever. As we shall see in Chapter 21, efficient coupling to a potential radiator generally requires antennas not very much shorter than (i.e., at least 1/5 to 1/10 of) a wavelength. Because wireless also often implies mobility, one frequently must consider antenna lengths compatible with portable form factors. If we arbitrarily set 10 cm as the upper limit of tolerable antenna length, then our requirement of efficient radiation forces us to a wavelength of no longer than roughly 0.5–1.0 m (or a frequency no lower than about 300–600 MHz; correspondingly, some early cellular telephone systems operated at 450 MHz).

As frequency increases, it becomes progressively easier to build efficient antennas. However, propagation becomes rapidly worse as radio waves take on more and more characteristics of light waves, experiencing greater attenuation and diffraction. Furthermore, biological effects become more prominent, limiting the amount of radiated power one might use to overcome path loss (remember: a typical microwave oven operates at 2.45 GHz). As a consequence, it's difficult to deploy a wide-area terrestrial network much above a few gigahertz. The 5-GHz band used by recently developed wireless data networks (e.g., IEEE 802.11a) probably represents a practical upper limit, within an octave or so. The finiteness of the useful available spectrum (roughly spanning the decade from 500 MHz to 5 GHz) accounts not only for the huge prices paid in spectrum auctions but also for the focus of this textbook: the design, construction, and measurement of discrete RF circuits up to the (fuzzy) upper limit of practical terrestrial wireless communications bands.

It would be wrong, however, to leave the reader with the impression that circuits and systems not within the scope of this book are somehow unworthy of consideration. For applications that do not need to operate over a large geographical area, or for which the increasing atmospheric absorption is actually an attribute, it is perhaps useful to know that water vapor absorbs strongly at around 22 GHz, and oxygen at around 63 GHz (see Figure 2.1). These absorption peaks are often exploited in covert communications, particularly between satellites, so that any ground-based eavesdroppers would detect little but noise. Similarly, automobile anticollision radar systems operate at frequencies high enough (e.g., 77 GHz) to provide the spatial resolutions desired while exploiting the high free-space attenuation to reduce interference among cars. Proposals for "piconets" operating in similar bands (e.g., 60 GHz) also seek to exploit the low spatial cross-talk of signals in these frequency ranges. The ability to use arrays of antennas to produce narrow beams of high-power density helps to offset the path loss at these higher frequencies. Explicit recognition of the potential of

2.3 LUMPED VERSUS DISTRIBUTED CIRCUITS

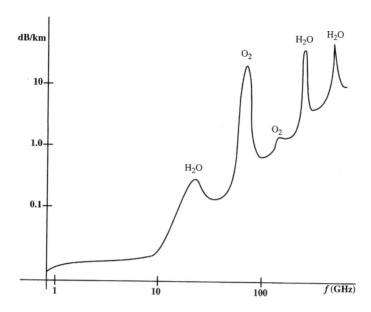

FIGURE 2.1. Approximate atmospheric attenuation vs. frequency at sea level, in dry air[3]

these bands for useful communications has been conferred recently with the designation of "E-band" spectrum by the FCC in the V- and W-band frequency range.

With that material as a background, we now turn to the main focus of this book, beginning with a more detailed examination of what is meant by lumped and distributed circuits.

2.3 LUMPED VERSUS DISTRIBUTED CIRCUITS

As we have argued, there are two important regimes of operating frequency, distinguished by whether one may treat circuit elements as lumped or distributed. The fuzzy boundary between these two regimes concerns the ratio of the physical dimensions of the circuit relative to the shortest wavelength of interest. At high enough frequencies, the size of the circuit elements becomes comparable to the wavelengths and so one cannot employ with impunity intuition derived from lumped circuit theory. Wires must then be treated as the transmission lines that they truly are, Kirchhoff's "laws" no longer hold generally, and identification of R, L, and C ceases to be obvious (or even possible).

In order to draw a proper boundary between lumped and distributed domains, we need to revisit (briefly) Maxwell's equations.

---

[3] After *Millimeter Wave Propagation: Spectrum Management Implications,* Federal Communications Commission, Bulletin no. 70, July 1997.

MAXWELL AND KIRCHHOFF

Many students (and a surprising number of practicing engineers, unfortunately) forget that Kirchhoff's voltage and current "laws" are *approximations* that hold only in the lumped regime (to which we have frequently alluded, but not yet defined). They are derivable from Maxwell's equations if we assume quasistatic behavior, thereby eliminating the coupling terms that give rise to the wave equation. To understand what all this means, let's review Maxwell's equations (for free space) in differential form:

$$\nabla \cdot \mu_0 \mathbf{H} = 0, \tag{1}$$

$$\nabla \cdot \varepsilon_0 \mathbf{E} = \rho, \tag{2}$$

$$\nabla \times \mathbf{H} = \mathbf{J} + \varepsilon_0 \frac{\partial \mathbf{E}}{\partial t}, \tag{3}$$

$$\nabla \times \mathbf{E} = -\mu_0 \frac{\partial \mathbf{H}}{\partial t}. \tag{4}$$

The first equation says that there is no net magnetic charge (i.e., there are no magnetic monopoles). If there were net magnetic charge, it would cause divergence in the magnetic field. We won't be using that equation at all.

The second equation (Gauss's law) acknowledges that there can be net electric charge and says that electric charge is the source of divergence of the electric field. We won't really use that equation either.

The third equation (Ampere's law, with Maxwell's famous modification) says that both "ordinary" current *and* a time-varying electric field produce the same effect: a magnetic field. The term that involves the derivative of the electric field is the famous displacement (capacitive) current term that Maxwell pulled out of thin air to produce the wave equation.

Finally, the fourth equation (Faraday's law) says that a changing magnetic field gives rise to (curl in) the electric field.

Wave behavior arises fundamentally because of the coupling terms in the last two equations: A change in **E** causes a change in **H**, which causes a change in **E**, and so on. If we were to set either μ_0 or ε_0 to zero, the coupling terms would disappear and no wave equation would result; circuit analysis could then proceed on a quasistatic (or even static) basis.

As a specific example, setting μ_0 to zero makes the electric field curl-free, allowing **E** to be expressed as the gradient of a potential (within a minus sign here or there). It then follows identically that the line integral of the **E**-field (which is the voltage) around any closed path is zero:

$$V = \oint \mathbf{E} \cdot dl = \oint (-\nabla \phi) \cdot dl = 0. \tag{5}$$

This last equation is merely the field-theoretical expression of Kirchhoff's voltage law (KVL).

To derive KCL (Kirchhoff's current law), we proceed in the same manner but now set ε_0 equal to zero. Then, the curl of **H** depends only on the current density **J**, allowing us to write:

$$\nabla \cdot \mathbf{J} = \nabla \cdot (\nabla \times \mathbf{H}) = 0. \tag{6}$$

That is, the divergence of **J** is identically zero. No divergence means no net current buildup (or loss) at a node.

Of course, neither μ_0 nor ε_0 is actually zero. To show that the foregoing is not hopelessly irrelevant as a consequence, recall that the speed of light can be expressed as[4]

$$c = 1/\sqrt{\mu_0 \varepsilon_0}. \tag{7}$$

Setting μ_0 or ε_0 to zero is therefore equivalent to setting the speed of light to infinity. Hence, KCL and KVL are actually the result of assuming infinitely fast propagation; we thus expect them to hold reasonably well as long as the physical dimensions of the circuit elements are small compared with a wavelength, so that the finiteness of the speed of light is not noticeable:

$$l \ll \lambda, \tag{8}$$

where l is the length of a circuit element and λ is the shortest wavelength of interest.

To develop a feel for what this constraint means numerically, consider a circuit element whose longest dimension is 1 cm. If we arbitrarily say that "much less than" means "at least about a factor of 10 smaller than," then such an element can be treated as lumped if the highest frequency signal has a wavelength greater than roughly 10 cm. In free space, this wavelength corresponds to a frequency of about 3 GHz. On typical circuit board materials, the frequency may be lower by about a factor of 2 or more. The calculation reveals a simple truth: People-sized objects can't be treated as lumped circuit elements in the multi-GHz frequency range.

Summarizing, the boundary between lumped circuit theory (where KVL and KCL hold, and where one can identify R, L, and C) and distributed systems (where KVL/KCL don't hold and where R, L, C can't always be localized) depends on the *size of the circuit element relative to the shortest wavelength of interest.* If the circuit element (and interconnect is certainly a circuit element in this context) is very short compared with a wavelength, we can use traditional lumped concepts and incur little error. If not, then use of lumped ideas is inappropriate. In this textbook, we will consider circuits whose dimensions can be comparable to a wavelength. Distributed effects must therefore be taken into account.

---

[4] Applying the "duck test" version of Occam's razor ("if it walks like a duck and quacks like a duck, it must be a duck"), Maxwell showed that light and electromagnetic waves are the same thing. After all, if it travels at the speed of light and reflects like light, it must be light. Most would agree that the derivation of Maxwell's equations represents the crowning intellectual achievement of the 19th century.

2.4 LINK BETWEEN LUMPED AND DISTRIBUTED REGIMES

We now turn to the problem of extending into the distributed regime design intuition developed in the lumped regime. The motivation is more than merely pedagogical for, as we shall see, extremely valuable design insights emerge from this exercise. One example is that *delay,* instead of gain, may be traded for bandwidth in amplifiers.

Interconnect is an example of a system that may be treated successfully at lower frequencies as a simple *RC* line, for instance. With that type of mindset, reduction of *RC* "parasitics" in order to increase bandwidth becomes a major preoccupation of the circuit and system designer (particularly of the IC designer). Unfortunately, reduction of parasitics below some minimum amount is practically impossible. Intuition from lumped circuit design would therefore (mis)lead us into thinking that the bandwidth is limited by these irreducible parasitics. Fortunately, a proper treatment of interconnect as a transmission line, rather than as a finite lumped *RC* network, reveals otherwise. We find that we may still convey signals with exceedingly large bandwidth as long as we acknowledge (indeed exploit) the true, distributed nature of the interconnect. By using, rather than fighting, the distributed capacitance (and inductance), we may therefore effect a decoupling of delay from bandwidth. This new insight is extremely valuable, and it applies to active as well as passive networks. Aside from giving us an appreciation for transmission line phenomena, such an understanding will lead us to amplifier topologies that relax significantly the gain–bandwidth trade-off that characterizes lumped systems of low order.

Given these important reasons, we now undertake a study of distributed systems by extension of lumped circuit analysis.

2.5 DRIVING-POINT IMPEDANCE OF ITERATED STRUCTURES

We begin by studying the driving-point impedance of uniform, iterated structures. It's important to note that certain nonuniform structures (e.g., exponentially tapered transmission lines[5]) have exceedingly useful properties, but we'll limit our present discussion to a consideration of uniform structures only.

Specifically, consider the infinite ladder network shown in Figure 2.2. Even though resistor symbols are used here, they represent arbitrary impedances.

To find the driving-point impedance of this network without summing an infinite series, note that the impedance to the right of node C is the same as that to the right of B and the same as that to the right of A.[6] Therefore, we may write

---

[5] For those of you who are curious, the exponentially tapered line allows one to achieve a broadband impedance match instead of the narrowband impedance match that a quarter-wave transformer provides. The transformation ratio can be controlled by choice of taper constants.

[6] This is an extremely useful technique for analyzing such structures, but a surprisingly large percentage of engineers have never heard of it or perhaps don't remember it. In any event, it certainly

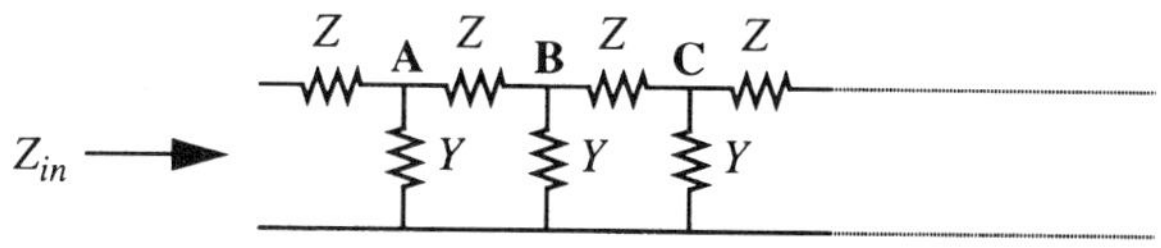

2.5 DRIVING-POINT IMPEDANCE OF ITERATED STRUCTURES

FIGURE 2.2. Ladder network

$$Z_{in} = Z + [(1/Y) \parallel Z_{in}], \tag{9}$$

which expands to

$$Z_{in} = Z + \frac{Z_{in}/Y}{1/Y + Z_{in}} \implies (Z_{in} - Z)\left(\frac{1}{Y} + Z_{in}\right) = \frac{Z_{in}}{Y}. \tag{10}$$

Solving for Z_{in} then yields

$$Z_{in} = \frac{Z \pm \sqrt{Z^2 + 4(Z/Y)}}{2} = \frac{Z}{2}\left[1 \pm \sqrt{1 + \frac{4}{ZY}}\right]. \tag{11}$$

In the special case where $Z = 1/Y = R$,

$$Z_{in} = \left(\frac{1 + \sqrt{5}}{2}\right) R \approx 1.618R. \tag{12}$$

This ratio of Z_{in} to R (or its reciprocal) is known as the *golden ratio* (or *golden mean* or *section*), and it shows up in contexts as diverse as the aesthetics of Greek geometers, Renaissance art and architecture, and solutions to several interesting (but largely useless) network theory problems.

IDEAL TRANSMISSION LINE AS INFINITE LADDER NETWORK

Let's now consider the more general case of the input impedance in the limit, where $|ZY| \ll 1$ and where we continue to disallow negative values of Z_{in}. In that case, we can simplify the result to

$$Z_{in} \approx \sqrt{Z/Y}. \tag{13}$$

We see that if Z/Y happens to be frequency-independent, then the input impedance will also be frequency-independent.[7] One important example of a network of this type is the model for an ideal transmission line. In the case of a lossless line, $Z = sL$ and $Y = sC$, where L and C represent differential (in the mathematical sense) circuit elements. The input impedance (called the *characteristic* impedance, Z_0) for an ideal, lossless, infinite transmission line is therefore

---

saves a tremendous amount of labor over a more straightforward approach, which would require summing various infinite series.

[7] Ladder networks with this property are called "constant-k" lines, since $Z/Y = k^2$ for a constant k.

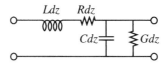

FIGURE 2.3. Lumped *RLC* model of infinitesimal transmission line segment

$$Z_{in} \approx \sqrt{Z/Y} = \sqrt{sL/sC} = \sqrt{L/C}. \qquad (14)$$

Because Y, the admittance of an infinitesimal capacitance, approaches zero as the length of the differential element approaches zero, while the reactance of the differential inductance element approaches zero at the same time, the ratio $1/YZ$ approaches infinity and so satisfies the inequality necessary to validate our derivation. The result – that we are left with a purely real input impedance for an infinitely long transmission line – should be a familiar one, but perhaps this particular path to it might not be.

An often-asked question concerns the fate of the energy we launch down a transmission line. If the impedance is purely real, then the line should behave as a resistor and should dissipate energy like a resistor. But the line is composed of purely reactive (and thus dissipationless) elements, so there would appear to be a paradox.

The resolution is that the energy doesn't end up as heat if the line is truly infinite. The energy just keeps traveling down the line forever, and so is lost to the external world just as if it had heated up a resistor and its environs; the line acts like a black hole for energy.

2.6 TRANSMISSION LINES IN MORE DETAIL

The previous section examined the impedance behavior of a lossless infinite line. We now extend our derivation of the characteristic impedance Z_0 to accommodate loss. We also introduce an additional descriptive parameter, the *propagation constant* γ.

2.6.1 LUMPED MODEL FOR LOSSY TRANSMISSION LINE

To derive the relevant parameters of a lossy line, consider an infinitesimally short piece of line, of length dz, as shown in Figure 2.3. Here, the elements L, R, C, and G are all quantities *per unit length* and simply represent a specific example of the more general case considered earlier.

The inductance accounts for the energy stored in the magnetic field around the line, while the series resistance accounts for the inevitable energy loss (such as due to skin effect; see Section 6.2) that all ordinary conductors exhibit. The shunt capacitance models the energy stored in the electric field surrounding the line, and the shunt conductance accounts for the loss due to such mechanisms as ordinary ohmic leakage and loss in the line's dielectric material.

2.6.2 CHARACTERISTIC IMPEDANCE OF A LOSSY TRANSMISSION LINE

To compute the impedance of a lossy line, we follow a method exactly analogous to that in Section 2.5:

$$Z_0 = Z\,dz + \left(\frac{1}{Y\,dz} \parallel Z_0\right) = Z\,dz + \frac{Z_0}{1 + (Y\,dz)Z_0}. \quad (15)$$

Since we will consider the limiting behavior of this expression as dz approaches zero, we may use the first-order binomial expansion of $1/(1+x)$:

$$Z_0 = Z\,dz + \frac{Z_0}{1 + (Y\,dz)Z_0} \approx Z\,dz + Z_0[1 - (Y\,dz)Z_0]$$
$$= Z_0 + dz(Z - YZ_0^2). \quad (16)$$

Cancelling Z_0 from both sides, we see that the final term in parentheses must equal zero. Hence, the characteristic impedance is

$$Z_0 = \sqrt{\frac{Z}{Y}} = \sqrt{\frac{R + j\omega L}{G + j\omega C}}. \quad (17)$$

If the resistive terms are negligible (or if RC just happens to equal GL), then the equation for Z_0 collapses to the result we derived earlier:

$$Z_0 = \sqrt{L/C}. \quad (18)$$

Because the impedance approaches $\sqrt{L/C}$ at sufficiently high frequency, independent of R or G, it is sometimes known as the *transient, surge,* or *pulse* impedance. It is this impedance that is used in the ratings of coaxial cables. The impedance of such cables at DC will be quite different in general, according to Eqn. 17, explaining why "50-Ω" cable doesn't measure anywhere near that value on an ohmmeter.

2.6.3 THE PROPAGATION CONSTANT

In addition to the characteristic impedance, one other important descriptive parameter is the *propagation constant,* usually denoted by γ. Whereas the characteristic impedance tells us the ratio of voltage to current *at* any one point on an infinitely long line, the propagation constant enables us to say something about the ratio of voltages (or currents) *between* any two points on such a line. That is, it quantifies the line's attenuation properties.

Consider the voltages at the two ports of a given subsection. The ratio of these voltages is readily computed from the ordinary voltage divider relationship:

$$V_{n+1} = V_n \left\{ \frac{Z_0 \parallel \frac{1}{Y\,dz}}{Z\,dz + \left[Z_0 \parallel \frac{1}{Y\,dz}\right]} \right\}. \quad (19)$$

Thus,
$$\frac{V_{n+1}}{V_n} = \frac{Z_0 \parallel \frac{1}{Y\,dz}}{Z\,dz + [Z_0 \parallel \frac{1}{Y\,dz}]} = \frac{Z_0}{Z_0 Z Y (dz)^2 + Z_0 + Z\,dz}. \tag{20}$$

Because we will use this expression in the limit of very small dz, we may discard the term that is proportional to $(dz)^2$ and again use the binomial expansion of $1/(1+x)$ to preserve only the first-order dependence on dz (remember, we're engineers – the whole universe is first-order to us!). This yields

$$\frac{V_{n+1}}{V_n} \approx \frac{Z_0}{Z_0 + Z\,dz} = \frac{1}{1 + \frac{Z}{Z_0}dz} \approx 1 - \frac{Z}{Z_0}dz = 1 - \sqrt{ZY}\,dz. \tag{21}$$

Despite our glibness, the net error in these approximations actually does converge to zero in the limit of zero dz.

Let us rewrite the previous equation as a difference equation:

$$V_{n+1} = V_n\left(1 - \sqrt{ZY}\,dz\right) \implies \frac{V_{n+1} - V_n}{dz} = -\sqrt{ZY}\,V_n. \tag{22}$$

In the limit of zero dz, the difference equation becomes a differential equation:

$$\frac{dV}{dz} = -\sqrt{ZY}\,V. \tag{23}$$

The solution to this first-order differential equation should be familiar:

$$V(z) = V_0 e^{-\sqrt{ZY}\,z}. \tag{24}$$

That is, the voltage at any position z is simply the voltage V_0 (the voltage at $z = 0$) multiplied by an exponential factor. The exponent is conventionally written as $-\gamma z$ so that, at last,

$$\gamma = \sqrt{ZY} = \sqrt{(R + j\omega L)(G + j\omega C)}. \tag{25}$$

To develop a better feel for the significance of the propagation constant, first note that γ will be complex in general. Hence, we may express γ explicitly as the sum of real and imaginary parts:

$$\gamma = \sqrt{(R + j\omega L)(G + j\omega C)} = \alpha + j\beta. \tag{26}$$

Then

$$V(z) = V_0 e^{-\gamma z} = V_0 e^{-(\alpha + j\beta)z} = V_0 e^{-\alpha z} e^{-j\beta z}. \tag{27}$$

The first exponential term gets smaller as distance increases; it represents the pure attenuation of the line. The second exponential factor has a unit magnitude and contributes only phase.

2.6.4 RELATIONSHIP OF γ TO LINE PARAMETERS

To relate the constants α and β explicitly to transmission line parameters, we make use of a couple of identities. First, recall that we may express a complex number in both exponential (polar) and rectangular form as follows:

$$Me^{j\phi} = M\cos\phi + jM\sin\phi. \tag{28}$$

Here, M is the magnitude of the complex number and ϕ is its phase. The polar form allows us to compute the square root of a complex number with ease (thanks to Euler):

$$\sqrt{Me^{j\phi}} = \sqrt{M}e^{j\phi/2} = \sqrt{M}\cos(\phi/2) + j\sqrt{M}\sin(\phi/2). \tag{29}$$

The last factoid we need to recall from undergraduate math is a pair of half-angle identities:

$$\cos(\phi/2) = \sqrt{\tfrac{1}{2}(1+\cos\phi)}; \tag{30}$$

$$\sin(\phi/2) = \sqrt{\tfrac{1}{2}(1-\cos\phi)}. \tag{31}$$

Now, γ is the square root of a complex number:

$$\begin{aligned}\gamma = \sqrt{ZY} &= \sqrt{(R+j\omega L)(G+j\omega C)} \\ &= \sqrt{(RG - \omega^2 LC) + j\omega(LG + RC)}.\end{aligned} \tag{32}$$

Making use of our identities and turning the crank a few revolutions, we obtain:

$$\alpha = \sqrt{\tfrac{1}{2}\left[\sqrt{\omega^4(LC)^2 + \omega^2[(LG)^2 + (RC)^2] + (RG)^2} + (RG - \omega^2 LC)\right]}; \tag{33}$$

$$\beta = \sqrt{\tfrac{1}{2}\left[\sqrt{\omega^4(LC)^2 + \omega^2[(LG)^2 + (RC)^2] + (RG)^2} - (RG - \omega^2 LC)\right]}. \tag{34}$$

These last two expressions may appear cumbersome, but that's only because they are. We may simplify them considerably if the product RG is small compared with the other terms. In such a case, the attenuation constant may be written as

$$\alpha \approx \sqrt{\tfrac{1}{2}\left[\sqrt{\omega^4(LC)^2 + \omega^2[(LG)^2 + (RC)^2]} - \omega^2 LC\right]}, \tag{35}$$

which, after a certain amount of bloodletting, further simplifies to

$$\alpha \approx \frac{R}{2}\sqrt{\frac{C}{L}} + \frac{G}{2}\sqrt{\frac{L}{C}}. \tag{36}$$

This, in turn, may be further approximated by

$$\alpha \approx \frac{R}{2}\sqrt{\frac{C}{L}} + \frac{G}{2}\sqrt{\frac{L}{C}} \approx \frac{R}{2Z_0} + \frac{GZ_0}{2}. \tag{37}$$

Thus, the attenuation per length will be small if the resistance per length is small compared with Z_0 and if the conductance per length is small compared with Y_0.

Turning our attention now to the equation for β, we have

$$\beta = \text{Im}[\gamma] \approx \omega\sqrt{LC}. \tag{38}$$

In the limit of zero loss (both G and $R = 0$), these expressions simplify to:

$$\alpha = \text{Re}[\gamma] = 0; \tag{39}$$

$$\beta = \text{Im}[\gamma] = \omega\sqrt{LC}. \tag{40}$$

Hence, a lossless line doesn't attenuate (no big surprise). Since the attenuation is the same (zero) at all frequencies, a lossless line *has no bandwidth limit*. In addition, the propagation constant has an imaginary part that is exactly proportional to frequency. Since the delay of a system is simply (minus) the derivative of phase with frequency, the delay of a lossless line is a constant, independent of frequency:

$$T_{delay} = -\frac{\partial}{\partial \omega}\Phi(\omega) = -\frac{\partial}{\partial \omega}(-\beta z) = \sqrt{LC}z. \tag{41}$$

We can now appreciate the remarkable property of distributed systems alluded to in the Introduction: The capacitance and inductance *do not directly cause a bandwidth reduction*. They result only in a propagation delay. If we were to increase the inductance or capacitance per unit length, the delay would increase but bandwidth (ideally) would not change. This behavior is quite different from what one observes in low-order lumped networks.

Also in stark contrast with low-order lumped networks, a transmission line may exhibit a *frequency-independent* delay, as seen here. This property is extremely desirable, for it implies that all Fourier components of a signal will be delayed by precisely the same amount of time; pulse shapes will be preserved. We have just seen that a lossless line has this property of zero dispersion. Since all real lines exhibit nonzero loss, though, must we accept dispersion (nonuniform delays) in practice? Fortunately, as Heaviside[8] first pointed out, the answer is no. If we exercise some control over the line constants, we can still obtain a uniform group delay *even with a lossy line* (at least in principle). In particular, Heaviside discovered that choosing RC equal to GL (or, equivalently, choosing the L/R time constant of the series impedance Z equal to the C/G time constant of the shunt admittance Y) leads to a constant group delay. There is nonzero attenuation, of course (can't get rid of that, unfortunately), but the constant group delay means that pulses only get smaller as they travel down the line; they don't smear out (disperse).

---

[8] By the way, he was the first to use vector calculus to cast Maxwell's equations in modern form and also the one who introduced the use of Laplace transforms to solve circuit problems.

Showing that Heaviside was correct isn't too hard. Setting RC and GL equal in our exact expressions for α and β yields:

$$\alpha = \text{Re}[\gamma] = \sqrt{RG}; \tag{42}$$

$$\beta = \text{Im}[\gamma] = \omega\sqrt{LC}. \tag{43}$$

Note that the expression for β is the same as that for a lossless line and thus also leads to the same frequency-independent delay.

Although the attenuation is no longer zero, it continues to be frequency-independent; the bandwidth is still infinite as long as we choose $L/R = C/G$. Furthermore, the characteristic impedance becomes exactly equal to $\sqrt{L/C}$ at all frequencies, rather than approaching this value asymptotically at high frequencies.

Setting $LG = RC$ is best accomplished by increasing either L or C, rather than by increasing R or G, because the latter strategy increases the attenuation (presumably an undesirable effect). Michael Pupin of Columbia University, following through on the implications of Heaviside's work, suggested the addition of lumped inductances periodically along telephone transmission lines to reduce signal dispersion. Such "Pupin coils" permitted significantly improved telephony in the 1920s and 1930s.[9]

2.7 BEHAVIOR OF FINITE-LENGTH TRANSMISSION LINES

Now that we've deduced a number of important properties of transmission lines of infinite length, it's time to consider what happens when we terminate finite-length lines in arbitrary impedances.

2.7.1 TRANSMISSION LINE WITH MATCHED TERMINATION

The driving-point impedance of an infinitely long line is simply Z_0. Suppose we cut the line somewhere, discard the infinitely long remainder, and replace it with a single lumped impedance of value Z_0. The driving-point impedance must remain Z_0; there's no way for the measurement apparatus to distinguish the lumped impedance from the line it replaces. Hence, a signal applied to the line simply travels down the finite segment of line, eventually gets to the resistor, heats it up, and contributes to global warming.

---

[9] Alas, the use of lumped inductances introduces a bandwidth limitation that true, distributed lines do not have. Since bandwidth and channel capacity are closely related, all of the Pupin coils (which had been installed at great expense) eventually had to be removed (at great expense) to permit an increase in the number of calls carried by each line. We will explore this idea further when we study filter design.

2.7.2 TRANSMISSION LINE WITH ARBITRARY TERMINATION

In general, a transmission line will not be terminated in precisely its characteristic impedance. Now, a signal traveling down the line maintains a ratio of voltage to current that is equal (of course) to Z_0 until it encounters the load impedance. The termination impedance imposes its own particular ratio of voltage to current, however, and the only way to reconcile the conflict is for some of the signal to reflect back toward the source.

To distinguish forward (incident) quantities from the reflected ones, we will use the subscripts i and r, respectively. If E_i and I_i are the incident voltage and current, then it's clear that

$$Z_0 = \frac{E_i}{I_i}. \qquad (44)$$

At the load end of things, the mismatch in impedances gives rise to a reflected voltage and current. We still have a linear system, so the total voltage at any point on the system is the superposition of the incident and reflected voltages. Similarly, the net current is also the superposition of the incident and reflected currents. Because the current components travel in opposite directions, the superposition here results in a subtraction. Thus, we have

$$Z_L = \frac{E_i + E_r}{I_i - I_r}. \qquad (45)$$

We may rewrite this last equation to show an explicit proportionality to Z_0 as follows:

$$Z_L = \frac{E_i + E_r}{I_i - I_r} = \frac{E_i}{I_i}\left[\frac{1 + E_r/E_i}{1 - I_r/I_i}\right] = Z_0\left[\frac{1 + E_r/E_i}{1 - I_r/I_i}\right]. \qquad (46)$$

The ratio of reflected to incident quantities at the load end of the line is called Γ_L and will generally be complex. Using Γ_L, the expression for Z_L becomes

$$Z_L = Z_0\left[\frac{1 + E_r/E_i}{1 - I_r/I_i}\right] = Z_0\left[\frac{1 + \Gamma_L}{1 - \Gamma_L}\right]. \qquad (47)$$

Solving for Γ_L then yields

$$\Gamma_L = \frac{Z_L - Z_0}{Z_L + Z_0}. \qquad (48)$$

If the load impedance equals the characteristic impedance of the line, then the reflection coefficient will be zero. If a line is terminated in either a short or an open, then the reflection coefficient will have a magnitude of unity; this value is the maximum magnitude it can have (for a purely passive system such as this one, anyway).

We may generalize the concept of the reflection coefficient so that it is the ratio of the reflected and incident quantities at any arbitrary point along the line:

$$\Gamma(z) = \frac{E_r e^{\gamma z}}{E_i e^{-\gamma z}} = \frac{E_r}{E_i} e^{2\gamma z} = \Gamma_L e^{2\gamma z}. \qquad (49)$$

Here we follow the convention of defining $z = 0$ at the load end of the line and locating the driving source at $z = -l$. With this convention, the voltage and current at any point z along the line may be expressed as:

$$V(z) = V_i e^{-\gamma z} + V_r e^{\gamma z}; \tag{50}$$

$$I(z) = I_i e^{-\gamma z} - I_r e^{\gamma z}. \tag{51}$$

As always, the impedance at any point z is simply the ratio of voltage to current:

$$Z(z) = \frac{V_i e^{-\gamma z} + V_r e^{\gamma z}}{I_i e^{-\gamma z} - I_r e^{\gamma z}} = Z_0 \left[\frac{1 + \Gamma_L e^{2\gamma z}}{1 - \Gamma_L e^{2\gamma z}} \right]. \tag{52}$$

Substituting for Γ_L and doing a whole heck of a lot of crunching yields

$$\frac{Z(z)}{Z_0} = \frac{\frac{Z_L}{Z_0}(e^{-\gamma z} + e^{\gamma z}) + (e^{-\gamma z} - e^{\gamma z})}{\frac{Z_L}{Z_0}(e^{-\gamma z} - e^{\gamma z}) + (e^{-\gamma z} + e^{\gamma z})}. \tag{53}$$

Writing this expression in a more compact form, we have

$$\frac{Z(z)}{Z_0} = \frac{\frac{Z_L}{Z_0} - \tanh \gamma z}{1 - \frac{Z_L}{Z_0} \tanh \gamma z}. \tag{54}$$

In the special case where the attenuation is negligible (as is commonly assumed, to permit tractable analysis), a considerable simplification results:

$$\frac{Z(z)}{Z_0} = \frac{\frac{Z_L}{Z_0} - j \tan \beta z}{1 - j\frac{Z_L}{Z_0} \tan \beta z} = \frac{Z_L \cos \beta z - jZ_0 \sin \beta z}{Z_0 \cos \beta z - jZ_L \sin \beta z}. \tag{55}$$

Here, z is the actual coordinate value and will always be zero or negative.

As a final comment, note that this expression is periodic. This behavior is strictly true only for lossless lines, of course, but practical lines will behave similarly as long as the loss is negligible. Periodicity implies that one need consider the impedance behavior only over some finite section (specifically, a *half*-wavelength) of line. This observation is exploited in the construction of the Smith chart, a brief study of which is taken up in Chapter 3.

2.8 SUMMARY OF TRANSMISSION LINE EQUATIONS

We've seen that both the characteristic impedance and the propagation constant are simple functions of the per-length series impedance and shunt admittance:

$$Z_0 = \sqrt{\frac{Z}{Y}} = \sqrt{\frac{R + j\omega L}{G + j\omega C}}; \tag{56}$$

$$\gamma = \sqrt{ZY} = \sqrt{(R + j\omega L)(G + j\omega C)}. \tag{57}$$

Using these parameters – in conjunction with the definition of reflection coefficient – allows us to develop an equation for the driving-point impedance of a lossy line terminated in an arbitrary impedance:

$$\frac{Z(z)}{Z_0} = \frac{\frac{Z_L}{Z_0} - \tanh \gamma z}{1 - \frac{Z_L}{Z_0} \tanh \gamma z}. \tag{58}$$

In the case of a lossless (or negligibly lossy) line, the expression for impedance takes on a reasonably simple and periodic form, setting the stage for discussion of the Smith chart:

$$\frac{Z(z)}{Z_0} = \frac{\frac{Z_L}{Z_0} - j\tan\beta z}{1 - j\frac{Z_L}{Z_0}\tan\beta z} = \frac{Z_L \cos\beta z - jZ_0 \sin\beta z}{Z_0 \cos\beta z - jZ_L \sin\beta z}. \tag{59}$$

2.9 ARTIFICIAL LINES

We've just seen that an infinite ladder network of infinitesimally small inductors and capacitors has a purely real input impedance over an infinite bandwidth. Although structures that are infinitely long are somewhat inconvenient to realize, we can always terminate a finite length of line in its characteristic impedance. Energy, being relatively easy to fool, cannot distinguish between real transmission line and a resistor equal to the characteristic impedance, and the driving-point impedance of the properly terminated finite line remains the same as that of the infinite line, and still over an infinite bandwidth.

There are instances when we might wish to approximate a continuous transmission line by a finite lumped network. Motivations for doing so may include convenience of realization or greater control over line constants. However, use of a finite lumped approximation guarantees that the characteristics of such an artificial line cannot match those of an ideal line over an infinite bandwidth.[10] The design of circuits that employ lumped lines must take this bandwidth limitation into account.

One important use of artificial lines is in the synthesis of delay lines; see Figure 2.4. Here, we use LC L-sections to synthesize our line. As in the continuous case, the driving-point impedance is just

$$Z_{in} = \sqrt{L/C}, \tag{60}$$

while the delay per section is just

$$T_D = \sqrt{LC}. \tag{61}$$

---

[10] One easy way to see this is to recognize that a true transmission line, being a delay element, provides unbounded phase shift as the frequency approaches infinity. A lumped line can provide only a finite phase shift (because of the finite number of energy storage elements) and hence a finite number of poles.

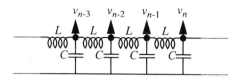

FIGURE 2.4. Lumped delay line

The value of a lumped delay line is that one may obtain large delays without having to use, say, a kilometer of coaxial cable.

2.9.1 CUTOFF FREQUENCY OF LUMPED LINES

Unlike the distributed line, the lumped line presents a real, constant impedance only over a finite bandwidth. Eventually, the input impedance becomes purely reactive,[11] indicating that real power can be delivered neither to the line nor to any load connected to the other end of the line. This behavior is the basis for a class of filters we consider in a separate chapter. Indeed, the birth of filter theory traces directly to the limited frequency response of this structure.

The frequency above which no power is delivered is known as the line's *cutoff* frequency, which is readily found by using the formula for the input impedance of an infinite (but lumped) *LC* line, reprised here from Section 2.5 for convenience:

$$Z_{in} = \frac{Z}{2}\left[1 \pm \sqrt{1 + \frac{4}{ZY}}\right]. \tag{62}$$

Here, let $Y = j\omega C$ and $Z = j\omega L$. Then the input impedance is

$$Z_{in} = \frac{j\omega L}{2}\left[1 \pm \sqrt{1 - \frac{4}{\omega^2 LC}}\right]. \tag{63}$$

At sufficiently low frequencies, the term under the radical has a net negative value. The resulting imaginary term, when multiplied by the $j\omega L/2$ factor, provides the real component of the input impedance.

As the frequency increases, however, the magnitude of the term under the radical sign eventually becomes zero. At and above this frequency, the input impedance is purely imaginary, and no power can be delivered to the line. The cutoff frequency is therefore given by

$$\omega_h = 2/\sqrt{LC}. \tag{64}$$

Since the lumped line's characteristics begin to degrade well below the cutoff frequency, one must usually select cutoff well above the highest frequency of interest. Satisfying this requirement is particularly important if good pulse fidelity is necessary.

---

[11] From inspection of the network, it should be clear that the driving-point impedance eventually collapses to that of the input inductor, since the capacitors act ultimately like shorts.

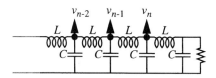

FIGURE 2.5. One choice for terminating lumped lines

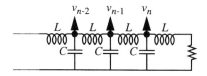

FIGURE 2.6. Alternative choice for terminating lumped lines

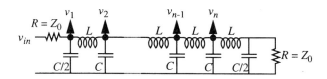

FIGURE 2.7. Half-sections for line termination

In designing artificial lines, the L/C ratio is chosen to provide the desired line impedance, while the LC product is chosen small enough to provide a high enough cutoff frequency to allow the line to approximate ideal behavior over the desired bandwidth. If a specified overall time delay is required, the first two requirements define the minimum number of sections that must be used.

2.9.2 TERMINATING LUMPED LINES

There's always a question as to how one terminates the circuit of Figure 2.4. One choice, shown in Figure 2.5, is to end in a capacitance and simply terminate across it. Another choice (Figure 2.6) is to end in an inductance. Though both of these choices will work after a fashion, a better alternative is to compromise by using a *half-section* at each end of the line, as shown in Figure 2.7.

Such a compromise extends the bandwidth over the circuit of Figure 2.5 or Figure 2.6. Each half-section contributes half the delay of a full section, so putting one on each end adds the delay of a full section. Furthermore, and more important, a half-section has twice the cutoff frequency of a full section, which is precisely why better bandwidth is obtained.

Many applications require delay elements that approximate a quarter-wavelength piece of transmission line, as we'll see when we discuss elements such as couplers

2.9 ARTIFICIAL LINES

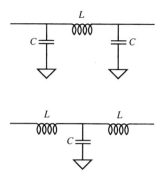

FIGURE 2.8. Simple π-section and T-section networks for approximation of $\lambda/4$ delay line

and combiners. Either π-section or T-section elements are suitable, with the most common versions using just three elements; see Figure 2.8. These two networks exhibit the same transfer function, so one may base a choice on practical considerations, such as accommodation of parasitics.

As usual, the ratio L/C is chosen as a function of the characteristic impedance of the line to be simulated:

$$\sqrt{L/C} = Z_0. \tag{65}$$

Furthermore, it is straightforward to demonstrate that, when the networks shown are driven and terminated in Z_0, they exhibit a quadrature phase lag at a frequency

$$\omega_0 = 1/\sqrt{LC}. \tag{66}$$

Solving for the element values then yields:

$$C = 1/\omega_0 Z_0; \tag{67}$$

$$L = Z_0/\omega_0. \tag{68}$$

2.9.3 *m*-DERIVED HALF SECTIONS

The port impedance of the LC half-section begins to increase significantly beginning at about 30–40% of the cutoff frequency owing to the parallel resonance formed by the output capacitance and the rest of the reactance it sees. This behavior can be moderated by the use of half-sections that are only marginally more elaborate than the single LC pair. Specifically, if the capacitor is replaced by a series LC branch, the frequency range over which the impedance stays roughly constant can be increased even further because the decreasing impedance of the series resonant branch helps offset the increasing impedance.

A simple network that achieves the desired result is shown in Figure 2.9. A more detailed derivation is presented in the chapter on filters. For now, accept that the element values are given by the following equations:

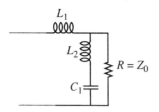

FIGURE 2.9. *m*-derived half-section for line termination

$$L_1 = \frac{mL}{2}; \tag{69}$$

$$L_2 = \frac{1-m^2}{2m}L; \tag{70}$$

$$C_1 = \frac{mC}{2}. \tag{71}$$

A network modified in this manner is called an *m*-derived half-section because, for any value of the parameter *m*, the nominal characteristic impedance remains the same as that of the simple *LC* half-section. This can be verified by direct substitution into Eqn. 62.

The impedance stays roughly constant up to about 85% of the cutoff frequency for a value of *m* equal to roughly 0.6. This choice is therefore a common one. The *m*-derived half-section may be used either as terminating sections on the ends of an artificial line or as elements for constructing the entire line. We will revisit many of these concepts in the context of filter design when we take up that important topic in Chapter 22.

2.10 SUMMARY

We have identified a fuzzy boundary between the lumped and distributed regimes, and we found that lumped concepts may be extended into the distributed regime. In carrying out that extension, we discovered that there are several (perhaps many) ways to trade gain for delay rather than bandwidth. As a final observation on this subject, perhaps it is worthwhile to reiterate that avoiding a straight gain–bandwidth trade-off requires a gross departure from single-pole dynamics. Hence, all the structures we've seen that trade gain for delay involve many energy storage elements. Another way to look at this issue is to recognize that, if we are to be able to trade delay for anything, we must have the ability to provide large delays. But large delays imply a large amount of phase change per unit frequency, and if we are to operate over a large bandwidth then the total phase change required is very large. Again, this need for large amounts of phase shift necessarily implies that many poles (and hence many inductors and capacitors) will be required, resulting in the relatively complicated networks we've seen.[12] Yet if one is pursuing operation over the largest possible bandwidth, use of these distributed concepts is all but mandatory. As we shall see in Chapter 12, distributed concepts may be applied to active circuits to allow the realization of amplifiers with exceptionally large bandwidth by trading delay in exchange for the improved bandwidth.

---

[12] An exception is the superregenerative amplifier. There, the time-varying nature of the system effectively causes weighted aliases of a single stage's response to combine in a way that produces a response similar to that of a cascade of such stages.

CHAPTER THREE

THE SMITH CHART AND S-PARAMETERS

3.1 INTRODUCTION

The design of microwave circuits and systems has its origins in an era where devices and interconnect were usually too large to allow a lumped description. Furthermore, the lack of suitably detailed models and compatible computational tools forced engineers to treat systems as two-port "black boxes" with graphical methods. The most powerful of these graphical aids, the Smith chart, dates from the 1930s, an age where slide rules dominated. Although Smith charts today are perhaps less relevant as a computational aid than they were then, RF instrumentation, for example, continues to present data in Smith-chart form. It also remains true that visualizing certain operations in terms of the Smith chart can inform design intuition in rich ways that modern computational aids may unfortunately bypass. This chapter thus provides a brief history and derivation of the Smith chart, along with an explanation of why a particular set of variables (S-parameters) won out over other parameter sets (e.g., impedance or admittance) to describe microwave two-ports.

3.2 THE SMITH CHART

Introductory presentations of the Smith chart are frequently devoid of any historical context, leaving the student with the impression that it sprang forth spontaneously and fully formed. This impression, in turn, makes many students feel mentally deficient if they are unable to appreciate instantly the subtle beauty, logic, and power that the chart must "obviously" possess. The real story, though, is that the Smith chart is the result of cumulative incremental refinements spanning about a decade. The Smith chart was evidently not quite obvious to Smith, so perhaps it should not be so immediately obvious to us. And besides, who was Smith, anyway?

Phillip Hagar Smith joined Bell Labs in the late 1920s, shortly before the big stock-market crash. He spent the early part of his career working on antenna systems where, naturally, the problem of impedance matching arises frequently. He found himself having to perform similar calculations over and over. The considerable tedium of those calculations led him to invent a succession of graphical aids (remember, there were no spreadsheets, only slipsticks). By the mid-1930s, he had devised rectangular and polar-plot representations of impedance that facilitated computation. By around 1937, he had learned enough mathematics (specifically, the art of conformal mapping) from his Bell Labs colleagues to complete the last steps and produce the version of the chart as it is basically known today.[1]

The reader may take comfort in the fact that few people besides Smith and his Bell Labs friends initially gave a rat's patootie about the chart.[2] It took nearly two years before he could get a publisher to print his article about it. Finally, in January of 1939, *Electronics* magazine published Smith's paper about his new chart.[3] For the most part, its publication reportedly evoked either a collective yawn or quizzical head scratching (sometimes both). This reaction is hardly surprising, for in that slide-rule age, engineers were always inventing charts, nomographs, and other graphical computational aids of one sort or another. The Smith chart had to compete with all of them for attention. The somewhat intimidating appearance of the chart undoubtedly also inhibited acceptance.

This situation changed upon the formation of the MIT Radiation Laboratory on the eve of the Second World War. The many physicists developing radar at the Rad Lab were unintimidated by (perhaps even attracted to) the mathematics underlying the Smith chart, having already encountered it in other contexts. Their comfort with the math, combined with the urgency of the task at hand, led to a rapid adoption of the Smith chart as an invaluable design tool. Before the war's end, a large community of engineers had become intimately familiar with the Smith chart. Spreading the word further was abetted by Smith himself in another publication in 1944.[4]

The mathematical basis for the Smith chart is the bilinear transformation (a ratio of two linear functions), which relates reflection coefficient to normalized load impedance:

$$\Gamma = \frac{\frac{Z_L}{Z_0} - 1}{\frac{Z_L}{Z_0} + 1} = \frac{Z_{nL} - 1}{Z_{nL} + 1}. \tag{1}$$

This bilinear relationship between the normalized load impedance and Γ is also biunique: knowing one is equivalent to knowing the other. This observation is important because the familiar curves of the Smith chart are simply a plotting, in the Γ-plane, of contours of constant resistance and reactance.

It's natural to ask why one should go to the trouble of plotting the equivalent of impedance in a nonrectilinear coordinate system, since it's certainly more straightforward to plot the real and imaginary parts of impedance directly in conventional

---

[1] R. Rhea, "Phillip H. Smith: A Brief Biography" (Introduction to reprint of Smith's book, *Electronic Applications of the Smith Chart: In Waveguide, Circuit, and Component Analysis,* by Noble Publishing, 2000).
[2] Louis Smullin, private communication.
[3] "A Transmission Line Calculator."
[4] "An Improved Transmission Line Calculator," *Electronics,* v. 17, January 1944, p. 130.

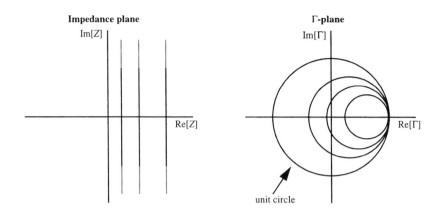

FIGURE 3.1. Mapping of constant-resistance lines in Z-plane to circles in Γ-plane

Cartesian coordinates. There are at least two good reasons for the seemingly non-obvious choice. One is that trying to plot an infinite impedance on a finite-sized piece of paper poses self-evident practical problems. Plotting Γ instead neatly handles impedances of arbitrary magnitude because |Γ| cannot exceed unity for passive loads. The other reason is that Γ repeats every half-wavelength when a lossless transmission line is terminated in a fixed impedance. Hence, plotting Γ is a natural and compact way to encode this periodic behavior. Much of the computational power of the Smith chart derives from this compact encoding, allowing engineers to determine rapidly the length of line needed to transform an impedance to a particular value, for example, or to read off the magnitude of the reflection coefficient by inspection.

The relationship between impedance and Γ given in Eqn. 1 may be considered a mapping of one complex number into another. In this case, we've already noted that it is a special type of mapping known as a bilinear transformation. Among the various properties of this transformation, a particularly relevant one is that circles remain circles when mapped. In this context, a line is considered a circle of infinite radius. Hence, circles and lines map into either circles or lines.

With the aid of Eqn. 1, it is straightforward to show that the imaginary axis of the Z-plane maps into the unit circle in the Γ-plane, while other lines of constant resistance in the Z-plane map into circles of varying diameter that are all tangent at the point $\Gamma = 1$; see Figure 3.1.

Lines of constant reactance are orthogonal to lines of constant of resistance in the Z-plane, and this orthogonality is preserved in the mapping (as they are in all conformal maps). Since lines map to lines or circles, we expect constant reactance lines to transform to the circular arcs shown in Figure 3.2. The Smith chart simply consists of both constant-resistance and constant-reactance contours in the Γ-plane without the explicit presence of the Γ-plane axes.

Because, as mentioned earlier, the primary role of the Smith chart these days is as a standard way to present impedance (or reflectance) data, it is worthwhile taking a

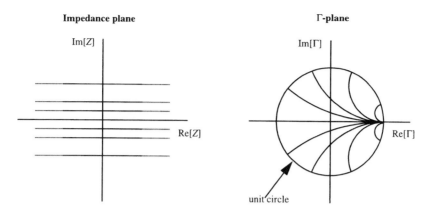

FIGURE 3.2. Mapping of constant-reactance lines in Z-plane to contours in Γ-plane

little time to develop a familiarity with it. The center of the Smith chart corresponds to zero reflection coefficient and hence to a resistance that is equal to the normalizing impedance.

The bottom half of the Z-plane maps into the bottom half of the unit circle in the Γ-plane, and thus capacitive impedances are always found there. Similarly, the top half of the Z-plane corresponds to the top half of the unit circle and inductive impedances. Progressively smaller circles of constant resistance correspond to progressively larger resistance values. The point $\Gamma = -1$ corresponds to zero resistance (or reactance), and the point $\Gamma = 1$ corresponds to infinite resistance (or reactance).

As a simple, but specific, example, let us plot the impedance of a series RC network in which the resistance is 100 Ω and the capacitance is 25 pF, all normalized to a 50-Ω system. Since the impedance is the sum of a real part (equal to the resistance) and an imaginary part (equal to the capacitive reactance), the corresponding locus in the Γ-plane must lie along the circle of constant resistance for which $R = 2$. The reactive part varies from minus infinity at DC to zero at infinite frequency. Since it is always negative in sign, the locus must be just the bottom half of that constant resistance circle, traversed clockwise from $\Gamma = 1$ as frequency increases, as seen in Figure 3.3.

Note that this curve is the impedance locus for *any* capacitance in series with 100 Ω. All that varies with capacitance value is the frequency that corresponds to a particular point on the curve. A corollary is that, for a fixed frequency, the addition (or subtraction) of capacitance merely shifts the position along this same semicircular arc. All such curves converge to a common point of infinite impedance, at zero frequency.

An inductor in series with that same 100-Ω resistor would trace out the upper half of the circle. Again, a clockwise traversal corresponds to increasing frequency. As with the RC example, a change in inductance simply produces motion along the semicircle at any given frequency. The impedance loci of all series RL networks share a common point of infinite impedance at infinite frequency.

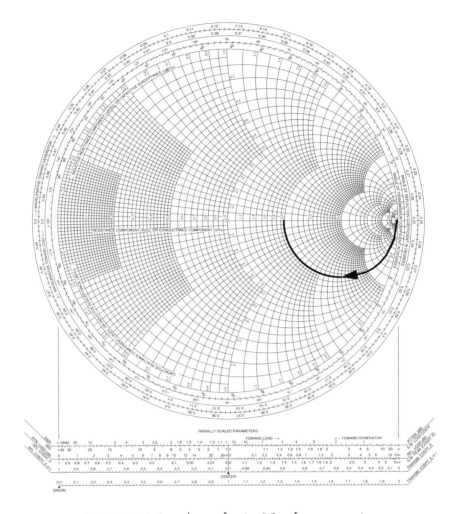

FIGURE 3.3. Impedance of series *RC* as frequency varies

It's not too great a leap beyond the foregoing examples to recognize that the impedance of a series RLC network traces out a circle as frequency varies. At the series resonant frequency, the impedance is purely real, corresponding to a point on the horizontal axis.

The Smith chart is also amenable to the evaluation of shunt impedances. We may express reflectance in terms of admittances or impedances with equal ease:

$$\frac{Z_{nL}-1}{Z_{nL}+1} = \frac{1/Y_{nL}-1}{1/Y_{nL}+1} = -\frac{Y_{nL}-1}{Y_{nL}+1}. \qquad (2)$$

Aside from the sign change, the equations for reflectance in terms of impedance and admittance are the same. Thus contours of constant conductance are also circles, and contours of constant susceptance are circular arcs orthogonal to the constant conductance circles. The minus sign is readily accommodated by simply flipping the chart

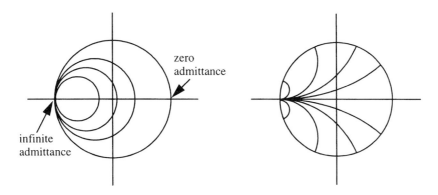

FIGURE 3.4. Constant conductance (left) and constant susceptance contours on admittance chart

over as in Figure 3.4. When oriented as shown, inductive susceptances are found in the lower half of the circle, and capacitive ones in the upper half. Some commercial versions of the Smith chart have both impedance and admittance contours (often rendered as different colors, or different shades of gray). The complexity of the combination can be a bit intimidating until familiarity breeds contentment.

A parallel GC network traces out a semicircle in the upper half of the admittance chart as frequency sweeps from zero to infinity. All such curves are tangent at the point of infinite admittance, converging there at infinite frequency. Similarly, a parallel GL network traces out a semicircle in the lower half of the chart. Again, all such semicircles are tangent at the point of infinite admittance, converging there at zero frequency. A parallel GLC network traces out a complete circle as frequency varies. These traversals are counterclockwise with increasing frequency, in contrast with the clockwise traversals of series impedance plots.

In addition to helping to visualize how the addition of shunt and series elements changes impedances or admittances, the Smith chart also enables us to evaluate rapidly how an impedance varies along a transmission line. Examination of commercially available Smith charts reveals the phrases "wavelengths toward generator" and "wavelengths toward load," coupled with orienting arrows disposed counterclockwise and clockwise, respectively. The meaning is simple enough: Imagine terminating a lossless line in some impedance. If we move "toward the generator" by some amount, we know that the reflectance changes in some manner. Specifically,

$$\Gamma(z) = \Gamma_L e^{2\gamma z}, \qquad (3)$$

so we conclude that the magnitude of the reflectance does not change; only the phase angle does. Now, the Smith chart is the result of converting impedance into reflectance (Γ). Thus, Γ is actually plotted with conventional Cartesian coordinates, but the corresponding axes are not printed on ordinary Smith charts (however, we have included these axes for tutorial purposes here). Just pretend that they are there, and you can see how Γ can be read off quickly. The center of the Smith chart corresponds to a

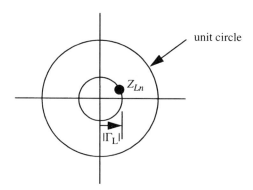

FIGURE 3.5. Single-frequency impedance locus

perfect match and is the origin of Γ's invisible rectangular coordinate system. Contours of constant Γ are thus simply circles centered about the origin, with mismatch increasing with radius. Therefore, as we move along the transmission line, we simply traverse some part of a circle that is concentric with the unit circle, allowing us to compute readily the impedance at any point along the line. See Figure 3.5.

Not only is the Smith chart useful analytically, it is also of inestimable value in deducing the length of line required to *produce* a specified impedance transformation, as is made clear in Chapter 4. A constant-reflectance magnitude circle corresponding to the maximum tolerable Γ may be overlaid on the Smith chart, facilitating the determination of whether matching objectives have been achieved.

As a further aid, a typical Smith chart has an additional set of rulers in a separate group below the chart proper. These are calibrated both in terms of reflectance and as translations of reflectance into other measures of impedance mismatch: standing-wave ratio (SWR) and return loss (RL).

There are numerous other properties of Smith charts, and the types of computations that may be performed graphically and rapidly with them are truly remarkable. We will study only one (but it's an important one) – impedance matching – in the next chapter. Since machine computation has diminished the role of Smith charts, we have presented only a truncated description and direct the interested reader to Smith's paper of 1944 for further applications.[5]

3.3 S-PARAMETERS

Engineers have devised many (perhaps too many) ways to describe systems. To simplify analysis and perhaps elucidate important design criteria, it is often valuable to

---

[5] Many books and papers have been written by others about the Smith chart, but you might as well learn directly from the man himself, starting with his 1944 paper, which describes the chart pretty much as we know it today. For a more thorough treatment, see his book, *Electronic Applications of the Smith Chart: In Waveguide, Circuit, and Component Analysis,* McGraw-Hill, New York, 1969 (reprinted by Noble Publishing, 2000).

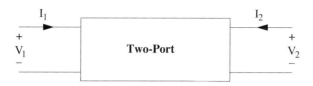

FIGURE 3.6. Port variable definitions

use macroscopic descriptions, which preserve input–output behavior but discard details of the internal structure of the system. At lower frequencies, the most common representations use impedance or admittance parameters, or perhaps a mixture of the two (called, sensibly enough, *hybrid* parameters).

The basis for linear two-port representations is simply that the voltage (or current) at one port may be expressed as a linear combination of the current (or voltage) at that port and the voltage or current at the other port. Depending on the particular choice of variables, the coefficients may be dimensionless or may have the dimensions of impedance or admittance. For example, in the impedance representation we express port voltages in terms of port currents. For the two-port shown in Figure 3.6, the relevant equations are:

$$V_1 = Z_{11}I_1 + Z_{12}I_2; \tag{4}$$

$$V_2 = Z_{22}I_2 + Z_{21}I_1. \tag{5}$$

It is most convenient to open-circuit the ports in succession to determine the various Z-parameters experimentally, because various terms then become zero. For instance, determination of Z_{11} is easiest when the output port is open-circuited because the second term in Eqn. 4 is zero under that condition. Driving the input port with a current source and measuring the resulting voltage at the input allows direct computation of Z_{11}. Similarly, open-circuiting the input port, driving the output with a current source, and measuring V_1 allows determination of Z_{12}.

Short-circuit conditions are used to determine admittance parameters, and a combination of open- and short-circuit conditions allow determination of hybrid parameters. The popularity of these representations to characterize systems at low frequencies traces directly to the ease with which one may determine the parameters experimentally.

At high frequencies, however, it is quite difficult to provide adequate shorts or opens, particularly over a broad frequency range. Furthermore, active high-frequency circuits are frequently rather fussy about the impedances into which they operate, and may oscillate or even expire when terminated in open or short circuits. A different set of two-port parameters is therefore required to evade these experimental problems. Called *scattering* parameters (or simply S-parameters), they exploit the fact that a line terminated in its characteristic impedance gives rise to no reflections.[6]

---

[6] K. Kurokawa, "Power Waves and the Scattering Matrix," *IEEE Trans. Microwave Theory and Tech.*, v. 13, March 1965, pp. 194–202.

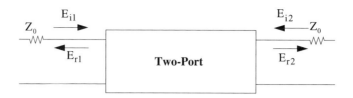

FIGURE 3.7. S-parameter port variable definitions

Interconnections between the instrumentation and the system under test therefore can be of a comfortable length since no short or open circuit needs to be provided, greatly simplifying fixturing.

As implied earlier, terminating ports in open or short circuits is a convenience for lower-frequency two-port descriptions because various terms then become zero, simplifying the math. Scattering parameters retain this desirable property by defining input and output variables in terms of incident and reflected (scattered) *voltage waves*, rather than port voltages or currents (which are difficult to define uniquely at high frequencies, anyway).

As can be seen in Figure 3.7, the source and load terminations are Z_0. With the input and output variables defined as shown, the two-port relations may be written as

$$b_1 = s_{11}a_1 + s_{12}a_2, \qquad (6)$$

$$b_2 = s_{22}a_2 + s_{21}a_1, \qquad (7)$$

where

$$a_1 = E_{i1}/\sqrt{Z_0}, \qquad (8)$$

$$a_2 = E_{i2}/\sqrt{Z_0}; \qquad (9)$$

$$b_1 = E_{r1}/\sqrt{Z_0}, \qquad (10)$$

$$b_2 = E_{r2}/\sqrt{Z_0}. \qquad (11)$$

The normalization by the square root of Z_0 is a convenience that makes the square of the magnitude of the various a_n and b_n equal to the power of the corresponding incident or reflected wave.

Driving the input port with the output port terminated in Z_0 sets a_2 equal to zero, which allows us to determine the following parameters:

$$s_{11} = \frac{b_1}{a_1} = \frac{E_{r1}}{E_{i1}} = \Gamma_1; \qquad (12)$$

$$s_{21} = \frac{b_2}{a_1} = \frac{E_{r2}}{E_{i1}}. \qquad (13)$$

Thus, s_{11} is simply the input reflection coefficient, while s_{21} is a sort of gain since it relates an output wave to an input wave. Its magnitude squared is a type of power gain known as the *transducer power gain,* with Z_0 as source and load impedance, but we defer a discussion of its precise definition to an appendix (see Section 3.4).

Similarly, terminating the input port and driving the output port yields:

$$s_{22} = \frac{b_2}{a_2} = \frac{E_{r2}}{E_{i2}} = \Gamma_2; \tag{14}$$

$$s_{12} = \frac{b_1}{a_2} = \frac{E_{r1}}{E_{i2}}. \tag{15}$$

Here, we see that s_{22} is the output reflection coefficient; s_{12} is the reverse transmission, whose magnitude squared is the reverse transducer power gain with Z_0 as source and load impedance.

Once a two-port has been characterized with S-parameters, direct design of systems may proceed in principle without knowing anything about the internal workings of the two-port. For example, gain equations, noise figure, and stability criteria can be recast in terms of S-parameters.[7] However, it is important to keep in mind that a macroscopic approach necessarily discards potentially important information, such as sensitivity to parameter or process variation. For this reason, S-parameter measurements are best used to derive element values for models whose topologies have been determined from first principles or physical reasoning.

To summarize, the reasons that S-parameters have become nearly universal in high-frequency work are that "zero"-length fixturing cables are unnecessary, there is no need to synthesize a short or open circuit, and terminating the two-port in Z_0 greatly reduces the potential for oscillation.

3.4 APPENDIX A: A SHORT NOTE ON UNITS

The inability to identify unique voltages and currents in distributed systems, coupled with the RF engineer's preoccupation with power gain, has made power the natural quantity on which to focus in RF circuits and systems. Power levels are expressed in watts, of course, but what can confuse and frustrate the uninitiated are the various decibel versions. For example, "dBm" is quite commonly used. The "m" signifies that the 0-dB reference is one milliwatt, while a "dBW" is referenced to one watt. If the reference impedance level is 50 Ω, then 0 dBm corresponds to a voltage of about 223 mV rms.

As clear as these definitions are, there are some who actively insist on confusing volts with watts, redefining 0 dBm to mean 223 mV rms *regardless of the impedance level*. Not only is this redefinition unnecessary, for one may always define a dBV, it is also dangerous. As we will see, critical performance measures – such as linearity and noise figure – intimately involve true power ratios, particularly in studying cascaded systems. Confusing power with voltage ratios leads to gross errors. Hence, throughout this text, 0 dBm truly means one milliwatt, and 0 dBV means one volt. Always.

---

[7] A representative reference that covers this topic is G. Gonzalez, *Microwave Transistor Amplifiers*, 2nd ed., Prentice-Hall, Englewood Cliffs, NJ, 1997.

With that out of the way, we return to some definitions. Common when discussing noise or distortion products in oscillators or power amplifiers is "dBc," where the "c" here signifies that the 0-dB reference is the power of the carrier.

Most engineers are familiar with the engineering prefixes ranging from that for 10^{-12} (pico) to that for 10^{12} (tera). Supplementing those below pico are femto (f), atto (a), zepto (z), and yocto (y), some of which sound like the names of lesser-known Marx brothers. Above tera are peta (P), exa (E), zetta (Z), and yotta (Y). You can see that abbreviations for prefixes associated with positive exponents are capitalized (with the exception of that for kilo, to avoid confusion with K, the unit of absolute temperature) and that those for negative exponents are rendered in lower case (with μ being used for "micro" to avoid confusion with the abbreviation for "milli"). With these additional prefixes, you may express quantities spanning an additional 24 orders of magnitude. Presumably, 48 orders of magnitude should suffice for most purposes.

By international convention, if a unit is named after a person then only the abbreviation is capitalized (W and watt, not Watt; V and volt, not Volt, etc.).[8] This choice assures that "two Watts" refers only to two members of the Watt family, and not to two joules per second, for example. Absolute temperature is measured in kelvins (not the redundant "degrees kelvin") and is abbreviated K. The liter (or litre) is an exception; its abbreviation may be l or L, the latter to avoid confusion with the numeral 1.

Finally, a good way to start a fistfight among microwave engineers is to argue over the pronunciation of the prefix "giga-". Given the Greek origin, both gs should be pronounced as in "giggle," the choice now advocated by both ANSI (American National Standards Institute) and the IEEE (Institute of Electrical and Electronics Engineers). However, there is still a sizable contingent who pronounce the first g as in "giant." The depth of emotion felt by some advocates of one or the other pronunciation is all out of proportion to the importance of the issue. Feel free to test this assertion at the next gathering of microwave engineers you attend. Ask them how they pronounce it, tell them they're wrong (even if they aren't), and watch the Smith charts fly.

DEFINITIONS OF POWER GAIN

While we're on the subject of definitions, we ought to discuss "power gain." You might think, quite understandably, that the phrase has an unambiguous meaning, but you would be wrong. There are four types of power gain that one frequently encounters in microwave work, and it's important to keep track of which is which.

Plain old *power gain* is defined as you'd expect: It's the power actually delivered to some load, divided by the power actually delivered by the source. However, because the impedance loading the source may not be known (particularly at high frequencies), measuring this quantity can be quite difficult in practice. As a consequence, other power gain definitions have evolved.

---

[8] The decibel is named after Alexander Graham Bell. Hence dB, not db, is the proper abbreviation.

Transducer power gain (a term we have already used in this chapter) is the power actually delivered to a load divided by the power *available* from the source. If the source and load impedances are some standardized value then computing the power delivered and the available source power is relatively straightforward, sidestepping the measurement difficulties alluded to in the previous paragraph. From the definition, you may also see that transducer gain and power gain will be equal if the input impedance of the system under consideration happens to be the complex conjugate of the source impedance.

Available power gain is the power available at the output of a system divided by the power available from the source. *Insertion power gain* is the power actually delivered to a load with the system under consideration *inserted*, divided by the power delivered to the load with the source connected directly to the load. Depending on context, any of these power gain definitions may be the appropriate one to consider.

Finally, note that if the input and output ports are all matched then the four definitions of power gain converge. Only in that instance would it be safe to say "power gain" without being more specific.

3.5 APPENDIX B: WHY 50 (OR 75) Ω?

Most RF instruments and coaxial cables have standardized impedances of either 50 or 75 ohms. It is easy to infer from the ubiquity of these impedances that there is something sacred about these values, and that they should therefore be used in all designs. In this appendix, we explain where these numbers came from in the first place in order to see when it does and does not make sense to use those impedances.

3.5.1 POWER-HANDLING CAPABILITY

Consider a coaxial cable with an air dielectric. There will be, of course, some voltage at which the dielectric breaks down. For a fixed inner conductor diameter, one could increase the outer diameter to increase this breakdown voltage. However, the characteristic impedance would then increase, which by itself would tend to reduce the power deliverable to a load. Because of these two competing effects, there is a well-defined ratio of conductor diameters that maximizes the power-handling capability of a coaxial cable.

Having established the possibility that a maximum exists, we need to dredge up a couple of equations to find the actual dimensions that lead to this maximum. Specifically, we need one equation for the peak electric field between the conductors and another for the characteristic impedance of a coaxial cable:

$$E_{max} = \frac{V}{a \ln(b/a)} \tag{16}$$

and

$$Z_0 = \sqrt{\frac{\mu}{\varepsilon}} \cdot \frac{\ln(b/a)}{2\pi} \approx \frac{60}{\sqrt{\varepsilon_r}} \cdot \ln\left(\frac{b}{a}\right), \tag{17}$$

where a and b are (respectively) the inner and outer radius and ε_r is the relative dielectric constant, which is essentially unity for our air-line case.

The next step is to recognize that the maximum power deliverable to a load is proportional to V^2/Z_0. Using our equations, that translates to:

$$P \propto \frac{V^2}{Z_0} = \frac{E_{max} \cdot a^2 \ln(b/a)^2}{(60/\sqrt{\varepsilon_r}) \cdot \ln(b/a)} = \frac{\sqrt{\varepsilon_r}[E_{max}^2 \cdot a^2 \ln(b/a)]}{60}. \tag{18}$$

Next, take the derivative, set it equal to zero, and pray for a maximum instead of a minimum:

$$\frac{dP}{da} = \frac{d}{da}\left[a^2 \ln\left(\frac{b}{a}\right)\right] = 0 \implies \frac{b}{a} = \sqrt{e}. \tag{19}$$

Plugging this ratio back into our equation for the characteristic impedance gives us a value of 30 Ω. That is, to maximize the power-handling capability of an air dielectric transmission line of a given outer diameter, we want to select the dimensions to give us a Z_0 of 30 Ω.[9]

But wait: 30 does not equal 50, even for relatively large values of 30. So it appears we have not yet answered our original question. We need to consider one more factor: cable attenuation.

3.5.2 ATTENUATION

It may be shown (but we won't show it) that the attenuation per length due to dielectric loss is practically independent of conductor dimensions. Using a simple equation that accounts only for the attenuation due to resistive loss, we have

$$\alpha \approx R/2Z_0, \tag{20}$$

where R is the series resistance per unit length. At sufficiently high frequencies (the regime we're concerned with at the moment), R is due mainly to the skin effect. To reduce R, we would want to increase the diameter of the inner conductor (to get more "skin"), but that would tend to reduce Z_0 at the same time, and it's not clear how to win. Again, we see a competition between two opposing effects, and we expect the optimum to occur once more at a specific value of b/a and hence at a specific Z_0.

Just as before, we invoke a couple of equations to get to an actual numerical result. The only new one we need here is an expression for the resistance R. If we make the usual assumption that the current flows uniformly in a thin cylinder of a thickness equal to the skin depth δ, we can write

$$R \approx \frac{1}{2\pi\delta\sigma}\left[\frac{1}{a} + \frac{1}{b}\right], \tag{21}$$

---

[9] Smith actually received a patent for this result. The single claim in the patent explicitly claims the ratio, $\sqrt{e}$ (about 1.6487).

where σ is the conductivity of the wire and δ is the same as always:

$$\delta = \sqrt{\frac{2}{\omega\mu\sigma}}. \tag{22}$$

With these equations, the attenuation constant may be expressed as:

$$\alpha = \frac{R}{2Z_0} \approx \frac{\frac{1}{2\pi\delta\sigma}\left[\frac{1}{a}+\frac{1}{b}\right]\sqrt{\varepsilon_r}}{2\left[60\ln\left(\frac{b}{a}\right)\right]}. \tag{23}$$

Taking the derivative, setting it equal to zero, and now praying for a minimum instead of a maximum yields:

$$\frac{d\alpha}{da} = 0 \implies \frac{d}{da}\frac{1/a + 1/b}{\ln(b/a)} = 0 \implies \ln\left(\frac{b}{a}\right) = 1 + \frac{a}{b}; \tag{24}$$

after iteration, this yields a value of about 3.6 for b/a. That value corresponds to a Z_0 of about 77 Ω. Now we have all the information we need.

First off, cable TV equipment is based on a 75-Ω world because it corresponds (nearly) to minimum loss. Power levels there are low, so power-handling capability is not an issue. So why is the standard there 75 and not 77 ohms? Simply because engineers like round numbers.

This affinity for round numbers is also ultimately the reason for 50 Ω (at last). Since 77 Ω gives us minimum loss and 30 Ω gives us maximum power-handling capability, a reasonable compromise is an average of some kind. Whether you use an arithmetic or geometric mean, the result after rounding is 50 Ω. And that's it.

3.5.3 SUMMARY

Now that we understand how the macroscopic universe came to choose 50 Ω, it should be clear that one should feel free to choose very different impedance levels if performance is limited neither by the power-handling nor the attenuation characteristics of the interconnect. As a result, IC engineers in particular have the luxury of selecting vastly different impedance levels than would be the norm in discrete design. Even in certain discrete designs, it is worth evaluating the trade-offs associated with using 50 Ω (or some other standard value) rather than simply using the standard values as a reflex.

CHAPTER FOUR

IMPEDANCE MATCHING

4.1 INTRODUCTION

Designers of low-frequency analog circuits are often puzzled by the seeming obsession of microwave engineers with impedance matching. Analog circuit design textbooks, for example, almost never have a chapter on this topic. Instead, engineers working at lower frequencies usually express specifications in terms of voltage gain, for example, with little or no reference to impedance matching. In striking contrast, RF engineers are indeed frequently preoccupied with the problem of impedance matching. The principal reason for the difference in philosophical outlook is that power gain is so abundant at low frequencies that designers there have the luxury of focusing on convenience, rather than necessity. For example, a textbook transformer could provide impedance matching to maximize power gain, but electronic voltage amplifiers are more readily realized (and are certainly more flexible) than are coils of wire wound around magnetic cores. The need for an impedance transformation is usually acknowledged only implicitly (if at all) and is frequently satisfied with a crude transformation to an unspecified low value with a voltage buffer, for example. On the other hand, RF power gain is often an extremely limited resource, so one must take care not to squander it. Impedance-transforming networks thus play a prominent role in the radio frequency domain.

There are many good reasons for seeking an impedance match at RF beyond simply maximizing gain. One is that a match makes a system insensitive to the lengths of interconnecting lines. Discrete RF and microwave components are generally designed for standardized impedances (such as 50 Ω) for precisely this reason. This standardization greatly facilitates assembly of such components into larger systems by eliminating the need for impedance transformers between them. Even so, coupling into and out of these standardized components may still require impedance transformations if the source or load impedances happen to differ from the standard values. This situation arises frequently in the case of coupling to antennas or other resonant structures, for example.

Maximizing power gain is not the only objective that requires impedance transformations. For every amplifier there is a particular value of impedance that minimizes its noise figure. As luck would have it, this magic value rarely coincides with the conditions that produce maximum gain or with standardized levels such as 50 Ω. Once again, impedance transformations are necessary.

Yet another motivation for providing a match is that filters are often the performance-limiting components in communications systems.[1] Important filter characteristics, such as passband and delay flatness, can be quite sensitive to the driving and terminating impedances, so impedance matching once again arises as a requirement.

Impedance matching is also valuable in high-power systems, where the presence of standing waves arising from mismatch may cause peak voltages to exceed breakdown limits at various points along a transmission line. Catastrophic breakdown is not the only possible outcome, for the output power and efficiency of a transmitter may depend critically on the impedance of the load it drives. An improper impedance may degrade efficiency (and thereby exacerbate thermal problems), cause unstable operation (including oscillation), prevent the attainment of a specified output power, or even damage the transmitter.

To study the important problem of impedance matching in greater detail, we start by presenting a theorem whose lessons provide much of the rationale for considering impedance matching at all.

4.2 THE MAXIMUM POWER TRANSFER THEOREM

Most undergraduate engineering students encounter the maximum power transfer theorem at some point in their coursework. Evidence is strong, however, that the full implications of the theorem are conveyed to students with less than perfect fidelity (debates rage about whether the problem lies in transmission or reception; we will not presume to settle the matter here). The question answered by the maximum power transfer theorem is this: Given a *fixed source* impedance Z_S, what load impedance Z_L maximizes the power delivered to that load? See Figure 4.1, which shows a network driven by a sinusoidal source.[2]

The power delivered to the load impedance by the source is entirely consumed by R_L, since reactive elements do not dissipate power. Hence, the power delivered to the load is just

$$\frac{|V_R|^2}{R_L} = \frac{R_L |V_S|^2}{(R_L + R_S)^2 + (X_L + X_S)^2}, \tag{1}$$

---

[1] This need not be the case in principle, but engineering practice frequently makes it so, for one commonly uses the least expensive filter that can meet design objectives.

[2] Where many students go awry is to forget the "fixed source impedance" part of the problem statement. If we were truly free to choose any *source* impedance, then of course we should choose to make it *zero*, not equal to the load! Then we could extract infinite power from the source.

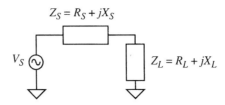

FIGURE 4.1. Network for maximum power transfer theorem

where the voltages V_R and V_S are the rms voltages across the load resistance and source, respectively.

In order to maximize the power delivered to R_L, it's clear that X_L and X_S should be algebraic inverses so that they sum to zero. That is, we want to have a net zero reactance. It's relatively easy to produce this condition at a single frequency or to approximate it over a narrow band. Further maximizing Eqn. 1 under that condition leads to the result that R_L should equal R_S. Hence, the maximum transfer of power occurs when the load and source impedances are complex conjugates.

Notice that this prescription implicitly tells us that the impedance transformation must be accomplished *losslessly*. The addition of, say, a series resistance to assure that the source impedance sees an equal resistance does not maximize power delivered to the *load*, because that added resistance is not part of the load to which we ultimately wish to deliver power.[3]

We can thus state a general strategy for maximizing power transfer: Using only lossless elements, null out any reactances while transforming the existing resistances until source and load match.

It is fortunate that there are many ways to implement that strategy, for it's true that no single impedance matching method is ideal in all cases. We'll therefore present a number of commonly used methods, which may be used alone or in combination.

MATCHING BENCHMARKS

All engineering involves trade-offs of one sort or another, and impedance matching is no exception to this rule. Because perfection is unattainable, we must set realistic goals for impedance matching. There are no hard and fast rules about what constitutes a good match, however; the required performance is very much dependent on the application. It is therefore important to develop fluency in describing mismatch in the several equivalent ways in which it is commonly reported: reflection coefficient Γ, which is the same as S_{11}; return loss (RL), which is simply the reciprocal of S_{11} (or its algebraic inverse when using dB quantities); and the standing-wave ratio (SWR), defined as

---

[3] Sometimes an added resistance is helpful for producing other results, such as taming parasitic oscillations in amplifiers. However, these ends are quite distinct from maximizing power gain.

Table 4.1. *Equivalent measures of mismatch*

$|\Gamma|$ ($|S_{11}|$)	RL (dB)	SWR
0	infinite	1.00
0.025	32.0	1.05
0.05	26.0	1.11
0.075	22.5	1.16
0.1	20.0	1.22
0.15	16.5	1.35
0.2	14.0	1.50
0.25	12.0	1.67
0.3	10.5	1.86
0.35	9.1	2.07
0.4	8.0	2.33
0.45	6.9	2.64
0.5	6.0	3.00

$$\text{SWR} = \frac{1+|\Gamma|}{1-|\Gamma|}. \qquad (2)$$

Recall that SWR is the ratio of the peak amplitude on the line to the minimum amplitude. A perfect match produces no reflections that would alternately interfere destructively and constructively with the forward wave, and thus it produces a 1:1 SWR. The worse the mismatch, the greater the disparity between the amplitudes resulting from constructive reinforcement and destructive interference, and the greater the SWR.

Because contours of constant reflectance magnitude are simply concentric circles, the distance from the center of the Smith chart is a direct measure of mismatch. As a convenience, the relationships among all of these metrics is taken care of on a typical (full) Smith chart. There is a set of auxiliary rulers below the chart proper that allows you to read off the mismatch in any of the equivalent ways in which mismatch is measured. Table 4.1 provides equivalent information.

As we've already mentioned, the quality of the match required is a function of the application. That said, it's generally the case that SWR values above 3 are rarely tolerated. A more typical specification might be expressed as a minimum return loss of 10 dB, with values in excess of 15 dB considered highly desirable. These goals correspond to SWR values of very roughly 2:1 and 1.5:1, respectively.

4.3 MATCHING METHODS

The maximum power transfer theorem tells us only that we wish to nullify the reactive part while simultaneously transforming the resistive part of the load to a value equal to that of the source. There are a great many ways of accomplishing these ends, all possessing different trade-offs among such dimensions as size, complexity, and bandwidth. We will first examine a collection of techniques that are relatively narrowband in nature; then we consider a number of methods for realizing broadband impedance matches.

4.3.1 CLASSIC LUMPED MATCHING METHODS

Review of Resonant Circuits

A great many of the narrowband impedance matching methods developed over the years are explainable with a unified treatment based on an understanding of lumped resonant circuits. Consequently, we first review briefly some properties of RLC resonant circuits.

The Q of a parallel RLC network at resonance is

$$Q = R/\sqrt{L/C}. \tag{3}$$

The quantity $\sqrt{L/C}$ has the dimensions of resistance and is sometimes called the *characteristic impedance* of the network.[4] It has the following significance: it is equal to the magnitude of the capacitive and inductive reactances at resonance. This is easily shown:

$$|Z_C| = |Z_L| = \omega_0 L = L/\sqrt{LC} = \sqrt{L/C}. \tag{4}$$

We will find that this quantity recurs with some frequency,[5] so keep it in mind.

Before we continue, let's see if our equation for Q makes sense. As the parallel resistance goes to infinity, Q does, too. This behavior seems reasonable since, in the limit of infinite resistance, the network degenerates to a pure LC system. With only purely reactive elements in the network, there is no way for energy to dissipate and so Q should go to infinity, just as the equation says it should. Plus, Q also increases as the impedance of the reactive elements decreases (by decreasing L/C), since the pure resistance becomes less significant compared with the reactive impedances.

For completeness, we may derive a couple of additional expressions for the Q of our parallel RLC network at resonance:

$$Q = \frac{R}{|Z_{L,C}|} = \frac{R}{\omega_0 L} = \omega_0 RC. \tag{5}$$

The ability of such networks to transform impedances is suggested by the behavior of the inductive and capacitive branch currents at resonance. We can readily compute how these currents may differ significantly from the overall network current (which

---

[4] This term is usually applied to transmission lines, but we see that it has a certain importance even in lumped networks.
[5] Recall that the characteristic impedance of a transmission line is given by the same expression, where L and C are there interpreted as the inductance and capacitance *per unit length*.

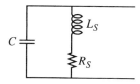

FIGURE 4.2. Not-quite-parallel RLC tank circuit

is simply due to the parallel resistance) and thereby extract some important insights about impedance transformation techniques.

At resonance, the voltage across the network is $I_{in}R$. Since the inductive and capacitive reactances are equal at resonance, the inductive and capacitive branch currents will be equal in magnitude:

$$|I_L| = |I_C| = \frac{|V|}{Z} = \frac{|I_{in}|R}{\omega_0 L} = \frac{|I_{in}|R\sqrt{LC}}{L} = |I_{in}|\frac{R}{\sqrt{L/C}} = Q|I_{in}|. \qquad (6)$$

That is, the current flowing in the inductive and capacitive branches is Q times as large as the net current into the whole network. Hence, if $Q = 1000$ and we drive the network at resonance with a one-ampere current source, then that one ampere will flow through the resistor but a thrilling *one thousand amperes* will flow through the inductor and capacitor (until they vaporize). From this simple example, you can well appreciate the incompleteness of simply stating that the inductor and capacitor cancel at resonance!

We might infer from these dramatic boosts in current that the network has somehow performed a downward impedance transformation. That is, the impedance in the high-current branches must be lower than the impedance across the entire combination. To pursue this idea further, consider the case illustrated by Figure 4.2. We've chosen to put a resistor in series with the inductor to reflect the practical truth that inductors are generally much lossier than capacitors. The model shown in the figure is therefore a realistic first approximation to typical parallel *RLC* circuits encountered in practice.

Since we've already analyzed the purely parallel *RLC* network in detail, it would be nice if we could re-use as much of that work as possible. So, let's convert the circuit of Figure 4.2 into a purely parallel *RLC* network by replacing the series *LR* section with a parallel one. Clearly, such a substitution cannot be valid in general, but over a suitably restricted frequency range (e.g., near resonance) the equivalence is pretty reasonable. To show this formally, let's equate the impedances of the series and parallel *LR* sections:

$$j\omega_0 L_S + R_S = (j\omega_0 L_P) \parallel R_P = \frac{(\omega_0 L_P)^2 R_P + j\omega_0 L_P R_P^2}{R_P^2 + (\omega_0 L_P)^2}. \qquad (7)$$

If we equate real parts and note that $Q = R_P/\omega_0 L_P = \omega_0 L_S/R_S$,[6] we obtain

---

[6] If the series and parallel sections are to be equivalent, then their Qs certainly must be equivalent.

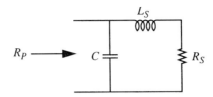

FIGURE 4.3. Upward impedance transformer

$$R_P = R_S(Q^2 + 1). \tag{8}$$

Similarly, equating imaginary parts yields

$$L_P = L_S\left(\frac{Q^2 + 1}{Q^2}\right). \tag{9}$$

We may also derive a similar set of equations for computing series and parallel RC equivalents:

$$R_P = R_S(Q^2 + 1); \tag{10}$$

$$C_P = C_S\left(\frac{Q^2}{Q^2 + 1}\right). \tag{11}$$

Let's pause for a moment and look at these transformation formulas. Upon closer examination, it's clear that we may express them – in a universal form that applies to both RC and LR networks – as

$$R_P = R_S(Q^2 + 1), \tag{12}$$

$$X_P = X_S\left(\frac{Q^2 + 1}{Q^2}\right), \tag{13}$$

where X is the imaginary part of the impedance. This way, one need only recall a single pair of "universal" formulas to convert any "impure" RLC network into a purely parallel (or series) one that is straightforward to analyze. Both parallel terms are always larger than their corresponding series terms. We must keep in mind that *the equivalences hold only over a narrow range of frequencies centered about ω_0.*

The L-Match

Now we're ready to reinterpret all of these results as descriptive of impedance transformers. As we've noted, the multiplication by Q of voltages or currents in resonant RLC networks hints at their impedance-modifying potential. Indeed, the series–parallel RC/LR network conversion formulas developed in the previous section actually show this property explicitly. To make this clearer, consider once again the circuit of Figure 4.2, now redrawn slightly as Figure 4.3.

4.3 MATCHING METHODS

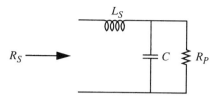

FIGURE 4.4. Downward impedance transformer

Here, we treat R_S as a load resistance for the network. When this resistance is viewed across the capacitor, it is transformed to an equivalent R_P according to the formulas developed in the previous section. From inspection of those "universal" equations, it is clear that R_P will always be larger than R_S, so the network of Figure 4.3 transforms resistances *upward*. If a downward impedance conversion is needed then we simply interchange ports, as shown in Figure 4.4.

There is a nice, intuitive way to keep track of which way the impedance transformation goes. For example, if the circuit in Figure 4.4 is driven by a test voltage source, the result is a parallel RLC network since the Thévenin resistance of the test source is zero. Now, the inductive current in a parallel RLC network at resonance is Q times as large as the current through R_P. This increase in current is seen by the source and may be interpreted as a reduction in resistance.

It should also be clear that interchanging the inductor and capacitor doesn't alter the transformation ratio, so you may accommodate other design considerations in deciding whether to use a high-pass or low-pass configuration. For example, if the source or load already possesses some reactance, it may be possible to absorb these undesired elements into the impedance transformer with the correct topological choice. Another consideration is the bandwidth over which a reasonable match is maintained. As it happens, the high-pass versions tend to be better in this regard. This is to be expected from Murphy's law, because the low-pass version is usually more useful (and thus used more often) for filtering undesired spectral components.

This circuit is known as an *L-match* because of its shape (you might have to contort yourself a bit to see this), not because it contains an inductor. Its chief attribute is its simplicity. However, there are only two degrees of freedom (one can choose only L and C). Hence, once the impedance transformation ratio and frequency have been specified, network Q (which might influence the frequency-selective properties of the network, including bandwidth) is automatically determined. If you want a different value of Q, you must use a network that offers additional degrees of freedom; we'll study some of these shortly.

As a final note on the L-match, the "universal" equations can be simplified if $Q^2 \gg 1$. If this inequality is satisfied, then the following approximate equations hold:

$$R_P \approx R_S Q^2 = R_S \left(\frac{1}{\omega_0 R_S C} \right)^2 = \frac{1}{R_S} \frac{L_S}{C}, \tag{14}$$

which may be rewritten as

$$R_P R_S \approx \frac{L_S}{C} = Z_0^2, \tag{15}$$

where Z_0 is the characteristic impedance of the network, as discussed in Section 4.3.1. Stated alternatively, select the characteristic impedance of the matching network equal to the geometric mean of the source and load impedances. When we consider the classic quarter-wavelength transmission line transformer, we will encounter a prescription that is highly reminiscent of this one.

One may also deduce that Q is approximately the square root of the transformation ratio:

$$Q \approx \sqrt{R_P/R_S}. \tag{16}$$

Finally, the reactances don't vary much in undergoing the transformation:

$$X_P \approx X_S. \tag{17}$$

As long as Q is greater than about 3 or 4, the error incurred will be under about 10%. If Q is greater than 10, the maximum error will be in the neighborhood of 1% or so. For quick, back-of-the-envelope calculations, at least, these simplified equations are adequate. Final design values can always be computed using the full "universal" equations if needed.

The L-match is also a nice, simple network with which to illustrate impedance matching graphically in terms of the Smith chart. We can think about the impedance-matching problem as a sort of game in which the goal is to return home from an arbitrary initial location, where "home" is located at the precise center of the Smith chart.[7] The rules of the game are such that motion is not along straight lines; that's what makes the game challenging and interesting. Indeed, the curved coordinate system that underlies the Smith chart means that each step we take will traverse a circular arc. The challenge is to assemble a sequence of arcs that will take us home. Developing a familiarity with how those trajectories correspond to the addition of shunt or series elements is the key to designing the matching networks we present explicitly – as well as to devising new ones on your own.

The addition of a series reactive element simply moves us along a circle of constant resistance. Similarly, the addition of a shunt susceptance simply moves us along a circle of constant conductance. Because impedance matching networks generally consist of a combination of series and shunt elements, it is helpful to use (or at least imagine) the complete ZY Smith chart when designing matching networks. Later we will add more trajectories to this mix, such as those that correspond to the addition of transmission line segments and those arising from changes in frequency.

These points are best illustrated with a specific example, so let's revisit the L-match network in Smith-chart terms. In Figure 4.5 we presume that the load resistance R_S

---

[7] A completely equivalent description is to seek to arrive at an arbitrary terminus from home as the initial position. The same rules of motion apply; one just runs the game in reverse. For an actual on-line impedance-matching "video game" played in this fashion, see ⟨http://contact.tm.agilent.com/Agilent/tmo/an-95-1/classes/imatch.html⟩.

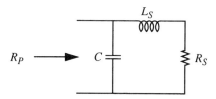

FIGURE 4.5. L-match example

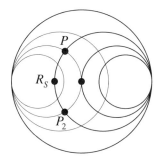

FIGURE 4.6. L-match example with ZY Smith chart
(thin lines are constant-conductance circles)

is too low, so that we want to transform it upward. Addition of a series inductance L_S to that load moves us along the circle of constant resistance corresponding to $R = R_S$, and the subsequent addition of a shunt capacitance C moves us along a circle of constant conductance. Ultimately, we want to end up at the center of the Smith chart, so this circle of constant conductance must include the terminal point $G = 1$. To design the match, then, simply draw two circles: one of constant conductance corresponding to $G = 1$, and the other of constant resistance corresponding to R_S. The intersection(s) of those two circles fixes both the inductance and the capacitance. A subsequent denormalization then yields the component values, completing the design.

As seen in Figure 4.6, we first move clockwise along a circle of constant resistance from the load resistance point, R_S, by adding a series inductance. We add just enough inductance to get to point P because it also lies on the (gray) circle of constant conductance for $G = 1$. That choice is significant because the subsequent addition of an appropriate shunt capacitance can move us clockwise along that circle down to the center of the Smith chart to complete the match. From reading off the shunt admittance or series reactance corresponding to point P, we can readily compute the capacitance and inductance, respectively.[8]

Note that there is another point of intersection of the two circles, at P_2, implying the existence of another (equivalent) matching network. In this case, a series capacitance takes us counterclockwise down to point P_2 along the constant-R_S circle, and

---

[8] For each point of intersection, we're talking about a single impedance from which both the capacitance and inductance may be found.

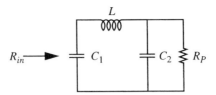

FIGURE 4.7. The π-match (low-pass version)

the addition of a shunt inductance takes us counterclockwise along the circle of constant conductance for $G = 1$. We add enough shunt inductance to end up again at the point $G = 1$. The shunt admittance and series reactance corresponding to point P_2 allow us to compute the shunt inductance and series capacitance to complete the design. The corresponding network corresponds to the high-pass version of the upward-transforming L-match.

From the Smith-chart constructions, we can also see that complex load impedances are readily accommodated as well. If, for example, the load already has some series inductance, then one doesn't have to add as much to get to point P, after which the design proceeds as before. Thus, we appreciate better how the correct choice of network can reduce the impact of parasitics by actually using them as circuit elements.

The Smith-chart constructions also help us remember that this network topology cannot transform downward. If the normalized load resistance exceeds unity, then the addition of a series reactance necessarily moves us along a constant-resistance circle that never intersects the $G = 1$ constant-conductance circle. Thus, there is no way for the addition of a shunt element to bring us to the center of the Smith chart. These same sorts of constructions allow us to deduce that merely exchanging ports will enable the desired impedance transformation.

The π-Match

As already discussed, one limitation of the L-match is that it allows us to specify only *two* of center frequency, impedance transformation ratio, and Q. To acquire a third degree of freedom, one can employ the network shown in Figure 4.7.

This circuit is known as a π-match, again because of its shape. The most expedient way to understand how this matching network functions is to view it as two L-matches connected in cascade, one that transforms down and one that transforms up; see Figure 4.8. Here, the load resistance R_P is transformed to a lower resistance (known as the *image* or *intermediate resistance,* here denoted R_I) at the junction of the two inductances. The image resistance is then transformed up to a value R_{in} by a second L-match section.

Now, it may feel suspiciously like a government-works project to use one L-section to go down and then another to go back up. However, we have gained an important additional degree of freedom. Recall that, for an L-match, Q is fixed at a value roughly equal to the square root of the impedance transformation ratio. Typically, the Q of an L-match isn't particularly high because huge transformation factors are

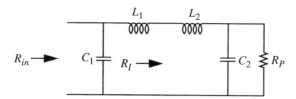

FIGURE 4.8. The π-match as a cascade of L-matches

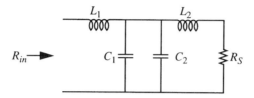

FIGURE 4.9. T-match

infrequently required. The π-match decouples Q from the transformation ratio by introducing an intermediate resistance value to transform to; this allows us to achieve much higher Q than is generally available from an L-match, even if the *overall* transformation ratio isn't particularly large.

Since we now have three degrees of freedom (the two capacitances and the sum of the two inductances), we can independently specify center frequency, Q (or bandwidth), and overall impedance transformation ratio. However, as with the L-match (or any other kind of match), impractical or inconvenient component values can result, so some creativity or compromise may be required to generate a sensible design. In many instances, cascading several matching networks may be helpful (e.g., extending the π-match concept with additional sections).

As a parting note, one final bit of trivia deserves mention. An additional reason that the π-match is popular is that the parasitic capacitances of whatever connects to it can be absorbed into the network design. This property is particularly valuable because capacitance is frequently the dominant parasitic element in many practical cases.

The T-Match

The π-match results from cascading two L-sections in one particular way. Connecting up the L-sections another way leads to the dual of the π-match: the T-match, shown in Figure 4.9. Here, what would be a single capacitor in a practical implementation has been decomposed explicitly into two separate ones to show clearly that this network consists of two conventional L-matches connected together at their high-impedance ports. The (parallel) image resistance is seen across these capacitors, either looking to the right or looking to the left as in the π-match.

The T-match is completely equivalent in performance to the π-match. It is the preferred choice when the source and termination parasitics are primarily inductive

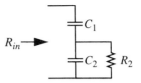

FIGURE 4.10. Tapped-capacitor matching network

in nature, allowing them to become part of the network, just as port capacitances are readily absorbed into the π-match.

Note that there remain other ways to cascade L-sections. They occasionally merit consideration, but are used infrequently enough that we simply call attention to their existence.

Tapped Capacitors and Inductors

A resistive voltage divider transforms impedances but is lossy. If we replace the resistances with pure reactances then the resulting network continues to transform impedance, but now without loss. There will also be residual reactance, but that can be tuned out separately. An example of such a circuit is shown in Figure 4.10.

The operation of this impedance transformer is best understood by noting that a voltage reduction in a perfectly lossless network must be accompanied by an impedance reduction proportional to the square of the voltage attenuation if power is to be conserved. This network is not perfectly lossless, but we do expect the impedance transformation ratio to be (roughly)

$$\frac{R_2}{R_{in}} \approx \left(\frac{1/sC_2}{1/sC_1 + 1/sC_2}\right)^2 = \left(\frac{C_1}{C_1 + C_2}\right)^2, \qquad (18)$$

so that the network either transforms a resistance R_{in} downward to a value R_2 or transforms a resistance R_2 upward to a value R_{in}.

To confirm this expectation, let us analyze the resistively loaded capacitive divider in isolation. The admittance of the combination is readily found after a little labor:

$$Y_{in} = \frac{j\omega C_1 - \omega^2 R_2 C_1 C_2}{j\omega R_2 (C_1 + C_2) + 1}. \qquad (19)$$

The real part is

$$G_{in} = \frac{\omega^2 R_2 C_1^2}{\omega^2 R_2^2 (C_1 + C_2)^2 + 1}. \qquad (20)$$

At sufficiently high frequencies, the equivalent shunt conductance indeed simplifies to

$$G_{in} \approx \frac{\omega^2 R_2 C_1^2}{\omega^2 R_2^2 (C_1 + C_2)^2} = G_2 \cdot \left[\frac{C_1}{C_1 + C_2}\right]^2 = \frac{G_2}{n^2}, \qquad (21)$$

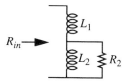

FIGURE 4.11. Tapped-inductor transformer

as anticipated. Equation 21 also defines a factor, n, which is the turns ratio of an ideal transformer that would yield the same resistance transformation as the capacitive divider. The concept of an equivalent turns ratio will prove particularly useful in unifying the treatment of various oscillators.

For the sake of completeness, we also compute the imaginary part of the admittance:

$$B_{in} = \frac{\omega C_1 + \omega^3 R_2^2 C_1 C_2 (C_1 + C_2)}{\omega^2 R_2^2 (C_1 + C_2)^2 + 1}; \tag{22}$$

at sufficiently high frequencies, this approaches a limiting value of

$$B_{in} \approx \omega \cdot \frac{C_1 C_2}{C_1 + C_2} = \omega \cdot C_{eq}, \tag{23}$$

where C_{eq} is the equivalent capacitance of the series combination of the two individual capacitances.

The foregoing series of equations serves well for analysis and particularly to develop design intuition. Equations 21 and 23 are also extremely useful for first-cut, back-of-the-envelope designs.

We now consider briefly the tapped *inductor* as a matching network (see Figure 4.11). As you might expect, its behavior is quite similar to that of its tapped-capacitor counterpart. We won't go through a detailed derivation of the design equations since they're completely analogous to those for the tapped-capacitor case, but we will make the following observation: R_2 must be less than R_{in} because, once again, we have a voltage divider.

Clearly, tapping as a method of impedance transformation applies more broadly than it does to pure capacitor or pure inductor networks. Indeed we shall see that coupling into and out of resonators is often accomplished through the use of impedance-transforming taps.

4.3.2 CLASSIC TRANSMISSION LINE IMPEDANCE TRANSFORMERS

As frequency increases, lumped implementations become progressively less appropriate. We therefore need to expand our palette of options to include several distributed matching techniques.

The most famous by far is the quarter-wave transmission line transformer. In terms of the Smith chart, recall that the addition of line to a load simply causes us to

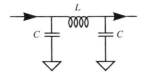

FIGURE 4.12. Lumped λ/4 impedance match

traverse a circle (again), this one representing a contour of constant reflectance, and that these contours are centered about the origin ($R = 1$). Also recall that impedance repeats every half-wavelength, representing one full rotation around the Smith chart. Thus, the addition of a piece of line that is a quarter-wavelength long causes us to travel precisely halfway around the constant-reflectance circle. Consequently, a given load impedance Z_1 at the end of the line transforms into its reciprocal (within a normalization constant) when viewed at the other end of the line. Specifically,

$$Z_2 = Z_0^2/Z_1. \tag{24}$$

This relationship tells us that we may use a quarter-wave piece of line as an impedance transformer by suitably selecting its characteristic impedance as follows:

$$Z_0 = \sqrt{Z_1 Z_2}. \tag{25}$$

This transformer is necessarily a narrowband element because its length meets the quarter-wave criterion at only one frequency. Another consideration is that there are limitations on the practical impedances one may realize, particularly in microstrip form. Thus, it may be difficult to realize the desired transformer in cases that require characteristic impedances that are below about 10–15 Ω or above about 150–200 Ω for typical microstrip.

If (as at low frequencies) the length of line required is impractically large or if the required characteristic impedance is inconvenient, then a lumped implementation may be preferable. The simplest of these is a π- or T-network approximation to a λ/4 line. See Figure 4.12.

The component values are given by these simple relationships:

$$L = \sqrt{R_1 R_2}/\omega_0; \tag{26}$$

$$C = 1/\omega_0 \sqrt{R_1 R_2}. \tag{27}$$

Here R_1 and R_2 are the source and load impedances. You can see that these formulas simply correspond to making the magnitudes of the inductive and capacitive reactances equal to the geometric mean of the source and load impedances.

The reader may have noticed that this network is actually a special case of a π-match. Its virtue is ease of design.

Quarter- and half-wave transmission lines thus have well-known special impedance transformation properties: the former reciprocates, and the latter replicates. Because complex loads thus remain complex, completing the design of an impedance match in general requires an additional element to cancel out any remaining imaginary part.

There is a third, less well-known, "magic" line length whose properties occasionally prove useful in the design of impedance transformers. Lines that are $\lambda/8$ in length have the ability to transform a complex load into a purely real one. To demonstrate this property, we first reprise the equation for the input impedance of a lossless line terminated in an arbitrary load:

$$Z(z) = Z_0 \frac{Z_L + jZ_0 \tan \beta z}{Z_0 + jZ_L \tan \beta z}, \tag{28}$$

where the coordinate z is zero at the load and takes on positive values elsewhere. Setting βz equal to $\pi/4$ yields

$$Z = Z_0 \frac{Z_L + jZ_0}{Z_0 + jZ_L} = Z_0 \frac{(R_L + jX_L) + jR_0}{R_0 + j(R_L + jX_L)} = Z_0 \frac{R_L + j(R_0 + X_L)}{(R_0 - X_L) + jR_L}. \tag{29}$$

Rationalizing the denominator and solving, we obtain

$$Z = Z_0 \frac{R_L(R_0 - X_L) + R_L(R_0 + X_L) + j[(R_0 + X_L)(R_0 - X_L) - R_L^2]}{(R_0 - X_L)^2 + R_L^2}. \tag{30}$$

Clearly, the impedance is not *automatically* purely real, but it is straightforward to identify the conditions under which the imaginary part disappears:

$$(R_0 + X_L)(R_0 - X_L) - R_L^2 = 0 \implies R_0^2 = X_L^2 + R_L^2. \tag{31}$$

Thus, if we select the characteristic impedance of a $\lambda/8$ line equal to the *magnitude* of the (complex) load impedance, then the transformed impedance will be purely real and with value

$$Z = Z_0 \frac{R_L(R_0 - X_L) + R_L(R_0 + X_L)}{(R_0 - X_L)^2 + R_L^2}, \tag{32}$$

which (after simplification) becomes

$$Z = Z_0 \frac{R_L}{R_0 - X_L} = Z_0 \frac{R_L}{|Z_L| - X_L}. \tag{33}$$

Note that the relationships among the variables guarantee a nonnegative resistance, satisfying necessary conditions on passivity. Further note that, as the load impedance approaches a pure reactance, the transformed resistance approaches infinity. Again, this result is to be expected on the basis of energy conservation alone, for the combination of a lossless line with a lossless reactance cannot result in a lossy impedance.

The $\lambda/8$ line, useful though it may be, rarely suffices to effect a match by itself, for merely converting a complex load into a real one is insufficient in general. The ultimate goal is to produce not merely a random real impedance but rather a particular value that is equal to Z_0. A classic microwave matching technique with more general utility is known as the *single-stub* match (see Figure 4.13). It exploits the

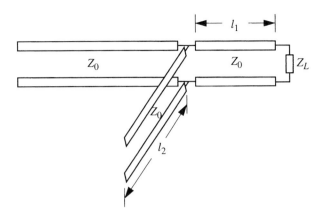

FIGURE 4.13. Illustration of single-stub impedance matching

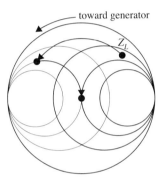

FIGURE 4.14. Single-stub match example with Smith chart (thin lines are admittance circles)

impedance variation along a transmission line with a mismatched load, and its action is best described in terms of Smith-chart trajectories. The addition of line moves the impedance around a circle of constant reflectance. Since we will ultimately add some shunt impedance (the stub), we want to add enough line to take us to a constant-conductance circle corresponding to $G = 1$. Then we add enough stub to take us to the center of the Smith chart.

As seen in Figure 4.14, adding some line moves us counterclockwise on a constant-reflectance circle whose radius, of course, is established by the value of Z_L (normalized). There are two values of line length l_1 that take us to the correct constant-conductance circle ($G = 1$), at which point we have obtained the correct real part. It is customary, but not obligatory, to select the shorter of the two possible lengths, as we have done here.[9] This choice reduces departures from expectations resulting

---

[9] Not only does this choice reduce the required length of line, it often leads to a somewhat broader band match. As a rule, it's generally a good idea to choose the option that minimizes length unless there's some other overriding consideration (e.g., bias convenience). Losses are smaller, the chances for radiation frequently diminish, and the structure occupies less space.

from the lossiness of real lines and also produces more compact structures. However, under some conditions (e.g., at very high frequencies), the required line length might be uncomfortably short. In such cases, it may be more convenient to select the longer length or occasionally even add integer multiples of a half-wavelength. Other practical considerations, such as ease of conveying bias to amplifiers, may also favor the other choice.

Once the value of l_1 is chosen, all that remains is to cancel any remanent imaginary part with a suitable shunt stub. In principle, an appropriate length of line terminated in either a short or an open would accomplish the desired end. Most recommendations found in the literature advocate the use of shorted lines as stubs for several quite sensible reasons. One is that it is possible to approximate a good short to a better degree than one may approximate a good open, since a fringing field always frustrates the production of a good RF open circuit. Another is that radiation from an open-circuited line may further perturb the impedance of the stub, to say nothing of the possibility of creating (or receiving) interference. Counterbalancing all of those good reasons is the inconvenience of implementing shorts in microstrip. Experience shows that the use of open-circuited stubs in microstrip form generally yields satisfactory results in spite of the concerns expressed, so we will assume from this point on that completing the impedance match will involve the shunt connection of an appropriate length of open-circuited line. In our example, note that selection of the shorter value of l_1 has already taken us to a point on the constant-conductance circle that requires the addition of a shunt capacitance to complete the match. Had we chosen the other line length, an inductive stub would have been required.

For completeness, we should also note that one could (again, in principle) complete the impedance match by using a stub in *series* with the load. However, this approach is more mechanically inconvenient than the shunt approach because it requires breaking existing connections. The shunt stub is therefore overwhelmingly more popular for microstrip.

The single-stub tuner, while perfectly effective as described, requires that the shunt stub connect to the line at a load-specific distance from the load. If the load impedance or distance is not known with infinite accuracy (or varies owing to manufacturing tolerances, etc.), then one must arrange for some adjustment of this point of attachment. Unfortunately, providing for a variable stub location is a nontrivial task, as is arranging for variation of the stub length. Commercial tuners are generally coaxial affairs and often use a stub with a sliding short (the combination is known colloquially as a trombone), but it is highly inconvenient to arrange for a sliding anything in microstrip form.

Accommodation of large changes in load impedance is possible with fewer mechanical engineering difficulties. Multiple-stub tuners consist of two or three (or occasionally more) single-stub tuners spaced apart by some fixed amount; see Figure 4.15. The added degree of freedom provided by the extra stub compensates partly for the degrees of freedom lost by fixing the stub locations, allowing the double-stub arrangement to provide a usefully wide tuning range. Of the many possible choices for the stub separation, most common multiple-stub tuners use a spacing of $\lambda/8$. One may

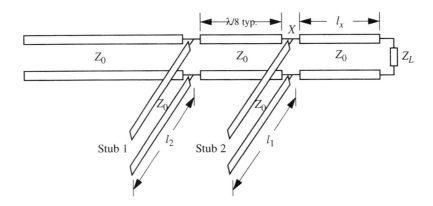

FIGURE 4.15. Double-stub tuner (with open-circuited stubs; mechanism for varying length not shown)

show that, with such a spacing, the tuner can produce a match to any load whose admittance contains a conductance less than $2G_0$ (as measured to the right of point X in Figure 4.15). Although we have drawn open-circuited stubs, they are typically implemented as sliding shorts. Again, the inconvenience of arranging for this type of short implies that you will almost certainly not encounter this structure in microstrip form.

As can be seen in Figure 4.16, the Smith-chart presentation for this tuner can be a bit intimidating. Nevertheless, we may still identify the arcs that together comprise a solution to the matching problem. In this particular case, we proceed to the origin in a series of zigzag arcs. As with other cases, the set of trajectories we show is not the only possible one that works.

The first arc corresponds to the transformation produced by the length of line l_x that lies between the load and stub 1. It is customary to place the tuner as close to the load as practical, making l_x considerably less than a wavelength, but this choice is not an absolute requirement. In Figure 4.16 we have chosen a completely arbitrary value of l_x. The arc traced out in moving from the load to the first stub lies along the circle of constant reflectance magnitude corresponding to the impedance mismatch between the load and the line.

The second arc corresponds to the action of stub 1. Here, we have assumed that the length is short enough that the open-circuited stub presents a capacitive susceptance. Therefore, the arc moves clockwise on a circle of constant conductance.

The third arc once again takes us along a circle of constant reflectance magnitude. We have assumed here that the length of stub 1 has been adjusted so that this third arc terminates on a constant-conductance circle corresponding to $G = 1$. From this point, the capacitive susceptance of stub 2 causes the admittance to traverse the $G = 1$ circle clockwise until a match is finally obtained.

We see that the double-stub tuner provides for an impedance match by spiraling the real part of the admittance stepwise, in zigzag fashion, toward the $G = 1$ circle. In practice, this tuning is almost always performed by trial and error – a method that is generally much faster than computing and then implementing an analytical solution.

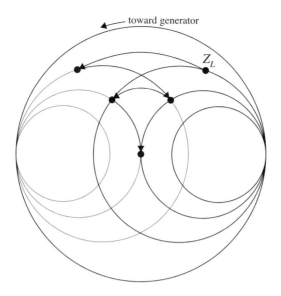

FIGURE 4.16. Double-stub match example with Smith chart (thin lines are admittance circles)

4.3.3 BROADBAND IMPEDANCE-MATCHING TECHNIQUES

The matching methods we've presented so far are fundamentally narrowband techniques, as the design procedures seek the desired correct impedance transformation at a single frequency only. One must therefore accept whatever impedance variation the design produces over the operational bandwidth. As luck would have it, the match is frequently acceptable over just a small fractional bandwidth. If a good match over a very wide frequency range is required then we must employ more sophisticated design methods, and the resulting transformation networks can quickly become rather elaborate. Before describing some of those methods, though, it's useful to consider an important bound initially derived by Hendrik Bode and extended by Robert Fano:[10]

$$\int_0^\infty \ln\left[\frac{1}{|\Gamma(\omega)|}\right] d\omega \leq \frac{\pi}{\tau}, \tag{34}$$

where τ is the time constant of the simple single-pole load impedance assumed in Bode's original derivation (Fano generalized the analysis to arbitrary loads). This inequality tells us that it is impossible to obtain a perfect match over a nonzero bandwidth (at best, we can obtain a perfect match at a finite number of frequencies within any finite frequency interval). It also tells us that we may pay for improved match

---

[10] H. W. Bode, *Network Analysis and Feedback Amplifier Design,* Van Nostrand, New York, 1945. Robert M. Fano extended Bode's work on matching theory for his Sc.D. thesis, excerpted in "Theoretical Limitations on the Broadband Matching of Arbitrary Impedances," *J. Franklin Institute,* January 1950, pp. 57–83 (part I) and February 1950, pp. 139–54 (part II).

over a given frequency range by worsening the match elsewhere. That is, designing for the best match over a given bandwidth will come at the expense of the largest possible mismatch at all other frequencies. This prescription is actually identical to that guiding the design of filters, where we wish to approach as closely as possible a zero transmission loss over some bandwidth coupled with infinite attenuation everywhere else. Consequently, the same approximation concepts that apply in the classical filter design problem apply here as well. One might consider absorbing the load into a structure reminiscent of a filter, for example, to produce or approximate Chebyshev-like reflectance behavior over some passband. We take up the subject of filter design in separate chapters, but for now it suffices to recognize that the broadband impedance-matching problem is the same as the bandpass filter design problem.

We can manipulate Eqn. 34 into a somewhat more intuitively useful form to provide an upper bound on the matching bandwidth for a specified reflectance (again, for an assumed single-pole load). Note that there is no contribution to the integral whenever the reflectance is unity. If we assume an (unattainable) ideal situation in which the reflectance is a uniform value Γ_{match} within a bandwidth $\Delta\omega$ and unity outside that bandwidth, then we can evaluate the integral directly:

$$\int_0^\infty \ln\left[\frac{1}{|\Gamma(\omega)|}\right] d\omega = -\int_0^{\Delta\omega} \ln[|\Gamma_{match}|] \, d\omega = -(\Delta\omega)\ln[|\Gamma_{match}|] \leq \frac{\pi}{\tau}. \quad (35)$$

Solving for Γ_{match} yields

$$|\Gamma_{match}| \leq \exp\left[-\frac{\pi}{(\Delta\omega)\tau}\right]. \quad (36)$$

It is also helpful to solve Eqn. 35 for the matching bandwidth:

$$\Delta\omega \leq \frac{\pi}{\tau \ln[|\Gamma_{match}|]}. \quad (37)$$

Or, expressing the bandwidth in hertz,

$$\Delta f \leq -\frac{1}{2\tau \ln[|\Gamma_{match}|]}. \quad (38)$$

We therefore see that the larger the permissible mismatch, the greater the bandwidth over which that match can be provided. To provide some quantitative orientation, accepting a Γ_{match} of 0.2 (corresponding to a 1.5 VSWR) in the passband means that the matching bandwidth will not exceed about $0.31/\tau$ Hz. In this case, the matching bandwidth is about double the -3-dB bandwidth of the load.

One extremely valuable broadband matching device is the multisection stepped-impedance line – and its continuous counterpart, the tapered transmission line. Although limited to the broadband transformation of purely real loads, these transformers are nonetheless extremely useful. The physical intuition motivating these designs is straightforward enough: If we need to connect two lines of differing impedance (and thus, presumably, of different dimensions), then using a sufficiently gradual and continuous transition between these two lines may avoid the generation of significant reflections.

We can place this idea on a more quantitative basis by examining how a single-stage quarter-wave transformer behaves away from the nominal center frequency. We start with

$$Z(z) = Z_m \frac{Z_L + jZ_m \tan \beta z}{Z_m + jZ_L \tan \beta z}, \tag{39}$$

where Z_m is the characteristic impedance of the matching section.

When the line is precisely $\lambda/4$ in length, we obtain the familiar impedance inversion:

$$Z(z) = \frac{Z_m^2}{Z_L}. \tag{40}$$

When used as an impedance matching device, we select the $\lambda/4$ line's characteristic impedance equal to the geometric mean of the source and load impedances:

$$Z_m = \sqrt{Z_0 Z_L}. \tag{41}$$

That's the ideal behavior, but now we'd like to consider what happens if the frequency is a bit above or below the nominal value. To do so in the most useful way, we first recast the impedance relationship in terms of reflectance:

$$\Gamma = \frac{\frac{Z(z)}{Z_0} - 1}{\frac{Z(z)}{Z_0} + 1} = \frac{Z_m(Z_L + jZ_m \tan \beta z) - Z_0(Z_m + jZ_L \tan \beta z)}{Z_m(Z_L + jZ_m \tan \beta z) + Z_0(Z_m + jZ_L \tan \beta z)}. \tag{42}$$

Collecting terms, simplifying, and then taking the magnitude yields

$$|\Gamma| = \left| \frac{Z_L - Z_0}{Z_L + Z_0 + j2\sqrt{Z_0 Z_L} \tan \beta z} \right| = \frac{|Z_L - Z_0|}{\sqrt{(Z_L + Z_0)^2 + 4Z_0 Z_L (\tan \beta z)^2}} \tag{43}$$

Now, at frequencies corresponding to wavelengths near $\lambda/4$ (βz near $\pi/2$), the tangent is very large. At the same time, we may approximate the sine as unity there. Consequently, we may approximate the reflection coefficient in that neighborhood as

$$|\Gamma| \approx \frac{|Z_L - Z_0|}{\sqrt{4Z_0 Z_L (\tan \beta z)^2}} = \frac{|Z_L - Z_0|}{2\sqrt{Z_0 Z_L} |\tan \beta z|}$$

$$\approx \frac{|Z_L - Z_0|}{2\sqrt{Z_0 Z_L}} |\cos \beta z| = \frac{\left|\frac{Z_L}{Z_0} - 1\right|}{2\sqrt{\frac{Z_L}{Z_0}}} |\cos \beta z|. \tag{44}$$

Since we are primarily concerned with how the reflectance behaves as one deviates a bit from center frequency, it is perhaps helpful to express the reflectance in terms of this deviation explicitly:

$$|\Gamma| \approx \frac{\left|\frac{Z_L}{Z_0} - 1\right|}{2\sqrt{\frac{Z_L}{Z_0}}} |\cos \beta z| = \frac{\left|\frac{Z_L}{Z_0} - 1\right|}{2\sqrt{\frac{Z_L}{Z_0}}} |\sin \theta|. \tag{45}$$

Here θ yields information about the deviation from center frequency, expressed as a phase angle:

$$\theta = \pi/2 - \beta z. \tag{46}$$

We see from Eqn. 45 that, for a fixed deviation from the center frequency, the reflectance magnitude gets worse (bigger) as the impedance transformation ratio grows. For very large mismatches the growth is approximately proportional to the square root of the impedance ratio. Thus, the larger the transformation ratio we demand of a single $\lambda/4$ section, the more dramatic the growth in mismatch away from the center frequency. For very small θ, the growth is approximately linear (as can be deduced from the small-angle approximation to the sine function), growing more rapidly (approximately as θ^3) for larger angles.

If Γ_{max} is the maximum reflectance that may be tolerated, then we may solve Eqn. 45 directly for the phase deviation limits:

$$|\Gamma_{max}| \approx \frac{\left|\frac{Z_L}{Z_0} - 1\right|}{2\sqrt{\frac{Z_L}{Z_0}}} |\sin \theta_{max}| \implies |\theta_{max}| \approx \sin^{-1}\left[\frac{2|\Gamma_{max}|\sqrt{\frac{Z_L}{Z_0}}}{\left|\frac{Z_L}{Z_0} - 1\right|}\right]. \tag{47}$$

The next step will be to bring frequency explicitly into this expression in order to compute the bandwidth over which the reflectance magnitude stays below the specified maximum.

First, recall that the phase constant β is

$$\beta = \omega/v_p. \tag{48}$$

Next, we know that the physical length of the line is $\lambda/4$ at the center frequency ω_0. So,

$$\theta = \frac{\pi}{2} - \beta z = \frac{\pi}{2} - \frac{\omega}{v_p}\frac{\lambda}{4} = \frac{\pi}{2} - \frac{\omega}{v_p}\frac{v_p}{4f_0} = \frac{\pi}{2}\left(1 - \frac{\omega}{\omega_0}\right). \tag{49}$$

We are interested in the two frequency limits that correspond to $\pm|\theta_{max}|$. If we call these two frequencies ω_1 and ω_2, then a little manipulation of quantities eventually allows us to write

$$\left(\frac{\omega_2 - \omega_1}{\omega_0}\right) = \frac{4}{\pi}|\theta_{max}| \approx \frac{4}{\pi}\sin^{-1}\left[\frac{2|\Gamma_{max}|\sqrt{\frac{Z_L}{Z_0}}}{\left|\frac{Z_L}{Z_0} - 1\right|}\right]. \tag{50}$$

Equation 50 is thus an approximation for the fractional bandwidth (multiply by 100 to get an expression in percent) for a specified reflectance tolerance and impedance transformation ratio. As a specific example, suppose that we may tolerate a reflectance magnitude of 0.2 (corresponding to a realistic SWR limit of 1.5), with an impedance ratio of 4. For this set of numbers, the normalized total match bandwidth is approximately 0.344. If we tighten up the specifications to permit a maximum reflectance magnitude half as large, 0.1 (SWR = 1.22), then the bandwidth also halves – from about 34% to 17%.

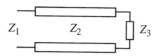

FIGURE 4.17. Circuit for derivation of small-reflection approximation

To supplement all of the foregoing mathematics, it's important to understand qualitatively why the frequency sensitivity of the match worsens as the transformation ratio increases. First, recognize that the ability to obtain a match at all is somewhat miraculous, for we have two interfaces (one at each end of the transforming line) at which definite impedance mismatches exist. Despite the reflections that are necessarily produced at those interfaces, the $\lambda/4$ transformer is nonetheless able to produce an impedance match at discrete frequencies. The transformer doesn't produce this result by somehow preventing reflections from occurring at the interfaces. Rather, it arranges for the ever-present reflections to *cancel*. The more dramatic the impedance mismatch at the interfaces, the more heroic the required cancellations. Thus the question isn't "Why does the match get worse?" but rather "Why is it ever good?" Consequently, it's not surprising that there should be a departure from the matched condition as frequency moves off center, with a steepness increasing with the mismatch.

Given this insight, it should seem reasonable that one might moderate the mismatch growth by moderating the impedance mismatches themselves, reducing thereby a dependency on miraculous cancellations. A more moderate impedance mismatch, however, necessarily implies an insufficient transformation ratio. To solve that problem, we could simply use, say, two $\lambda/4$ sections in cascade to provide the overall impedance transformation, with each section carrying only part of the total burden. Clearly, we may continue this process to any number of sections. The type of broadband impedance-matching device that results from this thinking is thus known as the (multisection) stepped-impedance transformer. The greater the number of sections, the better the bandwidth – though at the expense of increased total length. Questions that naturally arise concern the number of sections required (or total line length) and how the impedances of the individual sections should be chosen to meet design objectives. As one might imagine, some distributions of impedances are better than others. As with filter design (a topic we cover separately), one may seek a variety of passband behaviors, such as maximally flat or equiripple reflectance, with performance trading off with design complexity.

To explore some of these ideas rigorously involves mathematics of sufficient complexity to obscure most design insight. Therefore, we won't be rigorous. However, the approximations we will invoke are not so idealized as to render the results practically irrelevant. The demanding reader is invited to carry out an independent derivation without the simplifying assumptions.

Consider the two interfaces of a segment of line and also the reflections associated with the corresponding impedance mismatches (see Figure 4.17). An incident wave

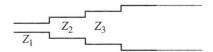

FIGURE 4.18. Multisection transformer structure

undergoes partial reflection at the Z_1–Z_2 interface. The portion that is not reflected continues on to the Z_2–Z_3 interface, where another reflection occurs. That reflected wave undergoes yet another partial reflection at the Z_2–Z_1 interface. The unreflected portion continues on its journey to superpose with the reflected portion of the incident wave, while the reflected portion heads back to the Z_2–Z_3 interface, and so on. If we neglect everything (specifically, everything involving the product of reflection coefficients) *except* for the two principal reflections, we obtain

$$\Gamma_{tot} \approx \Gamma_1 + \Gamma_2 \exp(-j2\theta), \tag{51}$$

where θ is the electrical length of the line expressed as a phase angle and where the reflection coefficients Γ_1 and Γ_2 correspond to those at the Z_1–Z_2 and Z_2–Z_3 interface, respectively. The exponential term merely expresses the fact that there is a round-trip delay of 2θ for the portion of the wave emerging from the first interface, traveling to the second interface, and being reflected back to the input of the line.

The reader may verify that this approximate expression for the overall reflectance yields the correct answer (zero) for a $\lambda/4$ transmission line transformer.

We now use Eqn. 51, the result of invoking a *small-reflection approximation* (because in ignoring the multiple reflections we implicitly assume that the product of the reflection coefficients is small compared to unity), as a computational atom in cascading sections to build up larger structures. See Figure 4.18. For simplicity, we will assume that all the sections are of the same length (*commensurate*), with an effective electrical length θ. This choice is made to simplify what comes but should not be taken to mean that it is necessarily optimal.

We now use the small-reflection approximation in a pairwise fashion to obtain

$$\Gamma_{tot} \approx \Gamma_1 + \Gamma_2 \exp(-j2\theta) + \Gamma_3 \exp(-j4\theta) + \cdots, \tag{52}$$

where the various Γ_n are the reflection coefficients at the discontinuities. If, in addition to the assumption of commensurate lines, we assume that the final load is real and that all impedances vary monotonically from one end of the structure to the other, then all of the reflection coefficients will be positive and real. Again, it is not necessary to satisfy these assumptions, but doing so enables the acquisition of valuable design insight (and also conforms to common engineering practice).[11]

Inspection of Eqn. 52 reveals that it is highly reminiscent of a Fourier series. In fact, it *is* a Fourier series. Recognizing the versatility of such a series for representing functions of a quite general character, one might imagine a synthesis procedure

---

[11] Relaxing the monotonicity condition, for example, allows the design of a certain class of filters (although the small reflection approximation usually no longer suffices for a good analysis).

by which one chooses the individual reflection coefficients to approximate a desired overall reflectance function.[12] Depending on the optimization objectives, one could select the coefficients to minimize the maximum reflectance within some specified passband or to guarantee monotonicity, for example. Again, the underlying philosophies of approximation are identical to those that guide the design of filters, a topic we treat later in a separate chapter. Consequently, one may select the coefficients to produce reflectance that is *maximally flat* (the so-called Butterworth condition, in which the maximum number of derivatives of the reflectance function is zero at center frequency) or of an equiripple nature (as in a Chebyshev approximation, where the error oscillates between specified limits), for example.[13]

We have seen that the performance of a stepped-impedance transformer improves as the number of sections increases, owing to the reduction in transformation ratio per section. It's a short leap from there to a continuous impedance variation. Just as there are many options for how to scale the impedance steps in the multisection transformer, there's a great body of literature on how one should "schedule" the impedance of a tapered matching device. By the Second World War, engineers had already translated the tapered acoustic horn (which performs an exactly analogous impedance transformation) into electromagnetic form.[14] The exponentially tapered line, in particular, has enjoyed enduring popularity because of the ease with which it is designed and understood. As its name suggests, the impedance of such a line has a simple exponential dependence on position:

$$Z(z) = Z_1 \exp(kz). \tag{53}$$

If the line is of some total length L and is to match an impedance Z_1 to Z_2, then the taper constant k is immediately determined from

$$k = \frac{1}{L} \ln \frac{Z_2}{Z_1}. \tag{54}$$

In order to carry out an actual design, one needs guidance on the choice of the total length. Intuitively, it seems reasonable that a better match would be obtained with more gradual tapers (larger L, or smaller k), and indeed this is the case. A useful criterion is to select the total length at least as large as half the wavelength of the

---

[12] Now you can appreciate why we choose to make the lines commensurate: it simply makes the reflectance function look like a Fourier series. However, this choice is not necessarily optimum, and the reader is invited to consider approximation methods that accommodate a distribution of segment lengths as well as segment impedances.

[13] Stepped transformers that produce the Butterworth maximally flat condition are sometimes known as *binomial* transformers because the coefficients are given by the binomial expansion.

[14] Harold A. Wheeler, "Transmission Lines with Exponential Taper," *Proc. IRE*, January 1939, pp. 65–71. Also see, e.g., George L. Ragan, *Microwave Transmission Circuits* (MIT Rad. Lab. Ser., vol. 9), McGraw-Hill, New York, 1948. We have Lord Rayleigh to thank for the acoustic horn. His work on acoustics also inspired him to consider electromagnetic analogies, leading him to analyze rectangular and circular waveguides before 1900. He was so far ahead of his time that this work had been largely forgotten by the time a use arose for it. Even today, many practicing microwave engineers are unaware of Rayleigh's contribution in this area.

lowest frequency component to which an impedance match is being provided. Thus,

$$k \leq \frac{2}{\lambda_{max}} \ln \frac{Z_2}{Z_1}. \tag{55}$$

Once the taper constant and line length have been chosen, the impedance as a function of position has been completely specified. All that remains is translating that impedance information into physical dimensions in order to build actual lines. Once again, however, we must be sure to take into account the practical limits on realizable impedances. As we've noted, impedance levels for most microstrip lines are typically constrained to lie within a factor of 3 or 4 of 50 Ω.

Subsequent work undertaken in the decade after the Second World War revealed that the exponential taper is not quite electrically optimum (although one may argue that it is certainly optimum in terms of performance obtained for the design effort expended). A design by Klopfenstein *is* optimum in the sense that, for a given permissible worst-case passband reflection, it is the shortest. Or, for a given total length, the Klopfenstein taper exhibits the lowest mismatch.[15] The Klopfenstein achieves this optimality by distributing the mismatch in a way that produces an equiripple reflectance over the passband. This egalitarian error distribution (known as a *minimax* optimum, because the maximum error is minimized within the passband) derives from extending to a continuum limit the design of a discrete Chebyshev multisegment transformer. Unfortunately, its design is not readily carried out with pen and paper, although computationally efficient algorithms have been available since 1968 or so.[16]

These various forms of tapered lines provide excellent matches to real loads over a broad frequency band. Achieving a broadband match with a complex load is more difficult, particularly as one seeks to approach the Bode–Fano limit. In applications requiring only moderate performance over a moderate bandwidth, it may suffice to absorb the imaginary part into the matching network, leaving a trivially solved, purely real impedance-matching problem. However, if the load is complicated (not just complex), then the reactance variation with frequency might be sufficiently extreme to preclude success with this strategy. We need not contrive a scenario with these characteristics, for providing a reasonably broadband match to a diode mixer is a commonly encountered challenge in terms of both theoretical design and practical realization. We will base an example on this problem because studying it, as well as its possible solutions, is highly instructive. Among others, it stimulates us to consider a few additional Smith-chart trajectories that will suggest relatively simple but powerful methods for greatly broadening the bandwidth of impedance matches. To understand such broadband methods, we need to extend our collection of Smith-chart patterns to include frequency variation, as the Smith loci we've presented up to this point all correspond to variation of parameters at a single frequency.

---

[15] R. W. Klopfenstein, "A Transmission Line Taper of Improved Design," *Proc. IRE*, v. 44, April 1956, pp. 539–48.

[16] M. A. Grossberg, "Extremely Rapid Computation of the Klopfenstein Impedance Taper," *Proc. IEEE*, v. 56, September 1968, pp. 1629–30.

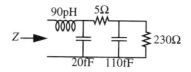

FIGURE 4.19. Equivalent linear circuit model for diode (absorbed power = 0 dBm)

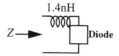

FIGURE 4.20. First pass: addition of series inductance

As a tutorial example, consider the problem of producing a reasonable match to a detector diode's impedance over a one-octave bandwidth. The first difficulty is that a diode is a nonlinear element and so its impedance depends on the bias point. It is common to characterize the small-signal impedance of mixer diodes when they are biased to dissipate one milliwatt, but it is important to perform the impedance characterization using the power levels that will prevail during use. In general, this power is not the same as the LO power; one must use the actual power level *absorbed* by, not incident on, the diode. Assume that a reasonable circuit model for a diode biased to the proper condition is as shown in Figure 4.19. Note the small value of series inductance (and low shunt capacitance). These numbers are typical for the beam-lead chip diode used in this example. Packaged devices generally have substantially worse parasitics.

The impedance of this network over a 6.5–13-GHz frequency span is shown in Figure 4.21 as curve A.[17] Noting that the impedance already happens to have about the right resistance near the middle of the band suggests the simple addition of a series inductance to rotate the impedance clockwise. This choice is made more attractive by the fact that there is already some inductance to begin with. To compute the required inductance, we only need to find the imaginary part of the impedance corresponding to a real part $R = 1$ (normalized). The magnitude of that capacitive reactance equals the magnitude of the desired inductive reactance needed to resonate it away. For this particular example, an inductance of about 1.4 nH produces a near match at the nominal center frequency of 10 GHz; see Figure 4.20.

---

[17] This example is inspired by two applications notes from Hewlett-Packard (now Agilent): AN-963, "Impedance Matching Techniques for Mixers and Detectors," and AN-976, "Broadband Microstrip Mixer Design – The Butterfly Mixer." We have modified the techniques somewhat in order to obtain a match over a full one-octave bandwidth, rather than the 1.5:1 frequency ratio in the applications notes.

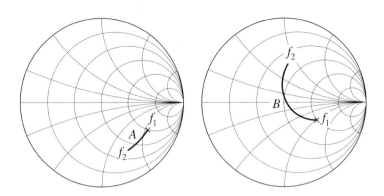

FIGURE 4.21. Rotation of impedance locus due to added series inductance

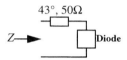

FIGURE 4.22. Second pass, step 1: addition of line

The rotation increases with frequency, however, so each point on curve A rotates clockwise by a frequency-dependent amount to produce curve B (see Figure 4.21). The point corresponding to the upper frequency limit f_2 naturally rotates much more than the impedance at the lower frequency limit. This behavior follows from the fact that, for a fixed inductance, the added impedance is directly proportional to frequency. Because we have chosen an inductance sufficient to produce a near-perfect match at the center frequency, the frequency-dependent rotation causes the match to degrade at the band edges. Such degradation is to be expected when simple, single-frequency matching methods are used. Here, the return loss is only about 7 dB at the band edges, corresponding to an SWR of about 2.7. This level of mismatch is generally considered undesirable.

Let us examine an alternative method for providing an impedance match. Instead of a series inductance, let's try using a shunt element. To produce a match at the nominal operating frequency, we simply add enough line to take us to the $G = 1$ normalized conductance circle. From there, a suitable shunt susceptance produces a match, as we've already seen. For this diode, we need to add about 43° worth of (50-Ω) line to produce the desired nominal rotation and complete the first step of a single-stub match; see Figure 4.22.

Once again, the line that we add produces a frequency-dependent rotation because its electrical length is proportional to frequency. Thus, if we add enough line to rotate the admittance curve to the $G = 1$ circle at the center frequency, there is necessarily an over-rotation at higher frequencies and an under-rotation at lower ones. This

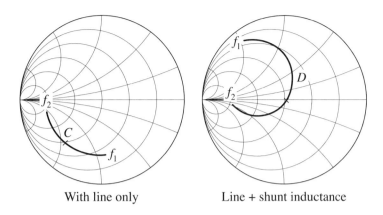

FIGURE 4.23. Frequency variation of match with single-stub tuner

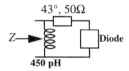

FIGURE 4.24. Second pass, step 2: addition of shunt inductance

behavior is apparent in the loci of Figure 4.23. Curve C corresponds to the clockwise rotation produced by adding a section of transmission line. Thanks to the frequency dependence of this rotation, the addition of line causes the locus to elongate (compared with curve A in Figure 4.21), in a manner similar to that observed when using a series inductance.

The final step is to add some shunt susceptance to produce a match at the nominal frequency. Once again, computing the necessary value begins with a measurement of the existing capacitive susceptance corresponding to the intersection with the $G = 1$ circle. The inductance is chosen so that its susceptance magnitude equals that of the capacitance. Here, we need to add a shunt inductance of about 450 pH to provide a match at the nominal center frequency (see Figure 4.24).

Again, the rotation is dependent on frequency, exacerbating the elongation we've already suffered from having added the short segment of line. In this case of a shunt inductance, the rotation is larger at low frequencies than at high, and the net effect is to produce the curve marked D. Thanks to the use of two elements that produce frequency-dependent rotations, we find that the match degrades much more dramatically for the shunt alternative than for the series example. In fact, the worst-case return loss (suffered at the lower band edge in this case) is less than 2 dB – a level of mismatch that would be intolerable in all but the most forgiving of applications. This result underscores the value of exploring the several options that typically exist before selecting the one that best satisfies all relevant optimization objectives. These

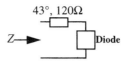

FIGURE 4.25. Third pass, step 1: addition of line

considerations may include not only the quality of the match but also realizability, size, and convenience.

Suppose that, despite our discouraging experience with that particular shunt network, we insist on pursuing an alternative that contains a shunt element. That insistence might be motivated, for example, by a need to provide a DC return path for the diode bias. Although it is not impossible to devise a bias circuit for the series-inductance example, a shunt inductance or transmission line segment naturally provides a DC path to ground and thus can also act as part of the bias network.

That choice still leaves us with the challenge of producing a reasonable match at the band edges. A large improvement is possible by using a resonant circuit as a matching element. The qualitative idea is simple: At the band edges, the mismatch may be (or can be made to be) attributable to reactances of opposite types. To promote a match, then, requires an inductance at one band edge and a capacitance at another. Neither single capacitors nor inductors exhibit the required behavior, but resonant networks do. To make the most effective use of this observation, we need to rotate the impedance locus if necessary so that the reactances at the band edges are indeed of opposite types. That loose specification still leaves open the question of what to do about the conductances. A good choice is to make the conductance at the band edges equal to the reciprocal of the conductance at the center of the band. We will see that this choice tends to result ultimately in an impedance locus roughly centered about a perfect match. Furthermore, the reflectance corresponding to these conductances provides a good estimate of the ultimate worst-case mismatch.

From inspection of Figure 4.23, it's clear that we need to rotate the curve D counterclockwise by some amount. Because the conductance already corresponds to a match at center frequency, performing the rotation by adding a piece of line whose characteristic impedance equals the Smith-chart normalization impedance cannot produce the desired reciprocal conductance relationship. Indeed, from the curvature of the admittance locus, it's clear that we need to supplement a rotation with a translation (actually, a scaling or renormalization) in order to increase the conductance at band center. This is readily achieved by selecting a line segment whose characteristic impedance is higher than the ordinary normalizing impedance. Selecting a line impedance of 120 Ω and electrical length of 43° at the 10-GHz center frequency produces the desired result (see Figure 4.25).

The effects of this combined rotation and renormalization are shown as curve E in Figure 4.26. Notice that the conductances at the band edges are indeed equal to each other and also roughly equal to the reciprocal of the conductance at the center

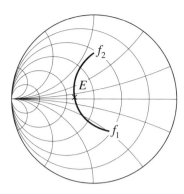

FIGURE 4.26. Locus after rotation and conductance equalization by added high-impedance line

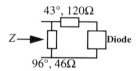

FIGURE 4.27. Third pass, step 2: addition of shunt line

frequency (very fussy engineers might choose a slightly higher line impedance to shift the locus to the left a bit more). Keeping track of the susceptances at the band edges, we note that curve E tells us that the network's admittance appears capacitive at the lower band edge and inductive at the higher one. Thus, a shunt resonator has the right general characteristics to offset the network susceptances at the band edges and thereby produce something close to a match. A too-high line impedance would result in a failure of the ends to meet, while a too-low one would overcompensate, causing the tails to cross. An improper electrical length would produce asymmetries in the shape.

Note that the network's inductive susceptance at the upper band edge is larger than the capacitive susceptance at the lower band edge. This asymmetry is compensated by introducing a complementary one, achieved by making the shunt line possess other than a 90° electrical length at the center frequency. Here, we need to provide a bit more capacitance at the upper band edge, meaning that we need to select the line a little longer than 90° (so that it acts a little capacitive) at the center frequency. We may produce the correct result by using a shunt 46-Ω line of 96° electrical length at the center frequency (see Figure 4.27).

The final locus appears in Figure 4.28. As is apparent from inspection of the final result, the admittance locus is now much more tightly distributed and also roughly centered about the perfect match condition. Consequently, the mismatch oscillates in some fashion. We know that if the locus were a perfect circle, the reflectance would

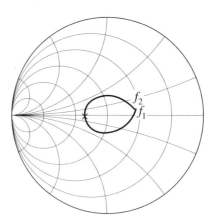

FIGURE 4.28. Admittance locus after addition of length of shorted line to produce convergence

then be perfectly constant. Other shapes are necessarily associated with ripple. If we could produce an equiripple response whose worst-case reflectance just meets our match criterion, we would have obtained a Chebyshev (minimax) optimum. Although we have not quite achieved that result, it is clear that the worst-case mismatch is substantially smaller than with either the series inductance or the single-stub matching network. For this particular example, the return loss in fact exceeds 10 dB over the entire octave range, corresponding to an SWR of better than 2:1. Although this quality of match is not excellent, it is well within the range of values commonly considered usable for many systems.

In the course of studying a variety of matching strategies for this example, we've built up a collection of methods ("macros") that have great general utility in producing matches over reasonably broad bandwidths. Although using them involves a somewhat ad hoc approach, this unstructured procedure is still commonly used over more complex, formal synthesis methods. Not all of these macros have official names, so we are going to invent some in summarizing these modules. The following is hardly an exhaustive list, but it contains enough operations to constitute a credible subset.

Rotation. Rotation can be produced by the addition of shunt susceptances or series impedances. These admittances or impedances may be produced by lumped components or suitable lengths of line, just as in the first step of a single-stub tuning operation. Because the electrical length is frequency-dependent, the rotation angle depends on frequency for all of these options.

Conductance (resistance) equalization. This operation makes the conductances (resistances) at the band edges equal to each other and may be provided by appropriate rotations.

Conductance (resistance) balance. This operation makes the band-edge conductances (resistances) equal to each other and equal to the reciprocal of the midband conductance (resistance) as well. Doing so sets the stage for *convergence,* as described in the next paragraph. Implementing rotations with line segments whose characteristic impedance differs from the system's normalization impedance is an expedient way to provide balancing at the same time.

Convergence. Once equalization has been achieved, the admittances (impedances) at the band edges can be brought close together by adding a shunt (series) resonator, as required. Because such a resonator appears inductive on one side of resonance and capacitive on the other, it has the ability to perform a minor miracle: bringing both ends of an equalized trajectory together. Because of the focus on equalizing reciprocal real parts, the imaginary parts may not have equal magnitudes at the edges of the frequency band. Hence the termination may have to be operated above or below its resonance at the center of the frequency band in order to produce convergence. Often convergence is the penultimate step, rather than the final one. Because of limitations on the range of line impedances that may be practically realized, it may be not be possible for a single operation to create a converged locus that is also centered about the origin of the Smith chart. A final rotation or translation (achieved, again, through the addition of a suitable length of line of the correct impedance) can complete the match.

As a final comment, note that the general procedures outlined here do not directly incorporate a specification on allowed mismatch in any formal way. Consequently, there is no guarantee of success. That said, carrying out this design procedure (and realizing the design in practical form) is simple enough that one often tries it first. Fortunately, it is not uncommon to find that it yields an acceptable result.

CHAPTER FIVE

CONNECTORS, CABLES, AND WAVEGUIDES

5.1 INTRODUCTION

Although the focus of this book is the implementation of discrete planar RF circuits, we consider here a number of important components that are fundamentally 3-D in nature: connectors, cables, and waveguides. We'll see that the useful frequency range of these components is bounded in part by the onset of *moding,* which (in turn) is a function of their physical dimensions. In addition, we'll examine the attenuation characteristics of these various ways to get RF energy from one place to another.

5.2 CONNECTORS

5.2.1 MODING AND ATTENUATION

For the flattest response over the largest possible bandwidth, an RF connector should exhibit a constant impedance throughout its length. This requirement is satisfied by maintaining constant dimensions throughout and by filling the intervening volume uniformly with a homogeneous dielectric. As straightforward and obvious as this requirement may seem, we will shortly see that there is at least one extremely popular connector that fails to meet it.

The best and most commonly used RF connectors are coaxial in structure. One important attribute of coaxial geometries is their self-shielding nature; radiation losses are therefore not an issue. One must always take care, however, to maintain transverse electromagnetic (TEM) propagation in which, you might recall from undergraduate electromagnetics courses, neither **E** nor **H** has a component in the direction of propagation. At sufficiently high frequencies, non-TEM propagation can occur, and the energy stored or propagated in higher-order modes can cause dramatic impedance changes. As the frequency increases, the tendency to excite these modes increases. Before actual propagation occurs, the energy stored in the modes necessarily represents reactive energy and correspondingly affects the impedance. Actual propagation results in a resistive perturbation.

5.2 CONNECTORS

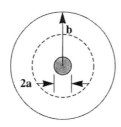

FIGURE 5.1. Cross-section of coaxial structure

Very loosely speaking (with due apologies to Professor Maxwell), waveguide propagation of higher-order modes occurs when the wavelengths are small enough that waves can "squeeze" their way through some critical dimension of the structure. Therefore, if the desire is to prevent such propagation, the conductor dimensions must be small enough to suppress all but the TEM mode.[1] For the coaxial case we are presently considering, the relevant critical dimension is approximately the average electrical circumference of the annular space between the inner and outer conductors. A non-TEM wave (specifically, the TE_{11} mode) can begin to propagate when its wavelength equals this average circumference.[2] The first such high-order mode thus has a wavelength given approximately by

$$\lambda_c \approx 2\pi[(a+b)/2]\sqrt{\varepsilon_r} = \pi(a+b)\sqrt{\varepsilon_r}, \tag{1}$$

where a and b are the inner and outer radii of the space between the conductors, and ε_r is the relative dielectric constant of the material that fills that space; see Figure 5.1. The error in this approximation is below 3% for b/a between 1 and 15.

It is the common desire for mode-free operation that explains why connectors and cables become progressively smaller as their intended operating frequency range increases. Some common coaxial connectors are the UHF, N, BNC, TNC, SMA, SMB, and SMC; see Figure 5.2.

The UHF connector is the oldest of these, having been developed in the 1930s by E. Clark Quackenbush of Amphenol Corporation for the broadcast radio industry.[3] It is commonly used in amateur radio gear. The socket version is SO-239, which mates to the PL-259 plug. Unfortunately, the bulky connector isn't actually suitable for use in the UHF range as it is now defined.[4] In fact, it has an important design flaw (a

---

[1] In the language of the profession, the word "mode" can be used as a verb. *To mode* means to launch higher-order modes. Thus, one worries about the frequency where a connector or transmission line begins "to mode." Yes, it may sound funny "to noun," but *c'est la mode*.

[2] If mode nomenclature is unfamiliar to you, don't worry. Section 5.4 discusses this topic in greater detail.

[3] At that time, Amphenol was still American Phenolic.

[4] The frequency range implied by *UHF* changed over time, but the name of the connector unfortunately did not.

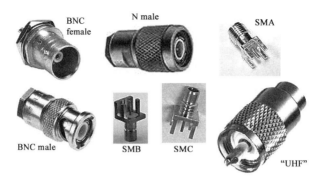

FIGURE 5.2. Assorted coaxial connectors (not shown to scale)

nonconstant impedance along the connector) that degrades its characteristics significantly toward the middle and upper end of the *VHF* range (30–300 MHz). The loss at 50 MHz can exceed 1 dB.[5] As a consequence, the connector really shouldn't be used much above about 100 MHz and, frankly, ought to be retired altogether. Nevertheless, it is used in a lot of 2-meter FM amateur radio equipment, and even occasionally in some 70-cm (440-MHz) radio gear, at the expense of significantly degraded performance. Its only attribute is that it is cheap. Please join in the crusade to eliminate this connector!

The N connector is named for its developer, Bell Labs researcher Paul Neill, and has been widely used in RF test equipment since WWII. Its dimensions are such that it is mode-free up to approximately 18 GHz.[6] However, the most common version has slots in the outer conductor, and this type is usually specified with an upper frequency limit of 11–12 GHz. The slotless version works better, with good characteristics up to 18 GHz. Be aware that not all cables equipped with N connectors are necessarily specified to operate over this entire range.

Another issue is that, by design, the center pin mates with its counterpart with a small gap (order of 75–100 μm) between shoulders of the conductors, as seen in Figure 5.3. The outer (ground) conductors (not shown) are designed to mate first, resulting in the space shown for the center conductors. The reason for the choice is that it is impractical to demand the tight machining tolerances and low temperature coefficient of expansion that would allow elimination of the buffer space. If the center pins were to connect without such a buffer space, tightening of the connector until the outer surfaces mated would almost guarantee destruction of the center pins. The small discontinuity associated with the effective change in dimensions causes the impedance to degrade as the frequency increases. As can be surmised from the figure, the reduction in conductor diameter is associated primarily with a parasitic series inductance.[7]

---

[5] Again, remember that a loss of "only" 1 dB is a 21% power loss!
[6] Some references claim that the *N* stands for "Navy."
[7] A more sophisticated model takes into account the capacitive fringing as well, resulting in a *CLC* π-model for the discontinuity. In any case, the discontinuity reduces the operating frequency range.

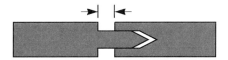

FIGURE 5.3. Detail of center pins when outer conductors are mated

As with *all* threaded connectors, one should never, repeat *never*, use pliers to tighten an N connector; otherwise the center pin can become damaged over time (and you'll scratch up the connector in any case). Use of a torque wrench to tighten to the specified 12 in-lb (135 N-cm) is highly recommended to maximize longevity. Barring that, a finger-tight connection is acceptable, provided you do not do this too often. While we're on the subject of caveats, you should never use lubricants (or other petroleum-based solvents) on connectors and cables. If the connectors are too tough to tighten properly because they've been cross-threaded, replace them. Spraying a lubricant won't repair the fundamental damage, but it could further harm the plastic portions of connectors and cables. If the connectors need cleaning, use compressed air or special solvents that are specifically designed for this purpose.

The BNC connector, named for developers Neill and (Amphenol engineer) Carl Concelman, is nearly ubiquitous in RF gear owing to its convenience and relatively small size. The use of a bayonet (the "B" in BNC)[8] ground connection in lieu of threads allows for rapid connection and disconnection. The internal dimensions are consistent with mode-free operation to at least 18 GHz, but the comparatively unstable bayonet connection begins to degrade performance above about 1 GHz. The upper useful frequency limit is conventionally taken to be approximately 3–4 GHz, though BNCs are occasionally used above this range – albeit at the expense of progressively worse (and erratic) behavior. Vibration can modulate the impedance, leading to a host of objectionable pathologies, particularly in mobile applications. This problem is solved in the otherwise identical threaded version of this connector, the TNC (threaded Neill–Concelman, developed in the late 1950s), which is suitable for use up to at least 11–12 GHz. The best can be used up to 18 GHz. The BNC shares with the N connector – and with the SMA/B/C (to follow) – a small gap between the shoulder of the inner pin and its receptacle, resulting in a reactive discontinuity that becomes more significant with increasing frequency.

The demand for still smaller connectors in the 1960s led to the development of the SMA (sub-miniature, type A). It is quite a bit smaller than the BNC and is usually specified to work up to at least approximately 18 GHz with semirigid coaxial

---

[8] A common belief, reinforced by many textbook authors, is that "BNC" stands for "baby N connector," but BNC really stands for "bayonet Neill–Concelman," to distinguish it from the threaded Neill–Concelman (TNC, not "tiny N connector," as still others claim). The misconception is evidently traceable to an otherwise reliable source: *Technique of Microwave Measurements* (MIT Rad. Lab. Ser., vol. 11), p. 10. Note that there also exists a less widely known C connector developed by (and named for) – you guessed it – Concelman. The C connector looks a lot like a BNC but is larger.

cable (often 12 GHz otherwise), although operation to a bit over 25 GHz is possible with care. Tightening to the proper specified torque of 56 N-cm (5 in-lb) ensures minimal degradation of the connector's characteristics over time. It is difficult to finger-tighten to sufficient torque yet too easy to over-tighten with an ordinary wrench.

Another small threaded connector is the SMC. It is less expensive than the SMA and somewhat smaller. The SMC typically functions well up to 10 GHz, and proper torque is in the range of 3–4 in-lb (34–45 N-cm). The SMB is a push-on/pull-off version of the SMC, and it can be used over a reduced frequency range of roughly 4 GHz (about the same as the BNC, and for similar reasons).

The insertion loss of connectors depends on whether they are straight-through or right-angle types as well as on whether they are threaded or snap-on. The precise values of attenuation are rarely of importance, however, since cable loss almost always dominates. However, we can offer some crude approximations. As a very rough rule of thumb, one can expect straight snap-on and threaded right-angle types to have twice the loss (on a dB basis) of an otherwise equivalent threaded straight-through version. Larger connectors have lower conductor losses per unit length but are physically longer, so insertion loss *per connector* tends to be narrowly distributed. The loss in decibels grows with frequency, generally exhibiting something between a square-root dependence and direct proportionality. In the low-gigahertz frequency range, N and SMA connectors typically exhibit losses under approximately 0.04 dB per root GHz (for straight-through versions) and under about 0.06 dB per root GHz (for right-angle types). Manufacturers frequently quote substantially worse values in their data sheets (either out of a sense of conservatism or so they won't have to test it in production). The BNC and TNC typically have losses that are under 0.1 dB and 0.15 dB for straight and right-angle types, respectively, per root GHz. The loss of a BNC is somewhat worse than for the TNC, increases faster with frequency, and is somewhat unstable owing to the flaky ground connection. Corresponding values for the SMC are similar, at about 0.15 dB and 0.2 dB per root GHz. Loss for the SMB can be as high as about twice (on a dB basis) as for the SMC. Because there is considerable variation among manufacturers, these numbers are to be taken only as rough approximations.

For instrumentation that must support frequent connection and disconnection, a different connector design is necessary. Ordinary connectors suffer from a gradual degradation of the center pins, which occurs as a result of fatiguing of the corresponding mating structure over time. And as mentioned earlier, manufacturing considerations demand a small buffer space in the mating structure of the center pin, guaranteeing increasing mismatch with increasing frequency. To avoid these problems, Amphenol and Hewlett-Packard modified a German design to produce a genderless connector in which the center and outer conductors make a butt contact (as opposed to the "pin and socket" model of the other connectors). The APC-7 (7-mm "Amphenol precision connector" or "a precision connector") is widely used in instrumentation up to at least 18 GHz.

To mate two APC connectors, first rotate the outer sleeve of one connector clockwise until the threads are exposed. This sleeve will be left loose. Then line up the two connectors and tighten the sleeve of the other connector on the exposed threads of the first. The sleeves should not butt against each other when you are finished; otherwise, they may push the mating surfaces apart. It is critically important to rotate only the sleeve while tightening. If the mating surfaces are allowed to rotate against one another then the resulting abrasion will damage the connectors. Of course, touching the mating surfaces is also taboo. Clearly, these types of connectors depend on maintenance of cleanliness and precise mechanical tolerances. Proper torque is therefore essential (135 N-cm, or 12 in-lb) both to avoid damage and to maintain flat impedance characteristics over frequency. In the absence of a torque wrench, ordinary finger-tightness is sufficient.

For operation at still higher frequencies, progressively smaller dimensions are necessary to avoid moding. The 3.5-mm, 2.92-mm (also known as the Wiltron K connector, for the highest frequency band covered), 2.4-mm, 1.85-mm (or V connector, again for the frequency band covered), and 1.0-mm (W) connectors are specified to operate up to about 33 GHz, 40 GHz, 50 GHz, 65 GHz, and 110 GHz, respectively. All of these should be tightened to 90 N-cm (8 in-lb), with the exception of the 1-mm connector, whose torque specification is 34 N-cm (3 in-lb). Again, use of a torque wrench is *absolutely* mandatory to maintain good characteristics, which include a typical loss on the order of 0.1 dB per root GHz. The higher the frequency, the fussier the requirements.

As a final note on these connectors, it should be mentioned that the SMA and the precision 3.5-mm and 2.92-mm connectors can mate to each other in the sense that the diameters and thread pitches are nominally compatible. However, the SMA has a teflon dielectric, whereas the precision connectors are air with small dielectric supports. To compensate, the SMA has a somewhat thinner outer sleeve. Further, SMA connectors from some vendors are not manufactured to sufficiently tight tolerances. It is altogether too possible to cause connector damage, even with a single connection–disconnection cycle. As a protective measure for instruments and other fixtures for which connector replacement is an arduous operation, it is prudent to attach a short extender (e.g., male on one end, female on the other) on 3.5-mm connectors to act as a sacrificial connector. This way, any damage is done to the extender, which may be removed and replaced with ease. "Connector savers" made explicitly for this purpose are available from a number of vendors.

The recent (and ongoing) rapid growth in compact consumer wireless devices – cellular telephones, GPS locators, wireless LAN cards, and the like – has driven the development of a new class of exceptionally small, low-cost connectors. Somewhat easing the design burden is that the physics of wireless propagation all but guarantees operating frequencies below about 5–10 GHz for mobile applications. Representative of connectors used in these applications are the MCX and MMCX, which are snap-on/snap-off types (similar to the SMB, but significantly smaller) and are rated for use up to approximately 6 GHz. One can generally expect insertion losses of better than 0.1 dB and 0.2 dB per root GHz for straight and right-angle versions, respectively.

Also popular with many wireless LAN access points are reverse versions of TNC and SMA connectors. These choices discourage the casual connection of antennas that may cause noncompliance with statutory limits on radiated power. Inevitably, the result has mainly been the creation of a cottage industry for the manufacture of suitable adapters.

75-Ω Connectors

The foregoing overview of connectors focuses on the standard types found in the 50-Ω world. There are fewer standard options in the 75-Ω universe. For example, the F connector is overwhelmingly the most common connector for 75-Ω applications (such as cable TV) and has no 50-Ω counterpart. Regrettably, specifications on the dimensions of the center pin are so loose that it is absurdly easy to produce a lousy connection or even cause damage: The center pin's diameter is permitted to range from about 0.5 mm to slightly over 1.6 mm! To make matters worse, it is occasionally possible to find cables (from some fly-by-night sources) whose center conductor diameter falls outside even this generous window. Needless to say, attempting to mate connectors at opposite ends of the range can cause connector damage or a failure to connect. Most F connectors are useful up to about 1 GHz.

Several other 75-Ω connectors are derived from standard 50-Ω prototypes, and one can find N, BNC, ... in the 75-Ω world, although they are rarer than the 50-Ω versions. It is important to note that the dimensions of the center pins are different for the two impedance levels, so attempting to mate a male 50-Ω with a female 75-Ω connector will almost certainly cause damage to at least one of the pair.

5.2.2 NONLINEAR EFFECTS

Given that only conductors and very linear dielectrics (e.g., air or Teflon) are nominally involved, it may come as a surprise that connectors can sometimes exhibit nonlinear behavior. It is actually possible for nonlinear effects in connectors to cause noticeable distortion (and even parasitic radiation of harmonic and intermod products), so if the engineer is unaware of this possibility, it can take a very long time to track down.

There are two principal sources of nonlinearity in connectors. One is due to simple corrosion. Many metallic oxides, chlorides, and sulfides are semiconductors, and corrosion can result in the formation of one or more of these types of compounds. The nonlinearities may be associated with Schottky or tunnel barriers formed with these materials. The obvious remedy is to protect all interfaces from corrosion. However, satisfying this simple dictum is not trivial, particularly in outdoor applications. Compounding the challenge is that conductors are often plated with silver to minimize high-frequency loss, and silver tarnishes quite readily.[9]

Another source of distortion can be magnetic materials that may saturate under large-signal conditions. One does not normally use magnetic materials on purpose,

---

[9] Contrary to popular belief, silver tarnish is primarily silver sulfide, not oxide. While we're at it, the green patina on copper is usually a chloride (near the ocean) or a sulfate (inland) rather than an oxide.

5.3 COAXIAL CABLES

5.3.1 WHY COAX?

The foregoing discussion on connectors for coaxial cables leaves unanswered a basic question: Why use coax? Maybe the answer is obvious, but bear with us nonetheless. At low frequencies, just about any old wires can be (and have been) used to get signals from one place to another with little loss. For example, twisted pairs are widely used in data communications (e.g., the now-ubiquitous "CAT-5" computer networking cable). The twisting ensures that both wires experience the same perturbations by stray fields and proximity to other conductors, thus maintaining balanced characteristics. Radiation losses are also reduced at the same time, permitting twisted-pair line to function well enough to enable 100-Mbps and 1-Gbps data rates. Ongoing work seeks to push those rates upward by another order of magnitude.

Ensuring that disturbances remain common-mode is certainly helpful but, as the frequency increases, such unshielded structures begin to behave as antennas. As explained in the chapter on antennas, radiation becomes noticeable when the conductor length is a reasonable fraction (i.e., a tenth) of a wavelength. Such radiation of energy implies that less signal makes it to the intended destination; attenuation has increased.

The coaxial structure, being shielded, does not suffer from radiation losses. Moding is always a concern, and we've already seen that the desire to suppress it explains the shrinking in connector and cable dimensions as the intended operating frequency increases. As for attenuation, the lack of radiation loss means that only conductor and dielectric losses need to be considered. These two mechanisms may be distinguished by their differing frequency dependence. Conductor losses, dominated by skin effect, vary with the square root of frequency, whereas dielectric losses generally vary directly (or faster) with frequency.

To verify these assertions and to study these mechanisms in greater detail, we begin with the following expression for the attenuation coefficient:[10]

$$\alpha \approx \frac{1}{2}\left(\frac{R}{Z_0} + \frac{G}{Y_0}\right), \qquad (2)$$

where α is the attenuation in nepers per unit length,[11] R is the series resistance per unit length, G is the shunt conductance per unit length, Z_0 is the characteristic impedance, and Y_0 is $1/Z_0$.

---

[10] See Chapter 2, Eqn. 37.
[11] Recall that the amplitude attenuation is a factor of e for every increase in distance of $1/\alpha$. Nepers and decibels are therefore proportional to each other, with one neper equal to $20 \log_{10} e$ or about 8.69 dB.

The first term accounts for the conductor loss. If we assume operation in a frequency regime where the skin depth is much smaller than the conductor thickness, then we may estimate R by making the usual assumption that current flows uniformly in a thin cylinder of a thickness equal to the skin depth δ:

$$R \approx \frac{1}{2\pi\delta\sigma}\left[\frac{1}{a} + \frac{1}{b}\right]. \quad (3)$$

Here σ is the conductivity of the wire and δ is the skin depth:

$$\delta = \sqrt{2/\omega\mu\sigma}. \quad (4)$$

Thus,

$$\alpha_c = \frac{[1/a + 1/b]\sqrt{\mu/32\sigma}}{\pi Z_0}\sqrt{\omega}. \quad (5)$$

We see that α_c, the attenuation due to conductor loss, indeed varies as the square root of frequency, as asserted earlier. We also see, sensibly enough, that larger conductor diameters and higher conductivity all help to reduce this attenuation term.

The frequency dependence of dielectric loss may be anticipated qualitatively by imagining the loss to arise from a sort of dipole "friction." If we postulate that one must expend a certain amount of energy to overcome this friction and thereby allow the reversal of dipole polarity, then the rate of energy loss (and thus the power dissipated) should be proportional to frequency. We therefore expect the attenuation due to this mechanism to vary directly with frequency. It is left as an exercise for the reader to show that α_d, the attenuation constant for dielectric loss, may be expressed as

$$\alpha_d = \frac{[\tan\delta]\sqrt{\mu_r\varepsilon_r\mu_0\varepsilon_0}}{2}\omega, \quad (6)$$

where $\tan\delta$ is the loss tangent of the dielectric material (it is perhaps unfortunate that the symbol δ is used in both the skin-depth and loss-tangent formulas). In any event, we see that dielectric loss does increase linearly with frequency on a neper and decibel basis, assuming that the loss tangent is independent of frequency.[12] Note that, unlike conductor loss, the attenuation due to dielectric loss is independent of conductor dimensions.

In accord with the preceding derivations, the attenuation of practical cables is approximately constant at very low frequencies, is subsequently rising as the square root of frequency as skin loss dominates, and changes to a direct proportionality as dielectric loss takes over. Some dielectric materials exhibit an increasing loss tangent with frequency, resulting in attenuation that rises faster than the first power of frequency. Upper limits on the operating frequency range of cables are often chosen somewhat near the point where dielectric loss begins to dominate.

From the foregoing analysis, we see that the lowest attenuation is provided by cables with air dielectric and the largest physical dimensions. Because of moding

---

[12] Most dielectric materials used in coaxial cables satisfy this assumption fairly well.

FIGURE 5.4. Typical coaxial cable

concerns, there is some upper bound on these dimensions and, as a result, there exists a lower bound on the achievable attenuation at any given frequency.

5.3.2 TYPES OF COAXIAL CABLE

For general-purpose use at lower frequencies (say, below 1 GHz or so), a common series of flexible coaxial cable is designated RG-n/U. The letters "RG" stand for "radio guide" and the "U" for "universal."[13] The designation originally included rigid lines as well as waveguide, but recent usage has focused on lower-frequency coaxial cable. These come in a variety of sizes with various dielectrics; see Figure 5.4. Numerous companies supply other types of cables featuring improved characteristics obtained via either better dielectrics (e.g., PTFE or polyethylene foam) or superior conductor and jacket properties. See Section 5.6 for characteristics of a representative sampling.

One problem with ordinary flexible coax is that repeated bending can cause degradation through fatiguing of both the conductors and the insulators. Additionally, it is relatively easy to bend the cable too sharply, resulting in dimensional distortions that degrade electrical characteristics, perhaps permanently if kinked. Semirigid and rigid coaxial cables were developed to solve these problems. Semirigid coax retains some measure of flexibility by employing a somewhat pliable outer conductor, allowing the cable to be bent into shapes it will retain. Use of special bending jigs is recommended to avoid kinking. As with ordinary flexible coax (or any kind of line, for that matter), too tight a turning radius will degrade electrical characteristics. Rigid coax (also known as hardline), as its name implies, is not meant to be bent at all, so manufacturers provide a variety of preformed shapes cut to various lengths.

Semirigid cable generally uses Teflon dielectric and comes in two common sizes: 0.141" and 0.085" diameters. There is also a 0.141" flexible cable with Teflon whose characteristics are intermediate between those of semirigid and larger-diameter flexible cable.

---

[13] G. L. Ragan, *Microwave Transmission Circuits* (MIT Rad. Lab. Ser., vol. 9), McGraw-Hill, New York, 1948, p. 244.

75-Ω Cables

Our focus is on 50-Ω connectors and cables because most RF instruments and systems use this impedance level. However, perhaps a few words about the 75-Ω world are in order.

As explained elsewhere,[14] 50 Ω arose as a standard impedance from trading off power-handling capability against attenuation. For coaxial air lines, an impedance of 30 Ω maximizes power handling capability, while 77 Ω results in minimum attenuation. The average of these two values (after rounding) was chosen as a compromise for general-purpose work.

In certain applications such as cable television, however, minimizing attenuation is extremely important for reducing the number of costly repeaters that must be deployed. In such applications the power levels are low, so the power-handling capability of the cable is unimportant. An impedance level of 75 Ω therefore arose as the standard for video applications.

Characteristics of several common 75-Ω cables are also presented in Table 5.1 (Section 5.6).

5.4 WAVEGUIDES

We have seen that the attenuation of coaxial cable is typically dominated by conductor loss at lower frequencies and by dielectric loss at higher frequencies. We've already noted that attenuation could be reduced if we were to use air as a dielectric and were free to increase cable dimensions without bound. However, moding considerations prevent us from doing the latter. Evading this constraint is one objective of waveguides.

Suppose that, rather than trying to prevent propagation of energy in higher-order modes by using ever-smaller geometries, we actually *exploit* moding. That is, we choose not to make use of TEM propagation. With such a choice, we are free to select much larger conductor dimensions. Furthermore, since the higher-order modes do not include DC, the center conductor of a coaxial cable is unnecessary. The resulting structure is called a waveguide. From the foregoing description, it is clear that waveguides are nothing more than hollow pipes. Waveguide propagation was demonstrated first in such circular hollow pipes (in the 1930s, independently and nearly simultaneously by Bell Labs engineer George C. Southworth and Wilmer L. Barrow of MIT), but elliptical, rectangular, and square cross-sections have also been used.[15] The math

---

[14] See Section 3.5.

[15] The first published theory of waveguide propagation was by Lord Rayleigh (John William Strutt) in 1897. He was too far ahead of his time, however. No one cared about waveguide propagation at a time when engineers could not practically generate signals that could propagate through pipes of reasonably small dimensions. Almost four decades later, Barrow and Southworth were working independently of each other (and without knowledge of Rayleigh's work) until they were close to publishing their own results.

is simplest (by far) for the rectangular case. Consequently, that is overwhelmingly the most common type of waveguide and is the only type we will consider in this brief treatment. For a comprehensive derivation of waveguide modes for other geometries, the interested reader is directed to the excellent treatments found in Ramo, Whinnery, and Van Duzer, *Waves and Fields in Modern Communications Systems* (3rd ed., Wiley, New York, 1994) or in *Electromagnetic Waves and Radiating Systems* by E. C. Jordan (Prentice-Hall, Englewood Cliffs, NJ, 1950).

5.4.1 MODE NOMENCLATURE

We've already used the term TEM, which refers to a mode in which both the electric and magnetic fields are transverse to the direction of propagation. Single conductor waveguides are unable to support TEM propagation, whose frequency range includes DC. Instead, allowed modes are either TE or TM, depending on how the waveguide is excited. In the standard system of nomenclature, the letter(s) following the "T" describes which field component is transverse to the direction of propagation. Subscripts complete the mode specification by identifying the number of half-waves at cutoff that fit across the x- and y-directions of a rectangular waveguide, or along the circumferential and diametrical directions for circular waveguide. Hence, $TE_{m,n}$ tells us that there are m half-waves of **E**-field across the x-dimension and n half-waves of **E**-field across the y-dimension for a rectangular waveguide.

Consider as a specific example the TE_{10} mode for a rectangular waveguide just at cutoff. This mode has a vertically oriented electric field that is zero at the left and right walls and is a maximum in the center (i.e., there is one half-wave in the x-direction). Additionally, the electric field exhibits zero variation in the vertical direction.

As another example, a TE_{01} designation means the following for a circular waveguide just at cutoff: The **E**-field exhibits no variation in the circumferential direction. It is purely circumferential in orientation and has a zero value at the surface of the waveguide. There is one peak – at the precise center of the waveguide.

Although waveguides can support an infinite number of modes, it is general practice to restrict the operating frequency range so that only one mode propagates. This recommendation derives from the same basic idea that applies in the coax case: as one approaches a mode cutoff frequency from below, reactive energy storage increases. Once a mode actually propagates, an additional real component results. Both effects cause impedance changes (or, in language more appropriate for waveguide work, SWR degrades). The lowest frequency mode is conventionally used, so waveguide dimensions must therefore be selected to span the desired operating frequency range. Just as with coaxial cables, then, waveguide dimensions shrink with increasing operating frequency. There is still a significant net improvement in attenuation because of the larger conductor surface area for the waveguide case.

At high frequencies, it may be impractical to fabricate waveguides of sufficiently small dimensions to suppress higher-order modes. In such cases, it is more sensible to make use of higher-order propagation. Such *overmoding* is of particular value in millimeter-wave work. Overmoded filters, waveguides, and other circuits are active research topics in near-terahertz electronics.

5.4.2 ATTENUATION PROPERTIES OF WAVEGUIDE

For both TE and TM modes in rectangular waveguides, the attenuation ultimately grows as the square root of frequency because of skin loss, and for the same reason that applies to TEM propagation in coax: increases in frequency cause a reduction in the effective volume through which the induced surface currents flow. This loss mechanism seems so fundamental that it may surprise you to learn that it is possible to evade this behavior (at least in theory). There's nothing that can be done about the reduction in *skin depth* with increasing frequency, but if the associated *surface currents* can be made to diminish as frequency increases then the net dissipation can actually decrease.

Remarkably, this is exactly what happens for all $TE_{0,m}$ modes in circular waveguides. The distinguishing feature of these modes (which cannot propagate in rectangular waveguides) is that there is no component of electric field normal to the conductor surface. In theory, the attenuation for such modes approaches *zero* as frequency increases toward infinity. In practice, bends, surface roughness, and any departure from perfect symmetry will ultimately excite other modes that are lossy. Nevertheless, with extraordinary care, it is possible to enjoy significantly smaller attenuation – over a usefully wide frequency range – than is suffered by the other modes. Until the development of optical fiber, circular waveguide was the lowest-loss way to convey a signal from one point to another.

Theoretical derivations are wonderful, but perhaps the best illustration that waveguide is indeed capable of exceptionally low attenuation is to examine some actual data. Off-the-shelf waveguide with loss on the order of 0.5 dB per 30 m at 10 GHz is readily available. Compare this value with the 13-dB loss at 5 GHz of the best flexible coaxial cable. One may also obtain waveguide that exhibits about 15-dB loss (again, per 30 m) at 100 GHz. As a final point of comparison, consider that optical fiber routinely exhibits losses of well under a decibel per kilometer at 400 THz.

5.5 SUMMARY

We've seen that connectors must be chosen and treated with care if proper operation is to be obtained and preserved over time. Both moding concerns and uniformity of mechanical dimensions affect the operational frequency range, explaining why BNC and TNC connectors, for example, have different upper frequency limits even though their critical dimensions are identical.

These same notions were applied to coaxial cable, with additional consideration given to the sources of attenuation. It was seen that larger dimensions reduce the prominence of conductor loss, but the desire for mode-free operation imposes hard limits on how far one may go in this direction.

Table 5.1. *Approximate attenuation and Z_0 of some flexible coaxial cables*

Type	Z_0	Loss (dB) per 30 m @			
		100 MHz	400 MHz	1 GHz	5 GHz
RG8/U	52	2	4	9	30
RG58/U	52	5	11	18	60
RG174/U	50	9	18	31	100
RG213/U	50	2	4	9	30
9913	50	1.4	2.7	4.5	13
RG6/U	75	2	4	6.5	–
RG11/U	75	1.3	2.6	4.3	–
RG59/U	75	2.5	5.1	8.2	–

One may evade these constraints by conveying power in structures whose dimensions are not required to be very small compared with a wavelength. In fact, for proper operation, we saw that waveguide dimensions must be at least a half wavelength in order to allow higher-order modes to propagate at all; they are inherently high-pass structures. This property explains why AM radio signals (with their ∼200–600-m wavelengths) are essentially unable to propagate into highway tunnels while FM signals (whose wavelengths are in the range of 3 m) can propagate reasonably well.

The combination of larger surface area and the absence of a lossy dielectric allows waveguides to exhibit extremely low loss. Higher breakdown voltages are also generally associated with the larger dimensions of waveguide, so the power-handling capability of waveguide is superior to that of coaxial cable.

5.6 APPENDIX: PROPERTIES OF COAXIAL CABLE

Table 5.1 lists the more popular types of coaxial cable, along with their characteristic impedance and typical attenuation of a 30-m length at four different frequencies. Note that the attenuation is in fact proportional to the square root of frequency at low frequencies for all of the cables listed (theory does occasionally work, after all). Skin loss dominates even up to 5 GHz, with dielectric loss becoming evident only as a modest increase in the slope of the loss.

The differing attenuation characteristics among the various cable types are easily explained. At only 0.1" (2.5 mm), the outer diameter of RG174 is the smallest of the cables listed. The attenuation is consequently the largest, being dominated by conductor loss. Because of its large attenuation, use of RG174 is restricted to rather short runs.

At the other extreme is 9913, whose exceptionally low loss is attributable to its large dimensions. The center wire itself is a rather hefty 9.5-gauge (0.405" diameter, about 1 cm) conductor (chosen as the largest that can fit within a UHF connector).

Table 5.2. *Approximate attenuation characteristics of some rigid lines*

Type	Loss (dB) per 30 m @ 1 GHz
0.141" semirigid	13
0.5" semirigid	4
0.5" hardline	2.5
0.625" hardline	1.5
1.625" hardline	1

Owing to its bulk, 9913 is neither easily bent nor should it be. This cable has the lowest loss of all conventionally available flexible cables, but it will begin to mode the soonest.

Less bulky is RG8, which is widely used for amateur radio applications. Its intermediate loss derives from its intermediate dimensions. One caveat is that the foam dielectric material is not physically robust. Repeated bending of the cable can cause displacement of the center conductor, resulting in impedance discontinuities. Extreme movement can even result in short circuits.

For general-purpose applications, RG58 is perhaps the most widely used coaxial cable up to the VHF range of frequencies because it is lighter than RG8 and has moderately low loss and reasonable power-handling capability. Note that RG58 ("thinnet") replaced RG8 in early ethernet implementations before it, in turn, was replaced by twisted pair.

Semirigid and rigid coaxial lines are used where one desires maximum mechanical stability combined with low loss. Table 5.2 presents representative characteristics for several types of semirigid coax and hardline (classified by outer dimension). Teflon-filled 0.141" semirigid line typically exhibits under 1.8 dB/m loss at 18 GHz. The low loss values of these cables make them particularly suitable for instrumentation fixtures and for long runs.

As a final note, putting a connector on the end of these lines usually involves stripping off a prescribed amount of shielding and then using the center conductor of the cable as the male pin of the connector.

CHAPTER SIX

PASSIVE COMPONENTS

6.1 INTRODUCTION

In this chapter we examine the properties of passive components commonly used in RF work. Because parasitic effects can easily dominate behavior at gigahertz frequencies, our focus is on the development of simple analytical models for parasitic inductance and capacitance of various discrete components.

6.2 INTERCONNECT AT RADIO FREQUENCIES: SKIN EFFECT

At low frequencies, the properties of interconnect we care about most are resistivity, current-handling ability, and perhaps capacitance. As frequency increases, we find that inductance might become important. Furthermore, we invariably discover that the resistance increases owing to the *skin effect* alluded to in Chapter 5.

Skin effect is usually described as the tendency of current to flow primarily on the surface (skin) of a conductor as frequency increases. Because the inner regions of the conductor are thus less effective at carrying current than at low frequencies, the useful cross-sectional area of a conductor is reduced, thereby producing a corresponding increase in resistance.

From that perfunctory and somewhat mysterious description, there is a risk of leaving the impression that all "skin" of a conductor will carry RF current equally well. To develop a deeper understanding of the phenomenon, we need to appreciate explicitly the role of the magnetic field in producing the skin effect. To do so qualitatively, let's consider a solid cylindrical conductor carrying a time-varying current, as depicted in Figure 6.1.

Assume for the time being that the return current (there must always be one in any real system) is far enough away that its influence may be neglected. A time-varying current I generates a time-varying magnetic field H. That time-varying field induces a voltage around the rectangular path shown, in accordance with Faraday's law. Ohm's law then tells us that the induced voltage in turn produces a current flow

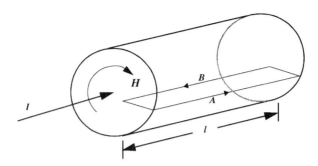

FIGURE 6.1. Illustration of skin effect with isolated cylindrical conductor

along that same rectangular path, as indicated by the arrows. Now here's the key observation: The direction of the induced current along path A is *opposite* that along B. The induced current thus adds to the current flowing along one side of the rectangle and subtracts from the other. Taking care to keep track of algebraic signs, we see that the current along the surface is the one that gets augmented and that the current below the surface is diminished. That is, current flow is strongest near the surface; that's the skin effect.

To develop this idea a little more quantitatively, let's apply Kirchhoff's voltage law (with proper accounting for the induced voltage term, both in magnitude and sign) around the rectangular path to obtain

$$J_B \rho l - J_A \rho l + \frac{d\phi}{dt} = 0, \tag{1}$$

where J is the current density, ρ is the resistivity, and the magnetic flux ϕ is perpendicular to the rectangle shown.

We see that, as deduced previously, the current density along path A is indeed larger than along B – by an amount that increases as either the depth, frequency, or magnetic field strength increases and as the resistivity decreases. Any of these mechanisms acts to exacerbate the skin effect. Furthermore, the presence of the derivative tells us that the current undergoes more than a simple decrease with increasing depth; there is a phase shift as well.

If we now increase the radius of curvature to infinity, we may convert the cylinder into the semi-infinite rectangular structure that is more commonly analyzed to introduce skin effect; see Figure 6.2. We will provide the barest outline of how to set up the problem and then simply present the solution.[1]

Computing the voltage induced by H around the rectangular contour proceeds with KVL as before:

---

[1] For a detailed derivation, consult any number of excellent texts on electromagnetic theory. See, e.g., S. Ramo, T. Van Duzer, and J. R. Whinnery, *Fields and Waves in Communications Electronics*, 3rd ed., Wiley, New York, 1994. Also see U. S. Inan and A. S. Inan, *Electromagnetic Waves*, Prentice-Hall, Englewood Cliffs, NJ, 2000.

6.2 INTERCONNECT AT RADIO FREQUENCIES: SKIN EFFECT

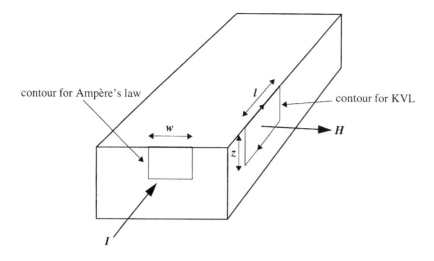

FIGURE 6.2. Subsection of semi-infinite conductive block

$$J\rho l - J_0 \rho l = \frac{d\phi}{dt} = -\frac{d}{dt}\int_0^z Bl\,dz, \qquad (2)$$

where the subscript 0 denotes the value at the surface of the conducting block. Now express J and H (and hence B) explicitly as sinusoidally time-varying quantities. For example, let

$$J_0 = J_{s0} e^{j\omega t}, \qquad (3)$$

where the subscript s is chosen arbitrarily to denote the magnitude of these skin-effect variables.

With these substitutions, the KVL equation allows us to write

$$\rho \frac{dJ_s}{dz} = -j\omega B_s = -j\omega \mu H_s, \qquad (4)$$

where we have used the relation between flux density B and magnetic field strength H,

$$B = \mu H; \qquad (5)$$

here μ is the permeability (equal to the free-space value in nearly all integrated circuits).

We need one more equation to finish setting up the differential equation. Ampère's law will give it to us:[2]

$$I_{encl} = w \int_0^z J\,dz = wH_0 - wH. \qquad (6)$$

Making the same substitutions as before yields

---

[2] Remember what Ampère's law says in words: The integral of the magnetic field around a closed path equals the total current enclosed by that path.

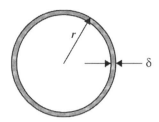

FIGURE 6.3. Application of skin depth concept to resistance calculation (cross-section shown)

$$-\frac{dH_s}{dz} = J_s. \tag{7}$$

Combining Eqn. 4 and Eqn. 7 yields a simple second-order differential equation for the current density:

$$\frac{d^2 J_s}{dz^2} = \frac{j\omega\mu}{\rho} J_s, \tag{8}$$

whose solution is

$$J_s = J_{s0} \exp(-z/\delta) \exp(-jz/\delta); \tag{9}$$

here

$$\delta = \sqrt{2\rho/\omega\mu} = \sqrt{2/\omega\mu\sigma} \tag{10}$$

is known as the skin depth. Notice that the current density decays exponentially from its surface value. Notice also (from the second exponential factor) that there is indeed a phase shift, as claimed earlier, with a 1-radian lag at a depth equal to δ.

For this case of an infinitely wide, infinitely long, and infinitely deep conductive block, the skin depth is the distance below the surface at which the current density has dropped by a factor of e. For copper at 1 GHz, the skin depth is approximately 2 μm. For aluminum, that number increases a little bit, to about 2.5 μm. What this exponential decay implies is that making a conductor much thicker than the skin depth doesn't help much, because the added material carries very little current. Furthermore, we may compute the effective resistance as that of a conductor of thickness δ in which the current density is uniform. This fact is often used to simplify computation of the AC resistance of conductors. To make sure that the result is valid, however, the boundary conditions must match those used in deriving our system of equations: The return currents must be infinitely far away, and the conductor must resemble a semi-infinite block. The latter criterion is satisfied reasonably well if all radii of curvature, and all thicknesses, are at least 3–4 skin depths.

As a specific example, let us estimate the AC resistance of an isolated wire. Assume that the wire's diameter is much larger than the skin depth. In that case, we may estimate the resistance by pretending that all of the current flows in an annulus of depth δ (see Figure 6.3). The resistance is readily computed as

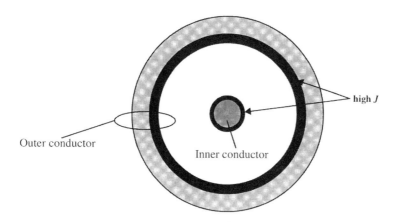

FIGURE 6.4. Coaxial cable cross-section

$$R = \frac{\rho l}{A} \approx \frac{\rho l}{2\pi r \delta}, \tag{11}$$

where l is the length of the wire and we have assumed that $r \gg \delta$ in constructing the last approximation.

In this case, calculations based on a simple skin-depth assumption yield excellent results. That's atypical; in many other cases, the results may be grossly in error. In evaluating the risk of making such a mistake, it's helpful to anticipate qualitatively where the currents will be flowing. To do so, recall that, in introducing skin depth, we invoked a qualitative argument that led us to several important insights. One of these is that skin effect is strongest where the magnetic fields are strongest. So, in determining which surfaces are likely to carry most of the current, we need to identify where the fields are the strongest.

Consider a coaxial system of conductors, as in a cable. There are three surfaces, but not all three exhibit skin effect. The outer cylindrical conductor conveys the return current for the central conductor. The coaxial structure is self-shielding in that both electrical and magnetic fields external to the cable are ideally zero, thanks to cancelling contributions by the two conductors (this attribute is why the coaxial structure is valued in the first place). The magnetic field is therefore the strongest in the space between them, and thus the skin effect is felt most acutely at the surface of the inner conductor and at the inner surface of the outer conductor. In Figure 6.4, the regions of high current density are indicated crudely in black. The outer surface of the outer conductor carries very little current (again, we're assuming conductor thicknesses that are very much larger than the skin depth). Therefore, computation of its resistance would consider only the black annulus at the inner surface of the outer conductor. *Not all skin exhibits the skin effect.*

To reinforce that last statement, consider another qualitative example. Specifically, consider what happens when we have two parallel cylindrical conductors in

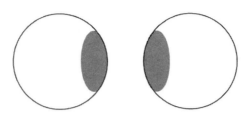

FIGURE 6.5. Cross-section of two-wire line

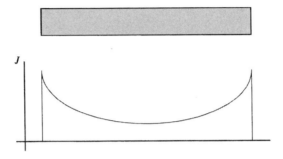

FIGURE 6.6. Thin, wide conductor cross-section and current density (approximate)

proximity, where the current in one serves as the return current for the other. As in the coaxial case, the magnetic field is strongest in the region between the two conductors. Therefore, the maximum current densities are found where the wire surfaces face each other; see Figure 6.5. For self-evident reasons, the phenomenon of current crowding at a surface because of current flowing in a nearby conductor is sometimes known as the *proximity effect*.

As one last example, consider a thin, wide conductor (again, with all other conductors very far away). The currents will distribute themselves roughly as shown in Figure 6.6. The current crowds toward the two ends (as a mnemonic aid, imagine this structure to be the central slice of a cylindrical conductor; the current-carrying ends of this slab correspond to the current-carrying outer surface of the cylinder). Because of the relatively low current density along the long edges, further widening of the conductor produces only modest resistance reductions. Thickening the conductor would have a much stronger effect.

The current distribution changes if we bring a conductor near this one, turning a complicated problem into a near-impossible one.

Computing the effective resistance for these last two structures is clearly not as straightforward as for the isolated conductor. Indeed, accurately computing the effective resistance of such a simple-seeming structure as a single-layer coil is virtually impossible because of all of the interactions among turns. This difficulty highlights the danger of automatically assuming that all surfaces are equally effective at carrying current.

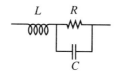

FIGURE 6.7. Simple lumped RF resistor model

Finally, for a collection of useful formulas, see Harold A. Wheeler, "Formulas for the Skin Effect," *Proc. IRE*, v. 30, September 1942, pp. 412–24.

6.3 RESISTORS

Even a component as simple as a resistor exhibits complex behavior at high frequencies. We may construct a very simple model by acknowledging first that current flows in both the connecting leads and the resistor proper. The energy stored in the magnetic field associated with that current implies the presence of some series inductance (typically about 0.5 nH/mm for leads in axial packages, as a rough approximation[3]). In addition, there is necessarily some capacitance that shunts the resistor as well, since we have a conductor pair separated by a distance. The simplest (but by no means unique) RF lumped circuit model for a physical resistor might then appear as shown in Figure 6.7.

The presence of parasitic inductance and capacitance causes the impedance to depart from a pure, frequency-independent resistance. Very low values of resistance suffer from an early impedance increase, starting approximately at a frequency where the reactance of the series inductance becomes significant compared with the resistance. Similarly, high resistances suffer a premature impedance decrease from the shunt capacitance. The frequency range over which the impedance remains roughly constant (at least for our simple model) is maximized for some intermediate (and definite) resistance value. As one might suspect from transmission line theory, this magic value is simply given by

$$R_{opt} = \sqrt{L/C} = Z_0. \tag{12}$$

The formal derivation of Eqn. 12, which we will not carry out here, begins by writing an expression for the impedance magnitude and then solving for the condition of maximal flatness by maximizing the number of derivatives whose value is zero at zero frequency.[4] We can construct an intuitive path to Eqn. 12 by considering two limiting case. If the inductance is neglected, the time constant of the network is simply RC. If the capacitance is neglected, the time constant is L/R. Setting these two

---

[3] For various equations for inductance and capacitance, see T. Lee, *The Design of CMOS Radio-Frequency Integrated Circuits*, 2nd ed., Cambridge University Press, 2004.
[4] Ibid. As mentioned there, it is often easier to carry out this procedure on the square of the magnitude.

Table 6.1. *Approximate element values for simple lumped RF resistor model*

Resistor type	L	C	Z_0	f_{max}
0.5-W axial-lead (3 cm total length)	15 nH	0.5 pF	170 Ω	1.8 GHz
0.5-W axial-lead, trimmed to 1 cm total length	5 nH	0.5 pF	100 Ω	3 GHz
0.25-W axial-lead (2 cm total length)	10 nH	0.25 pF	200 Ω	3 GHz
0.25-W axial-lead, trimmed to 0.7 cm total length	3 nH	0.25 pF	100 Ω	6 GHz
Type 0805 surface mount	<1 nH	0.02–0.1 pF (see text)	150 Ω	20 GHz

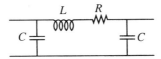

FIGURE 6.8. More elaborate model for surface-mount resistor in a microstrip environment

time constants equal and solving for R yields Eqn. 12. If the resistor value is smaller than Z_0 then the impedance of our model only rises, with a radian corner frequency given approximately by R/L. If the resistance exceeds Z_0 then the impedance initially drops, with a corner frequency of roughly $1/RC$. If the resistance equals Z_0, the bandwidth f_{max} over which the impedance remains approximately constant is given by the resonant frequency of the LC combination. The smaller the LC product, the greater the frequency range over which the resistor looks approximately resistive.

For reference, typical (and approximate) model parameter values for some representative resistors are given in Table 6.1. Note that the maximally flat impedance levels are in the range of 100–200 Ω. That is yet another reason why transmission line impedances tend not to be far removed from that range of values.

Again, the table entries are *highly* approximate and should be taken only as a rough guide. The shunt capacitance value for the surface-mount resistor in particular is hardly a reliable constant, depending as it does on the details of construction, composition of materials, and so forth. In most cases, what one measures has more to do with the fixturing than with the surface-mount resistor itself. Failure to accommodate this reality in extracting models from measurements is why published values for the capacitance of surface-mount resistors span an order of magnitude. As an example, a surface-mount resistor in a microstrip environment will generally behave differently from one mounted in some other type of fixture because the presence of the ground plane will change the electric field distribution. In this case, a more reasonable circuit model might be that appearing in Figure 6.8.

A typical set of model parameters for a type 0805 resistor might be $C = 80$ fF and $L = 0.7$ nH. Note that this model has no shunt capacitance across the resistor

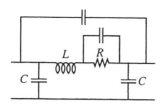

FIGURE 6.9. Still another model for a surface-mount resistor

itself. There probably is some, but it is so small that a good fit to measured behavior is possible without including it. It's important to note that there is no one "correct" model, and many different models may yield equally acceptable answers. For example, the models within many microwave CAD tools are often a sort of combination of that in Figures 6.7 and 6.8; this is shown in Figure 6.9.

Note that the values for optimum Z_0 are all in excess of typical line impedances (such as 50 Ω). If, for example, we are to provide terminations for a 50-Ω line, then the largest bandwidth is obtained with a parallel combination of devices rather than with a single 50-Ω resistor. Four 0.25-W resistors of 200 Ω each will do a reasonably good job of providing a 50-Ω termination over a bandwidth in excess of 1 GHz. Similarly, a parallel combination of two 0805 surface-mount resistors will provide an excellent termination over a bandwidth in excess of 10 GHz. The table also shows why conventional resistors (particularly the 0.5-W variety) are rarely used in microwave work. If higher-power terminations are required, it is preferable to make them out of parallel combinations of lower-power, higher-frequency resistors for operation over the largest bandwidth.

As an aside, the numerical identifiers for surface-mount components convey something about the physical dimensions. Within truncation errors, the first pair of numbers is four times the length in millimeters, while the second pair is four times the width.[5] Some common sizes are shown in Table 6.2 along with their power dissipation levels.

The parasitic capacitance depends to some extent on whether one terminal is grounded, the material comprising the body of the component, and whether the component is mounted flush over a grounded plane or on end with one terminal in the air. As a result, one cannot provide universally correct values of parasitic capacitance that are valid for all cases. Nevertheless, we can convey some rough idea of the parasitics associated with these packages. Consider the largest size listed in the table, the 2512, whose 6.4 × 3.2 × 3.2 dimensions are associated with ∼2.5-nH parasitic inductance and typically ∼0.18-pF shunt capacitance. If measured values are unavailable (as is usually the case), one may estimate the series inductance for the

---

[5] The original designations are actually in terms of mils (thousandths of an inch). Thus, each pair of digits represents a dimension that, when multiplied by ten, yields the approximate dimension in mils.

Table 6.2. *Some surface-mount resistor packages and characteristics*

Package	Approximate dimensions (mm × mm)	P_{diss} (mW)
0402	1.0 × 0.5	60
0603	1.6 × 0.8	60
0805	2.0 × 1.25	100
1206	3.2 × 1.6	125
1210	3.2 × 2.5	250
1812	4.5 × 3.2	500
2512	6.4 × 3.2	1000

other packages with the aid of the following equation for the inductance of an infinitesimally thin flat sheet:

$$L_{sheet} \approx \frac{\mu l}{2\pi}\left[0.5\ln\left(\frac{2l}{w}\right) + \frac{w}{3l}\right]. \quad (13)$$

This formula is appropriate because the resistive material is almost always a flat, thin layer deposited on the surface of a much thicker insulating substrate (even for "thick film" resistors). For the aspect ratios typically encountered in low-power surface-mount components, the inductance is usually in the range of 0.3–0.5 nH/mm.

One uses this formula twice to estimate the total parasitic inductance. One component of the total inductance is that of the main body, so its length and width are first plugged into the equation. To this (usually) dominant term, one must also add the inductance due to the flat vertical portions that contact the ends of component, with the height now replacing the width in Eqn. 13. The solder meniscus effectively thickens these vertical stubs, however, so it isn't quite fair to use the full inductance of each vertical section. As an arbitrary choice, 1/2 to 2/3 of the computed value of the vertical stubs is not an unreasonable factor. Using the former factor, we estimate an inductance of about 2.6–2.7 nH for the 2512 package, in good agreement with measurements.

Note that the inductance per length here is considerably lower than the 1-nH/mm rule of thumb that typically applies to thin round component leads. The reason is that the thick and wide shape of the surface mount components spreads out the magnetic field lines, thereby reducing flux density and hence inductance.

As alluded to previously, estimating the capacitance of this type of structure is somewhat complicated. Nevertheless, we can offer a crude lower bound based on the formula for the capacitance per length of a dipole antenna made out of a cylindrical conductor (see Chapter 21 for a derivation):

$$C \approx \left\{c^2\frac{\mu_0}{2\pi}\left[\ln\left(\frac{2l}{r}\right) - 0.75\right]\right\}^{-1} \approx \frac{2\pi\varepsilon_0}{\ln(2l/r) - 0.75} \approx \frac{5.56 \times 10^{-11}}{\ln(2l/r) - 0.75}. \quad (14)$$

This equation yields an estimate of the capacitance (per length) between ground and a conductor of length l and radius r, with values generally within a factor of 2 of 15 fF/mm for typical surface-mount component dimensions. The formula actually assumes a vertically oriented cylindrical conductor above a ground plane, with one terminal grounded.

The lower bound computed with this formula assumes (among other things) a unit dielectric constant and no additional metal pads, etc. The capacitances will be boosted by the dielectric constant of the package and also that of FR4. Furthermore, discontinuous transition between the PC board and the terminals of the resistors will give rise to shunt capacitances to ground from each terminal. Even though surface-mount components are hardly unique in this respect qualitatively, their small inherent capacitance means that this effect is more noticeable than for other discrete resistor types.

This approximate method may be used to estimate package parasitic capacitances of surface-mount inductors, as well as those of ordinary components of circular cross-section. Just remember that package parasitics may account for only a part of the total; some parasitics may arise internally (e.g., turn-to-turn winding capacitance in inductors). The computed package capacitance is therefore once again a lower bound estimate of the parasitic capacitance.

As a final comment, note that many axial-lead resistors are based on a carbon composition, which consists of a resistive powder formed into a cylindrical shape. Unfortunately such resistors can exhibit significant $1/f$ noise, with a power spectral density proportional to the DC bias current flowing through the resistor. Carbon film resistors are substantially better in this regard, and metal film resistors are even better. Although the $1/f$ corners are generally well below the RF range, one must be aware that oscillators can upconvert low-frequency noise into phase noise near the carrier.[6] Thus, even though $1/f$ noise is usually not an issue in circuits such as RF amplifiers, it cannot be completely neglected in all RF circuits. Fortunately, surface-mount resistors are generally of the film variety.

6.4 CAPACITORS

Many different dielectric materials are used in an effort to satisfy the numerous conflicting demands made on capacitor performance. Trade-offs among breakdown voltage, temperature coefficient, RF loss, and capacitance density inevitably lead to the many types of capacitors presently available. Space does not permit an encyclopedic review of all capacitor types, so we focus only on those that are commonly encountered in high-frequency circuits.

The lowest-loss capacitors are made with air (or vacuum) as the dielectric. Higher densities with low loss may be obtained with mica (a naturally occurring mineral) and

---

[6] Lee, op. cit. (see footnote 3).

Table 6.3. *Three-character capacitor codes (EIA)*

Temperature (°C)		Maximum percentage capacitance change over temperature range	
Minimum	Maximum		
X: −55	3: +45	A: ±1	P: ±10
Y: −30	4: +65	B: ±1.5	R: ±15
Z: +10	5: +85	C: ±2.2	S: ±22
	6: +105	D: ±3.3	T: −33, +22
	7: +125	E: ±4.7	U: −56, +22
		F: ±7.5	V: −82, +22

polystyrene. Although polystyrene has excellent electrical properties, it possesses an unfortunately low melting point, which limits use to temperatures below 85°C. One must consequently exercise care in soldering polystyrene capacitors.

Considerably more robust are capacitors made from PTFE, which (as noted earlier) is also an exceptionally low-loss dielectric. However, the expense of fabricating good thin films of PTFE has meant that capacitors made with it tend to have rather large dielectric thicknesses, leading to low capacitance densities (but very high breakdown voltages).

Ceramic capacitors themselves come in a number of varieties, distinguished by the characteristics of their dielectrics. To keep track of the many permutations, the Electronics Industry Association has settled on a three-character nomenclature. The first (second) character is a letter that indicates the minimum (maximum) operating temperature, and the third character is a letter that conveys the maximum capacitance change over the entire operating temperature range. The particulars are shown in Table 6.3. For example, a capacitor with a designation of X7R exhibits at most a ±15% capacitance variation over an operating temperature range of −55°C to +125°C.

Although a zero temperature coefficient is most commonly desired, there are important instances in which one wants instead a nonzero TC of a specified value. Oscillators are one example; inductors typically exhibit a positive TC,[7] so capacitors possessing a compensating negative TC are needed to produce an oscillation frequency with an overall zero TC. The characteristics of capacitors with controlled temperature coefficient are identified by the letter N (for "negative") or, more rarely, the letter P (yes, for "positive"), followed by the maximum TC magnitude in parts per million per degree C. A designation of N750 thus represents a capacitor with a −750-ppm/°C temperature coefficient. Just to make things confusing, however, there is an alternate system of codes that conveys the same information. Designed to save space for printing on small components, the three-digit EIA code unfortunately

---

[7] Consider that inductance is dimensionally proportional to length, and that most materials expand when heated. Thus, most physical inductors possess positive TCs.

Table 6.4. *Capacitor TC codes*

Older designation	Three-digit EIA	Older designation	Three-digit EIA
NP0	C0G	N330	S2H
N033	S1G	N470	T2H
N075	U1G	N750	U2J
N150	P2G	N1500	P3K
N220	R2G	N2200	R3L

does not directly convey numerical information about the actual TC, so Table 6.4 provides the necessary translation between the two labeling conventions.

The first letter in the three-digit TC convention conveys information about the TC's significant digits. The values are a subset of the values of standard resistors. For example, one can discern from the table that $P = 1.5$, $R = 2.2$, $S = 3.3$, $T = 4.7$, and $U = 7.5$. The middle digit of the code is the exponent. The NP0 designation (C0G) stands for "negative-positive-zero" and refers to the characteristics of a composite of negative- and positive-TC materials to yield a nominally zero TC (typically, a maximum of ± 30 ppm/°C).[8] The capacitance thus stays within approximately 0.15% of the nominal value over the military temperature range ($-55°C$ to $125°C$). Capacitance values of up to about 10 nF are available in the standard surface-mount package sizes. The loss of NP0/C0G is the lowest of the standard types, with peak Q-values in excess of 500–600 at low frequencies. This material also exhibits a low voltage coefficient.

Other commonly used materials include the somewhat less stable (but higher dielectric constant) X7R ceramic. Surface-mount types with values up to about 100 nF are available. As mentioned earlier, the capacitance might vary as much as $\pm 15\%$ over the military temperature range. Unlike C0G, the capacitance decreases (roughly linearly) with increasing DC bias, with up to an additional 30% drop at the rated voltage. This variation with voltage is associated with the piezoelectric nature of the dielectric, and the nonlinear behavior can generate significant distortion when these capacitors are used in the signal path. In addition, most X7R formulations are two orders of magnitude lossier than C0G materials.

High-K (high dielectric constant) ceramics, such as Y5V, give us capacitors that are physically the smallest but which suffer from extremely high TCs (e.g., up to an astounding 80% drop in capacitance at zero bias over a temperature range of $-30°C$ to $85°C$) and from losses that are a third of X7R. The voltage coefficient is also strongly negative, and one may expect a capacitance drop of up to 75% at the rated voltage. Such capacitances actually make effective mixers, so beware (or exploit this behavior). Furthermore, such dielectrics are piezoelectric to a surprising degree. It

---

[8] Note that these designations contain the numeral 0 and not the letter O.

Table 6.5. *Capacitor tolerance codes (EIA)*

Identifier	Tolerance (pF)	Tolerance (%)
B	±0.1	
C	±0.25	
D	±0.5	
E		±25
F	±1	±1
H		±2
J		±5
K		±10
M		±20

is not unusual for a sharp mechanical shock to generate spikes of volts (sometimes many tens of volts). Even if the spike does not cause direct damage to delicate circuitry, it should be obvious that the microphonic behavior of high-K capacitors can lead to a host of objectionable problems, especially if connected to sensitive circuit nodes and subjected to vibration (as in mobile applications). The most common use of these capacitors is therefore as supply bypasses, rather than in the signal path. Values up to about 1 μF are available in the standard surface-mount packages.

One should not overlook the option of making capacitors with the PC board as the dielectric. It is frequently convenient for trimming purposes to realize some part of a desired capacitance in PC board form to permit adjustment after fabrication. In any case, it's a good idea to be aware of how much capacitance is associated with a given area of conductor, if for no other reason than to estimate layout parasitics. With FR4, one can expect about 5 pF/cm^2 with a 1/32" (0.8-mm) thick substrate, or roughly 2.5 pF/cm^2 on a 1/16" (1.6-mm) substrate. The loss of FR4 is quite tolerable, being modestly better than that of X7R or Y5V. Of course, still lower loss (and somewhat lower capacitance) is obtained with a higher-quality board material, such as PTFE or RO4003. More discussion on the use of PC board traces for realizing capacitances and inductances may be found in Chapter 7.

Capacitor values are encoded as three digits stamped somewhere on the body (if the digits fit), followed by a letter that identifies the tolerance (see Table 6.5). The first two digits are a mantissa, and the third is an exponent. The implicitly understood unit is the picofarad. Hence, "221K" stands for a 220-pF capacitor with ±10% tolerance, and "105M" denotes a 1-μF, ±20% capacitor. Occasionally some other conventions are used, but this scheme is by far the most widespread. If in doubt, one can always verify a conjecture with an actual measurement.

Just as with resistors, parasitic effects cannot be ignored at radio frequencies. The simplest lumped RF model for real capacitors includes lead or terminal inductance (as before, this may be estimated as roughly 0.5–1.0 nH/mm for typical round wire

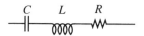

FIGURE 6.10. Simple lumped capacitor model

leads), and a resistive term to account for losses; see Figure 6.10. The inductance for surface-mount packages can be estimated using Eqn. 13, as before.

The resistive term of the model accounts for the effect of at least two distinct mechanisms. One is the loss of the dielectric, and the other is conductor loss (which is exacerbated at high frequencies by skin effect). Loss is often characterized by a dissipation factor D (or, equivalently, a loss tangent $\tan \delta$). Dissipation factor is simply the reciprocal of Q, while loss tangent is defined as the ratio of the imaginary and real parts of the dielectric constant. Strictly speaking, loss tangent applies only to the dielectric material, but it is often used to include all losses in a capacitor. In this latter case, loss tangent is the same as the capacitor dissipation factor. The reason for these multiple ways of describing loss is cultural. Power electronics folks tend to think in terms of power factor (the cosine of the phase angle between voltage and current, which angle is the same as that of the impedance), RF engineers generally think in terms of Q, and materials scientists tend to focus on loss tangent (dissipation factor, D).

The definition of power factor means that it is equal to the cosine of the arctangent of Q (the proof is left to you, because it is obvious that you don't have enough fun). For sufficiently large Q, the power and dissipation factors converge. For example, a Q-value in excess of 7 assures an error of less than about 1%. For all capacitors worth using in the signal path, Q will certainly be large enough that one may take loss tangent and power factor to be equal in practice.

Given these definitions, the component of effective series resistance (ESR) due to dielectric loss is

$$R \approx D/\omega C. \tag{15}$$

This formula is valid only at frequencies well below the series resonance. Clearly, ESR is a frequency-dependent quantity, especially when skin-effect conductor loss is considered as well.

At frequencies well above resonance, the resistance becomes proportional to frequency because the inductive reactance dominates, leading to the following approximation:

$$R \approx D\omega L. \tag{16}$$

We can deduce several important facts from the series RLC model. Above the resonant frequency of the network, the combination appears inductive and the impedance therefore increases with frequency. The minimum impedance is reached at the resonant frequency. If a capacitor is used, say, as a power supply bypass, then it is important to recognize that the quality of the bypassing will diminish at higher

Table 6.6. *Representative capacitors and lumped model parameters at 100 MHz*

Type	C	L	R	SRF
Ceramic disc (C0G/NP0)	10 nF	10 nH	0.5 Ω	17 MHz
0805 C0G/NP0	10 nF	~1 nH	0.08 Ω	50 MHz
0805 C0G/NP0	100 pF	~1 nH	0.25 Ω	500 MHz

frequencies because of series inductance. Simply exhibiting inductive behavior need not preclude use, however, since the most relevant quantity is the magnitude of the impedance. If this is sufficiently low, the capacitor can still be a satisfactory bypass element even when operating above the resonant frequency.

As a rough calibration on the magnitudes of these parasitic elements, consider the parameters (at 100 MHz) listed in Table 6.6. Here the disc capacitor is assumed to have a total length (measured from the tip of one lead, through the disc body, to the tip of the other lead) of about 10 mm. The 100-MHz test frequency considerably exceeds the 17-MHz self-resonant frequency in this case, so the effective series resistance is due more to the lead inductance than to the intrinsic capacitance. By exerting a little effort to shorten lead length, it is possible both to increase the self-resonant frequency and to reduce R by modest amounts.

It should be reiterated that loss is a strong function of both frequency and dielectric composition. Thus, the resistance values in Table 6.6 cannot be treated as universal constants. Your mileage may vary.

6.5 INDUCTORS

6.5.1 SURFACE-MOUNT INDUCTORS AND FERRITE BEADS

The never-ending drive to miniaturize circuits has resulted in the wide availability of tiny components, including inductors. Typical surface-mount inductors suitable for RF and microwave use are available in values ranging from about 10 nH on the low end to about 1 μH on the upper end. Manufacturers tend not to provide a great deal of detailed information about these inductors, so the user is obligated to perform the characterizations experimentally in all cases where it matters. That said, we can offer some crude generalizations that prove useful for initial back-of-the-envelope calculations. The shunt capacitances of most surface-mount inductors seem to be fairly narrowly distributed, with values typically around 0.1–0.2 pF; this corresponds to self-resonant frequencies of 4–5 GHz for the smallest inductances and up to several hundred megahertz for the largest. If higher self-resonant frequencies are required, you must make your own inductors. The same comment applies if you require a lower-loss inductor than is available commercially.

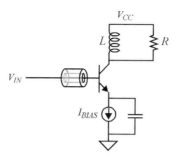

FIGURE 6.11. Typical use of ferrite bead

Occasionally, it proves essential to *increase* the RF loss of a circuit. Such a need may arise in connection with efforts to tame stubborn parasitic oscillations in an amplifier, for example, where the addition of "Q-spoiling" elements in series with the base (or gate) of a transistor has a long history of success (indeed, this trick dates back to the earliest days of the vacuum tube era). Often these annoying oscillations arise when the net collector load appears inductive at a frequency where the transistor still provides lots of gain. Thanks to feedback through the collector–base capacitance, this inductive load can produce the appearance of a negative input resistance. With a mixture of inductance, capacitance, and negative resistance, oscillation is all but assured. Adding a big enough resistor in series with the base is effective in ensuring a net positive base resistance, but it has the disadvantage of introducing a DC drop.

To introduce loss selectively at RF, one may use *ferrite beads*. These are nothing more than cylindrically shaped pieces of a lossy magnetic material. Threading a wire once or twice (or more, if necessary) through the center hole of a bead produces a transformer whose secondary is loaded by a resistance representing the lossy ferrite; see Figure 6.11. Because the coupling is effective only at RF, there is no DC loss.

Ferrite beads are routinely used in the suppression of electromagnetic interference (EMI, also called RFI, for radio-frequency interference). Their usefulness for such a wide variety of purposes has led many engineers to expect perhaps too much from these simple elements.

6.5.2 FORMULAS FOR INDUCTANCE

We have already incorporated inductance in many of the foregoing equations in a piecemeal manner. We now present a number of additional formulas for commonly encountered geometries. In all that follows, the equations strictly apply only at DC unless stated otherwise. At high frequencies, inductance drops because the shrinking of skin depth causes the contribution of internal flux to diminish. Fortunately, internal flux generally accounts for only a small percentage ($<5\%$) of the total, so its reduction does not cause dramatic changes in the overall inductance. Nonetheless, it is worthwhile avoiding unpleasant surprises by knowing explicitly what assumptions have gone into the derivations of formulas.

Flat Sheets

We've already presented a formula for the inductance of a current sheet. We repeat it here so that all the inductance formulas are in one place for easy reference:

$$L_{sheet} \approx \frac{\mu l}{2\pi}\left[0.5\ln\left(\frac{2l}{w}\right) + \frac{w}{3l}\right] = (2 \times 10^{-7})l\left[0.5\ln\left(\frac{2l}{w}\right) + \frac{w}{3l}\right]. \quad (17)$$

Wires

It is frequently desirable to know the inductance of lengths of conductor – either because parasitics need to be quantified or because one desires to use the inductance as a circuit element. If we may neglect the influence of nearby conductors (i.e., if we assume that the return currents are infinitely far away), then the DC inductance of a round wire is given by[9]

$$L \approx \frac{\mu_0 l}{2\pi}\left[\ln\left(\frac{2l}{r}\right) - 0.75\right] = (2 \times 10^{-7})l\left[\ln\left(\frac{2l}{r}\right) - 0.75\right]. \quad (18)$$

For a 2-mm–long standard IC bondwire, this formula yields 2.00 nH, leading to an oft-cited rule of thumb that the inductance of thin, round conductors is approximately 1 nH/mm. Notice that the inductance does grow faster than linearly with length because there is mutual coupling between parts of the wire (i.e., there is a weak transformer action) with a polarity that aids the inductance. From the logarithmic term, however, we see that this effect is minor. For example, going from 5 mm to 10 mm changes the DC inductance per millimeter from 1.19 nH to 1.33 nH (at least according to Eqn. 18). The inductance is similarly insensitive to the wire diameter, so even the larger conductors found in discrete circuits possess inductances of the same general order (e.g., 0.5 nH/mm).

If there is a conducting plane nearby – defined loosely as closer than a distance approximately equal to the length of the wire – then the inductance will be noticeably lower than that given by Eqn. 18. Intuitively, this reduction comes about as follows. Current flowing in the wire (which may be thought of, say, as positive charges moving in the x-direction) induces an image current in the ground plane (e.g., negative charges also moving in the x-direction). Opposite charges moving in the same direction are equivalent to two currents flowing in *opposite* directions, so their magnetic fields tend to cancel somewhat, leading to a reduction in magnetic flux. The closer the plane, the more dramatic the reduction in flux (and hence in inductance).

Air-Core Solenoids

Although our focus is on components that may be realized in a largely planar universe, more inductance per volume can be obtained with a classic 3-D textbook structure: the single-layer solenoid. See Figure 6.12.

---

[9] *The ARRL Handbook*, American Radio Relay League, 1992, pp. 2–18. The proximity of conducting planes may be ignored as long as they are located a distance away that is equal to one or two lengths, at minimum.

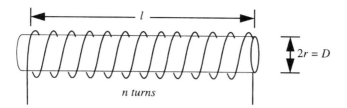

FIGURE 6.12. Single-layer solenoid

Assuming that the turns – unlike those in the figure – are tightly packed ("close-wound"), the inductance in *microhenries* is given by a famous formula presented by Wheeler in the late 1920s:

$$L \approx \frac{n^2 r^2}{9r + 10l}, \qquad (19)$$

where r and l are in *inches*.[10] In SI units, the formula is:

$$L \approx \frac{\mu_0 n^2 \pi r^2}{l + 0.9r}, \qquad (20)$$

where a free-space permeability is assumed. These formulas provide remarkable accuracy (typically better than 1%) for close-wound single-layer coils as long as the length exceeds two thirds the radius.[11] The formula underestimates the true inductance by about 4% for $l = 0.4r$, so the formula remains useful for moderately short coils.

For those interested in the origin of this famous and widely used expression, its derivation begins with the standard undergraduate physics equation for an infinitely long solenoid. For any segment of length l of this infinite structure, the inductance in SI units is given by

$$L = \mu_0 n^2 \frac{A}{l} = \frac{\mu_0 n^2 \pi r^2}{l}, \qquad (21)$$

where A is the cross-sectional area of the solenoid and n is the number of turns contained in the segment under consideration. The important thing to note is that the inductance drops as the solenoid lengthens, all other parameters held constant.

The magnetic field strength along a finite-length solenoid naturally diminishes near the ends. The inductance therefore drops; the solenoid acts electrically longer than its physical length. If the solenoid's length is very much greater than the radius, the finiteness is not felt as acutely. Thus, the correction for end effects is a function of the length-to-radius ratio. A famous paper by Nagaoka[12] provides a table and curves

---

[10] H. A. Wheeler, "Simple Inductance Formulas for Radio Coils," *Proc. IRE*, v. 16, no. 10, October 1928, pp. 1398–1400.

[11] As discussed later, the highest Q is generally obtained for particular ratios of wire diameter to wire spacing, with the precise value depending on the ratio of coil length to coil diameter.

[12] "The Inductance Coefficients of Solenoids," *J. College of Science* (Tokyo), v. 27, art. 6, 15 August 1909, p. 1.

for the correction factor, and later work by Grover provides an infinite series that may be truncated as necessary for a given level of accuracy. From these works one may discern that, as a first approximation, a simple estimate for the effective electrical length is the physical length, augmented by the radius. This ad hoc correction is similar to that for the fringing term in capacitors, and it turns out to be surprisingly good:

$$L = \frac{\mu_0 n^2 \pi r^2}{l + r}. \tag{22}$$

This formula is perfectly respectable (and relatively easy to remember, particularly if you work primarily in SI units), but a more rigorous analysis reveals that it does underestimate the inductance slightly, particularly for shorter coils. Wheeler's formula does a better job simply by adding 90%, rather than 100%, of the radius to the length.

The effective shunt capacitance across the inductor terminals depends on the boundary conditions to a significant degree. For example, if one terminal is grounded then the effective capacitance is relatively insensitive to the capacitance between adjacent turns and actually depends more on external fringing to ground. This latter capacitance is somewhat difficult to compute analytically. To the best of the author's knowledge, no correct general analytical solution has ever been published.[13] Consequently, the best we can offer here is a semi-empirical formula which assumes (a) that the wire insulation has a relative dielectric constant close to unity and (b) that one terminal is grounded. Within the validity of these assumptions, the effective shunt capacitance is approximately[14]

$$C_{coil} \approx \pi \varepsilon_0 \left[0.4(l/D + 1) + 0.9\sqrt{D/l} \right] D, \tag{23}$$

where l and D are the coil length and diameter, respectively. This equation matches Medhurst's data to better than 5% (roughly corresponding to the accuracy limits of his data) for l/D values ranging from 0.1 to 50.[15] Note that the primary dependence is on the coil diameter, with a weaker dependence on total length. Hence, for a given value of inductance, the highest self-resonant frequencies tend to be obtained with coils possessing the smallest radii.

To obtain accurate estimates of the total shunt capacitance, one must be careful to account also for the capacitance associated with any length of ungrounded lead. For this purpose, one may use the formula for the capacitance per length of an isolated wire, repeated here for convenience:

---

[13] Many have been offered, but close inspection reveals gross errors.
[14] This equation is based on data from Medhurst, *Wireless Engineer,* February 1947, pp. 35–43, and March 1947, pp. 80–92. The coefficients have been chosen to improve accuracy and reduce complexity over Medhurst's proposed formulas, as well as to employ SI units.
[15] Medhurst claims much better accuracy for his version of the formula, but in fact his maximum error is as large as 8%.

$$C_{wire} \approx \frac{2\pi\varepsilon_0}{\ln(2l/r) - 0.75} \approx \frac{5.56 \times 10^{-11}}{\ln(2l/r) - 0.75}. \qquad (24)$$

Recall that typical values are generally of the order of 10 pF/m.

In addition to the inductance value and parasitic shunt capacitance, effective series resistance is of great importance. To estimate it, one would be tempted quite naturally to make use of the skin-effect formula. Unfortunately, that formula assumes a uniformly illuminated semi-infinite block of conductor. In a solenoid, however, conditions are quite different: the magnetic field of one turn affects the current distribution of neighboring (and other) turns, so that the boundary conditions (and thus the effective cross-sectional areas) are considerably modified. Use of an unmodified skin-effect formula therefore usually leads to rather gross errors. Accounting analytically for the loss will be difficult if we want to handle the general case, but relatively simple expressions result if we focus specifically on the conditions that lead to maximum Q.

Here again, we are fortunate that Medhurst's extensive investigations allow us to express succinctly the conditions for maximizing the Q of air-core inductors. In what follows, it's to be understood that the optimum conditions are relatively flat, so the Q-value achieved is not overly sensitive to departures from the optimum conditions.

First, Medhurst expresses the coil Q with a deceptively simple formula as follows:

$$Q = 7.5 D \psi \sqrt{f}, \qquad (25)$$

where D is the coil diameter in meters, f is the frequency in hertz, and ψ is a complicated function of length l, coil diameter D, wire diameter d, and turns spacing s. The square-root frequency factor is to be expected on the basis of skin-effect considerations (operation well below self-resonance is assumed), but that's about as far as one can go with that knowledge. For metals other than copper, one must scale the constant of 7.5 linearly with the relative conductivity of the material.

The function ψ is too convoluted to express generally in a compact analytical form because it depends on both l/D and d/s. However, if our goal is to maximize Q, then it is not hard to provide an expression that yields the maximum value of ψ at each l/D. Such an expression is not helpful for analysis but is well suited to design, which (after all) is our aim:

$$\psi_{opt} \approx 0.96 \tanh\left(0.86 \sqrt{l/D}\right). \qquad (26)$$

Empirical as it is, this expression fits Medhurst's data to better than 3.6% for l/D up to 10 and to better than 4.5% for values greater than 10. Notice that the optimum ψ is a function only of l/D. Choosing l/D about 5 or so yields a value of ψ of about 0.88, which is close enough to the maximum value of ψ under all conditions that little is gained by using much larger l/D ratios.

Because Eqn. 26 implies that maximizing Q requires maximizing both D and ψ, one must choose the largest coil diameter consistent with achieving a self-resonant frequency well above (e.g., by a factor of at least 2–3) the desired operating frequency. One may use Eqn. 23 in conjunction with the desired inductance value to compute an upper bound on D.

Having effectively computed D and l (the latter from having chosen the l/D ratio), one may then use Wheeler's formula to compute the number of turns, n:

$$n = \frac{\sqrt{L(0.9r + l)/(\mu_0 \pi)}}{r}. \tag{27}$$

The winding pitch is thus n/l turns per meter.

The ratio of wire diameter to turns spacing (measured between adjacent wire centers) is roughly estimated as

$$\frac{d_{opt}}{s_{opt}} \approx \begin{cases} 0.5 + 0.07(l/D) & \text{for } l/D \lesssim 7, \tag{28} \\ 1 & \text{for } l/D > 7. \tag{29} \end{cases}$$

Again, these last two equations are derived from crude fitting to Medhurst's data and yield the d/s values leading to Eqn. 26.

The optimum wire diameter is therefore

$$d_{opt} \approx \frac{l}{n}\left[1 + \frac{s_{opt}}{d_{opt}}\right]^{-1}, \tag{30}$$

and the spacing between turns (measured from center to center of adjacent wires) is found from taking the ratio of Eqn. 30 to either Eqn. 28 or Eqn. 29.

The last bit of data that might be useful in designing these coils concerns the properties of wire. The conductivity of pure copper is about 5.7×10^7 S/m. The diameter of bare copper wire is usually presented in tabular form, but a simple (though approximate) formula is

$$D \approx \frac{0.32}{10^{(AWG)/20}}, \tag{31}$$

where the diameter D is in *inches* and AWG is the (American) wire gauge. This formula yields values correct to within about 2% for bare wires between 10 and 40 gauge, a range that spans the most commonly used sizes. Note that it implies a decrease in diameter by a factor of 10 for every wire gauge increase of 20, so the relative behavior of the wire gauge on diameter is the same as that of voltage expressed in decibels.

Solving for the wire gauge as a function of diameter (again, in inches) yields

$$AWG \approx 6.4 - 20 \log d. \tag{32}$$

There are no correspondingly simple formulas for enameled wire, but adding an arbitrary 0.0045" to the values for bare wire yields diameters that are typically correct to approximately 5% or better. It should be mentioned that insulator thicknesses vary somewhat from manufacturer to manufacturer, so values calculated from these equations must be verified in all cases where it matters. These formulas are presented mainly as guides for back-of-the-envelope types of calculations.

A Coil Design Example

To illustrate how one might use these equations, we now carry out the optimal design of a coil intended for use over AM radio frequencies (that is to say, below about 1.6 MHz).

1. Assume that the target inductance is 200 μH, based on the capacitance with which it is ultimately to be resonated.
2. We select $l/D = 5$, as suggested.
3. We desire the self-resonant frequency to exceed at least 3–5 MHz. Choosing the higher value for added margin, we find that the maximum allowed self-capacitance is about 5 pF. From Eqn. 23 we solve for D:

$$D_{max} \approx \frac{C_{coil}}{\pi\varepsilon_0\left[0.4(l/D+1)+0.9\sqrt{D/l}\right]}. \tag{33}$$

Here, we compute a maximum coil diameter of 6.4 cm (about 2.5"), making the coil about 32 cm long. This computation neglects the capacitance of any leads. This consideration is relevant here because the computed coil dimensions are somewhat large, and the lead wires may contribute noticeably to the total capacitance.
4. The number of turns required is

$$n = \frac{\sqrt{L(0.9r+l)/(\mu_0\pi)}}{r}, \tag{34}$$

or about 131 in this case.
5. Because the optimum d/s is about 0.85 for the chosen value of l/D, the optimum wire diameter is computed from Eqn. 30 as 1.1 mm and the spacing from turn to turn as 1.3 mm.
6. The required gauge of wire to use is found from Eqn. 32, and is here computed as 33.5. One may use wire from 32 to 36 gauge without a significant departure from expectations.
7. Evenly wind 131 turns of the chosen wire on an appropriate low-loss form to produce a winding with the design diameter and length.

The Q as predicted by Eqn. 25 is about 530. Note that this calculation neglects any loading or radiation losses, as well as any dielectric losses associated with wire insulation or coil form material. Loss due to eddy currents induced in nearby conductors is similarly neglected. The predicted Q is thus perhaps best regarded as a maximum to aspire to achieve, rather than a value that is to be routinely encountered. Nevertheless, the procedure outlined here does identify the optimum inductor design to a good approximation.

Coils Wound on Magnetic Cores

One problem with solenoidal structures is that they are not self-shielding. Unwanted and troublesome coupling can therefore occur between the inductor and other parts of a circuit, with attendant negative performance implications. Cylindrical shields are thus often placed over such inductors. However, such a shield is uncomfortably similar to (actually, the same as) a shorted single-turn secondary transformer winding. To avoid serious reduction of inductance and Q from induced image currents (eddy currents), the shield's diameter should be at least twice that of the coil (and preferably more) to place the image currents a reasonably large distance away and render their effects negligible.

An alternative is to use a toroidal inductor. Such a structure is magnetically (but not electrostatically) self-shielding if the core material is of sufficiently high permeability. The magnetic flux will then be concentrated in the core, leaving little to leak out. Sadly, all known magnetic core materials are rather lossy at high frequencies, so toroids are widely used only at lower frequencies (typically well below a few hundred megahertz).

Most manufacturers of toroids specify the core's "A_L" value, which they often cite as some number of millihenries per thousand turns. Unfortunately, that convention implies a linear dependence of inductance on the number of turns, and this often trips up the uninitiated (or the sleepy). A more rigorous unit would be nH/turns^2, which uses the same numerical value as A_L.

To maximize the self-resonant frequency of a toroidal inductor, spread the turns evenly around the entire circumference of the core, but leave a reasonable separation between the two terminals of the inductor.

Many RF transformers consist of coupled windings with equal numbers of turns (or simple integer ratios of turns). These often should be wound in multifilar fashion. That is, a multistrand (but individually insulated) bundle of wires are wound as a unit. This strategy minimizes differences among the turns in parasitics and mutual coupling coefficients.

Single Loop

Another useful formula is for the inductance of a single loop. Despite the simplicity of the structure, there is no exact, closed-form expression for its inductance (elliptic functions arise in the computation of the total flux). However, a useful "cocktail napkin" approximation is given by

$$L \approx \mu_0 \pi r. \tag{35}$$

This formula tells us that a loop of 1-mm radius has an inductance of approximately 4 nH.

In deriving this approximation, the flux density in the center of the loop is arbitrarily assumed to be half the average value in the plane of the loop; then the inductance is computed as simply the ratio of total flux to the current. In view of the rather coarse approximation involved, it is remarkable that the formula does as well as it typically does. Note that, for a single turn and in the limit of zero length, Wheeler's formula (Eqn. 22) converges to within about 10% of $\mu_0 \pi r$, providing some independent validation of Eqn. 35.

Much better accuracy is provided by the following expression, which takes into account a nonzero wire diameter as well as magnetic coupling among infinitesimal wire segments:[16]

$$L \approx \mu_0 r [\ln(16r/d) - 2], \tag{36}$$

where d is the diameter of the wire. With this equation, we see that Eqn. 35 strictly holds only for an r/d ratio of about 10.

To make a crude approximation even more so, Eqn. 35 can be extended to noncircular cases by arguing that all loops with equal area have about the same inductance, regardless of shape. Thus, we may also write:

$$L \approx \mu_0 \sqrt{\pi A}, \tag{37}$$

where A is the area of the loop. A closed contour of area one square centimeter has an inductance of about 7 nH, according to this formula. This equation, *very* approximate as it is, turns out to be quite handy in estimating the magnitude of various component and layout parasitics – as well as in evaluating the likely efficacy of proposed layout changes.

We can check the reasonableness of these expressions by considering the inductance of a loop of extremely large radius. Since we can treat any suitably short segment of such a loop as if it were straight, we can use the equation for the inductance of a loop to estimate the inductance of a straight piece of wire.

We've already computed that a circular loop of 1-mm radius has an inductance of 4 nH, so we have roughly 4 nH per 6.3 mm of length (circumference), which is in the same range as the value given by the more accurate formulas.

6.6 MAGNETICALLY COUPLED CONDUCTORS

6.6.1 TRANSFORMERS

It used to be that any electrical engineering graduate student would be familiar with the properties of an ideal transformer, at minimum. However, recent classroom evidence reveals that many schools omit material about transformers these days, so perhaps here is as good a place as any to plug that curricular hole (readers not in need of this refresher are invited to skip this section). We'll develop a model for ideal transformers first, and then patch it up to model real transformers.

A conventional transformer is a magnetically coupled system of inductors. Transformers get their name from their valuable ability to transform voltages, currents, and impedance levels over a relatively broad frequency range. In the simplest case, there are only two inductors, a primary and secondary. Just as the voltage across an

---

[16] Ramo, Whinnery, and Van Duzer, *Fields and Waves in Modern Radio*, Wiley, New York, 1965, p. 311.

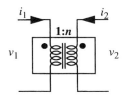

FIGURE 6.13. Ideal 1 : n transformer

isolated inductor is the result of a changing flux, a changing flux produced by the primary of a transformer can induce a voltage in the secondary, and vice versa.

For the ideal 1 : n transformer shown in Figure 6.13, n is the secondary-to-primary turns ratio. A changing magnetic flux common to both inductors thus generates n times the voltage at the secondary as at the primary (the polarity dots in the symbol identify which terminals are in phase). Energy conservation tells us that this voltage boost must be paid for by a corresponding current reduction of precisely the same factor. Because the ratio of voltage to current thus changes by n^2 in going from primary to secondary, an impedance transformation of that factor occurs at the same time. A turns ratio of 3, for example, corresponds to an impedance transformation ratio of 9. The ever-elusive ideal transformer would perform this function over an infinitely wide frequency range (including DC), and with zero loss. Even though such an element is physically unrealizable, it is nonetheless a useful starting point for constructing models of real transformers, as we'll soon see.

In the foregoing ideal example, we have implicitly assumed that all of the magnetic flux produced by, say, the primary couples to the secondary. The aim in most (but not all) transformer design is to approach this ideal as closely as possible. However, as with everything else in life, this aim is imperfectly met in practice, so our model must acknowledge a lack of perfect coupling or otherwise accommodate prescribed values besides unity.

Let us call L_1 the inductance of the primary alone (i.e., with the secondary open-circuited) and L_2 that of the secondary alone. From the physics of the arrangement, we expect the voltage at any port to be the superposition of a self- and mutual term. The V–I equations for the (still) lossless but imperfectly coupled transformer may therefore be expressed as follows:

$$v_1 = L_1 \frac{di_1}{dt} + M \frac{di_2}{dt}; \tag{38}$$

$$v_2 = M \frac{di_1}{dt} + L_2 \frac{di_2}{dt}. \tag{39}$$

Here M, the *mutual inductance* between the windings, enables us to model the degree of coupling between primary and secondary. Reciprocity (another concept emphasized less and less these days) is what permits us to use the same value of M in both primary and secondary voltage equations, even for asymmetrical transformers. Depending on the physical arrangement, the mutual inductance may take on positive or

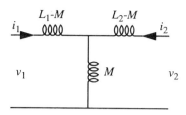

FIGURE 6.14. First-pass lossless transformer model (T version)

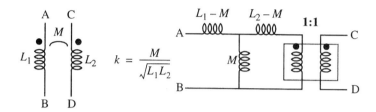

FIGURE 6.15. Transformer circuit model without common terminal

negative values, unlike isolated passive inductances. If the coupled flux adds to the self-flux, the mutual inductance is negative. If it opposes the self-flux, it is positive.

Although the total voltage across either the primary or secondary is the superposition of contributions from both the primary and secondary, the individual terms in the equations are isomorphic to that for an ordinary inductance. The corresponding circuit model for a transformer (Figure 6.14) thus contains only inductive elements. Here we have implicitly assumed a common connection between ports.

If the primary and secondary are very close to each other, then nearly all of the flux from one inductor will couple to the other. If far apart (or if their fields are orthogonally disposed), the coupling will be negligible and M will be very small. It is useful to describe the continuum of possibilities with a quantitative measure of coupling known, reasonably enough, as the coupling coefficient, defined as:

$$k \equiv \frac{M}{\sqrt{L_1 L_2}}. \tag{40}$$

Thus, the coupling coefficient is the ratio of the mutual inductance to the geometric mean of the individual inductances. For passive elements, the magnitude of the coupling coefficient may not exceed unity.

Our first-pass model is perfectly respectable, but it suffers from some deficiencies that occasionally motivate the development of alternatives. One specific limitation of the model shown is that it does not explicitly incorporate a turns ratio between primary and secondary; that information is buried inside of the various inductance parameters. A less important limitation is that the primary and secondary share a common terminal. That deficiency is readily repaired simply by cascading the model with an ideal 1:1 transformer, as shown in Figure 6.15.

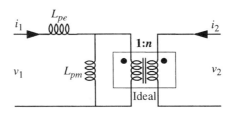

FIGURE 6.16. Alternative lossless transformer model

An alternative model that allows us to separate the ports completely and also explicitly incorporate an arbitrary turns ratio is depicted in Figure 6.16. Here

$$L_{pe} = L_1(1 - k^2), \qquad (41)$$

$$L_{pm} = k^2 L_1, \qquad (42)$$

and

$$n = L_2/M. \qquad (43)$$

This model contains an ideal transformer at its heart and then uses an isolated (uncoupled) *leakage inductance* L_{pe}, to account for the flux that doesn't participate in primary–secondary coupling. The *magnetizing inductance* L_{pm} models that portion of the primary inductance that does participate in coupling. It is therefore equal to the total primary inductance, diminished by an amount equal to the leakage inductance. The magnetizing inductance also properly accounts for a real transformer's failure to function at DC, and it explains why low-frequency transformers are generally bulkier than are high-frequency ones.

In cases where the coupling coefficient is close to unity, the magnetizing inductance is generally quite close in value to the primary inductance. For quick calculations of transformer circuits involving tight coupling, they may be treated as equal in most cases.

Having developed a lossless model that accommodates imperfect coupling as well as arbitrary turns ratios, we now need to account for a variety of parasitic elements that are always present. One potential source of significant parasitics is the material around or on which the inductor is wound. Although integrated circuit transformers almost never employ magnetic core materials (the transformers behave essentially as if they were wound on an air core), core materials are common for discrete circuits. All magnetic core materials exhibit loss of at least two types. *Hysteresis loss* arises from the inelasticity of magnetic domain walls. To support magnetic state changes, these walls must move. One may visualize a sort of friction as accompanying and inhibiting this wall movement. The energy lost per magnetic state transition is usually well modeled as constant for a fixed amplitude excitation, so the total power dissipated due to this mechanism is approximately proportional to frequency. We may account for this loss by adding a frequency-dependent resistance in shunt with the primary winding of our model.

Eddy current loss besets transformers as much as it does ordinary conductors. Currents may be induced in any nearby conductor, including electrically conductive core materials, adjacent windings, and conductive substrates. Because the induced voltage is proportional to frequency, eddy current losses are proportional to the square of frequency, as we've seen in the inductor case. The core losses augment those attributable to winding resistance, with due accounting for the skin effect.

In addition to the loss terms, the electric field surrounding and suffusing the windings stores electrostatic energy. Hence, a high-frequency model must also include capacitances to account for this additional energy storage mechanism. Further complications arise when attempting to model behavior at frequencies where the dimensions of the transformer are not very small relative to a wavelength. In those cases, a simple lumped description of the transformer will be inadequate, and we must treat the windings as coupled transmission lines.

Finally, to make matters even more complex than they already are, all core materials become noticeably nonlinear at sufficiently high flux density, and all parameters are generally functions of temperature as well. These factors explain the profusion of materials; no single core material satisfies all requirements of interest.

The most important implication of the other nonidealities is that the various parasitics limit both frequency response and efficiency. The magnetizing inductance shorts out the primary of the ideal transformer at DC, preventing transformer action there, while the winding capacitances perform a similar disservice at high frequencies. The net result of these nonidealities is to make classical transformers a rarity at gigahertz frequencies.

6.6.2 COUPLED WIRES

Not all transformers are intentionally realized. We need to appreciate that the magnetic fields surrounding conductors drop off relatively slowly with distance. As a result, there can be substantial magnetic coupling between adjacent (and even more remote) conductors. As with intentional transformers, a measure of this coupling is the mutual inductance between them. For two parallel, infinitesimally thin round wires of equal length, this inductance is given approximately by

$$M \approx \frac{\mu_0 l}{2\pi}\left[\ln\left(\frac{2l}{D}\right) - 1 + \frac{D}{l}\right], \qquad (44)$$

where l is the length of the wires and D is the distance between them.[17] For a 10-mm length and a spacing of 1 mm, the mutual inductance works out to about 4 nH. Since the inductance of each wire in isolation is about 10 nH, the 4-nH mutual inductance represents a coupling coefficient of 40%. In practice, one can expect coupling coefficients of this general order of magnitude between, say, adjacent pins of a typical integrated circuit package. Furthermore, the logarithmic dependence of M on spacing means that the coupling decreases slowly with distance, so the level of unwanted

---

[17] This formula is adapted from Frederick Terman, *Radio Engineers' Handbook*, McGraw-Hill, New York, 1943, Chap. 2.

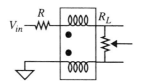

FIGURE 6.17. Guanella 1:1 "balun"

coupling between even nonadjacent pins can be troublesome. Obviously, the undesired cross-talk among pins can degrade signal integrity at high frequencies and otherwise cause a host of troublesome pathologies.

6.6.3 BROADBAND TRANSMISSION LINE TRANSFORMERS

Although this textbook's focus is on planar circuits, there are many practical instances where a planar circuit must use a distinctly nonplanar transformer somewhere. An example might be power amplifiers, where on-board transformers may not be up to the task. Other examples include broadband impedance transformers, where operation over several decades of frequency might be desired. Planar structures using ordinary PC board materials simply cannot provide this level of performance.

We've already examined conventional transformers and alluded to the limited bandwidth arising from winding capacitance and leakage inductance. We now briefly discuss *transmission line* transformers, which (because of their unique construction) suffer much less from these limitations, permitting operation over unusually large bandwidths.

The first description of this class of transformers was evidently by Guanella in an impossible-to-find reference in 1944.[18] Description of a related, but different, class of transformers by Ruthroff is much more frequently referenced because of its superior accessibility.[19] Both Guanella and Ruthroff's transformers are distinguished from their classic counterparts by having a DC path from input to output and for inducing nominally zero net flux in the transformer core material. This latter property greatly reduces the (frequency-dependent) hysteresis losses that often set the upper frequency of operation for many practical transformers.

The basic Guanella connection is shown in Figure 6.17. It is often called a balun in the literature, but because the transformer itself does not set the output common-mode voltage (AC-wise), it isn't quite a classic balun. Whatever is connected to the output determines the common-mode voltage. In the circuit of Figure 6.17, grounding the center tap of the output load resistor will cause the overall circuit to behave as a balun.

---

[18] G. Guanella, "Novel Matching Systems for High Frequencies," *Brown-Boverie Review*, v. 31, September 1944, pp. 327–9.

[19] C. L. Ruthroff, "Some Broad-band Transformers," *Proc. IRE*, v. 47, August 1959, pp. 1337–42.

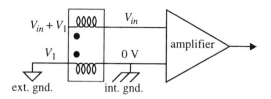

FIGURE 6.18. Guanella 1:1 balun as common-mode choke
(resistances not shown for clarity)

Broadband operation is enabled by exploiting the transmission line nature of the windings making up the primary and secondary. Indeed, Guanella's exposition of the subject explicitly uses transmission lines for the windings. Thus, the winding capacitances that limit the bandwidth of conventional transformers are absorbed into the fundamental operation of the transformer. If the characteristic impedance of the line equals that of the load, bandwidth will be maximized. The upper frequency range is therefore extended by two mechanisms working together: absorption of reactive parasitics into transmission line constants, and the reduction of hysteresis loss by keeping the core flux small.

The latter property is appreciated by considering what happens when the upper (primary) winding in Figure 6.17 is driven by an input voltage. The input current that flows in response causes a voltage to be induced in the secondary. Because of the 1:1 turns ratio, the winding voltages are the same and so are the currents. However, the primary and secondary currents flow in opposite directions, so the net magnetic flux they induce in the core is ideally zero. As a consequence, hysteresis-based core losses are correspondingly small. Indeed, for many practical realizations of the Guanella transformer, dielectric loss dominates over other core loss sources. The usual preoccupation in selecting transformer core material is low hysteresis and conductive losses, but dielectric loss is important in cores for transmission line transformers.

The ability of this circuit to reject common-mode signals over a wide bandwidth is routinely exploited in instrumentation. When used this way, the circuit is often called a common-mode choke (see Figure 6.18). It is valuable because the common-mode rejection provided by the transformer greatly relaxes the required common-mode rejection of subsequent circuits. The front-end vertical amplifiers of oscilloscopes, for example, routinely interpose common-mode chokes between the input connectors and the front ends of the amplifiers.

Imagine that a ground loop causes the internal and external grounds to be at different AC potentials. Just to keep things straight, let's arbitrarily assign the internal ground node a potential of zero volts, and let the external ground voltage be V_1. The voltage difference between these two nodes appears directly across the bottom winding, causing that same voltage to be induced across the top winding. Keeping track of signs, we see that this induced voltage is subtracted from the input voltage, causing the amplifier input to see V_{in}, exactly. The ground noise has been removed, thanks to the common-mode choke.

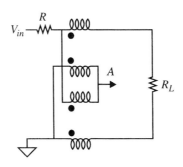

FIGURE 6.19. Guanella 1:4 balun

The purpose of the core material in these types of transformers is merely to provide sufficient common-mode reactance (because the aim of the circuit is to provide a purely differential output). The lower frequency limit of successful common-mode rejection is thus set by the common-mode inductance. The balun functions properly until the corresponding common-mode reactance ceases to be large relative to the characteristic impedance of the transmission line windings.

The 1:1 configuration can be extended to yield other impedance transformation ratios by an appropriate combination of series and parallel connections of windings. As just one example, consider driving two 1:1 transformers in parallel, but taking the output in series, to produce a 1:4 balun. See Figure 6.19.

Typically, the point A would be grounded. Alternatively, the midpoint of the output load could be grounded (as in the 1:1 case) to force a common-mode output voltage of zero. If wound with coaxial lines, the conductors should be sequenced as center and shield of one cable, then center and shield of a second cable, in that order, from top to bottom in Figure 6.19. Alternatively, the transmission line windings may be implemented with simple multifilar (in this case, quadrifilar) bundles.

The general idea of driving windings in parallel and taking outputs in series may be readily extended to create $1:n^2$ baluns, where n is an integer equal to the number of 1:1 building blocks used.

The Ruthroff configurations are somewhat different and, as we shall see, exhibit worse bandwidth as a consequence. We will consider just two basic examples of Ruthroff 1:4 transformers. One is a balun, and the other isn't.

In Figure 6.20, the configuration on the left has an unbalanced input and output. It is therefore sometimes called an "unun," but it sounds goofy no matter how many times you say it aloud (and pretty much no matter what your native language is). It looks funny in print, too. In any case, it works like this: An applied voltage impressed across the lower winding causes that same voltage to be induced in the upper winding. The series connection of the two windings causes the output voltage to double, producing the desired 1:4 impedance transformation. This configuration is also sometimes called a bootstrapped connection because of the series-aided voltage boost.

The configuration on the right of Figure 6.20 applies a voltage across the top winding, inducing the same voltage across the bottom winding. Again keeping track of

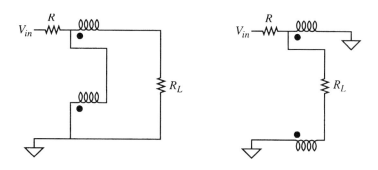

FIGURE 6.20. Ruthroff 1 : 4 "unun" (left) and balun

polarities, we see that the left side of the resistive load sees a voltage equal to the input voltage, while the right side sees its algebraic inverse. The voltage across the load is consequently once again twice the input voltage, producing a 1 : 4 transformation. Because neither end of the load resistor is grounded, it is considered a balanced configuration.

The Ruthroff versions have a somewhat more limited bandwidth because they sum an applied voltage with one that has propagated along a (transmission line) winding. The latter necessarily has a delay, which produces a fundamental upper frequency limit. The transformer functions well only as long as the delay along the winding is short relative to a period. Shortening the lines improves the upper frequency limit, but it degrades the lower frequency limit by reducing the common-mode inductance. Using cores with higher permeability can help compensate to an extent, but there's only a limited selection of suitable materials with low dielectric loss and high permeability.

In the Guanella version, the output is taken across two transmission lines. The delay is thus common-mode and therefore imposes no upper frequency limit. For much more information on this class of transformers, see Jerry Sevick's *Transmission Line Transformers,* 2nd ed., American Radio Relay League, Newington, CT, 1990.

6.6.4 NARROWBAND TRANSMISSION LINE TRANSFORMERS

If broadband operation is not required, impedance transformations may be implemented with simpler transmission line transformers. For example, a popular 1 : 4 narrowband balun is easily constructed out of a half-wavelength piece of transmission line.

In Figure 6.21, a resistor of value R is connected around a half-wavelength transmission line. The resistor is shown as a series connection of two resistors (of value $R/2$ each) to simplify analysis. Because of the line's length, a voltage V applied at the left end of the line undergoes an inversion when traveling to the right end of the line. The midpoint of the resistor is thus at ground potential. The right end of the line is therefore loaded by a resistance $R/2$, which – when reflected back to the left

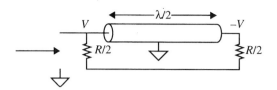

FIGURE 6.21. Narrowband 1 : 4 transmission line balun

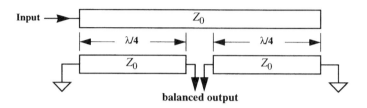

FIGURE 6.22. Classic Marchand balun

end of the line – remains $R/2$. This reflected load resistance is in parallel with the other $R/2$ resistance that is already at the left end of the line, producing a net input resistance of $R/4$. Thus, a balanced load of resistance R is transformed down to an unbalanced input resistance of $R/4$ (and vice versa).

To facilitate connection of the balanced load, the transmission line is typically bent into a U or O shape. This balun is extremely useful at frequencies high enough for practical realization in microstrip in particular.

A final balun we will consider is a translation into stripline form of a coaxial design originally due to Nathan Marchand.[20] It may be regarded as two couplers joined together and excited in antiphase. As seen in Figure 6.22, an unbalanced input drives an open-circuited $\lambda/2$ line. Signals at the two ends of that line are therefore precisely out of phase. These two out-of-phase signals drive a pair of couplers whose outputs are thus likewise out of phase. Because the coupled lines are each grounded on one end, the differential output is a balanced one, as desired.

Edge-coupled structures such as microstrip are not well suited to realizing the tight couplings normally desired in a Marchand balun. If additional layers of metal are available, broadside coupled layouts may be used to provide the required tight coupling without requiring absurdly small spacings.

Although the classic Marchand balun is a narrowband structure, it may be broadbanded significantly by using multisection lines in place of the single $\lambda/4$ segments shown (see Chapter 7). Operation over multi-octave or decade bandwidths is then possible. Finally, trimming of the output phase balance may be performed by adjusting the length of the uncoupled segment between the two coupled lines.

---

[20] N. Marchand, "Transmission Line Conversion Transformers," *Electronics,* v. 17, December 1944, pp. 142–5.

6.7 SUMMARY

We've seen that seemingly ordinary components must be modeled in progressively more sophisticated ways as frequency increases. Nominally simple components are seen to have important behaviors that may be ignored only at low frequencies. Even resistors, capacitors, and inductors must be treated as complicated impedances for proper design of microwave circuits. As an aid to developing appropriate models, this chapter has presented numerous equations and rules of thumb for estimating parasitic inductance and capacitance. Finally, we've considered various types of transformers, even though such components are rarely realized in purely planar form.

CHAPTER SEVEN

MICROSTRIP, STRIPLINE, AND PLANAR PASSIVE ELEMENTS

7.1 INTRODUCTION

A recurring theme in RF design is the need to pay careful attention to the electrical characteristics of *everything* along the signal path. This concern extends to printed circuit (PC) boards, so this chapter considers their high-frequency properties as well as those of numerous passive components made with PC board materials. We will focus on a particular type of transmission line known as *microstrip,* which is particularly suited to the realization of planar microwave circuits. In addition, numerous passive components can be made out of transmission lines, so capacitors, inductors, resonators, power combiners, and a variety of couplers – including baluns and hybrids – are presented as well.

7.2 GENERAL CHARACTERISTICS OF PC BOARDS

Just as with PC boards used at lower frequencies, those for RF applications consist of metal layers separated by a dielectric of some kind. By convention in the United States, the metal thickness is given indirectly as a certain weight of copper per square foot. Thus, "1 ounce" copper (a common value) is approximately 1.34 mil (35 μm) thick.[1] Half-ounce and 2-oz copper are other common values. The DC resistivity of bulk copper is approximately 1.8 $\mu\Omega$-cm, so the corresponding sheet resistivity of 1-oz copper is about 0.5 mΩ/square. The skin depth in copper is about 2.1 μm at 1 GHz, raising the sheet resistivity to roughly 8 mΩ/square at that frequency. Depending on how the copper is formed and deposited, the resistivity may be larger than the bulk value by as much as a factor of 2.[2] One additional factor that affects

---

[1] A mil is 0.001" and is not to be confused with a millimeter; 1 mil is actually about 25 μm.

[2] Because skin effect maximizes the current density on the surface facing the ground plane, the roughness of *that* surface is most relevant for resistance calculations. As it happens, *adhesion promoters* are often used to ensure that the conducting metals stick to the substrate. The interface between the promoters and the metal may be relatively rough, producing surprisingly high conductor loss at high frequencies.

the resistivity is the surface roughness; current must flow over a somewhat greater distance if the surface is rough. An estimate for the factor by which the resistance is boosted is given by[3]

$$F_{sr} \approx 1 + \frac{2}{\pi} \tan^{-1}\left[1.4\left(\frac{\Delta}{\delta}\right)\right],\tag{1}$$

where Δ is the rms surface roughness and δ is the skin depth. We see from this equation that surface roughness becomes a more prominent factor if its rms value is not small relative to skin depth.

Common dielectric thicknesses are 1/32" (0.8 mm), 1/16"(1.6 mm), and occasionally 1/8" (3.2 mm). In multilayer boards, materials 1/64" (0.4 mm) thick are also encountered. These are all approximate values, and actual thicknesses are often stated as multiples of 5 or 10 mils in the United States (and correspondingly round numbers in countries using the metric system).

For general-purpose work at lower frequencies, by far the most commonly used PC board dielectric is a fiberglass–epoxy composite called FR4 (for "flame-retardant formulation number 4"), or its more flammable counterpart, G10. The dielectric loss of FR4 dominates over conductor loss at microwave frequencies, with typical microstrip transmission lines (to be discussed in detail in Section 7.3) exhibiting a loss of 0.03 dB/cm/GHz for 1/16" (1.6-mm) material. Because of the dominance of dielectric loss, the attenuation really does increase linearly with frequency on a decibel scale over quite a wide frequency range (say, ~50 MHz to at least 5 GHz). This functional dependency implies a constant loss per wavelength at any frequency. In this case, the value is approximately 0.5 dB per wavelength.

In addition to somewhat high loss, FR4 is not manufactured to tight specifications, so it is normally considered unsuitable for the mass production of critical RF circuits. For example, 1.6-mm–thick FR4 boards purchased by Stanford University over the past several years from a number of vendors exhibit relative dielectric constants ranging from roughly 4.2 to 4.7 at 1 GHz. The distribution is not uniform, but to the extent that such a nonscientific, sparse sampling is at all representative, one might infer that the dielectric constant varies more than 5% from a nominal value of approximately 4.5.[4] Statistical theory being what it is, one should probably anticipate having to accommodate a greater variation from the mean, say ±10%. If such accommodation is possible and if the loss can be tolerated, FR4 can be used beyond 5 GHz (some intrepid folks have even used it at 10 GHz), despite the conventional wisdom that FR4 is unsuitable for applications above a few gigahertz.

It should also be noted that planar transmission lines are immersed in an inhomogeneous dielectric medium (e.g., part FR4 and part air). A weighted average of

---

[3] E. O. Hammerstad and F. Bekkadal, *A Microstrip Handbook,* ELAB Report, STF 44 A74169, N7034, University of Trondheim, Norway, 1975.

[4] To make matters more complex, a different ratio of epoxy to glass is frequently used for very thin FR4 substrates, causing the dielectric constant to be 5–10% lower for such material (used in laminates).

the dielectric constants accommodates this inhomogeneity to keep equations simple, at least in appearance. At high frequencies, however, the higher-dielectric constant substrate steals a progressively greater proportion of the flux, causing the effective dielectric constant to increase. High-frequency signals thus propagate more slowly than do lower-frequency ones; such transmission lines are consequently dispersive. One may typically expect a ~5% increase in FR4's effective dielectric constant as the frequency increases from 100 MHz to 5 GHz. The distortion of time waveforms arising from this dispersion may cause difficulty in broadband (e.g., high-speed digital) applications, where pulse fidelity is important.[5]

Because FR4 is made by binding glass fibers together within an epoxy matrix, anisotropies are possible if the fibers are not randomly oriented. It is not unheard of to encounter different dielectric constants in different directions, for example, so that transmission lines that are orthogonal to each other may have different characteristic impedances. Despite its various shortcomings, FR4's *extremely* low cost and wide availability continue to motivate engineers to devise ways to expand its use beyond noncritical, hobbyist, and low-volume prototyping RF applications. It is relevant here to note that the clock frequencies of digital systems have reached what previously had been known as the microwave realm.

In extremely low-cost consumer devices (e.g., toys, pocket radios, etc.), an even less expensive board material is not infrequently encountered. Phenolic is often a caramel brown, typically has an "organic chemical" odor, and is remarkably lossy. Although phenolic is occasionally used for RF toys up to 100 MHz, it is totally unsuitable for serious applications. It is mentioned here simply to answer the question: "What is that cheap, malodorous board made of?"

Microwave-compatible materials must be used in more demanding circumstances, of course. Many of the best soft substrate materials have historically been based on PTFE (polytetrafluoroethylene, better known by the DuPont trade name Teflon™). It is rather difficult to produce multilayer boards with such materials, however, and alternatives have been developed to solve this problem. A popular example is RO4003 from Rogers Corporation (⟨http://www.rogers-corp.com/mwu⟩). This material is based on a woven glass–reinforced hydrocarbon and ceramic thermoset plastic material. Transmission lines built on RO4003 exhibit approximately one fourth the loss of FR4 on a decibel basis. Its dielectric constant is 3.38, controlled to within 1.5%. Its low loss and stable, narrowly distributed characteristics – as well as its ease of manufacture – make such materials particularly suited for many applications.

Occasionally, one encounters polyphenylene oxide (PPO) as a PC board material. The lower loss of PPO (relative to FR4) led to its use in some high-frequency Tektronix oscilloscope plug-ins, for example. If you do come across it, take note that working with PPO requires extreme care because of its low melting temperature; it is the high-tech equivalent of butter. Manually soldering or de-soldering components

---

[5] This property has not prevented intrepid engineers from using equalizers to reduce the effects of dispersion. Through such means, exceptionally high bandwidth signals may be conveyed by FR4 lines.

Table 7.1. *Some characteristics of FR4 and RO4003 at 300 K*

Property	FR4	RO4003
Bulk ε_r @ 1 GHz	$4.5 \pm 10\%$, typ. (5.4 max @ 1 MHz, from IPC-4101)	$3.38 \pm 1.5\%$
TC of ε_r (ppm/°C)	?	+40
Dissipation factor	0.015 @ 1 GHz, typ. (0.035 max @ 1 MHz, from IPC-4101)	0.002 @ 10 GHz
Dimensional stability	<500 ppm	<300 ppm
TC of thermal expansion (ppm/°C)	15 in-plane, 100 in z-direction (typ.)	15 in-plane, 50 in z-direction

is just barely feasible and must be done with as much speed as you can muster. Simultaneously, one must be careful not to apply any forces while soldering, or else physical distortion is likely.

We'll focus mainly on FR4 (and, to a lesser extent, RO4003) for this rest of this book, so we summarize some of their relevant characteristics in Table 7.1.[6] Some entries for FR4 are guesses, since few parameters are actually controlled – as might be expected from the rather loose limits dictated by IPC-4101, the relevant standard for FR4.

As a final note on FR4, it is sometimes helpful to know that it is an excellent insulator. In addition to exhibiting low surface leakage (200-GΩ surface resistance is typical, *if* the surface is clean), its breakdown voltage is very high. A 1.6-mm–thick substrate will typically withstand in excess of 40 kV (sometimes above 60 kV). Excellent high-voltage capacitors may be handcrafted with FR4.

In more demanding applications, one may employ a variety of hard substrates, such as alumina, beryllia, quartz, and sapphire. Hard substrates have substantially better thermal conductivity than the soft substrates we've considered, but the former are *much* more expensive. Alumina is the most popular of the hard materials because it is the least expensive and has a number of reasonable electrical characteristics. Metals adhere to it well, so interconnect is not a problem. It also machines with relative ease. At the same time, it is very hard and strong, allowing it to survive its mounting on other substrates that have mismatched thermal coefficients. The material can also be polished to a fine degree, allowing deposited interconnect to have similarly low surface roughness, thus decreasing high-frequency conductor losses. Alumina's relatively high dielectric constant (∼10) is, however, both an advantage and a liability. At low frequencies it enables more compact circuits than are possible with FR4, yet at high frequencies it forces geometries that are uncomfortable or impossible to manufacture reliably. Furthermore, as we'll discuss shortly, the higher

---

[6] See Appendix C (Section 7.10) for a table of other dielectric materials and their corresponding dielectric constants.

dielectric constant implies that waves traveling along conductors on the surface suffer a greater inhomogeneity, with consequent negative implications for high-frequency performance.

The crystalline form of alumina, sapphire, can be polished even more finely, so it is occasionally used when cost is truly no object and where one is obsessed with reducing surface roughness to the absolute minimum. Paradoxically, its extreme smoothness inhibits metal adhesion, so some sort of adhesion-promoting layer must be interposed between the sapphire surface and the actual interconnecting metal. If the adhesion layer is itself not smooth, the advantages of sapphire can be quickly nullified.

Because it is desirable to minimize problems stemming from dielectric inhomogeneities, fused silica (quartz) is occasionally used instead of alumina or sapphire. With a dielectric constant of about 3.8, millimeter-wave circuits (and beyond) demand considerably relaxed machining tolerances. Its dielectric loss is also the lowest of all the hard substrates currently in use, remaining low well into the millimeter-wave bands. Unfortunately, quartz is incredibly brittle, so machining it is pretty much out of the question, and one must take extreme care in mating it to a mounting surface lest mismatches in thermal expansion coefficients result in cracking. Quartz is exceptionally smooth, so it possesses all of the advantages and disadvantages of sapphire in that respect. All in all, this material is an incredible pain to work with.

Recently, low-temperature co-fired ceramic (LTCC) substrates have become popular. These materials exhibit loss intermediate between FR4 and RO4003, for example, but are more easily manufactured than are traditional microwave substrates such as alumina.

We'll mention just one more material: Beryllia at one time was frequently found in high-power modules because it has good thermal conductivity and a thermal expansion coefficient that is well matched to that of copper. One *must not* machine it because breathing its toxic dust can induce a *fatal reaction* in some individuals. It's best to avoid the material altogether – but if you can't, *never* grind or crack BeO!

7.3 TRANSMISSION LINES ON PC BOARDS

The planar nature of PC boards effectively precludes the realization of coaxial structures. Furthermore, it is highly inconvenient in any case to make connections to intermediate points along a coaxial line. In 1951, Robert M. Barrett proposed the realization of planar versions of many classical microwave components by using PC board fabrication methods.[7] Three papers by researchers at the Federal Telecommunications Laboratory of ITT developed this general idea further (see the December

---

[7] R. M. Barrett and M. H. Barnes, "Microwave Printed Circuits," *Radio and TV News* (Radio-Electronic Engineering Section), v. 46, September 1951, pp. 16–31. Barrett credits V. H. Rumsey and H. W. Jamieson as having been the first to use a planar line (stripline) for a power divider, during WWII. Anecdotal reports also credit the prolific Harold Wheeler with having experimented with coplanar lines in the late 1930s.

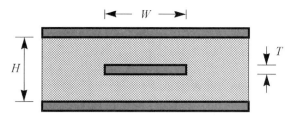

FIGURE 7.1. Stripline (cross-section)

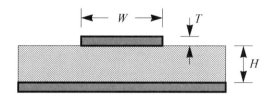

FIGURE 7.2. Microstrip (cross-section)

1952 *Proceedings of the IRE,* as well as the March 1955 *IRE Transactions on Microwave Theory and Techniques*).[8] These and subsequent efforts resulted in planar transmission lines now known as stripline and microstrip structures (see Figures 7.1 and 7.2). Although it took quite some time for these to catch on outside the laboratory, they are now by far the most common ways to build microwave circuits.

As seen in the figures, a stripline is constructed by sandwiching a conductor between two ground planes, whereas a microstrip is a conductor disposed above a single ground plane. The stripline structure is nearly self-shielding (leakage can occur only through the sides). It is therefore particularly useful in minimizing both losses and unwanted coupling due to radiation. Furthermore, the fact that the line is immersed in a uniform dielectric means that TEM propagation is supported. An important drawback, however, is the inconvenience of making connections to the center conductor.

In Figure 7.1, the stripline is shown as it is almost universally realized: a symmetrical structure in which the line proper is positioned midway between the two ground planes. The characteristic impedance of this line is given approximately by[9]

$$Z_0 \approx \frac{60}{\varepsilon_r} \cdot \ln\left[\frac{6H}{\pi W(0.8 + T/W)}\right]. \qquad (2)$$

---

[8] Perhaps most notable among these is by D. D. Grieg and H. F. Engelmann, "Microstrip – A New Transmission Technique for the Kilomegacycle Range," *Proc. IRE,* v. 40, December 1952, pp. 1644–50. This paper marks the debut of the term *microstrip* in the literature. By the way, ITT had bought Federal Telegraph (of arc technology fame), and it was that group (renamed Federal Telephone) that published this work on microstrip.

[9] This formula is a modification of one due originally to Seymour B. Cohn, "Characteristic Impedance of the Shielded-Strip Transmission Line," *IRE Trans. Microwave Theory and Tech.,* July 1954, pp. 52–7.

This formula is most accurate for narrow lines, defined by $W/(H-T) < 0.35$. For a total dielectric thickness of 1/8" (about 3.2 mm) using 1-oz copper, a 50-Ω line requires a conductor width of approximately 1.25 mm. Even though this width is not quite "narrow" in the sense of our inequality, it is close enough that Eqn. 2 is still reasonably accurate, at least for impedances near 50 Ω.

For a more accurate estimate, one may use Cohn's more elaborate original expressions:[10]

$$Z_0 = \frac{\eta K(k)}{4\sqrt{\varepsilon_r} K'(k)}, \qquad (3)$$

where

$$k = \left[\cosh\left(\frac{\pi W}{2H}\right)\right]^{-1} \qquad (4)$$

and η is the impedance of free space,

$$\eta = \sqrt{\mu_0/\varepsilon_0} \approx 120\pi. \qquad (5)$$

Furthermore, to such an excellent approximation that we will treat them as equalities,

$$\frac{K'(k)}{K(k)} = \begin{cases} \left[\frac{1}{\pi}\ln\left(2\frac{1+\sqrt{k'}}{1-\sqrt{k'}}\right)\right] & \text{for } 0 \leq k \leq 0.707, \qquad (6) \\ \left[\frac{1}{\pi}\ln\left(2\frac{1+\sqrt{k}}{1-\sqrt{k}}\right)\right]^{-1} & \text{for } 0.707 \leq k \leq 1, \qquad (7) \end{cases}$$

where $K(k)$ is the *complete elliptic integral* (of the first kind) and K' is its complementary function (in case you cared). Finally, it helps to know that

$$k' = \sqrt{1-k^2}. \qquad (8)$$

However, more widely used than stripline is microstrip – even though it is neither a shielded nor a homogeneous structure. It is important to note that propagation cannot be purely TEM because of this inhomogeneity: the dielectric constant of the material above the line (air) differs from that below it. Satisfying the boundary conditions on the electric field at the interface requires a field component in the direction of propagation. Loosely speaking, the portion of the wave below the line "wants" to propagate at a lower velocity than the portion above. Perhaps it is thus no surprise that rigorous theoretical treatments of this inhomogeneous structure are complex. The notion of "quasi-TEM" propagation is therefore usually invoked as a simplifying concept, in which the structure is treated as equivalent to a line surrounded by a *uniform* material whose dielectric constant has been suitably adjusted downward to yield an effective (averaged) dielectric constant.

---

[10] Also see K. C. Gupta, R. Garg, and R. Chadha, *Computer-Aided Design of Microwave Circuits*, Artech House, Dedham, MA, 1981. Cohn's formula, based on conformal mapping techniques, does not take into account nonzero conductor thickness.

Despite its shortcomings, microstrip is extremely popular because the line isn't buried, making it readily accessible. The standard way to make prototype boards involves etching a copper-clad piece of FR4 on which the desired pattern has been defined in a chemical resist. A faster way, and one that uses no toxic substances, is to lay down the conductor pattern with adhesive copper foil tape. A utility knife is ideal for trimming the foil to the desired dimensions. This construction technique is of course limited to use at frequencies low enough that the dimensional (in)stability may be tolerated.

Copper foil tape is usually a special-order item from electronics supply distributors; it is not commonly stocked by most chain electronic stores. However it is readily available from a surprising source: hardware stores and garden shops. It is sold as a barrier to snails. A representative brand is *SureFire™ Slug & Snail* copper barrier tape, sold by chains such as Ace Hardware. To use copper foil tape, first prepare the circuit board by *gently* abrading away any protective surface coating with a plastic scouring pad (a metal pad would work, but it's just a bad idea to use something that is guaranteed to leave conductive bits all over your circuit board and workbench). Then cut the copper tape to the desired dimensions and position it on the FR4. Smooth the tape with a plastic object such as the barrel of a pen, and trim any excess foil as needed. That's all there is to it. The beauty of this method, aside from its simplicity, is that modifications are trivial. If you make a mistake (such as cutting off too much material with the knife), no problem: remove the tape, clean the board again, and affix a new piece of tape. Whether formed in this way or with conventional PC board fabrication methods, components are easily connected to the line, facilitating construction, probing, adjustment, troubleshooting, and repair.

One unanticipated design challenge is the variety of published equations for the characteristic impedance of these lines. Fortunately, a comparison reveals that these equations typically differ in their predictions by only a few percent (at least, for impedances near 50 Ω; deviations can become large for other impedances). For the microstrip case, a representative set of equations is as follows:[11]

$$Z_0 \approx \sqrt{\frac{\mu_r \mu_0}{\varepsilon_r \varepsilon_0}} \cdot \frac{H}{W} \cdot \left[1 + 1.735 \varepsilon_r^{-0.0724} \cdot \left(\frac{W}{H}\right)^{-0.836} \right]^{-1}. \tag{9}$$

In nearly all cases, the permeability μ is that of free space, $4\pi \times 10^{-7}$ H/m, or approximately 1.257 μH/m.

---

[11] R. S. Carson, *High Frequency Amplifiers,* 2nd ed., Wiley, New York, 1982, p. 78. For much more accurate (and *infinitely* more complicated) equations, see E. Hammerstad and O. Jensen, "Accurate Models for Microstrip Computer-Aided Design," *IEEE MTT-S Digest,* June 1980, pp. 407–9; R. H. Jansen and M. Kirschning, "Arguments and an Accurate Model for the Power-Current Formulation of Microstrip Characteristic Impedance," *Archiv für Elektronik und Übertragungstechnik,* v. 37, no. 3/4, March/April 1983, pp. 108–12; and M. Kirschning and R. H. Jansen, "Accurate Wide-Range Design Equations for the Frequency-Dependent Characteristic of Coupled Microstrip Lines," *IEEE Trans. Microwave Theory and Tech.,* v. 32, no. 1, January 1984, pp. 83–90. These quasi-empirical equations result from fitting to a large number of field-solver simulations.

Table 7.2. *Representative physical widths of microstrip lines in FR4* ($\varepsilon_r = 4.4$)

H	W for 50 Ω	W for 100 Ω	f_c @ 50 Ω
1/16" (1.6 mm), 1-oz. Cu	3.0 mm	0.65 mm	26 GHz
1/32" (0.8 mm), 1-oz. Cu	1.5 mm	0.30 mm	48 GHz

Note that Eqn. 9 neglects any dependence on conductor thickness. As a result, it is somewhat in error, although generally by completely negligible amounts. To develop a first approximation, one may pretend that the vertical sides of the conductor contribute to the width as if they were folded flat. Thus, one may take the increase in effective width to be on the order of $2T$.

For fussier folks, the following equation may be used to compute a more accurate value for the effective width:

$$W_{eff} = W + \frac{T}{\pi} \cdot \left[\ln\left(\frac{2H}{T}\right) + 1\right]. \tag{10}$$

No matter how it's computed, the effective width should be used in Eqn. 9. As a specific numerical example, consider a board 1/32" (0.8 mm) thick with 1-oz copper cladding (which, as stated earlier, is ~35 μm thick). Assuming a bulk relative dielectric constant of 4.4, the mathematical correction given by Eqn. 10 in this case amounts to an effective electrical width increase of just ~$1.5T$, or 2.0 mil (50 μm). This correction is similar in magnitude to typical manufacturing tolerances (e.g., 3 mil) and is certainly smaller than what one can consistently achieve with the manual knife-and-copper tape method. Hence, it is frequently ignored. For this example, then, the physical width needs to be approximately 1.5 mm, or about 58 mil (i.e., 0.058") for a 50-Ω line (see Table 7.2). On 1/16" FR4 boards, it so happens that slicing 1/4"-wide copper foil tape (a standard width in the U.S.) straight down the middle yields conductors that produce a 50-Ω impedance to a very good approximation.

As mentioned earlier, the loss of a 50-Ω microstrip line over 1/16" material is typically about 0.07–0.09 dB/inch/GHz (~0.03 dB/cm/GHz) for FR4, while that for RO4003 is a fourth that value.[12] Figure 7.3 is a plot of typical transmission as a function of frequency for a 50-Ω microstrip line. In this particular case, the corresponding loss is approximately 0.03 dB/cm/GHz. As is evident from the plot, the attenuation (on a dB scale) does increase linearly with frequency over quite a wide frequency range (here, the plot extends over a two-decade range, from 50 MHz to 5.05 GHz).

As with connectors and any other conveyance for electromagnetic energy, one must be aware of the possibility of moding with transmission lines. In the case of microstrip (and stripline), moding begins to occur either when the frequency is high

---

[12] Continue to keep in mind, however, the variability of FR4. Also note again that this loss behavior implies a constant dB loss per *wavelength*.

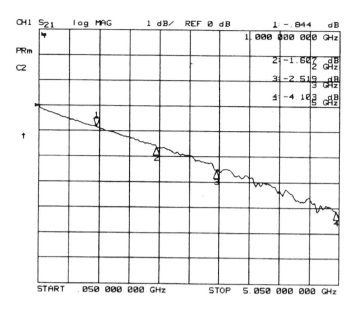

FIGURE 7.3. Typical transmission vs. frequency of 50-Ω microstrip line on 1.6-mm FR4

enough that a half-wavelength just fits across the width of the line, or approximately when a quarter-wavelength fits between the line and the ground plane.[13] The former condition describes a transverse resonance, whereas the latter corresponds to conditions favorable to the coupling of energy to a surface wave that may propagate along/within the dielectric.

The inhomogeneity of microstrip makes it difficult to derive these limits rigorously. However, one may derive a very crude estimate by glibly ignoring this inhomogeneity. Thus, the first transverse resonance possesses a wavelength given approximately by:

$$\lambda_c \approx 2W\sqrt{\varepsilon_r}. \tag{11}$$

The corresponding frequency is thus

$$f_c \approx c/2W\sqrt{\varepsilon_r}. \tag{12}$$

Fringing effects cause the line to act electrically as if it were somewhat wider than W.

Derivation of a formula for the onset of surface-wave moding is even more involved. Continuing with our crude rules of thumb, though, we may estimate this moding frequency *very* roughly as:

$$f_T \approx c/4H\sqrt{\varepsilon_r}. \tag{13}$$

---

[13] The distance between the line and its image is a half-wavelength under this condition, so the moding limits can all be expressed in terms of half-wavelength criteria.

A more rigorous derivation yields[14]

$$f_T = \frac{c}{2\pi H} \frac{1}{\sqrt{\varepsilon_r - 1}} \text{atan}(\varepsilon_r). \tag{14}$$

The approximate equation is seen to be somewhat conservative in that it predicts onset of surface-wave "launching" at a lower frequency than where it actually occurs. These two equations converge as ε_r approaches infinity.

For 50-Ω lines on FR4, the W/H ratio is quite close to 2, so the cutoff frequencies for these two conditions are similar. Since smaller dimensions are required to push up the mode-free bandwidth, maintenance of a constant line impedance requires that the substrate thickness shrink with the line width. The trade-off is that narrower lines have higher conductor losses, so line attenuation increases as one seeks to operate at ever higher frequencies. This unfortunate property is common to all transmission lines (e.g., microstrip, stripline, coaxial) and is one characteristic that motivates the use of waveguides at higher frequencies.

The last column of Table 7.2 gives approximate values for the frequency corresponding to the onset of transverse resonance. Good impedance characteristics are preserved for frequencies somewhat (e.g., 5–10%) below mode cutoff. Finally, note that even 1/16" FR4 has a high enough mode-free bandwidth to operate at any frequency that makes sense for this material.

7.3.1 COPLANAR WAVEGUIDE (CPW) AND COPLANAR STRIP (CPS)

Microstrip, as convenient as it is, still suffers from some deficiencies. One is that significant energy can be coupled into the substrate, leading to loss if the substrate is dissipative. Another is that probing tiny structures is exceedingly difficult because the ground plane is on the other side of the dielectric. And connections to ground can suffer because of the inductance of via connections. Populating the circuit with numerous vias to ground to solve these problems adds manufacturing complexity and expense, so that option does not represent a very practical solution.

An alternative is to use coplanar conductors, thus assuring that both ground and signal lines are accessible from the top surface. Needless to say, the coplanar arrangement greatly facilitates probing at millimeter wave frequencies and beyond. If three lines in a ground–signal–ground configuration are used, it is generally known as a coplanar waveguide (CPW) or coplanar transmission line. If a pair of lines is used in a ground–signal configuration, it is generally known as a coplanar strip (CPS). See Figure 7.4.

These structures have an additional advantage: different line widths and spacings can produce the same characteristic impedance. Thus lines may remain small

---

[14] G. Vendelin, "Limitations on Stripline," *Microwave Journal*, v. 13, no. 5, 1970, pp. 63–9.

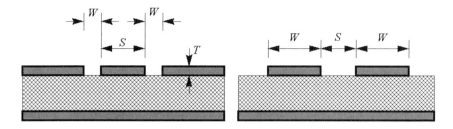

FIGURE 7.4. Coplanar waveguide (left) and coplanar strip

everywhere except at places that must interface to larger connectors, for example. Consequently, CPW provides excellent transitions to coaxial connectors. For these coplanar structures, performance generally improves as the ratio W/H shrinks because the increasing edge coupling implies less energy coupled into a potentially lossy substrate. At the same time, dispersion is also reduced for the same reason.

From Figure 7.4 it is apparent that these two structures are complementary, in a sense. That complementarity is formally acknowledged by using the same variable W for both the *gap* between CPW lines and the *width* of CPS conductors. The resulting equations for the characteristic impedance also reflect this complementarity, as we'll see. Thanks to the conformal mapping techniques used in the derivation, elliptic functions arise in the expressions, as they do for stripline. The terms "conformal mapping" and "elliptic functions" are actually codewords for "impossible to follow the derivation" and "hard to compute with." Any doubts about that statement are readily dispelled by examining the following equations.[15]

We have:

$$Z_0 = \begin{cases} \dfrac{\eta K'(k)}{4\sqrt{\varepsilon_e}K(k)} & \text{for CPW;} \quad (15) \\ \dfrac{\eta K(k)}{\sqrt{\varepsilon_e}K'(k)} & \text{for CPS.} \quad (16) \end{cases}$$

Those look simple, but the subexpressions are not quite trivial. First,

$$k = \frac{S/2}{S/2 + W}, \quad (17)$$

$$\varepsilon_e = 1 + \frac{\varepsilon_r - 1}{2} \frac{K'(k)K(k_1)}{K(k)K'(k_1)}. \quad (18)$$

In turn,

$$k_1 = \sinh\left(\pi \frac{S}{4H}\right) \Big/ \sinh\left(\pi \frac{S/2 + W}{2H}\right). \quad (19)$$

---

[15] We use the same notation and equations presented by I. Bahl and P. Bhartia, *Microwave Solid-State Circuit Design* (Wiley, New York, 1988), with simplifications and minor corrections. In turn they cite as the original source G. Ghione and C. Naldi, "Analytical Formulas for Coplanar Lines in Hybrid and Monolithic MICs," *Electronic Letters*, v. 20, 1984, pp. 179–81.

Then, we have again

$$\frac{K'(k)}{K(k)} = \begin{cases} \left[\dfrac{1}{\pi}\ln\left(2\dfrac{1+\sqrt{k'}}{1-\sqrt{k'}}\right)\right] & \text{for } 0 \leq k \leq 0.707, \quad (20) \\ \left[\dfrac{1}{\pi}\ln\left(2\dfrac{1+\sqrt{k}}{1-\sqrt{k}}\right)\right]^{-1} & \text{for } 0.707 \leq k \leq 1, \quad (21) \end{cases}$$

where $K(k)$ is again the complete elliptic integral of the first kind, and K' is again its complementary function.

Finally, it remains true that

$$K'(k) = k(k') \quad (22)$$

and that

$$k' = \sqrt{1-k^2}. \quad (23)$$

Although we see that the equations are hardly simple, they are nonetheless readily encoded in a spreadsheet or similar tool.

Because these edge-coupled structures concentrate the electric field in the lateral direction, the substrate becomes less important as the spacing/height ratio shrinks. Consequently, line attenuation may actually improve if the substrate is lossy. At the same time, reducing the amount of flux in the substrate diminishes the effective inhomogeneity, resulting in reduced dispersion.

7.3.2 LINE-TO-LINE DISCONTINUITIES

On occasion, one may wish to employ transmission lines of differing characteristic impedance. Even if one never uses more than one impedance, it is not always convenient (or possible) to run only perfectly straight lengths of transmission line. Consequently it's important to understand the nature of discontinuities that may exist between segments of transmission line. For example, if one connects two lines of unequal characteristic impedance (unequal width), the corresponding circuit model is not simply that of two constant-impedance lines in cascade – contrary to what you might expect. There is an additional complexity resulting from the distortion in field patterns that accompanies the distortion in geometry. Additional field components generally must be produced in the vicinity of the discontinuity in order to satisfy boundary conditions there, necessarily causing a departure from pure TEM propagation; higher-order modes must be excited. Recalling that deriving the characteristic impedance assumes TEM propagation, it should be no surprise that a more complex circuit model is required to describe line behavior in such cases. For similar reasons, complexities may arise if there are any bends, even if one intends a constant line impedance.

Thus an assumption of pure TEM propagation does not allow us to stitch the fields together properly at geometric discontinuities. If, as is common, these discontinuities excite modes that do not propagate then their effect is primarily to add a reactive

FIGURE 7.5. More realistic model of "open"-circuited microstrip line

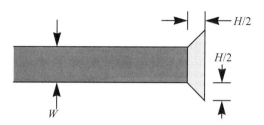

FIGURE 7.6. Approximate picture of end fringing (top view)

component to the impedance. If the energy is primarily stored in the electric field, the reactance is capacitive; if stored in the magnetic field, the reactance is inductive. If the physical extent of the discontinuity is short at all wavelengths of interest, then a single "lump" is an adequate description. Otherwise, a π- or T-model (or possibly cascades of such models) is needed.

Consider first the simple case of an open-circuited line. Because of fringing, the electric field does not drop abruptly to zero at the end of the line. This phenomenon is familiar; it's the reason parallel-plate capacitors (and antennas) act as if they are somewhat larger than their physical dimensions. As a consequence, a more accurate model for the "open"-circuited line is actually a capacitively loaded line, as shown in Figure 7.5.

An effective length extension of approximately $H/2$ is a reasonable first estimate, where H is the dielectric thickness as defined before (see Section 7.9 for a derivation, if you are really in the mood for lots of equations). From a top view (Figure 7.6), the fringing field also extends above and below the line, perhaps adding another effective area of approximately $(H/2)^2$. Thus, as a second crude approximation,

$$C_{eq} \approx \frac{\varepsilon}{H}\left(W\frac{H}{2} + \frac{H^2}{4}\right) = \varepsilon\left(\frac{W}{2} + \frac{H}{4}\right). \tag{24}$$

If a more accurate estimate of the effective length extension Δl is required, one may use an empirical formula due to Hammerstad and Bekkadal:[16]

$$\frac{\Delta l}{H} \approx 0.412\left(\frac{\varepsilon_{re} + 0.3}{\varepsilon_{re} - 0.258}\right)\left(\frac{W/H + 0.262}{W/H + 0.813}\right). \tag{25}$$

Here ε_{re} is the effective relative dielectric constant, which is given by

---

[16] Hammerstad and Bekkadal, *A Microstrip Handbook*.

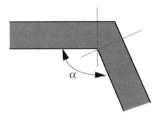

FIGURE 7.7. Sharp bend

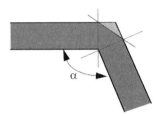

FIGURE 7.8. Excess capacitance of sharp bend

$$\varepsilon_{re} = \frac{\varepsilon_r + 1}{2} + \frac{\varepsilon_r - 1}{2}\left(1 + 10\frac{H}{W}\right)^{-1/2}. \tag{26}$$

Now consider a bend in a line of nominally fixed characteristic impedance. First, notice that this structure does not maintain constant conductor width throughout the bend (loosely identified as the region between the two intersecting boundary lines shown in Figure 7.7). The additional metal in the vicinity of the outer corner acts, to first order, as a shunt capacitance (a more sophisticated model would be a low-pass π-network with two shunt capacitances separated by some series inductance[17]). In noncritical applications, that capacitive loading may be ignored. More often, however, the reactive loading (and consequent low-pass filtering) is troublesome, and something must be done about it.

At minimum, we need some idea about how large a capacitance is produced by the corner so that we may evaluate its effect. As a very crude approximation, the shunt capacitance may be estimated as equal to that of the lighter triangle in Figure 7.8. The logical consequence of this identification is that compensation can be effected simply by slicing off the offending metal, as shown in Figure 7.9. The figure shows a more general situation in which the chamfer is of length a, because a detailed analysis reveals that optimum compensation results when an area somewhat greater than

---

[17] The series inductance models the fact that there is magnetic energy storage due to the current flowing in the bend. Alternatively, a T-network with two series inductances and a shunt capacitance is a perfectly acceptable model as well.

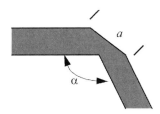

FIGURE 7.9. Mitered (chamfered) bend

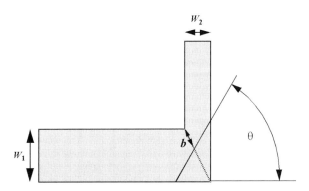

FIGURE 7.10. Mitering right-angle bend with lines of different widths

that highlighted in Figure 7.8 is removed. As a specific example, a value of approximately $1.8W$, rather than $\sqrt{2}W$, provides optimum compensation for a right-angle bend.[18] Fortunately, the optimum conditions are relatively flat, so small deviations from the optimum dimensions do not cause dramatic degradation.

To accommodate bends involving lines of different widths, one may use the approach suggested by Figure 7.10.[19] Here, the angle θ is given by

$$\theta = \tan^{-1}(W_1/W_2) \qquad (27)$$

and the inner corner-to-miter distance b is

$$b = 0.4\sqrt{W_1^2 + W_2^2}. \qquad (28)$$

The factor 0.4 is the default used by *Puff*, but it may be adjusted if necessary.

An alternative to a mitered transition is a smooth, circular bend with a radius of curvature that is at least several times the width of the line (a minimum value of 3× is often used as a rule of thumb). Parasitic reactances can be reduced to negligible levels with this type of bend, known variously as a swept or circular bend; see Figure 7.11. The mitered bend is much more popular because it works well enough for

---

[18] D. M. Pozar, *Microwave Engineering*, 2nd ed., Wiley, New York, 1998.
[19] This method is taken from the *Puff* 2.1 user's manual, chap. 7.

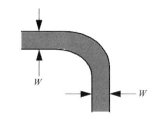

FIGURE 7.11. Circular (or swept) bend

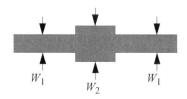

FIGURE 7.12. Step discontinuity

most applications, it is supported by more layout tools, and engineers are congenitally lazy (er, efficient).

Another commonly occurring case is a step transition from one impedance (width) to another. Again to first order, the step discontinuity can be modeled as a shunt capacitance; this is shown in Figure 7.12. As with all of these cases, exact equations for the parasitics associated with the step discontinuity are somewhat involved. However, an intuitively appealing (but admittedly crude) approximation can be developed by treating the wider portion of the line as simply producing a capacitive discontinuity. The capacitance may be computed to a first approximation as that of a parallel plate capacitor of area $l(W_2 - W_1)$, where l is the length of the wide portion. A small refinement in the estimate may be obtained by accounting for the small amount of fringing capacitance, but this is needed only for capacitors made with small l. In either case, a first-order model consists of two segments of transmission line of width W_1 and at whose junction there is a shunt capacitance to ground.

When connecting together two lines of different width, it is best to use a tapered transition rather than a step. A sufficiently gradual taper reduces the density of energy stored reactively in higher-order modes. A rough guideline as to what constitutes "sufficiently gradual" is to select the length of the tapered region at least as long as a quarter-wavelength of the lowest-frequency component for which a match must be preserved. Although a simple linear taper is shown in Figure 7.13 (because it's easy to draw, and the author is lazy), other taper shapes are better, as discussed in Chapter 4.

From the foregoing examples, it's easy to get the impression that every discontinuity is inherently capacitive in nature. Just to disabuse you of that notion, in Figure 7.14 we depict a case where one encounters an inductive parasitic. And now that we've given away the general answer, we can pretend that we could have anticipated it all

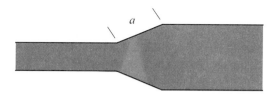

FIGURE 7.13. Tapered transition between two lines

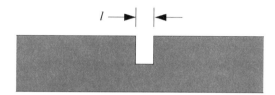

FIGURE 7.14. Notched line

along. The narrower section produces current crowding, increasing the magnetic energy density there. This increase may be interpreted as the action of an inductance.

Alternatively, consider the following: If the smaller-width section were of infinite length, then propagation would be TEM and its impedance would be higher than that of the wider section. Thus, for any suitably short segment of such a line terminated in a lower impedance, its impedance behavior will be similar to that of an inductance. Even though we know that TEM propagation is a poor assumption near the step change in conductor width, we will nonetheless calculate the inductance value as if propagation were TEM.

From the equation for the impedance of a transmission line segment (see Eqn. 31 in Section 7.4), we may derive the following expression for the inductance of a short stub, given our stated assumptions:

$$L \approx \frac{l Z_{0,n} \sqrt{\varepsilon_{r,eff,n}} \left[1 - \left(\frac{Z_{0,w}}{Z_{0,n}}\right)^2\right]}{c}, \qquad (29)$$

where the subscripts n and w stand for narrow and wide, respectively. At least this equation has the correct general behavior, with the inductance going to zero as the notch depth does. Yet because of the loosely justified assumptions used in its derivation, one should not rely on this equation except for a crude estimate.

More elaborate models take into account not only the shunt capacitance that arises from the same length-extension mechanism considered earlier (end-to-ground fringing) but also a capacitance in parallel with the inductance that arises from end-to-end fringing. These enhancements are left as "an exercise for the reader."

Another geometry that arises frequently is a T-connection of lines. In this case, the capacitive coupling between the two lines near the inner corners of the junction effectively produces a bypass path. See Figure 7.15. As suggested by the arrows,

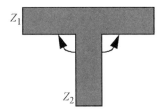

FIGURE 7.15. Illustration of T-junction shortening

capacitive coupling (for the most part) essentially short-circuits around the corner, making irrelevant a small segment of the vertical line where it meets the horizontal line. That is, that small segment behaves as if it is absent; the vertical line acts shorter than expected. For many situations, the effective shortening is not a problem. However, there are also certain configurations in which the lines must be some specified length. In those cases, one must lengthen the lines for them to possess the correct electrical length. A first-order estimate is that the line should be lengthened by approximately H, the dielectric thickness. Equation 61 (in Section 7.6) provides a more accurate estimate for the required lengthening.

As with the T-junction, shortening occurs at the junction of a wide and narrow line, and for the same reason. The amount of shortening, Δl, is approximately given by[20]

$$\frac{\Delta l}{H} = \frac{120\pi}{Z_1 \sqrt{\varepsilon_{re,ser}}} \left[0.5 - 0.16 \frac{Z_1}{Z_2} \left(1 - 2 \ln \left[\frac{Z_1}{Z_2} \right] \right) \right]. \tag{30}$$

7.3.3 TRANSITIONS BETWEEN CONNECTORS AND TRANSMISSION LINES

The transition from a coaxial connector to a planar transmission line can introduce serious impedance artifacts if badly handled. Just as with the line-to-line case, one important requirement is to avoid sharp bends in order to minimize the impact on impedance. Strictly satisfying this requirement, however, usually implies the mounting of connectors so that their conductor axes are in the plane of the circuit board (Figure 7.16). Needless to say, this electrically optimal arrangement is inconvenient in many cases, as it limits the position of connectors to the periphery of the PC board.

It is much more convenient to mount connectors on the surface of a PC board, but such an orientation necessarily forces signals to traverse a right-angle bend. A mitered or circular transition through the PC board is not practically realized, so it would seem that serious degradation of signal quality is simply inevitable. While some amount of degradation certainly can't be avoided, its magnitude can be reduced

---

[20] *Puff* 2.1 manual, p. 37. In turn, the manual cites the original source as Hammerstad and Bekkadal, *A Microstrip Handbook*.

FIGURE 7.16. Connector mounted in plane (side view)

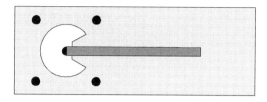

FIGURE 7.17. Improved SMA-to-microstrip transition (view from line side of PC board)

considerably by following a few simple layout rules. These compensation methods can be anticipated from our experience with the pure microstrip examples we've already studied. We expect the right-angle bend to introduce a shunt capacitance, just as in the earlier planar cases. Compensation then involves modification of the layout to reduce the capacitance in the vicinity of the bend. As a specific example, consider the mounting of an SMA connector. Much of the ground plane should be removed from around the center pin to reduce the shunt capacitance and thereby maintain an approximately constant impedance through the transition; see Figure 7.17. Versions of other connectors intended for PC board mounting should follow the same general layout strategy.

Note that the PC board versions of both the SMA and BNC normally require the drilling of five holes (one for the center conductor and four more for the grounded mounting flange). For rapid prototyping purposes, one may instead use a chassis-mount ("bulkhead") BNC, so that only one hole needs to be drilled. There is a challenge, though, because the threaded bushing of such a BNC is much longer than the thickness of any PC board. Thus, if one drills a hole large enough to accommodate the bushing, the center pin ends up sticking out a large distance from the surface of the board. Nevertheless, it is possible to use such BNCs if one employs a somewhat unorthodox mounting arrangement: Drill a *single* hole just large enough to allow only the BNC's Teflon dielectric – not the threads – to protrude through the board. Mount the BNC so that the threads may then be soldered to the ground-plane side (the heat required may exceed what many small soldering irons can provide), being careful to solder a nice, continuous bead around the full circumference.[21] Finish by soldering the microstrip line to the center pin, which should be nearly flush

---

[21] Make sure that the connectors are not coated with stainless steel; these won't solder! In these or other stubborn cases, sandpaper or a file can be used to remove surface coatings that prevent soldering.

FIGURE 7.18. Unorthodox mounting method for bulkhead BNC (side view)

with the board (use material 1/16" – or 1.6 mm – thick for a near-perfect fit). Then trim the connector's dielectric if necessary so that it ends up flush with the PC board. Clearly, such an unprofessional-looking mounting arrangement is completely unsuitable for a commercial product, but it is surprisingly effective for hobbyist and student applications in the low-gigahertz frequency range. In any case, it is certainly a good match to the qualities of FR4 and is by far the lowest-cost method for building decent circuits in that frequency range.

Despite its cheesiness, this arrangement is precisely the one used with excellent results in a microwave circuits laboratory course at Stanford for the past two decades. See Figure 7.18. If you build the coax-to-microstrip transition in the unorthodox manner described, you will invariably find that the discontinuity at the right-angle transition is a bit *inductive* in character, indicating that this mounting method actually overcompensates for the typically capacitive bend. The addition of a little capacitance in the vicinity of the transition (with a suitable amount of copper foil tape) can improve the quality of the transition significantly. Iteration with the guidance of a network analyzer or time-domain reflectometer (TDR) is highly recommended if the quality of the transition is important.

7.4 PASSIVES MADE FROM TRANSMISSION LINE SEGMENTS

In this section, we recognize explicitly that a line is in reality an impedance transformer. This transformation derives fundamentally from the delay in voltage and current associated with finite propagation velocity. The length-dependent phase shift between V and I causes a corresponding length-dependent change in impedance. The versatility of transmission lines as circuit elements ultimately derives from exploitation of this effect.

Recall that, for a lossless line terminated at some distance l in some arbitrary load impedance Z_L, the impedance looking into the end at $z = 0$ may be expressed as

$$\frac{Z(l)}{Z_0} = \frac{Z_{Ln} + j \tan \beta l}{1 + j Z_{Ln} \tan \beta l}, \tag{31}$$

where β is the imaginary part of the propagation constant, γ:

$$\beta = \text{Im}[\gamma] = \omega \sqrt{LC} = \omega/v, \tag{32}$$

7.4 PASSIVES MADE FROM TRANSMISSION LINE SEGMENTS

where v is the propagation velocity. Calculating the latter is a bit complicated in the case of microstrip because the line is not surrounded by a uniform dielectric. The effective dielectric constant is therefore intermediate between that of the board material and that of air. Additional approximate equations for the effective dielectric constant are as follows:[22]

$$\varepsilon_{r,eff} \approx \begin{cases} 1 + 0.63 \cdot (\varepsilon_r - 1) \cdot (W/H)^{0.1255} & \text{for } W/H \gtrsim 0.6; \quad (33) \\ 1 + 0.6 \cdot (\varepsilon_r - 1) \cdot (W/H)^{0.0297} & \text{for } W/H < 0.6. \quad (34) \end{cases}$$

For typical 50-Ω lines on FR4, $W/H \approx 2$ and so the effective dielectric constant is around 3.5 (if the actual bulk ε_r is 4.5). Wavelengths are thus a bit more than half that in free space.

We should expect, from physical arguments and intuition, that short segments of transmission line can be used to realize either inductances or capacitances. Specifically, let's review two extreme cases. In the first, assume that the transmission line is terminated in an open circuit. Then a short segment of such a line appears simply as a capacitor, since all we have are two conductors separated by a dielectric. Conversely, a short segment of line terminated in a short circuit will appear inductive, since we now have a current loop.

To derive an explicit formula for the capacitance, let the load impedance go to infinity in Eqn. 32. Then, the input impedance of a short piece of open-circuited line is approximately

$$Z \approx \frac{Z_0}{j\omega(l/v)}. \quad (35)$$

The capacitance is therefore

$$C = \frac{l}{vZ_0} = \frac{l\sqrt{\varepsilon_{r,eff}}}{cZ_0}. \quad (36)$$

A helpful mnemonic is that the time constant Z_0C is simply the one-way time of flight.

In applications where space is not a burdensome constraint, one should not overlook the option of making capacitors in this manner. Even if one does not realize all of a desired amount of capacitance this way, it is convenient for adjustment purposes to have at least part of the capacitance come from PC board traces that may be trimmed after fabrication. It is valuable in any case to know how much capacitance is associated with a given area of conductor, if only to facilitate estimation of layout parasitics. With FR4, for example, one can expect about 2.5 pF/cm² when using a 1/16" (1.6-mm)–thick substrate. Furthermore, the loss of FR4 is somewhat less than – but in the same general range as – that of X7R or Y5V ceramic dielectric materials used in discrete capacitors. Of course, lower loss (but somewhat lower

---

[22] Carson, *High Frequency Amplifiers*. These equations are allegedly inferior to the Hammerstad–Bekkadal equation (Eqn. 26) offered earlier.

capacitance per unit area) is obtained with a higher-quality board material, such as PTFE or RO4003.

An important consideration is that the linear dimensions must be small compared to a wavelength at the highest frequency of interest. Otherwise, distributed effects will alter the impedance behavior.

Similarly, for the inductance of a short line terminated in a short circuit we have

$$L = \frac{lZ_0}{v} = \frac{lZ_0\sqrt{\varepsilon_{r,eff}}}{c}. \tag{37}$$

Again, we find that the time constant (here, L/Z_0) is the one-way time of flight.

Additional formulas for inductances of various kinds are given in Appendix A (Section 7.8). Many of these have appeared earlier in Chapter 6 but are collected in this appendix for convenient reference.

Once a component has been designed, one must verify that the calculated line dimensions are indeed very small compared with an electrical wavelength (which may be computed using the given expressions for effective dielectric constant). If not, the approximations will be significantly in error.

In practice, these inductors and capacitors might not always be connected to open or short circuits. To validate the foregoing nonetheless, we can control the characteristic impedance of the lines so that their terminations may be considered as opens or shorts in comparison (see e.g. Eqn. 29). So, to make a capacitor, we would want to choose its Z_0 as low as possible (or practical). Conversely, we wish to choose Z_0 as high as possible to make an inductor, so that any impedance loading it appears approximately as a short circuit, relatively speaking.

Aside from the flexibility of being able to realize component values that span a continuum, using transmission line elements often permits us to build components featuring better characteristics than discrete ones. For example, building a 1-nH inductor is much easier in microstrip than in discrete form.

There are a few practical issues to consider when implementing transmission line components. First, one cannot specify arbitrarily high characteristic impedances for inductors because there is always a lower bound on the width of lines that may be fabricated reliably. Assuming a typical manufacturing tolerance of 2 mil, and supposing that this variation is allowed to represent at most 20% of the total width, one may assume a minimum practical linewidth of about 10 mil.[23] Hence, on 1/16" FR4, practical line impedances rarely exceed about 200 Ω. This value is nonetheless sufficiently high for many purposes. The effective series resistance tends to be somewhat large, however, so such inductors cannot be used where a high Q is required.

---

[23] These values are typical for FR4. On more rigid substrates, reliable fabrication of narrower lines becomes possible. Depending on the substrate type, linewidths of perhaps 1 mil (25 μm) represent the current state of the art, although such narrow lines are hardly common.

When realizing a capacitor, there are two practical constraints. If the linewidth gets too large, it may become an appreciable fraction of a wavelength, invalidating the approximations. The other limitation is simply one of board space.

In the next section, we relax the constraint on dimensions and also consider slightly lossy lines. In so doing we derive an explicit expression for the effective series resistance of elements constructed out of transmission line, allowing the computation of Q.

7.5 RESONATORS

The foregoing section imposes the constraint that all segments be very short compared with a wavelength. However, Eqn. 31 is not restricted to this condition; the constraint on length simply guarantees that only pure reactances are produced. In this section, we consider transmission line segments that purposefully violate our previous constraint in order to produce "impure" reactances. Specifically, let us study the input impedance of a shorted line that is nominally $\lambda/4$ in electrical length. At the frequency where the line is *precisely* $\lambda/4$ in extent, the input impedance is ideally infinite (assuming no loss). As the frequency increases from this condition, the line is somewhat longer than $\lambda/4$. As a consequence, the input impedance appears capacitive (as can be verified by examination of Eqn. 31). At frequencies where the effective line length is less than $\lambda/4$, the input impedance appears largely inductive. From this description, it should be clear that such a line behaves very much like a parallel RLC tank, even to providing a path at DC. One important difference, however, is that the transmission line version has multiple resonances whereas the lumped tank only has one.

A dual analysis of an open-circuited $\lambda/4$ line reveals that it behaves a great deal like a series RLC network. Whether open- or short-circuited, the resonant nature of $\lambda/4$ lines is widely exploited to make tanks for oscillators and filters (as discussed in detail in Chapters 15 and 23, respectively).

For this class of resonators, the Q-value is a function of conductor and dielectric losses, as well as of radiation. Because of its importance, we now undertake an approximate derivation of this critical parameter. In doing so, we will make liberal use of lessons learned from lumped RLC networks, having already drawn an analogy between them and resonant $\lambda/4$ lines.

For a lumped, series resonant tank, the Q-value at resonance may be expressed as the ratio of the inductive (or capacitive) reactance to the real part of the impedance. Now for a lossy transmission line, the normalized impedance is given by

$$\frac{Z(l)}{Z_0} = \frac{Z_{Ln} + \tanh \gamma l}{1 + Z_{Ln} \tanh \gamma l}, \qquad (38)$$

where the origin is again considered to be located at the load, and the coordinate l is the location of the port at which the impedance is measured.

In our case the load impedance is infinite, so the expression for the normalized input impedance simplifies to

$$\frac{Z(l)}{Z_0} = \frac{1}{\tanh \gamma l} = \frac{1}{\tanh(\alpha + j\beta)l}. \quad (39)$$

Because the loss cannot be too great for any resonator worth using, we will simplify our derivation by assuming that $\alpha l \ll 1$. When this condition is satisfied, the input resistance at resonance is approximately given by

$$R_{eff} \approx Z_0 \alpha l = Z_0(\alpha \lambda / 4). \quad (40)$$

We've already found an expression for the equivalent inductance of a short piece of line, but we cannot use it here because our line is not short in this instance. Instead, by equating the slope of the impedance of the line with that of a corresponding lumped LC network, one finds that there is precisely a factor-of-2 difference:

$$L = \frac{Z_0 \pi}{4\omega_0}. \quad (41)$$

That the effective resonator inductance is smaller by a factor of 2 may be anticipated from energy considerations. The current varies sinusoidally, rather than being constant, and thus the energy stored in the magnetic field is half what one would compute from an assumption of constant current.

With this equation for effective inductance, we may compute the quality factor of an open-circuited $\lambda/4$ line as simply

$$Q = \frac{\omega_0 L}{R_{eff}} = \frac{\pi}{\alpha \lambda} = \frac{\omega_0}{2\alpha v_p} = \frac{\omega_0 \sqrt{\varepsilon_{r,eff}}}{2\alpha c}. \quad (42)$$

These expressions for Q are quite general, and they allow us to deduce achievable resonator Q-values from measured or calculated transmission line attenuation and velocity factors (and vice versa).[24]

As a specific example, let's use the values cited earlier for FR4. The attenuation of a 50-Ω microstrip line might be approximately 0.07 dB/inch/GHz, corresponding to an α per GHz of about 0.35 Np/m.[25] At 1 GHz, a 0.35-Np/m attenuation value implies an unloaded resonator Q of about 50 (typical loaded Qs are half as large),

---

[24] A frequently asked question is: "Why is Q expressed only in terms of inductance?" Recall that, at resonance, the reactances of the capacitance and inductance are equal. So, you get the same answer if you use the capacitance to compute Q (don't take my word; try it). The focus on inductors is simply historical: discrete inductors generally have worse Q than capacitors, so engineers have always been obsessed with inductor Q.

[25] Remember that a neper (Np) is a factor of e. An amplitude change of one neper therefore corresponds to a power change of about 8.69 dB.

given a typical effective relative dielectric constant of 3.5.[26] Because the line attenuation for FR4 is roughly proportional to frequency over a broad range (implying a dominance by dielectric loss), the achievable resonator Q changes little with frequency when this board material is used. If improvements are needed but changing the substrate is not permitted (say, for cost reasons), then one may judiciously combine high-quality passive elements with microstrip ones. For example, if we need to improve a resonator implemented with an open-circuited microstrip line, we could shorten it (making it look more inductive) and then connect a capacitor from the end of the line to ground in order to compensate. Shortening the line reduces the peak voltage along the line and thereby reduces the dielectric loss. Excellent circuits are thus enabled by selectively using good discrete elements to supplement those made in inexpensive microstrip.

It is instructive to compare FR4's unloaded resonator Qs of ~50 (a usefully large value, to be sure) with what one may achieve with other materials and geometries. If RO4003 is used, a fourfold boost in Q is a reasonable expectation because of its superior dielectric loss properties. Use of coaxial cables with low loss dielectrics easily enables another tripling of Q, allowing the realization of resonators with Q-values of the order of 1000. *Cavity resonators,* made out of a closed conducting box or a solid block of special dielectric material, are capable of still another order of magnitude increase in Q. This progressive increase in Q may be understood qualitatively as follows. Exposed planar structures (such as microstrip) have the lowest Q-values because radiation losses add to conductor and dielectric loss. Coaxial structures are better because they can't radiate. Cavity resonators are even better – partly because of their lower surface-to-volume ratio and partly because of the absence of an additional dissipative conductor. The former consideration is important because, loosely speaking, energy is stored in volumes and dissipated on surfaces. A cube (or better, a sphere) has a more favorable surface-to-volume ratio than does a cylinder. The dielectric resonator has no dissipative conductors at all, so if a material with suitably low loss tangent is employed then exceptionally high Q-values (tens of thousands) are possible.

7.6 COMBINERS, SPLITTERS, AND COUPLERS

It is frequently necessary to combine signals from multiple sources to create a single output. An example is found in power amplifiers, in which the outputs of several lower-power stages are to be combined to produce a single high-power output. As with other linear passive elements, these combiners can be used in reverse to act as power splitters. When used as splitters, it is possible to obtain either differential or quadrature outputs, depending on the design details.

---

[26] Recall that the effective constant accounts for the inhomogeneous dielectric medium of microstrip lines.

184   CHAPTER 7 MICROSTRIP, STRIPLINE, AND PLANAR PASSIVE ELEMENTS

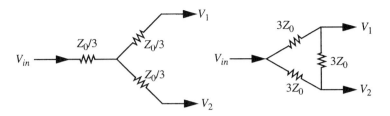

FIGURE 7.19. Lumped resistive 6-dB splitter/combiner
("Y" version on left, "Δ" version on right)

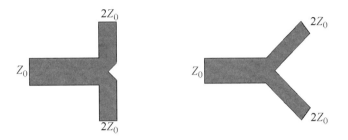

FIGURE 7.20. T- and Y-junction splitter/combiner

These elements are of such utility that engineers have expended a great deal of effort to devise numerous implementations. Space constraints force us to focus on the more commonly encountered types. We also present both lumped and distributed versions, partly to aid our understanding of one as a logical extension of the other and partly to expand our palette of options.

7.6.1 RESISTIVE COMBINERS

The simplest splitter/combiner consists of a network of three resistors, as illustrated by Figure 7.19. With the values shown, the reader may verify independently that the network provides a match if all ports are terminated in Z_0. At the same time, there is a 6-dB attenuation in this case. A compensating trait, however, is that the resistive splitter works well over a large bandwidth.

7.6.2 DISTRIBUTED COMBINERS

Alternative combiners are available for those cases where attenuation is unacceptable. The simplest lossless microstrip splitter/combiner consists of a junction of three lines, with the characteristic impedance of two of the lines set equal to twice that of the third. This is shown in Figure 7.20.

Because each arm must be terminated in its characteristic impedance, this particular type of combiner requires different terminations on the three ports. This property is one potential disadvantage. The T-combiner also suffers from a significant

7.6 COMBINERS, SPLITTERS, AND COUPLERS

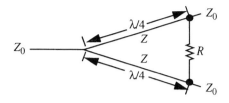

FIGURE 7.21. Wilkinson splitter/combiner (stylized)

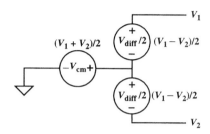

FIGURE 7.22. Decomposition of arbitrary voltages into common- and differential-mode components

impedance discontinuity at the junction of the three arms. As shown, a triangular notch is usually cut out of the vertical section opposite the Z_0 arm in order to provide a first-order compensation for the shunt capacitive loading caused by the discontinuity. The Y-combiner suffers less from this problem because of the less abrupt transition. One could use curved transitions (instead of the sharp ones shown) for still better performance.

A highly ingenious modification of the Y-junction by Ernest J. Wilkinson provides a simultaneous match on all ports as well as isolation between the output ports.[27] In the most general implementation, power may be split among an arbitrary number of ports. We first examine the simplest case of equal power splitting between two ports.

As seen in Figure 7.21, there are three explicit ports. We shall see that there is actually a fourth port, located at the midpoint of the resistor. Because this port is not normally used as an output, it is rarely identified explicitly. However, it is occasionally useful nonetheless to imagine its existence, as we shall see when we consider generalizations of Wilkinson's divider.

To determine the values of Z and R corresponding to the conditions of port isolation and equal power division, we may exploit the symmetry of the structure and explore the operation of the combiner with common-mode and differential-mode inputs. Recall from ordinary circuit theory that any pair of voltages (V_1, V_2) or currents may be expressed as the superposition of their average (common-mode) value with their difference, as seen in Figure 7.22. In microwave work, it is customary to refer to

---

[27] Wilkinson, then at Sylvania, describes this element in "*N*-Way Hybrid Power Combiner," *IRE Trans. Microwave Theory and Tech.*, v. 13, January 1960, pp. 116–18.

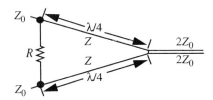

FIGURE 7.23. Wilkinson combiner (still stylized)

FIGURE 7.24. Wilkinson combiner after simplification for even-mode excitation

even- and *odd-*mode (instead of common- and differential-mode) excitations. They mean the same thing; it's just a matter of differing linguistic traditions. No matter what you call these decompositions, the response of the Wilkinson combiner to any arbitrary pair of input voltages may be computed simply by finding the responses to common-mode and differential inputs separately, and then summing those responses.

First consider the common-mode response. To do so, imagine driving both inputs of the combiner with identical voltages. In that case, we know by symmetry that the upper and lower halves of the combiner behave identically. To emphasize this symmetry, we may redraw the circuit slightly as shown in Figure 7.23.

All we have done is replace the single output line with a parallel combination of two double-impedance lines (note the shorting wire added at the apex). By symmetry, we know that no current flows across an imaginary horizontal line bisecting the complete structure. Consequently, we may separate the upper and lower half-circuits.[28] No current flows through the resistor, so we may remove it altogether. Similarly, no current flows through the tiny vertical bar that shorts the two double-impedance lines together. Carrying out a complete separation thus leaves us with a simplified circuit in which each of the two halves conveys complete information about the circuit's even-mode response. If we arbitrarily choose to eliminate redundancy by eliminating the bottom half, then the resulting circuit appears as shown in Figure 7.24. It is evident that an impedance match for the even mode is possible if we choose Z equal to the geometric mean of the input and output termination impedances:

$$Z = \sqrt{(Z_0)(2Z_0)} = \sqrt{2}Z_0. \tag{43}$$

Having used the even mode to discover the required characteristic impedance of the arms, we next use odd-mode analysis in hopes that it might tell us something

---

[28] An equally valid alternative to the path taken here would be to connect the upper and lower half-circuits in parallel, because corresponding mirror-image points are of equal voltage under common-mode excitation. This approach leads to the same answer.

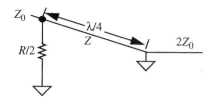

FIGURE 7.25. Wilkinson combiner after simplification for odd-mode excitations

about the resistor R. In this case, a purely differential (odd) excitation produces antisymmetric responses, again enabling simplifications. Because the voltages along the upper and lower halves are now mirror images of each other, a horizontal line midway through the structure represents a constant-voltage contour of zero value; it's a virtual ground. After simplification, the combiner reduces to the half-circuit shown in Figure 7.25. Here we see that the apex of the high-impedance arms is grounded. When transformed to the input by the quarter-wavelength arm, this short becomes an open. Thus, the only loading seen by the odd-mode input is provided by the resistor. To provide a match to the odd-mode component, we need only satisfy

$$\left(\frac{R}{2} = Z_0\right) \implies R = 2Z_0. \tag{44}$$

Just as with any other symmetrical circuit (such as a differential amplifier, for example), we see that use of even- and odd-mode decomposition enables great simplifications in analysis and design. In this specific case, it's allowed us to identify the conditions for providing matches separately to even and odd components and then arrange for the satisfaction of these conditions. Finally, this analysis method also reveals that the Wilkinson combiner indeed provides the isolation claimed at the outset. We've already seen that, for even excitations, the upper and lower half-circuits are decoupled from each other because of an effective open-circuiting of all connections between them. For odd excitations, the upper and lower half-circuits are again decoupled, this time because of the virtual short circuit that forms at the midpoint of the bridging resistance and at the apex of the arms. Because arbitrary excitations are always expressible as the sum of even and odd mode excitations, the decoupling holds generally. Thanks to all of these attributes, the Wilkinson splitter/combiner is an extremely popular microwave element.

Our analysis has revealed that the bridging resistance acts as the termination for odd-mode components only. In normal operation, the Wilkinson splitter/combiner is operated symmetrically (i.e., in a purely even mode). Hence, the bridging resistance normally has only to dissipate incidental odd-mode energy that might arise from imperfect matching between loads connected to the two halves. Nevertheless, *incidental* doesn't guarantee *small*, so it is important to compute the worst-case dissipation and then select the resistor's power-handling capability accordingly. A common error is to omit such a calculation until olfactory cues call attention to the insufficiency of the resistor's power ratings.

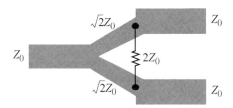

FIGURE 7.26. Wilkinson splitter/combiner (stylized, and definitely not optimum)

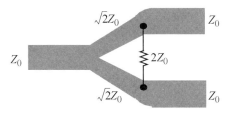

FIGURE 7.27. Wilkinson splitter/combiner (more representative of real layouts)

Practical layouts for the Wilkinson splitter/combiner improve upon the suboptimal initial arrangement shown in Figure 7.26. For reasons that should be clear by now, the transition to the output lines is usually smoother in practice. In fact, practical Wilkinson combiners often use curved transitions throughout its entire structure (see Figure 7.27).

Identical Wilkinson combiners may be cascaded to provide additional outputs. Such binary (or *corporate*) arrays are sometimes used in power amplifier modules to combine the outputs of several low-power amplifiers. Wilkinson's original paper describes another generalization of his technique to accommodate splitting into an arbitrary number of paths (not just binary values). That technique in turn may be extended to permit unequal power splitting as well. These possibilities are best appreciated with a visual aid like Figure 7.28.

As with the two-way case, seeking an impedance match for the even mode permits us to deduce that the required impedance of the $\lambda/4$ arms is given by

$$Z = \sqrt{(NZ_0)(Z_0)} = Z_0\sqrt{N}. \tag{45}$$

Providing a match to odd components requires the connection of a resistor of value $R = Z_0$ from each line to a common point (the virtual port to which we've already alluded) – a point to which no other connections are normally made.

If N exceeds about 10, the required line impedance can grow to values that are difficult to realize practically in microstrip form. Another practical consideration is that connecting the bridging resistors gets complex for $N > 2$, since a simple series connection no longer suffices. To eliminate or alleviate both of these difficulties, it

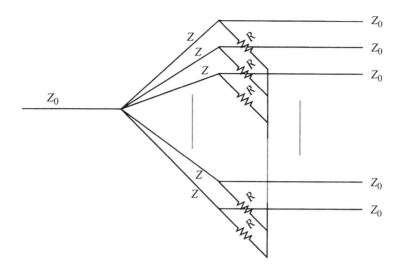

FIGURE 7.28. Generalized N-way Wilkinson splitter/combiner

is sometimes preferable to use a binary tree version, either alone or in combination with N-way implementation.

One may also use the generalized equal divider as a starting point for designing a divider that provides unequal power division. Suppose, for example, we wish to split the output power in a ratio of $M:1$. We begin consideration of this case with that of the generalized equal divider in which $N = (M + 1)$; see Figure 7.29.

In the first step of the evolutionary sequence shown in the figure, we simply tie together M of the $(M + 1)$ output lines. To preserve matching on that output, the load impedance must be Z_0/M there. After appropriate simplifications by way of computing the parallel combination of the rest of the higher-power branch, we obtain the second of the three designs shown in Figure 7.29.

Next, we scale the result so that the geometric mean of the load impedances equals Z_0. This normalization reduces the likelihood that the impedance of the narrower line is too high for practical realization. The output terminations are thus scaled upward by a factor K, and the impedances of the arms are scaled appropriately as well. The power ratio M thus equals K^2. Finally, the two separate resistances for terminating the odd mode are combined into a single resistor.

Clearly, power may be divided unequally among additional outputs as well. One need only begin with the generalized N-way divider and then proceed in an analogous fashion.

The Wilkinson combiner may also be realized in lumped form (Figure 7.30). This alternative version is somewhat more amenable to tuning (as long as the frequency of operation isn't too high) and can be made quite compact, particularly at lower frequencies. It shares a limited bandwidth with its microstrip counterpart. As is evident from the figure, the lumped version simply replaces each $\lambda/4$ arm of the microstrip combiner with a simple π-network transmission line approximation. We adapt here the relevant design equations from Chapter 2.

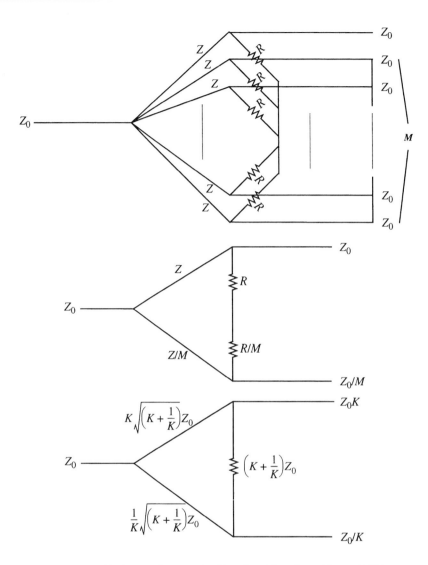

FIGURE 7.29. Evolution of unequal splitter/combiner from N-way Wilkinson

First, choose the L/C ratio to produce the correct arm impedance:

$$\sqrt{L/C} = \sqrt{2}Z_0. \tag{46}$$

Next choose the LC product to produce the correct center frequency:

$$1/\sqrt{LC} = \omega_0. \tag{47}$$

Solving for each element yields:

$$C = 1/\sqrt{2}\omega_0 Z_0, \tag{48}$$

$$L = \sqrt{2}Z_0/\omega_0. \tag{49}$$

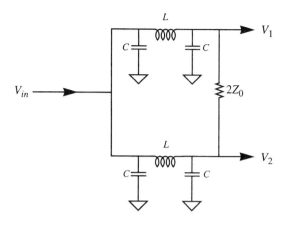

FIGURE 7.30. Lumped two-way Wilkinson splitter/combiner

Thus, the impedance magnitude of each element is simply the characteristic impedance of the corresponding line. Note also that the two input capacitors may be combined into a single one. Thus, only three capacitors would be used in practical implementations. Four are shown in the figure for purely pedagogical reasons.

Finally, the inductors must be spaced far enough apart that we may neglect magnetic coupling between them. The required separation is smaller if the inductors are orthogonal to each other or if they are shielded (e.g., wound as toroids).

7.6.3 HYBRIDS AND BALUNS[29]

Hybrids are extremely versatile elements. The first hybrid was invented to solve a difficult problem in telephony, that of enabling duplex communication over a single wire pair. The most straightforward ways to provide duplex capability require three or four wires (one each for transmit and receive, as well as ground returns for each). Thanks to hybrids, a single wire pair suffices to support full duplex communications. This feat relies on the ability to decompose signals into common-mode (even-mode) and differential (odd-mode) components. By using, say, the even mode to carry transmissions and the odd mode to receive, one can use a single wire pair to achieve duplex operation, as seen in Figure 7.31.

A voltage applied to port 1 generates equal secondary voltages at transformer secondaries A and B. The voltage across A drives a current flow into port D. In turn, the voltage across port B drives port C with the opposite polarity. The voltage across port D is the same as that across C, but of the opposite polarity. If you've managed to follow all that and keep track of the polarities, we see that port 1's voltage ultimately appears across the resistor bridging port 3. That is, port 1 couples to port 3. At the same time, that the voltages across ports D and C are equal but opposite means that no net voltage appears across port 2; ports 1 and 2 are isolated from each other.

---

[29] *Balun* derives from "balanced-to-unbalanced" and thus rhymes with *gallon*.

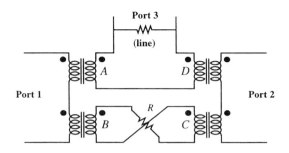

FIGURE 7.31. Classic telephone hybrid[30]

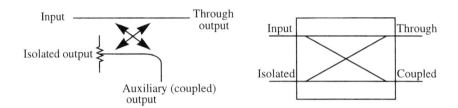

FIGURE 7.32. Some symbols for a hybrid coupler

Thus, this connection of transformers remarkably allows a microphone and speaker connected to ports 1 and 2 to share a common wire pair at port 3 without creating an unwanted internal feedback loop.[31]

As one might gather from this description, there are a great many ways to implement the general idea of discrimination on the basis of symmetry, because there are many ways to realize circuits that are selectively responsive to either differential- or common-mode signals. In the case of the classic telephone hybrid, a broadband multi-tap transformer wound on a soft iron core provides this discrimination. At microwave frequencies, the nonideality of classical transformers motivates alternative realizations, but the underlying principles are the same.

Before discussing hybrids any further, it's probably a good idea to provide a definition. The meaning of *hybrid* has changed over time, but today it generally refers to any four-port device that possesses the following properties: matched impedances at all ports; at least one isolated output port (i.e., a port that produces zero output under certain input conditions); and equal power division. Thus hybrids are often differentiated by the particular phase relationships among the output port signals.

Two of many symbols for hybrids are shown in Figure 7.32. The figure on the left shows an internal termination for the isolated port. Many hybrids are constructed this

---

[30] This balun is but one of several types used in dial telephones.
[31] In actual telephones, the port isolation is deliberately made a little imperfect (by selecting a cancellation impedance R different from the line impedance) so that a person may hear his or her own voice. The acoustic feedback enabled by this *sidetone* helps individuals choose an appropriate speaking amplitude.

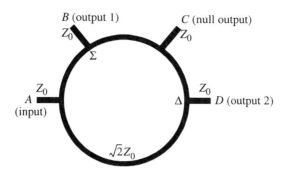

FIGURE 7.33. Ring (or rat-race) hybrid

way as a convenience. In the simplified version on the right, the diagonal lines identify which port pairs are coupled together, leaving open the question of termination on the isolated port.

As we will see in this section, there are many different types of hybrids, distinguished by implementation details and phase relationships among the port signals. A 180° hybrid, named for its ability to provide two antiphase outputs, is useful for performing single-ended-to-differential conversion (and vice versa); it can serve as a splitter or combiner (but not all splitters or combiners are hybrids). In planar microwave form, a narrowband 180° hybrid consists of a closed path (e.g., a circular loop) of microstrip whose electrical circumference is $3\lambda/2$. Three taps (labeled A, B, C, and D in Figure 7.33) are separated from each other by $\lambda/4$. A signal supplied to A splits between the clockwise and counterclockwise paths. In traveling to point B, the clockwise-going signal has been shifted by $\lambda/4$ and the counterclockwise signal by $5\lambda/4$, so they add in phase. In traveling to C, the signals undergo phase shifts of $\lambda/2$ and λ, respectively, leading to cancellation; no signal emerges from that tap. Finally at tap D, the signals have shifted $3\lambda/4$ in each direction, leading once again to in-phase addition. Note that the signals emerging from B and D are shifted in phase from each other by $\lambda/2$. Thus a single-ended input at A becomes a differential signal between B and D. And by reciprocity, a differential input supplied at B and D combine to produce a single-ended output at A.

The rat-race hybrid is quite versatile, for with different port assignments the hybrid performs different functions. For example, a signal supplied to port B will split into two equal in-phase signals at ports A and C (by symmetry, a signal supplied to port C will also produce in-phase outputs at B and D). In this case, port D is the isolated port. Running the argument in reverse, we conclude that equal signals supplied to ports A and C will sum to produce an output at port B. For this reason, port B is sometimes called the sum port, denoted by Σ, even if it used as an input.

A signal supplied to port D produces a differential output between ports A and C, with port B now acting as the isolated port. Consequently, a differential signal supplied between ports A and C produces an output at port D. Therefore, port D is sometimes called the difference port, denoted by Δ. By superposition, we may conclude that the application of any two arbitrary signals to ports A and C results in the appearance of their sum at port B and their difference at port D.[32]

It is important to note that the characteristic impedance of the ring proper must differ from that of each of the taps if we are to provide an impedance match. The precise value can be derived by recognizing that the power splits evenly between the two identically loaded output taps. Therefore, the source driving point A sees an equivalent load impedance of $Z_0/2$ that is effectively driven in parallel through two paths, each of which has an impedance we'll call Z_{ring}. Recall that a quarter-wave line can be used as an impedance matcher if its characteristic impedance is chosen as the geometric mean of the source and load impedances. Here, the source impedance is Z_0 and the effective load impedance is $Z_0/2$, so a match results when

$$\frac{Z_{ring}}{2} = \sqrt{Z_0 \cdot \frac{Z_0}{2}} = \frac{Z_0}{\sqrt{2}} \implies Z_{ring} = \sqrt{2} Z_0. \tag{50}$$

A hybrid of this type is known as a ring (or *rat-race*) hybrid.[33] It should be clear that a circular shape is not strictly necessary. The only requirements are that the total perimeter and tap locations satisfy the various wavelength criteria, and that the impedance of the ring proper be $\sqrt{2}$ times that of the taps. Note that these wavelength-based criteria imply that such a hybrid is of necessity a narrowband element. The precise value of bandwidth depends on the amplitude or phase mismatch (or port reflectance) that may be tolerated. That said, typical useful bandwidths are generally on the order of 15% for relatively flat amplitude and phase (again, the precise bandwidth is entirely a function of what is meant by "relatively flat"). If reflectance is more important than amplitude flatness, the bandwidth is considerably larger (e.g., 50%).

When the hybrid is used to produce differential outputs from a single, ground-referenced one (or vice versa), it is also known as a *balun* (for balanced-to-unbalanced converter). It is often mispronounced "bail-un." Occasionally you may see the term *unbal* used to connote single-ended-to-differential conversion, but *balun* is much more common.

Because the classic ring hybrid's diameter is a function of wavelength (roughly on the order of $\lambda/2$), there may be practical implementation problems at both very high and very low frequencies. At extremely high frequencies, the diameter of the ring may be so small that it is of the same order as the feedline width. Aside from the simple layout challenges that such crowding implies, the mutual proximity of the feedlines may significantly perturb operation of the hybrid. To fix things up a bit, one may increase the circumference of the ring by integer multiples of $\lambda/2$, taking care to reposition the ports as necessary to maintain the proper phase relationships.

---

[32] Here, "arbitrary" doesn't really mean *arbitrary*, of course. We are limited to signals within the narrow bandwidth over which the hybrid functions as it should.

[33] It is sometimes also called a hybrid ring, but in this context *hybrid* is the noun and *ring* is the adjective, so "ring hybrid" is more grammatically correct.

7.6 COMBINERS, SPLITTERS, AND COUPLERS

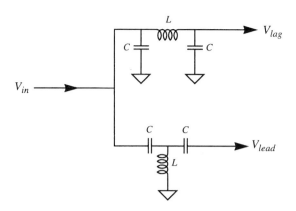

FIGURE 7.34. Simple lumped 180° splitter/combiner (diplexer version)

At very low frequencies, the problem is simply one of excessive size. In such cases it is often preferable to consider a lumped alternative, where a microstrip hybrid might occupy an unacceptably large area. The network shown in Figure 7.34 is actually a diplexer (i.e., a network that splits signals into two frequency bands). The low-pass and high-pass filters produce frequency-dependent lagging and leading phase, respectively. The low-pass filter acts as a lumped approximation to a λ/4 line, while the high-pass filter is its dual. Each filter's corner frequency is thus chosen to produce a phase shift magnitude of 90° at the center frequency of operation. Even though the phase shift of each filter certainly varies with frequency, the *difference* between the output phases is a constant 180° over a broad frequency range (although the amplitudes are strictly equal at only one frequency). This lumped implementation shares with its microstrip cousin the limitation of relatively narrowband operation. Even though it lacks an isolated output, it's frequently called a hybrid anyway.[34]

If Z_0 is the source resistance driving the input port, then we wish the input impedance of each filter to equal $2Z_0$. If a load of value $2Z_L$ connects the two outputs to each other (equivalent to a single Z_L connected to ground from each output), then we need to choose the characteristic impedance of each filter arm equal to the geometric mean of $2Z_0$ and Z_L:

$$\left(\sqrt{L/C} = \sqrt{2Z_0 Z_L}\right) \implies L/C = 2Z_0 Z_L. \tag{51}$$

The other equation needed to complete the design derives from choosing the corner frequencies of each filter to produce the necessary quadrature phase shift at the center frequency of operation for the hybrid:

$$\omega_0 = 1/\sqrt{LC}. \tag{52}$$

---

[34] Occasionally the argument is made that – since the two antiphase outputs can be isolated from each other if the ports are driven with equal amplitudes – the use of *hybrid* for this circuit represents no misuse of the term.

Solving for each element yields:

$$C = 1/\omega_0\sqrt{2Z_0Z_L}; \qquad (53)$$

$$L = \sqrt{2Z_0Z_L}/\omega_0. \qquad (54)$$

Thus, for a 1-GHz hybrid driven by 50 Ω and terminated in 100 Ω (50 Ω from each output to ground), the component values are about 11.3 nH and 2.25 pF. Either discrete components or microstrip equivalents (or some combination of these) may be used to implement this hybrid.

If, as is frequently the case, there is some parasitic capacitance in parallel with the load resistance, then one may resonate it away with a suitable inductance (in principle) if necessary. In a similar fashion, parasitics in series with the load may also be removed (again, over a narrow frequency band).

If one desires a more exact lumped analogue to the distributed coupler, the individual $\lambda/4$ segments may be replaced by low-pass π-sections. For somewhat improved bandwidth, the $\lambda/2$ section is best implemented by a high-pass T-network.[35] The reason is, once again, that the phase shifts behave in a complementary fashion with frequency, leading to a more constant phase shift overall, even though the amplitude is hardly constant with frequency. The overall coupler then appears as shown in Figure 7.35.

The element values are chosen so that all reactances equal the ring impedance at the center frequency of operation:

$$C_1 = 1/\omega_0 Z_0\sqrt{2}; \qquad (55)$$

$$L_1 = Z_0\sqrt{2}/\omega_0. \qquad (56)$$

Lumped implementations have about the same narrow bandwidth (order of 10–15%) as the distributed versions. They have become increasingly popular because of their potential for realization in a smaller space than their classical distributed counterparts, particularly at low frequencies. Even at higher frequencies, the greater amenability to trimming makes the lumped implementations attractive in certain cases.

In still other applications, it is desirable to generate (or combine) two signals that are in quadrature, rather than in antiphase, with each other. Examples of such applications include phase shifters (since combining variable proportions of in-phase and quadrature components yields arbitrary phase shifts); single-sideband generation and detection; and quadrature modulation (including the important case of driving diode mixers with quadrature signals to perform this modulation). A popular microstrip element for performing these functions is the 90° (quadrature) hybrid, also known as a *branchline* coupler or hybrid. See Figure 7.36.

---

[35] S. Parisi, "A Lumped-Element Rat-Race Coupler," *Applied Microwaves*, August/September 1989, pp. 84–93.

7.6 COMBINERS, SPLITTERS, AND COUPLERS

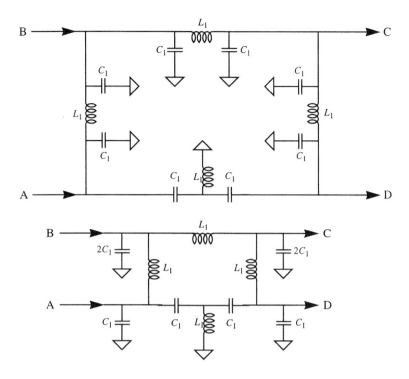

FIGURE 7.35. One possible lumped implementation of a 180° hybrid

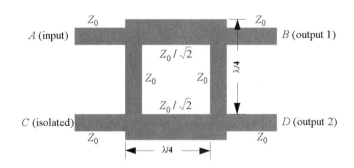

FIGURE 7.36. Branchline hybrid (unmitered 3-dB version shown)

Each arm making up the sides of the central box is nominally $\lambda/4$ in length, so it's not too hard to see how there could be a quadrature relationship between signals at adjacent ports. Beyond that observation, however, it is difficult to figure out the branch impedances by inspection (that is, if you don't already know the answer). Again, even- and odd-mode analyses per Figure 7.37 allow us to complete the design with a minimum of fuss. In order to maximize the similarity with our analysis of the Wilkinson divider, let us perform even- and odd-mode analysis on adjacent ports A

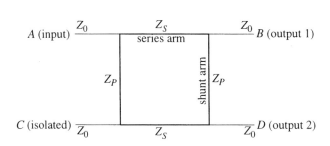

FIGURE 7.37. Stylized branchline hybrid for even- and odd-mode analysis

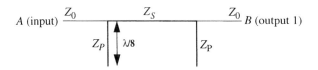

FIGURE 7.38. Equivalent half-circuit for even mode

and C. If we use current source drives, then the superposition of even and odd excitations allows to analyze what happens when we drive only the main input port.

A common-mode drive allows us to bisect the circuit, just as in the Wilkinson case; see Figure 7.38. Because a common-mode excitation guarantees the absence of current flow between the upper and lower half-circuits, bisection results in two open-circuited stubs, each $\lambda/8$ in length. Each stub consequently behaves as a shunt capacitance of admittance jG_P. It is straightforward to use the circuit of Figure 7.38 to show that, if the input impedance is to equal Z_0 when the output is terminated in Z_0, we need to satisfy

$$G_P^2 - G_S^2 = Y_0^2, \qquad (57)$$

where $G_S = 1/Z_S$. Inspection of the corresponding equations for the odd-mode case reveals that the same matching constraint applies to both even- and odd-mode excitations. Producing a match for one component thus automatically produces a match for the other component.

We see that there are infinitely many combinations of arm impedances that provide a match. We need to impose an additional constraint to fix the individual values of the characteristic impedances. That additional constraint is a specification of the ratio of powers to be delivered to the two outputs. A superposition of even- and odd-mode responses allows us to compute the ratio of powers delivered to the output ports:

$$\frac{P_{out1}}{P_{out2}} = \left(\frac{Z_0}{Z_P}\right)^2. \qquad (58)$$

If (as is almost always the case) we desire to split power evenly, then solving for the corresponding impedances yields:

7.6 COMBINERS, SPLITTERS, AND COUPLERS

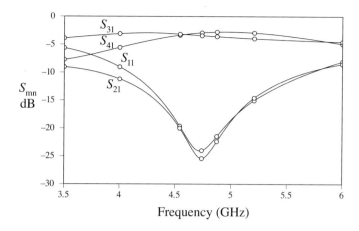

FIGURE 7.39. Simulated performance of a representative 4.7-GHz branchline coupler

$$Z_P = Z_0; \tag{59}$$

$$Z_S = Z_0/\sqrt{2}. \tag{60}$$

These are the impedances shown in Figure 7.36. The equal power split is often emphasized by referring to the coupler as a 3-dB quadrature hybrid.

Although sharp corners are shown in the crude drawing of the branchline coupler, it is common to miter the outer corners of the $Z_0/\sqrt{2}$ arms. One may also use a circular layout instead of a square, although the square version is much more common.

Like its rat-race cousin, the classic branchline hybrid is a relatively narrowband device. As the frequency varies, so does the phase shift. Again, the useful bandwidth depends on one's definition of "useful," but as a general rule, the fractional bandwidth is of the order of 10–15%, just as with the ring hybrid (and for the same reasons). This small value demands high accuracy in modeling and fabrication. This necessity is well illustrated by a simulation (in this case, by *Sonnet Lite* 9.51) of a simple 4.7-GHz branchline coupler. Even though this particular coupler's design uses unmitered corners (as in Figure 7.36) and thus is not optimum, its general performance limits are typical. As can be seen in Figure 7.39, the coupling magnitude quickly deviates significantly from the 3-dB target as the frequency varies more than about 5% away from the design center frequency. At the same time, return loss and isolation degrade rapidly as well.

The asymmetry in the simulated response is due mainly to the lack of any discontinuity correction. As mentioned previously, sharp corners can produce deviations from ideal behavior. Outer corners add shunt capacitance (to first order), and inner corners cause the series and shunt arms of the coupler to possess electrical lengths that are no longer equal to each other. Mitering corrects for the former effect, as we have seen numerous times. A first-order correction for the latter effect may be accomplished

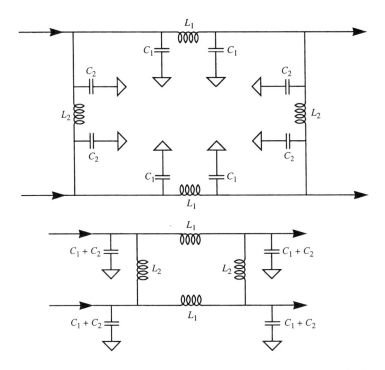

FIGURE 7.40. One possible lumped implementation of a 90° hybrid

by lengthening the shunt arms by an amount d_2, which can be derived from an expression we've already given (as Eqn. 30) but reprise here for convenience:[36]

$$\frac{d_2}{H} = \frac{120\pi}{Z_{series}\sqrt{\varepsilon_{re,ser}}}\left[0.5 - 0.16\frac{Z_{series}}{Z_{shunt}}\left(1 - 2\ln\left[\frac{Z_{series}}{Z_{shunt}}\right]\right)\right]. \quad (61)$$

The required lengthening is typically about $1.3H$ for FR4 branchline couplers.

Just as with earlier examples, one may realize a lumped analogue of the microstrip version by replacing each $\lambda/4$ section with a suitable approximation, as seen in the upper schematic of Figure 7.40. After combining elements, the complete lumped network consists of the four capacitors and four inductors shown in the lower circuit diagram. The component values are given by these familiar relationships:

$$L_1 = Z_0/\sqrt{2}\omega_0, \quad (62)$$

$$C_1 = \sqrt{2}/\omega_0 Z_0, \quad (63)$$

$$L_2 = Z_0/\omega_0, \quad (64)$$

$$C_2 = 1/\omega_0 Z_0. \quad (65)$$

There are other possible lumped implementations, and some of these may have better characteristics. For example, one may alternate high- and low-pass sections

---

[36] See footnote 20.

to improve bandwidth, exploiting their complementary phase change with frequency to provide first-order insensitivity of phase shift to frequency.[37] This property is the same as that exploited in the simple lumped approximation to the ring hybrid presented earlier.

We'll later see that, among the great many uses for a quadrature coupler, an extremely valuable one is in mitigating the problem of poor impedance matching in power amplifiers (see Chapter 20).

7.6.4 DIRECTIONAL COUPLERS

Directional couplers, which may be hybrids, are a subset of power splitters in that they route input power to separate destinations. As they are typically constructed, directional couplers are four-port devices capable of splitting power by prescribed amounts that generally differ considerably from 1:1. One common application for directional couplers is as signal sampling devices. Feedback from the output of a power amplifier, for example, is often provided by a directional coupler. This feedback signal might be used simply as an output power indication, or it may be used to close a negative feedback loop for the reduction of distortion.

Directional couplers are valuable for several reasons. One is that they are able to resolve signals into their separate forward- and reverse-signal components. It is from this latter property that directional couplers get their name. The modern network analyzer depends critically on this ability to make precise measurements of impedance, for example. We will explore this theme more fully in Chapter 8.

Another attribute is the presence of an isolated port when implemented by hybrids. Power supplied to the input port ideally does not couple at all to the isolated port. When used as an *input,* the isolation reduces the interaction between the sources driving the two ports. This isolation is valuable when carrying out two-tone intermodulation tests on receivers, for example, where a lack of isolation between two signal generators might cause one or the other instrument to misbehave and thereby corrupt the measurement.

As with many other examples we have seen, directional couplers may be realized in lumped or distributed form.

Although it is fundamentally a four- (or more) port device, a directional coupler is frequently used as a three-port element. As with all cases, each port, including any unused port (the isolated output, usually), must be terminated in its characteristic impedance. Most commercially available three-port couplers have an internal broadband terminator.

Figures of merit for a directional coupler include coupling factor, isolation, and directivity. The *coupling factor* is defined as the ratio of input power to the power delivered to the coupled (auxiliary) port:

---

[37] See e.g. K. Ali and A. Podell, "A Wide-band GaAs Monolithic Spiral Quadrature Hybrid and Its Circuit Applications," *IEEE J. Solid-State Circuits,* v. 26, no. 10, October 1991.

$$C_F \equiv \left. \frac{P_{IN}}{P_{AUX}} \right|_{forward}. \tag{66}$$

Most sources simply use C to denote the coupling factor, but we want to avoid confusion with the symbol for capacitance and so append the subscript F. However, we will be sloppy and let C_F denote both a power ratio and its decibel version, leaving it to the reader to determine from the context and other clues which one is being discussed at any given time.

Typical values of coupling factor might range from 3 dB up to over 20 dB. A larger coupling factor means that more power is coupled to the main (through) output, not to the auxiliary (coupled) output. The nomenclature can get a bit confusing, because a higher coupling factor corresponds to *less* power coupled to the coupled output.[38] By the same token, the lower the coupling factor, the greater the attenuation in going from the input to the through output. Just to keep you disoriented, some data sheets and textbooks define the coupling factor more rigorously as the reciprocal of Eqn. 66, so that the decibel versions will have negative signs. Though such a definition is more correct (because the fractional power coupled to the coupled output is indeed subunity), convention has dropped the minus sign from common usage.

If the directional coupler is operated in reverse, with power now supplied to the through output with the input terminated, ideally no signal should be measured at the auxiliary output (that's the reason for the "directional" nomenclature). Inevitably, though, some reverse power will leak through to the auxiliary output. A measure of how well the reverse leakage is suppressed is the *isolation factor,* defined as

$$I \equiv \left. \frac{P_{IN}}{P_{AUX}} \right|_{reverse}. \tag{67}$$

Isolation factors of 30–60 dB are not uncommon for commercially available units.

The two quantities are often combined to yield a figure of merit called the *directivity, D*:

$$D \equiv \frac{I}{C_F} = \frac{P_{AUX}|_{forward}}{P_{AUX}|_{reverse}}. \tag{68}$$

Thus directivity is a measure of how well the coupler discriminates between forward and reverse components. We desire an infinite directivity, but all real couplers fall short of the ideal. Microstrip couplers in particular tend to have relatively low directivity (e.g., a common range might be 10–15 dB), owing to a significant difference between odd- and even-mode phase velocities, as we shall explain shortly. For this reason, stripline (or coaxial) geometries are favored over microstrip in commercial couplers.

For completeness, note that the insertion loss of a coupler depends on the degree of coupling. The more power is conveyed to the coupled output, the less remains to

---

[38] This is simply an artifact of eliminating the minus sign from the decibel version.

7.6 COMBINERS, SPLITTERS, AND COUPLERS

Table 7.3. *Ideal insertion loss versus coupling factor*

Coupling factor (dB)	Insertion loss (dB)
3	3
6	1.3
10	0.46
15	0.14
20	0.044

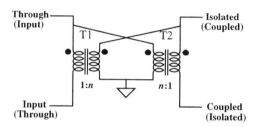

FIGURE 7.41. Broadband directional coupler (transformer version)

appear at the through output. Assuming an otherwise lossless coupler, the insertion loss may be related to the coupling factor as follows:

$$(P_{out} = P_{in} - P_{coupled}) \implies \frac{P_{out}}{P_{in}} = 1 - \frac{P_{coupled}}{P_{in}} = 1 - 10^{-C_F/10}, \quad (69)$$

where the coupling factor C_F is in decibels. Table 7.3 provides a few numbers to clarify the relationship between insertion loss and coupling. For coupling factors exceeding about 10 or 15 dB, the insertion loss in most practical couplers is dominated by parasitic mechanisms (e.g., skin effect or dielectric loss), rather than by the power diverted to the coupled port.

Directional couplers may be implemented in distributed or lumped form. Of the latter class, a widely used broadband implementation is the transformer-based coupler shown in Figure 7.41. As seen in the figure, this coupler offers a choice of port assignments (differentiated by parentheses). As we'll see shortly, the coupled output signal of one choice is inverted relative to the input, and that of the other choice is in phase with the input voltage.

Of the many variations described by Sontheimer and Frederick around 1969, this particular choice has enjoyed enduring popularity because of its good performance and the ease with which it is constructed.[39] Good directivity is routinely obtained

---

[39] See Carl G. Sontheimer and Raymond E. Frederick, "Broadband Directional Coupler," U.S. Patent #3,426,298, granted 4 February 1969.

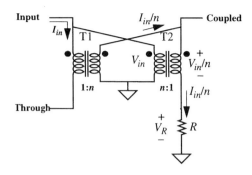

FIGURE 7.42. Broadband 0° directional coupler (approximate analysis)

over two or so decades of frequency. With the addition of a few elements to improve performance at the band edges and with the use of calibration to characterize and compensate algorithmically for the remaining deficiencies, this type of coupler can function over several decades of frequency.

Depending on port assignments, this coupler can provide a coupled output that is either in or out of phase with the main input. In Figure 7.41, the port assignments corresponding to the in-phase version are given in parentheses. We will analyze the in-phase version first. Because an exact analysis would obscure the operational principles, we present a simplified derivation. We'll ultimately provide the results of a complete derivation that yields more accurate design equations.

In Figure 7.42, we assume that the coupling factor is large enough (transformer turns ratio is high enough) that we may neglect the current flowing in the n-turn winding of T2. With this assumption, one may regard this coupler as using one transformer (T1) to sense the current flowing into the input port, and the other (T2) to sense the input voltage. The former statement is only approximately true, but the latter involves no approximations. If we may neglect any current flowing into a load (not shown) connected to the coupled port, then a scaled version of the input current flows through the secondary of T2 and also through the resistive termination. The voltage across the resistor is therefore proportional to this current. If we may arrange for the input port's impedance to equal R, then

$$I_{in} = \frac{V_{in}}{R}, \tag{70}$$

and the voltage dropped across the resistor shown is

$$V_R = \frac{I_{in}}{n} R = \frac{V_{in}}{nR} R = \frac{V_{in}}{n}. \tag{71}$$

Now, the voltage across the secondary of T2 is a scaled version of the input voltage (specifically, it's V_{in}/n). The voltage at the coupled output is therefore simply

$$V_{coupled} = V_R + \frac{V_{in}}{n} = 2\frac{V_{in}}{n}. \tag{72}$$

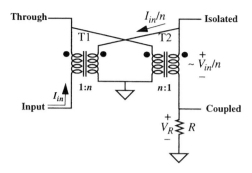

FIGURE 7.43. Broadband 180° directional coupler (approximate analysis)

Note that if the current I_{in} were negative in sign (corresponding to a reverse current, flowing from the through port to the input port), then the voltage developed across the resistor would be negative as well, leading to a cancellation at the coupled port. Thus, we see that the output at the coupled port is preferentially responsive to the forward component and discriminates against the reverse one.

The transformer turns ratio, n, determines the coupling factor. The greater the value of n, the smaller the amount of power fed to the coupled port. When the coupled port is terminated with a resistor of value R, the voltage drops to half the open-circuit value given by Eqn. 72. The coupling factor (in dB) is thus given by

$$C_F \approx 20 \log n. \tag{73}$$

Carrying out an exact analysis isn't difficult, but it is tedious. Because of its minimal intuitive value, we omit the details of that analysis and simply present the results as the following set of equations:

$$V_{thru} = V_{in}\left(\frac{2n^2 - 1}{2n^2 + 1}\right), \tag{74}$$

$$V_{coupled} = V_{in}\left(\frac{2n}{2n^2 + 1}\right), \tag{75}$$

$$V_{iso} = -V_{in}\left(\frac{1}{2n^3 + n}\right). \tag{76}$$

We see that, as the turns ratio increases, the voltage at the through output converges to the input voltage. At the same time, the voltage at the coupled port approaches V_{in}/n, as anticipated. Finally, the signal at the isolated port drops rapidly with increasing n. Also note the inversion at the isolated port.

An alternative set of port assignments yields a coupler that possesses an isolated port and produces a 180° phase inversion between input and coupled ports (see Figure 7.43). In this implementation, current into the input port is sampled exactly by T1. If we are permitted to assume that the coupling factor is large, then the voltages at the through and input ports are nearly equal. Thus, a reasonable approximation of the input voltage appears across the n-turn primary of T2, and a scaled version of that approximation appears across the secondary of T2. Assuming that the current flowing into any load connected to the isolated port may be neglected, then the current flowing through the resistor R is simply I_{in}/n. Consequently

$$V_R = -\frac{I_{in}}{n}R = -\frac{V_{in}}{nR}R = -\frac{V_{in}}{n}, \tag{77}$$

where we have once again assumed that the input port's impedance equals R, so that

$$I_{in} = \frac{V_{in}}{R}. \tag{78}$$

Note that the voltage V_R is equal in magnitude to that developed across the secondary of T2, but opposite in polarity. Thus, these voltages add destructively to produce zero output at the isolated port.

The coupling factor (again, in dB) is approximately

$$C_F \approx 20 \log n. \tag{79}$$

A reverse current produces constructive interference at the isolated port, thus providing discrimination between forward and reverse components of input current.

An exact analysis yields the following expressions for the through, coupled, and isolated port outputs:

$$V_{thru} = V_{in}\left(\frac{4n^4 - 2n^2}{4n^4 + 1}\right), \tag{80}$$

$$V_{coupled} = -V_{in}\left(\frac{4n^3}{4n^4 + 1}\right), \tag{81}$$

$$V_{iso} = -V_{in}\left(\frac{2n}{4n^4 + 1}\right). \tag{82}$$

From these expressions we discern that the approximations used earlier are reasonable. Imperfect cancellations result in not-quite-zero output at the isolated port, but increases in the turns ratio rapidly reduce that port's output to tiny values. Similarly, the magnitude of the voltage at the coupled port so rapidly converges to V_{in}/n as n increases that one may use the approximation (Eqn. 79) for the coupling factor in nearly all practical cases.

Aside from its broadband nature, the transformer-based coupler accommodates a wide range of termination impedances.[40] Because the transformer itself imposes no fundamental limit on the allowable impedances, the same coupler may be used for 50-Ω or 75-Ω systems, for example. The broadband transformer coupler is commonly wound on a *binocular* core (a solid block of magnetic material with two adjacent cylindrical holes), or on two parallel stacks of toroidal ferrite cores, or simply on a pair of toroids.

---

[40] Actually, these two attributes are related here.

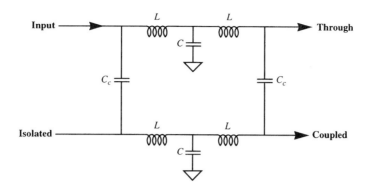

FIGURE 7.44. Lumped narrowband directional coupler

If broadband operation is not essential, then one may use alternative realizations to avoid what many engineers consider to be the inconvenience of winding transformers. One of these employs a pair of lumped quadrature phase shifters (T-network approximations of $\lambda/4$ transmission lines are used in the particular version shown in Figure 7.44), coupled together with capacitors.

To understand how this type of coupler functions, consider driving the input with a test signal (through a resistance equal to Z_0). Ignoring for now the left-hand coupling capacitor of value C_c, most of the input power proceeds directly to the through port, experiencing a quadrature delay in the process. The remainder of the power traversing the top delay line flows downward through the right-hand C_c. Half of *that* power in turn is delivered to the load terminating the coupled output, and the other half flows leftward through the bottom quadrature delay line to the isolated port. Note that this signal has now experienced a total of two quadrature phase shifts since originating from the input port and is therefore 180° out of phase with the input signal. We'll call that the reverse signal.

There is also a direct path to the isolated port from the input, through the left-hand coupling capacitor. The signal flowing through that path is equal in magnitude, but precisely out of phase, with the reverse signal. They therefore cancel when they sum at the isolated port, which explains how that port got its name.

Finally, there is one more path between input and coupled ports. A forward signal may propagate from the input, down through the left-hand coupling capacitor to the bottom quadrature delay line, and finally to the coupled port. This signal adds in phase with that coming from the input through the upper delay line and down through the right-hand coupling capacitor (both signals have experienced a quadrature delay and traversed a coupling capacitor). Thus, the coupled output is nonzero by some prescribed amount controlled by the value of C_c, while the isolated output ideally produces no output at all.

Notice that this analysis implicitly assumes that all ports are properly terminated (otherwise, the power splittings will not be as described).

If we assume that the reactance of the coupling capacitance is much greater than Z_0 within the band of interest (meaning that we limit ourselves to coupling factors no smaller than, say, 15 to 20 dB), then we can offer a simplified design procedure in which the T-section delay line element values are given by familiar relationships. At minimum, it provides a serviceable first pass from which a final design may evolve after a very few iterations. In a great many cases, this procedure yields a fully practical design by itself.

We again choose the L/C ratio according to the desired characteristic impedance:

$$\sqrt{L/C} = Z_0. \tag{83}$$

We then select the LC product as a function of the desired center frequency:

$$1/\sqrt{LC} = \omega_0. \tag{84}$$

The element values are thus chosen so that the reactances equal the characteristic impedance:

$$C = 1/\omega_0 Z_0; \tag{85}$$

$$L = Z_0/\omega_0. \tag{86}$$

To finish the design, choose the capacitance C_c to produce the desired level of coupling. The ratio of coupled to through power is simply the square of the voltage divide factor associated with C_c and Z_0:

$$\frac{(\omega Z_0 C_c)^2}{1 + (\omega Z_0 C_c)^2} = 10^{-C_F/10}, \tag{87}$$

where C_F is the coupling factor expressed in decibels.

In the case of the very weak coupling we are considering, $(\omega Z_0 C_c)^2$ is negligibly small compared to unity. In that regime, solving Eqn. 87 for the coupling capacitance yields the following approximation:

$$C_c \approx (1/\omega Z_0)[10^{-C_F/20}]. \tag{88}$$

It is also straightforward to use Eqn. 85 to relate the coupling capacitance to the delay line capacitance:

$$C_c \approx (1/\omega Z_0)[10^{-C_F/20}] = C[10^{-C_F/20}]. \tag{89}$$

According to these formulas, a 1-GHz, 15-dB, 50-Ω coupler would use 3.18-pF capacitors and 7.96-nH inductors to synthesize the lines, along with coupling capacitances of approximately 0.565 pF. Again, this latter value may be somewhat more approximate than would be implied by (falsely) reporting three significant digits, but it serves as a good starting point for subsequent refinement based on accurate simulations or experimental data. For this particular example, simulations reveal that the coupling factor is extremely close to the design target: about 14.9 dB at 1 GHz (15 dB at 970 MHz). At 1 GHz, the isolated output provides 30 dB of attenuation (33 dB at 970 MHz) relative to the main input. Thus the directivity is just a bit in excess of 15 dB at the design center frequency and rises to 18 dB at 970 MHz. Perhaps it is not too surprising that maximum directivity occurs somewhat below the

7.6 COMBINERS, SPLITTERS, AND COUPLERS

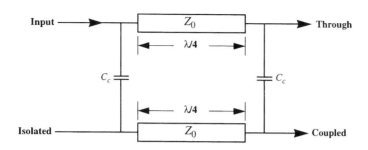

FIGURE 7.45. Microstrip directional coupler with lumped capacitive coupling

design target frequency, for the coupling capacitance adds some capacitive loading (another side effect of which is a tiny reduction in characteristic impedance). If absolutely necessary, compensation for this effect may be provided by reducing the line capacitance. However, the small amounts of correction required generally fall within standard component tolerances, usually rendering such fine adjustments unnecessary in practice. Finally, the input return loss exceeds 15 dB up to 1.24 GHz for this particular design.

A variation on this theme combines true $\lambda/4$ transmission lines with lumped coupling capacitors; see Figure 7.45. This combination is sometimes favored because it eliminates a collection of passive components while retaining the relative ease with which the coupling is varied through suitable adjustment of the coupling capacitances. This configuration is most practical when the frequency is high enough to enable realization of the transmission line in a reasonably small space. For good predictability, the two lines need to be separated from each other by an amount sufficient to assure that essentially all of the coupling is due to the lumped capacitances. Generally, "sufficient separation" corresponds to a line-to-line spacing that is at least several times (e.g., 5×) the dielectric thickness. In practice the separation is frequently constrained by the physical size of available coupling capacitors. The coupling capacitor value is computed using the same formula that applies for the purely lumped implementation (Eqn. 88). The capacitors' parasitics must be kept small enough to ignore or must otherwise be accommodated explicitly in simulations, as with all things microwave. Within those constraints, this particular implementation is capable of excellent performance, as it is least sensitive to the most serious deficiencies of FR4.

The previous coupler design results from replacing some lumped components with distributed ones. We may continue this process to devise a coupler that uses no lumped elements at all. The necessary coupling is provided by a suitable choice of spacing between two microstrip lines, as seen in Figure 7.46.

At first glance, the coupler in this figure looks like a trivial extension of that in Figure 7.45, with the necessary coupling capacitance now provided by lateral proximity. However, note that the isolated and coupled ports have exchanged positions in the figure. To understand why this is not a labeling error, recognize that the coupling between the two lines now could be due to a combined action of electric and

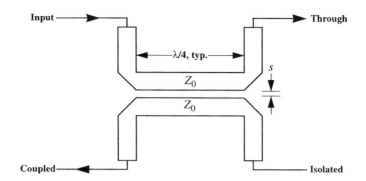

FIGURE 7.46. Classical microstrip directional coupler

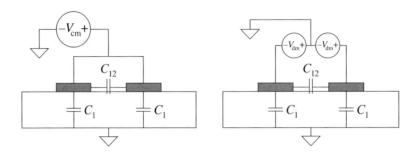

FIGURE 7.47. Determination of even- and odd-mode impedances (not drawn to scale)

magnetic fields. We therefore need to analyze how this coupler functions without preconceptions based on the previous example. To do so, we must first expand upon the concepts of even- and odd-mode excitation introduced earlier.

The concept of a characteristic impedance for a single, isolated microstrip line is straightforward enough. The (approximate) formula for this quantity is similarly simple and familiar:

$$Z_0 = \sqrt{L/C}, \qquad (90)$$

where L and C are the inductance and capacitance per unit length. However, when we bring a second line into proximity with the first, additional degrees of freedom are introduced into the system and hence we can no longer speak of a single characteristic impedance. To take some of the mystery out of that statement, let's consider what happens when we excite a pair of lines with common- and differential-mode signals. For simplicity's sake, let's examine the capacitances that couple each line to ground (and to each other), as shown in the cross-section of Figure 7.47.

For the even mode illustrated in the left-hand figure, the equality of line voltages assures the irrelevance of interline coupling capacitance C_{12}. The common-mode capacitance (per line) is thus smaller than for a single isolated line (Z_0). By convention, we define the *even-mode impedance*, Z_{0e}, as twice the common-mode impedance.

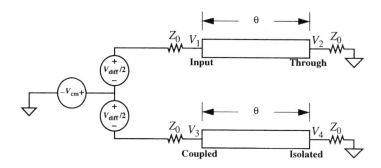

FIGURE 7.48. Port and signal definitions with generalized excitation

This way, Z_{0e} approaches Z_0 as the coupling strength goes to zero. Note that Z_{0e} is higher than Z_0.[41]

Under odd-mode (antisymmetric) voltage excitation, the coupling capacitance no longer has zero voltage across it and therefore now plays a role. If we consider that capacitance to consist of two capacitances of value $2C_{12}$ in series, then the midpoint of those two capacitors is at ground potential. The effective capacitance on each line is therefore larger ($C_1 + 2C_{12}$). If we define the odd-mode impedance Z_{0o} as half the differential impedance, then it approaches Z_0 in the limit of zero coupling. We see that Z_{0o} is less than Z_0.

From the picture, it should be evident that the even- and odd-mode impedances deviate from each other by increasing amounts as the lines are brought closer together. That is, the more closely coupled the lines, the greater the separation between even- and odd-mode impedances. Because of this relationship, these impedances convey the same knowledge as does the coupling coefficient, and design for a specified degree of coupling may be cast alternatively in terms of mode impedances.

Let us now derive explicit relationships among the port voltages as a function of mode impedances. To simplify what is to come, we label the ports and identify signal variables as in Figure 7.48. We wish to discover the port voltage relationships for the specific case when the input line alone is driven through its termination by a source of value V_{in}; the other ports are all grounded through their respective terminations. Note that we may treat the coupled port as *driven* (through a termination) with a voltage source of zero value. As with any pair of voltages, we may decompose these two input signals into the sum of a common- and differential-mode excitation, as shown in the figure. The common-mode value is simply the average of the two input voltages, or just $V_{in}/2$. Similarly, the total differential-mode value is the difference between the two, or V_{in}, half of which is given to each of the two differential sources in the decomposition of Figure 7.48.

---

[41] With currents flowing in the same direction in the two lines, the inductance increases. This increase also contributes a boost to the even-mode impedance. Similarly, the opposing current flows under odd-mode excitation reduce the inductance, contributing to a reduction in odd-mode impedance.

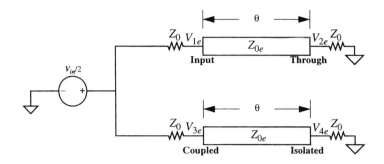

FIGURE 7.49. Coupler with purely common-mode (even-mode) excitation

In order to find the port voltages corresponding to the original problem, we will compute the various node voltages for even- and odd-mode excitations separately, then sum the results to obtain the complete solution. The symmetry of the structure itself simplifies the derivation considerably, as we'll see momentarily.

First, we reprise the equation for the input impedance of a loaded (but lossless) line:

$$Z(\theta) = Z_n \frac{Z_{Ln} + j \tan \theta}{1 + j Z_{Ln} \tan \theta}, \tag{91}$$

where Z_{Ln} is the load impedance, normalized to Z_n, and θ is the effective electrical length of the lines (to accommodate explicitly the possibility of lengths other than $\lambda/4$ or, equivalently, behavior at other than the nominal frequency), expressed as a phase angle. We'll be using this equation quite a bit.

Let us first consider the even mode. In this case, we set the differential voltage to zero, so that both lines are driven (through their termination resistors) by the common-mode voltage. See Figure 7.49. By symmetry, we know immediately that $V_{1e} = V_{3e}$ and $V_{2e} = V_{4e}$. In turn, a voltage divider equation allows us to relate V_{1e} to V_{in}:

$$\frac{V_{1e}}{V_{in}/2} = \frac{Z_{0e} \dfrac{Z_0/Z_{0e} + j \tan \theta}{1 + j(Z_0/Z_{0e}) \tan \theta}}{Z_0 + Z_{0e} \dfrac{Z_0/Z_{0e} + j \tan \theta}{1 + j(Z_0/Z_{0e}) \tan \theta}}$$

$$= \frac{Z_0 + j Z_{0e} \tan \theta}{Z_0(1 + j(Z_0/Z_{0e}) \tan \theta) + (Z_0 + j Z_{0e} \tan \theta)}, \tag{92}$$

which simplifies a bit to

$$\frac{V_{1e}}{V_{in}/2} = \frac{Z_0 + j Z_{0e} \tan \theta}{2 Z_0 + (j \tan \theta)(Z_0^2/Z_{0e} + Z_{0e})}. \tag{93}$$

Now consider the odd-mode behavior, as depicted in Figure 7.50. By (anti)symmetry, we know that $V_{1o} = -V_{3o}$ and $V_{2o} = -V_{4o}$. Again, a simple voltage divider equation allows us to relate V_{1o} to V_{in}:

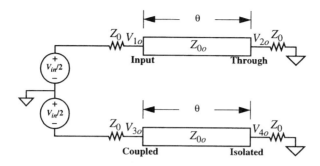

FIGURE 7.50. Coupler with purely differential-mode (odd-mode) excitation

$$\frac{V_{1o}}{V_{in}/2} = \frac{Z_0 + jZ_{0o}\tan\theta}{2Z_0 + (j\tan\theta)(Z_0^2/Z_{0o} + Z_{0o})}. \quad (94)$$

Now we can solve for the voltage at the input of the line, V_1, by summing the even and odd contributions:

$$V_1 = V_{1e} + V_{1o}$$
$$= \frac{V_{in}}{2}\left[\frac{Z_0 + jZ_{0e}\tan\theta}{2Z_0 + (j\tan\theta)(Z_0^2/Z_{0e} + Z_{0e})} + \frac{Z_0 + jZ_{0o}\tan\theta}{2Z_0 + (j\tan\theta)(Z_0^2/Z_{0o} + Z_{0o})}\right]. \quad (95)$$

If the main line is to present a match to the source, the term in brackets must be unity. Imposing this requirement and solving, we ultimately find that

$$Z_0 = \sqrt{Z_{0e}Z_{0o}}. \quad (96)$$

Thanks to symmetry and reciprocity, satisfying Eqn. 96 produces an impedance match at all four ports – not only at the input to the main line. However it is important to note that such an impedance does not properly terminate the even- and odd-mode waves individually. Indeed, we see that the odd-mode impedance is less than Z_0 while the even-mode impedance is greater than Z_0. The mere fact of their inequality implies that it is impossible to provide a perfect match to both components with a single termination. Each component thus suffers a reflection upon reaching the end of the line. It just so happens that the reflections cancel if the impedances satisfy Eqn. 96, leading to a net match overall.

The voltage at the coupled port is just as readily computed, again by summing the even and odd contributions:

$$V_3 = V_{3e} + V_{3o}$$
$$= \frac{V_{in}}{2}\left[\frac{Z_0 + jZ_{0e}\tan\theta}{2Z_0 + (j\tan\theta)(Z_0^2/Z_{0e} + Z_{0e})} - \frac{Z_0 + jZ_{0o}\tan\theta}{2Z_0 + (j\tan\theta)(Z_0^2/Z_{0o} + Z_{0o})}\right]. \quad (97)$$

Simplifying things a bit yields

$$V_3 = \frac{V_{in}}{2}\left[\frac{(j\tan\theta)(Z_{0e}-Z_{0o})}{2Z_0 + (j\tan\theta)(Z_{0e}+Z_{0o})}\right]. \quad (98)$$

Here, the magnitude of the term in brackets is the (subunity) coupling factor, $|V_3/V_1|$. It is a maximum when θ is an odd multiple of $\pi/2$, corresponding to lengths that are odd multiples of $\lambda/4$, which explains the popularity of quarter-wavelength couplers.[42] The maximum value of coupling is readily found from Eqn. 98 as

$$C_F = \frac{Z_{0e}-Z_{0o}}{Z_{0e}+Z_{0o}}. \quad (99)$$

A little rearranging yields

$$\frac{Z_{0e}}{Z_{0o}} = \frac{1+C_F}{1-C_F}, \quad (100)$$

which – when combined with Eqn. 96 – yields:

$$Z_{0e} = Z_0\sqrt{\frac{1+C_F}{1-C_F}}; \quad (101)$$

$$Z_{0o} = Z_0\sqrt{\frac{1-C_F}{1+C_F}}. \quad (102)$$

We thus see that, as asserted earlier, the (normalized) even- and odd-mode impedances indeed convey the same information as the coupling factor. They may therefore be used interchangeably.

With these relationships, we may now write

$$V_3 = \frac{V_{in}}{2}\left[\frac{j\tan\theta\frac{Z_{0e}-Z_{0o}}{Z_{0e}+Z_{0o}}}{\frac{2Z_0}{Z_{0e}+Z_{0o}}+j\tan\theta}\right] = \frac{V_{in}}{2}\left[\frac{(j\tan\theta)C_F}{\frac{2Z_0}{Z_{0e}+Z_{0o}}+j\tan\theta}\right]$$

$$= \frac{V_{in}}{2}\left[\frac{(j\tan\theta)C_F}{\sqrt{1-C_F^2}+j\tan\theta}\right]. \quad (103)$$

The coupling is thus seen to differ little from the maximum value as long as the tangent term overwhelms the other term in the denominator. Thanks to this easily fulfilled requirement, the coupling factor is roughly constant over a relatively broad bandwidth (e.g., an octave), centered about the nominal $\lambda/4$ condition.

Sometimes thinking about a result raises more questions than it answers. Depending on your philosophical bent, you might decide that it's therefore best not to think. That might be the case here. It's certainly true that most published descriptions of this coupler are almost purely mathematical and thus bypass a conundrum: In deriving the expression for the signal developed at the coupled port, we use even- and odd-mode

---

[42] See e.g. B. M. Oliver, "Directional Electromagnetic Couplers," *Proc. IRE*, v. 42, November 1954, pp. 1686–92.

excitations that are precise algebraic inverses of each other. Consequently, we might expect their superposition to result in a near cancellation and thereby always produce a negligible voltage at the coupled port. Yet that isn't what happens. The resolution to this seeming paradox is that, again, the even- and odd-mode components generate reflections upon reaching the end of the line. As with the main line, the two components undergo reflections of opposite polarity (and of identical magnitude). Unlike the main line, there is already an inverse relationship between the excitations to begin with (leading to a broadband destructive interference at the isolated port). Consequently, the two polarity reversals produce reflections that add constructively upon arriving at the coupled port. The more tightly coupled the lines, the greater the difference between the mode impedances and the stronger these reflections. Because the coupled output is thus generated by the action of reflected waves, this device is sometimes called a *backward-wave coupler*. We'll soon see why we have spent this much time on a verbal explanation of how this coupler operates.

We can proceed in like fashion to discover that the voltage at the isolated port is ideally zero. This result is independent of frequency, at least in principle, for it arises from an exact cancellation of even- and odd-mode waves. For microstrip in particular, however, isolation is imperfect in practice because the required miraculous cancellations by destructive interference do not quite materialize, for reasons that will be discussed shortly.

Finally, a little additional work shows that the signal at the through port is given by

$$\frac{V_2}{V_1} = \frac{\sqrt{1-C_F^2}}{\sqrt{1-C_F^2}\cos\theta + j\sin\theta}. \tag{104}$$

In the limit of very weak coupling, Eqn. 104 is well approximated by

$$\frac{V_2}{V_1} = \frac{\sqrt{1-C_F^2}}{\sqrt{1-C_F^2}\cos\theta + j\sin\theta} \approx \frac{1}{\cos\theta + j\sin\theta} = e^{-j\theta}, \tag{105}$$

reflecting the fact that – in the weak-coupling regime – nearly all of the power is transmitted to the through port, experiencing a phase lag of θ in the process.

We've derived the coupling factor as a function of mode impedances and electrical line length. To carry out an actual design, though, we need to relate those quantities to layout dimensions. Regrettably, there do not seem to be any simple, generally applicable formulas for microstrip that yield accurate values for the necessary spacing as a function of desired coupling.[43] The simplest ones of any useful accuracy are

---

[43] Quite accurate, but complicated, analytical expressions are offered by M. Kirschning and R. H. Jansen in "Accurate Wide-Range Design Equations for the Frequency-Dependent Characteristics of Coupled Microstrip Lines," *IEEE Trans. Microwave Theory and Tech.*, v. 32, no. 1, January 1984, pp. 83–90. These equations yield answers that are generally in close agreement (e.g., within a couple of percent) with the results of field-solver simulations and experimental measurements. The equations accommodate dispersion, loss, unequal mode velocities, and other practical effects. The commercial simulator *APLAC* is an example of one that uses these equations.

perhaps the equations of Akhtarzad et al., which unfortunately must be solved implicitly for the width and spacing of the lines:[44]

$$\frac{W_e}{H} = \frac{2}{\pi} \cosh^{-1}\left(\frac{2d - g + 1}{g + 1}\right); \qquad (106)$$

$$\frac{W_o}{H} = \begin{cases} \frac{2}{\pi} \cosh^{-1}\left(\frac{2d - g - 1}{g - 1}\right) + \frac{4}{\pi(1 + \varepsilon_r/2)} \cosh^{-1}\left(1 + 2\frac{W/H}{S/H}\right) \\ \qquad \qquad \qquad \qquad \qquad \qquad \qquad \text{if } \varepsilon_r < 6, \quad (107) \\ \frac{2}{\pi} \cosh^{-1}\left(\frac{2d - g - 1}{g - 1}\right) + \frac{1}{\pi} \cosh^{-1}\left(1 + 2\frac{W/H}{S/H}\right) \quad \text{if } \varepsilon_r \geq 6. \quad (108) \end{cases}$$

For these equations, note that:

$$g = \cosh\left(\frac{\pi S}{2H}\right); \qquad (109)$$

$$d = \cosh\left[\pi\left(\frac{W}{H} + \frac{S}{2H}\right)\right]. \qquad (110)$$

The ratios W_e/H and W_o/H are those of single isolated microstrip lines, whose characteristic impedances are $Z_{0e}/2$ and $Z_{0o}/2$ (respectively) and thus may be found using the equations presented earlier for ordinary microstrip lines. After completing that step, Akhtarzad's equations are solved iteratively for the actual line width and spacing for the coupled lines.

The foregoing equations are the ones used by the simulator *Puff* for its ideal coupled-line simulations. The difference in mode velocities (and in conductor and substrate losses) are not comprehended in these equations. Fortunately, *Puff* also offers a more comprehensive simulation that does take these effects into account. Thus, one could perform an initial design with the equations of Akhtarzad et al. to find some initial line dimensions, then evaluate the design with the more accurate simulations. Because the ideal equations are good enough to generate a reasonable first-pass design, usually only one or two iterations are required to converge on an acceptable final design.

If we accept the need for iteration then it may be acceptable for the initial design to be somewhat inexact. If so, then perhaps the complexity of the foregoing equations can be avoided altogether. If we neglect the second term in Akhtarzad's equations for the odd-mode width/height ratios, then we may obtain a closed-form approximation for the spacing/height ratio:

---

[44] Sina Akhtarzad, Thomas R. Rowbotham, and Peter B. Johns, "The Design of Coupled Microstrip Lines," *IEEE Trans. Microwave Theory and Tech.*, v. 23, June 1975, pp. 486–92. Also see R. Garg and I. J. Bahl, "Characteristics of Coupled Microstriplines," *IEEE Trans. Microwave Theory and Tech.*, v. 27, June 1979, pp. 700–5, with corrections in v. 28, March 1980, p. 272.

$$\frac{S}{H} \approx \frac{2}{\pi}\cosh^{-1}\left\{\frac{\cosh\left(\frac{\pi}{2}\frac{W_o}{H}\right) + \cosh\left(\frac{\pi}{2}\frac{W_e}{H}\right) - 2}{\cosh\left(\frac{\pi}{2}\frac{W_o}{H}\right) - \cosh\left(\frac{\pi}{2}\frac{W_e}{H}\right)}\right\}. \qquad (111)$$

Once we have found S/H, the next step is finding W/H with the aid of Eqn. 106 or Eqn. 107. The design is completed by specifying the length of the coupled section (e.g., as $\lambda/4$). Again, because the odd- and even-mode velocities differ, it's traditional to select the length using the average velocity of the two modes.

Having a closed-form equation for S/H certainly simplifies the procedure enough for design to proceed rapidly with the aid of a spreadsheet, for instance, but the procedure is still not quite simple. We therefore offer the following alternative, quasi-empirical formula that applies for loosely coupled (10-dB or more) lines for 50-Ω systems. It's simple enough to require only a few keystrokes on an ordinary hand calculator and is valuable for computing rapid, crude estimates:[45]

$$\frac{S}{H} \approx 1.11 \ln \frac{0.32}{1 - \sqrt{\frac{1-C_F}{1+C_F}}}, \qquad (112)$$

where the coupling factor C_F is a pure ratio here, not the decibel equivalent. For very small values of C_F, the formula simplifies to the following approximation:

$$\frac{S}{H} \approx 1.11 \ln \frac{0.32}{C_F}. \qquad (113)$$

These formulas result from combining the analytical relationship between odd-mode impedance and coupling factor,

$$\frac{Z_{0o}}{Z_0} = \sqrt{\frac{1-C_F}{1+C_F}}, \qquad (114)$$

together with an empirical formula for the normalized odd-mode impedance as a function of S/H:

$$\frac{Z_{0o}}{Z_0} \approx 1 - 0.32 \exp\left(-1.11\frac{S}{H}\right). \qquad (115)$$

Finally, we assume that – in the weak-coupling regime – the lines are of the same width as uncoupled microstrip lines. In practice, the width needs to be reduced as coupling gets tighter, but the amount of narrowing is small for weakly coupled lines. We consequently neglect any width adjustments.

Our crude formula (Eqn. 112) is satisfactory only for noncritical designs or for generating a reasonable starting point for further refinement. The formula misbehaves for coupling values tighter than about 10 dB, as it has been optimized for the

---

[45] This formula is based largely on comparisons between simulations from *RFSim99* for 1.6-mm FR4 and the results of *Sonnet Lite* field-solver simulations.

Table 7.4. *Comparison of empirical formula and RFSim99* (FR4, $H = 1.6$ mm, $\varepsilon = 4.6$)

C_F, dB (target)	S, mm (Puff)	W, mm (Puff)	S, mm (formula)	C_F, dB (formula)	C_F, dB (RFSim99)
10	0.21	2.49	0.24	10.9	9.2
15	0.76	2.88	1.183	13.6	13.9
20	1.70	3.02	2.15	18.2	18.9
25	3.39	3.10	3.10	26.5	25.8
30	9.73	3.19	4.14	57.5	41.7

range most commonly encountered in practice for edge-coupled structures (10–30-dB coupling factors). Table 7.4 compares the predictions of Eqn. 112 with values as computed by the simulation tools *RFSim99* and *Puff* over that range of coupling factors.

The second and third columns of the table show the spacing and width as computed by *Puff*, which uses the Akhtarzad equations. The spacings necessary to achieve the coupling factors of column 1 as computed by the crude formula (Eqn. 112) are given in the fourth column. As a comparison, the last two columns summarize the coupling factors computed by the crude formula and by *RFSim99* for the spacing (and width, for *RFSim99*) given by *Puff*. The algorithm used by *RFSim99* is undocumented; we present its results simply to get a rough idea of how well our crude formula does relative to those used in available tools. As can be seen, the crude formula agrees reasonably well both with *Puff* and with *RFSim99* over a 10–25-dB coupling range. All three disagree rather significantly for the 30-dB coupling case. Fortunately, such a weak value of coupling is rarely desired. Note also the very small spacing required to achieve a 10-dB coupling factor. Achieving still tighter coupling is thus effectively precluded from practical implementation with this structure. Finally, note that *Puff*'s computation of linewidth shows only a small variation for coupling factors of 15 to 30 dB. Thus, it may suffice to use lines whose widths are computed as for an isolated microstrip line, at least for initial designs.

When built in microstrip form as shown, such a coupler invariably exhibits relatively low directivity. The reason is that the velocities of the even and odd modes are not the same, thanks to the inhomogeneity of microstrip. To understand how this inhomogeneity produces unequal mode velocities, contrast the capacitive fringing field for a common-mode excitation with that for a differential-mode excitation. For the odd mode, fringing is much stronger than for the even mode. Because there is substantially more line-to-line fringing, there is more fringing field above the substrate than for the even mode. The effective dielectric constant is correspondingly lower for the odd mode, and the phase velocity is higher. Since the phase velocities for the two modes differ, the line length can only be $\lambda/4$ for at most one of those modes. It is customary to compute a line length that is $\lambda/4$ for the average of the two modes. The

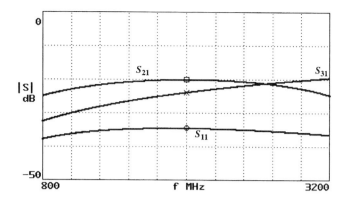

FIGURE 7.51. Results of *Puff* simulation of a 2-GHz, 20-dB coupled-line coupler (1.6-mm FR4)

inequality of mode velocities implies imperfect phase shift. This, in turn, degrades isolation and therefore directivity in particular. For this reason, commercial couplers are almost never built in microstrip, using instead stripline or coaxial geometries.

The results of a *Puff* simulation of a 20-dB coupler (Figure 7.51) underscore the extent to which unequal mode velocities can degrade isolation for microstrip. Instead of infinite isolation (and therefore infinite directivity), the signal at the isolated port is not much below that at the coupled port. The directivity barely exceeds a pathetic 4 dB here, whereas a typical stripline implementation could be expected to provide ~30-dB directivity. This particular microstrip example provides such poor performance, in fact, that the output power at the isolated port actually *exceeds* that at the coupled port above about 2.6 GHz. A related phenomenon is the lack of quadrature between the signal at the isolated port and the input. The isolated port's output lags the input by nearly 180 degrees at the nominal center frequency. On the other hand, neither the coupled output nor the input match suffers significant degradation, as expected. The coupling bandwidth remains about an octave, for example.

From our qualitative argument we know more than simply that the odd-mode phase velocity is too fast; we also know why. Consequently, we can propose some methods for reducing the velocity disparity. Of the several methods that have been devised over the years, the most popular at moderate frequencies is simply to add some capacitance across the gap. Capacitances added there do not affect the even-mode velocity (in theory), so they represent a simple means by which one may equalize the two phase velocities. One way to realize the necessary capacitance is with serpentine or meander-type structures (e.g., sawtooth-like boundaries) to increase the line-to-line edge capacitance. This method works best for tightly coupled lines, but the lack of any simple descriptive equations makes designing such compensated couplers a decidedly nontrivial task. It's easier to use discrete capacitors, placed at the ends of an ordinary coupled line. Adjustment to maximize directivity is often best effected by symmetrically sliding a pair of fixed capacitors along the line, rather than by using variable capacitors at the ends; see Figure 7.52. With care, use of these capacitors

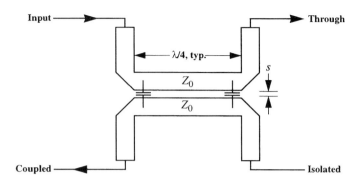

FIGURE 7.52. Compensated microstrip directional coupler

can improve directivity by ~10 dB, typically. At high frequencies, however, it may be difficult to find or fabricate capacitors of sufficient quality and of the right physical size. Also, avoiding parasitics in the mere act of connecting the capacitors to the lines becomes progressively more difficult as frequencies increase.

At frequencies where adding capacitors is impractical, one may deposit dielectric materials directly over the line (effectively making the structure appear more stripline-like) or bring a grounded shield plate down over the lines (again, making the structure appear more stripline-like). These approaches work well, but they considerably increase manufacturing complexity, offsetting the chief advantage of microstrip construction.

A completely different approach to improving directivity is to avoid a dependency on two different modes altogether. By eliminating a reliance on interference to produce the coupled and isolated outputs, the bandwidth over which high directivity is obtained can be extended considerably.

A structure with these attributes is the *forward-wave* or *codirectional* coupler.[46] It superficially resembles the conventional backward-wave coupler, but it suppresses the generation of backward waves by tapering the end sections to provide a good broadband termination for both even- and odd-mode components simultaneously.[47] When reflections from both modes are suppressed, there are no longer two components to interfere constructively at the port nearest the input to create the coupled signal. That port consequently now becomes the isolated output, and what was formerly the isolated output becomes the coupled output.

Despite its attributes, the codirectional coupler is rarely used primarily because there are no simple design equations. For the most part, the codirectional coupler has therefore remained an academic curiosity. Its days as an example of microwave

---

[46] Pertti K. Ikäläinen and George L. Matthaei, "Wide-band, Forward-Coupling Microstrip Hybrids with High Directivity," *IEEE Trans. Microwave Theory and Tech.*, v. 35, August 1987, pp. 719–25.

[47] The even- and odd-mode impedances must not differ too much if the tapered lines are to provide a good match to both modes. Hence, the coupling must be relatively weak in order for the codirectional coupler to provide high directivity.

7.6 COMBINERS, SPLITTERS, AND COUPLERS

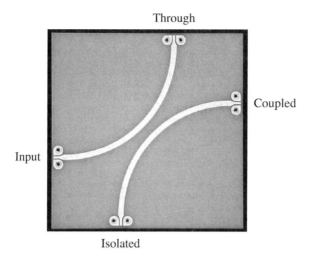

FIGURE 7.53. Codirectional arc coupler [*courtesy of Matthew Morgan and Sander Weinreb*]

exotica may finally be drawing to an end, however, as a simplified and elegant form of the codirectional coupler has been reported recently.[48] As seen in Figure 7.53, this version of a forward-wave coupler simply consists of two quarter-circle arcs, separated by a gap. The two design degrees of freedom represented by the arc radius and minimum arc-to-arc gap constrains the design space enough to permit reasonably rapid convergence to an acceptable design with the aid of an electromagnetic field solver. A 12-dB codirectional coupler designed in this way exhibits a 20-dB directivity over an octave spanning 50 GHz to 110 GHz.[49] It would be difficult to obtain this level of performance with a standard uncompensated backward-wave coupler in standard microstrip.

Making accurate measurements at these frequencies is another challenge. One difficulty is simply providing good terminations. The best are on chip, to avoid the inevitable degradations that attend transitions. However, even on-chip terminations can have poor performance at millimeter-wave frequencies, owing to discontinuities associated with vias, for example. To solve these problems, one may use a *disk terminator* made out of material whose sheet resistivity is 50 Ω per square (see Figure 7.54). Thanks to its symmetry, any moding that arises in the disk still excites only a square's worth of material, keeping the resistance constant. The diameter of the disk determines the *lower* frequency limit, which should be set roughly equal to at least $\lambda/4$ at the lowest frequency of interest. The upper frequency limit is set by

---

[48] Matthew Morgan and Sander Weinreb, "Octave-Bandwidth High-Directivity Microstrip Codirectional Couplers," *MTT-S Intl. Microwave Symposium,* Philadelphia, 2003. These types of couplers can practically provide only relatively loose coupling values because it is extremely difficult to suppress the backward wave under tight coupling.

[49] Ibid.

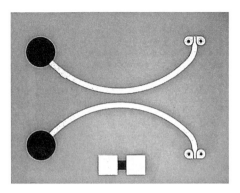

FIGURE 7.54. Illustration of disk terminators [*courtesy of Matthew Morgan and Sander Weinreb*]

the onset of moding in the feedline itself or by the excitation of asymmetrical modes owing to unwanted (but ever-present) imperfections.

Neither the codirectional nor ordinary backward-wave coupler is suitable for the production of very tight coupling, however.[50] As we've seen in the case of the quadrature hybrid, *equal* power splitting is often precisely what is needed, but 3-dB coupling is practically unattainable with ordinary edge-coupled lines, even with substrates featuring a high dielectric constant. In some cases, the version shown in Figure 7.45 can provide somewhat higher couplings without requiring heroic fabrication techniques and thus might be preferable (for manual prototyping, if nothing else).[51] Naturally, such an option only makes sense at frequencies that are amenable to the use (and mounting) of available discrete capacitors – and for coupling that is still not too tight.

Noting that a single coupled pair makes use of only a single edge per line, one might wonder if using additional lines might be an alternative method for obtaining greater coupling for a given interline spacing. For example, one might propose the use of two *pairs* of lines in an interdigitated arrangement, as shown in Figure 7.55.

The two pairs of coupled lines possess three pairs of coupled edges, instead of just one. Consequently, a given interline spacing produces much tighter coupling. Or, a larger spacing may be used to produce a given degree of coupling, relaxing dimensional tolerances. Clearly, this process of interdigitation may be continued indefinitely in principle. However, the law of diminishing returns – along with the practical difficulties of fabricating ever-narrower lines (made necessary by the parallel connections) and implementing the connections implied by the flying bondwires in Figure 7.55 – drives most engineers to use just two pairs, although more are used on occasion.

---

[50] If additional conductor layers are available then one may place one line above the other; such area-coupled (*broadside*-coupled) lines can have much tighter coupling than edge-coupled ones. This theme will recur when we examine various ways to implement filters.

[51] This statement presumes that the operational frequencies are low enough that one may safely neglect the parasitics of the discrete components.

7.6 COMBINERS, SPLITTERS, AND COUPLERS

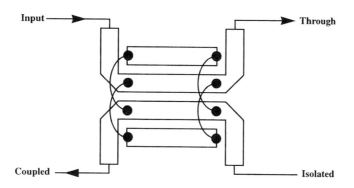

FIGURE 7.55. Coupler with additional lines for tighter coupling (linewidths not to scale)

The solution of the coupling problem frequently forces us to solve another. If we assume that the reason for seeking tighter coupling is to make practical the implementation of a 3-dB coupler, then it is not much of a leap to assume that it is important to preserve symmetry to avoid degrading the equality of outputs that a 3-dB coupler provides. Regrettably, the through and coupled outputs emerge from diagonally opposed corners instead of from the same side. Consequently, it is nigh impossible to avoid some degradation as a result of this fundamental lack of output symmetry.

In 1969 an engineer at Texas Instruments, Julius Lange, described an ingenious solution to both the coupling and symmetry problems.[52] His coupler design augments interdigitation with a clever splitting of one of the lines into two pieces. This way, a simple rearrangement allows both the through and coupled outputs to emerge from the same side, enabling a symmetrical feed to whatever follows the coupler itself. In its simplest form, the Lange coupler appears as shown in Figure 7.56. If additional layers of metal are available, they may be used instead of the bondwires shown (in conjunction with vias). In all cases, one must take care to ensure that the parasitics of the interconnect don't degrade performance.

As mentioned before, the use of paralleled lines implies that the individual lines themselves must possess characteristic impedances that are higher than the termination impedances. The coupled lines will thus be narrower than the port feedlines (despite the equal relative dimensions shown in Figure 7.56). More typically, an actual layout might appear as shown in Figure 7.57. From the typical layout, one may better appreciate the fabrication complexities involved (particularly if more fingers are used), as well as the challenge of making a good transition between the coupled sections and the much wider feedlines that connect to the four ports.

Aside from those considerations, the design of a Lange coupler involves somewhat more complex analytical formulas. Even with that additional complexity, final

---

[52] J. Lange, "Interdigitated Stripline Quadrature Hybrid," *IEEE Trans. Microwave Theory and Tech.*, v. 20, December 1969, pp. 1150–1.

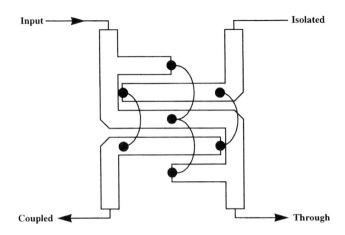

FIGURE 7.56. Lange coupler (linewidths not to scale)

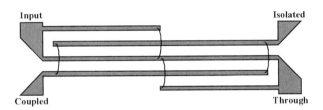

FIGURE 7.57. More representative layout of a microstrip Lange coupler (bondwire version)

refinement with the aid of electromagnetic field solvers is still almost always necessary. With that disclaimer out of the way, we offer a sequence of equations that may be used to generate a credible first-pass design.[53]

First, from Ou[54] we have

$$Z_0^2 = \frac{Z_{0e}Z_{0o}(Z_{0e} + Z_{0o})^2}{[Z_{0e} + (k-1)Z_{0o}][Z_{0o} + (k-1)Z_{0e}]} \quad (116)$$

and

$$C = \frac{(k-1)Z_{0e}^2 - (k-1)Z_{0o}^2}{(k-1)(Z_{0e}^2 + Z_{0o}^2) + 2Z_{0e}Z_{0o}}, \quad (117)$$

---

[53] A coherent presentation of this procedure is found in T. C. Edwards and M. B. Steer, *Foundations of Interconnect and Microstrip Design*, 3rd ed., Wiley, New York, 2000, and also in V. Fusco, *Microwave Circuits*, Prentice-Hall, Reading, MA, 1987.

[54] W. P. Ou, "Design Equations for Interdigitated Directional Coupler," *IEEE Trans. Microwave Theory and Tech.*, v. 26, October 1978, pp. 801–5.

where k is the number of fingers. Here C is the direct coupling factor, not the dB version.

Osmani then combined Ou's derivations to yield a series of equations you would actually use for designing a Lange coupler.[55] First define a factor q as

$$q = [C^2 + (1 - C^2)(k - 1)^2]; \qquad (118)$$

then compute the odd-mode impedance from

$$Z_{0o} = Z_0 \sqrt{\frac{1-C}{1+C} \frac{(k-1)(1+q)}{(C+q) + (k-1)(1-C)}}. \qquad (119)$$

The even-mode impedance is computed next, using

$$Z_{0e} = Z_{0o} \frac{C+q}{(k-1)(1-C)}. \qquad (120)$$

Finally, use the method of Akhtarzad to determine actual conductor dimensions and spacings to complete the design.

Like the branchline coupler, the 3-dB Lange coupler may be used as a quadrature combiner or splitter. One practical limitation is that the coupling lines can get rather narrow, to say nothing of the difficulty of implementing the connections suggested by the bondwires shown in Figure 7.56 (even if vias in conjunction with an additional metal layer are used instead). A compensating advantage is that the Lange coupler operates over a much broader bandwidth than the 10% or 15% bandwidth provided by a branchline hybrid. As with ordinary coupled-line couplers, the useful bandwidth of a Lange coupler typically exceeds an octave.[56] This attribute explains the Lange's popularity in spite of the fabrication challenges.

The following simulations (by *Sonnet Lite* 9.51) of a 12-GHz coupler highlight the performance of a conventional Lange coupler in greater detail. As can be seen in Figure 7.58, the coupling is close to 3 dB over a frequency range that extends from below 8 GHz to above 16 GHz. Over that same frequency range, the input return loss exceeds 18 dB.

In light of all this attention given to tight coupling, it's easy to get the false impression that achieving maximum coupling is always an overriding goal. It's therefore important to note that there are many instances – such as SWR measurement, or sampling an amplifier's output to measure power or to close a feedback loop – that do not always call for maximum coupling. Consequently one should feel free to consider a conventional backward-wave coupler with lines that are shorter than $\lambda/4$, especially where space is an important engineering consideration.

---

[55] R. M. Osmani, "Synthesis of Lange Couplers," *IEEE Trans. Microwave Theory and Tech.*, v. 29, February 1981, pp. 168–70.

[56] This statement applies to the coupling magnitude, not to maintenance of quadrature.

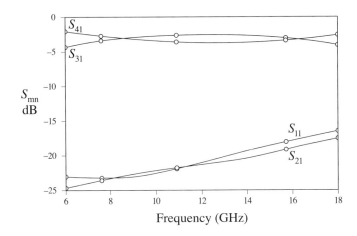

FIGURE 7.58. Simulated performance of a 12-GHz Lange coupler

FIGURE 7.59. Multisection coupler (symmetrical version shown; relative dimensions not to scale)

7.6.5 BROADBAND AND SIZE REDUCTION TECHNIQUES

Many of the couplers we've studied make extensive use of line segments whose lengths are expressed as some fraction of a wavelength, generally implying narrowband operation. This property is sometimes an attribute instead of a limitation owing to the incidental filtering of out-of-band noise and interference that might be provided as a by-product of frequency-dependent coupling. However, there are also many instances (particularly in instrumentation) where broadband operation is desirable. It is therefore valuable to consider methods for extending the frequency range over which these couplers may operate.

The most common broadbanding technique – use more sections – is simple in concept but somewhat difficult in execution. Just as in the stepped-impedance transformer, broader band operation is enabled by cascading sections, each of which carries a smaller burden for the overall performance. See Figure 7.59. This idea may be applied to the coupled line to provide good coupling and directivity over decade bandwidths.

Consider, as we did with the classical single-section backward-wave coupler, a wave flowing from the input port to the through port. That wave induces a backward wave in a coupled line. In this specific instance, it induces a succession of backward waves as it traverses the various coupled sections. For example, a wave traveling along the zeroth section induces a backward wave in the coupled line with a particular coupling coefficient C_0, then another with a coupling coefficient C_1 upon entering the next section, and so on. As all of the induced waves flow backward toward the coupled port, they superpose to yield the overall coupled signal.

Now we've already seen that the backward-wave coupler actually depends on reflections for operation. Weak coupling produces small differences in mode impedances, and hence it generates small reflections. If we additionally assume that cascading such weak couplers does not appreciably alter this general property of weak reflections, we may again invoke the small-reflection approximation and thereby estimate the overall coupling as a simple weighted sum of delayed couplings.

We begin by generating an approximation for the coupling factor that is valid in this limit of weak coupling. Starting with Eqn. 103, we may develop the following expression:

$$\frac{V_3}{V_1} = \frac{(j\tan\theta)C_F}{\sqrt{1-C_F^2}+j\tan\theta}$$

$$\approx \frac{(j\tan\theta)C_F}{1+j\tan\theta} = \frac{(j\sin\theta)C_F}{\cos\theta+j\sin\theta} = [(j\sin\theta)C_F]e^{-j\theta}; \qquad (121)$$

here, as before, C_F is the maximum coupling. Note that the only approximation made stems from the weak coupling assumption (C_F small compared to unity). Aside from that approximation, Eqn. 121 remains valid over an arbitrarily large bandwidth (again, in principle).

Next, note that the contribution of the Nth section to the coupled output is delayed by $2N\theta$, because there is a delay of $N\theta$ incurred while the forward wave travels to the Nth section, and another $N\theta$ delay during which the induced backward wave travels back to the coupled port. Summing all of these backward waves yields

$$\frac{V_3}{V_1} \approx j\sin\theta[C_0 e^{-j\theta} + C_1 e^{-j(\theta+2\theta)} + \cdots + C_N e^{-j(\theta+2N\theta)}], \qquad (122)$$

where the total number of coupled sections is $N+1$.

Recognizing that the summed term in the brackets of Eqn. 122 is of the form of a Fourier series, we can see that (as in the other examples in which we've invoked the small-reflection approximation) the coefficients C_n may be chosen to produce an array of useful behaviors. Here, it is the coupling that may be made maximally flat or equiripple, for example.[57]

The transitions between sections are generally mitered, rather than abrupt as shown in Figure 7.59. As with the single-section coupler, it is common to implement each

---

[57] There is naturally an extremely close connection between this structure and the multisection impedance transformer. For a good summary, see Inder Bahl and Prakash Bhartia, *Microwave Solid State Circuit Design*, Wiley, New York, 1988. Also see Matthaei, Young, and Jones, *Microwave Filters, Impedance Matching Networks, and Coupling Structures*, reprinted by Artech House, Dedham, MA, 1980. Known as *MYJ* or "the (big) black book," this classic volume comprehensively describes the state of the art of its title subject as of the mid-1960s.

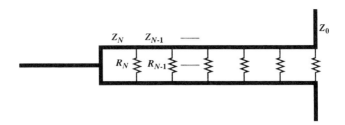

FIGURE 7.60. Multisection Wilkinson divider (stylized)

section as nominally $\lambda/4$ in length (measured at the center of the frequency band). And as with the single-section coupler, unequal even- and odd-mode phase velocities (as well as dispersion) degrade the multisection coupler's performance when implemented in microstrip. Compensation for these effects (e.g., with lumped capacitors, as described earlier) is essential if reasonable broadband performance is to be achieved with microstrip implementations.

The Wilkinson divider can be made to operate over similarly broad frequency spans through cascading, also achieving decade or more bandwidths instead of the octave-range values commonly provided by a single stage.[58] It may be viewed as a combination of a Wilkinson divider with impedance transformers. As with the multisection impedance transformer, the small-reflection approximation allows us to understand that the greater the number of sections, the greater the bandwidth.

In the stylized circuit of Figure 7.60, each section is $\lambda/4$ in length. In practical layouts, the arms are almost never straight and parallel segments as shown in the stylized figure. Instead, the arms are generally implemented as semicircular loops. At least two considerations motivate this choice. One is that such a geometry improves the aspect ratio, especially in the case of many sections. The other is that parallel lines are also coupled lines, and the coupling may perturb the operation of the divider.

In the case of the branchline coupler, the underlying idea is that coupling resonant systems together results in the creation of resonances above and below the original, uncoupled resonances. A proper choice of initial resonant frequencies and coupling strength can lead to broader band operation. In fact, this mechanism is responsible for the surprisingly large bandwidth of the Lange coupler. Even though it is most familiar to engineers in the context of amplifier and filter design, the same insights apply to the coupling problem. For example, one may cascade branchline couplers (Figure 7.61) to broaden considerably the frequency range over which they function well. Regrettably, there are no simple formulas for the general design of such branchline couplers, only algorithms.[59]

The arms are again $\lambda/4$ in length at the center frequency of operation, and the distribution of line impedances and choice of number of sections determines both the

---

[58] Seymour B. Cohn, "A Class of Broadband 3-Port TEM Hybrids," *IEEE Trans. Microwave Theory and Tech.*, v. 16, February 1968, pp. 110–18.

[59] O. Maraguchi et al., "Optimum Design of 3-dB Branchline Coupler Using Microstrip Lines," *IEEE Trans. Microwave Theory and Tech.*, v. 31, August 1983, pp. 674–8.

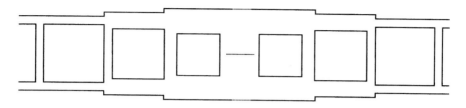

FIGURE 7.61. Multisection branchline coupler

FIGURE 7.62. Schiffman "alligator clip" reduced-size broadband coupler

bandwidth and the passband ripple. As with the single branchline coupler and many other examples we've presented, one may replace each $\lambda/4$ arm with a π-section *CLC* network to realize a lumped version of the broadband multisection quadrature hybrid.

One trade-off produced by cascading is an increase in size. To reduce the cost of that trade-off, one may use structures such as shown in Figure 7.62.[60] The idea is to compress the length of the coupled sections through the use of zigzag (or serpentine or meander) geometries. The effective electrical length lies somewhere *between* the horizontal dimension occupied by the line and the total length along the teeth of the zigzag edges. As one might expect, no analytical formulas exist for the precise design of such couplers. However, if one begins with a three-section stepped-impedance prototype (whose design itself is nontrivial, but not impossible), then an initial zigzag design is achievable using a simple heuristic procedure.[61] For the segment with the closest spacing, choose the section length so that the total zigzag length is the same as that of the prototype. For the third segment (the one with the widest spacing), assume that the edge-to-edge spacing is measured between midpoints of the teeth. Then, simply choose the line spacing and length reduction of the middle section as the geometric means of the first and third section values. Simulate and refine as necessary to converge on an acceptable design. Over a 1–8-GHz frequency range, a

---

[60] As reported by S. Uysal, *Nonuniform Line Microstrip Directional Couplers and Filters*, Artech House, Dedham, MA, 1993.

[61] See Dana Brady, "The Design, Fabrication and Measurement of Microstrip Filter and Coupler Circuits," *High Frequency Electronics*, July 2002, pp. 22–30.

prototype design exhibits a ~10-dB directivity, a coupling of 19 dB with 1.5-dB ripple, and a worst-case return loss of 16 dB.[62] Although this performance level by itself is not quite up to commercial standards, the design represents a credible first pass from which a suitable coupler might emerge after several iterations.

Finally, it's important to note that many of these solutions (interdigitation, meander lines, etc.) are necessitated to a large extent by an insistence on simple planar realizations. If additional levels of metal are available, then broadside-coupled (rather than edge-coupled) structures will provide tight coupling much more simply. Indeed, significant size reductions may then be enabled by using broadside-coupled spirals, for example.

7.7 SUMMARY

This chapter has presented the basic characteristics of microstrip transmission lines, along with numerous approximate equations and rules of thumb for estimating circuit constants and parasitics. Many of these approximations facilitate the characterization and mitigation of reactive discontinuities formed at bends and other transitions between segments of line, or between connectors and lines. We considered the use of suitably designed segments of transmission line as capacitors, inductors, and resonators, and we presented an extended discussion of couplers, splitters, and combiners.

7.8 APPENDIX A: RANDOM USEFUL INDUCTANCE FORMULAS

We have already presented some formulas for inductance in a previous chapter in a piecemeal manner. We now present a number of additional formulas for commonly encountered geometries. In all that follows, the equations strictly apply only at DC unless stated otherwise. At high frequencies, inductance drops somewhat because the shrinking of skin depth causes the contribution of internal flux to diminish. Fortunately, internal flux generally accounts for only a small percentage (e.g., below 5%) of the total inductance at DC, so its reduction at high frequency does not cause dramatic changes in the overall inductance. Nonetheless, it is worthwhile avoiding unpleasant surprises by knowing explicitly what assumptions have gone into the derivations.

In the sections that follow here, transmission line effects are neglected, so the formulas apply only when dimensions are short compared with a wavelength.

7.8.1 FLAT SHEETS AND ROUND WIRES

In Chapter 6 we presented a formula for the inductance of a current sheet. It is repeated here so that all the inductance formulas are in one place for easy reference:

---

[62] Brady, ibid.

7.8 APPENDIX A: RANDOM USEFUL INDUCTANCE FORMULAS

$$L_{sheet} \approx \frac{\mu_0 l}{2\pi}\left[0.5\ln\left(\frac{2l}{w}\right) + \frac{w}{3l}\right] = (2 \times 10^{-7})l\left[0.5\ln\left(\frac{2l}{w}\right) + \frac{w}{3l}\right]. \quad (123)$$

See Section 6.5.2 for details.

The DC inductance of a round wire is given by[63]

$$L \approx \frac{\mu_0 l}{2\pi}\left[\ln\left(\frac{2l}{r}\right) - 0.75\right] = (2 \times 10^{-7})l\left[\ln\left(\frac{2l}{r}\right) - 0.75\right]. \quad (124)$$

7.8.2 SINGLE LOOP

A useful approximation for a circular loop is given by:

$$L \approx \mu_0 \pi r. \quad (125)$$

This formula tells us that a loop with radius of 1 mm has an inductance of about 4 nH.

Better accuracy is provided by the following equation, which takes into account a nonzero wire diameter:[64]

$$L \approx \mu_0 r[\ln(8r/a) - 2], \quad (126)$$

where a is the radius of the wire. Given Eqn. 126, we see that Eqn. 125 strictly holds only for an r/a ratio of about 20.

Making a crude approximation even more so, we could extend Eqn. 125 to noncircular cases by arguing that all loops with equal area have about the same inductance, irrespective of shape. Then we could write

$$L \approx \mu_0\sqrt{\pi A}, \quad (127)$$

where A denotes area of the loop. See Section 6.5.2 for details.

7.8.3 PLANAR SPIRALS

In keeping with the planar world view that dominates this text, we now consider a popular geometry for realizing small-valued inductances in a PC board (or IC) context. Planar spirals with circular, octagonal, hexagonal, and square shapes have all been used. The inductance and Q-values attainable are very much second-order functions of shape (despite much lore to the contrary), so engineers should feel free to use their favorite shape with impunity. A square spiral is the simplest to lay out and is thus the overwhelming favorite of lazy engineers (a set of which the author is a proud member).

---

[63] *The ARRL Handbook,* American Radio Relay League, 1992, pp. 2–18. The proximity of conducting planes may be ignored as long as they are located a distance away that is equal to one or two lengths, at minimum.

[64] Ramo, Whinnery, and Van Duzer, *Fields and Waves in Modern Radio,* Wiley, New York, 1965, p. 311.

Table 7.5. *Coefficients for inductance formula*

Shape	c_1	c_2	c_3	c_4
Square	1.27	2.07	0.18	0.13
Hexagon	1.09	2.23	0.00	0.17
Octagon	1.07	2.29	0.00	0.19
Circle	1.00	2.46	0.00	0.20

The formulas for all of these shapes can be cast in a unified form as

$$L = \frac{\mu n^2 d_{avg} c_1}{2}\left[\ln\left(\frac{c_2}{\rho}\right) + c_3\rho + c_4\rho^2\right], \quad (128)$$

where n is the number of turns, d_{avg} is the average of the inner and outer diameters, and ρ is a *fill factor*, defined as

$$\rho \equiv \frac{d_{out} - d_{in}}{d_{out} + d_{in}}. \quad (129)$$

From this last equation, you can see why the term "fill factor" is appropriate: ρ approaches unity as the inductor windings fill the entire space, and it approaches zero as the inductor becomes more and more hollow.

The various c_n coefficients are a function of geometry and are given in Table 7.5 for four representative shapes.[65] To an excellent approximation, the coefficient c_1 is the area for a given outer dimension, normalized to the area of the largest circle that can be inscribed within the layout. The factor c_2 is the primary term, while c_3 and c_4 may be considered first- and second-order correction factors, respectively. When all four factors are used, the equations are typically accurate to within a couple of percent (and almost never in error by more than 5%), thus generally obviating the need for a full electromagnetic field solver to evaluate the inductance of such structures.

On those rare occasions where other regular polygons are of interest, one may use the following analytical formula:

$$L \approx \frac{\mu n^2 d_{avg} A_{out}}{\pi d_{out}^2}\left[\ln\left(\frac{2.46 - 1.56/N}{\rho}\right) + \left(0.20 - \frac{1.12}{N^2}\right)\rho^2\right], \quad (130)$$

where A_{out} is the outer area and N is the number of sides of the polygon. This formula is simply a restatement of Eqn. 128 with analytical approximations used for the coefficients c_1, c_2, and c_4. The coefficient c_3 is set to zero, which is a good approximation for all regular polygons with more than four sides. This analytical formula is only one or two percent more inaccurate than the tabulated one.

The Q of a planar spiral inductor may be estimated roughly by using the skin-effect formula to compute an approximation of the effective resistance. This formula isn't

---

[65] S. S. Mohan et al., "Simple Accurate Inductance Formulas," *IEEE J. Solid-State Circuits*, February 2000.

quite adequate because it neglects the influence of a given turn's field on the current distribution in adjacent turns. Hence, one should expect the estimate to be rather crude at best.

Generally speaking, somewhat hollow inductors have the highest Q because the innermost turns tend not to contribute much magnetic flux but do contribute significant resistance. Hence, removing them is a good idea in general. Although there is no simple rule for what is optimum in all cases, a reasonable guideline is to have a 3:1 ratio between the outer and inner diameters. Fortunately, the optimum conditions are relatively flat, so this guideline is satisfactory for most practical cases.

In addition to series resistance, one is also generally interested in the self-resonant frequency. The self-capacitance gives rise to this resonance, which is due mainly to the overlap between the line that makes connection to the center of the spiral and the rest of the turns of the inductor. That overlap capacitance can be estimated from a simple parallel-plate formula. The turn-to-turn capacitance is usually negligible because the individual terms all appear in series.

7.9 APPENDIX B: DERIVATION OF FRINGING CORRECTION

DANGER, WILL ROBINSON – INTEGRALS, CHEESE AND A BREEZE AHEAD!

A rigorous calculation of fringing capacitance is rather difficult. When precise answers are needed, or if the geometry is complex, often the best practical choice is to employ numerical methods. Unfortunately, such an approach often obscures design insight. As a complement to those valuable numerical approaches, we offer here an analytical expression whose inaccuracy perhaps can be forgiven in view of its simplicity and near universality. And although its derivation may not exactly fit on a cocktail napkin, the final result certainly does (as is made clear by its incorporation into Eqn. 12).

Field theorists have devised many ingenious strategies for accomplishing what we seek. The approach we'll take is inspired by one of the many wonderful chapters in Feynman's *Lectures on Physics,* in particular, "The Principle of Least Action."[66] There, Feynman points out that powerful minimum principles can frame novel and elegant solutions to old problems. For example, if you were to forget the current divider law for two parallel resistors, you could derive it using the principle that currents will distribute themselves in a way that minimizes the total power dissipation. Any other current distribution would result in a higher total dissipation (try it!). Satisfying the same minimum principle is the voltage divider law.

Similarly, if our task is to deduce the electric field between two conductors, it is valuable to know that charges will distribute themselves to minimize the total energy stored in the system – and also to remember that a unique potential distribution is linked to the charge distribution. Because (for a given voltage) energy is proportional

---

[66] Volume II, Chapter 19 (Addison-Wesley, Reading, MA, 1964).

to capacitance, we may infer from this minimum principle that the correct potential distribution is the one among all possible distributions that minimizes the computed capacitance.[67] We use this observation by proposing a "reasonable" functional form for the potential distribution, computing the capacitance it implies, and then choosing parameters (if any) to minimize that capacitance. Feynman's minimum principle then says that we will have generated the best possible approximation to the truth for *that* particular guess (even if it is wrong). Furthermore, we will know that our approximation error will always be positive (since our approximate formula will necessarily overestimate the true capacitance).

To start, we equate two different formulas for the energy stored in a capacitor:

$$\frac{1}{2}CV_0^2 = \frac{1}{2}\varepsilon \int_{\text{Vol}} |\nabla V|^2 \, d\,\text{Vol}, \tag{131}$$

where ∇V is the gradient of the potential V (recall that the electric field is equal to minus this gradient). The term on the left comes from ordinary circuit theory, and that on the right is from field theory.

Next (and this is the tricky part), guess a "reasonable" form for the potential. To aid in guessing, first look at our structure, which is shown in Figure 7.63. The electric field lines are idealized as perfectly vertical until the very end of the line is reached, and then as progressively curving outward more and more until they are perfectly circular at a distance H beyond the end. Along the radial line shown in the figure – and at an angle θ with respect to the ground plane – assume that the potential increases in some fashion as the radius r increases from 0 to H. Further assume (rather fancifully) that, at a given r, the potential increases linearly from 0 as the angle θ varies from 0 to $\pi/2$. Assume also that negligible energy is stored in the electric field for $r > H$. This latter assumption avoids an embarrassing prediction of potentials in excess of the applied voltage, V_0, as the radius approaches infinity. It also causes us to underestimate the energy stored. This error is at least in the right direction to offset the systematic overestimation inherent in the method when used with any incorrect potential distribution (although there is the possibility of overcompensation). Finally, assume that the plates are infinitesimally thin.

Given these assumptions (and they are just that), we may postulate an approximate potential function of the form

$$\tilde{V}(r,\theta) = V_0 \left(\frac{r}{H}\right)^k \left(\frac{2\theta}{\pi}\right), \tag{132}$$

where k is some parameter whose value is to be determined later. The tilde denotes that it is a postulated – and approximate – potential. You can verify that this equation satisfies the conditions stated previously (but not necessarily *all* relevant boundary conditions; if it did then it would have to be the correct solution). Note that we neglect variations in the z-direction (i.e., out of the plane of the page).

---

[67] The same minimum principle can be used to derive formulas for inductance.

7.9 APPENDIX B: DERIVATION OF FRINGING CORRECTION

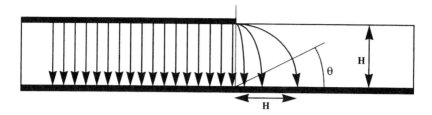

FIGURE 7.63. Very approximate field distribution for fringing capacitance estimation (side view)

With that potential function in hand, the rest is just plugging and chugging:

$$\tilde{C} = \frac{4\varepsilon}{\pi^2 H^{2k}} \int_{\text{Vol}} |\nabla \tilde{V}|^2 \, d\,\text{Vol} = \frac{4\varepsilon W}{\pi^2 H^{2k}} \int_{\text{Area}} |\nabla \tilde{V}|^2 r \, dr \, d\theta; \tag{133}$$

$$\frac{\nabla \tilde{V}}{\tilde{V}} = \frac{k}{r} \mathbf{r} + \frac{1}{r\theta} \boldsymbol{\theta} \quad \left(\text{or } \nabla \tilde{V} = \frac{\tilde{V}}{r}\left(k\mathbf{r} + \frac{1}{\theta}\boldsymbol{\theta}\right)\right), \tag{134}$$

where $\mathbf{r}$ and $\boldsymbol{\theta}$ are the r- and θ-directed unit vectors, respectively. After combining these equations and evaluating the double integral as r ranges from 0 to H and as θ varies from 0 to $\pi/2$, we obtain

$$\frac{\tilde{C}}{W} = \frac{\pi \varepsilon k}{12} + \frac{\varepsilon}{\pi k}, \tag{135}$$

where W is the width of the line.

Now, we want to select k to minimize the estimated capacitance. Setting the first derivative of Eqn. 135 to zero yields

$$k = \sqrt{12}/\pi, \tag{136}$$

which – given the approximate nature of this entire endeavor – may be treated as essentially unity (meaning that we could have just started with $k = 1$ and ended up with pretty much the same answer).

Substitution of this value of k into Eqn. 135 finally gives us an approximate equation for the per-width fringing capacitance:

$$\frac{\tilde{C}}{W} = \frac{\varepsilon}{\sqrt{3}}. \tag{137}$$

Note that this capacitance is *independent* of H. To the extent that the approximations leading to its derivation are valid, it is therefore a *universal* fringing correction whose value is very roughly 5 fF/mm. That is, any open structure will contribute approximately this capacitance per length of edge, at least to cocktail-napkin accuracy.[68] The

---

[68] This statement is true for a finite-length conductor over an infinite ground plane. For two equal-size plates, the universal correction is precisely half the value, or about 2.5 fF/mm. But the length extension remains $H/\sqrt{3}$ (or, rounding, $H/2$) per edge. Also remember that we have ignored field variations in the z-direction, so the correction becomes increasingly dubious as W/H diminishes.

Table 7.6. *Circular parallel-plate capacitance*

H/D	"Exact" correction factor	Cheesy correction factor (using $H/2$)	Residual cheese error (%)	Cheesy correction factor (using $H/\sqrt{3}$)	Residual cheese error (%)
0.005	1.023	1.010	1.2	1.012	1.1
0.01	1.042	1.020	2.1	1.023	1.6
0.025	1.094	1.051	4.0	1.059	3.2
0.05	1.167	1.102	5.5	1.119	4.1
0.10	1.286	1.210	5.9	1.244	3.3

foregoing assumes a vacuum dielectric, but we'll pretend that the correction remains valid even for microstrip.

Because capacitance is proportional to the ratio of effective area WL_{eff} to plate spacing H, the fringing capacitance is equivalent to an ideal fringing-free parallel-plate capacitor whose dimensions are W by $H/\sqrt{3}$.

But wait, you say: Equation 24 contains $H/2$, not $H/\sqrt{3}$. Here's how we get $H/2$: First, we know that our proposed functional form is wrong (consider its behavior at large radii). Thus, by the minimum principle we know that our estimate is probably too high ($\sqrt{3}$ is too low) if the field for $r > H$ were truly negligible. But by how *much* is our estimate high? We don't know (if we did, we could remove the error altogether). But under the assumption that the estimate isn't too horribly wrong, we arbitrarily round the denominator upward a little bit to get to the closest convenient number, 2. Cheesy? You bet. Can we evaluate the extent of cheesiness? Consider Table 7.6, which presents correction factors for the capacitance between circular parallel plates of diameter D and spacing H. We define a correction factor as the ratio of actual (or estimated) capacitance to the value given by the simple fringing-free undergraduate physics formula. Values in the second column are obtained from numerical field solutions, and in the third from assuming that the capacitors act as if we extended the radius by an amount $H/2$ (so that the effective diameter is $D + H$).

As expected, the correction factors are very close to unity for small spacings, so all three formulas yield answers that differ negligibly from each other. As H/D ratios grow, however, the fringing-free parallel-plate formula underestimates the true capacitance by increasing amounts. The true capacitance is nearly 30% larger than the value computed by the fringing-free formula when the H/D ratio is 0.1. Application of the cheesy correction factor results in a residual error that is under 6% at that same spacing. This close tracking is encouraging, because the correction factor was derived for a rectangular structure but was applied successfully to a circular one. Hence, our assertion of a universal fringing correction seems not so unreasonable.

As a final comment, if we use the $H/\sqrt{3}$ factor actually derived instead of the $H/2$ arbitrarily substituted for it, the error improves a little bit in the particular case

7.10 APPENDIX C: DIELECTRIC CONSTANTS OF OTHER MATERIALS

Table 7.7. *Other dielectrics*

Material	ε_r	Material	ε_r
Al_2O_3 (96%) (alumina)	9.5	Mica	5.4
Al_2O_3 (99.5%)	9.8	Mylar	3
AlN	8.7	Paper	2.7 typ.
$BaTiO3$	~600	Plexiglas	3.45
BeO (99.5%); toxic!	6.6	Polyethylene	2.25
Diamond	5.5	Polystyrene	2.55
Fused silica	3.82	PTFE (Teflon)	2.1
GaAs	13	TiO_2	~100
Ge	16	RT/duroid	2.5
Glass (borosilicate)	4.8	Si	11.7

Source: Primarily *AppCAD* (Agilent Technologies).

of Table 7.6, as can be seen in the last two columns.[69] At a normalized spacing of 0.1, the correction factor becomes 1.24; this reduces the error to almost a full order of magnitude below the fringing-free estimate, to a little bit above 3%. So, which to use? Fortunately, the contribution by fringing constitutes a second-order correction to first-order formulas, so small errors in those corrections result in very small overall errors. The choice of whether to use $H/2$ or $H/\sqrt{3}$ (or some other value) is thus not of critical import, and the selection can be made on the basis of other criteria. The slothful author uses the simpler value of $H/2$ most of the time, since it minimizes another kind of energy – his own.

7.10 APPENDIX C: DIELECTRIC CONSTANTS OF OTHER MATERIALS

It's important not to allow this book's focus on FR4 to convey the impression that other materials are not used in microwave circuits. To the contrary; FR4 is rarely used in "serious" microwave work. Its primary virtue is its low cost, making it increasingly popular for high-volume consumer applications. This book's choice of FR4 as a tutorial medium is an acknowledgment of this growing popularity as well as a response to the general absence of information on FR4 in most of the microwave literature. When seeking the best performance, however, other dielectric materials are strongly favored over FR4. In Table 7.7 we provide an incomplete sampling of additional dielectric materials. Not all of them are necessarily good dielectrics for microwave work, but they may be frequently encountered.

---

[69] A value of $2H/3$ is even better for this particular data set, with almost no error at $H/D = 0.1$.

CHAPTER EIGHT

IMPEDANCE MEASUREMENT

8.1 INTRODUCTION

Both time- and frequency-domain characterizations provide comprehensive information about a system. The latter require the ability to generate and measure sinusoidal voltages and currents over a broad frequency range. The network analyzer, in either scalar or vector incarnations, is an example of such an instrument.

An alternative is to use time-domain methods to characterize a system. The principal tool of this type is the time-domain reflectometer (TDR), which is in essence a miniature radar system. The TDR launches a pulse ("the main bang") into the device under test and then observes any echoes. The timing of a reflection with respect to the main bang indicates the location of a discontinuity, and the shape of the reflected pulse conveys important information about its nature. With a reflectometer, then, one can quickly locate and characterize both resistive and reactive discontinuities (and evaluate their fixes). A network analyzer can also provide this information but requires considerably more labor to do so.

We consider both network analyzers and TDRs in this chapter, beginning with the latter.

8.2 THE TIME-DOMAIN REFLECTOMETER

There are two primary applications of TDRs: finding and characterizing impedance discontinuities. These capabilities translate directly into the ability to correct defects and evaluate the quality of any compensation performed.

8.2.1 LOCATING DISCONTINUITIES

As shown in Figure 8.1, a TDR consists of just two main modules: a pulse generator and an oscilloscope. A pulse generator applies a fast risetime step to the device under test (DUT). A portion of the signal is tapped off and fed to an oscilloscope, whose sweep is synchronized with the step. The synchronizing signal is timed to allow the display of the voltage both a bit before the rising edge and well after.

8.2 THE TIME-DOMAIN REFLECTOMETER

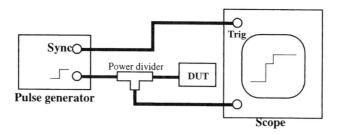

FIGURE 8.1. Time-domain reflectometer

A pulse's risetime determines its spectral content and, hence, the bandwidth over which the TDR can perform a useful characterization. Similarly, the oscilloscope's bandwidth must be consistent with the desired characterization bandwidth. A common rule of thumb is that the -3-dB bandwidth of a step is related inversely to the 10–90% risetime as follows:

$$f_{-3\text{dB}} t_r \approx 0.35. \qquad (1)$$

This relationship, although strictly correct only for single-pole systems, allows us to estimate the performance requirements of a TDR system. For example, suppose we wish to characterize a transmission line up to 10 GHz. Using our rule of thumb, we find that the TDR's risetime must be shorter than about 35 ps. The fastest commercially available TDRs are capable of characterizing systems beyond 50 GHz, implying risetimes of under 7 ps.[1]

The risetime of the incident pulse determines not only the bandwidth over which the system is characterized but also the spatial resolution of the characterization. If a pulse reflects off of a discontinuity some distance x_d from the source, then the total time taken in the round trip back to the source is

$$t_{prop} = \frac{2x_d}{v_{prop}}, \qquad (2)$$

so that

$$x_d = \frac{t_{prop} v_{prop}}{2}, \qquad (3)$$

where v_{prop} is the propagation velocity. Clearly, if the pulse's risetime is too slow then reflections will be obscured during the rising edge. Roughly speaking, the spatial resolution is approximately equal to the distance traveled during the risetime. The first reflectometers were developed to locate faults in very long cables, where the ability to pin down the location of an open or short to within 100 meters or so suffices. Given that the speed of light along a typical cable is about 60–80% of the free-space value, the corresponding delay is about 4 ns per meter. Risetimes in the range of hundreds

---

[1] Here we are excluding systems that employ cryogenics and superconductors. Laboratory demonstrations of pulse generators with sub-picosecond risetimes suggest that bandwidth will continue to increase.

of nanoseconds – implying bandwidths in the low-megahertz range – can therefore be satisfactory for such cable fault–finding applications. The far faster sub–10-ps risetimes cited earlier for today's leading-edge gear correspond to the ability to locate discontinuities to a resolution of a few millimeters in free space. Such risetimes and their corresponding spatial resolutions are much more compatible with the size of typical microwave circuit elements and modules.

It is not necessary to know the propagation velocity to locate a discontinuity, despite the seeming implications of Eqn. 3. With microstrip, for example, just run a finger along the line while observing the TDR trace. When the bump produced by your finger coincides with the bump produced by the discontinuity you're trying to investigate, you've found it: the discontinuity will be right underneath your finger. (Of course this method should not be used if the TDR pulse is of an unusually high power!)

8.2.2 CHARACTERIZING DISCONTINUITIES

One reason that the TDR is so valuable is that it conveys much more information than merely the location of discontinuities. This is most directly understood from the relationship between the reflection coefficient and the termination impedance:

$$\Gamma = \frac{Z_{Ln} - 1}{Z_{Ln} + 1}, \tag{4}$$

where Z_{Ln} is the normalized load impedance,

$$Z_{Ln} \equiv \frac{Z_L}{Z_0}. \tag{5}$$

Note that the reflection coefficient is a complex quantity in general, possessing both a magnitude and phase (or real and imaginary part). It thus contains information about how the spectral components of the step response are modified in reflecting off of the discontinuity. Note also that Γ contains similarly complete information about the load impedance (Eqn. 4 may be solved for Z_{Ln} in terms of Γ). Adding the assumption of linearity allows us to bring to bear on the problem all of the powerful tools of linear system theory. In particular, finding the step response is a moldy staple of system theory, and that is precisely what the TDR displays. Even though we'll start with a formal mathematical approach, we'll quickly examine a few representative cases to extract physical insight on how one might guess the correct answer for these and many other cases of practical relevance.

The response to any input is the sum of the input excitation as well as any reflection that arises. The reflection is merely Γ times the incident signal. Hence the transfer function that relates the total output to the input is

$$H(s) = 1 + \Gamma = \left(\frac{2Z_{Ln}}{Z_{Ln} + 1}\right). \tag{6}$$

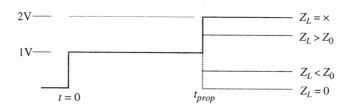

FIGURE 8.2. Idealized TDR trace for open, shorted, and resistive loads

FIGURE 8.3. Idealized TDR trace of capacitively terminated transmission line

When using this equation, it's important to keep track of the fact that the inverse Laplace transform of Eqn. 6 is valid only for times that are greater than the round-trip time of flight,

$$t_{prop} = \frac{2x_d}{v_{prop}}. \tag{7}$$

Before this time, the response is just the value of the input alone (e.g., one volt if we have assumed a unit step excitation).

Using these relationships, it is straightforward to determine the TDR traces for several commonly encountered cases. For example, consider open- and short-circuit loads. In those two cases, the normalized load impedances are infinite and zero, respectively, with corresponding values of 2 and 0 for $H(s)$. Keeping in mind that these values apply only after the time-of-flight delay, the unit step responses thus appear as shown in Figure 8.2. For resistive loads in between these two extremes, the step response will jump to some level between zero and 2 V. If the load resistance is less than the characteristic impedance, the final value will be below 1 V. If greater, the final value will lie between 1 V and 2 V; and if equal to Z_0, no discontinuity will be observed.

Now consider the step response when reactive loads terminate a line. If the load element is a capacitance, then

$$\frac{2Z_{Ln}}{Z_{Ln}+1} = \frac{2}{1+1/Z_{Ln}} = \frac{2}{1+sZ_0C}. \tag{8}$$

This is simply the transfer function of a single-pole low-pass filter, whose step response should be familiar; see Figure 8.3. In like manner, the step response for any number of discontinuities can be readily determined. Without providing detailed derivations (which are left as a pleasant exercise for the reader), Figure 8.4 presents a

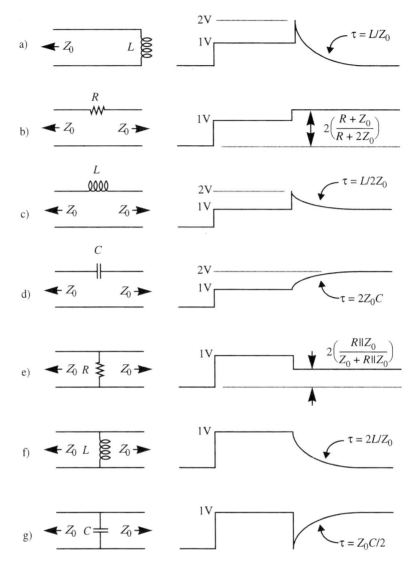

FIGURE 8.4. Idealized TDR traces for several discontinuities (incident amplitude = 1 V in all cases)

short catalog of simple yet practically relevant discontinuities and their corresponding TDR traces.

The shapes of the TDR traces can be anticipated from purely physical arguments with a minimum of mathematics. In all of the reactive examples, there is only one time constant because we have considered only single-reactance loads. A single time constant implies a single exponential factor. An inductive termination (case a) appears initially as an open circuit, but ultimately acts as a short. The time constant of the exponential transition between these two conditions is the ratio of inductance to the effective resistance it sees (here, Z_0). In case c, the inductance sees a total resistance of $2Z_0$ (one Z_0 each to the left and right), and the final value is 1 V.

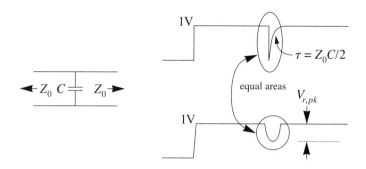

FIGURE 8.5. Ideal and more realistic TDR traces for shunt capacitance

In case b, that of a series resistive discontinuity, the step response must jump up because the effective load resistance is the resistance as viewed from the discontinuity to the right. Here, that is the sum of R and Z_0. A simple voltage divider equation yields the result shown in the figure (just remember that the open-circuit step amplitude is 2 V).

Arguments similar to the foregoing can be used to sketch the TDR traces for the rest of the examples given in Figure 8.4.

In practice, the observed TDR traces will differ somewhat from the idealized ones shown in the figure. The main difference is due to the finite risetime of the step excitation. If one considers a practical step to be the result of low-pass filtering an infinitely fast one, then the actual TDR traces may be deduced by low-pass filtering the ideal traces through a filter whose step response has the same risetime as that of the actual step. This filter will slow down rising edges and cause a rounding of sharp corners.

8.2.3 PARAMETER EXTRACTION

Using our catalog of TDR traces, it is often possible to measure small inductances and capacitances – or even to extract a more complex circuit model – from a measured step response. Doing so requires that we consider explicitly how the limited bandwidth of all real systems affects the shape of the waveform. As a specific example, consider the shunt capacitive discontinuity described by Figure 8.5.

The reflection coefficient is

$$\Gamma = \frac{Z_{Ln} - 1}{Z_{Ln} + 1} = \frac{1/(sCZ_0 + 1) - 1}{1/(sCZ_0 + 1) + 1} = \frac{-sCZ_0}{sCZ_0 + 2}, \quad (9)$$

and Γ is seen to have a pole at a frequency given by

$$\omega = 2/CZ_0. \quad (10)$$

Spectral components above this pole frequency are attenuated by the low-pass filter that is effectively formed with the capacitor. This filtering is the reason for the change in shape shown in Figure 8.5. The sharp edge gets smeared out, resulting in the smooth bump shown in the bottom trace. Despite the rounding, the areas of the

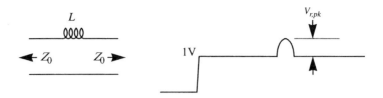

FIGURE 8.6. Somewhat more realistic TDR trace for series inductance

bumps are the same (we will exploit this invariance later in an alternative measurement method). If the capacitive discontinuity is small enough that this filtering effect may be neglected, then the reflection coefficient may be approximated by

$$\Gamma \approx \frac{-sCZ_0}{2}. \tag{11}$$

The incident and reflected signals thus may be related approximately by a derivative:

$$V_r = \Gamma V_i \approx \frac{-CZ_0}{2}\frac{dV_i}{dt}. \tag{12}$$

The peak value of V_r is proportional to the peak value of the input slope, so

$$C \approx \frac{-2V_r}{Z_0}\left(\frac{dV_i}{dt}\right)^{-1} \approx \frac{-2V_{r,pk}}{Z_0}\left(\frac{V_i}{\tau}\right)^{-1}, \tag{13}$$

where $V_{r,pk}$ is as shown in Figure 8.5, V_i is the amplitude of the input step, and τ is the time constant of the input step. (We have used the fact that, for a single-pole system, the maximum slope of the step response is simply the amplitude of the step, divided by the time constant.) The 10–90% risetime of a step is approximately equal to 2.2τ, so we could also write

$$C \approx \frac{|V_{r,peak}|}{1.1Z_0}\left(\frac{t_{rise}}{V_i}\right). \tag{14}$$

An analogous derivation for the case of a series inductive discontinuity yields the following estimate:

$$L \approx 2Z_0V_r\left(\frac{dV_i}{dt}\right)^{-1} \approx \frac{Z_0V_{r,pk}}{1.1}\left(\frac{t_{rise}}{V_i}\right), \tag{15}$$

where a typical waveform is as shown in Figure 8.6. This method can provide remarkable measurement resolution. Suppose, for example, that a given TDR system possessed the ability to resolve a voltage as small as 1 mV along with a 15-ps–risetime, 1-V step. The smallest measurable capacitance and inductance would be about 0.3 fF and 0.7 pH! Needless to say, it would be exceedingly difficult to make measurements of such small values using other methods. From this calculation, it is clear that even relatively insensitive and slow TDR systems are capable of impressive measurements of inductance and capacitance.

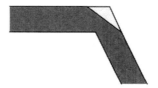

FIGURE 8.7. Mitered bend

An Alternative Measurement Approach

The foregoing method is based on derivatives. Because various imperfections along a line can distort the waveshapes considerably, it is usually difficult to apply the method in as straightforward a fashion as implied. In such cases, a measurement based on an integral formulation will almost always yield more satisfactory results.

Beginning again with

$$V_r = \Gamma V_i \approx \frac{-CZ_0}{2} \frac{dV_i}{dt}, \qquad (16)$$

we can say that

$$\int V_r \, dt = \frac{-CZ_0}{2} \implies C = -\frac{2}{Z_0} \int V_r \, dt. \qquad (17)$$

Similarly,

$$\int V_r \, dt = \frac{L}{2Z_0} \implies L = 2Z_0 \int V_r \, dt. \qquad (18)$$

Estimating the area of a bump seems to be considerably easier for most people than estimating slopes and the like. As a consequence, the area method is preferable whenever the bumps are significantly rounded.

8.2.4 COMPENSATION

By identifying the location and type of discontinuity, the TDR enables you to design compensators (should that prove necessary). Because of the speed with which TDR characterizations may be performed, the efficacy of any compensation scheme is rapidly evaluated. As a specific example, consider the mitered bend pictured in Figure 8.7.

The optimum amount of mitering is easily determined experimentally with a TDR. Pieces of the corner are sliced off until the reflections are minimized. To achieve this same result with, say, a vector network analyzer or a slotted line SWR measurement would require more (and perhaps considerably more) work.

An important consideration is that a given discontinuity may mask the existence or size of other discontinuities further down the line. For example, a large series inductance (or a large shunt capacitance) may reduce the bandwidth of the TDR pulse downstream of the discontinuity, reducing the ability to characterize other discontinuities past the inductor. Therefore, the proper method is to fix the discontinuity nearest the source first, retest with the TDR, attend to the next discontinuity, and so forth until all problems are fixed.

8.2.5 SUMMARY OF TDR

The TDR is an indispensable complement to traditional frequency-domain equipment, permitting the characterization of microwave systems over a broad frequency range in a remarkably short time. The ability to locate discontinuities is a particularly valuable capability of TDRs, as is the related ability to evaluate expediently the quality of any compensation methods over a broad band of frequencies.

An excellent applications note on the use of the TDR may be found in the February 1964 issue of the *Hewlett-Packard Journal* (v. 15, no. 6). Though four decades old, the principles have not changed.

8.3 THE SLOTTED LINE

8.3.1 INTRODUCTION

The development of the automatic vector network analyzer (VNA, or simply *network analyzer*) has revolutionized the characterization of microwave circuits. By computing all of the S-parameters of a network over a broad frequency range, the network analyzer provides the designer with a comprehensive overview of circuit behavior that would be extremely cumbersome to obtain manually.

Before describing the VNA, we begin with a study of one instrument that the VNA has largely displaced: the slotted line. There are several motives for this retrospection. One is pedagogical, for the slotted line affords us an opportunity to investigate directly the quintessential wave phenomena of reflection and interference. Another is practical, because the slotted line exploits these phenomena to measure impedance at high frequencies with comparatively inexpensive equipment. Yet another is that important calibration issues that also apply to the VNA are quite naturally introduced with the slotted line. Finally, the labor involved in making accurate measurements with a slotted line is large enough to explain what motivated development of the network analyzer.

A detailed description of the VNA then follows, including illustrative examples of how it is used. There is a focus on identifying and mitigating sources of error – along with a comprehensive description of calibration techniques – because much of the modern VNA's power derives from its ability to characterize and remove its own errors.

The chapter concludes with instructions on how to build an inexpensive slotted line system capable of measuring impedance over a 1–5-GHz range.

8.3.2 THE OLD DAYS: SLOTTED-LINE IMPEDANCE MEASUREMENT

Prior to the development of the network analyzer, characterization of microwave systems was a cumbersome process. Consider first the basic problem of measuring impedance. At low frequencies, it is a relatively simple matter to use a bridge measurement technique or to excite a network with, say, a voltage and then measure the current that flows in response. Finding the ratio of voltage to current is straightforward, even if one must keep track of the relative phase between them in order to compute both the real and imaginary parts of the impedance:

$$Z = \frac{V}{I} = |Z|e^{j\phi}. \tag{19}$$

As frequency increases, however, the situation gets progressively more complicated. Adding to the usual difficulties associated with making instruments operate at high frequencies are the serious problems of fixturing: the impedance of a given length of conductor perturbs the measurement more and more significantly as frequency increases. To guarantee that one is truly characterizing the device under test – rather than a combination of the device and the interconnect – requires extraordinary care.

A recurring theme in good engineering is the conversion of a liability into an asset ("it's not a bug, it's a *feature*"). In this case we acknowledge a priori the futility of trying to reduce fixturing impedances to insignificant levels. Rather than attempt to quantify and remove the effect of the fixturing on the measured impedance, we consider instead the effect of the load impedance on the fixturing. To understand why this change in viewpoint is so valuable, recall that the amplitude of a sinusoidal voltage is independent of position only along a properly terminated transmission line (or waveguide). Any mistermination gives rise to a reflection that periodically interferes constructively and destructively with the incident wave, producing standing waves along the line. The amplitude and phase of the standing waves depend uniquely on the mismatch between the load impedance and the line's characteristic impedance, Z_0. Measurement of the standing waves, coupled with knowledge of Z_0, thus allows computation of the load impedance Z_L. The core of this impedance measurement method is therefore the bi-unique relationship between impedance and reflection coefficient:

$$Z_{Ln} = \frac{1+\Gamma}{1-\Gamma}; \tag{20}$$

here Z_{Ln} is the normalized load impedance,

$$Z_{Ln} \equiv Z_L/Z_0, \tag{21}$$

and the reflection coefficient is a complex quantity,

$$\Gamma = |\Gamma|e^{j\phi}. \tag{22}$$

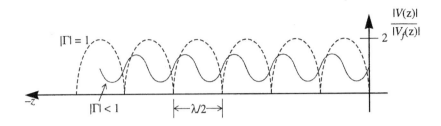

FIGURE 8.8. Typical plot of amplitude vs. position for two values of Γ

The mathematical basis for the measurement technique becomes clear by first expressing the voltage along a transmission line as the sum of forward and reflected components:

$$V(z) = V_f + V_r = V_f(e^{-j\beta z} + \Gamma e^{j\beta z}) = V_f e^{-j\beta z}(1 + \Gamma e^{j2\beta z}). \tag{23}$$

Here $z = 0$ is defined as the location of the load, with z increasingly negative as one approaches the source, and β is the phase constant, $2\pi/\lambda$.

The magnitude of the line voltage as a function of position is

$$|V(z)| = |V_f e^{-j\beta z}||(1 + \Gamma e^{j2\beta z})|$$
$$= |V_f||(1 + \Gamma e^{j2\beta z})| = |V_f||(1 + |\Gamma|e^{j(\phi+2\beta z)})|, \tag{24}$$

where ϕ is the phase angle of the reflection coefficient. Note from Eqn. 24 that the voltage magnitude is periodic. These standing waves have a periodicity of $\lambda/2$, so the distance between (say) minima corresponds to π radians of phase; see Figure 8.8. The minimum and maximum amplitudes occur when the exponential factor is -1 and $+1$:

$$V_{min} = |V_f|(1 - |\Gamma|); \tag{25}$$

$$V_{max} = |V_f|(1 + |\Gamma|). \tag{26}$$

Recall that the *standing wave ratio* (SWR) is defined as the ratio of maximum to minimum amplitudes:

$$\text{SWR} = \frac{V_{max}}{V_{min}} = \frac{1 + |\Gamma|}{1 - |\Gamma|}. \tag{27}$$

From Eqn. 27 it is clear that measurement of SWR allows the computation of $|\Gamma|$.

To complete the measurement we need ϕ, the phase of Γ. The key is to note that the minimum amplitude occurs when

$$1 + |\Gamma|e^{j(\phi+2\beta z)} = 1 - |\Gamma| \tag{28}$$

or (equivalently)

$$\phi + 2\beta z = (2n + 1)\pi, \tag{29}$$

where n is any integer. Therefore, the phase of Γ can be computed from

8.3 THE SLOTTED LINE

$$\phi = (2n+1)\pi - 2\beta z, \qquad (30)$$

where z is the location of the minimum (again, z is a negative quantity in our coordinate system).[2]

A practical consideration is that the precise location of the *electrical* reference plane $z = 0$ is not always obvious. As a consequence, an additional experiment is generally required to determine this piece of information. The traditional (and simplest) method is simply to terminate the line in as good a short circuit as possible. Clearly, the minima will be nulls (ideally, anyway), again periodically disposed along the line. Any of these nulls may be taken as the location of the reference plane $z = 0$, although it is customary to choose the one closest to the physical short. Pick one and record its position. Also note the spacing between successive minima (this is equal to $\lambda/2$), so that you can readily compute the phase constant β. Then replace the short with the impedance to be measured. Measure SWR to enable a calculation of $|\Gamma|$, and note the shift in the position of the minima relative to the zero reference established with the shorted load, counting shifts away from the load as having a negative sign. Plug that value into Eqn. 30 and solve for ϕ. Then use Eqn. 22 in Eqn. 20 to find the (normalized) load impedance. The actual load impedance is found simply by multiplying this value by Z_0.

The beauty of this technique is that the fixturing does not need to be short compared with a wavelength. In fact, the fixture's length actually must exceed one half-wavelength (and preferably be several half-wavelengths) in order for standing waves to be characterized.

The measurement requires knowledge of the voltage as it varies along the line. In turn this requires that we have physical access to the line. A slotted-line system therefore consists of an air-dielectric transmission line (or waveguide) that is slit open to admit a probe (which is generally a simple high-impedance diode detector capacitively coupled to the line). The slit and probe are carefully designed to minimize disturbance of the fields. In the case of a coaxial line, a lengthwise slit in the outer conductor has a minimal effect because no currents flow circumferentially. The primary effect of the slit is a small reduction in capacitance per unit length and a consequent small increase in characteristic impedance. A suitably narrow slit minimizes this effect to negligible levels and also ensures a minimum of radiation and its attendant losses.[3] Similarly, the coupling to the diode detector may be adjusted to minimize perturbation of the measurement by detector loading.

---

[2] In principle, one could also use the maxima in the measurement. However, the minima are sharper, so that a given amplitude measurement uncertainty translates into a smaller (perhaps much smaller) timing (phase) uncertainty than if the maxima were used.

[3] As long as the slit is comparable to (or narrower than) the wall thickness, it will act much like a waveguide far beyond cutoff. As a result, it is almost always the case that radiation can be considered truly negligible, and many texts consequently don't even both to mention the possibility of radiation at all.

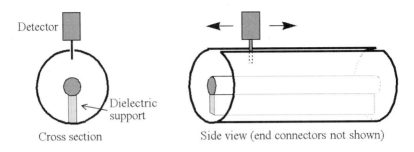

FIGURE 8.9. Coaxial slotted line

The probe is mounted on a slider with a calibrated ruler so that its position along the line can be measured (a coaxial line is shown in Figure 8.9, but a slotted waveguide also works). In most slotted lines, the probe's depth into the line is adjustable, allowing a trade-off between detector sensitivity and disturbance of the field pattern. Fortunately, the probe's presence does not affect the location of nulls (because the electric field is zero there), so the probe may be adjusted to high sensitivity for determining that data accurately. However, the probe will affect the shape of the standing waves – with distortion increasing with amplitude – leading to errors in measuring the value (and location) of the peaks. The amount of asymmetry in amplitude-vs.-position provides a qualitative assessment of probe disturbance.

An Example

Having described the method and equipment, it's helpful to go through an actual numerical example to elucidate the procedure.

Step 1. *Establish the location of the reference plane.* Connect a short-circuit load and note the location of the minima (which should be nearly nulls if the short is reasonably good and if line losses are negligible). Feel free to increase the probe depth for greater detector output to allow a more accurate pinpointing of the null locations.

Assume for our example that these minima occur at $z = -1$ mm, -121 mm, and -241 mm. Note also that the wavelength is twice the distance between nulls, or 240 mm.

Step 2. *Replace the short with the impedance to be measured.* Withdraw the probe enough to reduce distortion of the pattern (as evaluated by symmetry) and also verify that the probe output is small enough to lie within its calibrated range. Readjust the probe position if necessary to satisfy both requirements. Note both the voltage SWR and the new locations of the minima. If, as is generally the case, the probe produces an output voltage proportional to power, don't forget to compute the SWR by taking the square root of the ratio of the probe output at the maxima and minima.

Assume for our example that the measured SWR is 1.6 and that these new minima are located at $z = -41$ mm, -161 mm, and -281 mm.

Step 3. *Choose one of the null positions* from step 1 as the origin, and calculate the difference between this coordinate and the corresponding minimum observed with the load connected.

Here, choose $z = -1$ mm as the origin (it's the closest to the load). Then the displacement we use in the calculation of ϕ is -41 mm $- (-1$ mm$) = -40$ mm. Given a wavelength of 240 mm, we compute ϕ as

$$\phi = (2n+1)\pi - 2\beta z = \pi - \frac{4\pi}{240 \text{ mm}}(-40 \text{ mm}) = \frac{5\pi}{3}, \quad (31)$$

where we have arbitrarily chosen $n = 0$.

Step 4. *Compute* Γ.

First find $|\Gamma|$ from the SWR measurement:

$$|\Gamma| = \frac{\text{SWR} - 1}{\text{SWR} + 1} = \frac{0.6}{2.6} \approx 0.23. \quad (32)$$

Next, use the phase angle calculated in step 3 to complete the calculation of Γ:

$$\Gamma = |\Gamma|e^{j\phi} \approx 0.23 e^{j5\pi/3} = 0.23\left[\cos\frac{5\pi}{3} + j\sin\frac{5\pi}{3}\right] \approx 0.115 - j0.2. \quad (33)$$

Step 5. *Compute the normalized impedance* using Eqn. 20:

$$Z_{Ln} \approx \frac{1 + (0.115 - j0.2)}{1 - (0.115 - j0.2)} = \frac{(1.115 - j0.2)(0.885 - j0.2)}{0.885^2 + 0.2^2} \approx 1.15 - j0.486. \quad (34)$$

Then multiply by Z_0 (here assumed to be 50 Ω) to find the load impedance at last:

$$Z_L \approx 57.5 - j24.3. \quad (35)$$

We see that the load impedance (at this frequency) is equivalent to a resistance in series with a moderate capacitance.

And that's all there is to it (more or less).

From this example, it should be clear that the slotted-line method involves a fair amount of effort to characterize impedance at a single frequency, let alone over a broad frequency range. This is one reason that the method is used less frequently today, although it continues to live on in millimeter-wave work, where VNAs are either prohibitively expensive or simply unavailable, or where fixturing discontinuities may obscure measurement. It remains without question the best option for hobbyists or labs on a budget, since slotted-line gear is readily available on the surplus market at low cost. As an even lower-cost alternative, instructions on how to build a simple microstrip-based "slotted" line instrument are given in Section 8.9.

Error Sources (and Their Mitigation)

Mechanical imperfections are one source of error. For example, if the center and outer conductors are not perfectly cylindrical and truly concentric, then the impedance of

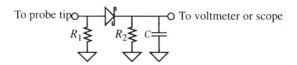

FIGURE 8.10. Schematic of typical probe

the line won't be independent of position. Similarly, if the probe carriage assembly does not maintain a constant distance from the center conductor, a position-dependent error will arise. Finally, the dielectric supports that are necessary for mechanical stability inevitably disturb the field patterns as well. In Figure 8.9 the support is shown as continuous along the bottom, but periodically distributed posts, spaced no closer than is necessary to provide adequate mechanical support, are also frequently used to minimize perturbations. In any case, the best slotted lines are superb examples of mechanical engineering, with near perfect concentricity. Many are equipped with low-backlash verniers to allow position measurement to a precision of better than 25 μm.

Another (and generally dominant) source of error is associated with the characteristics of the probe. Most probes are simple diode circuits intended to behave approximately as square-law detectors. They thus generate an output voltage roughly proportional to power. See Figure 8.10. The resistor R_1 is not a physical component of slotted-line probes. It is shown in the schematic simply to remind us that the voltage being sampled by the probe is that of a transmission line, whose impedance is about Z_0 (assuming that mismatches are small).

To gain a crude understanding of the attributes and limitations of a diode as a power detector, assume that the load capacitor in Figure 8.10 appears as such a low impedance at RF that negligible voltage appears across it. Further assume that the diode continues to exhibit an exponential relationship between current and voltage, even in the RF regime:

$$i_D = I_S\left(\exp\left[\frac{qv_D}{kT}\right] - 1\right)$$
$$= I_S\left[\frac{qv_D}{kT} + \frac{1}{2!}\left(\frac{qv_D}{kT}\right)^2 + \frac{1}{3!}\left(\frac{qv_D}{kT}\right)^3 + \frac{1}{4!}\left(\frac{qv_D}{kT}\right)^4 + \cdots\right]. \quad (36)$$

Next, let the diode voltage (which is equal to the probe voltage with the given assumptions) be a sinusoid:

$$v_D = V_p \sin \omega t. \quad (37)$$

Because of the nonlinearity, the diode current will consist of even and odd harmonics of the input frequency. All of these harmonics have a zero time average, so the only contribution to a DC diode current is from the zero-frequency component. Only the even-order terms in the expansion of Eqn. 36 can produce DC components, so

$$\langle i_D \rangle = \left\langle I_S\left[\frac{1}{2!}\left(\frac{qv_D}{kT}\right)^2 + \frac{1}{4!}\left(\frac{qv_D}{kT}\right)^4 + \cdots\right]\right\rangle. \quad (38)$$

Clearly, the quadratic term is the one that provides an average diode current proportional to the square of the voltage, or proportional to power. All other terms contribute error (as judged by conformance with a linear power law), with an increasing prominence as the voltage increases. If we are arbitrarily willing to tolerate, say, a contribution from the fourth-order term as large as 5% as that from the quadratic, then we must satisfy

$$\frac{1}{4!}\left(\frac{qv_D}{kT}\right)^4 < \frac{1}{20}\left[\frac{1}{2!}\left(\frac{qv_D}{kT}\right)^2\right] \qquad (39)$$

or, equivalently,

$$\left(\frac{qv_D}{kT}\right)^2 < 0.6. \qquad (40)$$

Therefore, as a very crude approximation we must limit peak diode voltage to values

$$v_D < \sqrt{0.6}(kT/q) \approx 20 \text{ mV}. \qquad (41)$$

Although practical diode detectors vary considerably in their characteristics, it is generally the case that one should probably distrust output voltages readings when they exceed about 5–10 mV (perhaps corresponding to input powers on the order of -20 dBm or thereabouts). Deviations from ideal behavior increase rapidly as the output voltage increases because the higher-order even terms rapidly increase in significance. The useful range can be extended by incorporating a resistive load, though at the expense of reduced output level. To understand why the simple trick of resistive loading should be effective, note that the peak *open*-circuit output voltage approaches the input voltage at high amplitudes, behavior that is linear (and therefore clearly subquadratic). Loading the circuit with a resistor produces a condition intermediate between the short-circuit case (where the current grows too fast with large input amplitudes) and the open-circuit case (where the current doesn't grow fast enough), leading to a significant range extension. Best results are typically found when using a load within a factor of 2 of 470 Ω, with the optimum found by experiment. With the proper load, the acceptable input power range can be extended another 10 dB or more. The reduction of output level, however, produces a trade-off between sensitivity and accuracy.

From the foregoing, it is clear that minimizing the peak voltages applied to the detector improves accuracy. However, for a given level of sensitivity, reducing the peak level implies a desire to minimize the voltage ratio to be measured by the detector. If we measure both the minimum line voltage and some voltage other than the maximum (as well as the position at which this other voltage is measured), then we can accomplish precisely this reduction in the dynamic range required of the detector. This other voltage can be related mathematically to the maximum because the precise shape of the standing wave is known. Specifically, it can be shown that the following relationship holds:[4]

---

[4] See e.g. Terman and Pettit, *Electronic Measurements,* 2nd ed., McGraw-Hill, New York, 1952, p. 140. This method is described also in *Microwave Measurements* (MIT Rad. Lab. Ser., vol. 20), McGraw-Hill, New York, 1948.

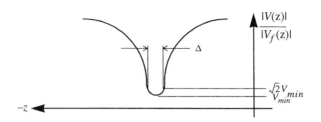

FIGURE 8.11. Alternate measurement method for high SWR

$$\text{SWR} = \lambda/\pi\Delta, \tag{42}$$

where the quantity Δ is as defined in Figure 8.11. This method requires only that the probe accurately measure a voltage ratio of about 1.41 : 1 (corresponding to a power ratio of 2 : 1). It is especially useful when attempting to measure very high SWR values, where the maximum-to-minimum voltage ratios are large.

One obvious way to improve accuracy is simply to calibrate the probe to determine explicitly the actual relationship between input and output. However engineers, being the lazy (oops, *efficient*) lot they are, have devised clever workarounds that completely bypass the need to calibrate a probe altogether. Because SWR is dimensionless, a purely ratiometric measurement is all that is required. Consider interposing a calibrated attenuator between the signal generator and the slotted line. The attenuation is set to its minimum value (say), the probe is slid along the line until a minimum is found, and the voltage there is noted. The probe is then moved to find the maximum, and the attenuation factor increased until the output is the same as at the minimum. This attenuation factor is precisely the desired ratio V_{max}/V_{min}. Note that this measurement places essentially no demands on the probe at all, having replaced with a readily realized linear attenuator the need to characterize a nonlinear probe.

Finally, a considerable improvement in sensitivity is possible if the signal generator produces a modulated output. Rather than measuring the DC output of the probe, a demodulator, followed by a bandpass amplifier, provides the output. Using a modulating frequency well above the $1/f$ noise corner of the system improves SNR, allowing the use of higher postdetector gain and a consequent reduction in the required level of coupling of the probe to the line. This reduction in perturbation improves accuracy.

8.4 THE VECTOR NETWORK ANALYZER

8.4.1 BACKGROUND

Each data point with a slotted line requires setting the frequency to the desired value, locating the new reference null with a shorted load, and then measuring SWR and locating the minima with the device under test (DUT) connected as load. The vector network analyzer (VNA) automates this process and adds greater functionality as

8.4 THE VECTOR NETWORK ANALYZER

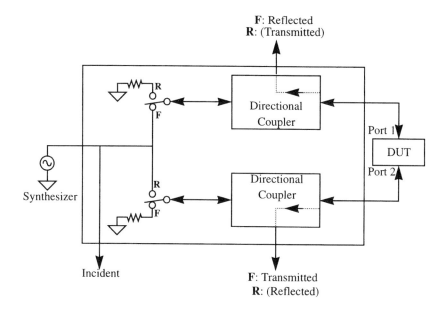

FIGURE 8.12. Typical VNA core block diagram (simplified)

well, permitting a rapid and complete characterization of all of the S-parameters of a microwave system over an exceptionally broad frequency range (e.g., from 50 MHz to 110 GHz in one instrument!).

At the heart of the VNA is a device (e.g., a directional coupler) that miraculously resolves signals along a line into its forward and reflected components. This decomposition into the two components is valuable because the measurement of impedance can be reduced to measurement of reflection coefficient, as we've seen with the slotted-line measurement method. Similarly, measurement of power gain involves ratios of forward components, and so on. Thus a VNA can characterize the full set of S-parameters for a two-port.

A representative block diagram of a VNA reveals the central role of the directional coupler (or equivalent). As shown in Figure 8.12, a frequency synthesizer provides the input to the network analyzer. Both the output power and frequency are controllable. A part of the synthesizer output is sampled as the incident signal, and the rest is steered by a pair of SPDT switches. When the switches are in the position marked "F," the DUT is driven in the forward direction, and the top directional coupler provides an auxiliary output that corresponds to the signal reflected from port 1 of the DUT. At the same time, the lower directional coupler provides an output corresponding to the power coming out of port 2 of the DUT.

To make measurements of reverse characteristics, the switches are moved to position "R," reversing the roles of ports 1 and 2 of the DUT. The incident, reflected, and transmitted signals are sent to a receiver/detector (not shown) whose job is to measure the magnitude and phase of these signals, followed by processing of the data and presentation in a display.

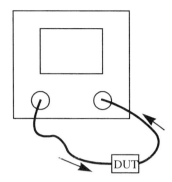

FIGURE 8.13. Transmission measurement

It is clear that a VNA comprises all of the building blocks of a complete transceiver, and more. Not only does the VNA have to cover an exceptionally wide range of frequencies (e.g., 50 MHz to 110 GHz in one instrument, albeit with degraded characteristics toward the limits of this range), it must also make accurate measurements of signals spanning a wide dynamic range of amplitudes at the same time. Operation over such a wide range requires identifying and correcting as many sources of error, both external and internal, as possible. The modern VNA employs sophisticated computational means to accomplish this error reduction, but it requires a knowledgeable operator to ensure that the calibration is performed correctly. Mistakes in calibration are an all-too-common source of anomalous results, so we will spend considerable time examining the error sources associated with the VNA's various measurement modes.

8.4.2 BASIC MEASUREMENT MODES AND ERROR SOURCES

First consider making a transmission measurement (either in the forward or reverse direction); see Figure 8.13. There will generally be some cable and fixturing external to the VNA. The total electrical length and loss of these external elements are variable. More to the point, they are beyond the control of the VNA because fixturing is a prerogative of the user.

One basic calibration step is therefore the measurement of fixturing loss and delay so that these can be subtracted from a subsequent measurement performed with the DUT in place. This step, called the *through* (often abbreviated as "thru") measurement, involves removing the DUT and connecting the rest of the fixturing together directly. The VNA then measures the fixture's phase shift and loss over the user-specified frequency range, storing the data for later subtraction.

After a through calibration, the DUT is inserted and the VNA is ready to measure its insertion loss and phase shift. In many cases, one is interested in the time delay rather than phase. Since delay is simply (minus) the derivative of phase with respect to frequency, the VNA can readily compute the delay from phase data. There are

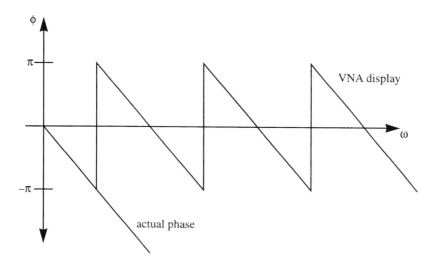

FIGURE 8.14. Phase shift vs. frequency, ideal versus VNA display

some subtleties, however, that one must appreciate if correct measurements are to be made. One such consideration is that the instrument measures phase at a discrete set of frequencies, rather than continuously over the entire band. Hence, the derivative must be approximated by a ratio of finite differences:

$$\frac{d\phi}{d\omega} \approx \frac{\Delta\phi}{\Delta\omega}. \tag{43}$$

The frequency interval in the denominator of Eqn. 43 is known as the frequency *aperture* and is controllable by the user. A narrow aperture provides fine resolution but may be sensitive to noise in the data. A wide aperture is less sensitive to noise because it effectively performs an averaging over the frequency interval, but it can miss fine structure precisely because of this averaging. Modern instruments default to an aperture that is satisfactory for most applications but which may be overridden by the user if desired.

Another subtlety is that the phase detector within a VNA functions over a finite interval, modulo some phase. A typical detector range is $\pm\pi$ radians, so the VNA cannot distinguish phase shifts outside of this range from those lying within it. Hence, a pure time delay's phase appears as a periodic sawtooth when plotted against frequency in linear coordinates; this is shown in Figure 8.14. The user must employ physical arguments or other knowledge to splice the various regions together properly; the VNA fundamentally lacks the information necessary to do so. However, there is an advantage to the sawtooth-like display when plotting: it reduces the total vertical height for a given resolution.

A related consequence of the modulo-ϕ behavior is that, if a VNA's computation of delay uses an aperture value that corresponds to a phase step in excess of π radians, then the displayed delay will be in error. To guard against these types of problems,

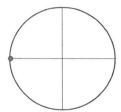

FIGURE 8.15. Idealized VNA display after calibration with shorted load

it is good practice to examine both the phase and delay curves and not just the delay. Simple checks of reasonableness are carried out rapidly, so there is hardly an excuse for not performing them.

In addition to transmission measurements, the other basic VNA operating mode is measurement of reflection. Just as with the slotted line, it is necessary to establish a reference plane. And, just as with the slotted line, a short-circuit load is best used for this purpose.[5] Therefore, in the simplest calibration for a reflection measurement, the best available short-circuit load is connected to the test port in place of the DUT. The VNA measures the magnitude and phase of the reflection over the specified frequency range and stores this data, using it to locate the reference plane and correct for fixturing losses (the VNA cannot use the information about fixturing losses from the through measurement because the latter does not identify the loss over the relevant fixturing path length). After this calibration step, a display of S_{11} with the short-circuit load should consist of data points tightly clustered about the -1 point, as shown in Figure 8.15.

If other than a tight distribution (e.g., an arc) is observed, carefully check the fixturing (particularly the connectors), correct any problems, and repeat the calibration. After re-verification, the VNA is ready to perform one-port reflectance measurements. As we saw with the slotted line, such a measurement is equivalent to an impedance measurement. Depending on context, the user may wish the data to be displayed as reflection coefficient or impedance. The modern VNA can provide a display either of Γ in polar form or of impedance on a Smith chart. The format is deliberately left unspecified in Figure 8.15 because, for the special case of a shorted load, the data are located in the same spot.

A subtle issue is that the calibration with the shorted load establishes the reference plane at the physical location of the short within the calibration standard. The fixturing may add some physical length beyond that plane. One could correct for this by performing the calibration with a short at the actual DUT terminals. Another option is to make use of the "port extension" feature of modern VNAs in which the instrument algorithmically adds length, effectively moving the reference plane further

---

[5] In principle, an open circuit would serve as well. However, physical approximations to a short are better than those to an open.

8.4 THE VECTOR NETWORK ANALYZER

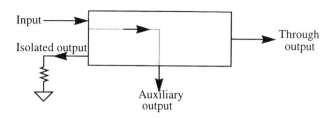

FIGURE 8.16. Directional coupler port definitions

away from the VNA connectors. The correct extension is determined by producing the best possible short at the DUT terminals and then varying the extension value to achieve the best distribution near the -1 point on the Smith chart.

Another consideration is that, after calibration, an open-circuit load ordinarily should *not* produce a dot. Because of fringing, there will always be some capacitive load shunting the otherwise open-circuited port. There may also be radiation out of that port. The net result is that the display will generally be an arc if the VNA has been calibrated *properly*. A common error is to attempt to "correct" this calibration "problem" by making ad hoc adjustments to port extension or time-delay settings, thus actually degrading calibration.

The foregoing description focuses on how external fixturing errors can be removed. Using the through and short calibrations, the VNA can reduce by large amounts the errors in transmission and reflection measurements. For even greater accuracy, the VNA is capable of characterizing its own internal errors with the aid of additional calibration steps. To appreciate how the VNA performs these additional corrections, it is necessary to identify the errors corrected in these various calibrations.

The VNA depends on directional couplers to decompose signals into incident and reflected components. As with everything else, practical directional couplers are imperfect. To quantify these imperfections, we need to define the various figures of merit that apply to the directional coupler as it is configured for use in a VNA (see Figure 8.16). It should be noted that Figures 8.12 and 8.16 both use a simplified symbol for the directional coupler. It is clear that the directional coupler is generally a four-port device, but a VNA typically uses only three of them, terminating the fourth (the isolated output) internally with a matched load.

Recall that a directional coupler is characterized by parameters such as coupling factor, isolation, and directivity. For a VNA in particular we desire an infinite directivity, but all real couplers fall short of the ideal. The lack of infinite directivity is a significant error source, and correcting for this deficiency is a major aim of VNA calibration.

To illustrate how directivity errors can corrupt measurements, first examine Figures 8.12 and 8.16 to review how directional couplers are hooked up inside a VNA. Notice that the main input of each directional coupler is connected to a port of the DUT. Thus, when performing a reflectance measurement, the synthesizer drives the through output. Power flows from the synthesizer, "backwards" through the coupler, to the DUT. Any power reflected by the DUT feeds back into the main input of the directional coupler, and a portion of the reflected power exits the auxiliary port for sampling and measurement.

If the directivity were infinite then the auxiliary port signal would be due entirely to the power reflected from the DUT, allowing direct measurement of the reflected power. However, a finite directivity implies that some of the power flowing from the synthesizer to the main input leaks out of the auxiliary port as well. The VNA would then measure an auxiliary port signal that is a weighted sum of both the forward and reflected power. Because these may add both in and out of phase over frequency, typical manifestations of imperfect directivity are ripples in, say, the measured reflection coefficient as a function of frequency.

A representative calculation illustrates the magnitude of the problem. Suppose we have a 10-dB coupler with 30-dB directivity and then attempt to measure the impedance of a load that has a 20-dB return loss. That is, $C = 10$ dB, $D = 30$ dB, $I = C + D = 40$ dB, and $RL = 20$ dB. The signal reflected by the DUT is $RL = 20$ dB below the incident power level, and the amount of the reflected signal surviving to the auxiliary output is $C = 10$ dB below that, for a total of 30 dB below incident. The unwanted signal at the auxiliary port is $I = 40$ dB below incident. Thus we see that the error power is an unacceptable 10% of the signal power in this example. If we were to attempt to measure a return loss of 30 dB, the situation would be even worse, for the error power would then equal the desired signal power.

Another error that behaves much like directivity error arises from reflections at interfaces with adapters, cables, and fixturing. These reflections necessarily produce signals at the auxiliary output of the coupler and, as in the previous example, such parasitic signals can obscure the component of signal that is due to the actual reflection from the DUT.

Source mismatch is yet another potential source of error. Consider the flow of power from the source, through the coupler, and to the load. Some power reflects off of the load and returns to the source. If there is a mismatch in source impedance, there will be a subsequent re-reflection from the source back through the coupler. Some of that power reflects off the DUT and finds its way out of the auxiliary port. From the qualitative description of this process, this error term is clearly most significant when the load has a high reflection coefficient.

A third type of error is related to one we've already examined: frequency response. The couplers, cables, and adapters – as well as the part of the system that actually measures magnitude and phase – may all have frequency-dependent characteristics.

The three types of one-port errors – directivity, source mismatch, and frequency response – can be removed by performing three experiments. For example, consider attaching a perfectly resistive matched load as the DUT. In this case, the auxiliary output of the directional coupler should have no signal. Any deviation from that condition indicates an effective directivity error, which can be measured and stored for later removal. The extent to which directivity errors are nulled out depends critically on the quality of the "perfect" load used in this step of the calibration sequence.

A common choice for the other two experiments is to use both a shorted and open load. As with the perfect load, the ultimate accuracy of VNA measurements depends on how close the impedance of the loads are to zero and infinity. It is particularly hard to implement a good open circuit at high frequencies because stray capacitance is difficult to control. To underscore the difficulty involved, note that a 0.1-pF stray capacitance (which is about the right value for an open-circuited APC-7 connector) has an impedance of only about 160 Ω at 10 GHz. Also, radiation from the open end is an increasing problem as frequency increases, and this loss produces a real component of impedance in parallel with the parasitic shunt capacitance. This problem is mitigated by sliding a short along a line until it is positioned a quarter wavelength away from the reference plane. The shorted line is a closed structure that prevents radiation.

When calibrating for two-port operation, the three experiments are augmented with a fourth: a through measurement to characterize the frequency response of the fixturing, as described previously. Hence, the quartet of experiments is often known as the "short/open/load/thru" (SOLT) two-port calibration method.

There are several minor variations on the SOLT technique, all aimed at solving the problem of imperfect impedance standards. One of these replaces the fixed matched load with a sliding load. Here, a movable load with a near-perfect match is slid along an air line, and the whole assembly is used in place of the fixed load. As the load slides along the line, the small reflections combine with the incident signals in a periodic manner – alternately adding and subtracting – leading to a data set that is distributed in a circle in the complex plane. The directivity vector is the center of that circle. Three points uniquely determine a circle, so in principle only three measurements are needed. In practice, a larger number is used to improve the error estimation.

Yet another variation replaces the sliding load with an offset load, which may be thought of as a sliding load in which the load no longer slides. If two points and the angle of the offset are known, the center of the circle can again be determined. These two points are obtained with two loads of different length. The sliding or offset loads are popular at millimeter-wave frequencies, where good approximations to ideal loads are simply not available. At these frequencies, a shim of known thickness is inserted between mounting flanges to produce the second measurement.

An alternative calibration suite is known as the thru-reflect-line (TRL) method.[6] This method corrects for the same errors as the SOLT method, but it depends less on the perfection of the impedance standards used as calibration loads. As the name of the method suggests, the first step in the calibration is to connect the two ports of the external fixture together in a low-reflectivity "through" configuration to characterize the fixturing. The next step is to connect a grossly mismatched load to each port of the fixture separately (hence the name "reflect"). The precise nature of the

---

[6] This calibration suite was first described by G. Engen and C. Hoer in "Thru-Reflect-Line: An Improved Technique for Calibrating the Dual Six-Port Automatic Network Analyzer," *IEEE Trans. Microwave Theory and Tech.*, v. 27, December 1979, pp. 987–93.

mismatch is not too important (although the phase of the reflection coefficient should be known to within approximately 90° because the response of most phase detectors is periodic), and its magnitude need not be known; it just has to have the same high reflectance at both ports (a nominal short is frequently used). Finally, the two ports are again connected together through the low-reflectivity fixturing, but now with a different cable (or other fixturing) length than was used in the first thru measurement. Again in a concession to the limitations of practical phase detectors, this line should be nominally a quarter-wavelength longer than the thru at the band center. For best accuracy, it's advisable to limit the frequency span to a value that assures that the length difference corresponds to phase angles not too different from 90°. The precise limits vary from instrument to instrument and also as a function of required accuracy. That said, a typical value for the lower end is 15–25° while that for the upper end is 155–165°, implying that a given line is useful for calibration over roughly an 8:1 range. If a larger frequency range must be spanned, then additional lines with different lengths should be used.

The TRL method is particularly attractive for noncoaxial systems such as microstrip, where good impedance standards are difficult to realize (or are simply unavailable commercially). It is even attractive for coaxial media, because impedance standards are expensive and the TRL procedure requires no expensive elements (in principle, anyway).

An alternative calibration suite corrects for the same errors as does the TRL suite, but it facilitates on-wafer measurements in particular. Called LRM for *line-reflect-match,* this calibration method does not use lines of different length.[7] This consideration is especially important for on-wafer measurements, because it is highly inconvenient to move probes around to perform calibration. Aside from the inconvenience, it is certainly true that repeatability in electrical measurements is limited by the repeatability of contacting. Obviously, moving probes around is antithetical to that repeatability. And in many cases, it is not possible to change the probe-to-probe distance at all, effectively ruling out a TRL calibration altogether.

To permit a TRL-like calibration result without requiring variable-length or excessively long calibration standards, the LRM suite relies on the availability of a good broadband match (the *M* in LRM). As such, the ultimate quality of the calibration is dependent on the quality of this matched load. A particular vulnerability is the series inductance typically encountered when contacting any structure with a probe, and the aim of many practical implementations of LRM calibration is to determine this inductance and remove it algorithmically.

As its name implies, the other standards are a line and a high-reflectance load. The latter is usually an open circuit, so that the LRM standards are a subset of those used

---

[7] D. F. Williams and R. B. Marks, "LRM Probe-Tip Calibrations Using Nonideal Standards," *IEEE Trans. Microwave Theory and Tech.,* v. 43, February 1995, pp. 466–9. This paper is not the first description of LRM, but it summarizes well the set of practical considerations related to LRM's use.

for SOLT. The dependence on a broadband match effectively forces a dependence on standards purchased from a vendor, such as Cascade Microtech, just as one must generally purchase calibration standards for the other methods.

8.4.3 SPECIAL CONSIDERATIONS FOR MICROSTRIP

As implied in the preceding paragraph, microstrip environments pose some challenges for calibration. Consider, for example, a microstrip fixture for measuring the S-parameters of a transistor. It is not entirely obvious how to carry out, say, an SOLT calibration sequence for such a noncoaxial structure. Since one important aim of calibration is to null out fixturing artifacts, we evidently wish to implement and use a short, open, matched load as well as a through line at various stages of the calibration, all at the physical location where the DUT (here, a transistor) would be placed. An open circuit sounds easy enough (but it isn't really, because of ever-present fringing capacitance), and so does a thru. The former can be approximated simply by not installing a DUT, and the latter can be approximated by a second fixture that is identical to the first but in which the microstrip line extends all the way across. Implementing a reasonable approximation to a short is similarly straightforward, with a third fixture (again, otherwise identical to the first) in which the ports are shorted to ground (e.g., through a short, wide piece of copper foil). The tough one is implementing a good matched termination. A surface-mount resistor at the end of the microstrip line, for example, might suffice for crude prototyping, but it is unsatisfactory for accurate characterization because its series inductance and shunt capacitance cause the impedance of the "matched" load to vary with frequency.

A reasonable solution to this problem becomes apparent when we re-examine what errors are being calibrated out with the matched load connected to the VNA. For the most part, internal VNA directivity errors are being nulled out at this step of a SOLT calibration, so there is no need for the rest of the fixturing to be involved at all. Hence, the ordinary coaxial matched impedance standard may be connected directly to the VNA port without worrying at all about whether a microstrip test environment will eventually be used. We will call this method the modified SOLT technique.

The modified SOLT calibration unfortunately will not fix errors in effective directivity caused by a mismatch past the APC port. Hence, if that transition is poor or if precise answers are necessary, then a TRL calibration should be performed. For exacting work, then, the TRL calibration method is best, but the modified SOLT is often a good enough compromise.

A final consideration is that the "open" condition with a microstrip is imperfect (as it is with all open structures) because of fringing capacitance (on the order of 50 fF). A partial correction for this is possible through the use of software connector subtraction. Many VNAs have the ability to remove the effect of a connector by means of software. Alas, microstrip is not one of the ordinary options. Of the options that are typically available, the best approximation is the APC, whose ∼100-fF fringing capacitance is reasonably close to that of a typical open line on FR4. A residual error remains, however, and one may use the "variable port extension" feature of many VNAs to reduce this residual error substantially.

Finally, one must always observe power limits to avoid nonlinear operation or even damage. In general, one should apply less than about 20 dBm (100 mW) to avoid damage and less than about 0 dBm to avoid nonlinearity. These are only rough rules of thumb, so be sure to consult the actual documentation for a given instrument for the correct values.

8.5 SUMMARY OF CALIBRATION METHODS

The previous section describes so many permutations that it is easy to get a bit confused (things would have been even more confusing had we covered all that exist). Here is a summary of the calibration methods along with some comments to remind you what their relative attributes and weaknesses are, allowing you to make an informed choice of calibration technique.

The simplest is the short-thru one- and two-port calibration, which only corrects for external fixturing and detector frequency response errors. Errors from finite directivity and source mismatch are not corrected. This technique is also known as *response calibration*.

Better for one-port measurements is the short-open-load (SOL) suite of calibrations. As long as the impedance standards are perfect, this method is capable of nulling out errors from finite directivity as well as source mismatch and detector frequency response errors. A thru measurement may be added to yield a SOLT calibration, which is a two-port method that corrects for all of the errors corrected by SOL and also corrects for the remaining cabling of a two-port fixture.

Variations on the basic SOLT theme include replacing the load measurement with either a sliding load or a fixed offset load. The modified SOLT method employs an SOT sequence with a microstrip (or other noncoaxial) fixture to null out all but VNA directivity errors. Use of a standard coaxial matched load without the fixture completes the calibration by (almost) zeroing out directivity errors.

The thru-reflect-line (TRL) method eliminates the need for using perfect impedance standards as calibration loads and corrects the same errors as the SOLT technique. The TRL method is particularly attractive in characterizing noncoaxial systems such as microstrip. With TRL calibration, it is possible to reduce directivity and source mismatch errors to levels as low as -60 dB at 18 GHz and also to essentially eliminate frequency response errors. These values should be compared to the -40-dB directivity and -35-dB source mismatch errors typically achieved with the SOLT method (fixed load). The sliding and offset load options improve the SOLT errors to levels in between those of fixed SOLT and TRL methods.

Finally, the LRM calibration suite corrects for the same errors as does TRL but has the additional advantage of not requiring lines of different lengths. By thus not requiring a change in probe distance, fixturing complexities are reduced. Furthermore, repeatability is improved because fixturing artifacts remain more constant throughout the calibration cycle. The drawback is the need to purchase expensive (and fragile) calibration substrates from an external vendor.

8.6 OTHER VNA MEASUREMENT CAPABILITIES

Thanks to the extensive use of computation, the modern VNA is capable of more than a complete measurement of S-parameters. For example, once the S-parameters are measured over a broad frequency range, the frequency response data can be transformed to time response data. Step responses and TDR traces can be generated from VNA data. Although the time taken to perform all of the measurements and computations is substantially larger than it would take for a "real" TDR, this additional functionality is nonetheless welcome.

Because of the algorithmic nature of the transform, it is possible to perform a little mathematical magic that would be impractical to carry out with actual time-domain instrumentation. For example, consider a case where a TDR trace contains reflections from multiple sources. The early discontinuities can mask the effect of subsequent ones in a real TDR measurement. A VNA, however, can remove the first discontinuity, allowing examination of the previously masked reflections. The extent to which a VNA can perform this removal (called *gated impedance* or *gated TDR* measurements) depends on the accuracy and noise of the S-parameter measurements.

8.7 REFERENCES

Aside from the sources cited in footnotes, the reader may also find useful "Vector Measurement of High Frequency Networks" (Hewlett-Packard High Frequency Vector Measurement Seminar Notes, April 1989). These notes contain an excellent high-level summary of how a VNA is used, with a concise discussion of error sources and calibration methods. Another useful reference is the user manual of almost any VNA, such as the HP8720C (a 130-MHz–20-GHz instrument) or the HP8510C (capable of operating from 50 MHz to over 100 GHz).

8.8 APPENDIX A: OTHER IMPEDANCE MEASUREMENT DEVICES

8.8.1 THE SWR METER

A classic device for rapidly assessing impedance mismatches is the standing-wave ratio meter. In its simplest form, it is a directional coupler (with weak coupling) combined with a means to measure amplitude. The device under test is connected to the through port of the main line, as shown in Figure 8.17.

In the least expensive incarnations, a diode-based peak detector and meter provide a relative power indication. The detector is first connected to measure the coupled

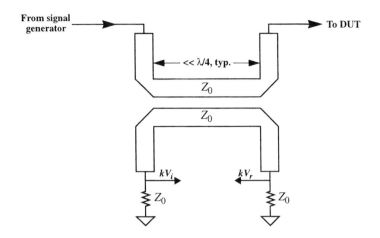

FIGURE 8.17. Classic SWR meter

signal corresponding to the incident power. Typically, a potentiometer is adjusted to normalize the indicated reading on the meter to full scale. Then, the detector is connected to the other coupled port to measure the reflected power. The display is calibrated to provide the SWR directly.

It should be apparent that other directional couplers may be used, particularly in view of the inferior directivity of microstrip coupled-line couplers. The edge-coupled version shown in Figure 8.17 is merely illustrative (and certainly easy for the weekend experimenter to cobble together).

8.8.2 THE GRID-DIP OSCILLATOR

A staple of ham shacks in days gone by was the grid-dip oscillator (GDO). Although it is not typically used at gigahertz frequencies (commercial units typically top out at the lower end of the UHF range), there's no fundamental reason why it couldn't function satisfactorily there, physical constraints permitting (we could even imagine integrating one into a probe-station probe). In any case, it gives us an opportunity to integrate multiple RF and microwave principles.

The idea behind a classic GDO is simple. If a network is simply brought near an oscillator's own tank, the oscillator's losses will increase owing to the energy coupled into the external network. This increase in loss will be particularly acute if the external network is resonant at the same frequency as the oscillator's own tank. The energy given up to the external load produces a consequent reduction in bias, and this dip is measured with a simple meter. The GDO thus consists of an oscillator that may be tuned over some frequency range, a probe that allows the necessary coupling to an external network, and a meter for indicating the dip.

The grid-dip oscillator is exceptionally versatile. If the GDO is well calibrated then it can readily determine the resonant frequency of external tanks, facilitating their tuning to within at least a coarse accuracy (say, a couple of percent). If one

8.8 APPENDIX A: OTHER IMPEDANCE MEASUREMENT DEVICES

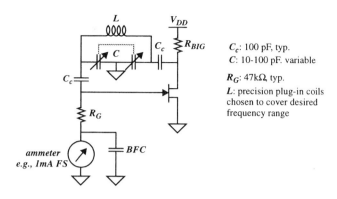

FIGURE 8.18. JFET-based "grid"-dip oscillator

connects an unknown inductor to a capacitor whose value is accurately known, the GDO's measurement of the resonant frequency can be used to calculate the value of the unknown inductor. A known inductor similarly allows determination of an unknown capacitor.

A simple modification allows GDOs to measure frequency as well, by using the instrument as both a local oscillator and mixer. Coupling the GDO to an external oscillator results in mixing action due to the inherent nonlinearity of vacuum tubes. By simply putting headphones into the plate circuit of the GDO's oscillator tube, one can listen to the beat frequency vary as the GDO is tuned. When the audible beat frequency diminishes to zero, the GDO's frequency has been adjusted to that of the external oscillator, and the value may be read off the calibrated dial directly. Frequency determinations on the order of a percent or so are generally possible, depending on the quality of the dial calibration and GDO stability.

A rough translation of a representative GDO circuit into JFET form appears as Figure 8.18.[8] Commercial dippers come with a set of precisely wound plug-in coils whose inductance values are typically known to better than a percent. The user selects an inductance whose corresponding GDO frequency span is roughly centered about the value of interest. The plug-in coils serve not only as a component in the GDO's tank but also as the probe for coupling to the DUT.

To use a GDO properly, it's important to avoid the temptation to over-couple the instrument to the DUT. The reason is that coupling changes the resonant mode frequencies. Thus, one should couple the GDO to the DUT only as tightly as necessary to attain a reliable dip. Typically, one initially brings the instrument so near the DUT that overcoupling is likely, just to get an indication of a large magnitude. Then, while backing the instrument away from the DUT, the GDO is constantly readjusted to track the dip.

---

[8] Variations on this basic circuit have appeared in countless publications, kits, and commercial instruments through the decades. The translation we offer here is most closely based on a circuit from *The Radio Amateur's Handbook,* 29th ed., Rumford Press, Concord, NH, 1952.

8.9 APPENDIX B: PROJECTS

8.9.1 MICROSTRIP "SLOTTED"-LINE PROJECT

As we've seen, the modern network analyzer is a truly remarkable instrument that is capable of extremely accurate characterizations of microwave networks over a broad frequency range. Unfortunately, this capability comes at a price: A typical gigahertz VNA costs more than the average sports utility vehicle, and the few that are available on the surplus market are rarely discounted much. Clearly, a VNA is generally priced out of the reach of most hobbyists (and even out of the reach of many academic laboratories), so the slotted line is the device of choice for those on a budget. On top of that, the slotted line is a superb pedagogical tool for teaching the principles of Smith-chart manipulations (e.g., providing explicit explanations for phrases such as "wavelengths toward the generator" and so forth). As mentioned earlier, many slotted lines are available on the surplus market for quite reasonable prices, at least for lines designed for use in the low-gigahertz frequency range.

This section describes a much cheaper (and much cheesier) alternative: a microstrip "slotted"-line system capable of measuring impedance and frequency to 5 GHz and beyond. This instrument (and that is a loose use of the term, to be sure) has important attributes: it costs very little (the total parts and materials cost should not exceed $5–$10) and is extremely easy to make using ordinary tools and materials. The trade-off is that the instrument's accuracy is not particularly good, and the shaky mechanicals are not terribly robust. However the performance is adequate for virtually any home project in the low-gigahertz range of frequencies. With care, the instrument can be made to perform well enough for a great many student laboratory projects as well.

The design presented here is based on FR4 material and right-angle mounted bulkhead BNCs of the type described in the chapter on microstrip, in keeping with a focus on minimizing cost. In particular, a microstrip line is much easier to make than a slotted coaxial airline. Of course, better performance can be obtained by using lower-loss PC board material in tandem with better connectors, and the reader is certainly invited to improvise variations on the basic design as budget, patience, and performance requirements dictate.

The first step is to get a piece of FR4 longer than the largest electrical wavelength of interest – but not so long that the loss is excessive over the desired operating frequency range. For a minimum operating frequency of 1 GHz, a good compromise is about 25 cm. A line of this length typically exhibits about 0.8 dB of loss at 1 GHz and perhaps 4 dB of loss at 5 GHz (compared with a worst-case loss of less than 1 dB for a true coaxial slotted line). If you are going to use the instrument only at the higher frequencies, performance will improve by shortening the line to reduce the loss (we only need the line to be long enough to contain a couple of minima, and the loss *per wavelength* is roughly constant, with a value of a bit under 1 dB per λ).

Mount BNCs at the two ends, and then construct a 50-Ω microstrip line using copper foil tape. It is important that the foil be as smooth as possible. Next, affix a

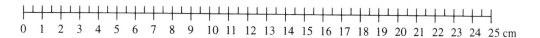

FIGURE 8.19. Metric ruler for microstrip line (drawn at half size)

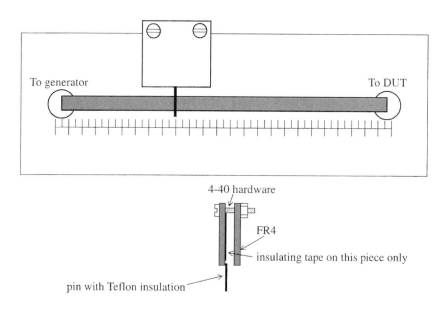

FIGURE 8.20. Bottom view of slotted line and side view of probe assembly (not to scale)

nonconductive metric ruler just below the line (if you don't have a suitable ruler then use a photocopier to duplicate, at 200%, the metric ruler shown as Figure 8.19). Because photocopier accuracy varies considerably, verify that the enlargement hasn't distorted the scale factor of the ruler. Careful interpolation between the 5-mm markings should allow a precision of ∼1 mm, though accuracy is a different matter!

The next step is to construct the detector. Here we use a Schottky diode–based detector circuit capacitively coupled to the microstrip line. It's a simple circuit, and the biggest challenge you'll face is mechanical: to construct the slider while guaranteeing proper and consistent coupling of the detector to the line.

The probe is a common needle (such as the kind that come with new shirts) carefully jammed into the bottom side of one part of the slider and cemented with a little epoxy, then clipped to length (see Figure 8.20). The probe is surrounded by a short length of insulation (preferably Teflon taken from a piece of hookup wire) to act as the dielectric between the probe and the line and also to provide a smooth rolling action. The probe motion – and the line itself – must be as smooth as possible in order to maintain a constant coupling as the probe slides along the line.

The slider assembly is made out of two pieces of FR4 that are bolted together. Teflon tape (or very smooth copper foil tape) may be affixed to the inner surface of

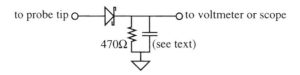

FIGURE 8.21. Schematic of probe assembly

the piece that's on the line side of the unit to reduce sliding friction and abrasion. Its thickness needs to be carefully controlled to ensure that the probe makes good contact with the line.

The foil side of both pieces faces the main board. The foil that contacts the ground plane of the main board provides the ground contact for the probe circuitry. The slider contains the probe circuitry, which consists of a diode detector and a resistive load (see Figure 8.21).

The Schottky can be any low-capacitance, high-frequency unit (such as the HP 5082-2835 or -2860), whose anode lead is connected to the actual probe tip. The input signal amplitudes must be small enough that the diode acts approximately as a square-law detector, making the output voltage roughly proportional to power. The proportionality constant, as long as it truly is a constant, is fortunately irrelevant since only ratios are used in computing SWR. However, the square-law behavior necessitates taking the square root of the probe output voltages in order to compute SWR.

The resistive load is shown as 470 Ω, but you are encouraged to experiment with its value in order to maximize the detector's useful range. A surface-mount chip resistor is used to keep parasitics small, although the output side of the detector is not terribly sensitive to parasitics of typical magnitude.

The capacitance of the slider, plus the input capacitance of most meters and scopes to which the output of the probe is connected, will generally be large enough not to require any additional capacitance. To maximize usefulness, the ability to measure sub-millivolt signals is desirable. Some amplification may be necessary to boost detector outputs to levels that are conveniently measurable with inexpensive instruments.

So, when all is said and done, how good is the microstrip slotted line? The lossiness of FR4 and the right-angle mounted BNCs, probe coupling irregularity, lack of probe calibration, and the hand-built nature of the line itself all conspire to make this impedance measurement tool a rather crude one. As a rough rule of thumb, one can expect reasonable accuracy for impedances between about $Z_0/5$ and $5Z_0$. For the most common case of producing a good match to Z_0, the tool works extremely well and allows the attainment of S_{11} values below -15 dB with ease. Results at 1 GHz generally are surprisingly good, with progressive degradation as the frequency increases to 5 GHz and beyond.

If the instrument is to be used only at the higher frequencies, there are several necessary refinements. Replace the right-angle BNCs with inline SMA connectors, use RO4003 instead of FR4, and abandon the idea of handcrafting the line out of copper foil tape; it is insufficiently uniform, so conventional PC board manufacturing techniques are required. Finally, a better diode may have to be used (e.g., the M/A-COM MA4E2054, which is specified beyond 10 GHz). If the line is shortened by a factor of 5 or so, satisfactory operation between 5 and 10 GHz is possible when all of these refinements are combined.

Measuring Frequency with the Slotted Line

If frequency measurements to ~10% accuracy are acceptable, then the slotted line is a remarkably economical alternative to a gigahertz frequency counter. Since standing-wave minima are separated by half a wavelength, knowledge of the propagation velocity is the only additional information needed to allow computation of the frequency. For 1.6-mm–thick FR4, the effective dielectric constant is typically within 5–10% of 3.5, leading to an on-board wavelength that is a factor of 0.535 times the free-space wavelength. This value corresponds to a 160-mm wavelength at 1 GHz. To make frequency measurements, then, simply open- or short-circuit the line to guarantee maximum SWR, and measure the distance between minima. Accuracy is enhanced by measuring the distance between several minima (if present) and dividing by the number of half-wavelengths. Double the measurement to compute the wavelength, then divide that value (in millimeters) into 160 mm to find the frequency in gigahertz.

As a numerical example, suppose we measure a distance of 25 mm between nulls. Then the wavelength is 50 mm, so dividing that value into 160 mm yields a frequency of 3.2 GHz.

Without calibration of any kind, ~10% absolute accuracy is possible. With calibration, the error can be reduced by a factor of 5 to 10, depending on the care taken in construction and calibration. Furthermore, the line can be used to measure frequency well beyond where it is useful as an impedance measurement instrument, because only the distance between nulls is important for frequency measurement; characteristics such as loss and the like are completely irrelevant.

8.9.2 HOMEGROWN SUB-NANOSECOND PULSE GENERATORS

The art of fast pulse and step generation is highly specialized, and it is unrealistic to expect that we could generate pulses with risetimes competitive with state-of-the-art instruments by using only what's available in the typical home laboratory. However, you may be pleasantly surprised to find that it isn't difficult to generate pulses with risetimes in the neighborhood of ~200 ps using components readily available to hobbyists. Such a pulse generator is especially valuable for evaluating the quality of oscilloscopes and particularly of scope probes.

A trivial modification to the pulse generator converts it into a triggerable generator, also with 200-ps risetime. Of the possible ways to generate pulses of this speed, the most economical for hobbyists is unquestionably to make use of an abnormal mode

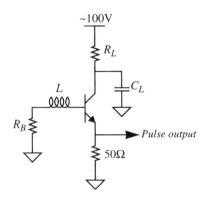

FIGURE 8.22. Free-running avalanche mode pulse generator

of transistor operation: *avalanche breakdown*. In this type of breakdown, the collector voltage is high enough to rip electrons from their orbits, creating hole–electron pairs. The electrons accelerate toward the positive terminal (here assumed to be the collector), while the holes accelerate toward the base. As the freed carriers accelerate, some bash into other silicon atoms, creating still more hole–electron pairs, and so on, causing a rapid increase in collector current.

The simple pulse generator circuit shown in Figure 8.22 exploits this avalanching. In this circuit, the collector supply voltage is chosen well above the transistor's breakdown voltage, and its precise value is not at all critical. However, *under no circumstances should you derive this voltage directly from the mains*; it is simply too dangerous to do so! Rather, a battery-operated circuit is highly recommended. A particularly handy source of high voltage is the xenon flash circuitry of disposable cameras. These may often be obtained at low cost (or even free) from neighborhood photo labs. Typically 200–300 V may be found across the one large capacitor in such units, and this level of voltage is more than adequate to avalanche almost any transistor of interest. Exercise caution when removing the board from the camera case and (certainly) while wiring it up. The main capacitor can store a dangerous amount of charge for quite some time.

The capacitor C_L may be made out of copper foil tape over ground plane on 1.6-mm FR4. A good starting value is a square strip approximately 0.75 cm on a side. Foil may be added or trimmed as necessary to adjust pulse duration and amplitude. It's important to minimize the length of the connection to the collector. Any inductance between the collector and C_L will slow down the pulse. It only takes a few nanohenries to slow down the circuit by a factor of 2 or more.

Together with C_L, the load resistor R_L determines the pulse repetition rate. A reasonable target for the pulse repetition frequency is in the neighborhood of very roughly 100 kHz, typically corresponding to R_L in the range of 100 kΩ to a few megohms. Slow repetition rates are typically associated with worsening jitter, and very high repetition rates result in large dissipation.

Once avalanching begins, the collector current increases rapidly for two reasons: the direct effect of avalanche electron multiplication in the collector, and the increase in base current produced by the avalanching holes. The increased base current increases the collector current through ordinary transistor action. This positive feedback mechanism is enhanced by biasing the base through a relatively large impedance to allow the hole current that comes *out* of the base to raise the base voltage significantly. A typical value for R_B is in the neighborhood of kilohms to tens of kilohms. The precise value is not critical, but if you are trying to optimize any given implementation, adjustment of this value is one place to start. Similarly, series parasitic inductance is not a problem (in fact, it is somewhat beneficial). To underscore that the inductance may be useful, it is shown explicitly in the schematic of Figure 8.22.

The pulse width depends on the size of the collector capacitor (larger capacitances lead to taller and wider pulses) and the characteristics of the transistor. Low collector–base capacitance is favored to allow the base and collector voltages to move rapidly in opposite directions. A more critical parameter is the ratio of BV_{CBO} to BV_{CEO}. The former is a measure of collector–base breakdown voltage with the emitter open-circuited, while the latter is the breakdown voltage with the base open-circuited. The latter is always smaller than the former precisely because of the same internal positive feedback mechanism already described. For most small-signal transistors, the ratio of these two breakdown voltages falls within the range 1.5–2.0, but a few (such as the 2N2369 or Zetex FMMT-417) have ratios that exceed 2.5 or so. Those few are the ones that are particularly well suited for making avalanche pulsers.

It is important to underscore that transistors are almost never specified by manufacturers for avalanche mode operation. Even if you do find a transistor that avalanches well, you should not expect all transistors of a given type to avalanche similarly. Consequently some hand selection will generally be necessary. That said, typically more than 75% of a given batch of 2N2369 transistors will avalanche well enough to provide a 5–10-V peak pulse into 50 Ω with rise and fall times close to 200 ps – speeds that were state of the art for quite expensive laboratory instruments in the mid-1960s. The ∼1-A/ns current slew rate is difficult to achieve through conventional means, so having to try a handful of transistors seems a modest price to pay indeed. This high a slew rate also underscores the importance of assiduously reducing parasitic inductance in series with the emitter circuit: a single nanohenry of stray inductance drops an entire volt!

The pulse generator is a versatile instrument with multiple uses in high-speed work. As one specific example, the bandwidth of a scope–probe combination can be rapidly evaluated with such a generator by observing the displayed rise and fall times. Aberrations introduced by defective or improperly calibrated instruments and cables are also readily observed. Given that the typical alternative is to measure frequency response by sweeping a sine-wave generator over a gigahertz range, the pulse generator is clearly an extremely inexpensive option.

A simple modification turns the free-running pulse generator into a triggered device. See Figure 8.23. Here, the supply voltage is adjusted to a value just below

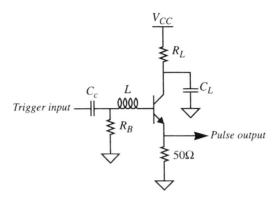

FIGURE 8.23. Triggered avalanche mode pulse generator

where free-running operation is enabled. Then a trigger pulse, coupled into the base circuit through a small-value (e.g., 1–10-pF, depending on the amplitude and risetime of the trigger pulse), low-inductance capacitor will push the transistor over the edge and into avalanche breakdown.

CHAPTER NINE

MICROWAVE DIODES

9.1 INTRODUCTION

The modern microwave diode owes its existence to the demands of military radar during the Second World War. Vacuum tubes of the time were simply unable to operate in the multi-GHz radar frequency bands. Fortunately, the seeds of a breakthrough had been planted in the mid-1930s by Bell Labs scientist George C. Southworth during his early work with cylindrical waveguides. Proving the old adage that "necessity is the mother of invention," an inspired bit of thinking by his colleague, Russell Ohl, led him to try out crystal detectors (then nearly obsolete) as power sensors, hoping that the low capacitance associated with point contacts would permit operation at the high frequencies he was using (within an octave of 1 GHz).[1] Promising results of tests on silicon confirmed that such diodes do indeed succeed where vacuum tubes fail. A crash development program by the MIT Radiation Laboratory and others successfully delivered reliable point-contact silicon diodes capable of operating at frequencies in excess of 30 GHz by the end of the war.[2]

In this chapter, we examine diodes of this type. However, we also broaden the term "diode" to include many other two-terminal semiconductor elements that find use in microwave circuits. So, in addition to ordinary junction and Schottky diodes, we'll consider varactors (parametric diodes), tunnel diodes (including backward diodes), PIN diodes, noise diodes, snap-off (step recovery) diodes, Gunn diodes, MIM diodes, and IMPATT diodes. Space constraints force us to leave out a few of the less widely used yet very interesting types (TRAPATT, LSA, etc.).[3]

---

[1] Acting on Ohl's suggestion, Southworth obtained a few old-style silicon catwhisker detectors from a local surplus shop and, after some experimentation and refinement, enjoyed success.

[2] Torrey and Whitmer describe this remarkable program in volume 15 of the *MIT Radiation Laboratory Series,* McGraw-Hill, New York, 1948.

[3] For a comprehensive review of other diode types, see Sze's *The Physics of Semiconductor Devices* (Wiley, New York, 1981) as well as a volume he edited, *High-Speed Semiconductor Devices* (Wiley, New York, 1990).

This chapter also contains an appendix (Section 9.13) describing how to make a Schottky diode by baking pocket change in a home oven, and how to use it in a crystal radio that operates in the distinctly macrowave frequency band of ~1 MHz.

9.2 JUNCTION DIODES

The workhorse of diodes is the "plain old" junction type. Most textbooks on device physics fail to mention that the junction diode was discovered before it was invented. The first was made accidentally at Bell Laboratories during an experiment with silicon crystal growth that inadvertently left opposite ends of an ingot doped with opposite polarities. In early 1940 Russell Ohl examined a wafer that had been cut from a particular part of that ingot that happened to straddle the transition from n-type to p-type. He discovered first its photovoltaic properties while trying to understand the cause of erratic resistance measurements. Generating nearly half a volt, his solar cell astounded his colleagues, who were accustomed to seeing an order of magnitude less from the cuprous oxide and selenium cells of the day. Although the junction diode's rectification properties were also noted at the time, the far superior frequency response of point-contact devices engaged the attention of engineers throughout the war years. Serious development of junction diodes was thus deferred until the end of the 1940s, when dissatisfaction with the mechanical instability of point contacts encouraged engineers to pay closer attention to junction devices.

As do point-contact diodes, junction diodes exhibit an exponential relationship between voltage and current:

$$i_D = I_S \exp\left(\frac{v_D}{nV_T}\right), \tag{1}$$

where I_S is a temperature-dependent current, V_T is the thermal voltage (kT/q), and n is the ideality factor (ideally unity). For most practical diodes, the ideality factor generally lies somewhere between 1 and 2.[4] At room temperature, the forward diode current doubles every $18n$ millivolts and increases by a factor of 10 every $60n$ millivolts.

At radio frequencies, one must also consider parasitic and junction capacitances, as well as parasitic resistance and inductance. Under reverse bias, the small-signal capacitance is well approximated by

$$C_j = \frac{C_{j0}}{(1 - V_j/\psi_0)^m}, \tag{2}$$

where C_{j0} is the capacitance at zero bias, V_j is the junction voltage (positive in forward bias), ψ_0 is nominally the contact potential (related to the bandgap of the material, and typically on the order of 1 V for silicon), and m is a constant whose value depends

---

[4] The ideality factor approaches unity only when surface leakage and recombination in the depletion layer may be ignored.

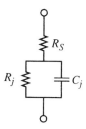

FIGURE 9.1. Simple small-signal diode model

on the doping profile. For linearly graded junctions, m is $1/3$; for abrupt junctions, it is $1/2$. Values of m in excess of $1/2$ are possible through the use of *hyperabrupt* junctions (described in greater detail in Section 9.4).

The junction capacitance of real diodes conforms surprisingly well to this equation for reverse bias and also for weak forward bias. In deep forward bias, the capacitance typically limits at a value about 2–3 times C_{j0}, rather than growing without bound as the junction voltage approaches ψ_0.

The voltage dependency of this capacitance is both an asset and a liability. When an electronically controllable capacitance is desired, as in a varactor (more on this in Section 9.4), one wishes to enhance this characteristic. In other cases, the capacitance variation is an unwelcome source of distortion.

One basic requirement for a diode is that its impedance in the forward direction be much lower than that in reverse. In forward bias at high frequencies, the impedance is dominated by the parasitic series resistance. In the blocking direction, the diode's impedance is well approximated as a series combination of a resistance and the junction capacitance; see Figure 9.1. In this model, R_j represents the fundamental bias-dependent diode resistance whose value in forward bias is

$$R_j = \frac{nV_T}{I_{BIAS}}, \qquad (3)$$

where nV_T is the same as in Eqn. 1 and I_{BIAS} is the bias current through the diode. A range of ideality factors between 1 and 2 implies R_j values of 25–50 Ω at 1-mA of bias current at room temperature. In reverse bias, we may usually assume that R_j is so much larger than the reactance of C_j at RF that it may be treated as infinite.

The back-to-front impedance ratio is approximately

$$\frac{Z_R}{Z_F} \approx \frac{R_S + 1/j\omega C_j}{R_S} = 1 + \frac{1}{j\omega R_S C_j}. \qquad (4)$$

The reciprocal of $R_S C_j$ is known as the diode cutoff frequency. Maximizing the reverse-to-forward impedance ratio is equivalent to saying that the cutoff frequency should be much higher than the intended operating frequency. For standard junction diodes, the series resistance and diode capacitance are opposing functions of

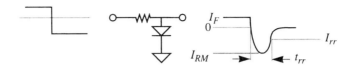

FIGURE 9.2. Representative test circuit and typical waveforms for reverse recovery

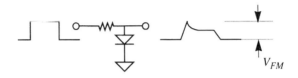

FIGURE 9.3. Representative test circuit and typical waveforms for forward recovery

cross-sectional area, so that their product is independent of area to first order. This lack of scaling constrains the operational frequency range of junction diodes.

An additional consideration is that a junction diode cannot turn off abruptly because, being a minority-carrier device, it remains on until the stored minority charge is actively removed or dies off through recombination. The time that the diode remains in the conductive state is known as the reverse recovery interval; see Figure 9.2. The value of t_{rr} is measured from the time the diode current reverses polarity to the time it has decayed to some specified value I_{rr}, and thus it depends on the latter. It is also a function of the forward current that precedes the transition to off, increasing as I_F increases. In addition, it is a function of the peak reverse current I_{RM}, decreasing as I_{RM} increases.

Some diodes are designed to produce a very rapid transition to the nonconductive state at the end of reverse recovery. The rapid shift from on to off generates high-frequency harmonics of an imposed fundamental signal. Uses of such snap-off (or step recovery) diodes are discussed further in Section 9.8.

Less commonly discussed is a phenomenon known as forward recovery. Just as diodes can't make an instantaneous transition from conduction to blocking, they also take nonzero time to turn on. As a consequence, it is possible for the forward voltage of a diode to exhibit considerable overshoot before settling down to its steady-state value of, say, 0.7 V. Alas, the forward recovery time is rarely specified. Instead, data concerning forward recovery is usually given in terms of the maximum forward drop, V_{FM}, under some prescribed test conditions. An example is the simple circuit shown in Figure 9.3. The amplitude of the step is adjusted to produce some specified value of steady-state current, and V_{FM} is then measured. It is not unusual for V_{FM} to be 3–4 times as large as the steady-state forward drop. Diodes of the same nominal type may exhibit considerable spread in forward recovery behavior.

In addition to the parasitic series resistance already described, packaged diodes will always have additional parasitics. At minimum, one may expect a certain amount

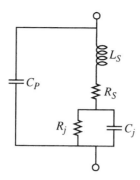

FIGURE 9.4. Simple high-frequency small-signal model of packaged diode

of series inductance and shunt capacitance. A reasonable high-frequency model for a packaged diode thus appears something like Figure 9.4. This model is fairly general and applies to virtually all packaged diodes.

Series inductance L_S accounts for the parasitic inductance associated with the physical length of the diode, and shunt capacitance C_P models the unavoidable capacitive coupling between the two ends of the diodes. The package parasitics are, of course, dependent on the particulars of the package. That said, typical values for the series parasitic inductance are usually within a factor of 2 of 1 nH, and the shunt capacitance typically within a factor of 2 of 100 fF, for most packages currently used in microwave work. As a calibration point, perhaps it is useful to know that the 4-mm–long type-PP glass package commonly used by lower-frequency diodes, such as the popular 1N914 and 1N4148, exhibits approximately 3-nH inductance (assuming zero additional lead length beyond the ends of the glass envelope) and ~150-fF shunt capacitance.

More complex models may include additional inductance in series with the model of Figure 9.4 to account for additional inductance (such as that associated with leads) beyond that of the device package itself.

9.3 SCHOTTKY DIODES

Schottky diodes (also known as hot carrier diodes) actually predate the junction variety. Ferdinand Braun, inventor of an early form of oscilloscope, published the first paper on crystal rectifiers, in 1874.[5] Exactly 25 years later he returned to that earlier work in an effort to improve the nascent radio art, but apparently he did not apply for any patents. That distinction belongs to the remarkable Jagadish Chandra Bose, who

---

[5] "Ueber die Stromleitung durch Schwefelmetalle," *Poggendorf's Annalen der Physik und Chemie*, v. 153 (now v. 229), pp. 556–63. In recognition of his many important contributions, he shared the 1909 Nobel Prize with Guglielmo Marconi.

needed detectors for the ~60-GHz radiation he was generating with special spark-gap apparatus.[6] He applied in 1901 for a patent on a galena-based point-contact diode acting as a bolometer, not as a rectifier (it was granted in 1904). That the operating principles would not be fully understood for decades did not inhibit researchers from patenting numerous related detectors in the early years of the century. Significant among them was a rectifying silicon detector patented in 1906 by the prolific Greenleaf Whittier Pickard.[7]

Pickard's silicon detector played an important role in stimulating the development of modern semiconductors for, as mentioned, it was just such a detector that was tried, almost in desperation, by Southworth. As he had hoped, a point contact does indeed guarantee exceptionally low device capacitance, far lower than that of a vacuum tube. More important, in rare defiance of Murphy's law, the resistance associated with a point contact scales inversely with the contact radius, rather than the area:

$$R_S = 1/4\sigma r, \qquad (5)$$

where σ is the conductivity and r is the contact radius. This equation for spreading resistance holds as long as the point contact's radius is much smaller than (a) the thickness of the semiconductor and (b) the radius of the other contact. Now, the parasitic capacitance does scale directly with area. Since the capacitance therefore shrinks faster than the resistance grows, the parasitic RC product drops approximately linearly with the radius, rather than remaining constant. This scaling behavior differs greatly from that of planar junction devices, and it accounts for the superior frequency response of point-contact devices.

The key to good high-frequency performance, then, is to produce as small a contact area as is consistent with providing a reliable contact. Methods for producing exceptionally sharp tips (typically of tungsten) – as well as the development of thin, highly polished wafers of pure silicon – enabled the achievement of K-band compatible point-contact Schottky diodes by the early 1940s.[8] It should be noted that the estimated contact diameters of ~2 μm were not matched by integrated circuit lithography until the 1980s.

The differences between those early point-contact diodes and modern ones are actually relatively minor. Packages have improved, resulting in a reduction of parasitic inductance in particular. A more important difference is in the quality of the semiconductors. Monocrystalline germanium and silicon ingots were not available until the 1950s, so early diodes were made from polycrystalline wafers. Carrier transport is inferior in polycrystalline material, so resistivity is significantly higher for a given doping level. The poly wafers were broken into shards, with much of the breakage

---

[6] This is not a typographical error. Bose was working with millimeter-wave radiation in the 19th century!

[7] Alan Douglas, in "The Crystal Detector" (*IEEE Spectrum,* April 1981), tells us that Pickard had tested over 31,000 combinations of minerals and wires by 1920.

[8] The 1N26 has a maximum specified conversion loss of 8.5 dB at 24 GHz.

9.4 VARACTORS

FIGURE 9.5. Schottky diode symbol

occurring along grain boundaries. Efforts were made to select material from within the larger grains in hopes of obtaining approximations to monocrystalline samples. The selected pieces were mounted on threaded studs, then the edges ground. After assembly into a cartridge and adjustment of the catwhisker (usually of tungsten), diodes were tested and sorted according to quality. A considerable spread in characteristics was common, and good diodes were zealously guarded by individual radar engineers. Today, of course, semiconductors are uniformly excellent, so yield loss and series resistance are now much lower. Recent work has resulted in GaAs point-contact diodes capable of operating above 170 GHz by using tungsten catwhiskers possessing 100-nm–diameter tips.[9] Modern deep sub-micron photolithography has advanced enough to produce exceptionally small contacts, so most Schottkys made today exploit this ability by forming an array of tiny contacts all over the surface of the diode. The catwhisker is then pressed against the surface with a reasonable pressure (for good stability), making contact with one of the tiny contact dots. Experimental "dot matrix" devices with cutoff frequencies in the range of 10 THz have been demonstrated, with much of the performance due to the etching away of parasitic material.[10]

Since Schottky diodes do not function by minority carrier injection, there is no reverse recovery delay. Also, many modern Schottky devices exhibit ideality factors n that are very close to unity, often within a range of 1.05 to 1.1:

$$i_D = I_S \exp\left(\frac{v_d}{nV_T}\right). \qquad (6)$$

For a 1-mA forward current, the dynamic resistance is therefore typically not much greater than 25 Ω at room temperature. Figure 9.5 displays the symbol for a Schottky diode.

9.4 VARACTORS

The voltage-dependent capacitance exhibited by both Schottky and ordinary junction diodes is extremely useful for making voltage-tuned circuits, and varactors consequently find wide use in voltage-controlled oscillators (VCOs), for example.

We've already seen that a linear doping profile leads to a relatively lazy voltage dependence, whereas abrupt doping produces a more dramatic capacitance variation with voltage. Diodes with this latter type of doping can provide a typical capacitance

---

[9] "Die MIM oder MIS Dioden," ⟨http://mste.laser.physik.uni-muenchen.de/~mst/mim.html⟩.
[10] ⟨http://info.iaee.tuwien.ac.at/gme/jb97/97_07.htm⟩.

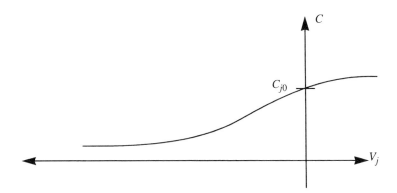

FIGURE 9.6. Typical varactor V–C curve

tuning ratio of about 2:1. Although it might initially appear that one could not improve upon the results of abrupt doping, an even greater sensitivity to voltage may be obtained by using hyperabrupt doping profiles. These are accomplished by approximating an exponential doping profile.

The voltage dependency of diodes made with all of these profiles is still reasonably well approximated by the relationship given earlier, repeated here for convenience:

$$C_j = \frac{C_{j0}}{(1 - V_j/\psi_0)^m}. \tag{7}$$

We've already observed that the exponent m is $1/3$ and $1/2$ for linearly graded and abrupt junctions, respectively. For certain hyperabrupt profiles, m can even exceed unity (values in excess of 2 are readily available commercially). As a consequence, hyperabrupt doping profiles can produce diodes whose capacitance varies over a decade range. Such diodes permit the tuning of LC oscillators over a 3:1 frequency range, enabling the realization of electronically tuned AM radios, for example.

A typical voltage-vs.-capacitance curve for a hyperabrupt varactor is shown in Figure 9.6.

PARAMETRIC AMPLIFICATION

As the Second World War drew to a close, researchers at the MIT Radiation Laboratory discovered a puzzling phenomenon. Under certain conditions, an ordinary point contact diode could provide power gain. No violation of energy conservation is implied, since both a DC bias and a local oscillator were present as potential power sources, but it was unclear exactly how energy from either or both of these sources could end up at the signal frequency. Subsequent investigation with an analog computer revealed that the presence of a nonlinear junction capacitance is essential to producing this effect. Beyond this key observation, though, no other explanation was offered.[11]

---

[11] Torrey and Whitmer, op. cit. (see footnote 2).

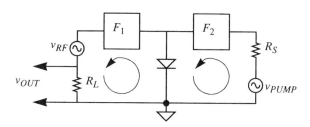

FIGURE 9.7. Simplified parametric amplifier

Today, this phenomenon is known as parametric amplification. The nomenclature derives from the purposeful variation of a circuit parameter in some prescribed manner. In the case of the parametric diode, the essential features of how amplification may arise can be explained qualitatively by considering a simple LC tank in which the capacitance is a conventional parallel-plate capacitor, except with a variable spacing. At any fixed value of plate spacing, and given some initial charge, the tank will oscillate at its natural frequency. If any loss were present, of course, it would cause the amplitude to decay exponentially with time.

Now consider what happens if we suddenly pull the plates of the capacitor apart some distance just as the capacitor voltage reaches its peak. At that instant, the attractive electrostatic force between the two plates is at a maximum, so we have to work *against* the field in order to pull the plates apart. Where does the energy of that work go? With no dissipative mechanisms in the tank, that energy is clearly injected into the LC tank, causing the amplitude to increase. To restore the capacitance without perturbing the energy, we can quickly push the plates together again 90° later, when the tank voltage is zero. Another 90° later, we pull the plates apart once again, and so on. By injecting energy twice every cycle in this manner, we can continue to increase the amplitude of the oscillation; we have achieved amplification.

Note that the frequency at which we vary the plate spacing is twice that of the tank resonance. The former, known as the pump frequency, is supplied by a separate oscillator in practical realizations. The pump oscillator periodically modulates the diode capacitance, causing a negative resistance to appear at the tank's resonant frequency.

So, if we're not violating energy conservation, where does the energy come from in a practical paramp? In a standard amplifier, the DC power supply is the ultimate source of the added signal energy. In a parametric amplifier, the energy comes from the pump oscillator. Just as we had to do work to separate the plates in our thought experiment, the pump has to do work in the actual parametric amplifier. The parametric amplifier can therefore be thought of as a system that violates linear time invariance, allowing energy at one frequency to show up as energy at a different frequency.

The simplified paramp schematic of Figure 9.7 shows in more detail how the pump oscillator and main input are applied to the diode, and where the output port is located. The RF input, v_{RF}, is fed to the diode through a bandpass filter F_1, which is tuned to the input frequency. At the same time, the pump input (nominally at twice the input frequency) is also fed to the diode through another bandpass filter, F_2, which is tuned to the pump frequency. The pump modulates the diode capacitance, causing it to produce a negative effective resistance at the RF input frequency. The left-hand loop is then capable of producing a net gain, resulting in an amplified output available across R_L.

The valuable attribute of a paramp is that only reactances are directly involved in the amplification process, so very low noise figures are possible. In fact, until the invention of the maser, the lowest-noise amplifiers in existence were paramps. Today, they are rarely used because conventional amplifiers frequently perform well enough without requiring a pump at twice the signal frequency. Nonetheless, it is instructive that amplification can occur parametrically. At the very minimum, it shows us that it is possible to obtain gain from a two-terminal element, an observation that has relevance for the section that follows.

As a final note on this subject, distributed amplifiers (also known as traveling wave amplifiers) have been constructed with parametric diodes.

9.5 TUNNEL DIODES

In the summer of 1957, a young Leo Esaki was working on his Ph.D. thesis at a young Sony Corporation. The topic was an investigation of how the breakdown voltage of germanium diodes varies with doping concentration. Since it was "well known" what to expect, the thesis topic probably appeared rather unpromising at first, something in danger of consignment to the dustbins of science (to paraphrase a familiar saying). Esaki himself perhaps sensed this danger, so he chose to fill in gaps in the literature, employing doping levels so extreme that no one else had bothered to try them, and observing the temperature dependence over a very wide range.

First, he confirmed that heavier doping leads to reduced breakdown voltages. That wasn't new. What *was* novel was his surprising observation that, with incredibly heavy (degenerate) doping, it is possible to reduce the apparent breakdown voltage to less than zero. That is, in some cases, it is necessary to apply an increasing *forward* bias to *reduce* the current. Over that regime of operation, such a heavily doped diode exhibits a negative incremental resistance. With still more forward bias, the diode characteristics eventually converge to those of a standard diode.

As seen in Figure 9.8, a tunnel diode's forward current increases to a peak value (I_P) and then decreases to a valley current (I_V) as the forward voltage increases from zero to V_P to V_V. Convergence with ordinary junction diode behavior takes place beyond V_V. Esaki did more than simply report these observations; he explained them as consequences of quantum mechanical tunneling.[12] Up to that time, tunneling was a theoretical construct that physicists found useful for considering various phenomena, but that no one had observed directly. Esaki's direct demonstration of tunneling – in

---

[12] L. Esaki, "New Phenomenon in Narrow Germanium p–n Junctions," *Phys. Rev.*, v. 109, 1958, p. 603.

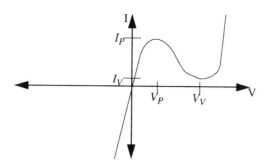

FIGURE 9.8. Typical tunnel diode characteristics

so simple a structure, no less – was so dramatic that he was awarded a Nobel Prize in physics in 1975.

Aside from its purely scientific significance, the negative resistance exhibited by a tunnel diode is potentially useful as a microwave gain element. Consider, for example, an amplifier formed from a standard resistive voltage divider in which one of the resistances has a negative value:

$$\frac{v_O}{v_{IN}} = \frac{R_1}{R_1 + R_2}. \tag{8}$$

With $R_2 < 0$, the output voltage exceeds the input voltage, so we have made an amplifier.

Commercially available Ge tunnel diodes exhibit typical valley currents ranging from a few hundred microamps to several milliamps (a few are as high as 20 mA, but these are unusual), at a forward bias of about 350 mV. Peak currents are typically 5–10 times higher than the valley current and occur at roughly 65-mV forward bias. Typical peak negative conductances range from several microsiemens to a few hundred millisiemens.

An important figure of merit concerns the frequency range over which the diode produces a net negative resistance. Consider again our general diode model (Figure 9.9). The shunt capacitance C_P and series inductance L_S can be neutralized at any given frequency (at least in principle) by external tuning elements, so the only model elements that need to be considered are R_S, R_j (which can be negative), and C_j. The parallel $R_j C_j$ network can be transformed into a series equivalent RC, whose resistance is

$$R = \frac{R_j}{1 + (\omega R_j C_j)^2}. \tag{9}$$

The net resistance is therefore $R_S + R$, which will be negative up to a frequency limit (also known as the resistive cutoff frequency) given by

$$\omega_{max} = \frac{1}{R_j C_j} \sqrt{1 - \frac{R_j}{R_S}}. \tag{10}$$

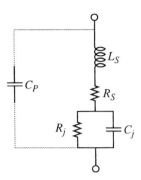

FIGURE 9.9. Simple high-frequency model of tunnel diode

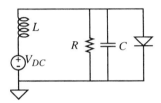

FIGURE 9.10. Simplified tunnel diode oscillator

Good tunnel diodes minimize the junction (and stray) capacitance and series resistance.

Another obvious use of a negative resistance is in realizing oscillators. Connecting a tunnel diode to a resonator makes an oscillator, provided that the diode's negative resistance exceeds the positive resistance that models the parallel tank loss. See Figure 9.10. In this simplified oscillator, the RLC circuit (with the inductor connected through the DC power supply) is a lossy resonator. The tunnel diode is placed in parallel with the tank and then biased to its negative resistance region. Oscillation occurs as long as the tank loss (including loading) can be negated by the diode's effective resistance. In many high-frequency applications, the tank is not a lumped RLC but rather a resonant cavity (a class that includes resonant transmission lines).

Because the extreme doping levels implied that neither very pure materials nor extreme cleanliness during fabrication was necessary, many were optimistic that tunnel diodes – with their simple structure, high operational frequency, and ease of manufacture – would displace transistors in many applications. Work on microwave systems and even computers based on tunnel diodes was undertaken in laboratories throughout the world.

Nevertheless, the reader may be aware of a total lack of dominance by the tunnel diode in modern microwave technology. One reason that the tunnel diode never lived up to its early promise is that, being a two-terminal element, it is very difficult

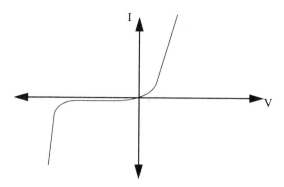

FIGURE 9.11. Typical backward diode characteristics

to construct unilateral amplifiers. Hence, tunnel diode circuits have the unfortunate quality that changes in the load impedance cause changes all the way up to the input. This property makes cascading quite difficult. Hence, except for very specialized or simple stand-alone circuits, the tunnel diode was found unsuitable. Another factor is that the negative resistance region spans a relatively small voltage range, so that useful signal levels are constrained to a fraction of a volt, leading to typical signal output powers below 1 mW. Finally, stabilizing the gain against variations in bias and temperature is generally necessary. Manufacturers have thus largely abandoned production, and the few tunnel diodes available today are remarkably expensive, reflecting industry's reluctance to make them. By one estimate, fewer than 10,000 tunnel diodes were sold worldwide in 1999. Nevertheless, tunnel diodes have remained a favorite of academic research, and publications continue to appear.

BACKWARD DIODE

The backward diode is simply a tunnel diode whose anode and cathode have been interchanged and which employs doping levels that cause near-equal values of peak and valley current. The resulting V–I characteristic looks approximately as shown in Figure 9.11. By interchanging the roles of anode and cathode, a diode with very low "forward" drop results, thereby approximating more closely the characteristics of an ideal diode, at least in the forward direction. However, note that the "reverse" breakdown of a backward diode is under a volt, so this diode has limited utility as a power rectifier. It finds occasional use as a small-signal detector or mixer.

9.6 PIN DIODES

The minority-carrier storage problem that afflicts standard junction diodes is exploited in the PIN (for p-intrinsic-n) diode. (There is an important theme here: A problem in one context can often be turned into a critical advantage in another, if one just thinks about it for a little bit.) As its name implies, PIN diodes interpose a

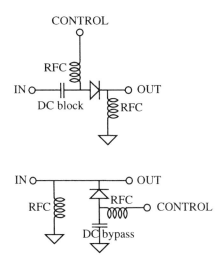

FIGURE 9.12. PIN diode attenuator/switch circuits

very lightly doped (i.e., intrinsic or near-intrinsic) semiconductor layer between the standard junction layers. Because of the high resistivity of this interposed layer, its resistance dominates that of the whole diode. When the diode is forward biased, injected charge raises the conductivity of the intrinsic region.

At low frequencies, the PIN diode simply behaves pretty much as an ordinary diode, since recombination processes are fast enough to kill off the injected charge over the period of a slowly varying terminal voltage. However, carrier lifetimes are particularly large when the doping is light, so the diode is unable to keep up with quick variations in applied voltage (e.g., at MHz rates and above). The charge density therefore remains roughly constant, causing the PIN diode to appear like a resistor. Furthermore, the value of resistance is a function of the DC bias current through the diode.

Because the PIN diode thus behaves as an electronically controllable resistance, it has found wide application in the RF domain as a modulator, variable attenuator, and switch. Since it is amenable to fabrication by standard microelectronic processing techniques, a PIN can be made physically small, with correspondingly small parasitics. Filter components can then be electronically connected to (or removed from) a circuit to effect band switching, for example. Automatic gain control (AGC) is another application in which PIN diodes excel. The circuits depicted in Figure 9.12 show typical methods for using a PIN as attenuators or switches.

The top circuit in the figure places the PIN diode in series between the input and load. The bias current for the diode is supplied through the control input with the aid of RF chokes whose impedance is large at all signal frequencies of interest. A DC blocking capacitor prevents this DC current from flowing into the input source. No blocking capacitor is generally needed on the output because the ground-returned RFC forces the DC output voltage to be zero. At low bias currents, the diode is a high impedance, and little signal propagates from input to output. In the limit of zero bias, the feedthrough is due entirely to the total shunt capacitance in parallel with the diode. Typical values of maximum attenuation range from perhaps 20 to 40 dB, with increasing degradation from feedthrough at higher frequencies. At high bias currents, the conductance of the diode is very high, connecting input to output. Typical insertion losses are below 1 dB. In this limit, the PIN therefore may be considered a switch.

The second circuit uses a shunt arrangement in which the PIN diode places an additional resistance in parallel with the load. Here, a high bias current is associated with maximum attenuation whereas a low bias current provides the minimum attenuation.

The two circuits are frequently combined into a T-arrangement, with a shunt branch in the middle that provides a path to short out any feedthrough component. High levels of isolation (e.g., >70–80 dB) are possible with a T-switch, but at the expense of somewhat higher minimum values of loss.

It should be clear that permutations of the basic configurations can be used to implement more complex switching arrangements, such as single-pole, double-throw (SPDT) and others. These are left as an "exercise for the reader."

Finally, because the PIN is fundamentally an electronically controlled resistor, it may be used in applications wherever a potentiometer might be used at lower frequencies. Examples include (but are certainly not limited to) voltage-controlled filters and phase shifters.

9.7 NOISE DIODES

When a sufficiently large reverse bias is applied across a junction, the electric field may be able to rip bound electrons from their host atoms. If the field and mean free path are large enough, the freed electrons may accelerate sufficiently to liberate still more electrons during subsequent collisions. These secondary electrons themselves may accelerate and liberate even more electrons, and so forth. This avalanche breakdown is therefore associated with an extremely rapid rise in current once the breakdown voltage is reached. Stated another way, the voltage is roughly insensitive to current in the breakdown region. For this reason, such diodes (often called Zener diodes, even though Zener's breakdown mechanism isn't the same as avalanching) are widely used as voltage references or voltage regulators.

It was discovered early on that Zener diodes are extremely noisy (actually, this noisiness was first observed in gas-filled "vacuum" tube predecessors, which also function by avalanche breakdown).[13] Engineers must work hard to ensure that this noise doesn't enter critical signal paths and thereby degrade signal-to-noise ratio.

In keeping with the idea that "if life gives you lemons, make lemonade," this noisiness has been exploited to make broadband white noise sources whose effective noise

---

[13] It should be mentioned that avalanching gas-filled waveguide noise sources have been used as national laboratory noise standards to which diode noise sources are traced.

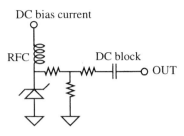

FIGURE 9.13. Typical noise diode

temperatures are in the many thousands of degrees – but without requiring the actual heating of a resistor to such temperatures. Such diodes are therefore indispensable as safe, compact, and low-power noise sources for noise figure measurement, to name one prominent example.

The noise fundamentally arises from the stochastic nature of avalanching: an electron goes some random distance, acquires a random energy, bashes into an atom, and frees a random number of electrons. The spectrum of avalanching is consequently extremely broad, so as long as the device parasitics are small, the output noise extends over an exceptionally wide frequency range.

One problem is that there is considerable variation in noise from diode to diode, so each one must be calibrated against some primary standard if the diodes are themselves to be used as secondary standards in noise measurement. Other characteristics, such as changes over time or with ambient temperature variations, must be accommodated or controlled. Subsurface or guard-ring–stabilized zeners have been developed to reduce instability over time. Partly as a result of the somewhat nonstandard structure and partly because of the labor involved in calibration, noise diodes that are traceable to an NIST (or other laboratory) standard are quite expensive (of the order of $2000 at the time of this writing).

Figure 9.13 shows a typical connection of a noise diode. The diode is biased with a current chosen to provide a stable, flat noise spectrum (the resistive T-network aids in impedance matching). The RF choke prevents shunting of high-frequency components to ground, and the DC blocking capacitor (not present in all designs) prevents the DC diode voltage from upsetting whatever load receives the output of the noise diode.

More elaborate circuits include some additional components to extend the frequency range over which the spectrum is white. The added circuitry is often hand-tailored for each diode.

9.8 SNAP DIODES

Again in keeping with the lemonade philosophy, *snap* diodes (also known as *snap-off* or *step recovery* diodes) are p–n junction diodes carefully designed to exploit the minority carrier storage inherent in such diodes. Recall that applying a reverse bias to

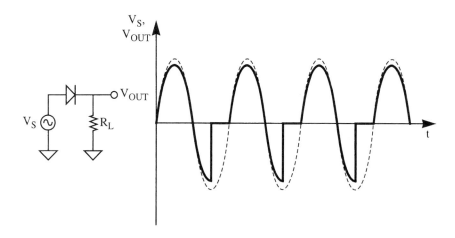

FIGURE 9.14. Illustration of reverse recovery

a diode that was previously in forward bias does not instantly cause a p–n diode to turn off; the injected minority carriers must be removed first. Until then, the diode remains on, conducting current with a low voltage drop (e.g., 0.7 V). This departure from ideal behavior has implications for RF circuits and even for switching power supplies, where conduction in the reverse direction can result in significant loss in efficiency. In extreme cases, it can even cause component failure.

The snap diode converts this drawback into an advantage. More precisely, it makes use of the rather abrupt transition from on to off that occurs once the stored charge has been removed. It is therefore possible to produce signals with extremely fast risetimes with snap diodes. By the principle of time-frequency duality, we recognize that fast risetimes imply high-frequency spectral content. Snap diodes have been used as harmonic multipliers and generators of control pulses for sampling gates and other very fast circuits. Hewlett-Packard, the company that pioneered snap diodes,[14] still uses them in front ends for multi-GHz sampling oscilloscopes. It is possible to obtain snap diodes capable of producing sub–10-ps risetimes.

To understand better why snap diodes are useful, consider the circuit – and its associated waveforms – shown in Figure 9.14. In the figure, the dashed curve is the source voltage V_S and the solid curve is the output voltage V_{OUT}. If the diode had infinitely fast reverse recovery then the output voltage would never go negative, instead going to zero as soon as the input voltage goes negative. However, a nonzero recovery time causes the diode to continue to remain on for some time, with a flow of significant reverse current until the stored minority charge has been removed. The diode current rapidly goes to zero once this charge is gone, leading to a sharp transition from on to off. In the figure, the reverse recovery interval just happens to occupy about 90° of the applied sinusoid.

[14] A. F. Borr, J. Moll, and R. Shen, "A New High-Speed Effect in Solid-State Diodes," *ISSCC Digest of Technical Papers,* February 1960.

From inspection of the output waveform, the spectrum clearly contains significant harmonic energy. With a periodic input, the output spectrum consists of harmonics of the input fundamental, resulting in a comblike spectrum with significant energy out to approximately the inverse risetime of the snap recovery.[15] Filters can be used to select the desired component(s), enabling use of the snap diode as a frequency multiplier. This way, a 100-MHz input signal can be used to generate gigahertz outputs, for example. On the other hand, this frequency multiplication can be troublesome if it occurs unexpectedly. An all-too-common example is an ordinary power supply rectifier, where a 60-Hz input may be multiplied upward in frequency into the high AF and low RF range, causing interference. This phenomenon is particularly objectionable in audio amplifiers and AM radios, where the interference may result in audible artifacts.

An alternative description of snap diode operation is that it produces a sharpening up of slow waveforms. It is in this context that snap diodes are often employed in high-speed samplers, where a slower pulse is sped up by a snap diode circuit to produce the exceptionally fast pulses needed for operating a sampling switch.

The snap-off speed is dependent on doping because the rate at which minority carriers are swept out is related to the internal electric field. Heavier doping leads to higher built-in fields and hence faster snap-off speeds. Furthermore, doping profiles that maximize electric field are also helpful. The exponential grading used by hyperabrupt varactors is particularly suitable for making snap diodes. Generally speaking, the larger the value of grading coefficient m, the snappier the diode.

Since breakdown voltage is also related to doping, one expects to find lower breakdown voltages associated with faster turn-off, all other factors held equal, and this correlation is indeed observed in practice. The fastest snap diodes are low–breakdown-voltage varactors.

The reverse recovery time is a function of the forward current (more stored charge implies longer recovery), and the spectral content is therefore also a function of the bias. In comb generators or frequency multipliers, provisions for adjusting the bias are frequently added to the simple test circuit shown in Figure 9.14. The bias and input amplitude may be varied to optimize some performance parameter (e.g., amount of tenth harmonic). Using these degrees of freedom in conjunction with filters allows many snap diode frequency multipliers to provide conversion gains as high as -15 dB or so, with the conversion gain changing little until the bandwidth limit (which may be estimated using the bandwidth–risetime rule of thumb) is reached.

As a final note on snap diodes, it is important to distrust most *Spice* simulations of reverse recovery. The diode models provided by most vendors fail to capture the subtle effects needed for accurate simulations. In particular, the equations for charge storage seem to be fundamentally in error, with incomplete accommodation of true carrier transit time.

---

[15] The old rule of thumb relating bandwidth and risetime ($f_{-3dB} t_r = 0.35$) may be taken as a first approximation here.

9.9 GUNN DIODES

One way of thinking about how an oscillator works is to consider it a combination of a negative resistance and a resonator. The negative resistance is often realized at the lower microwave frequencies with a transistor connected in a positive feedback configuration – as in a Colpitts oscillator, for example. At extremes of frequency, however, the phase shift through the active element itself is large enough to cause the oscillation frequency to depend less on the resonator and more on the transistor. In the limit, the oscillation frequency is constrained by factors such as f_{max}.

An alternative is to employ a two-terminal element that is capable of generating a negative resistance through some fancy physics. Recall that the effective mass of an electron in a solid generally differs from that in free space because of interaction with all of the other charges in the material. These other charges include those that flow as a result of an applied field. Thus, there exists a possibility of a field-dependent effective electron mass.

Now, gallium arsenide (GaAs) is a material whose band structure includes two conduction band valleys separated by a relatively small energy difference (indium phosphide, InP, is another). Electrons in these two valleys have two different effective masses, with heavier electrons in the higher energy valley. At low electric fields, conduction involves only electrons in the lower energy valley. As the field strength increases, that valley fills up. Eventually, the field strength reaches a threshold (around 300–350 kV/m for GaAs) beyond which the upper valley begins to fill. Because of the higher electron mass, the drift velocity decreases as the electric field increases across this transition. Because the current density is proportional to velocity, we have the interesting situation that there is a range of field strengths over which current actually decreases as the applied voltage increases. See Figure 9.15. The behavior we have just described is an incremental negative resistance. It was first observed experimentally by J. B. Gunn, for whom the effect and device have been named.[16] However, theoretical predictions of the Gunn effect were actually made several years earlier, by Ridley and Watkins[17] and by Hilsum.[18] Since it is much easier to say "Gunn" than it is to say "Ridley, Watkins, and Hilsum," the former name has stuck.

The attractive property of a Gunn diode is that it is a bulk device; calling it a diode is to use the term in its most general sense: a two-terminal element. A Gunn diode contains no junctions. No exotic processing or expensive lithography is required. "Simply" package it properly and bias the material to the negative resistance region, and it's ready to go.

---

[16] "Microwave Oscillation of Current in III-V Semiconductors," *Solid-State Commun.,* v. 1, 1963, p. 88.
[17] "The Possibility of Negative Resistance Effects in Semiconductors," *Proc. Phys. Soc. London,* v. 78, 1961, p. 293.
[18] "Transferred Electron Amplifiers and Oscillators," *Proc. IRE,* v. 50, 1962, p. 185.

FIGURE 9.15. Typical V–I plot of Gunn diode

When used in CW oscillator applications, the bias current is typically adjusted to a value 10–20% below the value that produces the maximum output power, corresponding to applied voltages on the order of three times the threshold voltage V_{th}. For pulsed applications, the voltage may be an additional factor of 3 higher for larger output power on a transient basis.

Gunn oscillators for frequencies ranging from approximately 5 GHz to 150 GHz are commercially available,[19] with CW output power spanning a few milliwatts to several hundred milliwatts or so and with DC-to-RF conversion efficiencies in the range of ~10%. Pulsed power output values are a couple of hundred times higher at any given frequency, and the largest Gunns are capable of pulsed output power in the kilowatt range.

An important consideration in all amplifiers and oscillators is their noise. A commonly used device noise parameter is called the noise measure, M, which for a two-terminal element is simply the factor by which the effective device resistance is noisier than a passive resistor of the same absolute value:

$$M \equiv \frac{\overline{v_n^2}}{4kT|R|\Delta f}. \tag{11}$$

Gallium arsenide Gunn diodes typically exhibit noise measures of the order of 15 dB at 10 GHz.

There is a trade-off between frequency and power capability in device design that leads to a bound on the product of output power and (at least) the *square* of frequency. An approximate but intuitively appealing derivation begins with the observation that some critical device dimension usually limits frequency response. Hence, this dimension must scale directly with wavelength (or inversely with frequency). But as the critical dimension scales, so does the breakdown voltage (assuming a constant critical breakdown field). Power is proportional to the square of voltage and thus inversely proportional to the square of frequency. For CW operation, this maximum is about 500 W-GHz^2 and about 10^5 W-GHz^2 in pulsed mode.[20] The inverse square

---

[19] The tuning range of any given Gunn oscillator is typically under 10–20% and is generally limited by the resonator technology.

[20] D. M. Pozar, *Microwave Engineering*, 2nd ed., Wiley, New York, 1998, p. 590.

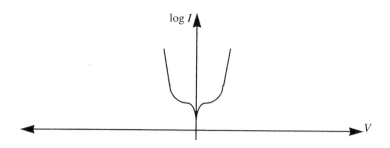

FIGURE 9.16. Representative V–I characteristics of symmetrical MIM diode

frequency dependence worsens to an inverse cubic relationship as one moves deeper and deeper into the millimeter-wave bands.

The bias voltage affects the phase delay between applied voltage and the current that flows, and hence it affects the effective reactance of the Gunn diode. This reactance forms part of whatever resonant structure is used to set the oscillation frequency. As a consequence, there is a nonzero supply pushing effect, and one must stabilize bias voltages if this is to be minimized. Alternatively, the sensitivity to supply voltage provides a means of fine-tuning the frequency of a Gunn oscillator.

Temperature also has a strong effect, with both the threshold and peak output power voltages decreasing as temperature rises. Typical values for the temperature coefficient of those voltages are around −5000 ppm/°C. Prevention of supply push is therefore complicated if the temperature varies.

9.10 MIM DIODES

Another structure that depends on tunneling is the metal-insulator-metal diode. If the insulator is very thin (on the order of a few nanometers or less, just as for the tunnel diode, and for the same reasons), significant tunnel current may flow. Because of the nonlinear dependence on voltage, MIM structures act as symmetrical nonlinearities. See Figure 9.16. With a suitable bias (which may include zero), MIM diodes are capable of detection, just as are more conventional diodes. Since current transport through the insulator takes place through tunneling and since the other electrodes are of very conductive metal, such diodes are capable of operation at exceptionally high frequencies. Operation up to near-optical frequencies is not out of the question for integrated implementations of MIMs.

9.11 IMPATT DIODES

Around 1956, Read[21] proposed an avalanching diode structure that would generate a negative resistance. A greatly simplified version of his reasoning is as follows:

---

[21] W. T. Read, "A Proposed High-Frequency, Negative Resistance Diode," *Bell System Tech. J.*, v. 37, March 1958, p. 401. This paper was circulated within Bell Labs two years before publication.

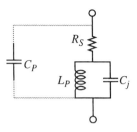

FIGURE 9.17. Simple high-frequency model of IMPATT diode

Operate a diode in avalanche breakdown, but lengthen it enough so that there is a nonnegligible carrier transit delay across the device.[22] Because of this delay, it may be possible for the current density to lag behind an applied sinusoidal voltage. If this lag can be made large enough, it may be possible to arrange for the current to decrease while the voltage increases. If this happens, we have created a negative resistor, which may be used in all of the applications we've cited for the other negative resistors (tunnel and Gunn diodes).

Inspired by Read, several groups sought to reduce his theoretical ideas to practice. Some early work turned out to be surprisingly easy to carry out. By coincidence, ordinary manufacturing practice of the 1950s just happened to produce diodes with a structure suitable for creating a negative resistance. As one Bell Labs engineer noted upon testing a junk box full of ten-year-old diodes, they "just seemed to want to work."[23] In relatively short order, oscillators at tens of gigahertz were demonstrated with remarkably simple circuits (essentially consisting of just the diode and a resonator tuned to the diode's resonant frequency).

As can be seen from Figure 9.17, an IMPATT's avalanche and drift region is modeled as a parallel LC resonator, reflecting the tight coupling between the carrier transit delay and the operational frequency. Variations on Read's original structure provide differing trade-offs among efficiency, tuning range, and operating frequency, but it is generally true that IMPATTs are narrowband devices with typical tuning ranges measured in the low tens of percent.[24] The IMPATT must therefore be selected to match the resonant cavity in which it is to be mounted. Coarse tuning can be performed by adjusting the dimensions of the cavity, and fine tuning (including modulation) can be effected by varying the bias voltage across the diode.

By the late 1960s, IMPATTs capable of operating up to 340 GHz had been demonstrated. Devices with 5 W of CW power at 14 GHz and 30 W pulsed at 8.5 GHz further underscored the promise of these devices. Peak efficiencies of 10–20% are

---

[22] Avalanching by itself involves significant delay, to which the delay along the rest of the structure must be added when computing the total delay.

[23] B. C. De Loach, Jr., "The IMPATT Story," *IEEE Trans. Electron Devices*, v. 23, no. 7, July 1976.

[24] For an excellent treatment of these and other microwave diodes, see S. Sze, *Physics of Semiconductor Devices*, Wiley, New York, 1981, pp. 567–613.

comparable to values achieved by Gunn diodes. Limits on performance are significantly better than for Gunns, with CW silicon sources constrained to below 3 kW-GHz^2 and pulsed ones three times better than that.[25]

Because of the need for avalanching and also because output power is related to the output voltage swing, rather high voltages are needed to operate IMPATT devices. Typical values range around 50–100 V, and IMPATTs may be connected in series for higher output power.

The high output powers at high frequencies would seem to make IMPATT sources attractive alternatives to traditional vacuum tube sources (such as magnetrons) for many applications. Unfortunately, as we've already seen, avalanching is fundamentally a noisy process, so oscillators made with IMPATTs have high levels of phase noise. This drawback limits their use to the small set of applications that do not require low levels of noise, and it also explains why Gunn diodes dominate even with their somewhat lower output power. Noise measures for IMPATTs are typically in excess of 30 dB and can approach 50–60 dB under some conditions, particularly as the voltage is increased. It is common practice to operate IMPATT oscillators at bias voltages ~15% below values that produce the maximum output power as a compromise between noise measure and output power.

As a closing comment on avalanche mode devices, it should be mentioned that experimenters stumbled upon an "anomalous" mode of IMPATT operation, later termed TRAPATT (for trapped plasma avalanche transit time).[26] This mode is operative at lower frequencies and enables remarkably high efficiencies. Pulsed output power of 400 W at 1 GHz and an impressive efficiency of 60% are mentioned by De Loach (see footnote 23). These devices share with their IMPATT cousins the problem of high noise, and commercial application of TRAPATTs is relatively rare.

9.12 SUMMARY

We've seen numerous ways in which two-terminal devices can be designed to provide a rich variety of functions that are useful at RF and microwave frequencies. High-frequency amplifiers, attenuators, modulators, switches, detectors, oscillators, and short-pulse generators are readily constructed with remarkably simple circuits.

9.13 APPENDIX: HOMEGROWN "PENNY" DIODES AND CRYSTAL RADIOS

Although silicon is today the dominant semiconductor by a huge margin, we shouldn't forget that there are *many* other semiconducting materials, both organic and inorganic. We've already noted that Braun first described quintessential semiconducting behavior – rectification – in 1874, long before the word "semiconductor" was

---

[25] Pozar, *Microwave Engineering.*
[26] De Loach, "The IMPATT Story."

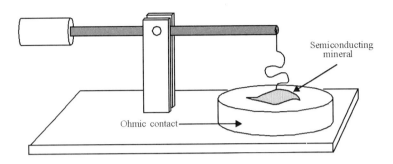

FIGURE 9.18. Typical crystal detector

coined and well before the invention of radio. He found that many naturally occurring minerals, such as galena (PbS), iron pyrites (FeS_2, more commonly known as fools' gold), tennantite/tetrahedrite (Cu_3AsS_{3-4}/Cu_3SbS_{3-4}), and chalcopyrites ($CuFeS_2$/$Cu_2S \cdot Fe_2S_3$) disobeyed Ohm's "law" by conducting current unequally in the two polarities. An understanding of solid-state quantum physics lay more than fifty years in the future, so Braun was unable to provide a satisfactory theoretical explanation for his puzzling experimental observations. Even today, quantitative descriptions of many compound semiconductors are somewhat elusive, owing to the difficulty of making sufficiently pure and perfect samples of these complicated substances.

Three decades later, as the age of wireless dawned, simpler group-IV semiconductors finally made their debut, again by accident. Around the turn of the century Acheson inadvertently created carborundum (SiC) during attempts to make synthetic diamonds. Not nearly as precious yet almost as hard as diamond, it found almost immediate and enduring application as an abrasive. It was left to Gen. Henry Harrison Chase Dunwoody to discover and exploit its semiconducting nature to make radio detectors in 1907.

At nearly the same time, Greenleaf Whittier Pickard discovered that silicon was an excellent semiconductor and patented its use as a radio detector. This achievement represents the first recognition and use of an elemental semiconductor. Pickard went on to study the field exhaustively, eventually testing over 30,000 combinations of minerals and contacting wires (whimsically known as catwhiskers), patenting many of these in the process. See Figure 9.18. One, consisting of a contact between the minerals zincite (ZnO) and bornite (Cu_5FeS_4), found fairly wide use. Pickard named it the Perikon detector, for PERfect pIcKard cONtact. Pickard was clearly a better scientist than a product marketeer.

All of these mineral-based materials were used to make what became known as crystal[27] radios. Remarkably, these typically operate without any power source

---

[27] The term *crystal rectifier* was coined by Prof. G. W. Pierce of Harvard University around 1909, who studied these materials exhaustively (but ultimately in vain) in search of an explanation for why they behaved as they did. At least he showed that all contemporary theories were wrong, leaving the path clear for the quantum theorists.

9.13 APPENDIX: HOMEGROWN DIODES AND CRYSTAL RADIOS

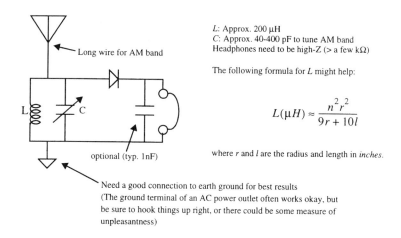

FIGURE 9.19. Simple crystal radio (not optimized)

beyond what is contained in the incoming radio wave. In its simplest form, a crystal radio consists of an antenna, a diode, and a headphone; see Figure 9.19. The optional capacitor is rarely needed with most practical headphones, but it is shown for completeness and for consistency with circuit diagrams from the era.

To understand how such a simple radio could possibly work, it is important to know that the incoming wave is amplitude modulated (AM), meaning that speech or music is encoded as variations in the amplitude of the radio-frequency carrier wave that actually propagates through space. The modulated wave has zero average value, and our ears are too slow to respond to anything but the average. The diode rectifies the symmetrical AM signal, producing an asymmetrical signal whose time-varying average is proportional to the modulation.

The reason that no batteries are needed is that the human auditory system is truly amazing. The threshold of hearing at 1 kHz corresponds to an eardrum displacement on the order of the diameter of a hydrogen atom! So, the magic here isn't in the radio, it's in the biology. Of course, the absence of any electrical amplification means that crystal radios require rather good antennas and reasonable proximity to radio stations.

The standard crystal for these radios consists of a lump of galena (the best being argentiferous, which – although silver-bearing – is known as steel galena because of its appearance), to which one contact was historically made through immersion in a low–melting-point alloy of lead, cadmium, and bismuth known as Wood's metal. This contact is not at all critical, the aim being merely to make a low-resistance contact to the crystal. Wood's metal is actually unnecessary, and simple clamping is sufficient. The jaws of an alligator clip work just fine.

The other contact, though, is tricky, because it is the region near the interface between the catwhisker and the surface of the crystal that is the seat of rectification. If the contact pressure is too great, a short circuit results; if too light, the resistance is excessive. Consisting of a very fine wire (e.g., 2 mil in diameter), the catwhisker must make only the lightest of contacts to the surface of the galena. One must also hunt around the surface to find a suitably sensitive spot. Because of the tenuous contact, it is easily jarred, requiring frequent adjustment. These days, it is much easier to make a good crystal radio because excellent diodes that do not require adjustment are available commercially. The 1N34A germanium diode (available from a decreasing number of sources) is particularly suitable for crystal radio work because it is a sealed unit, but it lacks the charm of a catwhisker to fiddle with.

An alternative to galena was frequently used by soldiers during the Second World War: razor blades. A rusty one works okay, since it turns out that Fe_2O_3, the dominant oxide in ordinary rust, is a semiconductor (Pickard discovered this in 1906). Better still are the "Blue Pal" blades once made by the American Safety Razor Company, because of their thin oxide coating of high quality.[28] An improvised catwhisker, made out of a sharpened pencil lead or safety pin, was generally used in these "foxhole" radios.

It turns out that one can make a remarkably good diode out of an ordinary penny. In fact, this diode is much better than those made of razor blades and almost as good as store-bought diodes.

In the early 1920s, two fellows named Grondahl and Geiger (not *that* Geiger) discovered accidentally that Cu_2O (cuprous oxide) is a semiconductor (as it happens, it is always p-type). What set this discovery apart from all the earlier semiconductor work was that copper oxide devices have sufficiently uniform characteristics over the surface that a large-area contact (instead of a point contact) is practical. Although a large area is undesirable for high-frequency operation, it is precisely what is needed for high-current applications. Power supply rectifiers thus became the first application for this material, displacing the less reliable, far bulkier, and power-consumptive vacuum tubes then in use. The $\sim 1010°C$ ($\sim 1850°F$) required processing temperature, however, precludes making this oxide safely at home. Furthermore, a quick look at the phase diagram for the Cu/O system reveals that, at atmospheric pressure, the temperature must be controlled within fairly narrow limits (the window is only about 20–30°C wide) to guarantee formation of Cu_2O. If the temperature deviates from this window, one forms cupric oxide, CuO, instead. The literature of the day (much of it authored by Walter Brattain, who would later win the Nobel Prize in physics for co-inventing the transistor, and his colleague Joseph Becker) states repeatedly that CuO is crud[29] and describes various methods for getting rid of it (e.g., sandblasting, grinding, and electrolytic treatments).

Nevertheless, a little experimentation by the author has shown that CuO, crud though it may be, is indeed another semiconducting oxide of copper (in fact, it appears always to be n-type), and one that can be made easily at home. At atmospheric

---

[28] Other manufacturers also used an oxide coating, but the Blue Pal is the razor blade most often mentioned in reminiscences. The blue color is evidently due to the tempering process during manufacture.

[29] This is a perfectly valid scientific term, an acronym for "cupric, random useless debris."

pressure and temperatures below 1000°C, one almost can't help but form CuO. You don't need a clean room. It can even be a dirty room.

The recipe for making a penny diode is therefore as follows.

1. Obtain a penny dated no earlier than 1983. This is a requirement because pennies made before then are a homogeneous alloy of copper and zinc (in 95/5 ratio), and the zinc apparently greatly weakens the action. Newer pennies are actually almost entirely zinc on which has been electroplated a 10–15-μm coating of very pure copper. (The change actually occurred in mid-1982, but it's not always easy to tell just by looking which of the two types a given penny is, so it's safest to choose one from 1983 and later.)
2. Clean the penny thoroughly but gently using a copper pot cleaner and an old toothbrush until it shines with copper's characteristic gleam. Do not use abrasives (remember, the copper coating is exceedingly thin). It's not a bad idea to handle the penny only by the edges to avoid contaminating the surface with skin oils. Better still, wear gloves. Or don't. Who knows? Maybe perspiration is a key dopant.
3. After rinsing and drying (it's okay to hasten the process by daubing with a paper towel), place the penny on a cookie sheet (a folded sheet of aluminum foil is fine), and into an oven that has been preheated to at least 500°F (use the maximum your oven will allow), and bake for 15–30 minutes.
4. Turn off the oven and let it cool. Verify that the pennies are now covered uniformly with a nice, dark brownish film. These pennies are still legal tender, by the way. All you've done is accelerate their natural oxidation.[30]

You may wish to bake many pennies at the same time, rather than just one. They're cheap (extensive calculations with a supercomputer show that they cost approximately one cent each), it's the best use of the energy needed to run the oven, and it increases your chances of finding "lively" specimens.

For a catwhisker, a tiny phosphor-bronze wire is the canonical choice. However, one can also use a bent safety pin, although one must be extremely careful to use the *lightest* touch. A pencil lead is even better because it has a little more "give" to it, making the structure less critically sensitive to contact pressure. A 0.5-mm HB grade lead from Pentel is known to work quite well; others may work fine also. When using such a lead, make contact with the surface at an angle, rather than head-on, so that the edge of the lead touches the penny surface. A tungsten filament salvaged from a light bulb may also be suitable.

---

[30] This topic presents an opportunity to clear up two common misconceptions. Contrary to widespread belief, it is not illegal to destroy a penny. Also, many people are taught incorrectly that copper turns green when it oxidizes (as in the Statue of Liberty, or roofing copper). However, most of the time the green patina is actually either hydrated copper sulfate (if inland) or hydrated copper chloride (if near a body of salt water). Cuprous oxide is a pleasant red color, and cupric oxide is dark brown or black.

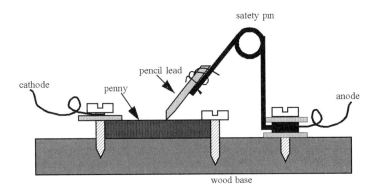

FIGURE 9.20. Penny diode (side view)

To make the ohmic contact, just clamp the penny to a piece of scrap wood under the head of an ordinary wood screw or two. The pressure of a tightened screw is more than sufficient to make an excellent conducting contact to the penny. Connect one wire to this screw. Since CuO made using the recipe given is apparently always n-type, the screw connection is the cathode, while the catwhisker is the anode terminal.

As seen in Figure 9.20, a short segment of pencil lead can be secured to a safety pin with a few turns of stiff wire. If two washers are used to allow some rotation of the safety pin (as shown), then there are enough degrees of freedom to provide considerable adjustment of the location and pressure of the pencil lead. As suggested earlier, one may dispense with the pencil lead altogether and use the safety pin as the catwhisker. However, one must then take great care to use only the lightest pressure. Even with such care, you will probably find adjustment tricky and unstable.

Now, to use this diode in a crystal radio, a good idea is to tune in a strong station initially using a 1N34A germanium diode, because it may be difficult or frustrating to have to deal with too many variables simultaneously. Once a station has been tuned in, disconnect the germanium diode and replace it with the penny diode, leaving all other settings the same. Now begins the fun part, that of hunting around the surface for a good spot, and fiddling with the pressure of the catwhisker to obtain the maximum sensitivity. Then, just as everything is working great, the slightest motion might make it stop working altogether (or it will quit on you for no apparent reason), forcing you to do it all over again. Depending on your personality, this behavior is either delightful and charming or maddening to distraction. If the latter holds, then simply use the 1N34A all the time.

If, on the other hand, you are of the former bent, you may be interested in trying other semiconductors once you've enjoyed the penny diode to the fullest. Visit your local mineral shop, and use the partial list given here as Table 9.1 (as well as those mentioned in Braun's research) as a guide in your quest for diodes made from rocks.

Of those in the table, galena is by far the best. It is widely available, relatively inexpensive (I recently bought a golfball-sized hunk for about $5, but your mileage

Table 9.1. *Some natural semiconductors*

Name	Formula	Polarity
Galena	PbS	varies
Iron pyrite	Fe_2S	
Iron oxide	Fe_2O_3	p-type
Carborundum	SiC	varies
Chalcopyrite	CuFeS	
Cuprous oxide	Cu_2O	p-type
Cupric oxide	CuO	n-type
Copper pyrite	Cu_2S	
Zincite	ZnO	n-type
Psilomelane	MnO_2	

may vary), and generally full of lively spots of great sensitivity. In general, though, a real catwhisker made of very fine wire will have to be used if the best results are to be obtained. The best catwhiskers tend to be of phosphor-bronze wire, not because of their chemical composition but simply because they have the right degree of springiness to maintain a given contact pressure. However, fine wire salvaged from a small junked motor (say, from a microprocessor fan) will work well enough as long as it stays clean.

If you'd like to hunt for other materials, it is helpful to note that researchers in the early days were guided in their search by a vague notion that imperfect contacts were important. So, look for materials that are metallic compounds. Chances are high that you can get rectification out of them. Good choices are various metallic oxides and sulfides (as you can tell by examining Table 9.1). Or try to make them yourself. Take a nail, and make it rust. Grab a hunk of copper that's turned green from exposure to the elements. If your family objects to your experimenting with a valuable patina-covered antique, drop some pennies in salt water and try to make your own copper chloride.

Try baking copper-bearing coins from other countries. The best results I've ever obtained, in fact, have been with the 10-won coin from South Korea. The quality and number of active sites greatly exceeds that of a penny. A 5-won coin is of the same composition, so it presumably would work as well (as of this writing, a 5-won coin is worth just under a half a cent, making it the least expensive substrate material of all). Other coins I've tried include a New Taiwan dollar, a German Pfennig, a Singapore dollar, and a Danish 25-øre coin. None of these works nearly as well as a penny, but functioning sites could still be found.

Finally, a special headphone or earphone must be used with this setup. Ordinary headphones, such as those from personal stereos, are intended to be used with amplifiers and typically are 32 Ω in impedance. Other, much higher-impedance headphones are needed here. Luckily, these (and germanium diodes) are readily available at low cost from the following source:

Antique Electronic Supply
6221 S. Maple Ave.
Tempe, Arizona 85283
ph: 480-820-5411
fax: 800-706-6789 from the US and Canada, 480-820-4643 from elsewhere
website URL: http://www.tubesandmore.com
e-mail: info@tubesandmore.com

They have a small piezoelectric earphone for $2.50 (catalog item P-A480) and a headset for $15.25 (part number P-A466). Either will work fine with the crystal radio circuit described in this document. Both the earphone and headset have enough internal capacitance that no additional parallel capacitance is required.

CHAPTER TEN

MIXERS

10.1 INTRODUCTION

Electrical engineering education focuses so much on the study of linear, time-invariant (LTI) systems that it's easy to conclude that there's no other kind. Violations of the LTI assumption are usually treated as undesirable, if acknowledged at all. Small-signal analysis, for example, exists precisely to avoid the complexities that non-linearities inevitably bring with them. However, the high performance of modern communications equipment actually depends critically on the presence of at least one element that fails to satisfy linear time invariance: the mixer. The superheterodyne[1] receiver uses a mixer to perform an important frequency translation of signals. This invention of Armstrong has been the dominant architecture for 75 years because frequency translation solves many problems simultaneously.[2]

In the architecture shown in Figure 10.1, the mixer heterodynes an incoming RF signal to a lower frequency,[3] known as the intermediate frequency (IF). Although Armstrong originally sought this frequency lowering simply to make it easier to obtain the requisite gain, other significant advantages accrue as well. As one example, tuning is now accomplished by varying the frequency of a local oscillator rather than by varying the center frequency of a multipole bandpass filter. Thus, instead of adjusting several *LC* networks in tandem to tune to a desired signal, one simply varies a single *LC* combination to change the frequency of a local oscillator (LO). The intermediate frequency stages can then use *fixed* bandpass filters.[4] Selectivity is therefore

---

[1] Why "super"heterodyne? The reason is that Reginald Fessenden had already invented something called the "heterodyne," so Armstrong had to name it something different.

[2] Proving once again that success has many fathers (while failure is an orphan, to complete John F. Kennedy's saying), Lucien Lévy and Walter Schottky, among others, laid claim to the superheterodyne. While it is certainly true that Armstrong was not the first to conceive of the heterodyne principle, he was the first to recognize how neatly it solves so many thorny problems and was certainly the first to pursue its development vigorously. Schottky eventually conceded his historical claim, but Lévy went to his grave embittered (and largely forgotten).

[3] Actually, one may also translate to a *higher* frequency.

[4] Recognition that one could use a fixed IF distinguished Armstrong's version from Lévy's.

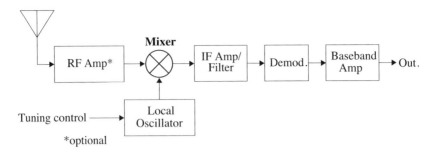

FIGURE 10.1. Superheterodyne receiver block diagram

determined by these fixed-frequency IF filters, which are much easier to realize than variable-frequency filters. Additionally, the overall gain of the system is distributed over a number of different frequency bands (RF, IF, and baseband), so that the required receiver gain (typically 120–140 dB on a power basis) can be obtained without much worry about potential oscillations arising from parasitic feedback loops. These important attributes explain why the superheterodyne architecture still dominates nearly a century after its invention.

10.2 MIXER FUNDAMENTALS

Since linear, time-invariant systems cannot produce outputs with spectral components not present at the input, mixers must be nonlinear or have time-varying elements to provide frequency translation. Historically, many devices (e.g., electrolytic cells, magnetic ribbons, rusty scissors, vacuum tubes, transistors, and brain tissue) operating on a host of diverse principles have been used, demonstrating that virtually *any* nonlinear element can be used as a mixer.[5]

At the core of all mixers presently in use is a multiplication of two signals in the time domain. The fundamental usefulness of multiplication may be understood from examination of the following trigonometric identity:

$$(A\cos\omega_1 t)(B\cos\omega_2 t) = \frac{AB}{2}[\cos(\omega_1-\omega_2)t + \cos(\omega_1+\omega_2)t]. \quad (1)$$

Multiplication thus results in equal-power output signals at the sum and difference frequencies of the input, signals whose amplitudes are proportional to the product of the RF and LO amplitudes. Hence, if the LO amplitude is constant (as it usually is), any amplitude modulation in the RF signal is transferred to the IF signal. By a similar mechanism, an *undesired* transfer of modulation from one signal to another can also occur through nonlinear interaction in both mixers and amplifiers. In that context it is called *cross*-modulation, and its suppression through improved linearity is an important design consideration.

---

[5] Of course, some nonlinearities work better than others; we will focus on the more practical types.

Having recognized the fundamental role of multiplication, we now enumerate and define the most significant characteristics of mixers.

10.2.1 CONVERSION GAIN

One important mixer characteristic is conversion gain (or loss), which is defined as the ratio of the desired IF output to the value of the RF input. For the multiplier described by Eqn. 1, the conversion gain is therefore the IF output, $AB/2$, divided by A (if that is the amplitude of the RF input). Hence, the conversion gain in this example is $B/2$, or half the LO amplitude.

Conversion gain, if expressed as a power ratio, can be greater than unity in active mixers, while passive mixers are generally capable only of voltage or current gain at best.[6] Conversion gain in excess of unity is often convenient because the mixer then provides amplification along with the frequency translation. However, it does not necessarily follow that sensitivity improves, since noise figure must also be considered. For this reason, passive mixers may offer superior performance in some cases despite their conversion loss.

10.2.2 NOISE FIGURE: SSB VERSUS DSB

Noise figure is defined as one might expect: it's the signal-to-noise ratio (SNR) at the input (RF) port divided by the SNR at the output (IF) port. There's an important subtlety, however, that often trips up both the uninitiated and a substantial fraction of practicing engineers. To appreciate this difficulty, we first need to make an important observation: In a typical mixer, there are actually *two* input frequencies that will generate a given intermediate frequency. One is the desired RF signal, and the other is called the *image* signal. In the context of mixers, these two signals are frequently referred to collectively as *sidebands*.

The reason that two such frequencies exist is that the IF is simply the *magnitude* of the difference between the RF and LO frequencies. Hence, signals both above and below ω_{LO} by an amount equal to the IF will produce IF outputs of the same frequency. The two input frequencies are therefore separated by $2\omega_{IF}$. As a specific numerical example, suppose that our system's IF is 100 MHz and we wish to tune to a signal at 900 MHz by selecting an LO frequency of 1 GHz. Aside from the desired 900-MHz RF input, a 1.1-GHz image signal will also produce a difference-frequency component at the IF of 100 MHz.

The existence of an image frequency complicates noise figure computations because noise originating in both the desired and image frequencies therefore become

---

[6] Sometimes cited as an exception is a class of systems known as *parametric* converters or amplifiers, in which power from the LO is transferred to the IF through reactive nonlinear interaction (typically with varactors), thus making power gain possible. This fact tells us that such systems are more properly classified as active.

IF noise, while there is generally no desired signal at the image frequency. In the usual case where the desired signal exists at only one frequency, the noise figure that one measures is called the *single-sideband* noise figure (SSB NF), while the rarer case where both the "main" RF and image signals contain useful information leads to a double-sideband (DSB) noise figure. Clearly, the SSB noise figure will be greater than for the DSB case since both have the same IF noise, while the former has signal power in only a single sideband. Hence, the SSB NF will normally be 3 dB higher than the DSB NF.[7] Unfortunately, DSB NF is reported much more often because it is numerically smaller and thus (falsely) conveys the impression of better performance, even though there are few communications systems for which DSB NF is an appropriate figure of merit.[8] Frequently, a noise figure is stated without any indication as to whether it is a DSB or SSB value. In such cases, one may usually assume that a DSB figure is being quoted.

Noise figures for mixers tend to be considerably higher than those for amplifiers because noise from frequencies other than at the desired RF can mix down to the IF. Representative values for SSB noise figures range from 10 to 15 dB or more. It is mainly because of this larger mixer noise that one uses LNAs in a receiver. If the LNA has sufficient gain then the signal will be amplified to levels well above the noise of the mixer and subsequent stages, so the overall receiver NF will be dominated by the LNA instead of the mixer. If mixers were not as noisy as they are, then the need for LNAs would diminish considerably.

10.2.3 LINEARITY AND ISOLATION

Dynamic range requirements in modern, high-performance telecommunications systems are quite severe, frequently exceeding 80 dB and approaching 100 dB in many instances. As discussed in Chapter 13, the floor is established by the noise figure, which conveys something about how small a signal may be processed, while the ceiling is set by the onset of severe nonlinearities that accompany large input signals.

As with amplifiers, the compression point is one measure of this dynamic range ceiling and is defined the same way. Ideally, we would like the IF output to be proportional to the RF input signal amplitude; this is the sense in which we interpret the term "linearity" in the context of mixers. As with amplifiers (and virtually any other physical system), however, real mixers have some limit beyond which the output has a sublinear dependence on the input. The *compression point* is the value of RF signal[9]

---

[7] The 3-dB difference assumes that the conversion gain to two equal sidebands is the same. Although this assumption is usually well satisfied, it need not be.

[8] Two important exceptions in which both sidebands contain useful information are radioastronomy (as in the measurements of the cosmic background radiation – the echoes of the Big Bang), and direct-conversion receivers. We revisit this DSB-vs.-SSB explanation in Chapter 14.

[9] Some manufacturers (and authors) report an *output* compression point. If the conversion gain at that point is known, the figure can be reflected back to the input point. Sadly, many insist on burying that bit of information, making it extremely difficult to perform fair comparisons of mixer performance. We will always state explicitly whether the figure is an input or output parameter.

10.2 MIXER FUNDAMENTALS

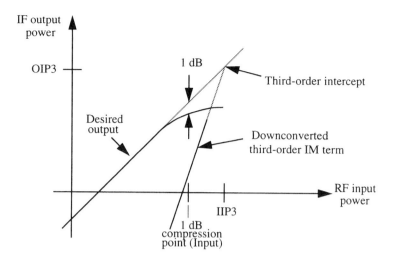

FIGURE 10.2. Definition of mixer linearity parameters

at which a calibrated departure from the ideal linear curve occurs. Usually, a 1-dB (or, more rarely, a 3-dB) compression value is specified. One should specify either the input or output signal strength at which this compression occurs, together with the conversion gain, to allow fair comparisons among different mixers.

The two-tone third-order intercept is also used to characterize mixer linearity. A two-tone intermodulation test is a relevant way to evaluate mixer performance because it mimics the real-world scenario in which both a desired signal and a potential interferer (perhaps at a frequency just one channel away) feed a mixer input. Ideally, each of two superposed RF inputs would be translated in frequency without interacting with each other. Of course, practical mixers will always exhibit some intermodulation effects and the output of the mixer will thus contain frequency-translated versions of third-order IM components whose frequencies are $2\omega_{RF1} \pm \omega_{RF2}$ and $2\omega_{RF2} \pm \omega_{RF1}$. The difference-frequency terms may heterodyne into components that lie within the IF passband and are therefore generally the troublesome ones, while the sum-frequency signals can usually be filtered out.

As a measure of the degree of departure from linear mixing behavior, one can plot the desired output and the third-order IM output as a function of input RF level. The third-order intercept is the extrapolated intersection of these two curves. In general, the higher the intercept, the more linear the mixer. Again, one ought to specify whether the intercept is input- or output-referred, as well as the conversion gain, to permit fair comparisons among mixers. Additionally, it is customary to abbreviate the intercept as IP3, or perhaps IIP3 or OIP3 (for input and output third-order intercept, respectively). These definitions are summarized in Figure 10.2.

Cubic nonlinearity can also cause trouble with a *single* RF input. As a specific example, consider building a low-cost AM radio. The standard IF for AM radios happens unfortunately to be 455 kHz (mainly for historical reasons). Tuning in a station at 910 kHz (a legitimate AM radio frequency) requires that the LO be set to

1365 kHz.[10] The cubic nonlinearity could generate a component at $2\omega_{RF} - \omega_{LO}$, which in this case happens to coincide with our IF of 455 kHz.

One might be tempted to assert that such a component is not a problem since it adds to the desired output. One therefore might even be tempted to consider this an asset. However, the third-order IM products have amplitudes that are no longer proportional to the input signal amplitude. Hence, they represent amplitude distortion that can corrupt the "correct" output (we're talking about an amplitude-modulated signal, after all).

Even if the exact numerological coincidence of the foregoing example does not occur, various third-order IM terms can possess frequencies within the passband of the IF amplifier, ultimately degrading signal-to-noise or signal-to-distortion.

Another parameter of great practical importance is isolation. It is generally desirable to minimize interaction among the RF, IF, and LO ports. For instance, since the LO signal power is generally quite large compared with that of the RF signal, any LO feedthrough to the IF output might cause problems at subsequent stages in the signal processing chain. This problem is exacerbated if the IF and LO frequencies are similar, so that filtering is ineffective. Even reverse isolation is important in many instances, since poor reverse isolation might permit the strong LO signal (or its harmonics) to work its way back to the antenna, where it can radiate and cause interference to other receivers.

10.2.4 SPURS

Mixers, by their nature, may heterodyne a variety of frequency components that you never intended to mix. For example, harmonics of some signal (desired or not) could lie (or be generated) within the passband of the mixer system and subsequently beat against the local oscillator (and its harmonics). Some of the resulting components may end up within the IF passband. The signals that do ultimately emerge from the output of the IF system are known as spurious responses, or just spurs. Evaluation of mixer spurs is straightforward in principle but *highly* tedious in practice. The availability of software tools to take care of this task has taken the tedium out of the calculations, but it's instructive to describe the process, just the same.[11]

Let m and n be the harmonic numbers of the RF input and LO frequencies, respectively. Then the spur products present at the output of the mixer (prior to any filtering) are given by

$$f_{spur} = mf_{RF} + nf_{LO}. \tag{2}$$

---

[10] A local oscillator frequency of 455 kHz also works, but it is a less practical choice because such "low-side injection" requires the local oscillator to tune over a larger range than if the LO frequencies were above the desired RF.

[11] An excellent (and free) program that performs this calculation (and many others of great value to the RF/microwave designer) is *AppCAD,* originally from HP, now from Agilent. Both older DOS and newer Windows versions are available. The program *RFSim99* also has a spur search tool.

10.2 MIXER FUNDAMENTALS

Table 10.1. *Spur table for FM radio example*

m	f_{RF}, low	f_{RF}, high	n
−3	73.8	93.9	3
−2	72.0	92.1	2
1	*88.0*	*108.2*	*−1*
2	82.7	102.8	−2
3	80.9	101.0	−3

The apparent simplicity of that equation is misleading: The calculation must be repeated for all combinations and *signs* of m and n, ranging up to the maximum harmonic order you care to consider. To make a laborious procedure even more so, one must actually consider RF signals of frequencies below the nominal input passband – at least down to the lower passband edge frequency divided by the maximum value of m. One must also consider input frequencies above the upper passband edge. Because no input filter is perfect, harmonics of the LO can still heterodyne with RF signals above the input passband to produce in-band spurs at the output of the mixer.

For each (m, n) pair, examine the spur frequency to determine whether it lies within the IF passband (or sufficiently close to it) and so merits further consideration. For each spur that does, work backward to the implied RF input frequency and evaluate the likelihood that there will be a signal at that frequency of sufficient strength to be a source of trouble. Then make appropriate modifications to the input filtering, if necessary, to avoid those troubles.

This exercise is sometimes performed with the worst-case assumption that there is no filtering of any kind at the RF input port. In that case, the number of calculations grows very large quite quickly. If one is patient enough, however, the information generated can be used to guide the design of the input filter.

As a specific example, suppose we wish to design a mixer for an FM receiver whose nominal input passband is to accommodate signals spanning 88.1 MHz to 108.1 MHz. With a conventional 10.7-MHz IF, the LO needs to tune from 77.4 MHz to 97.4 MHz (assuming low-side injection). To keep the numbers easy, assume a bit unrealistically that the IF system possesses a nominal bandwidth of approximately 200 kHz. Further assume that we need not consider harmonic orders higher than 3.[12] With these assumptions, we can construct Table 10.1.

Examining just the first entry in the table, we see that the third harmonic of RF signals in the 73.8–93.9-MHz frequency band may heterodyne with the third harmonic

---

[12] Remember, the spectrum of a signal rolls off approximately as $1/n$, where n is the number of derivatives needed to produce impulses from the time-domain representation of the signal. Most signals of practical interest have spectra that roll off fast enough that consideration of orders higher than about 5 or 7 is probably overkill for typical situations. Your mileage may vary, however.

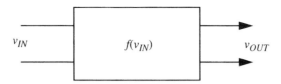

FIGURE 10.3. General two-port nonlinearity

of the LO to produce signals within the 10.6–10.8 MHz IF passband. Notice that input filtering would only be partially effective at best, because much of the spurious input band overlaps the desired FM radio band. If there were indeed significant interferers within this spurious band, our only choice would be to improve the spectral purity of the LO in order to minimize its third harmonic content. Such an improvement also benefits the spur problems implied by the last row of the spur table. Similarly, reduction in second harmonic content is the only practical way to avoid the problems implied in the second and fourth rows. The third row is actually not undesired: it describes the intended mode of operation for the receiver and is included simply for completeness.

By carrying out this laborious procedure for any contemplated system, it is possible to assess the sensitivity to various imperfections and thus to evaluate the need for remediation.

10.3 NONLINEARITY, TIME VARIATION, AND MIXING

We now consider how to implement the multiplication that is the heart of mixing action. Some mixers directly implement a multiplication, while others provide it incidentally through a nonlinearity. We follow a historical path and first examine a general two-port nonlinearity[13] (see Figure 10.3), since mixers of that type predate those designed specifically to behave as multipliers.

If the nonlinearity is "well-behaved" (in the mathematical sense), we can describe the input–output relationship with a series expansion:

$$v_{OUT} = \sum_{n=0}^{N} c_n (v_{IN})^n. \tag{3}$$

To use such an Nth-order nonlinearity as a mixer, the signal v_{IN} would be the sum of the RF input and the local oscillator signals. In general, the output will consist of three types of products: DC terms, harmonics of the inputs, and intermodulation products of those harmonics.[14] Not all of these spectral components are desirable, so part of the challenge in mixer design is to devise topologies that inherently generate few undesired terms.

---

[13] We will shortly see the advantages of three-port mixers.
[14] Keep in mind that fundamentals are harmonics.

10.3 NONLINEARITY, TIME VARIATION, AND MIXING

Even-order nonlinear factors in Eqn. 3 contribute DC terms; these are readily filtered out by AC coupling, if desired. Harmonic terms, at $m\omega_{LO}$ and $m\omega_{RF}$, extend from the fundamental ($m = 1$) all the way up to the Nth harmonic. As with the DC terms, they are also often relatively easy to filter out because their frequencies are usually well away from the desired IF.

The intermodulation (IM) products are the various sum- and difference-frequency terms. These have frequencies expressible as $p\omega_{RF} \pm q\omega_{LO}$, where integers p and q are greater than zero and sum to values up to N. Only the second-order intermodulation term ($p = q = 1$) is normally desired.[15] Unfortunately, other IM products might have frequencies close to the desired IF, making them difficult to remove, as we shall see. Since it is generally true that high-order nonlinearities (i.e., large values of N in the power series expansion) tend to generate more of these undesirable terms,[16] mixers should approximate square-law behavior (the lowest-order nonlinearity) if they only have one input port (as shown in Figure 10.3). We now consider specifically the properties of a square-law mixer to identify its advantages over higher-order nonlinear mixers.

TWO-PORT EXAMPLE: SQUARE-LAW MIXER

To see explicitly where the desired multiplication arises in a square-law mixer, note that the only nonzero coefficients in the series expansion are the c_1 and c_2 terms.[17] If we assume that the input signal v_{IN} is the sum of two sinusoids,

$$v_{IN} = v_{RF} \cos(\omega_{RF} t) + v_{LO} \cos(\omega_{LO} t), \tag{4}$$

then the output of this mixer may be expressed as the sum of three distinct components:

$$v_{OUT} = v_{fund} + v_{square} + v_{cross}, \tag{5}$$

where

$$v_{fund} = c_1 [v_{RF} \cos(\omega_{RF} t) + v_{LO} \cos(\omega_{LO} t)], \tag{6}$$

$$v_{square} = c_2 \{[v_{RF} \cos(\omega_{RF} t)]^2 + [v_{LO} \cos(\omega_{LO} t)]^2\}, \tag{7}$$

$$v_{cross} = 2c_2 v_{RF} v_{LO} [\cos(\omega_{RF} t)][\cos(\omega_{LO} t)]. \tag{8}$$

The fundamental terms are simply scaled versions of the original inputs and therefore represent no useful mixer output; they must be removed by filtering. The v_{square}

---

[15] The order of a given IM term is the sum of p and q, so a second-order IM product arises from the quadratic term in the series expansion.

[16] As with most sweeping generalities, there are exceptions to this one. In building frequency multipliers, for example, high-order harmonic nonlinearities are extremely useful. However, in mixer design it is usually true that higher-order nonlinearities are undesirable.

[17] There may also be a nonzero DC term (i.e., c_0 may be nonzero), but this component is easily removed by filtering, so we will ignore it at the outset to reduce equation clutter.

components similarly represent no useful mixer output, as is evident from the following special case of Eqn. 1:

$$[\cos \omega t]^2 = \tfrac{1}{2}[1 + \cos 2\omega t]. \tag{9}$$

Thus, we see that the v_{square} components contribute a DC offset (as well as second harmonics) of the input signals. These also must generally be removed by filtering.

The useful output comes from the v_{cross} components because of the multiplication evident in Eqn. 8. Using Eqn. 1, we may rewrite v_{cross} in a form that shows the mixing action more clearly:

$$v_{cross} = c_2 v_{RF} v_{LO}[\cos(\omega_{RF} - \omega_{LO})t + \cos(\omega_{RF} + \omega_{LO})t]. \tag{10}$$

For a fixed LO amplitude, the IF output amplitude is linearly proportional to the RF input amplitude. That is, this nonlinearity implements a linear mixing, since the output is proportional to the input.

The conversion gain for this nonlinearity is readily found from Eqn. 10:

$$G_c = \frac{c_2 v_{RF} v_{LO}}{v_{RF}} = c_2 v_{LO}. \tag{11}$$

Just as any other gain parameter, conversion gain may be a dimensionless quantity (or a transconductance, transresistance, etc.). It is customary in discrete designs to express conversion gain as a power ratio (or its decibel equivalent), but the unequal input and output impedance levels in typical IC mixers also makes a voltage or current conversion gain appropriate. To avoid confusion, of course, it is essential to state explicitly the type of gain.[18]

As asserted earlier, the square-law mixer's advantages are that the undesired spectral components are usually at a frequency quite different from the intermediate frequency and are thus readily removed. For this reason, two-port mixers are frequently designed to conform to square-law behavior to the maximum practical extent.

Square-law mixers may be implemented with virtually any common nonlinearity, because the quadratic term typically dominates. In the simplified schematic of Figure 10.4, the bias, RF, and LO terms are shown as driving a bipolar base in series. The bias voltage V_{BIAS} is chosen as necessary to place the active device in the desired operating region. For example, a JFET or MESFET would require a negative bias, and an ordinary MOSFET and bipolar transistor would need a positive bias. Refinements would be needed to provide a stable bias, of course.

The discrete tank in the output circuit may be implemented with a $\lambda/4$ transmission line resonator at frequencies where the size may be tolerated. The summation of RF and LO signals can be accomplished in practical circuits with resistive or reactive

---

[18] All too frequently, published "power" gain figures for IC implementations are essentially voltage gain measurements and are therefore grossly in error if the input and output impedance levels differ significantly, as they often do. It seems necessary to emphasize that watts and volts are not the same.

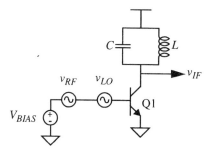

FIGURE 10.4. Square-law mixer (simplified)

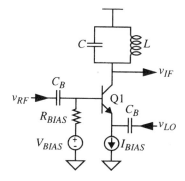

FIGURE 10.5. Square-law mixer (alternative configuration)

combiners. Because the RF and LO signals are in series, there is poor isolation between them. An alternative (but functionally equivalent) arrangement that reduces the effect of the relatively large LO signal on the RF port is as shown in Figure 10.5.

The RF signal drives the base directly (through a DC-blocking capacitor), while the LO drives the source terminal. This way, the base-to-emitter voltage is the sum of ground-referenced LO and RF signals. The bias current is set directly with a current source, while the DC base voltage is determined by the value of V_{BIAS}. Resistor R_{BIAS} is chosen large enough to avoid excessive loading and also to minimize its noise contribution.

Perfect square-law device behavior is not at all necessary to obtain mixing action. The bipolar circuit in the figure functions as a result of the quadratic term in the series expansion for the exponential i_C–v_{BE} relationship that dominates over a range of input amplitudes. Precisely because many nonlinearities are well approximated by a square-law shape over some suitably restricted interval, one can estimate the conversion gain for other nonlinear devices used as mixers once the value of the quadratic coefficient (c_2) is found. To underscore this point, let's estimate the conversion gain for a bipolar transistor. To simplify the calculation, we continue to ignore dynamic effects. Then we can use the exponential v_{BE} law:

$$i_C \approx I_S e^{v_{BE}/V_T}, \tag{12}$$

where V_T is the thermal voltage kT/q, not the threshold voltage.

Expansion of this familiar relationship up to the second-order term yields[19]

$$i_C \approx I_C \left[1 + \frac{v_{IN}}{V_T} + \frac{1}{2}\left(\frac{v_{IN}}{V_T}\right)^2 \right]. \tag{13}$$

By inspection (well, almost),

$$c_2 = \frac{g_m}{2V_T}, \tag{14}$$

so that an estimate of the conversion gain is

$$G_c = c_2 v_{LO} = g_m \cdot \frac{v_{LO}}{2V_T}. \tag{15}$$

The conversion gain here is a transconductance that is proportional both to (a) the standard incremental transconductance and (b) the ratio of the local oscillator drive amplitude to the thermal voltage. The conversion gain for a bipolar transistor is therefore dependent on bias current, LO amplitude, and temperature.

Let us now consider an ideal square-law long-channel FET, for which

$$i_D = \frac{\mu C_{ox} W}{2L}(V_{GS} - V_T)^2. \tag{16}$$

Short-channel (high-field) devices are more linear as a result of velocity saturation, and thus they are generally inferior to long devices as mixers.[20]

If the gate–source voltage V_{GS} is the sum of RF, LO, and bias terms then we may write

$$i_D = \frac{\mu C_{ox} W}{2L}\{[V_{BIAS} + v_{RF}\cos(\omega_{RF}t) + v_{LO}\cos(\omega_{LO}t)] - V_T\}^2, \tag{17}$$

from which one may readily find that the conversion gain (here a transconductance) is simply

$$G_C = \frac{\mu C_{ox} W}{2L} v_{LO} = \frac{I_D}{V_{OD}^2} v_{LO} = g_m \frac{2v_{LO}}{V_{OD}}, \tag{18}$$

where V_{OD} is the DC bias value of the gate overdrive ($V_{GS} - V_T$). Note that this ideal square-law FET has a conversion transconductance that is similar in form to that of a bipolar transistor, with the overdrive voltage playing the role of the thermal voltage.

---

[19] We have implicitly assumed that the base–emitter drive contains a DC component as well as the RF and LO components, so that I_C is nonzero.

[20] The reader is reminded once again that "short-channel" actually means "high-field." Hence, even "short" devices may still behave quadratically for suitably small drain–source voltages.

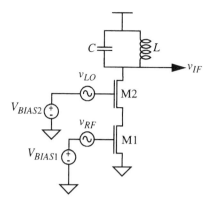

FIGURE 10.6. FET dual-gate mixer (simplified)

Note also that this ideal device has a conversion transconductance that is independent of bias.[21] It is still dependent on temperature (through mobility variation) and LO drive amplitude, however.

As in the corresponding derivation for a MOSFET, the foregoing computation ignores parasitic series base and emitter resistances. These resistances can linearize the transistor and hence weaken mixer action. Proper selection of device type is therefore necessary to maximize performance.

Another method for improving RF–LO isolation is the dual-gate FET mixer. As its name implies, this mixer uses a FET that possesses two gates. For all practical purposes, this structure may be regarded as equivalent to two FETs in a cascode-like configuration, as seen in Figure 10.6.

In this circuit, M1 functions as a transconductor with deliberately poor output resistance; it is generally operated in the triode region. With that choice of bias, the LO drive applied to M2 causes the drain current, and resistance, of M1 to vary at the LO rate. The drain current of M1 (and hence of M2) thus contains components at the sum and difference frequencies, as in any other mixer.

As with the other circuits, the LC tank can be replaced by its transmission line equivalent at frequencies where it makes sense to do so.

10.4 MULTIPLIER-BASED MIXERS

We've seen that nonlinearities produce mixing incidentally through the multiplications they provide. Precisely because the multiplication is only incidental, these nonlinearities usually generate a host of undesired spectral components. Furthermore,

---

[21] This independence of bias holds only in the square-law regime. Enough bias must therefore be supplied to guarantee this condition. Hence, V_{BIAS} is not permitted to equal zero. In fact, it must be chosen large enough to guarantee that the gate–source voltage always exceeds the threshold voltage, since a MOSFET behaves exponentially in weak inversion.

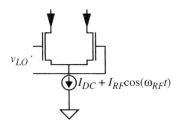

FIGURE 10.7. Single-balanced mixer

since two-port mixers have only one input port, the RF and LO signals are generally not well isolated from each other. This lack of isolation can cause the problems mentioned earlier, such as overloading of IF amplifiers, as well as radiation of the LO signal (or its harmonics) back out through the antenna.

Mixers based directly on multiplication generally exhibit superior performance because they ideally generate only the desired intermodulation product. Furthermore, because the inputs to a multiplier enter at separate ports, there can be a high degree of isolation among all three signals (RF, LO, IF). Finally, FET technologies provide excellent switches, with which one can implement outstanding multipliers.

10.4.1 SINGLE-BALANCED MIXER

One extremely common family of multipliers first converts the incoming RF voltage into a current, then performs a multiplication in the current domain. The simplest multiplier cell of this type is shown in Figure 10.7.[22] Again, the MOSFET symbol is simply a proxy for any active device. This topology, with suitable accommodation of differing bias requirements, functions with MESFETs, MOSFETs, JFETs, and bipolars.

In this mixer, v_{LO} is chosen large enough so that the transistors alternately switch (commutate) all of the tail current from one side to the other at the LO frequency.[23] The tail current is therefore effectively multiplied by a square wave whose frequency is that of the local oscillator:

$$i_{out}(t) \approx (\text{sgn}[\cos \omega_{LO} t])[I_{BIAS} + I_{RF} \cos \omega_{RF} t]. \tag{19}$$

Because a square wave consists of odd harmonics of the fundamental, multiplication of the tail current by the square wave results in an output spectrum that appears

---

[22] Mixers of this general kind are often lumped together and called Gilbert mixers, but only some actually are. True Gilbert multipliers function entirely in the current domain, deferring the problem of V–I conversion by assuming that all variables are already available in the form of currents. See Barrie Gilbert's landmark paper, "A Precise Four-Quadrant Multiplier with Subnanosecond Response," *IEEE J. Solid-State Circuits,* December 1968, pp. 365–73.

[23] One may also interchange the roles of LO and RF input, but the resulting mixer has lower conversion gain and worse noise performance (among other deficiencies).

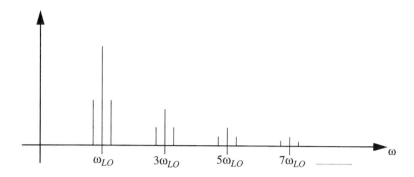

FIGURE 10.8. Representative output spectrum of single-balanced mixer

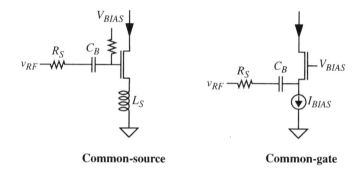

FIGURE 10.9. RF transconductors for mixers

as shown in Figure 10.8 (ω_{RF} is here chosen atypically low compared with ω_{LO} to reduce clutter in the graph).

The output thus consists of sum and difference components, each the result of an odd harmonic of the LO mixing with the RF signal. In addition, odd harmonics of the LO appear directly in the output as a consequence of the DC bias current multiplying with the LO signal. Because of the presence of the LO in the output spectrum, this type of mixer is known as a *single-balanced* mixer. Double-balanced mixers, which we'll study shortly, exploit symmetry to remove the undesired output LO component through cancellation.

Although the current source of Figure 10.7 includes a component that is perfectly proportional to the RF input signal, V–I converters of all real mixers are imperfect. Hence, an important design challenge is to maximize the linearity of the RF transconductance. Linearity is most commonly enhanced through some type of source degeneration, in both common-gate and common-source (or common-base and common-emitter) transconductors; see Figure 10.9. The common-gate circuit uses the source resistance R_S to linearize the transfer characteristic. This linearization is most effective if the admittance looking into the source terminal of the transistor is much larger than the conductance of R_S. In that case, the transconductance of the stage approaches $1/R_S$.

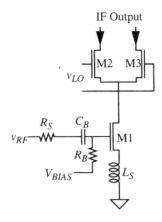

FIGURE 10.10. Single-balanced mixer with linearized transconductance

Inductive degeneration is usually preferred over resistive degeneration for several reasons.[24] An inductance has no thermal noise to degrade noise figure and no DC voltage drop to diminish supply headroom. This last consideration is particularly relevant for low-voltage/low-power applications. Finally, the increasing reactance of an inductor with increasing frequency helps to attenuate high-frequency harmonic and intermodulation components.

A more complete single-balanced mixer that incorporates a linearized transconductance is shown in Figure 10.10. The value of V_{BIAS} establishes the bias current of the cell, while R_B is chosen large enough not to load down the gate circuit (and also to reduce its noise contribution). The RF signal is applied to the gate through a DC blocking capacitor C_B. In practice, a filter would be used to remove the LO and other undesired spectral components from the output.

The conversion transconductance of this mixer can be estimated by assuming that the LO-driven transistors behave as perfect switches. Then the differential output current may be regarded as the result of multiplying the drain current of M1 by a unit-amplitude square wave. Since the amplitude of the fundamental component of a square wave is $4/\pi$ times the amplitude of the square wave, we may write:

$$G_C = \frac{2}{\pi} g_m, \tag{20}$$

where g_m is the transconductance of the V–I converter and G_C is itself a transconductance. The coefficient is $2/\pi$ (-3.92 dB) rather than $4/\pi$ because the IF signal is divided evenly between sum and difference components.

---

[24] Capacitive degeneration is sometimes suggested but is markedly inferior to inductive degeneration because it increases noise and distortion at high frequencies. It also provides no DC path for biasing, so additional circuitry must be added in any case. An inductive element solves all of these problems.

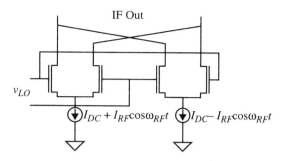

FIGURE 10.11. Active double-balanced mixer

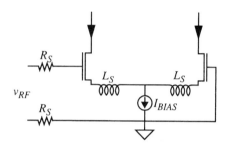

FIGURE 10.12. Linearized differential RF transconductor for double-balanced mixer

10.4.2 ACTIVE DOUBLE-BALANCED MIXER

To prevent the LO products from getting to the output in the first place, two single-balanced circuits may be combined to produce a double-balanced mixer; see Figure 10.11. We assume once again that the LO drive is large enough to make the differential pairs act like current-steering switches. Note that the two single-balanced mixers are connected in antiparallel as far as the LO is concerned but in parallel for the RF signal. Therefore, the LO terms sum to zero in the output, and the converted RF signal is doubled in the output. This mixer thus provides a high degree of LO–IF isolation, easing filtering requirements at the output. With care, this circuit routinely provides 40 dB of LO–IF isolation, with values in excess of 60 dB possible.

As in the single-balanced active mixer, the dynamic range is limited in part by the linearity of the V–I converter at the RF port of the mixer. So, most of the design effort is spent attempting to find better ways of providing this V–I conversion. The basic linearizing techniques used in the single-balanced mixer may be adapted to the double-balanced case, as shown in Figure 10.12.

In low-voltage applications, the DC current source can be replaced by a parallel LC tank to create a zero-headroom AC current source. The resonant frequency of the tank should be chosen to provide rejection of whatever common-mode component is most objectionable. If several such components exist, one may use series

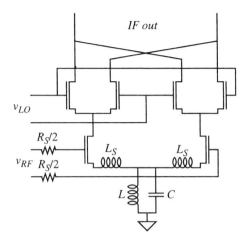

FIGURE 10.13. Minimum supply–headroom double-balanced mixer

combinations of parallel LC tanks. With such a choice, a complete double-balanced mixer appears as shown in Figure 10.13. The expression for the conversion transconductance is the same as for the single-balanced case.

These mixers may also be modified to act as low-noise mixers by the simple expedient of adding suitable gate inductances to the inductively degenerated pair that receives the RF input. By following a prescription essentially identical to that for stand-alone LNAs, it is possible to construct a low-headroom, low-noise mixer that may obviate the need for a separate LNA in some applications. Adjustment of the tuning of the input loop allows a variable trade-off among conversion gain, noise figure, and distortion.

Noise Figure of Gilbert-Type Mixers

Computing the noise figure of mixers is difficult because of the cyclostationary nature of the noise sources. One simulation-based technique involves characterization of the time-varying impulse response, arguing that a mixer is at least linear, if not time-invariant.[25] Although the method is accurate and quite suitable for analysis, its complexity does inhibit acquisition of design insight. Nonetheless, we can identify several important noise sources and make general recommendations about how to minimize noise figure.

One noise source is certainly the transconductor itself, so that its noise figure establishes a lower bound on the mixer noise figure. The same approach used in computing LNA noise figure may be used to compute the transconductor noise figure.

The differential pair also degrades noise performance in a number of ways. One noise figure contribution arises from imperfect switching, which causes attenuation

---

[25] C. D. Hull and R. G. Meyer, "A Systematic Approach to the Analysis of Noise in Mixers," *IEEE Trans. Circuits and Systems I,* v. 40, no. 12, December 1993, pp. 909–19.

of the signal current. Hence, one challenge in such mixers is to design the switches (and associated LO drive) to provide as little attenuation as possible.

Another noise figure contribution of the switching transistors arises from the interval of time in which both transistors conduct current and hence generate noise. Additionally, any noise in the LO is also magnified during this active gain interval. Minimizing this simultaneous conduction interval reduces this degradation, so sufficient LO drive must be supplied to make the differential pair approximate ideal, infinitely fast switches to the maximum practical extent. Finally, the 3-dB attenuation inherent in ignoring either the sum or difference signal automatically degrades noise figure (by 3 dB), because the noise cannot be discarded so readily. As a result, practical current-mode mixers typically exhibit SSB noise figures of at least 10 dB, with values more frequently in the neighborhood of 15 dB.

Linearity of Gilbert-Type Mixers

The IP3 of this type of mixer is bounded by that of the transconductor, so the three-point method used to estimate the IP3 of ordinary amplifiers may also be used here to estimate the IP3 of the transconductor. If the LO-driven transistors act as good switches, then the overall mixer IP3 generally differs little from that of the transconductor. To guarantee good switching it is important to note that, while sufficient LO drive is necessary, excessive LO drive should be avoided. To understand why excessive LO drive is a liability rather than an asset, consider the effect of ever-present capacitive parasitic loading on the common source connection of a differential pair. As each gate is driven far beyond what's necessary for good switching, the common source voltage is similarly overdriven. A spike in current results. In extreme cases, this spike can cause transistors to leave the saturation region. Even if that does not occur, the output spectrum can become dominated by the components arising from the spikes, rather than the downconverted RF. Hence, one should use only enough LO drive to guarantee reliable switching, and no more.

A Short Note on Simulation of Mixer IP3 with Time-Domain Simulators

Just as we noted with simulations of intermodulation distortion in amplifiers, common circuit simulators (such as *Spice*) provide accurate mixer simulations only reluctantly, if at all. The problem stems from two fundamental sources: the wide dynamic range of signals in a mixer forces the use of far tighter numerical tolerances than are adequate for "normal" circuit simulations; and the large span of frequencies of important spectral components forces long simulation times. Hence, obtaining an accurate value for IP3 from a transient simulation, for example, is usually quite challenging. The reader is therefore cautioned to treat mixer simulation results with a healthy degree of skepticism.

Because even the "accurate" options available in some simulation tools are orders of magnitude too loose to be useful for IP3 simulations, one specific action that mitigates some of these problems is to tighten tolerances progressively until the simulation results stop changing significantly. In particular, the behavior of the IM3 component in an IP3 simulation is an extremely sensitive indicator of whether the tolerances are sufficiently tight. If the IM3 terms do not exhibit a $+3$ slope (on a dB scale), chances are good that the tolerances are too loose. One must also make sure that the amplitudes of the two input tones are chosen small enough (i.e., well below either the compression or intercept points) to guarantee quasilinear operation of the mixer; otherwise, higher-order terms in the nonlinearity will contribute significantly to the output and confound the results. In the early phases of design, the three-point method may be applied to the transconductor to estimate its IP3 without having to suffer the agony of a transient simulation.

Another subtle consideration is to guarantee equal time spacing in the transient simulation, since FFT algorithms generally assume uniform sampling. Some simulators use adaptive time stepping to speed up convergence, so significant spectral artifacts can arise when computing the FFT. One may set the time step to a tiny fraction of the fastest time interval of interest to assure convergence without resort to adaptive time stepping. As an example, one might have to use a time step that is three orders of magnitude smaller than the period of the RF signal. Hence, for a 1-GHz RF input, one might have to use a 1-ps time step. It is this combination of iteration, tight time step, and numerical tolerance problems that causes IP3 simulations to execute so slowly. As with the case of amplifiers, alternatives to time-domain simulators have evolved in response to these problems.[26]

Additional Linearization Techniques

Because the linearity of these current-mode mixers is controlled primarily by the quality of the transconductance, it is worthwhile to consider additional ways to extend linearity. Philosophically, there are four methods for doing so: predistortion, feedback, feedforward, and piecewise approximation. These techniques can be used alone or in combination. What follows is a representative (but hardly exhaustive) set of examples of these methods.

Predistortion cascades two nonlinearities that are inverses of each other, and it shares with feedforward the need for careful matching. Predistortion is actually nearly ubiquitous, as it is the principle underlying the operation of current mirrors. In a mirror, an input current is converted to a gate-to-source voltage through some nonlinear function that is then undone to produce an output current exactly proportional to the input. Predistortion is also fundamental to the operation of true Gilbert mixers, where a pair of junctions computes the inverse hyperbolic tangent of an input differential current, and a differential pair subsequently undoes that nonlinearity.

Negative feedback computes an estimate of error, inverts it, and adds it back to the input, thereby helping to cancel the errors that distortion represents. The reduction in distortion is large as long as the loop transmission magnitude is large. Because

---

[26] Measuring these quantities in the laboratory also requires some care. As with the simulation, the amplitudes of the two input tones must be low enough to avoid excitation of higher-order nonlinearities (which would cause a slope of other than $+3$) yet sufficiently larger than the noise floor.

10.4 MULTIPLIER-BASED MIXERS

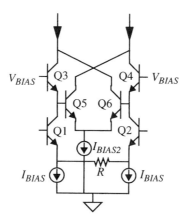

FIGURE 10.14. Classic bipolar cascomp

a negative feedback system computes the error a posteriori, the overall closed-loop bandwidth must be kept a small fraction of the inherent bandwidth capabilities of the elements comprising the system – else the error estimate will be irrelevant at best and destabilizing at worst. The series feedback examples of this chapter are popular methods for linearizing high-frequency transconductors.

Feedforward is another linearization technique; it computes an estimate of the error at the same time the system processes the signal, thereby evading the bandwidth and stability problems of negative feedback. However, the error computation and cancellation then depend on matching, so the maximum practical reduction in distortion tends to be substantially less than generally attainable with negative feedback. Feedforward is most attractive at high frequencies, where negative feedback becomes less effective owing to the insufficiency of loop transmission.

An example of feedforward correction applied to a transconductor is Pat Quinn's "cascomp" circuit, originally implemented with bipolar transistors.[27] As Figure 10.14 shows, this transconductor consists of a cascoded differential pair to which an additional differential pair has been added. Some linearization is provided by the source degeneration resistor R, but significant nonlinearity remains in the transconductance of inner differential pair Q1–Q2. To see this explicitly, consider that the voltage across the resistor is the input voltage minus the difference in gate-to-source voltages of Q1 and Q2:

$$V_R = v_{IN} - (v_{BE1} - v_{BE2}) = v_{IN} - \Delta v_{BE1}. \tag{21}$$

The goal is to have a differential output current precisely proportional to v_{IN}, so any nonzero Δv_{BE} represents an error. The cascoding pair possesses the same Δv_{BE} as the input pair, which is measured by the inner differential pair. A current proportional to this error is subtracted from the main current to linearize the transconductance.

---

[27] "Feedforward Amplifier," U.S. Patent #4,146,844, issued 27 March 1979, reissued 1984.

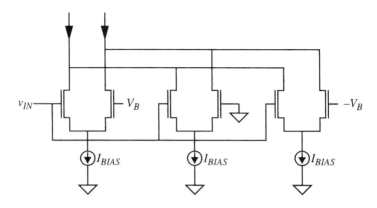

FIGURE 10.15. High-linearity g_m cell

The name "cascomp" derives from this combination of a cascode and error compensation. Although the inner pair is shown as an ordinary differential pair for simplicity, it is frequently advantageous to linearize it and thus relax the requirements on the correction, thereby increasing the range over which the transconductance remains constant.

Another nonfeedback approach is piecewise approximation, which exploits the observation that virtually any system is linear over some sufficiently small range. It divides responsibility for linearity among several systems, each of which is active only over a small enough range so that the composite exhibits linearity over an extended range.

Gilbert's bipolar "multi-tanh" arrangement is an example of piecewise approximation. In MOS form (just to keep you disoriented), it appears as shown in Figure 10.15.[28] Each of the three differential pairs behaves as a reasonably linear transconductance over an input voltage range centered about V_B, zero, and $-V_B$, respectively. For input voltages near zero, the transconductance is provided by the middle pair and is roughly constant for small enough v_{IN}. As the input voltage deviates significantly from zero, the tail current eventually steers almost completely to one side of the middle pair but, with an appropriate selection of bias voltage V_B, one of the outer pairs takes over and continues to contribute an increase in output current; see Figure 10.16.

The overall transconductance is the sum of the individual offset transconductances and can be made roughly constant over an almost arbitrarily large range by using a sufficient number of additional differential pairs, each offset appropriately. The tradeoff is an increase in power dissipation and input capacitance.

10.4.3 PASSIVE DOUBLE-BALANCED MIXER

So far, we've examined active mixers, which have the attribute of providing conversion gain. However, active devices with sufficient gain may simply be unavailable

---

[28] The name derives from the fact that the transfer characteristic of a bipolar differential pair is a hyperbolic tangent.

10.4 MULTIPLIER-BASED MIXERS

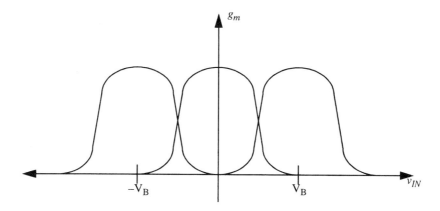

FIGURE 10.16. Illustration of linearization by piecewise approximation

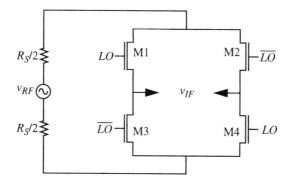

FIGURE 10.17. Simple double-balanced passive MOS mixer

at very high frequencies, so we need to consider passive mixers as well. Aside from their potential for very high-frequency operation, such mixers may also have the potential for low-power operation at very low supply voltages.

In the active mixers considered so far, representations of the RF signal in the form of currents, rather than the RF voltages themselves, are effectively multiplied by a square-wave version of the local oscillator. An alternative that avoids the V–I conversion problem is to switch the RF signal directly in the voltage domain. This option is considerably easier to exercise in FET than in bipolar form, which is why bipolar mixers are almost exclusively of the active, current-mode type.

The simplest passive commutating FET mixer consists of four switches in a bridge configuration; see Figure 10.17. Unlike most of the other circuits in this chapter, bipolar devices cannot substitute gracefully for the FET devices shown. Bipolar transistors are excellent current-mode switches, but they are not good voltage-mode switches.

The switches are driven by local oscillator signals in antiphase so that only one diagonal pair of transistors is conducting at any given time. When M1 and M4 are on, v_{IF} equals v_{RF}; when M2 and M3 are conducting, v_{IF} equals $-v_{RF}$. A fully equivalent description is that this mixer multiplies the incoming RF signal by a unit-amplitude square wave whose frequency is that of the local oscillator. Hence, the output contains many mixing products that result from the odd-harmonic Fourier components of the square wave.[29] Luckily, these are often readily filtered out, as discussed previously.

The voltage conversion gain of this basic cell is easy to compute from the foregoing description. Assuming multiplication by a unit-amplitude square wave, we may immediately write

$$G_c = 2/\pi, \quad (22)$$

where (again) the $2/\pi$ factor results from splitting the IF energy evenly between the sum and difference components.[30]

In practice, the actual voltage conversion gain may differ somewhat from $2/\pi$ because real transistors do not switch in zero time. Hence, the incoming RF signal is not multiplied by a pure square-wave signal in general. Perhaps contrary to intuition, however, the effect of this departure from ideal assumptions is usually to *increase* the voltage conversion gain above $2/\pi$.

A more general expression for the voltage conversion gain is somewhat cumbersome to derive, so we will only state the relevant insights here.[31] The output of the mixer may be treated as the product of three time-varying components and a scaling factor:

$$V_{IF}(t) = v_{RF}(t) \cdot \left[\frac{g_T(t)}{g_{Tmax}} \cdot m(t)\right] \cdot \left[\frac{g_{Tmax}}{\overline{g_T}}\right]. \quad (23)$$

The function $g_T(t)$ is the time-varying Thévenin-equivalent conductance as viewed from the IF port, while g_{Tmax} and $\overline{g_T}$ are the maximum and average values (respectively) of $g_T(t)$. The *mixing function, m(t)*, is defined as

$$m(t) = \frac{g(t) - g(t - T_{LO}/2)}{g(t) + g(t - T_{LO}/2)}, \quad (24)$$

where $g(t)$ is the time-varying conductance of each switch and T_{LO} is the period of the LO drive. The mixing function has no DC component, is periodic in T_{LO}, and has only odd harmonic content if the LO signal has perfect half-wave symmetry.

The Fourier transform of the first bracketed term in Eqn. 23 has a value of $2/\pi$ at the LO frequency for a square-wave drive (as asserted earlier) and a value of $1/2$ for a sinusoidal drive, so the effective mixing function indeed contributes a higher conversion gain for a square-wave drive. However, the second bracketed term is unity for

---

[29] This situation is the same as with the current-mode mixers, however. Also, even harmonics of the LO terms may be nonzero if the duty cycle of the square wave is not exactly 50%.

[30] If we assume equal source and load terminations, then this gain corresponds to a 3.92-dB voltage and power loss. Many practical implementations (such as the discrete passive mixers discussed in Sections 10.4.4–6) typically exhibit a somewhat greater conversion loss than this theoretical limit because of additional sources of attenuation (e.g., nonzero switch drop, skin-effect loss, etc.). Common conversion losses for mixers of this type are in the neighborhood of 5–6 dB.

[31] For a detailed derivation, see A. Shahani et al., "A 12mW Wide Dynamic Range CMOS Front-End for a Portable GPS Receiver," *IEEE J. Solid-State Circuits,* December 1997.

a square-wave drive (because the peak and average conductances are equal) but $\pi/2$ for a sinusoidal drive. The overall conversion gain is greater with a sinusoidal drive because the second term more than compensates for the smaller contribution by the (effective) mixing function. The difference is not particularly large, however. With a sinusoidal drive, the conversion gain is $\pi/4$ (-2.1 dB), compared with the $2/\pi$ gain (-3.92 dB) obtained with a square-wave drive.

Owing to the spectrum of the (effective) mixing function, undesirable products can appear at the IF port of this type of mixer. The subject of filtering therefore deserves careful consideration, especially in connection with the issue of input and output terminations. In virtually all discrete designs, the source and load impedances are real and well-defined at 50 Ω. In other cases, such as integrated circuit implementations, the load at the IF port of the mixer might not be terminated in this fashion. In MOS forms, for example, the load at the IF port is frequently capacitive to an excellent approximation. In such cases, the capacitive loading can be exploited to form a simple low-pass filter in conjunction with the resistance of the switches. A thorough analysis[32] reveals that the transfer function of this filter is simply

$$H(s) = \left[s \frac{C_L}{g_T} + 1 \right]^{-1}. \tag{25}$$

We see that the pole frequency is simply the ratio of the average conductance (again, as viewed from the IF port, back through the switches) to the load capacitance. This inherent filtering action may be tailored to provide a much-desired attenuation of unwanted mixer products.

Both noise figure and IP3 are strong functions of the LO drive, since the resistance of the switches in the on state must be kept low and constant to optimize both parameters. The IP3 is also a function of the amount of voltage boost provided by the L-match. This boost may be adjusted downward to trade conversion gain for improved IP3 and, in some cases, it may be appropriate to remove the L-match altogether. Typical SSB noise figures of 10 dB and input IP3 of 10 dBm are readily achievable with an LO drive amplitude of 300 mV.[33] As a crude estimate, the SSB noise figure of this type of mixer is approximately equal to the power conversion loss.

As a final note on the noise performance of this type of mixer, one might expect the absence of DC bias current to imply the absence of $1/f$ noise. However, because a mixer is a periodically time-varying system, it's still possible for spectral components centered at integer multiples of the local oscillator to fold down to DC, for example. Thus, $1/f$ noise may still appear at the output of the mixer without requiring any DC bias in the mixer itself. That is, a DC current may nonetheless arise. In cases where it is important to minimize $1/f$ noise in the mixer output, it is generally helpful (a) to reduce the LO drive to the minimum value consistent with acceptable mixing action and (b) to design the local oscillator carefully to minimize its close-in

---

[32] Ibid.
[33] This value applies to a sine-wave LO.

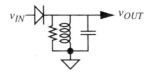

FIGURE 10.18. Simple diode mixer

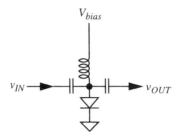

FIGURE 10.19. Simple single-diode mixer

phase noise in particular (see Chapter 18). Finally, terminating the mixer in a relatively high resistance keeps induced DC currents small and thus reduces $1/f$ noise. These considerations are particularly important in the design of receivers that are sensitive to $1/f$ noise, such as the direct-conversion (also known as the homodyne or zero-IF) receiver and low-IF receiver.

To reduce the power consumed by the LO drivers, the gate capacitance of the switches may be resonated with an inductor (for narrowband applications), resulting in a power reduction by a factor of Q^2. It is trivial to reduce the power to the order of a milliwatt or less, even at gigahertz frequencies.

10.4.4 SINGLE-DIODE MIXER

The simplest and oldest passive mixer uses a single diode. In the circuit of Figure 10.18, the output RLC tank is tuned to the desired IF, and v_{IN} is the sum of RF, LO, and DC bias components. The nonlinear V–I characteristic of the diode provides diode currents at a number of harmonic and intermodulation frequencies, and the tank selects only those at the IF.

It is tempting to reject this circuit as hopelessly unsophisticated, since it provides neither conversion gain nor isolation, for example. However, at the very highest frequencies, it may be difficult to implement mixers any other way. Indeed, all of the radars used in the Second World War featured single-diode mixers. Much of the modern work in the millimeter-wave bands simply would not be possible without these types of mixers. A more detailed schematic of such a mixer is shown in Figure 10.19.

Here, the input is the sum of RF and LO components only; the DC bias term is provided through an RF choke (or quarter-wave line). Capacitive coupling is used throughout to prevent this bias from upsetting circuits driving the input and driven

10.4 MULTIPLIER-BASED MIXERS

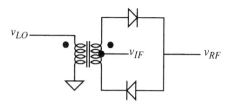

FIGURE 10.20. Classic transformer-based single-balanced diode mixer

by the output. The necessary summation at the input may be provided by numerous means, such as resistive combiners or various types of hybrids. The latter are often preferred because of the isolation they offer.

As another note on this circuit, we remind you that it can be used as a crude demodulator if the input signal is an AM signal (at either RF or IF). When used in this manner, the output inductor is removed entirely, no LO is used, and a simple *RC* network provides the output filtering. Millions of "crystal" radio sets used this type of detector (known in this context as an *envelope* detector), and nearly all AM superheterodyne radios built today use a single-diode demodulator.

10.4.5 TWO-DIODE MIXERS

There are seemingly an infinite number of ways to use diodes as mixers. As we'll see, it will appear that a diode bridge can be used as just about anything, depending on which terminals are defined as input and output and which way the diodes point.[34]

With two diodes, it's possible to construct a single-balanced mixer. In this case, one may obtain isolation between LO and IF, but there is poor RF–IF isolation; see Figure 10.20. The transformer-based version works very well at low to moderate frequencies (say, up to 1–2 GHz). In this range, the transformer in many commercial implementations is typically a small trifilar-wound toroid. Two of the windings are connected in series to create the secondary winding. This method of construction ensures good symmetry in the secondary windings as well good primary–secondary coupling.

Assume that the LO drive is sufficient to make the diodes act as switches, regardless of the magnitude of the RF input. With a positive value for v_{LO}, both diodes will conduct (note the reference dots on the transformer windings) and effectively

---

[34] The nonlinear junction capacitance of diodes can be used to build a parametric amplifier. The nonlinearity can be used to transfer energy from a local oscillator (known as the *pump* in paramp parlance) to the signal, instead of the more conventional transfer of power from a DC source to the signal frequency. Parametric amplifiers can be extremely low-noise devices, since only pure reactances are needed to make them work. Prior to invention of the maser, such amplifiers exhibited the lowest noise at microwave frequencies, enabling important achievements in radioastronomy. Chapter 9 discusses parametric amplifiers in a little more detail.

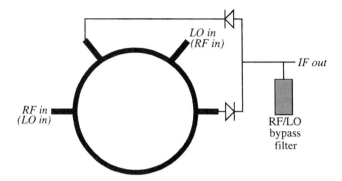

FIGURE 10.21. Typical single-balanced diode mixer with rat-race hybrid (filters not shown)

connect v_{RF} to the IF output. When v_{LO} goes negative, the diodes open-circuit and disconnect v_{RF}.

The poor RF–IF isolation should be self-evident from the comment that the diodes connect the RF and IF ports together whenever the diodes are on. Similarly, it should be evident that symmetry guarantees excellent RF–LO isolation. Whenever the diodes are on, the RF voltage can develop only a common-mode voltage across the transformer windings, so no voltage can be induced at the LO port (at least ideally; asymmetries always exist in practice to spoil perfection).

At the highest frequencies, the transformer is almost always replaced by a hybrid of some sort. A typical example uses a ring hybrid to provide the antiphase diode drive; see Figure 10.21. For this configuration to provide good performance, the LO and RF frequencies must both lie within the narrow bandwidth of the hybrid. The maximum permitted frequency separation is therefore about 10%. Fortunately, that constraint does not prevent sensible frequency plans. Another requirement is for the IF port to present a very low impedance termination at the frequency of the RF (and LO) input signal. A lumped resonant circuit (at lower frequencies) or transmission line stub (at higher frequencies, as shown in the figure) can be designed to have the desired behavior. One must ensure that component parasitics do not cause the filter to attenuate signals at the desired IF. A second requirement is that the IF port must present a low-reflectance termination at the desired (e.g., difference) IF frequency and at the image (e.g., sum) IF frequency as well. Failure to terminate the image component is a common oversight and often leads to pathological mixer behavior.[35]

Note that, as suggested by Figure 10.21, the input ports can be driven two ways. The conversion loss is identical for the two possibilities, and isolation doesn't depend on the choice, either. However, the spurs generated by the two options are not the same. Depending on system details, there may be an advantage to one connection or the other.

---

[35] These pathologies are not always undesired, but dependence on them is risky.

10.4 MULTIPLIER-BASED MIXERS

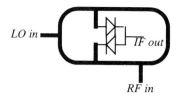

FIGURE 10.22. Mixer layout with noncircular ring hybrid (IF return and RF/LO bypass not shown)

The lines connecting the hybrid to the diodes also need to be of matched length (more precisely, they must possess a small phase angle difference). If this requirement is not well satisfied then the conversion loss will increase, matching may degrade, and linearity may suffer as well. The lines also need to be transmission lines for as much of the distance to the diodes as possible. Furthermore, the diode packages should be chosen so that the parasitics are consistent with operation at the desired frequency. Since the best package is no package, beam-lead devices are frequently used at the highest frequencies.

A more subtle consideration is that the feed to the diodes needs to be modified with a path to (an AC) ground for the IF signal. At high frequencies, a high-Z_0 grounded stub that is a quarter-wavelength at the RF (LO) is preferable. Two such stubs are required here, one from each diode to ground. If the diodes are to operate with some DC bias (as is occasionally, but infrequently, the case), the stubs may be returned to the DC bias supply. The connection must be well bypassed to ground at the RF/LO frequencies.

As a final remark on this single-balanced case, it should be noted that the area within the hybrid is permitted to contain circuitry – if care is exercised to avoid unwanted coupling. A reasonable guideline is to space the diodes and other elements away from the ring proper by a distance equal to at least 4–5 dielectric thicknesses. Frequently, the ring itself is also deformed into a rectangular or elliptical shape in order to accommodate layout constraints. See Figure 10.22. An unfortunate by-product of this type of layout is an increased need for electromagnetic field-solver simulations to verify that the design functions as desired.

Don't Try This at Home (or Anywhere Else)

In homage to the "bad circuit ideas" sections that populate *The Art of Electronics* of Horowitz and Hill, we will mention that an all-too-common variation on the single-balanced mixer replaces the rat-race hybrid with a branchline coupler. Because the latter provides a quadrature phase shift rather than an inversion, it is surprising that such a replacement can work at all. Even more surprising is its enduring popularity. To accept seriously degraded performance just to save a half-wavelength of interconnect seems an injudicious trade-off. The slimmest justification for using this approach might be that broadband branchline couplers could be used. However, the utility of a bad broadband mixer is questionable.

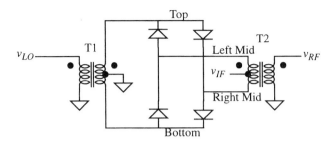

FIGURE 10.23. Classic double-balanced diode ring mixer

10.4.6 DOUBLE-BALANCED DIODE MIXER

By adding two more diodes and one more transformer, we can construct a double-balanced mixer to provide isolation among all ports (see Figure 10.23). We initially show conventional transformers, but this structure may also be translated into quasi-planar form by using, for example, Marchand-based baluns, as we'll see later.[36]

As with the transformer-based single-balanced version, this mixer is popular up to the low-gigahertz frequency range. To understand how it operates, once again assume that the LO drive is sufficient to cause the diodes to act as switches. In the circuit shown, the left pair of diodes conducts whenever the LO drive is negative whereas the right pair of diodes conducts whenever the LO drive is positive.

With the LO drive positive, the voltage at "Right Mid" must be zero by symmetry, since the center tap of the input transformer is tied to ground. Thus, v_{IF} equals v_{RF} (again, note the polarity dots). With the LO drive negative, it is "Left Mid" that has a zero potential, and v_{IF} equals $-v_{RF}$. Hence, this mixer multiplies v_{RF} approximately by a unit-amplitude square wave whose frequency is that of the LO.

Isolation is guaranteed by the symmetry of the circuit. The LO drive forces a zero potential at either the top or bottom terminal of the output transformer, as noted previously. If the RF input is zero, there will be no IF output. Hence, this configuration provides LO–IF isolation. Similarly, we can show LO–RF isolation by considering a zero IF input. Since again there is a zero potential at either the top or bottom terminal of the output transformer, there will be no primary voltage and hence no secondary voltage.

To an excellent approximation, the impedance seen at the IF port is the source impedance that drives the RF port – as long as the LO drive power is high enough to place the diodes in a highly conductive state, and assuming a 1:1 turns ratio between the RF port and any two adjacent taps of T2.

These passive mixers are available commercially in discrete form and perform exceptionally well. The upper limit on the dynamic range is typically constrained by

---

[36] See Stephen Maas, "Harmonic Balance Simulation Speeds RF Mixer Design," ⟨http://www.planetanalog.com/story/OEG20020328S0107⟩, 28 March 2002. Also see his book, *Microwave Mixers*, 2nd ed., Artech House, Norwood, MA, 1993. Maas has written extensively on the subject, and anyone interested in microwave mixers would profit enormously from reading his many publications on the topic.

diode breakdown, while isolation is a function of the matching levels achieved (both in diode characteristics and transformer winding parasitics).

The conversion gain of the double-balanced passive mixer is readily computed by noting that the RF signal is effectively multiplied by a close approximation to a unit-amplitude square wave, assuming the LO drive is large enough. The amplitude of the fundamental is $4/\pi$, and multiplication results in sum and difference components with half that amplitude. Consequently, the conversion gain ideally should be

$$20\log(2/\pi) \approx -3.92 \text{ dB}. \tag{26}$$

In practice, mixers with a single quad of diodes typically exhibit minimum conversion losses in the neighborhood of 5–6 dB for LO drives above about 5–7 dBm. The difference is partly due to losses in the diodes and transformers and also to the fact that the LO drive does not cause the diodes to switch on and off instantaneously. If LO drive power is limited, the mixer can still function, albeit with degraded performance. For example, a 0-dBm LO power will typically produce a 9-dB conversion loss. The compression point of such mixers tracks the LO power and is typically about 6 dB below the nominal LO power (on an input-referred basis). Reduced LO levels therefore degrade SFDR (spurious-free dynamic range) rapidly, since the noise floor rises (owing to attenuation) and the intercept point diminishes.

Higher RF levels can be accommodated if series connections of diodes are used in place of each diode in Figure 10.23, the drawback being an increased LO drive requirement to guarantee complete switching of the diodes. There are mixers that possess 1-dB compression points in excess of 21 dBm while requiring 27 dBm (!) of LO drive.

Isolation between RF and LO ports becomes increasingly important with these high-level mixers, but it rarely exceeds 30 dB. The potential for LO signals to leak back through the RF port for re-radiation by the antenna is also a factor that must be taken into account in any careful design. Filters or unilateral amplifiers may be used – in conjunction with an appropriate frequency plan – to mitigate the re-radiation problem.

10.4.7 IMAGE TERMINATION

When actually using such mixers, one should be aware that it is important to terminate all ports in the proper characteristic impedance: not only at the RF, IF, and desired LO frequencies, but at the image frequencies as well. If only narrowband terminations are used, it is possible for reflections of various intermodulation products to degrade performance seriously (or at least alter it in ways that may be difficult to anticipate). Hence, it is generally insufficient merely to use a standard RLC tank as an output bandpass filter without a diplexer[37] or an intermediate buffering stage to guarantee a proper resistive termination at the sum and difference frequencies. Failure to

---

[37] Recall that a diplexer splits an input signal into high- and low-frequency bands. A diplexer can also be used as part of a duplexer (a device that permits simultaneous transmit and receive). The terms are so similar that they are often confused with each other.

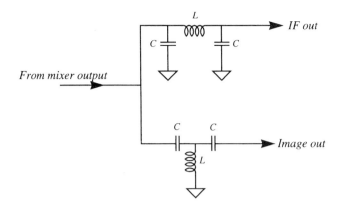

FIGURE 10.24. Diplexer for simultaneous termination of IF and image

satisfy this condition can be the source of many perplexing phenomena. Normally, poor image termination produces degraded SFDR as expected, but occasionally it improves SFDR. To avoid such unpredictable behavior, attention to termination at all frequencies is essential.

As is evident from the schematic of Figure 10.24, a simple diplexer consists of low-pass and high-pass branches. Assuming that the desired IF is the difference component, the low-pass branch provides the desired output while the high-pass branch provides the image signal. Terminating both outputs guarantees that the mixer's IF output port sees a broadband resistive termination. The particular diplexer shown is considered relatively elaborate. If filtering requirements are more modest, one may use a simpler implementation consisting of a single LC pair in each branch.

At higher frequencies, transformers with adequate characteristics may be unavailable. In those cases, various stripline couplers may be adapted to replace them. One in particular, based on the Marchand balun, has been used successfully to implement broadband mixers into the lower-frequency end of the millimeter-wave bands. The tight coupling required is virtually impossible to provide with edge-coupled structures, so microstrip implementations are effectively precluded. Instead, broadside-coupled versions in stripline are more practical. To keep the figures simple, however, we will draw them in a fashion that does not explicitly reflect a stripline arrangement. Just keep this in mind as you view the drawings.

The classic Marchand (Figure 10.25) by itself isn't quite suitable, however; there aren't enough ports! That slight problem is readily solved by slicing the upper line in half. Then both the leftmost and rightmost ends may serve as RF (LO) and LO (RF) input ports, and the newly created terminals in the center of the top line serve to drive the diode quad (Figure 10.26).[38] We see from the figure that the left pair of coupled lines provides a balanced drive (say, of RF) to one opposing pair of terminals of the diode quad, while the right pair of coupled lines provides a balanced LO

---

[38] See the Maas citations in footnote 36.

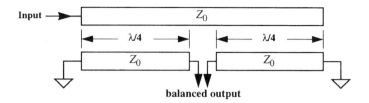

FIGURE 10.25. Classic Marchand balun

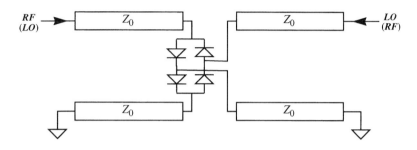

FIGURE 10.26. Double-balanced mixer based on Marchand balun
(core only shown; see text)

drive to the other. Thus, the diodes see a drive arrangement that is exactly the same as in the classic transformer-based version.

A moderately tricky part is arranging for the IF signal to find a way out of this structure. By examining the classic prototype of Figure 10.23, we can conclude that we need to synthesize the equivalent of a center tap between the two RF-driven terminals of the diode quad. An expedient way of doing so is to connect a line that is $\lambda/4$ at the center of the RF–LO span to each of those terminals and then tie the other ends together. However, the IF signal appears at the RF port as well. To prevent this loss of isolation, one could capacitively couple the RF drive to the diode quad. If the capacitor is small enough, it will attenuate the IF signal without materially affecting the RF signal. Replacing the capacitor with a parallel-resonant trap (tuned to the IF) will improve RF–IF isolation significantly but is considerably harder to implement.

We still need to provide a return path for the IF signal. As in the single-balanced case, tying a pair of grounded $\lambda/4$ lines to the other two diode quad terminals solves that problem neatly.

In yet another variation, neither IF-blocking capacitors nor resonant traps are used. Instead, the RF signal drives the structure through the high-pass port of a diplexer. The common port of the diplexer feeds the mixer core, and the IF signal is extracted from the lowpass port of the diplexer.

The baluns that drive the mixer core, and the various stubs that connect to it, need to be designed carefully if satisfactory operation is to be obtained over a broad band. The lengths of the coupled sections are chosen $\lambda/4$ at the center of the operating range. Operation over a 2–3 : 1 ratio above and below that center frequency is then

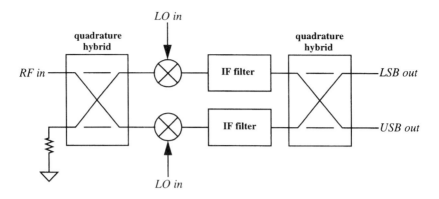

FIGURE 10.27. Complex mixer example

possible. Thus, the overall operating frequency range of the couplers can span 5:1 to 10:1.

The circuit's bandwidth may be considerably less, however, because of the $\lambda/4$ stubs. For example, the shorted stubs will appear as short circuits instead of opens at double the nominal frequency, and thus a 2:1 frequency range becomes the limit. If the stubs are implemented with very narrow (very high-impedance) lines, however, the increase in losses at higher frequencies can prevent short circuits; hence the bandwidth may be better than might seem reasonable on the basis of lossless line assumptions.

10.4.8 OTHER MIXER CONFIGURATIONS

Complex Mixers

In many types of modulation, one must effectively keep track of the image frequencies. In some cases, the purpose is to cancel the undesired image and so relax requirements on external image filters. In other cases, the image is simply another modulation sideband. The quadrature (or complex) mixer lies at the heart of receivers of this type.

In Figure 10.27, a quadrature hybrid (typically a Lange or branchline coupler) splits an incoming RF signal into in-phase and quadrature components. These two signals feed a pair of mixers driven by in-phase LO signals. The outputs are then filtered, and the surviving IF components are fed to another quadrature hybrid, which performs phase-shifting, summing, and differencing operations on the quadrature IF components. The upper and lower image signals (sidebands) are available as separate outputs.

Subharmonic Mixers

At the extremes of frequency, it may not be practical or possible to generate the LO drives required to implement an ordinary superheterodyne stage. This situation arises

10.4 MULTIPLIER-BASED MIXERS

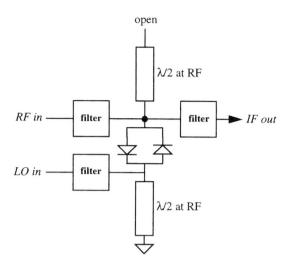

FIGURE 10.28. Example of subharmonic mixer ($N = 2$)

commonly in radioastronomy, for example, where one often wishes to detect radiation of sub-millimeter (or shorter) wavelength. We don't have to solve an RF signal generation problem, since the signal is provided by nature, but we are left with a difficult LO generation problem.

One solution is take advantage of nonlinearities. After all, they're hard to avoid, so one might as well exploit them. For example, suppose we need an LO drive with a frequency f_1, but it turns out to be impractical or inconvenient to acquire this. One could use a nonlinear element to generate harmonics of that frequency, filter to select the harmonic desired, and use that signal to drive a mixer.

An alternative is to take advantage of the spur modes that are generally present in practical mixer circuits. In this case, we drive the LO port at a frequency f_1/N and select the spur mode corresponding to the LO's Nth harmonic. Because we drive the LO port with a signal whose fundamental is a subharmonic of the actual LO desired, such a mixer is known as a subharmonic mixer.

The usual goal in mixer design is to suppress spur modes, but here we wish to enhance the effect – at least for one particular spur mode. A common circuit for doing so is shown in Figure 10.28. Here we assume that the RF input signal and the desired LO frequency are so close that they may be treated as equal. Furthermore, we assume that the LO actually supplied is at half this frequency, so that the subharmonic order $N = 2$. The open-circuited transmission line in the upper portion of the circuit thus presents an *open* circuit to the desired RF signal but a *short* circuit to the LO drive that is supplied, because the line is only a quarter-wavelength long there. At the same time, the short-circuited line presents a short circuit to the RF signal and an open circuit to the LO drive that is supplied. The voltage across the diodes is thus the sum of the RF input signal and LO drive. The use of two diodes connected in inverse parallel assures an enhancement of even-mode spurs. The low-pass IF filter does its best to remove undesired hash, but its ability to do so is limited precisely because of the spur-mode enhancement inherent in this architecture. Nevertheless, this circuit is often the only option available at extremely high frequencies.

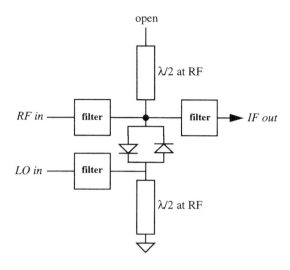

FIGURE 10.28. Example of subharmonic mixer ($N = 2$)

commonly in radioastronomy, for example, where one often wishes to detect radiation of sub-millimeter (or shorter) wavelength. We don't have to solve an RF signal generation problem, since the signal is provided by nature, but we are left with a difficult LO generation problem.

One solution is take advantage of nonlinearities. After all, they're hard to avoid, so one might as well exploit them. For example, suppose we need an LO drive with a frequency f_1, but it turns out to be impractical or inconvenient to acquire this. One could use a nonlinear element to generate harmonics of that frequency, filter to select the harmonic desired, and use that signal to drive a mixer.

An alternative is to take advantage of the spur modes that are generally present in practical mixer circuits. In this case, we drive the LO port at a frequency f_1/N and select the spur mode corresponding to the LO's Nth harmonic. Because we drive the LO port with a signal whose fundamental is a subharmonic of the actual LO desired, such a mixer is known as a subharmonic mixer.

The usual goal in mixer design is to suppress spur modes, but here we wish to enhance the effect – at least for one particular spur mode. A common circuit for doing so is shown in Figure 10.28. Here we assume that the RF input signal and the desired LO frequency are so close that they may be treated as equal. Furthermore, we assume that the LO actually supplied is at half this frequency, so that the subharmonic order $N = 2$. The open-circuited transmission line in the upper portion of the circuit thus presents an *open* circuit to the desired RF signal but a *short* circuit to the LO drive that is supplied, because the line is only a quarter-wavelength long there. At the same time, the short-circuited line presents a short circuit to the RF signal and an open circuit to the LO drive that is supplied. The voltage across the diodes is thus the sum of the RF input signal and LO drive. The use of two diodes connected in inverse parallel assures an enhancement of even-mode spurs. The low-pass IF filter does its best to remove undesired hash, but its ability to do so is limited precisely because of the spur-mode enhancement inherent in this architecture. Nevertheless, this circuit is often the only option available at extremely high frequencies.

CHAPTER ELEVEN

TRANSISTORS

11.1 HISTORY AND OVERVIEW

With the growing sophistication in semiconductor device fabrication has come a rapid expansion in the number and types of transistors suitable for use at microwave frequencies. At one time, the RF engineer's choices were a bipolar or possibly a junction field-effect transistor. The palette of options has since grown to a dizzying collection of MOSFETs, VMOS, UMOS, LDMOS, MESFETs, pseudomorphic and metamorphic HEMTs (MODFETs), and HBTs, all offered in an ever-expanding variety of materials systems. We'll attempt to provide a description of these types of devices, starting with a deciphering of their abbreviations. Then we'll focus on a small subset of these devices in an expanded discussion of modeling.

The bipolar transistor was discovered – not invented – in December of 1947 while the Bell Labs duo of John Bardeen and Walter Brattain was attempting to build a MOS field-effect transistor at the behest of their boss, William Shockley.[1] Their repeated failures led them to suspect that the problem lay with the surface, where the neat periodicity of the bulk terminates abruptly, leaving unsatisfied bonds to latch onto contaminants. To verify this "surface state" hypothesis, they undertook a detailed study of semiconductor surface phenomena. One of their experiments, designed to modulate the postulated surface states, itself happened to exhibit power gain. It wasn't the MOSFET they had been trying to build; it was a germanium point-contact bipolar transistor. Its behavior was never quantitatively understood, and repeatability of characteristics was only a fantasy. But it was good enough to earn the team – including Shockley – the 1956 Nobel Prize in physics. Although the point-contact transistor had minimal commercial impact, at least it gave us the names for the electrodes of a transistor. From Figure 11.1, you can appreciate where the base terminal got its name – it's the mechanical base for a point-contact device.

---

[1] Michael Riordan and Lillian Hoddeson, *Crystal Fire,* Norton, New York, 1997. This superb book is a must-read for anyone who is even remotely interested in the history of semiconductors. Also see the excellent review article by W. Brinkman et al., "The History of the Transistor and Where It Will Lead Us," *IEEE J. Solid-State Circuits,* v. 32, no. 12, December 1997, pp. 1858–65.

CHAPTER 11 TRANSISTORS

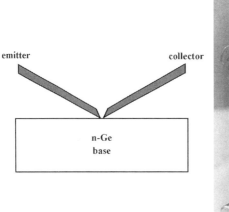
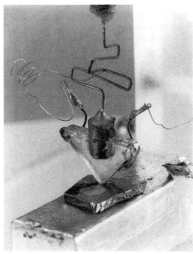

FIGURE 11.1. Point-contact bipolar transistor
[*photo courtesy of and copyright Lucent Bell Labs*]

Shockley disliked point-contact transistors, partly because it was a messy three-dimensional structure and partly because he hadn't been one of its direct inventors.[2] Working diligently in secret to develop a transistor he could call his own, he invented the junction bipolar transistor as it is known today: a sandwich of three semiconductor layers of alternating polarity, whose essential behavior is understandable with relatively simple one-dimensional analyses. Unlike its point-contact predecessor, the junction transistor was a true invention. In a scientific *tour de force,* Shockley correctly predicted its fundamental electrical characteristics in 1948 – well before the first crude one was demonstrated in mid-1950.[3] The modern bipolar device is considerably more complicated than depicted in Figure 11.2. We will consider enhancements to this picture a bit later, when we derive models for the bipolar transistor.

Field-effect transistors finally debuted in 1953, but in the form of junction FETs (JFETs), not MOSFETs.[4] See Figure 11.3. Shockley had invented the device a year earlier. Varying the voltage on the reverse-biased gate junction varies the extent of a depletion layer, adjusting the effective cross-section of – and therefore current through – a semiconducting bar. As conventionally built, JFETs are depletion-mode devices, meaning that their default state is one of conduction. One must actively turn them off. Because their gate electrode is one terminal of a reverse-biased diode, JFETs exhibit very large DC resistance, and their low-frequency power gain is correspondingly high.

---

[2] Shockley says so himself in "The Path to the Conception of the Junction Transistor," *IEEE Trans. Electron Devices,* v. 23, no. 7, July 1976, pp. 597–620.

[3] William Shockley, Morgan Sparks, and Gordon K. Teal, "P–N Junction Transistors," *Phys. Rev.,* v. 83, 1 July 1951, pp. 151–62.

[4] George C. Dacey and Ian M. Ross, "Unipolar Field-Effect Transistor," *Proc. IRE,* August 1953, pp. 970–9.

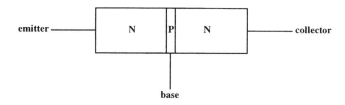

FIGURE 11.2. Junction bipolar transistor (vastly simplified, but similar to first prototype)

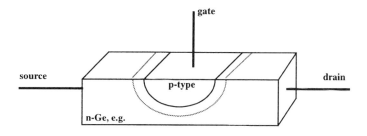

FIGURE 11.3. n-channel junction FET (simplified; most practical devices have two gate diffusions)

The first transistors were made of germanium, whose high leakage motivated a shift to silicon. Unfortunately, silicon's much higher melting point made the production of high-quality monocrystalline ingots almost impossible. Gordon Teal of Texas Instruments succeeded in building the first ones in early 1954, using a grown-junction process.[5] Thanks to Teal, TI's monopoly on silicon transistors was not broken until four years later, when Fairchild Semiconductor introduced their superior silicon "mesa" devices (which used layers defined by diffusion), followed soon after by planar silicon devices in 1959 (the first year of Moore's law).[6]

Research into MOSFETs continued, and Dawon Kahng and Martin Atalla of Bell Labs finally succeeded in building a working silicon MOSFET in 1960.[7] They exploited the surprising discovery that silicon's own oxide tames the troublesome

---

[5] Teal is an unsung hero of the early semiconductor age. He had chosen the study of germanium for his doctoral thesis in part because "its utter uselessness intrigued and challenged" him, as he later wrote (see the July 1976 issue of the *IEEE Transactions on Electron Devices*). At Bell Labs, he stubbornly insisted on developing monocrystalline material, despite initial opposition from Shockley, who thought polycrystalline material would be forever adequate. Without Teal's high-quality crystals of uniform characteristics, progress in the field would have been infinitely more difficult than it was.

Teal left Bell Labs for his home state of Texas for family reasons. That's how he came to present a paper on 10 May 1954 at the IRE National Conference on Airborne Electronics. He spoke after a succession of presenters had affirmed the hopelessness of any short-term success with silicon. Teal wowed the audience by pulling some silicon transistors out of his pocket and announcing their imminent availability from TI.

[6] J. A. Hoerni, "Planar Silicon Transistors and Diodes," *IRE Electron Devices Meeting*, October 1960.

[7] D. Kahng and M. M. Atalla, *IRE Solid-State Devices Research Conference*, Carnegie Institute of Technology, 1960. Also see D. Kahng, U.S. Patent #3,102,230, filed 1960, issued 1963.

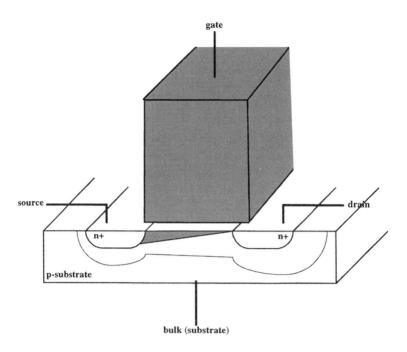

FIGURE 11.4. Typical MOSFET (greatly simplified, but with realistic relative gate electrode size)

surface states that had set Bardeen and Brattain on the road to discovering the transistor a dozen years earlier. Despite this success, however, the MOSFET's relatively poor device physics would limit its use to applications where low cost and high circuit density were prized above performance. Not until the late 1980s would Moore's law finally work its magic to the point where microwave MOSFETs became available. Thanks to its physical structure, it is relatively simple to tailor a MOSFET to operate as a depletion- or enhancement-mode device.

As with our simple rendering of a bipolar transistor, the drawing of Figure 11.4 leaves out a great many details. The modern MOSFET is a highly sophisticated, barely recognizable descendant of the Kahng–Atalla device of 1960. This evolution has been forced by the continuing drive toward ever-smaller geometries while maintaining acceptable device reliability in the face of increasing stresses (electrical, thermal, and mechanical). We will shortly consider a slightly more detailed picture of a MOSFET when we take up the issue of modeling.

A variant of the MOSFET, known as LDMOS (for laterally diffused MOS), has enjoyed wide deployment in cellular base–station power amplifiers. Its combination of reasonable gain, good linearity, moderately high output power, and low cost has made it particularly attractive for those applications. See Figure 11.5. As its name suggests, it employs an additional diffused (or implanted) layer. The purpose of the added n-layer is essentially to provide a large series resistance to reduce the peak electric field near the drain. The drain depletion layer extends into this region (rather than punching through to the source) and thus permits the device to operate with

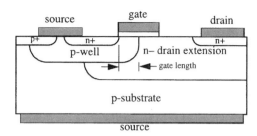

FIGURE 11.5. Typical LDMOS cross-section (layers not drawn to scale)

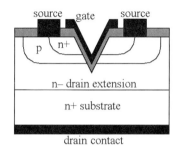 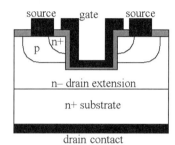

FIGURE 11.6. VMOS and UMOS cross-sections (layers not drawn to scale)

higher supply voltages than would otherwise be possible. The trade-off is a higher resistance in the "on" state. Breakdown voltages of 70–100 V allow robust operation with 28–50-V supplies. At the same time, the effective gate length is defined by the position of the n+ source diffusion, relative to the p-well. As such, the gate length can be substantially smaller than ordinary lithographic tools would usually support (observe that, in Figure 11.5, the gate electrode extends into the drain region). In the case of typical LDMOS devices used for base stations, the effective gate length is of the order of 0.25 μm, enabling good performance out to 3–5 GHz.

To simplify packaging, the source connection may be made at the bottom of the die. In those cases, a deep p+ "sinker" diffusion is commonly used to provide a low-resistance connection from the top surface (where the actual source terminal resides) to the contact on the underside of the die.

One reason that LDMOS is attractive is that it may be fabricated using rather simple equipment, as its structure differs only slightly from that of conventional low-power MOSFETs. However, it remains a lateral device because its current still flows only along the surface, in contrast with the vertical flow of bipolar transistors. Because the density of current per surface area is thus not particularly large, power MOS devices consume significant die area.

A modest modification reduces the total device area needed to support a particular current. Depending on the particular geometry used, such transistors are known as VMOS or UMOS power FETs. As is evident from Figure 11.6, the current flow is primarily vertical. In effect, VMOS and UMOS devices provide double the number of channels for the same number of processing steps. Furthermore, and in common with LDMOS, the channel length is not determined by photolithography. The gate length is controlled by adjusting the depth of the n+ source diffusion into the p-region. This control allowed VMOS devices with submicron channel lengths to be constructed in an era when lithography was limited to resolutions of several microns. In fact, VMOS devices were the first commercially available MOSFETs for which carrier velocity saturation affected electrical characteristics in a first-order fashion. The combination of short channel length (for high speed), an electric-field–absorbing drain extension (for high breakdown voltage), and vertical current flow (for high current density per die area) has made VMOS and UMOS devices popular for many RF power applications.

Silicon has come to dominate the industry because it is easily understood (being an elemental semiconductor), straightforward to process, and readily forms a superb insulating oxide – not because any one electrical parameter is outstanding. For example, the mobility of electrons is much higher in germanium than in silicon. Unfortunately, germanium's oxide is water soluble. That fact, plus its high leakage current (arising from its lower bandgap), has made germanium all but disappear as a commercially important semiconductor.

Gallium arsenide's far superior low-field electron mobility had been appreciated in the mid-1950s, but the difficulties of mastering the liquid-phase epitaxy required to grow the binary semiconductor limited the availability of the raw material. Point-contact GaAs diodes capable of operating well beyond Ka band were available by the late 1950s, but making transistors was another thing altogether. Doping to make bipolar transistors proved even trickier than it had with silicon (in particular, doped p-type GaAs is very troublesome), and the lack of a native oxide makes GaAs MOSFETs much more of an oxymoron than in silicon. Carver Mead of Caltech finally succeeded in making a crude GaAs transistor in 1965, three years after the first GaAs-based LEDs had been demonstrated.[8] By using a Schottky (metal-to-semiconductor) diode to form a gate, the metal–semiconductor FET (MESFET) sidesteps the MOSFET's need for a gate oxide as well as the JFET's and bipolar's need for forming a conventional junction. Unlike modern silicon MOSFETs, the MESFET uses actual metal for the gate (e.g., TiWN cladded with gold), making the parasitic series gate resistance much lower for MESFETs than for MOSFETs of the same dimensions. Like its ordinary JFET cousin, the MESFET is normally a depletion-mode device (enhancement-mode MESFETs do exist, though they are much less common). After considerable refinement, the GaAs MESFET came to dominate power amplifiers for cell phones through the 1980s and mid-1990s. See Figure 11.7. Note that, in addition to a gate–source Schottky diode, there is also gate–drain diode. Under normal operation, both of these diodes are in reverse bias.

For modern MOSFETs and MESFETs both, the gate–source capacitance is of the order of roughly 1 pF/mm of gate width, and the intrinsic voltage gains ($g_m r_{ds}$)

---

[8] See Schottky Barrier Gate Field Effect Transistor," *Proc. IEEE*, v. 54, 1966, p. 307. A more practical MESFET is described by W. W. Hooper and W. I. Lehrer, *Proc. IEEE*, v. 55, 1967, p. 1237.

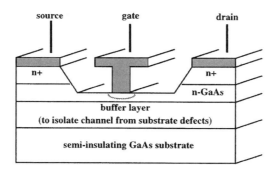

FIGURE 11.7. Typical n-channel GaAs MESFET

are generally in the range of 10–20. Although dependent on process technology, the maximum transconductance is generally around 1–2 mS/mm, and the maximum drain current is very roughly 1–2 mA/mm. These numbers are provided simply to convey a rough order-of-magnitude calibration. They are hardly constants of nature and thus should only be used for the coarsest back-of-the-cocktail napkin calculations.

For cell-phone handset power amplifiers, the GaAs MESFET now has been largely supplanted by another of Shockley's forward-looking ideas: the *heterojunction bipolar transistor* (HBT). Ordinary semiconductor devices use a single material for all layers; the only electronic differences among the layers are related to doping. All junctions are therefore *homojunctions* in these classic devices. Shockley had recognized about 1950 that combining semiconductors with different bandgaps represents powerful degrees of freedom in device design.[9] In a bipolar transistor, for example, heavier base doping reduces parasitic base resistance and increases breakdown voltage and output resistance, but it degrades the current gain. At the same time, heavier emitter doping improves current gain but increases emitter capacitance. These tradeoffs can be eased considerably by using a wide-bandgap material for the emitter. By suppressing base-to-emitter hole injection, such a heterojunction permits heavier base doping and lighter emitter doping without seriously degrading current gain. High-frequency performance improves as a direct result of reducing the parasitic base resistance and emitter capacitance. The heavier base doping also reduces the extent to which the collector–base depletion layer extends into the base region. That, in turn, reduces basewidth modulation (the Early effect) and thereby increases the output resistance.

Although this basic idea had been appreciated for decades, it wasn't until the 1970s that semiconductor technology had advanced enough to enable construction of devices. An important constraint is that the different semiconductors must have substantially equal lattice constants, for otherwise their interface will be full of defects. The most widely used heterojunction bipolar transistor is made with AlGaAs for the

---

[9] U.S. Patent #2,569,347, issued 25 September 1951. The first open publication on the concept is apparently by Herbert Kroemer ("Theory of a Wide-Gap Emitter for Transistors," *Proc. IRE*, November 1957, pp. 1535–7), who evidently learned of Shockley's patent from the paper's reviewers.

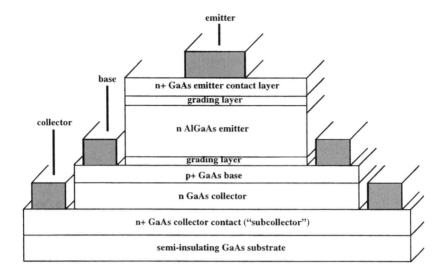

FIGURE 11.8. Typical GaAs HBT (idealized)[11]

emitter and GaAs for the base (and collector).[10] Also popular is the silicon germanium (SiGe) HBT, although it is not quite a true heterojunction device in Shockley's original sense of the term. For material stability reasons, the SiGe HBT is unable to use a large enough mole fraction of germanium in the base region to effect a significant shift in bandgap. Rather, a germanium concentration gradient in the base serves mainly to enhance the classic dopant-induced electric field already present there. This electric field speeds up carrier transport across the base, increasing high-frequency current gain. The popularity of SiGe is due to its compatibility with mainstream IC fabrication process technology, since the SiGe transistor is not very different from a conventional bipolar device.

Grading layers (see Figure 11.8) serve to moderate stresses at the GaAs–AlGaAs interfaces arising from the small mismatch in lattice constants. They are made as thin as possible to reduce any parasitic series resistance they might contribute.

More recently, indium phosphide HBTs have become available, partly driven by the material's compatibility with infrared generation and detection. Many InP HBTs used in RF applications are double-heterojunction devices, with InAlAs emitters, InGaAs bases, and InP collectors. The additional base–collector heterojunction increases breakdown voltage while reducing leakage current and output capacitance. The InGaAs base material has 50% higher mobility than GaAs.

---

[10] It is customary to refer to the technology according to the material comprising the substrate. Thus, an HBT with an AlGaAs emitter and GaAs substrate would generally be called a GaAs HBT. In those instances where it is important to identify the emitter explicitly, one might speak of a GaAs/AlGaAs HBT.

[11] S. M. Sze (ed.), *High-Speed Semiconductor Devices,* Wiley, New York, 1990.

Note that HBTs are vertical-flow devices, as opposed to the lateral-flow MOSFET and MESFET. The fabrication techniques currently in use (e.g., MOCVD, MBE) invariably produce the mesa-like structures of Figure 11.8. Silicon mesa devices were displaced by planar transistors because mesa structures have exposed junctions, resulting in poor device lifetimes and highly variable characteristics. As of yet, no similar revolution seems to be in the offing for HBTs.

The tailoring of device properties through the use of heterojunctions is not limited to realization of HBTs. In 1980, the high electron mobility transistor (HEMT) made its debut, based on pioneering work throughout the 1970s by tunnel diode inventor Leo Esaki.[12] This device relies on the insight that carriers drift fastest in dopant-free regions. Unfortunately, an ordinary MESFET's threshold voltage depends on the doping level. Thus, threshold voltage and mobility are effectively coupled in a classic MESFET. One cannot adjust the threshold voltage to arbitrary values without degrading high-frequency performance. Conversely, maximizing high-frequency performance might correspond to an inconvenient or unusable threshold voltage.

To acquire the additional degree of freedom necessary to decouple these two properties, one may use an appropriately doped wide-bandgap material (e.g., AlGaAs) with which the Schottky gate contact is formed together with a narrower-bandgap intrinsic semiconductor (e.g., GaAs) below it. Thanks to the bandgap difference, carrier flow can be well confined to the intrinsic layer (creating a "two-dimensional electron gas"), while other key device characteristics (such as threshold voltage) are controlled by the doping levels in the wide-bandgap layer. Thanks to superior carrier transport in the undoped channel, mobilities are readily doubled or tripled, improving device transconductance and high-frequency performance by similar factors.

The spacer layer above the undoped channel (see Figure 11.9) helps isolate the channel from the donors in the n+ AlGaAs layer. Its presence is necessary to maintain high mobility in the channel, but it must be made as thin as possible to maximize the coupling of charge into the channel.

The HEMT is also known by several other names, such as the modulation-doped FET (MODFET) or heterostructure FET (HFET). Less common abbreviations you may encounter include TEGFET or 2DEGFET (both for two-dimensional electron gas FET) and SDFET (selectively doped FET). An informal survey suggests that HEMT is the most popular term, with HFET and MODFET lagging well behind.

Subspecies of HEMTs include the pseudomorphic and metamorphic varieties. *Pseudomorphic* refers to the exploitation of useful electrical effects that may attend intentional strain caused by (slight) mismatches in the lattice constant (as long as the mismatch is not so great as to generate serious defects). These electrical effects may include increased bandgap differences and enhanced mobility. In a typical pseudomorphic HEMT (pHEMT), the undoped high-mobility GaAs channel of a conventional HEMT is replaced with a still higher–mobility undoped InGaAs

---

[12] In 1973, Esaki won the Nobel Prize in physics for his discovery of tunneling in heavily doped p–n junctions.

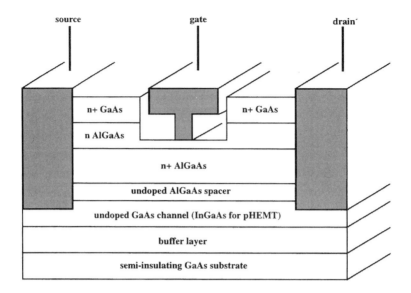

FIGURE 11.9. Typical GaAs HEMT[13]

channel for even better high-frequency performance. Other versions are based on an InAlAs/InGaAs/InP material combination.

Metamorphic HEMTs (mHEMTs) expand the number of suitable material combinations by accommodating lattice mismatch with compositionally graded layers. Such strain-absorbing buffer layers permit larger mole fractions of indium to be used in the undoped channel, for example. Even if defects result from the lattice mismatch, they can be confined to the buffer layers, where they are electrically innocuous for the most part.

Because HEMTs are modified MESFETs, they are commonly depletion-mode devices. As such, biasing them can be somewhat inconvenient if only a single supply is available. In fact, this inconvenience has been a significant factor in driving the shift toward HBTs in handset power amplifiers. By eliminating the need for a negative supply voltage generator, the HBT reduces handset size and cost and also improves battery life.

Another characteristic worth mentioning is that GaAs and InP have significantly worse thermal conductivity than silicon. Silicon's thermal conductivity is triple that of GaAs and double that of InP. These differences are particularly important for power amplifiers, yet even for allegedly low-power systems there may be instantaneous power densities large enough for thermal effects to manifest themselves on occasion. Engineers accustomed to working only with silicon-based devices are often surprised by the extent to which thermally related phenomena must be accommodated in a design.

[13] Sze, op. cit.

Finally, it must be mentioned that device engineers have by no means exhausted their considerable creativity. Thus, the incomplete sampling we have provided here will necessarily grow more incomplete with time. The use of silicon carbide (SiC) and gallium nitride (GaN), for example, is being actively studied, and one can expect the list to extend well beyond those materials in the future.

11.2 MODELING

Active devices are generally quite nonlinear, and they contain many parameters whose values may vary considerably as a function of frequency, temperature, and bias. Accommodating nonlinearities in circuit analysis is challenging enough, so having to consider reactances as well further complicates an already difficult task. Although the advent of cheap computation has made it feasible to include "everything" in a device model, doing so is best for analysis, not for design.

An alternative approach largely abandons hope of understanding the underlying physics and instead bases design primarily on measured parameters (such as S-parameters). This approach can yield excellent results because it is based on the "truth," but the lack of physically based compact models means that extensive data sets may be required for a robust design, and prediction of performance beyond the data set (or even for a different device of the same nominal type) may not be possible.

A common complaint is that one or another model is "wrong." It's important to recognize that *all* models are wrong at some level, so a search for a correct model will never terminate. Rather, the proper engineering philosophy is to seek models that are capable of answering the questions that are asked. If the questions are simple, then simple models may be perfectly acceptable. Complexity beyond that required to perform the task at hand just makes the acquisition of insight less likely.

A recurring theme in modeling *for design* is the reduction of complexity through selective and conscious neglect of phenomena that may be considered of second order. Part of the art, of course, is to discern which phenomena truly are second order *for a particular situation*. It is also important to know the model's limits of validity, and then to apply the model only within those boundaries.

The purpose of this chapter is to review models that are physically based and comparatively simple. As such, they are appropriate for first-pass designs of circuits. These models are also indispensable for subsequent refinement of a design, should more sophisticated simulations or experimental measurements reveal the need for modifications. Without guidance from simple models, it can be difficult to discern which parameters should be adjusted, in what direction, and by (approximately) how much.

We will present small-signal models for FETs and bipolars. For the latter, we will assume that the collector–base junction is never forward biased (i.e., the transistor never saturates). We will also spend a little time discussing how to extract key model parameters from often cryptic data-sheet information.

11.3 SMALL-SIGNAL MODELS FOR BIPOLAR TRANSISTORS

11.3.1 A SIMPLE DC MODEL

At frequencies low enough that reactances may be neglected and for a fixed collector–emitter voltage, the relationship between collector current and base–emitter voltage is well approximated by an exponential:

$$i_C \approx I_S \left[\exp\left(\frac{v_{BE}}{V_T} \right) \right], \tag{1}$$

where V_T, the thermal voltage, is equal to kT/q, which has a value of approximately 25 mV at room temperature. Remarkably, practical silicon devices conform to this exponential behavior within a few percent over 6–8 orders of magnitude of collector current. It is occasionally useful to note that Eqn. 1 implies a doubling of collector current for every 18-mV increase in v_{BE}, and a decade increase for every 60 mV.

The parameter I_S is the saturation current, and the way it is described in some texts encourages many to infer (incorrectly) that it is some sort of a constant. In fact, its behavior is more subtle:

$$I_S = I_0 \exp\left(-\frac{V_{G0}}{V_T} \right), \tag{2}$$

where V_{G0} is the bandgap voltage (extrapolated to 0 kelvins, and equal to about 1.2 V for silicon). A quasi-empirical expression for I_0 is

$$I_0 = A_e B T^r, \tag{3}$$

where A_e is the emitter area, B is a process-dependent constant, T continues to be the absolute temperature, and r is a process-dependent quantity we'll call the curvature coefficient. For the relatively deep, diffused emitters of older bipolar processes, r typically has a value between 2 and 3, whereas for the shallow, implanted (and very heavily doped[14]) diffusions that are common in modern high-speed processes, r is typically closer to 4. Clearly I_S is a temperature- and fabrication-dependent "constant."

Equation 1 assumes operation of the transistor in the forward active region, in which only the base–emitter junction is in forward bias. Furthermore, it assumes values of collector current far in excess of I_S, an assumption that is certainly well satisfied for all microwave devices under normal operation.

From Eqns. 1–3 it may be discerned that the base–emitter voltage drops about 2 mV for every degree increase in temperature, at constant collector current. See Figure 11.10. Use of Eqn. 1 alone would lead one to predict the wrong sign for this temperature coefficient, underscoring the importance of using a detailed expression

---

[14] Bandgap narrowing and nonlinearity in the heavily doped emitters are probably responsible for the high values of r.

11.3 SMALL-SIGNAL MODELS FOR BIPOLAR TRANSISTORS

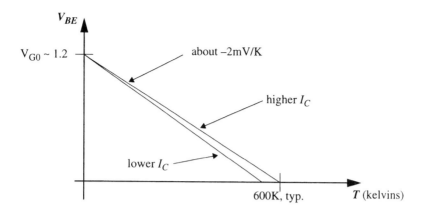

FIGURE 11.10. Approximate behavior of V_{BE} vs. temperature at constant collector current

for I_S. Not shown in the figure is a small curvature in the temperature behavior. This curvature is small enough that it may be neglected except when constructing certain types of voltage references (so-called bandgap references).

Another key relationship is that between collector and base current:

$$I_C = \beta_F I_B. \tag{4}$$

Trade-offs made in device design for good high-frequency performance usually result in lower β-values than exhibited by lower-frequency transistors. Nominal values are frequently below 100 and may be as low as 10. Keep in mind that β may vary considerably from device to device and even for a single device (as a function of current and temperature).

Clearly, the relationship between i_C and v_{BE} is nonlinear. Analysis of circuits containing a single nonlinearity can be complicated; analysis of those with more than one can be hopeless. As a consequence, linearized models (also known as incremental or small-signal models) have evolved to permit tractable analysis, albeit at the expense of a restricted range of validity.

In the case of the bipolar transistor, development of a suitable model begins by considering the base–emitter voltage as the sum of a DC term and a small signal superimposed on it:

$$v_{BE} = V_{BE} + v_{be}. \tag{5}$$

Inserting Eqn. 5 into Eqn. 1 and expanding in a Taylor series yields

$$i_C \approx I_S \left[\exp\left(\frac{V_{BE} + v_{be}}{V_T}\right) \right]$$

$$= I_S \left[\exp\left(\frac{V_{BE}}{V_T}\right) \right] \left[1 + \left(\frac{v_{be}}{V_T}\right) + \frac{1}{2!}\left(\frac{v_{be}}{V_T}\right)^2 + \cdots \right], \tag{6}$$

which yields a linear relationship if we throw away all the terms that make it nonlinear (remember, we're efficient engineers!):

$$i_C \approx I_C\left[1 + \frac{v_{be}}{V_T}\right] = I_C + \left(\frac{I_C}{V_T}\right)v_{be} = I_C + i_c. \qquad (7)$$

That is, the total collector current is the sum of the DC term and a term that is approximately proportional to the small-signal base–emitter voltage:

$$i_c = \left(\frac{I_C}{V_T}\right)v_{be} = g_m v_{be}. \qquad (8)$$

The proportionality constant, g_m, is known as the transconductance (*conductance* because it is a ratio of current to voltage, and *trans* because the current and voltage are measured at two different ports). The transconductance is therefore modeled by a voltage-dependent current source.

There is also a relationship between collector and base currents. Thus, we have a relationship between base current and base–emitter voltage:

$$i_b = \frac{i_c}{\beta_0} = \left(\frac{g_m}{\beta_0}\right)v_{be} = \frac{v_{be}}{r_\pi}. \qquad (9)$$

The symbol β_0 is used to underscore that the small-signal ratio of collector and base currents may differ from that at DC. That said, one usually takes it to equal β_F in most circumstances. Such a substitution is generally justified because β is so variable that it makes little sense to be overly fussy.

The linear small-signal relationship between base current and base–emitter voltage is represented by a simple resistor between the base and emitter terminals.

In all real bipolar transistors, the collector current is also a weak function of the collector–emitter voltage, rather than solely a function of base–emitter voltage. The reason is that the thickness of the collector–base depletion layer increases with increasing collector voltage and so the effective electrical width of the base region is similarly dependent on the collector voltage. Because the collector current depends on the basewidth, increases in collector voltage cause increases in collector current. This phenomenon, known as *basewidth modulation* (or the *Early effect*, after Bell Labs engineer James Early, who first identified and described this mechanism), is modeled by a collector–emitter resistor, r_o.[15] If one plots curves of collector current as a function of collector–emitter voltage, the effects of basewidth modulation manifest themselves in an upward tilt. Extrapolating all such curves back to the V_{CE}-axis results in a near common intercept. That intercept is (minus) the *Early voltage*, V_A. At a given collector bias current, the collector–emitter resistor has a value

$$r_o = V_A/I_C. \qquad (10)$$

---

[15] There is also a reduction in base current, which may be accounted for by adding another resistor between collector and base. Under the conditions that typically prevail in microwave circuits, that resistor can almost always be ignored. Those interested in more sophisticated modeling should consult Ian Getreu's classic *Modeling the Bipolar Transistor* (Tektronix, Beaverton, OR, 1979) or *The Design and Analysis of Analog Integrated Circuits* by P. Gray and R. Meyer (Wiley, New York, 1996).

11.3 SMALL-SIGNAL MODELS FOR BIPOLAR TRANSISTORS

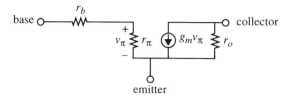

FIGURE 11.11. Low-frequency hybrid-π model for bipolar transistor

Typical Early voltages are of the order of 10–50 V for most microwave bipolar transistors.

To complete the DC portion of the model, recognize that a transistor is made of semiconducting materials whose resistance is nonzero. Thus, there is always some resistance in series with each terminal. For the most part, base resistance is most significant, so we generally ignore resistances in series with the other two terminals (however, the validity of this neglect should be checked in all cases where it might matter). The small-signal model that results from all of these considerations then appears as in Figure 11.11.

A little bit about the nomenclature: The "π" in "hybrid-π" comes from the shape of the model (yes, a little imagination is required), while *hybrid* refers to the fact that not all of the parameters have the same dimensions – here, we have resistances and a (trans)conductance.

As we'll see when we consider FETs, all transistor models have the same basic topology. The reason is simple: All transistors in use today are fundamentally voltage-controlled current sources. That is, they are transconductors.[16] The small-signal models naturally reflect this universal truth. Differences lie primarily in the nature of the impedances connected to the various terminals of the transconductor.

Now, what about the "small signal" part? That is, how small is small? The answer is readily obtained from recognizing that truncation of the Taylor series expansion is fundamentally at the heart of the linearization. We might therefore define the limits of "small" as corresponding to where the neglected terms become some significant percentage of the main linear term. As a specific and arbitrary choice, suppose we are satisfied if the second-order term remains no larger than 10% of the first-order one. In that case, we desire

$$\frac{1}{2!}\left(\frac{v_{be}}{V_T}\right)^2 < \frac{1}{10}\left(\frac{v_{be}}{V_T}\right) \implies \left(\frac{v_{be}}{V_T}\right) < \frac{1}{5}. \tag{11}$$

---

[16] Many textbooks describe bipolar transistors as current-controlled devices. That point of view, while potentially leading to mathematically correct descriptions, is one step removed from the actual physics. It is the base–emitter voltage that fundamentally lowers the barrier height at the base–emitter junction, injecting carriers into the base region, where they diffuse (and drift) toward the collector. Any base current that flows is a parasitic current whose ideal value is zero. That's hardly the description of a fundamental control variable!

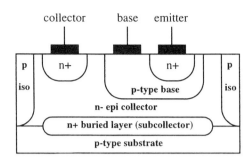

FIGURE 11.12. More typical IC bipolar transistor structure (layers not drawn to scale)

Thus, small-signal analysis becomes progressively more unreliable as the variation in base–emitter voltage exceeds about 5 mV at room temperature. Note that the boundary between small and large is somewhat fuzzy, with the allowable error very much a function of the problem being solved. Nevertheless, it remains true that *small* is never very many millivolts for a bipolar transistor.

11.3.2 A SIMPLE HIGH-FREQUENCY MODEL

To develop a simple model, it's helpful to consider a cross-section that's more representative of actual devices; see Figure 11.12. Even though this cross-section is itself a simplified representation, it is considerably more elaborate than the simple NPN sandwich presented earlier. Here, the current flow is principally vertical, although there is a component of lateral flow before carriers head upward to the collector contact. Often, to reduce the corresponding parasitic series collector resistance, an extra collector contact is provided to the right of the emitter in Figure 11.12. The function of the heavily doped buried layer is likewise to reduce this resistance – by placing a low-resistance path in parallel with the high-resistance epitaxial collector proper. A second base contact to the right of the emitter contact will reduce parasitic base resistance.

At high frequencies, we have to account for at least the most basic frequency-dependent device behavior. To do so, first recognize that a bipolar transistor consists of two junctions. The base–emitter junction is normally in forward bias, while the collector–base junction is normally reverse-biased. Hence there must be, at minimum, two junction capacitances (one each for the base–emitter and collector–base terminal pairs).

The small-signal capacitance is well approximated by

$$C_j = \frac{C_{j0}}{(1 - V_j/\psi_0)^m}, \tag{12}$$

where C_{j0} is the junction capacitance at zero bias, V_j is the junction voltage (positive in forward bias), ψ_0 is nominally the contact potential (dependent on the bandgap

11.3 SMALL-SIGNAL MODELS FOR BIPOLAR TRANSISTORS

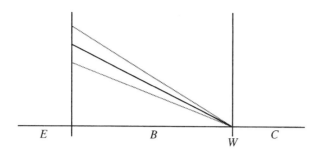

FIGURE 11.13. Excess carrier concentration in the base region at different base–emitter voltages

of the material, and typically of the order of 1 V for silicon), and m is a constant that depends on the doping profile. For linearly graded junctions, m is $1/3$, and for abrupt junctions, it is $1/2$. Values of m in excess of $1/2$ are possible through the use of *hyperabrupt* junctions (as discussed in greater detail in Chapter 9).

Junction capacitance behavior conforms surprisingly well to this equation for reverse bias, and even for weak forward bias. In deep forward bias, the capacitance typically limits at a value about 2–3 times C_{j0}, rather than growing without bound as the junction voltage approaches ψ_0. In amplifiers, the voltage dependency of this capacitance is usually considered a liability because it is a source of distortion. In special circumstances, nonlinearities can be exploited for a number of purposes, such as harmonic generation.

For the reverse-biased collector–base junction, the parallel-plate capacitance is all there is. For the forward-biased base–emitter junction, however, we have to do a little more work. To understand how there could be a capacitance besides a parallel-plate term (which is, essentially, what is being modeled by Eqn. 12), consider Figure 11.13: a simple (and highly approximate) plot of excess carrier concentration as a function of position in the base.

The concentration of injected carriers at the emitter edge is exponentially related to the total base–emitter voltage. Shown in the figure are hypothetical distributions for three different values of v_{BE}. Neglecting recombination leads to the perfect straight-line distributions shown. The concentrations all go to zero at the collector edge because that junction is in reverse bias. Carriers flow principally by diffusion from the region of high concentration (at the emitter edge) to the collector edge, where the concentration is zero.[17]

The total charge in the base region is proportional to the area of the triangle for any given v_{BE}. If v_{BE} increases then extra charge must be supplied by the base terminal;

---

[17] In virtually all modern transistors, the base doping varies with position, creating a built-in electric field that greatly enhances transport of carriers across the base. This field opposes drift in the opposite direction, which is one reason why a transistor has much poorer frequency response when operated with collector and emitter reversed.

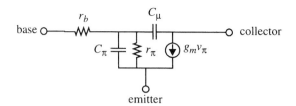

FIGURE 11.14. Higher-frequency hybrid-π model for bipolar transistor

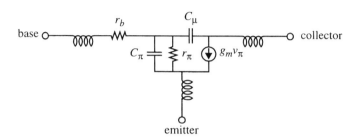

FIGURE 11.15. High-frequency hybrid-π model for bipolar transistor including some parasitics

if v_{BE} decreases, charge must be extracted. This behavior is modeled by a capacitance. Unlike an ordinary junction capacitor, however, this one has a value that is proportional to v_{BE} (the bigger the value of v_{BE}, the bigger the triangles swept out for a given change in v_{BE}). With our linearization, this proportionality implies that the capacitance, called the diffusion capacitance C_b, is proportional to the DC collector bias current. Thus, the base–emitter capacitance is the sum of the ordinary (voltage-dependent) junction term and this (current-dependent) diffusion term:

$$C_\pi = C_{je} + C_b. \tag{13}$$

Including all of these capacitances thus leads to the model shown in Figure 11.14.[18]

Packaged devices have additional parasitics. The most significant are inductances in series with each terminal; see Figure 11.15. A challenge is to minimize these unwanted inductances. Excessive parasitic emitter inductance is particularly unwelcome. For example, a 1-nH emitter inductance represents an impedance of more than 6 Ω at 1 GHz. This built-in impedance can cause a significant decrease in the gain of a stage and also threaten gain flatness by introducing the possibility of resonance. Many high-frequency packages offer two emitter leads, for example, precisely to lower the effective series inductance.

---

[18] Note that we are neglecting the output resistance, r_o. Although such an omission usually produces negligible errors, one should check the validity of that assumption in any case where it might matter. Examples include high-gain circuits as well as low-noise amplifiers, where the input impedance may be a sensitive function of such a resistor.

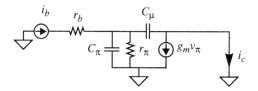

FIGURE 11.16. Circuit for f_T determination

A more elaborate model of package parasitics might include various resistances (perhaps modeling lead losses) and capacitances (e.g. between all terminal pairs) as well. However Figure 11.14 captures most phenomena of interest to the RF and microwave engineer. Again, keep in mind that we are looking for models that give us design insight, not necessarily the correct answers. Simulators, using much more elaborate models, take care of the analysis problem for us.

11.3.3 HIGH-FREQUENCY FIGURES OF MERIT

A common figure of merit is the frequency at which the short-circuit common–emitter current gain is *extrapolated* to drop to unity. For the circuit shown in Figure 11.16,[19]

$$\omega_T = \frac{g_m}{C_\pi + C_\mu}. \tag{14}$$

Notice that this figure of merit is completely insensitive to any impedance in series with the base terminal. Since the base is driven by a current source, any such impedance is totally irrelevant for ω_T.[20] However, actual circuit performance is certainly sensitive to these impedances, revealing one serious deficiency of this particular (and overused) measure. All too often, this parameter is maximized at the expense of more practically relevant ones. Another problem is that any shunt impedance across the collector–emitter terminals is similarly ignored, here because they are shorted out.

Because there is nothing fundamental about current gain (for example, in principle one could always use an ideal transformer to obtain any current gain without consuming any power) and because ω_T is otherwise an incomplete figure of merit, a different measure of device performance is often offered to supplement ω_T. The maximum frequency of oscillation, ω_{max}, tells us when the *power* gain of the transistor has dropped (or is extrapolated to drop) to unity. For our simple model,

$$\omega_{max} = \tfrac{1}{2}\sqrt{\omega_T/r_b C_\mu}. \tag{15}$$

This measure is more satisfactory because it explicitly shows that the series base resistance has a definite impact on high-frequency performance.

---

[19] In deriving this directly from the circuit, the contribution to the output current by C_μ is neglected but its contribution to the input current isn't.

[20] "A current source in series with a Buick is still a current source."

For measurement of both ω_T and ω_{max}, the various gains are plotted as a function of frequency until they drop with a clear single-pole trend.[21] Each trend is extrapolated to the unity gain frequency. This extrapolation is necessitated by the appearance of more complex behaviors as frequency increases. In the case of ω_T, it is possible with our model never to reach the unity gain frequency because of feedthrough via C_μ.

Although the preceding discussion is based on the silicon bipolar transistor, the basic model applies to HBTs as well – at least for the purpose they are meant to serve (acquisition of design insight). Model element parameters, of course, will differ.

11.3.4 EXTRACTION OF MODEL PARAMETERS FROM DATA SHEETS

Despite (or because of) the large amount of data contained in data sheets, it can be challenging to extract model parameters. Again keeping in mind that everything we say and do in modeling is wrong to some degree, it is nevertheless possible to outline a parameter extraction method that is sufficiently accurate for back-of-the-envelope calculations. The model elements of interest are the capacitances C_μ and C_π and the parasitic base resistance r_b. We may be interested in the inductances in series with the terminals as well. Occasionally, we may also care about mutual inductances between terminals (RF power transistors for base stations are an example, since they are typically made of multiple devices wired in parallel).

The value of C_μ is usually given directly on data sheets but often labeled C_{ob}, which derives its name from its measurement in a common-*b*ase test circuit with the emitter terminal *o*pen. A quick look at our transistor model reveals that the capacitance actually measured in such a test is what we have been calling C_μ.

The junction portions of C_π and C_{je} are typically given as well. However, they are almost always measured at reverse bias. Since the base–emitter junction is operated in forward bias, the C_{je} values that are given must be converted somehow into forward-bias values. The easiest way is simply to find the zero bias value and double it. (Eqn. 12 cannot be used as a basis for this conversion because it is invalid for ordinary forward bias.) Fortunately, C_{je} is rarely dominant, so a somewhat large uncertainty in its value is acceptable.

The diffusion capacitance term, C_b, is best inferred from a plot of ω_T as a function of collector current (see Eqn. 14). As a rough approximation, one may extract an estimate of the diffusion capacitance per current from data-sheet parameters as

$$\frac{C_b}{I_C} \approx \frac{1}{V_T \omega_{T,pk}}, \tag{16}$$

where $\omega_{T,pk}$ is the maximum value of ω_T, V_T is the thermal voltage (kT/q), and the collector current corresponds to the value that produces $\omega_{T,pk}$.

---

[21] The current gain thus falls off as $1/f$, whereas the power gain falls off as $1/f^2$.

At low currents, where the diffusion term is negligible, the capacitances are fixed and ω_T increases linearly with g_m and, hence, with collector current. Ultimately, the diffusion term dominates the denominator, leading to a constant ω_T. Then, to a reasonable approximation, C_b is simply g_m/ω_T in that regime of constant ω_T. In real devices, ω_T ultimately drops with increasing I_C because of high-level injection effects that cause the base to widen and thereby increase the charge stored at a superlinear rate, so one should make the determination in the flattest possible region. In true HBTs, high-level injection effects are suppressed to such an extent that it is easier to make this determination for those devices than for ordinary homojunction bipolar transistors.

Deducing r_b is the most difficult of all, and one must generally accept a fairly high uncertainty in its inferred value. As shown in Chapter 13, the input resistance will have a real part that is the sum of r_b and an induced resistance created by the interaction of emitter inductance and the transistor:

$$r_{in} \approx r_b + \omega_T L_e. \qquad (17)$$

Most data sheets for microwave transistors provide a plot of input impedance as a function of frequency. Find where the input resonance occurs, and note the resistance at resonance; this is r_{in}. Use the resonant frequency and knowledge of device capacitances to deduce the total package inductance. Since the total inductance is the sum of base and emitter inductance, you need to make some judgment or calculation about how the total is partitioned between the two leads. For many packages, an even split is a good choice. Finally, use knowledge of ω_T to complete the computation of r_b. Unfortunately, $\omega_T L_e$ is in many cases a significant fraction of the measured r_{in}, so that estimation of r_b is associated with a large uncertainty.

It's also helpful to note that r_b varies with bias point. At low collector current, the base current is correspondingly low. The voltage dropped along the series resistance in the base is therefore small, and the entire base–emitter interface is effective at injecting carriers into the base region. At high currents, the drop along the base resistance is large. Thanks to this drop, the voltage varies along the base–emitter interface, being larger near the surface than deeper in the structure because of the shorter path between base and emitter contacts. The base–emitter interface thus injects current nonuniformly, with greater emission near the surface. The mean distance between the base contact and the effective emitting regions thus diminishes, producing a corresponding reduction in r_b. Typically, the ratio of maximum to minimum r_b is a factor of about 2.

11.4 FET MODELS

Fortunately, MOSFETs, MESFETs, and HEMTs share a common model (to zeroth order). As asserted earlier, it is quite similar to that for a bipolar transistor, displayed in Figure 11.17. We consciously avoid presenting any detailed large-signal V–I equations because, for modern MOSFETs and MESFETs, the number of second-order

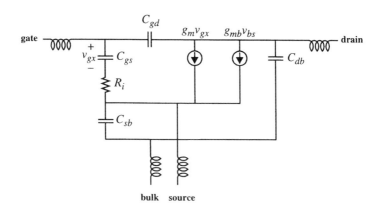

FIGURE 11.17. Simple incremental FET model

effects is so large as to make any simple set of equations quantitatively useless. It's not even qualitatively correct to make statements such as "the current varies quadratically with the gate–source voltage." Therefore, we'll offer a few qualitative and quasi-quantitative observations.

First, all FETs work by modulating the conductivity of a semiconductor. In the case of depletion-mode devices such as a JFET or a typical MESFET/HEMT, the default state is one of conduction and so we apply a control voltage to reduce the conductivity. In the case of an n-channel device, the gate–source voltage will be negative.[22] A parameter known as the *pinchoff* voltage corresponds to the gate–source voltage that produces an extrapolated drain current of zero. In enhancement-mode devices, the default state is one of nonconduction and so we must induce conductivity by the application of a sufficient gate–source voltage. Conduction is extrapolated to begin once the gate–source voltage exceeds a certain *threshold* voltage. From these descriptions, you see that the threshold voltage and the pinchoff voltage are actually describing the same point, just approached from opposite default conditions.

There are two distinct regimes of operation for a FET. In *triode,* by analogy with its vacuum tube ancestor, the FET acts like a voltage-controlled (albeit nonlinear) resistor. For small drain–source voltages, the resistance is inversely proportional to the gate *overdrive* (the gate–source voltage beyond threshold):

$$r_{ds} \propto 1/(v_{GS} - v_T). \tag{18}$$

The FET remains in this resistive region of operation as long as the drain–source voltage is smaller than the gate–source voltage. For higher drain–source voltages, the

---

[22] Nearly all microwave devices are designed to have electrons carry the current. Thus, microwave bipolar transistors are generally NPN devices, and microwave FETs are usually n-channel transistors. The reason is that electron mobility is always greater than hole mobility. In silicon, there is a 2:1 disparity, making PNP or p-channel FETs reasonably practical. For other materials, however, the disparity can exceed an order of magnitude. Thus commercial p-channel GaAs MESFETs, for example, do not exist.

gate voltage at the drain end of the device is insufficient to induce carriers. The conductive channel is then said to be pinched off. Nonetheless, current still flows. This statement should bother you, for it's based on a nonphysical sleight of hand. The density of charges goes to zero, but these charges flow infinitely fast (got that?). By the magical properties of "infinity times zero," the product can be finite (or anything that you want). In reality, true pinchoff does not occur; the charge density goes down near the drain end, but not to zero. At the same time, the carrier velocity does not go to infinity but rather to a maximum velocity determined by material properties: the saturation velocity. It's the ultimate speed limit for a given material.[23] Gallium arsenide's saturation velocity is about double that of silicon, explaining its popularity for RF and microwave applications.

Once pinchoff (sort of) occurs, the drain current saturates (because the velocity and charge density both do). This regime of operation is thus called, reasonably enough, *saturation*. In the ever-elusive classical, textbook FET, the drain current in saturation depends quadratically on gate overdrive:

$$i_D = K(v_{GS} - v_T)^2, \tag{19}$$

where K is whatever it needs to be to make the drain current correct. For JFETs, it's traditional to express K as

$$K = I_{DSS}/V_P^2, \tag{20}$$

where I_{DSS} is the drain current at zero gate–source voltage. In this case, the threshold voltage would be replaced by the pinchoff voltage in Eqn. 19.

As with bipolar transistors, the current doesn't really saturate perfectly; a dependence on drain–source voltage remains. This effect is modeled in the same way, by the addition of a drain–source resistance r_o.

From this point on, we focus instead on the small-signal picture, leaving the large-signal details to more elaborate simulation models. Perhaps it is sufficient to note that in MESFETs there are gate–source and gate–drain Schottky diodes, both operating in reverse bias during normal operation. Either or both diodes can be forward biased under unusual operating conditions, however. In contrast, MOSFETs have no junctions tied directly to the gate.

Note that there are two transconductances in our small-signal model. A typical MOSFET actually has two gates. One is the nominal gate, and the other is the substrate. Because variations in potential on both terminals can modulate the drain current, two separate transconductances are needed in the model. If the source and substrate terminals are tied together (as in all discrete devices), the model simplifies because the back-gate effect and its corresponding model elements disappear.

---

[23] This statement is true only as long as the charges flow along a path that is at least several mean-free paths (the average distance between collisions or other scattering events) in extent. If the path is short enough, then carriers will flow ballistically and the concept of a saturation velocity breaks down.

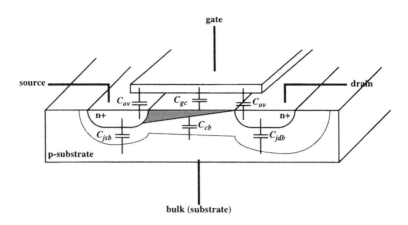

FIGURE 11.18. MOSFET capacitances

The transconductance may be expressed in a number of ways, but a compact one is

$$g_m = B \frac{I_D}{V_{OD}}, \qquad (21)$$

where I_D is the drain current bias value and V_{OD} is the gate overdrive voltage. The equation is valid only for V_{OD} greater than several kT/q, so the transconductance per unit current is guaranteed by nature to be worse for FETs (of any kind) than for bipolars. The parameter B is ideally 2 for square-law devices, and it approaches unity for real devices.

Not shown in the simple model of Figure 11.17 are resistances in series with the terminals. Because the gate electrode is usually made of thin polysilicon in MOSFETs, the series gate resistance may not be neglected in general. Device designers go to great pains to subdivide silicon MOSFETs into several paralleled pieces to keep the total gate resistance low. Because MESFET gates are made out of much lower-resistivity metal, the series gate resistance tends to be correspondingly lower than for MOSFETs.

In the case of a MESFET (and its cousins), the gate capacitance charging resistance, R_i, is sometimes important. The main transconductance depends on the voltage across the gate–source capacitance, so the presence of R_i produces a low-pass filter. A typical range of values for R_i is 5–10 Ω.

11.4.1 DYNAMIC ELEMENTS

Capacitances limit the high-frequency performance of FET circuits, just as they limit the performance of bipolar ones, so we need to understand where they come from and roughly how big they are.

First, since the source and drain regions form reverse-biased junctions with the substrate, one expects the standard junction capacitance from each of those regions to the substrate. These capacitances are denoted C_{jsb} and C_{jdb}, as shown in Figure 11.18, where the extent of the depletion region has been greatly exaggerated.

There are also various parallel-plate capacitance terms in addition to the junction capacitances. The capacitors shown as C_{ov} in Figure 11.18 represent gate–source and gate–drain *overlap* capacitances; these are highly undesirable, but unavoidable. During manufacture, the source and drain regions may diffuse laterally by an amount similar to the depth that they diffuse. Hence, they bloat out a bit during processing and extend underneath the gate electrode by some amount. As a crude approximation, one may take the amount of overlap, L_D, as 2/3 to 3/4 of the depth of the source–drain diffusions. Hence,

$$C_{ov} \approx \frac{\varepsilon_{ox}}{t_{ox}} W L_D = 0.7 C_{ox} W x_j, \tag{22}$$

where x_j is the depth of the source–drain diffusions, ε_{ox} is the oxide's dielectric constant (about $3.9\varepsilon_0$), and t_{ox} is the oxide thickness.

The parallel-plate overlap terms are augmented by fringing, and thus the "overlap" capacitance would be nonzero even in the absence of physical overlap. In this context, one should keep in mind that, in modern devices, the gate electrode is actually considerably thicker than the channel is long, so the relative dimensions of Figure 11.18 are misleading. Think of a practical gate electrode as a tall oak tree instead of a thin plate. In addition, the interconnecting wires to the source and drain are hardly of negligible dimensions. Because the thickness of the gate electrode now scales little (if at all), the "overlap" capacitance now changes somewhat slowly from generation to generation.

Another parallel-plate capacitance is the gate-to-channel capacitance, C_{gc}. Since both the source and drain regions extend into the region underneath the gate, the effective channel length decreases by twice the bloat, L_D. Hence, the total value of C_{gc} is

$$C_{gc} = C_{ox} W (L - 2L_D). \tag{23}$$

In strong inversion, the charge carriers at the surface and those in the bulk are of the opposite type. In between there is a depletion region. As a result, there is also a capacitance between the channel and the bulk, C_{cb}, that behaves as a junction capacitance. Its value is approximately

$$C_{cb} \approx \frac{\varepsilon_{Si}}{x_d} W (L - 2L_D); \tag{24}$$

here x_d is the depth of the depletion layer, whose value is given by

$$x_d = \sqrt{\frac{2\varepsilon_{Si}}{q N_{sub}} |\phi_s - \phi_F|}. \tag{25}$$

The quantity within the absolute-value bars is the difference between the surface potential and the Fermi level in the substrate. In strong inversion (for both triode and saturation regions), this quantity has a magnitude of twice the Fermi level.

Now, the channel is not an explicitly accessible terminal of the device, so finding how the various capacitive terms contribute to the terminal capacitances requires

Table 11.1. *Approximate MOSFET terminal capacitances*

	Off	Triode	Saturation
C_{gs}	C_{ov}	$C_{gc}/2 + C_{ov}$	$2C_{gc}/3 + C_{ov}$
C_{gd}	C_{ov}	$C_{gc}/2 + C_{ov}$	C_{ov}
C_{gb}	$C_{gc}C_{cb}/(C_{gc}+C_{cb})$ $<C_{gb}<C_{gc}$	0	0
C_{sb}	C_{jsb}	$C_{jsb}+C_{cb}/2$	$C_{jsb}+2C_{cb}/3$
C_{db}	C_{jdb}	$C_{jdb}+C_{cb}/2$	C_{jdb}

knowledge of how the channel charge divides between the source and drain. In general, the values of the terminal capacitances depend on the operating regime because bias conditions affect this partitioning of charge. For example, when there is *no* inversion charge (the device is "off"), the gate–source and gate–drain capacitances are just the overlap terms to a good approximation.

When the device is in the linear region there is an inversion layer, and one may assume that the source and drain share the channel charge equally. Hence, half of C_{gc} adds to the overlap terms. Similarly, the C_{jsb} and C_{jdb} junction terms are each augmented by half of C_{cb} in the linear region.

In the saturation regime, potential variations at the drain region don't influence the channel charge. Hence, there is no contribution to C_{gd} by C_{gc}; the overlap term is all there is. The gate–source capacitance *is* affected by C_{gc}, but "detailed considerations"[24] show that only about 2/3 of C_{gc} should be added to the overlap term. Similarly, C_{cb} contributes nothing to C_{db} in saturation but does contribute 2/3 of its value to C_{sb}.

The gate–bulk capacitance may be taken as zero in strong inversion (both in triode and in saturation, as the channel charge essentially shields the bulk from what's happening at the gate). When the device is off, however, there is a gate-voltage–dependent capacitance whose value varies in a roughly linear manner between C_{gc} and the series combination of C_{gc} and C_{cb}. Below (but near) threshold, the value is closer to the series combination and approaches a limiting value of C_{gc} in deep accumulation, where the surface majority carrier concentration increases above that of the bulk, owing to the positive charge induced by the strong negative gate bias. In deep accumulation, the surface is strongly conducting and may therefore be treated as essentially a metal, leading to a gate–bulk capacitance that is the full parallel-plate value.

The variation of this capacitance with bias presents one additional option for realizing varactors. To avoid the need for negative supply voltages, the capacitor may be built in an n-well using n+ source and drain regions to form an accumulation-mode MOSFET capacitor. The terminal capacitances are summarized in Table 11.1.

---

[24] The 2/3 factor arises from the calculation of channel charge, and inherently comes from integrating the triangular distribution assumed in Figure 11.18 in the square-law regime.

As a final comment, you may take as a *rough* rule of thumb (if no other information is given to you) that the gate–source capacitance of all FETs is of the order of 1 pF per millimeter of gate width. Scaling trends have curiously led to a rough constancy of this factor, despite orders-of-magnitude changes in device feature sizes. However, we cannot guarantee that this constancy will extend into the future. This number is offered simply as a crude guide to aid zeroth-order, back-of-the-cocktail napkin calculations with which to impress your party guests.

Transit Time Effects (Nonquasistatic Behavior)

The lumped models of this chapter clearly cannot apply over an arbitrarily large frequency range. As a rough rule of thumb, one may usually ignore with impunity the true distributed nature of transistors up to roughly a tenth or fifth of ω_T. As frequencies increase, however, crude lumped models become progressively inadequate. The most conspicuous shortcomings may be traced to a neglect of transit time ("nonquasistatic" or NQS) effects.

To understand qualitatively the most important implications of transit time effects, consider applying a step in gate-to-source voltage. Charge is induced in the channel and drifts toward the drain, arriving some time later owing to the finite carrier velocity. Hence, the transconductance has a phase delay associated with it.

A side effect of this delayed transconductance is a change in the input impedance: the delayed feedback from the channel back through the gate capacitance necessarily prevents a pure quadrature relationship between gate voltage and gate current. As a consequence, the applied gate voltage performs work on the channel charge. This dissipation must be accounted for in any correct circuit model. Van der Ziel[25] has shown that, at least for long-channel devices, the transit delay causes the gate admittance to have a real part that grows as the square of frequency:

$$g_g = \frac{\omega^2 C_{gs}^2}{5 g_{d0}}. \tag{26}$$

To get roughly calibrated on the magnitudes implied by Eqn. 26, assume that g_{d0} is approximately equal to g_m. Then, to a crude approximation,

$$g_g \approx \frac{g_m}{5} \left(\frac{\omega}{\omega_T} \right)^2. \tag{27}$$

Hence, this shunt conductance is negligible as long as operation well below ω_T is maintained. However, for accurate computations of power gain at high frequencies and *thermal noise,* this conductance must be included. We shall see later that a proper noise figure calculation must take this noise source into account, for example. Finally, because the derivation of the extrapolated unity–power gain frequency presented earlier neglects nonquasistatic effects, it overestimates the true value of ω_{max}.

---

[25] *Noise in Solid State Devices and Circuits,* Wiley, New York, 1986.

As a final note on NQS effects, we note that a series resistance could be used instead of a shunt conductance by transforming the parallel $g_g C_{gs}$ circuit into a series $R_i C_{gs}$. In that case, the model parameter R_i is given by

$$R_i \approx 1/5g_{d0}. \tag{28}$$

A common error is to define the gate-to-source voltage as across the capacitor (in the MESFET, the transconductor's current indeed depends on that capacitor voltage). After the transformation to a series representation, the control voltage for the transconductor is the voltage across the series combination for a MOSFET.

11.4.2 DIFFERENCES IN THE MESFET AND HEMT

The basic small-signal models for the MESFET and HEMT are essentially the same as that for the MOSFET, although the particular functional dependencies are different. In most of the literature on MESFETs, nonquasistatic effects are rarely mentioned, even though they must be present.

In the case of the HEMT, gate current leakage is often nonnegligible. In those cases, simply adding a shunt resistance across the gate capacitance suffices to model the effect.

11.5 SUMMARY

We've seen that it is possible to develop simple small-signal transistor models that are nonetheless useful enough both for first-pass designs and as a guide to subsequent iteration, even if analysis is performed with more sophisticated computer models. A method for extracting model parameters from data sheets allows the engineer to make progress even if suitable models are not offered by the manufacturer.

CHAPTER TWELVE

AMPLIFIERS

12.1 INTRODUCTION

The design of amplifiers for signal frequencies in the microwave bands involves more detailed considerations than at lower frequencies. One simply has to work harder to obtain the requisite performance when approaching the inherent limitations of the devices themselves. Additionally, the effect of ever-present parasitic capacitances and inductances can impose serious constraints on achievable performance. Indeed, parasitics are so prominent at RF that an important engineering philosophy is to treat parasitics as circuit elements to be exploited, rather than fought.

Having evolved during an era where modeling and simulation capabilities were primitive, traditional microwave amplifier design largely ignores the underlying details of device behavior. Instead, S-parameter sets describe the transistor's macroscopic behavior over frequency. In doing so, vast simplifications can result, but at a cost. By effectively insulating the engineer from the device physics, it is difficult to extrapolate beyond the given data set. Furthermore, real transistors are nonlinear, so the S-parameter characterizations are strictly relevant only for the bias conditions used in their generation.

Because simulation and modeling tools have advanced considerably since that time, we will consider the design of both broadband and narrowband amplifiers from a device-level point of view, rather than with the more traditional Smith-chart–based approach. Thus, we will not spend time examining stability and gain circles, for example.[1] Readers interested in the classical approach are directed to any of a number of representative texts that cover the topic in detail.[2]

We'll assume that the reader is already familiar with basic amplifier configurations (such as common-emitter, cascode, etc.), so we will instead focus on several techniques for extending bandwidth. We'll also study a collection of "strange impedance behaviors" that can afflict amplifiers, as well as their cures.

---

[1] This neglect should not be interpreted as a condemnation. The traditional approach allows one to proceed with design in the absence of models. That ability can be a powerful asset.
[2] See e.g. G. Gonzalez, *Microwave Transistor Amplifiers,* 2nd ed., Prentice-Hall, New York, 1996.

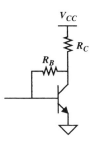

FIGURE 12.1. "Foolproof" bipolar bias circuit

12.2 MICROWAVE BIASING 101

We begin, though, with a consideration of practical ways to bias transistors. Compared with design practice at lower frequencies, appropriate biasing idioms are more limited at RF because it is difficult to keep the bias circuit's parasitics from interacting with the signal path at microwave frequencies. In some cases, you might be able to arrange for the bias circuit to serve double duty (e.g., act as part of the matching network). In all cases, you *must* keep in mind the physical details of implementing any proposed bias network. Developing this habit is a challenge for newcomers to RF and microwave circuits. Many designs that appear promising on paper are spoiled by inattention to bias circuit implementation details.

As it is impossible to examine an exhaustive collection of bias circuits, we will present just a few of the more commonly used types most appropriate for use in discrete circuits. The selected examples and their analyses should be sufficient, though, to give the reader a fairly comprehensive understanding of the subject.

12.2.1 BIPOLAR TRANSISTOR BIASING

In noncritical applications, it may be acceptable to use the bias circuit shown in Figure 12.1. We assume that the base is AC coupled to whatever drives this stage. Thus, the bias is established only by the elements explicitly shown in the figure. The circuit is "foolproof" in the sense that the transistor is *guaranteed* to be biased somewhere in the forward active region for any supply voltage high enough to forward bias a diode.

Perhaps the easiest proof is by *reductio ad absurdum*. First, recall that a bipolar transistor can find itself in only one of three possible regions of operation: cutoff, saturation, and active. In cutoff, there is essentially no collector current. In saturation, there is significant forward bias on both junctions. The collector–emitter voltage is then the difference between two junction voltages and is necessarily smaller than either. In the forward active region, only the base–emitter junction is in forward bias.

Now we have all we need. First assume that the transistor is in cutoff. Then there is no current through any branch and hence no voltage drop across the two resistors.

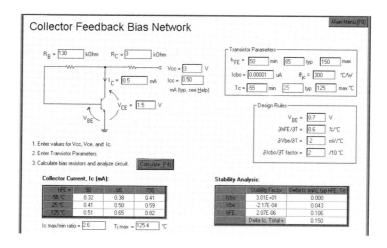

FIGURE 12.2. AppCAD simulation results for simple foolproof bias circuit

The base–emitter voltage is thus V_{CC}, which contradicts the assumption that the transistor is in cutoff.

As a second try, assume that the transistor is in saturation. Then, the collector–emitter voltage is less than the base–emitter voltage. If that's the case, the current through the feedback resistor R_B must flow out of the base. That condition is incompatible with the assumption of forward bias on the base–emitter junction, so the transistor can't be in saturation. All that's left is the active region. *QED.*

An additional virtue of this circuit is that the emitter is truly grounded (not just bypassed to ground). By not interposing any elements (like a resistor) between the emitter and ground, parasitic inductance is minimized. This consideration is important because series inductance can degrade RF gain, produce unwanted resonances, and otherwise alter the input impedance in ways that may be undesired.

Offsetting those attributes is relatively poor bias stability. The voltage on the collector equals the base–emitter voltage, augmented by the drop across R_B. In turn, the latter depends on the base current, and is thus sensitive to β:

$$V_C = V_{BE} + \frac{I_C}{\beta} R_B. \tag{1}$$

Regrettably, β increases with temperature at the rate of about 0.6%/K. Therefore, the collector voltage decreases – and the collector current increases – with increasing temperature.

In the particular bias design analyzed in Figure 12.2, we see that the collector current varies over a 2.6 : 1 ratio over the military temperature range. A smaller variation results if the supply voltage is increased, because a given absolute change in V_C represents a decreasing percentage of the voltage dropped across the collector load resistor as V_{CC} increases. Similarly, bias stability improves if the target collector voltage is decreased because, again, the total drop across the load increases.

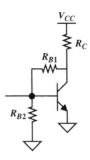

FIGURE 12.3. More stable "foolproof" bipolar bias circuit (V_{BE}-based feedback bias)

If neither of those options is acceptable, the addition of a single resistor can provide improvements by reducing the dependency on β. See Figure 12.3. Suppose that the current through the two bias resistors is large compared to the base current. Now, the voltage across R_{B2} is simply V_{BE}. If the base current can indeed be neglected, then the voltage across R_{B1} is $V_{BE}(R_{B1}/R_{B2})$ to a good approximation. Thus, the collector voltage is approximately

$$V_C \approx V_{BE}(1 + R_{B1}/R_{B2}). \tag{2}$$

The output voltage, and thus the current, is dependent primarily on V_{BE}. Because V_{BE} typically drops about 2 mV/K, the collector voltage will drop with increasing temperature as well. Over the military temperature range, the change in V_{BE} can be expected to be about 350 mV. With a fixed supply voltage, that behavior implies an increase in collector current with increasing temperature, just as in our original foolproof bias circuit. *AppCAD*'s analysis of this circuit reveals that the current variation is now 2.1:1 over temperature, for the same conditions as before. If the supply voltage is raised to 5 V, the variation drops to 1.5:1 over temperature.

These feedback bias methods allow the emitter to be tied to ground. Bias stability results from negative feedback's reduction of DC gain. Unfortunately, this negative feedback is implemented with broadband elements and is thus equally effective at reducing signal gain, too. One option for fixing this problem is to interpose a choke between the base and the resistive feedback network to decouple the action of the DC bias network from the signal path at the input; see Figure 12.4. If R_{B1} and R_{B2} are sufficiently large compared with R_C, then the loading of the output node by the bias network can be neglected.

It's important to resist the urge to make the choke infinitely large. Remember that every inductor has parasitic capacitance. Above the corresponding self-resonant frequency, the inductor actually looks capacitive. Thus, it's best to adopt a philosophy of moderation, choosing a choke that is no larger than absolutely necessary. A rough rule of thumb is to choose the reactance of the choke equal to 5–10 times the impedance seen at the base at the lowest signal frequency of interest.

12.2 MICROWAVE BIASING 101

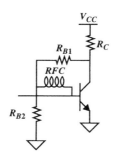

FIGURE 12.4. Use of choke to decouple
DC and AC signal paths

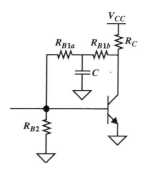

FIGURE 12.5. Use of a capacitor to decouple
DC and AC signal paths

An alternative that becomes increasingly attractive as frequency increases is to replace the choke with a high-impedance $\lambda/4$ line. If the junction of R_{B1} and R_{B2} is well bypassed to ground with a suitable capacitance, the other end of the line will present a very high impedance. Clearly, use of a $\lambda/4$ bias feed is limited to relatively narrowband systems. Within that constraint, however, the technique is extremely useful.

Yet another alternative is to break the feedback resistor R_{B1} into two pieces and capacitively bypass the midpoint to ground; this is shown in Figure 12.5. When using this method, it's important to bear in mind that the resistor R_{B1b} needs to be large enough compared with RC not to load the output too much. At the same time, the parallel combination of R_{B1a} and R_{B2} needs to be large enough not to load the input excessively. Satisfying these requirements is somewhat at odds with achieving good bias stability, however.

As with the choke, one must resist the temptation to specify an excessively large value for C. A good rule of thumb is to select it large enough so that its reactance is 5–10 times smaller than the effective resistance it faces at the lowest frequency at which the amplifier is to operate.

374   CHAPTER 12 AMPLIFIERS

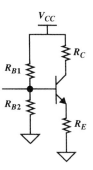

FIGURE 12.6. Bias for β- and V_{BE}-independence

Just to round out this collection of common bipolar bias methods, in Figure 12.6 we present a classic topology that is popular at lower frequencies. Because of the emitter degeneration, it isn't quite as popular at microwave frequencies – except in applications that are tolerant of the probable gain reduction that accompanies the presence of irreducible parasitics.

There are many ways to analyze this bias configuration, but the rigorous ones tend to obscure the principles of operation. Instead, we'll provide an approximate analysis that is more useful for design. As always, one should then verify that design objectives have in fact been met, using more rigorous analyses. That said, the method that we'll present will yield conservative results in the vast majority of practical cases.

Bias current variation arises from changes in β and V_{BE} over temperature, and from device to device. The circuit of Figure 12.6 suppresses both of these sources. To understand most directly how, let's consider the individual contributions of each to the overall variation in bias point. In effect, we'll invoke a small-signal assumption and simply add the two variations together to estimate the total.

To estimate the change in bias caused by V_{BE}-variation over temperature, first recall that, at constant collector current, V_{BE} decreases by about 2 mV/K, so that V_{BE} might change by a total of about 350 mV over the full military temperature range. If we neglect β-variation for this calculation, then we may assume that the base voltage is fixed. Then, the variation in collector current is due entirely to changes in the emitter voltage caused by the change in V_{BE}:

$$\Delta I_{C,VBE} = \frac{\Delta V_{BE}}{R_E}. \tag{3}$$

Expressing this as a fraction of the nominal collector current, we can write

$$\frac{\Delta I_{C,VBE}}{I_{C,NOM}} = \frac{\Delta V_{BE}}{I_{C,NOM} R_E}. \tag{4}$$

Thus, we can suppress changes in collector current to any desired degree by choosing a sufficiently large nominal voltage drop across R_E. For example, by choosing

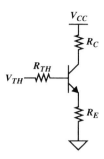

FIGURE 12.7. Equivalent circuit for β- and V_{BE}-independent bias

the nominal emitter voltage equal to 3.6 V, we can keep the variation in collector current (due to V_{BE} variation) down to about 10% over the entire temperature range. This result shouldn't be surprising, because the mechanism that reduces the sensitivity to this variation is precisely the same as that which reduces the gain to desired signals. That is, the circuit can't really distinguish between desired inputs and undesired voltage variations, so whatever is effective at suppressing one will be equally effective at reducing the other.

We now consider how to reduce sensitivity to variations in β. It's important to understand that β varies considerably from device to device (even of the same nominal type) and also over temperature. Generally speaking, the variation in β is about 0.6%/K. Over the military temperature range, it's not unusual to see a 2:1 or 3:1 change in β. Compounding that variation is the normal manufacturing spread.[3] As a result, a conservative bias design needs to accommodate perhaps a 5:1 (or greater) variation in β. To see how best to do so, let's first replace the input circuit with its Thévenin equivalent; see Figure 12.7. Here,

$$V_{TH} = V_{CC} \frac{R_{B2}}{R_{B1} + R_{B2}} \quad (5)$$

and

$$R_{TH} = R_{B1} \parallel R_{B2}. \quad (6)$$

In this part of the analysis, we assume that V_{BE} is fixed at its nominal value. Thus, the variation in collector current results from the change in base voltage that accompanies changes in base current arising from β-variation:

$$V_B = V_{TH} - \frac{I_C}{\beta} R_{TH}. \quad (7)$$

---

[3] Few manufacturers' data sheets specify the variation, often opting for a minimum β without a maximum or simply stating a "typical" value (whatever that means). If you take their data too literally, you may overdesign the bias network.

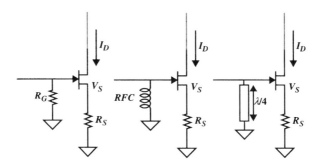

FIGURE 12.8. *N*-channel depletion-mode FET self-bias
(bypass across degeneration not shown)

It's clear that choosing a sufficiently small R_{TH} will make the base voltage insensitive to β to any desired degree. That, in turn, will make variations in collector current as insensitive to β as needed. As a crude rule of thumb, selecting the current through the base bias resistors equal to 5–10 times the maximum base current will yield satisfactory stability. For a more quantitative guide, first note that the maximum collector current coincides with maximum β. Then we may write

$$\Delta V_B = R_{TH}\left(\frac{I_{C,max}}{\beta_{max}} - \frac{I_{C,min}}{\beta_{min}}\right). \tag{8}$$

With the aid of Eqn. 8, you can select a specific R_{TH} to produce as small a ΔV_B as your budget allows.

In summary: First select the voltage drop across the emitter degeneration sufficiently large to suppress base–emitter voltage variations, and then select the base bias resistors sufficiently small to suppress β-variations. Finally, bypass the degeneration resistor to restore the stage's AC gain to the value desired. Owing to parasitics, this last step becomes increasingly difficult as frequency increases. Again, a numerically small parasitic inductance becomes a large impedance at high frequencies. Just a single nanohenry is 63 Ω at 10 GHz.

12.2.2 DEPLETION-MODE FET BIASING

All junction FETs are depletion-mode devices, as are most MESFETs and HEMTs. Thus, in normal operation, the gate–source voltage will be negative for n-channel devices. In configurations that do not demand maximum efficiency, it may be acceptable to use a simple source-degeneration resistor for biasing. With that choice, it is possible to return the gate terminal to a DC ground. The resulting configuration is often known as *self-bias* (see Figure 12.8), and it dates back to the earliest days of the vacuum tube era. In particular, n-channel depletion-mode FETs are close (in fact, the closest) semiconductor analogues to vacuum tubes. One may usually translate classical vacuum tube circuits into JFET versions with ease, for example, with only minor changes necessitated by supply voltage differences.

Even though the gate current is typically very small, the gate terminal must not be allowed to float to a random potential. An explicit DC path must always be provided to set the gate voltage, or else many pathologies will result. Here, we assume that the gate is AC-coupled to the previous stage, so that R_G sets the DC gate potential to zero volts. Fortunately, the value of R_G may be chosen within wide limits. Its minimum acceptable value is set by a desire to avoid attenuation, and so it only has to be 5–10 times the impedance driving the gate, for example. The upper value is set by the worst-case gate leakage current (generally occurring at the upper temperature limit) and the maximum tolerable shift in gate voltage. As seen in Figure 12.8, the gate resistor can be replaced by a choke or, in narrowband applications, a $\lambda/4$ piece of high-impedance line.

The value of R_S is readily computed from knowing the desired drain current and the V_{GS} that corresponds to that current. Thus,

$$R_S = \frac{|V_{GS}|@I_D}{I_D}. \tag{9}$$

Analytical equations that allow determination of V_{GS} for the target drain current are convenient, but not required; one may use experimental V–I data if device equations are unavailable.

Because the resulting drain current is controlled by the device's particular V–I characteristics, this bias circuit is device-dependent. If the transistor is replaced, a different value of R_S may be required if the drain current is to remain within tolerance. Clearly, this device-specific dependency must be accommodated if mass reproducibility is a requirement.

It's also important to note that the temperature variation of the resulting drain current is not directly controlled in this simple circuit, so one must accept whatever temperature drift results. Two competing effects produce temperature variation. One is a decrease in carrier mobility with increasing temperature. This mechanism by itself causes drain current to decrease. The other is the -2-mV/K typical change in the threshold voltage (traditionally called the pinchoff voltage, V_P, in JFET terminology) with temperature.[4] By itself, that mechanism tends to increase the drain current. For each JFET, there is a particular drain current for which these two competing effects cancel to first order, leading to a substantially constant drain current.

To facilitate the design and characterization of self-bias networks, *AppCAD* offers a handy tool with which one can determine the value of the bias resistor and also assess the variation in drain current over temperature. See Figure 12.9. Of course, a

---

[4] The difference in nomenclature reflects a difference in viewpoint arising from how enhancement and depletion devices differ. A depletion-mode device's default state is one of conduction. Thus, the gate voltage is regarded as the means whereby current is reduced. Eventually, you can shut off the current by pinching off the channel. An enhancement-mode device's default state, however, is one of nonconduction. A sufficient gate voltage – the threshold voltage – must be applied before it turns on.

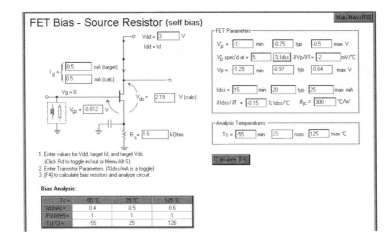

FIGURE 12.9. Sample JFET self-bias design spreadsheet from *AppCAD*

full-featured circuit simulator will also provide equivalent information, but the overhead in running *AppCAD*'s bias design tool is much smaller. Certainly, its tutorial value is large, for one can rapidly develop design insights that one can't always get quickly from staring at equations.

That said, here's an equation anyway:

$$I_D = K(V_{GS} - V_P)^n, \tag{10}$$

where $n = 2$ for classic JFETs. The constant K is sometimes called the FET *perveance,* by analogy with a similar constant in equations that describe *V–I* behavior of vacuum tubes. For classic square-law FETs, the constant K has the dimensions of current per squared voltage.

When deciphering data sheets, it may be helpful to know that the parameter I_{DSS} is the drain current that flows with zero V_{GS}. Also, the pinchoff voltage is usually given by the manufacturer as the value of V_{GS} corresponding to a drain current that is some small percentage of I_{DSS}, rather than the theoretical (but experimentally problematic) value of zero. The bias design tool allows the user to specify that percentage or, alternatively, an absolute drain current. In the example shown, the value is 5% of I_{DSS}. To distinguish such an arbitrary pinchoff definition from the rigorous one, the data-sheet pinchoff voltage is identified as a primed variable. The tool then extracts the "true" V_P by extrapolating the $V_{GS}-I_D$ behavior to zero current. The reader will note that this latter pinchoff voltage is indeed more negative than the data-sheet pinchoff voltage, as expected.

From the numbers used in the example (which are representative of practical discrete devices), you can see that the drain current happens to increase with temperature in this case. The current at the maximum temperature is 50% larger than that at the minimum temperature for this particular device with this particular set of numbers.

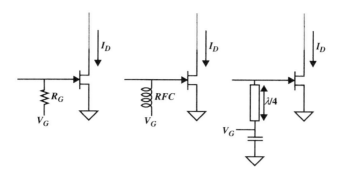

FIGURE 12.10. Some MESFET biasing options if extra supply is available ($V_G < 0$)

Finally, in applications that are intolerant of the efficiency or headroom reductions caused by the voltage dropped across a degeneration resistor, one must provide a negative voltage to bias the gate. The bias voltage can be applied through an inductor or resistor, as shown in Figure 12.10. In narrowband applications, another option is to use a λ/4 line to feed the gate, just as in the degenerated case. The necessity for an extra supply is an inconvenience one tolerates only reluctantly. Gallium arsenide MESFETs once dominated handset power amplifiers, yet their displacement in increasing numbers by HBTs is due, in no small measure, to the latter's ability to operate with a single supply voltage.

Finally, we should mention that many of the bias topologies used for bipolars may be adapted to work with FETs as well. As just one example, the circuit of Figure 12.6 functions fine with a FET, with due allowance for the negative V_{GS}. The reader is encouraged to consider replacements of this type if one of the topologies we've explicitly presented proves unsuitable for some reason.

12.2.3 ACTIVE BIAS

All of the foregoing examples use passive elements to establish bias. As we've seen, it is not trivial to realize a stable bias using only passive elements, especially if we want to avoid interfering with the signal path. To round out our survey of biasing ideas, we now consider the addition of active elements that can provide an extra degree of freedom, which may be exploited to control bias to a much finer degree than is otherwise possible.

To validate what is to come, we assume that the spectrum of signals we wish to amplify does not extend all the way down to DC. That way, the bias circuitry may again be treated separately from the signal path. As a philosophical approach, then, we wish to compare our bias current or voltage with some sort of reference, and then close a bias feedback loop around the amplifier to enforce a negligible difference between the bias point and that reference. As might be expected, there are a great many

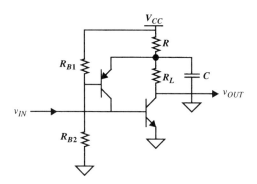

FIGURE 12.11. Example of active bias

ways to implement this general idea. We will consider just one to illustrate the basic principle.

A brute-force method would be to use an op-amp as the active element. In some cases, that choice might even be a good one. For most commonly encountered instances, however, use of an op-amp for this purpose is vast overkill. Often, a single additional transistor suffices. As a specific example, consider the simple circuit displayed in Figure 12.11.

As with all of the bias circuits here, there are many variations on this basic theme. The PNP transistor supplies just enough base current to the NPN device to keep the emitter potential of the PNP transistor a diode drop above its own base potential. The latter is established by the bias divider formed by R_{B1} and R_{B2}. By thus setting the voltage dropped across R, the current through that resistor is similarly set. Since the current through the PNP device is only the base current of the NPN, we may neglect it in concluding that the current through R is essentially the collector current through the NPN transistor:

$$I_C \approx \frac{V_{CC} - \left[V_{CC}\left(\frac{R_{B2}}{R_{B1} + R_{B2}}\right) + V_{EB,PNP}\right]}{R}. \tag{11}$$

From inspection of Eqn. 11, it's clear that this bias method forces a collector current that does not depend on the undependable. However, it's not entirely foolproof. The capacitor C must be chosen large enough (and of a sufficient quality) to act as a good RF short at all frequencies of interest. Also, we have created a feedback loop with two active devices, so the potential for unstable behavior has increased. In cases where that potential is realized, it may be beneficial to kill the gain of the feedback loop by adding some resistance in series with the emitter, or by inserting resistance between the collector of the PNP and the base of the NPN. When doing the latter, it may be helpful to bypass the collector to ground with a capacitor (large enough to form the dominant pole) or with a resistor–capacitor series combination (to create a lag compensator to improve stability). Not directly connecting the PNP's collector

12.3 BANDWIDTH EXTENSION TECHNIQUES

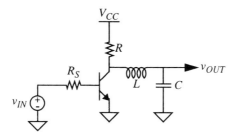

FIGURE 12.12. Amplifier with series peaking
(bias details omitted)

to the NPN's base is desirable for another reason, too, since it is best not to add the output capacitance of the PNP to the total capacitance on the base of the NPN.

The bias methods described here are by no means exhaustive, but they are sufficiently representative that the reader can probably generate many others based on the underlying ideas.

12.3 BANDWIDTH EXTENSION TECHNIQUES

12.3.1 SERIES AND SHUNT PEAKING

Back in the 1930s, in the early days of television development, one problem of critical importance was that of designing amplifiers with a reasonably flat response over the astounding video bandwidth of ∼4 MHz. Though obtaining this bandwidth is trivial today, it was challenging with the vacuum tubes available at the time. In an era where the number of vacuum tubes primarily determined the cost of a circuit, engineers adopted the philosophy of allowing any number of passive components while restricting to an absolute minimum the number of vacuum tubes.

One of the earliest examples of following that prescription is the series-peaked amplifier, shown in Figure 12.12. Here, we assume that the inductor value is the only degree of freedom. The load resistance is set by the gain sought, and the capacitance represents the irreducible parasitic load that would otherwise limit bandwidth to $1/RC$.

If we assume that the transistor itself has negligible parasitics, then we can treat it as a pure transconductor. Since all we're after is insight about bandwidth, we may neglect the transistor (within the assumptions stated) and focus solely on the transresistance of the output network,

$$\frac{v_{out}}{i_c}\left[R \parallel \left(sL + \frac{1}{sC}\right)\right]\left[\frac{1/sC}{sL + 1/sC}\right] = \frac{R}{s^2LC + sRC + 1}. \tag{12}$$

To facilitate derivations, express the inductance as

$$L = R^2C/m \tag{13}$$

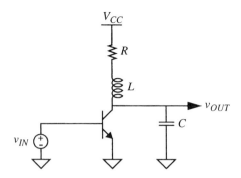

FIGURE 12.13. Simplified shunt-peaked amplifier (biasing not shown)

and let $RC = \tau$, the reciprocal of the bandwidth without the peaking network. Then, we will explore what happens to the transresistance as we vary m:

$$\frac{v_{out}}{i_c} = R_L \left[\frac{s^2 \tau^2}{m} + s\tau + 1 \right]^{-1}. \tag{14}$$

With this equation it is straightforward to show that (a) the maximum bandwidth is $\sqrt{2}$ times that of the uncompensated case ($L = 0$) and (b) it occurs for $m = 2$. A choice of $m = 3$ leads to maximally flat delay, with a corresponding bandwidth boost factor of about 1.36. For both cases, the bandwidth boost comes entirely from the resonant peaking provided by complex poles.

Series peaking requires only the addition of a single inductor yet increases bandwidth by ~40%. Although this improvement is impressive considering the trivial modification that produces it, it is possible to do better still simply by moving the inductor to a different position. Such *shunt peaking* was used in countless television sets at least up to the 1970s. It continues to make intermittent appearances, having been used in the video display circuits of the original compact Macintosh computers as well as in a host of broadband amplifiers for optical communications.

Figure 12.13 depicts a shunt-peaked amplifier in its simplest form. This amplifier is a standard common-emitter configuration, with the addition of the inductance in series with the collector load resistor. The term *shunt peaking* comes from the observation that the added inductance appears in a branch that is in parallel with (in shunt with) the load capacitance. Note that, unlike the series-peaked case, any device output capacitance may be absorbed into the load, allowing the theoretical benefits to be enjoyed more fully in practice.

Given our assumptions, we may model the amplifier for small signals as shown in Figure 12.14. But before launching into a detailed derivation, it's helpful to think about why repositioning the inductor this way should provide a greater bandwidth extension than for the series case.

First, we know that the gain of a purely resistively loaded common-emitter amplifier is proportional to $g_m R_L$. We also know that, when a capacitive load is added, the

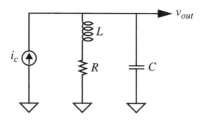

FIGURE 12.14. Model of shunt-peaked amplifier

gain eventually falls off as frequency increases because the capacitor's impedance diminishes. The addition of an inductance in series with the load resistor provides an impedance component that increases with frequency (i.e., it introduces a zero), which helps offset the decreasing impedance of the capacitance – leaving a net impedance that remains roughly constant over a broader frequency range than that of the original RC network.

An equivalent time-domain interpretation may be provided by considering the step response. The inductor delays current flow through the branch containing the resistor, making more current available for charging the capacitor, reducing the risetime. To the extent that a faster risetime implies a greater bandwidth, an appropriate choice of inductor therefore increases the bandwidth.

Formally, the impedance of the RLC network may be written as

$$Z(s) = (sL + R) \parallel \frac{1}{sC} = \frac{R[s(L/R) + 1]}{s^2LC + sRC + 1}. \tag{15}$$

We recognize that there are two poles (possibly complex), just as in the series peaking case. In addition, there is indeed a zero, as argued earlier.

Since the gain of the amplifier is the product of g_m and the magnitude of $Z(s)$, let's now compute the latter as a function of frequency:

$$|Z(j\omega)| = R\sqrt{\frac{[\omega(L/R)]^2 + 1}{(1 - \omega^2 LC)^2 + (\omega RC)^2}}. \tag{16}$$

As with the series-peaked amplifier, we now introduce a factor m, defined as the ratio of the RC and L/R time constants, to facilitate subsequent derivations:

$$m = \frac{RC}{L/R}. \tag{17}$$

Then, our transfer function becomes

$$Z(s) = (sL + R) \parallel \frac{1}{sC} = R\frac{\tau s + 1}{s^2 \tau^2 m + s\tau m + 1}, \tag{18}$$

where $\tau = L/R$.

We did not show the details of how to derive the optimum conditions for the series-peaked case because shunt peaking is superior (as we will see momentarily). It is

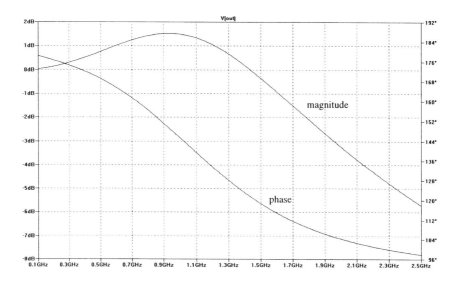

FIGURE 12.15. Frequency response of shunt-peaked amplifier (maximum bandwidth m)

instructive to study the procedures involved, so let's now examine the shunt-peaked amplifier more closely.

The magnitude of the impedance, normalized to the DC value ($= R$) as a function of frequency, is

$$\frac{|Z(j\omega)|}{R} = \sqrt{\frac{(\omega\tau)^2 + 1}{(1 - \omega^2\tau^2 m)^2 + (\omega\tau m)^2}}, \quad (19)$$

so that

$$\frac{\omega}{\omega_1} = \sqrt{(-m^2/2 + m + 1) + \sqrt{(-m^2/2 + m + 1)^2 + m^2}}, \quad (20)$$

where ω_1 is the uncompensated -3-dB frequency ($= 1/RC$).

The goal is to determine values of m that lead to desired behaviors. As with the series case, there is more than one choice. Maximizing the bandwidth is one obvious possibility. Taking the derivative of Eqn. 20, setting it to zero, and enduring some pain, one finds that this maximum occurs at a value of

$$m = \sqrt{2} \approx 1.41, \quad (21)$$

which extends the bandwidth to a value about 1.85 times as large as the uncompensated bandwidth. Anyone who has labored to meet a tough bandwidth specification can well appreciate the value of nearly doubling bandwidth through the addition of a single inductance at no increase in power.

Unfortunately, however, this choice of m leads to nearly a 20% peak in the frequency response, a value often considered undesirably high. See Figure 12.15. To moderate the peaking, one might seek a bandwidth other than the absolute maximum by increasing m. One common choice is to set the magnitude of the impedance equal to R at a frequency equal to the uncompensated bandwidth. Solving for this condition yields a value of 2 for m, with a corresponding bandwidth

$$\omega = \omega_1 \sqrt{1 + \sqrt{5}} \approx 1.8\omega_1. \tag{22}$$

Hence, the bandwidth in this case is still quite close to the maximum. Further calculation shows that the peaking is substantially reduced, to about 3%.

The arbitrary choice that leads to this result is often used because it yields such a significant bandwidth enhancement without excessive frequency response peaking. However, there are many cases where one desires the frequency response to be completely free of peaking. Thus, perhaps one might seek the value of m that maximizes the bandwidth without producing any peaking.

The conditions for such maximal flatness may be found through the following general technique: Form an expression for the frequency response magnitude (or, as is often more convenient, the square of the magnitude); then maximize the number of derivatives whose value is zero at DC.

Carrying out this method manually is frequently labor-intensive, but in this particular example, a straightforward calculation reveals that the magic value of m is:

$$m = 1 + \sqrt{2} \approx 2.41, \tag{23}$$

which leads to a bandwidth that is about 1.72 times as large as the unpeaked case. Hence, at least for the shunt-peaked amplifier, both a maximally flat response and a substantial bandwidth extension can be obtained simultaneously.

In other situations, there may be a specification on the time response of the amplifier rather than on its frequency response. Examples of practical interest are ultrawideband (UWB) systems, many of which require good pulse fidelity. Another is an oscilloscope vertical deflection amplifier, whose time response (characterized, say, by the step or pulse response) must be similarly "well behaved." That is, not only must we amplify uniformly the various spectral components of the signal over as large a bandwidth as practical, but the *phase relationships* among its Fourier components must be preserved as well. If the spectral components do not experience equal delay (measured in *absolute time*, not degrees), potentially severe distortion of the waveshape can occur. Such "phase distortion" is objectionable for the bit errors it can cause in digital systems and for its obvious negative implications for the fidelity of such analog instrumentation as oscilloscopes.

To quantify this type of distortion, first consider the phase behavior of a pure time delay. If all frequencies are delayed by an equal amount of *time,* then this fixed amount of time delay must represent a linearly increasing amount of *phase shift* as frequency increases. Phase distortion will be minimized if the deviation from this ideal linear phase shift is minimized.

Evidently, then, we wish to examine the *delay* as a function of frequency. If this delay is the same for all frequencies, we will have no phase distortion (other than the change in shape that results from the ordinary filtering any bandlimited amplifier provides). Formally, the delay is defined as follows:

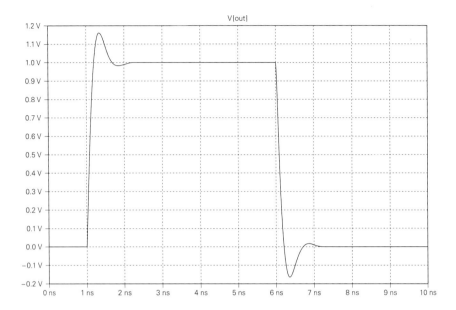

FIGURE 12.16. Pulse response for shunt-peaked amplifier (maximum bandwidth m)

$$T_D(\omega) \equiv -\frac{d\phi}{d\omega}, \qquad (24)$$

where ϕ is the phase shift of the amplifier at frequency ω.

Unfortunately, it is impossible for a network of finite order to provide a constant time delay over an infinite bandwidth: infinite phase shift would ultimately be required, but poles and zeros contribute only bounded amounts of phase shift. All we can do in practice, then, is to provide an approximation to a constant delay over some finite bandwidth.

By analogy with the frequency response case, we see that a maximally flat time delay will result if we maximize the number of derivatives of $T_D(\omega)$ whose value is zero at DC. Again, this method is general.

Because arctangents arise in expressing the phase shift due to poles and zeros, computing the relevant derivatives is generally quite a bit more unpleasant than in the magnitude case. Even for our shunt-peaked amplifier, which is only second order, the amount of labor is substantial. Ultimately, however, one may derive the following cubic equation for m (computational aids are of tremendous benefit here):

$$m^3 - 3m^2 - 1 = 0, \qquad (25)$$

whose relevant root is:

$$m = 1 + \left[(3 + \sqrt{5})/2\right]^{1/3} + \left[(3 - \sqrt{5})/2\right]^{1/3} \approx 3.104, \qquad (26)$$

corresponding to a bandwidth improvement factor of a bit under 1.6.

The two plots that appear as Figures 12.16 and 12.17 allow us to assess the value of maximally flat delay by comparing the pulse response of a shunt-peaked amplifier designed for absolute maximum bandwidth with one designed for maximally linear

12.3 BANDWIDTH EXTENSION TECHNIQUES

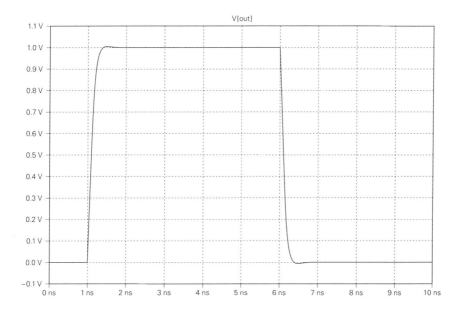

FIGURE 12.17. Pulse response for shunt-peaked amplifier (maximally flat delay)

Table 12.1. *Shunt peaking summary*

Condition	$m = R^2C/L$ (approx.)	Bandwidth boost factor	Normalized peak freq. response
Maximum bandwidth	1.41	1.85	1.19
$|Z| = R$ @ $\omega = 1/RC$	2	1.8	1.03
Best magnitude flatness	2.41	1.72	1
Best delay flatness	3.1	1.6	1
No shunt peaking	infinite	1	1

phase. As is apparent, the two pulse responses have roughly similar rise and fall times, indicating roughly similar bandwidths. However, the pulse response is much better behaved for the linear phase case.

Since the conditions for maximally flat frequency response and maximally flat time delay do not coincide, one must compromise, as is the case for the series-peaked amplifier. We therefore see that, depending on requirements, there is a range of useful inductance values; see Table 12.1. A larger L (smaller m) gives a larger bandwidth extension but poorer pulse fidelity, while a smaller L yields less bandwidth improvement but better pulse response.

Shunt Peaking Example

Even though shunt peaking traces its origins to video amplifiers from the 1930s, it is relevant and useful even in the modern era for the same reasons it was originally valued: it allows one to squeeze extra performance from a given technology. It is useful

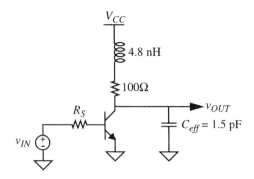

FIGURE 12.18. Shunt-peaked amplifier example

to note that the technique does not require a high-Q peaking inductor and is therefore easily implemented. To underscore this point, consider the problem of designing a 1.5-GHz common-emitter broadband amplifier. Let us assume that phase linearity is important for this application, so let's select $m = 3.1$ for best group delay uniformity.

Let the total capacitive loading on the collector be 1.5 pF (from both the transistor and loading by interconnect and subsequent stages), and assume that the load resistance cannot be made smaller than 100 Ω without increasing by an unacceptable amount the power consumed in keeping the gain constant. If the bandwidth is entirely controlled by the output node, then the bandwidth of the amplifier is just a bit over 1 GHz – somewhat shy of the 1.5-GHz goal.

If we assume that the minimum acceptable collector load resistance is used, then the required shunt peaking inductor is readily calculated as

$$L \approx R^2 C/3.1 = 4.8 \text{ nH}. \qquad (27)$$

This value is readily implemented in many technologies, both discrete and integrated; see Figure 12.18. The estimated bandwidth increases to approximately 1.7 GHz, comfortably in excess of the requirement. Again, shunt peaking provides this improvement without increasing the power consumed by the stage. Finally, note that the Q of the collector network is of the order of unity at the band edge, so inductors with truly modest Q (such as IC spiral inductors) suffice. Any resistance in series with the inductor can form part of the resistive load, making inductor loss even less important.

From the simulation results presented in Figure 12.19, it's easy to see that the bandwidth has been boosted to just below 1.7 GHz. The linear frequency axis makes it easier to see that the phase is reasonably linear until frequencies approach the bandwidth of the amplifier, implying a relatively constant delay over the amplifier passband, as intended.

12.3.2 MORE ON ZEROS AS BANDWIDTH ENHANCERS

The shunt-peaked amplifier highlights the utility of zeros. Thinking about a zero as an "antipole" helps to understand why it can extend bandwidth. One problem, though,

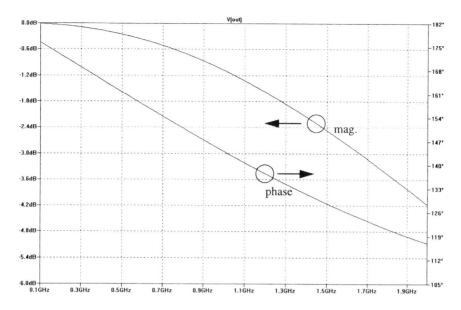

FIGURE 12.19. *LTSpice* simulation of shunt-peaked amplifier ($m = 3.1$ for best delay)

is that it is impossible to assure that the zero is at precisely the same frequency as the pole it is to cancel, so we ought to evaluate the effect of this inexactness with the aid of the following transfer function:

$$H(s) = \frac{\alpha \tau s + 1}{\tau s + 1}. \tag{28}$$

For simplicity's sake, note that no gain or attenuation factor is included in this expression. Thus, the ideal value of the constant α is unity, so that $H(s)$ is ideally unity at all frequencies.

Let's now consider the step response of this combination (often known as a *pole–zero doublet*). The initial- and final-value theorems tell us that the initial value is α and the final value is unity. Because the system state evolves exponentially with the time constant of the pole,[5] we can rapidly sketch a couple of possible step responses: one with $\alpha_1 < 1$, and one with $\alpha_2 > 1$. See Figure 12.20.

We see that the response jumps immediately to α but then settles down (or up, as the case may be) to the final value, with a time constant of the pole. If α happens to equal unity then the response reaches final value in zero time. In all practical circuits, of course, additional poles force a nonzero risetime, but the general idea should be clear from this example. Certainly we can appreciate that the bandwidth extension does not come entirely for free, if we care about the details of the time response.

---

[5] For some inexplicable reason, there seems to be a fair amount of confusion about this point. The presence of the zero merely alters the initial error, but this error always settles to the final value with a time constant of just the pole.

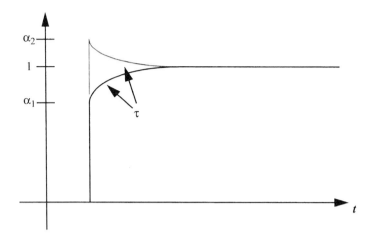

FIGURE 12.20. Possible step responses of pole–zero doublet

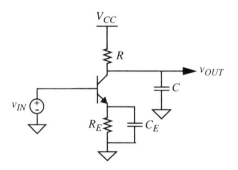

FIGURE 12.21. Zero-peaked common-emitter amplifier

This type of cancellation may be implemented as in the degenerated common-emitter amplifier shown in Figure 12.21. Here, C_E is *not* chosen large enough to behave as a short at all frequencies of interest. Instead, it is chosen just large enough to begin shorting out R_E when C begins to short out R. It is therefore relatively straightforward to understand that ideal compensation should result when $R_E C_E \approx RC$. Proper adjustment is necessary to obtain the best response.

12.3.3 TWO-PORT BANDWIDTH ENHANCEMENT

Shunt peaking is a form of bandwidth enhancement in which a one-port network is connected across the amplifier and load. Series peaking interposes a two-port network between amplifier and load. Although shunt peaking provides a larger bandwidth enhancement than does series peaking, it would be improper to conclude that two-port enhancement is therefore inferior (especially since a one-port is a degenerate subset of a two-port).

12.3 BANDWIDTH EXTENSION TECHNIQUES

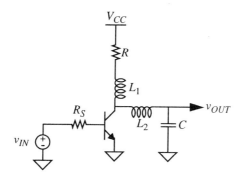

FIGURE 12.22. Amplifier with shunt and series peaking

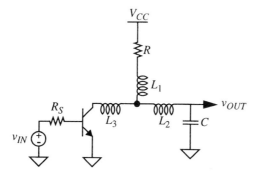

FIGURE 12.23. Shunt and (double) series peaking

One may augment shunt peaking by additionally separating the load capacitance from the output capacitance of the device. If a series inductor is used to perform this separation, the overall result is a combination of shunt and series peaking; see Figure 12.22. We won't spend any time analyzing this combination because it is an intermediate step on the way to a much better bandwidth extension method. The next (also intermediate) evolutionary step is to add an inductance between the device and the rest of the network, as shown in Figure 12.23.

This combination is functionally identical to the previous iteration if we ignore the transistor's output capacitance (because L_3 is then in series with a current source). So let's now consider it qualitatively as follows. Just as in the step response of an ordinary shunt-peaked amplifier, the flow of current into the load resistor continues to be deferred by the action of L_1. This action alone speeds up the charging of the load capacitance. In addition to that mechanism, the transistor initially has to drive only its own output capacitance (not shown) for some time because L_3 delays the diversion of current into the rest of the network. Hence, risetime at the collector improves, which we again interpret as implying an improved bandwidth. Some time after the collector voltage has risen significantly, the voltage across the load capacitance begins to rise as current finally starts to flow through L_2. Hence, such a network charges the capacitances *serially in time*, rather than in parallel. The trade-off is an increased delay in exchange for the improved bandwidth. We will see that this bandwidth–*delay* trade-off is a recurrent theme.

To save space, the combination of three inductors can be realized conveniently as a single pair of magnetically coupled inductors (i.e., a transformer), since the equivalent circuit model of such a connection is precisely the arrangement we seek (consider the T-model of a transformer in particular). That is, inductance L_2 comes for free (and may take on positive or negative values, depending on the relative current directions through the two windings; this network requires a negative value). The inductors may be implemented as a pair of coils that have been placed proximate enough to produce the desired amount of coupling. In the 500-series oscilloscopes of Tektronix, for example, the inductors are realized as single-layer coils wound around a plastic rod. Merely spacing the two coils from each other by the correct distance produces the correct coupling factor.

In the special case where $L_1 = L_3 = L$, one may derive the transresistance of the load network as

$$\frac{v_{out}}{i_c} = R\frac{s(L/R)+1}{s^2(L+L_2)C + sRC + 1}. \quad (29)$$

Note that the transfer function of the T-coil network is essentially identical to that for shunt peaking in that there are two complex poles and one zero. As mentioned previously, we therefore cannot identify any possible advantage of this network if we continue to neglect the output capacitance of the transistor.

Adding a small bridging capacitance across the inductors to create a parallel resonance provides further improvement. The increased circulating currents associated with the resonance help to push the bandwidth out even further. Derivation of the transresistance for this case is straightforward but most tedious, so we defer the details to the Appendix (Section 12.8). As shown there, the result of that tedium is

$$\frac{v_{out}}{i_c} = R\frac{2s^2LC + s(L/R) + 1}{s^4L(L+2L_2)CC_B + s^3 2RCLC_B + s^2[C_B(L+L_2) + 2LC] + sRC + 1}. \quad (30)$$

After considerably more agony than suffered in deriving the equations for the shunt-peaking case, one may show that the coupled inductances should each have a value given by

$$L_1 = L_3 = L = \frac{R^2C}{2(1+|k|)}, \quad (31)$$

where L is interpreted as the primary or secondary inductance with the other winding open-circuited (see the Appendix for a complete derivation). Hence, this is the value of inductance used in designing and laying out each spiral in an IC implementation, for example, or each single-layer solenoid in a lower-frequency discrete realization.

The bridging capacitance should have a value of

$$C_B = \frac{C}{4}\left[\frac{1-|k|}{1+|k|}\right]. \quad (32)$$

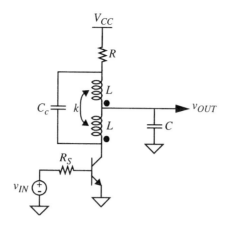

FIGURE 12.24. Amplifier with bridged T-coil bandwidth enhancement

It it also shown in the Appendix that a coupling coefficient of 1/3 yields a Butterworth-type (maximally flat magnitude) response, while a k of 1/2 leads to maximally flat group delay. These coupling coefficients are not particularly large and thus are readily obtained in practice. Two adjacent bondwires typically have coupling coefficients in this range, for example.

Applying these conditions then leads to the circuit pictured in Figure 12.24. The resulting network is called a bridged T-coil and has been used for over forty years in oscilloscope circuitry.[6] The bridged T-coil is capable of almost tripling the bandwidth (the practical maximum is a $2\sqrt{2}$ improvement, or about 2.83×, obtained with the Butterworth condition) if the output capacitance of the device is negligibly small compared with the load capacitance.

It may be shown that the bandwidth is maximized if the junction of the two inductors drives the higher-capacitance node. In Figure 12.24, we have assumed that the load capacitance is larger than the output capacitance of the transistor (remember, we have frequently assumed that the transistor's output capacitance is zero). The collector and load capacitance connections may be reversed if the output capacitance happens to exceed the load capacitance.

As a final refinement, some additional compensation for the output capacitance of the transistor may be provided by adding more inductance in series with it, effectively providing more series peaking. A nearly equivalent result may be obtained merely by tapping the inductors at other than their midpoint (in this case, closer to the load resistor end).

---

[6] The design equations, however, remained a trade secret for the entire time that the 500- and 7000-series oscilloscopes of Tektronix were in production. These networks are presented without design equations – and with only the sketchiest qualitative descriptions – in Bob Orwiler's *Oscilloscope Vertical Amplifiers,* Tektronix Circuit Concepts Series, Tektronix, Beaverton, OR, 1969.

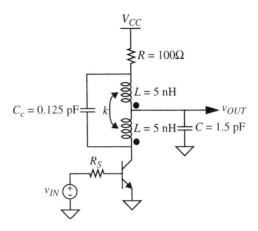

FIGURE 12.25. Bridged T-coil bandwidth enhancement example

Modifying our earlier shunt-peaked design example yields the circuit of Figure 12.25. Here, to keep the comparison fair, we have continued to assume that we desire a maximally flat group delay. As can be seen, the total inductance has doubled. However, since the two inductors are situated on top of each other with a small offset, the additional area is modest (on the order of 50%). In addition, the 125-fF bridging capacitance might be provided as an inherent by-product of merely placing the inductors near each other. The theoretical bandwidth improvement factor provided by this circuit is about 2.7.[7] Hence, roughly a 2.9-GHz bandwidth can be expected, substantially better than the 1.7-GHz bandwidth of the shunt-peaked case. It is important to underscore that this improvement is obtained without an increase in power and without requiring any advances in device technology.

The simulation results of Figure 12.26 show that, indeed, the bridged T-coil network provides the expected bandwidth boost. In addition, the linear frequency axis allows us to see from the phase plot that linear phase behavior (constant delay) is well approximated up to about the bandwidth of the amplifier.

We will later appreciate the structure displayed in Figure 12.25 as itself an intermediate evolutionary step on the way to a completely "distributed amplifier" (to be discussed shortly), in which parasitic capacitances are absorbed into structures that trade gain for delay, not for bandwidth. (Consider a transmission line, for example – it consists of inductance and capacitance, but these elements impose no limit on bandwidth because the capacitances are charged serially in time.) Meanwhile, the circuit may be considered simply as a more sophisticated way to divert current away from the load resistor and into the load capacitance.

---

[7] Again, this value assumes that the output capacitance of the transistor is negligibly small compared with the load capacitor. If this inequality is not well satisfied, additional series compensation will be required to achieve bandwidth boosts of this order.

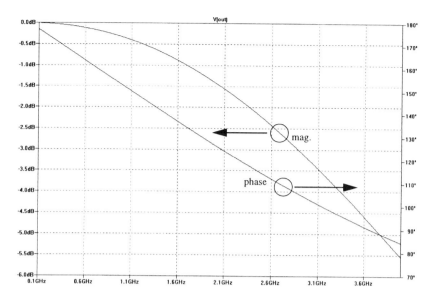

FIGURE 12.26. *LTSpice* simulation of bridged T-coil amplifier (linear phase)

12.4 THE SHUNT-SERIES AMPLIFIER

In contrast with the open-loop architectures we've studied so far, an alternative approach to the design of broadband amplifiers is to use negative feedback. One particularly useful broadband circuit that employs negative feedback is the shunt-series amplifier. Its name derives from the use of a combination of shunt and series feedback, and its utility derives from the relative constancy of input and output impedances over a broad frequency range (which makes cascading much less complicated) as well as from its ease of design. In addition, the dual feedback loops confer the usual benefits normally associated with negative feedback: a reduced dependency on device parameters, improved distortion, broader bandwidth, and rosier cheeks.

Stripped of biasing details, the shunt-series amplifier appears as shown in Figure 12.27. Here R_S is the resistance of the input source and R_L is the load resistance. Thus, the amplifier core consists of just R_F, R_E, and the transistor. To understand how this amplifier works, initially assume that R_E is large enough (relative to the reciprocal of the transistor's g_m) that it degenerates the overall transconductance to approximately $1/R_E$. Since R_E is in series with the input and output loops, the degeneration by R_E is the "series" contribution to the name of this amplifier. To continue the analysis, assume also that R_F is large enough so that its loading on the output node may be neglected. With these assumptions, the voltage gain of the amplifier from the base to the collector is approximately $-R_L/R_E$.

Although we have assumed that R_F has but a minor effect on gain, it has a controlling influence on the input and output resistance. Specifically, it reduces both

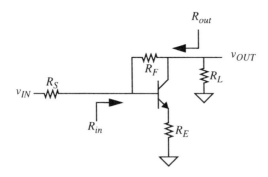

FIGURE 12.27. Shunt-series amplifier (biasing not shown)

quantities through the (shunt) feedback it provides. Additionally, the reduction of input and output resistances helps to increase the bandwidth still further by reducing the sum of open-circuit time constants.[8]

To compute the input resistance R_{in}, we use the fact that the gain from base to collector is approximately $-R_L/R_E$. If (as seems reasonable) we may neglect base current, then the input resistance is due entirely to current flowing through R_F. Applying a test voltage source at the base terminal allows us to compute the effective resistance in the usual way. Just as in the classic Miller effect, connecting an impedance across two nodes that have between them an inverting gain results in a reduction of impedance. Formally, R_{in} is given by

$$R_{in} = \frac{R_F}{1 - A_V} \approx \frac{R_F}{1 + R_L/R_E}, \qquad (33)$$

where A_V is the voltage gain from base to collector.

Now, to compute the output resistance, apply a test voltage source to the collector node and again take the ratio of v_{test} to i_{test}; this yields

$$R_{out} = \frac{R_F + R_S}{1 + R_S/R_1} \approx \frac{R_F}{1 + R_S/R_E}. \qquad (34)$$

If the source and load resistances are equal (a particularly common situation in discrete realizations), then the denominators of Eqns. 33 and 34 are approximately equal. Since the numerators are also approximately equal, it follows that R_{in} and R_{out} are themselves nearly equal. If $R_S = R_L = R$, then we may write

$$R_{out} \approx R_{in} \approx \frac{R_F}{1 + R/R_E} \approx \frac{R_F}{1 - A_V}. \qquad (35)$$

The ease with which this amplifier provides a simultaneous impedance match at both input and output ports accounts in part for its popularity. Once the impedance

---

[8] See T. H. Lee, *The Design of CMOS Radio-Frequency Integrated Circuits,* 2nd ed., Cambridge University Press, 2004, Chap. 8.

12.4 THE SHUNT-SERIES AMPLIFIER

level and gain are chosen, the required value of the feedback resistor is easily determined. Combining knowledge of the load resistance with the required gain leads quickly to the necessary value of R_E. To complete the design, a suitable device width and bias point must be chosen. Generally, these choices are made to ensure sufficient g_m to validate the assumptions used in developing this set of equations.

12.4.1 DETAILED DESIGN OF SHUNT-SERIES AMPLIFIER

The foregoing presentation outlines the first-order behavior of the shunt-series amplifier in order to help develop design intuition. To carry out a more detailed design, however, we now consider some of the second-order factors neglected in the previous section.

Low-Frequency Gain and Input–Output Resistances

We start by computing the gain from base to collector since it allows us to find the input and output resistances easily. Once the base-to-collector gain and input resistance are known, the overall gain is trivially found from the voltage divider relationship.

First, recall that the effective transconductance of a common-emitter amplifier with emitter degeneration is:

$$g_{m,eff} = \frac{g_m}{1 + g_m R_E}. \quad (36)$$

Note from Eqn. 36 that the effective transconductance is approximately $1/R_E$ as long as $g_m R_E$ is much larger than unity.

Applying a test voltage from base to ground causes a collector current to flow through both the load and feedback resistors. Some fraction of the test voltage also feeds forward directly to the output. Superposition allows us to treat each of these contributions to the output voltage separately:

$$v_{out} = -g_{m,eff} v_{test} \frac{R_F R_L}{R_F + R_L} + v_{test} \frac{R_L}{R_F + R_L}. \quad (37)$$

Solving for the gain yields

$$A_V = \frac{v_{out}}{v_{test}} = -\frac{R_L}{R_E} \cdot \left[\frac{1}{1 + 1/g_m R_E}\right] \cdot \left[\frac{1}{1 + R_L/R_F}\right] \cdot \left[1 - \frac{1}{g_{m,eff} R_F}\right]. \quad (38)$$

Although not the most compact expression, Eqn. 38 shows the gain derived earlier from first-order theory, multiplied by three factors (in brackets), each of which is ideally unity.

The first "nonideal" factor reflects the influence of finite g_m on the effective transconductance. While $g_{m,eff}$ approaches $1/R_E$ in the limit of large $g_m R_E$, this first factor shows quantitatively the effect of finite $g_m R_E$. The second term is the result of the loading by R_F on the output node. As long as R_F is substantially larger than the load resistance R_L, the gain reduction is small.

The final gain reduction factor is due to feedforward of the input signal to the output. This feedforward reduces the gain because the ordinary gain path inverts, while the feedforward path does not. Hence, the feedforward term partially cancels the desired output. The transconductance of the feedforward term is $1/R_F$, so as long as this parasitic transconductance is small compared with the desired transconductance $g_{m,eff}$, the gain loss is negligible.

Having examined the complete gain equation term by term, we now present a much more compact (but still exact) expression, useful for calculations to follow:

$$A_V = -\frac{R_L}{R_{eff}} \cdot \left[\frac{R_F - R_{eff}}{R_F + R_L} \right], \qquad (39)$$

where R_{eff} is simply the reciprocal of the effective transconductance. The upshot is simply that, in order to obtain the desired gain, one must choose a value of R_E (or R_{eff}) that is somewhat smaller than would be anticipated on the basis of the first-order equations.

Now that we have a complete expression (two, even) for the low-frequency gain, we can obtain a more accurate value for the resistance between base and ground:

$$R_{in} = \frac{R_F}{1 - A_V}, \qquad (40)$$

which (after using Eqn. 39) becomes

$$R_{in} = \frac{R_F}{1 + \frac{R_L}{R_{eff}} \left(\frac{R_F - R_{eff}}{R_F + R_L} \right)} = \frac{R_{eff}(R_F + R_L)}{R_{eff} + R_L}. \qquad (41)$$

In general, one designs specifically for a particular value of gain. Assuming success at achieving that goal, the value of feedback resistance necessary to produce a desired input resistance is readily found simply from Eqn. 40.

The output resistance (i.e., as seen by R_L) is also simple to find. Again, we apply a test voltage source to the collector node and compute the ratio of test voltage to test current. Performing this exercise yields

$$R_{out} = \frac{v_{test}}{i_{test}} = \frac{R_F + R_S}{1 + g_{m,eff} R_S} = \frac{R_F + R_S}{1 + R_S/R_{eff}} = \frac{R_{eff}(R_F + R_S)}{R_{eff} + R_S}. \qquad (42)$$

Comparing the expressions for input and output resistance, we see that if R_S and R_L are equal (as is commonly the case) then R_{in} and R_{out} will also be precisely equal. This happy coincidence is one reason for the tremendous popularity of this topology.[9]

---

[9] It should be noted that there is a minor difference in that finite β causes the input and output resistance to be somewhat unequal, although the error is small for typical values of β. The input resistance is smaller by a factor of approximately $1 - 1/(2\beta)$, while the output resistance is higher by a factor of about $1 + 1/(2\beta)$. The gain is also slightly lower, by a factor of about $1 - 2\beta$.

It should be emphasized that, when carrying out a design (as opposed to analysis), the desired gain is known. Hence, if the input and output resistances are to be equal, selection of the feedback resistor is trivial from Eqn. 40. The value of R_E is then chosen to provide the correct gain, completing the design.

Bandwidth and Input–Output Impedances

Having presented exact expressions for various low-frequency quantities (gain and input–output resistances), we now derive approximate expressions for the bandwidth as well as the input and output *impedances* of this amplifier.

Before plowing through a slew of equations, let's see if we can anticipate the qualitative behavior of these quantities. Because this amplifier is a low-order system, we expect gain and bandwidth to trade off more or less linearly. Furthermore, precisely because it is a low-order system, an open-circuit time constant estimate of bandwidth should be reasonably accurate.

We also expect the input impedance to possess a capacitive component, partly because of the presence of C_{gs} but also because of the augmentation of C_{gd} by the Miller effect. The output impedance, on the other hand, could behave differently because the shunt feedback that reduces the output resistance becomes less effective as frequency increases. As a result, the output impedance could actually rise with frequency, leading to an inductive component in the output impedance.

Having made those predictions, let us proceed with a calculation of the open-circuit time constant sum. To simplify the development, assume that the only device capacitances are C_{gs} and C_{gd}. Furthermore, neglect the series base resistance. Finally, assume that the source and load resistance are equal to each other and to a value R.

The effective resistance facing C_{gd} is clearly R_F in parallel with a resistance given by

$$r_{left} + r_{right} + g_{m,eff} r_{left} r_{right}, \tag{43}$$

so the resistance is

$$R_F \parallel (R_S + R_L + g_{m,eff} R_S R_L) = R_F \parallel R(2 + g_{m,eff}R); \tag{44}$$

after substitution for R_F, this may be rewritten as

$$R(1 - A_V) \parallel R(2 + g_{m,eff}R). \tag{45}$$

Note that, in the limit of large gain, the resistance facing C_{gd} approaches

$$|A_V|(R/2), \tag{46}$$

as might be anticipated from considering the Miller effect.

Computing the resistance facing C_{gd} is somewhat more involved, but ultimately one may derive the following expression:

$$\frac{R(R_F + R + 2R_E) + R_E R_F}{(2R + R_F)(1 + g_m R_E) + g_m R^2}. \tag{47}$$

In the limit of large gain, this quantity simplifies to

$$\frac{R}{R_E} \frac{1}{g_m}. \tag{48}$$

Note that the ratio R/R_E is approximately the magnitude of the gain (from base to collector). Because both open-circuit resistances are then roughly proportional to gain, the gain–bandwidth product of the shunt-series amplifier is approximately constant.

The estimated bandwidth of the amplifier in this limit is therefore

$$(BW) \approx \left[|A_V| \left(\frac{C_{gs}}{g_m} + \frac{RC_{gd}}{2} \right) \right]^{-1}. \tag{49}$$

Having derived an approximate expression for the bandwidth, we now consider the input impedance. As stated earlier, the input impedance should possess a capacitive component because of C_{gs} and the Miller-multiplied C_{gd}. A crude approximation to the total capacitance may be obtained simply by assuming that the impedance at the base controls the bandwidth of the amplifier. That is, assume that the time constant of the amplifier's pole is the product of the source resistance R_S ($= R$) and the capacitance at that node. With this assumption, the effective input capacitance is just the bracketed portion of Eqn. 49 divided by R:

$$C_{in} \approx \frac{C_{gs}}{g_m R_E} + C_{gd} \frac{|A_V|}{2}. \tag{50}$$

In almost all practical cases, the Miller-augmented C_{gd} dominates.

Note that the presence of this capacitance, which effectively appears between base and ground, makes it impossible to achieve a perfect input impedance match at all frequencies. Furthermore, as the frequency increases, C_{gs} progressively shorts out, connecting the emitter-degeneration resistance R_E to the base node. Hence, even the input resistance tends to degrade as well, diminishing as the frequency increases.

These effects can be mitigated to a certain extent by using some simple techniques. First, an L-match can be used to transform up the resistive part to the desired level, such as 50 Ω, at some nominal frequency (generally a little beyond where the quality of the match has begun to degrade noticeably). Of the possible types of L-matches, the best choice is usually the one that places an inductance in series with the base and a shunt capacitance across the amplifier input, because such a network becomes transparent at low frequencies (where no correction is required).

The series inductance of the L-match generally leaves a residual inductive component. This inductance is easily compensated by simply augmenting the shunt capacitance of the L-network. With this compensation, the frequency range over which a reasonably good input match is obtained can often be doubled.

To compute the output impedance, apply a test voltage source to the collector and calculate the ratio of the test voltage to the current that it supplies. In the limit of high gain, one finds that the output impedance includes an inductive component whose value is approximately

$$L_{out} \approx ARC_{gs}/g_m, \qquad (51)$$

where C_{gd} has been neglected.

To develop a deeper understanding of the origins of this inductance, note that the base voltage is some fraction of the test voltage applied to the collector. Specifically, the base voltage is an attenuated and low-pass–filtered version of the applied collector voltage attributable to the capacitance at the base. Hence, the base voltage lags behind the voltage at the collector. The transistor then converts the lagging base voltage into a lagging collector current. From the viewpoint of the test source, it must supply a current with a component that lags the applied voltage. This phase relationship between voltage and current is characteristic of an inductance.

From this insight, we can assess the effect of neglecting C_{gd}. Since C_{gd} supplies a *leading* component of voltage at the base, it tends to offset the inductive effect. As a result, the output inductance actually observed can be considerably smaller than the upper bound estimated by Eqn. 51 if C_{gd} is not negligibly small.

Design Example

Let's now consider a design example, which will reinforce important concepts and also force us to contend with the challenges introduced by the need to provide bias. Suppose that we want to design a general-purpose 10-dB–gain block, with 50-Ω input and output resistances to facilitate cascading. In that case, we are seeking a voltage gain magnitude of about 3.2. From our first-order equations, we would want an emitter degeneration resistance of approximately $(50\ \Omega)/(1+3.2) = 12\ \Omega$. However, we know that the first-order gain equation is optimistic in that it neglects the transistor's own (emitter) output resistance. Thus, let's arbitrarily choose $R_E = 10\ \Omega$, with the understanding that we may have to revise the value a bit.

Next, we select the value of the feedback resistor:

$$R_F \approx Z_0(1 - A_V) = (50\ \Omega)(1 + 3.2) \approx 2 = 210\ \Omega. \qquad (52)$$

We will set it at 220 Ω, the nearest standard 10% tolerance value. As with R_E, we understand that the calculated value for R_F is subject to adjustment.

Now that we've completed the core amplifier design, we need to select a bias point for the transistor and then figure out how to establish that bias. One consideration is that we would like to validate (at least reasonably well) our use of the approximate gain equation. That, in turn, means that we need to select a sufficiently high collector bias current so that $1/g_m$ added to $R_E = 10\ \Omega$ is no larger than the first-order estimate of 12 Ω. If we choose a 10-mA I_C, then $1/g_m$ is about 2.5 Ω at room temperature, which is reasonably close. Even though it's over the limit, we'll stay with that value for now. Note that the bias current also sets a bound on the output swing (and hence on output power). Depending on the design objectives, it may be necessary to increase the bias current even further. For this design example, let us assume that our output swing requirements are sufficiently modest that 10 mA suffices.

Next, we need to select a transistor with acceptable characteristics (e.g., sufficient cutoff frequency, low parasitics, etc.) at this collector current. For the remainder of

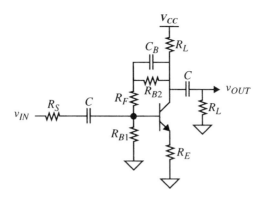

FIGURE 12.28. Complete shunt-series amplifier

this example, assume that we use a 2SC3302 microwave transistor, whose characteristics are satisfactory.

Now that we've selected the bias condition, we must design a network that establishes it stably. Suppose we select the topology shown in Figure 12.28. We've added several elements to produce a stable bias, but the AC circuit remains essentially unchanged. The drop across resistor R_{B1} is multiplied by the sum of R_F and R_{B2} to produce the collector voltage. To validate that statement and to ensure minimal dependency on transistor β, we require the current through those resistors to be large relative to the base current. At the same time, we don't wish to load down the input too much, or else gain and input match will suffer. To quantify the range of acceptable values, suppose the nominal base current is 100 μA for this transistor at the target 10-mA collector bias current and that the nominal base voltage is roughly 1 V. Thus, R_{B1} must be no larger than about 1 kΩ in order for the current through it to be at least 10 times the base current. To avoid excessive loading, we would like R_{B1} to be no smaller than 10 times the source resistance, or 500 Ω. We arbitrarily select a value of 680 Ω, since it lies roughly in the middle of those two limits and is a standard 10% tolerance resistor value. This value corresponds to a current through R_{B1} of very roughly 1.2 mA.

Next, assume that the supply voltage is 10 V. Because junction capacitances diminish as the reverse bias across them increases, we might feel motivated to drop as much of that 10 V across the transistor as possible. However, we need to leave some drop across the collector load resistor to enable some output swing. In the absence of any more detailed design goals, let us arbitrarily choose to drop half the supply voltage across R_L. Thus, we select R_L equal to (5 V)/(10 mA) = 500 Ω.

To finish choosing the bias elements, we wish to drop about 4 V across the sum of R_F and R_{B2}, making the total resistance about 3.3 kΩ when given a 1.2-mA current through the bias network. Subtracting R_F from that sum leaves us with a bit over 3 kΩ. We will round it up to 3.3 kΩ, the nearest 10% value, in effect neglecting the DC drop across R_F.

12.4 THE SHUNT-SERIES AMPLIFIER

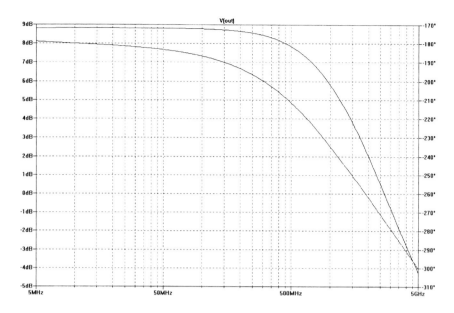

FIGURE 12.29. *LTSpice* simulation of first-pass design

Finally, the coupling capacitors, C, simply need to be large enough so that their reactance is small relative to 50 Ω at the lowest signal frequency of interest. If we violate our philosophy of moderation and use too large a capacitor, the risks of parasitics at high frequencies (e.g., series resistance and inductance) are too large. It may be necessary to use parallel combinations of capacitors if the amplifier must span a very large frequency range. Paralleled combinations may be desired in any case to keep the physical width of signal paths constant and thereby avoid introducing impedance discontinuities.

As always, we also need to be somewhat concerned about the parasitics associated with the feedback bypass capacitor C_B. Fortunately, as we'll see shortly, we do not require heroic characteristics. For now, it suffices to specify a value such that the reactance is at least 10 times smaller than R_{B2}.

The simulation graphed in Figure 12.29 shows that this preliminary design is reasonably close to meeting the design objectives. The bias current is 9.5 mA, and the gain is just shy of 9 dB. The former is close enough to the target not to warrant any adjustment. If desired, the 1-dB gain discrepancy may be eliminated by reducing the emitter degeneration value by a couple of ohms.

We did not specify any bandwidth target for this design, but note that simulations reveal a bandwidth of almost exactly 1 GHz. By coincidence, this value happens to be the same as that of the uncompensated open-loop amplifiers studied in connection with a variety of bandwidth extension methods. If a greater bandwidth were necessary, we could apply many of those same bandwidth extension methods here. One particularly straightforward one is a simple inductance in series with the shunt

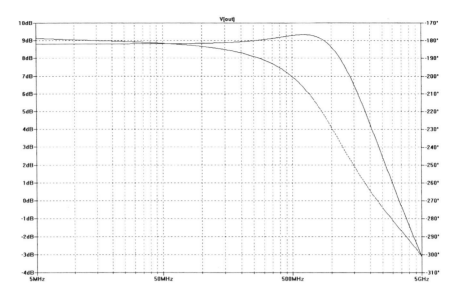

FIGURE 12.30. *LTSpice* simulation of shunt-series amplifier with feedback inductance

feedback path to create a zero, just as in a shunt-peaked amplifier. If that zero is placed properly, it can offset the effect of the most dominant pole, pushing bandwidth out. In this case, we would want to choose a zero in the general vicinity of the 1-GHz bandwidth we presently observe. Because our feedback resistor is 220 Ω, we would initially try an inductor of about

$$L \approx \frac{R_F}{\omega_{zero}} = \frac{220\ \Omega}{2\pi 10^9} \approx 35\text{ nH}. \tag{53}$$

The simulation in Figure 12.30 shows the result of including this inductance. As can be seen, the bandwidth is now about 1.7 GHz. Note that the boost factor is similar to what shunt peaking provides. The ∼0.5-dB gain peaking is not normally considered too objectionable but can be eliminated, if necessary, by reducing the value of the feedback inductance.

Because inductance in the feedback path is helpful in extending bandwidth, we need not be terribly concerned about the inductive parasitics of C_B. And because of the relatively large feedback resistance, we probably do not need to worry about resistive parasitics, either. However, one must always verify these suppositions.

This first-pass design is certainly incomplete, but fairly straightforward optimizations are all that are required to finalize it. These we leave "as an exercise for the reader."

We close our examination of this topology with a consideration of how to improve the input match. Although the shunt-series feedback topology is highly effective at providing a broadband match at both ports, the match is never perfect over an arbitrarily broad frequency range. At the input port, for example, the capacitive input impedance of the transistor itself produces an ever-worsening mismatch as frequency

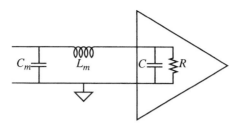

FIGURE 12.31. One proper method for providing an improved match at high frequencies

increases. We typically find that the real part reduces as well. These effects are natural consequences of the gain reduction as the band edge is approached. As the gain goes down, the feedback diminishes in effectiveness and so the input impedance approaches that of an open-loop stage.

If we wish to extend the frequency range over which a good match is obtained, we need either to redesign the shunt-series amplifier (e.g., modify the feedback networks in some ways) or simply to cascade a matching network with the amplifier as it stands. If we select the latter strategy, we must constrain our search to those matching idioms that leave untouched the good match that the shunt-series amplifier already provides at lower frequencies. Furthermore, we seek those matching networks that do not complicate biasing.

As a specific example, suppose that measurements of the input impedance near the 1-GHz bandwidth allow us to deduce that a reasonable model for the amplifier input around that frequency is a 25-Ω resistance shunted by a 1-pF capacitance. There are several matching networks that would theoretically transform us to a purely real 50 Ω, but not all of them satisfy the criteria we have established. As an example of an unsatisfactory solution, suppose we attempt to resonate out the capacitance with a suitable shunt inductance, then transform the remaining real part upward. There are at least two problems with that proposed solution. One is that a shunt inductance produces a DC short between the amplifier input and ground. Fixing that problem requires the inconvenience of adding a large capacitance in series with that inductance, forcing us now to worry about its parasitics. A more serious problem is that this network completely destroys whatever match might have existed at lower frequencies. Thus, in exchange for fixing the match at 1 GHz, we end up ruining the amplifier's match just about everywhere else.

A much better choice is to use a simple low-pass L-match (see Figure 12.31). The series inductance L_m and shunt capacitance C_m facilitate biasing. Plus, such a network has no effect at low frequencies, and thus we might be able to leave largely untouched the good match already provided there.

In terms of Smith-chart loci, the series inductance is chosen to rotate the impedance enough to touch the $G = 1$ circle. Addition of the shunt capacitance then rotates us to the center of the Smith chart, completing the match. Depending on objectives,

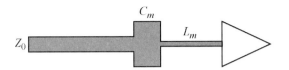

FIGURE 12.32. Matching network layout for amplifier

one may target a perfect match at a frequency intermediate between (a) where the uncompensated amplifier exhibits a poor match and (b) the upper frequency limit where the compensated amplifier is to exhibit an acceptable match. When seeking moderate improvements in matching bandwidth (say, tens of percent), simple iteration usually suffices to converge on an acceptable design.

The matching network can be realized at lower frequencies with lumped elements and at higher frequencies by microstrip equivalents. A representative layout might appear as shown in Figure 12.32. If a broadband improvement in the match is required (say, around an octave or so), one may employ the more sophisticated resonant networks described in Chapter 4. When seeking an exceptionally broadband match improvement (e.g., approaching a decade or more), the best strategy is to form, for example, a Chebyshev-like filter in which the last stage is the input impedance of the amplifier. This way, it is possible to improve the bandwidth of the match to a good fraction of the Bode–Fano limit.

12.4.2 THE DISTRIBUTED AMPLIFIER (TRAVELING-WAVE AMPLIFIER)

Without question, the most elegant exploitation of distributed concepts is the 1936 distributed amplifier of U.K. inventor William S. Percival.[10] He apparently didn't talk about it very much, though, and widespread awareness of this scheme had to await the publication in 1948 of a landmark paper by Ginzton, Hewlett, Jasberg, and Noe.[11]

In the abstract to their paper, the authors note that "the ordinary concept of 'maximum bandwidth–gain product' does not apply to this distributed amplifier." Let's see how this structure achieves a gain-for-delay trade-off without affecting bandwidth.

As is evident in Figure 12.33, inputs to the transistors are supplied by a tapped delay line, and the outputs of the transistors are fed into another tapped delay line. Although simple sections are shown, the best performance is obtained when m-derived or (bridged) T-coil sections are employed, as discussed earlier. The distributed amplifier is thus the logical continuation of a progression of using complexity to enable delay–gain–bandwidth trade-offs.

---

[10] W. S. Percival, "Thermionic Valve Circuits," British Patent Specification no. 460,562, filed 24 July 1936, granted January 1937.

[11] E. L. Ginzton, W. R. Hewlett, J. H. Jasberg, and J. D. Noe, "Distributed Amplification," *Proc. IRE*, August 1948, pp. 956–69.

12.4 THE SHUNT-SERIES AMPLIFIER

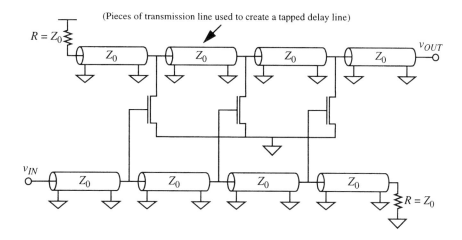

FIGURE 12.33. Distributed amplifier (FET version)

A voltage step applied to the input propagates down the input line, causing a step to appear at each transistor in succession. Each transistor generates a current equal to its g_m multiplied by the value of the input step, and the currents of all the transistors ultimately sum in time coherence if the delays of the input and output lines are matched.

Since each tap on the output line presents an impedance of $Z_0/2$, the overall voltage gain (neglecting losses) is

$$A_V = n g_m Z_0/2. \tag{54}$$

In contrast with ordinary amplifier cascades, we see that this amplifier has an ideal gain that depends linearly on the number of stages. It is important in this connection to observe that this amplifier architecture is essentially an additive one: the overall gain is the result of summing, not multiplying, individual stage gains. As a consequence, the overall amplifier may provide gains in excess of unity even at frequencies where the individual stages do not. A conventional cascade of amplifiers cannot perform this miracle. Consequently, the distributed amplifier can operate at substantially higher frequencies than can conventional amplifiers. Furthermore, note that the delay is similarly proportional to the number of stages; this amplifier does trade gain for delay, since bandwidth does not factor into the trade-off in any direct or obvious way. In fact, bandwidth limitations primarily result from line attenuation that increases with frequency. In turn, this attenuation is partly due to the transistor's dissipative parasitics.

Another way to understand this amplifier's advantages is to recognize that one source of limited bandwidth in conventional amplifiers is the drop in input impedance with increasing frequency that accompanies input capacitance. Unlike dissipative attenuation, which increases with frequency, a lumped amplifier's limited bandwidth results primarily from worsening reflections as frequency increases. The

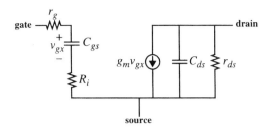

FIGURE 12.34. Simple incremental FET model

traveling-wave amplifier evades this problem by absorbing the device's input capacitance into the constants of the tapped delay line.[12] Thus, by using that capacitance to produce delay, the amplifier's input impedance remains substantially constant and equal to Z_0, until the cutoff frequency of the line itself is approached. In designing such an amplifier, then, it is important to use enough sections to ensure that the bandwidth is not limited by the cutoff frequency of the lines. Although the lines in this case are a hybrid mix of lumped and distributed components, a worst-case estimate for the line cutoff frequency (sometimes called the *Bragg* cutoff) may be computed on the basis of lumped assumptions. That is,

$$\omega_c \approx 2/\sqrt{LC}, \tag{55}$$

where the inductance L is the total inductance of the line segment between transistors and where C is the sum of the line segment capacitance and the device capacitance. We can always make the line cutoff frequency as high as needed by subdivision into sufficiently small segments.

Several factors conspire to make traveling-wave amplifiers (TWAs) function imperfectly. The most important of these may be appreciated by first considering a simple model for a transistor (in this case, a MESFET, but it could model a MOSFET as well, with a suitable adjustment of parameters): see Figure 12.34.

Note that the model neglects C_{gd}, because its inclusion would inhibit the extraction of useful design insights. For similar reasons, we have neglected inductive parasitics in series with the three terminals. We have also neglected the gate–source shunt conductance arising from nonquasistatic (NQS) effects. This neglect can be quite serious for distributed amplifier analysis in particular because the extreme bandwidths typically associated with TWAs almost certainly extend to regions where NQS loading is significant.

It is important to note from this model that the transistor loads the input and output lines differently. The input capacitances are larger than the output capacitances, so significant adjustment is necessary to guarantee matched delays and matched

---

[12] We are assuming that the input impedance of the device looks capacitive at high frequencies. However, this assumption is not always satisfied, and the departure from this assumption must be taken into account in practical designs if good results are to be achieved.

impedances. Satisfying these requirements requires that the drain lines be longer and wider than the gate lines in order to equalize delay and characteristic impedance, respectively. The specific conditions that we must satisfy are approximately as follows:

$$Z_{0,in} \approx \sqrt{L_{in}/(C_{in} + C_{gs})}, \tag{56}$$

$$Z_{0,out} \approx \sqrt{L_{out}/(C_{out} + C_{ds})}; \tag{57}$$

$$T_{d,in} \approx \sqrt{L_{in}(C_{in} + C_{gs})}(l_{in}), \tag{58}$$

$$T_{d,out} \approx \sqrt{L_{out}(C_{out} + C_{ds})}(l_{out}). \tag{59}$$

In these equations, L_{in} and C_{in} represent the total inductance and capacitance of a segment of input line of length l_{in} separating the input taps (we are thus assuming that the segments are so short relative to a wavelength at the highest frequency of interest that we may still treat them as lumped elements). The delay between those taps is $T_{d,in}$. The corresponding quantities for the output lines are identified with the subscript *out*. Normally, we desire both equal impedances and delays.

From the equations we see that both the input and output lines must be designed to have a characteristic impedance that is higher than Z_0 without the transistors. Furthermore, the input line must be designed for a higher native impedance than the output line because the transistor's output capacitance loads the output line less heavily than the input capacitance loads the input line. If we assume, as is usually the case, that one design objective is for both the input and output ports to have the same characteristic impedance Z_0, then we may derive a specific condition on the line constants as follows:

$$Z_0 \approx \sqrt{\frac{L_{in}}{C_{in} + C_{gs}}} \implies T_{d,in} \approx \frac{L_{in}}{Z_0}(l_{in}); \tag{60}$$

$$Z_0 \approx \sqrt{\frac{L_{out}}{C_{out} + C_{ds}}} \implies T_{d,out} \approx \frac{L_{out}}{Z_0}(l_{out}). \tag{61}$$

Now setting the delays equal yields

$$(L_{in})(l_{in}) = (L_{out})(l_{out}) \implies \frac{L_{in}}{L_{out}} = \frac{l_{out}}{l_{in}}. \tag{62}$$

Thus, the segment lengths must be chosen in an inverse ratio to the segment inductances. The input line segments must have higher inductance than the output line segments in order to equalize impedances, and the output lines need to be longer to equalize delays.

We still don't have quite enough information to proceed with an actual design; we need specific design goals and constraints. For example, we could seek to maximize the gain with no constraints; maximize the gain with fixed segment lengths; maximize the bandwidth for a given gain; or maximize the gain–bandwidth product for a fixed power consumption. Aiming for each of these objectives will lead to distinct designs. We will be able to provide only an outline of how to implement strategies for achieving some of these goals.

If one wishes to maximize gain with fixed transistors and fixed segment lengths without power constraints, then the optimization will in fact yield the number of stages corresponding to this condition. As we increase the number of stages in an effort to increase the gain, the total line length increases as well. But thanks to line attenuation (which is exponential), we eventually reach a point where adding one more stage introduces more attenuation than gain. The precise number of stages that produces this maximum gain is a function of device technology, bias point, device dimensions, and line constants. For this academic condition, one may show that the optimum number of stages is approximately

$$N_{opt} \approx \frac{\ln(\alpha_{in}l_{in}/\alpha_{out}l_{out})}{\alpha_{in}l_{in} - \alpha_{out}l_{out}}, \qquad (63)$$

where α_{in} and α_{out} are the attenuation constants for the input and output lines (including the effect of transistor loading).[13] Note that this equation does not provide guidance about how to choose the optimum transistor or its bias. Instead, it presumes that we are simply handed a fixed device and are given its bias (from which we compute the transistor's contribution to line loadings). In integrated circuit implementations, we have total control over the device width. Even in a discrete design context, we often have at least some selection of device sizes, but this particular optimization procedure doesn't take advantage of that important degree of freedom. Further note that the design procedure does not accommodate any specific bandwidth target.

As a more realistic (but considerably more difficult) alternative, we would generally fix the bias voltages on the transistors, as well as their total width. These constraints are equivalent to fixing the total power consumption and gain. Thus, identifying the optimum number of stages in this case yields a design whose bandwidth is maximized within a fixed power budget.

In seeking any of these optima, it is important to acknowledge that the resistive device parasitics alter the characteristics of the input and output lines in troublesome ways. For example, accounting for nonquasistatic transistor dynamics (neglected in Figure 12.34) reveals a shunt conductance that grows as the square of frequency. That NQS loss compounds the other device losses, as well as the loss inherent in all real lines. To complicate matters, the line loss is also frequency dependent. If dominated by skin effect, the loss will grow as the square root of frequency. If dielectric loss dominates, then the growth will be linear in frequency. If both mechanisms are significant, the loss will exhibit some mixture of those two behaviors.

Whatever their origins, the loss mechanisms produce attenuation and dispersion. We also see from the transistor model that the input and output lines are coupled through the drain–gate capacitor. All of these second-order phenomena combine to

---

[13] See David M. Pozar, *Microwave Engineering*, 2nd ed., Wiley, New York, 1998.

degrade performance. They certainly make the design somewhat more challenging.[14] In general, iteration with a combination of circuit simulators and electromagnetic field solvers are required to converge on a good design. That these challenges are not insurmountable is clear from the record-setting gain–bandwidth products that TWAs exhibit.

To underscore that observation, perhaps it helps to know that *vacuum tube* distributed amplifiers were successfully used in many Tektronix oscilloscopes for many years (their model 513 was their first to use this type of amplifier). The amplifiers were used in the final vertical deflection stage and typically involved six or seven "matched" pairs of vacuum tubes. Bandwidths of roughly $\omega_T/2$ were routinely achieved, so that 100-MHz general-purpose oscilloscopes were available by around 1960.[15]

Given these attributes, one may reasonably ask why this type of amplifier is not ubiquitous today. A significant reason is that it is rather power hungry, since many stages are required to provide a given gain. Another is that TWAs are rather noisy, as the transistors are typically driven by nothing approaching a noise match. Yet another is that the active devices that first supplanted vacuum tubes – bipolar transistors – have several characteristics that make them unsuitable for use in distributed amplifiers. The biggest offender is a significant parasitic base resistance, r_b, which spoils line Q and therefore degrades the line. Bipolar distributed amplifiers consequently acquired an unsavory reputation. Finally, the lumped lines could not be integrated until very recently, when devices improved enough that frequencies of operation increased to a range where fully integrated lines become practical. Distributed amplifiers all but disappeared as a consequence.

They finally made their reappearance in about 1980, when workers in GaAs technology rediscovered the principle. Since that time, distributed amplifiers have been constructed in a variety of compound semiconductor technologies, with InP versions achieving 100-GHz bandwidths.

12.4.3 INTERMODULATION DISTORTION IN BROADBAND AMPLIFIERS

The narrowband focus of the chapter on low-noise amplifiers (Chapter 13) extends to the way nonlinearities are characterized. The value of third-order intercept as a linearity measure depends on the ability to neglect harmonic and intermodulation distortion products that lie outside the bandwidth of the system.

---

[14] One may always use a cascode cell, of course, to reduce the coupling problem – at the expense of a reduction in headroom. Having to accommodate the additional delay through the common-gate device also adds complexity to the design process.

[15] The distributed amplifiers in the Tektronix 585A 100-MHz oscilloscope used 6DJ8 duo-triodes, which have f_T values of roughly 300 MHz. The delay lines were composed of T-coils, which provide better bandwidth than ordinary *m*-derived lumped approximations. It is likely that the proximity of the coils also provided some bridging capacitance, further boosting bandwidth.

By contrast, broadband amplifiers accept, amplify, and distort signals over a wide range of frequencies. Consequently, distortion measures that are appropriate for relatively narrowband systems may convey an inadequate assessment of linearity for broadband systems, such as cable television amplifiers and emerging ultrawideband (UWB) systems. In those cases, third-order intercept needs to be replaced by, or at least supplemented with, alternative linearity measures.

The case of cable television serves as a practical and illustrative example. There could be as many as 91 channels, each 6 MHz wide, transmitted to the subscriber from the "head end." With a total bandwidth exceeding half a gigahertz, there is little hope of filtering out the many distortion products that are inevitably generated by ever-present nonlinearities. As it happens, the human eye is exquisitely sensitive to certain types of artifacts. Thus, even though the signal-to-distortion ratio may be numerically large (e.g., comfortably in excess of what any digital system would require for a completely negligible bit-error rate), that alone may not guarantee acceptable visual quality. As cable television systems came to be deployed in large numbers beginning in the 1970s, engineers became aware of an important source of these visibly objectionable distortions. The most important of these are composite triple beat (CTB) and composite second-order (CSO) distortions. As their names imply, CTB is caused by the intermodulation of three carriers and CSO by the intermodulation of two carriers.

Consider three carriers of frequencies f_1, f_2, f_3. If they are near each other, the intermodulation product $(f_1 + f_2 - f_3)$, as just one example, will also lie near the original carrier frequencies. If this triple-beat product is not sufficiently small, it will produce a visible "crawling" effect that is quite noticeable until large carrier-to-CTB ratios are approached. Note that there are other intermodulation products of those same three frequencies that will lie near the original triplet. In fact, any combination involving one sum and one difference will produce a triple-beat component near the original frequencies.

Extensive experimentation has revealed that these products are judged by most people to be of negligible consequence when they lie below the carrier by about 50 dB. The FCC consequently requires that cable TV equipment satisfy a 51-dB carrier-to-CTB ratio. Most commercially available headend amplifiers typically exceed that minimum requirement by 10 dB or more.

Because CTB is a third-order distortion, it theoretically drops 3 dB for every 1-dB drop in power. Thus, for every decibel reduction in power, the C-CTB ratio improves by 2 dB while the SNR degrades 1 dB. Similarly, the C-CSO ratio improves 1 dB for every 1 dB of reduction in power. As with IP3, these theoretical expectations are not always met in practice; one can always construct an amplifier that does not conform to these relationships. That said, the rules of thumb stated apply much more often than not.

Measuring CTB is similar to measuring third-order intercept. This time, three signal generators feed the device under test. As with measuring IP3, the signal generator outputs should be isolated from each other (or individual attenuators used ahead of the summing point) to prevent one generator from upsetting another.

12.5 TUNED AMPLIFIERS

Commercial instrumentation uses unmodulated carriers (CW) for the CTB measurement. Such a condition is considered the most demanding, because modulated carriers have smaller average carrier power and hence produce lower levels of CTB on average. Typically, one finds about a 10-dB improvement in subjective CTB in going from the CW measurement to tests with actual television signals.

12.5 TUNED AMPLIFIERS

12.5.1 INTRODUCTION

We've already seen that the design of broadband amplifiers can be aided by such bandwidth extension tricks as shunt peaking. However, it is not always necessary (or even desirable) to provide gain over a large frequency range. Often, all that is needed is gain over a narrow frequency range centered about some high frequency.

Such tuned amplifiers are used extensively in communications circuits to provide selective amplification of wanted signals and a degree of filtering out of unwanted signals. As we'll see shortly, eliminating the requirement for broadband operation allows one to obtain substantial gain at relatively high frequencies. That is, to zeroth order, the effort required to achieve a gain of 100 over a bandwidth of 1 MHz is roughly independent of the center frequency about which that bandwidth is obtained; the difficulty of obtaining a specified gain–bandwidth product is approximately constant and independent of center frequency (within certain limits). Furthermore, the power required to obtain this gain can be considerably less for a narrowband implementation. This last consideration is particularly important when designing portable equipment, where battery life is a major concern.

12.5.2 COMMON-EMITTER AMPLIFIER WITH SINGLE TUNED LOAD

To understand why the gain–bandwidth product should be roughly independent of center frequency, consider the amplifier shown in Figure 12.35. If we drive from a zero-impedance source (as shown) and if we can neglect series base resistance, then the collector-base capacitance C_{gd} may be absorbed into the capacitance C. In that case, we can model the circuit as an ideal transconductor driving a parallel RLC tank. At low frequencies, the inductor is a short and the incremental gain is zero; while at high frequencies, the gain goes to zero because the capacitor acts as a short. At the resonant frequency of the tank, the gain becomes simply $g_m R$ since the inductor and capacitor cancel.

For this circuit, the total -3-dB bandwidth is (as usual) simply $1/RC$. Hence, the product of gain (measured at resonance) and bandwidth is just

$$G(BW) = g_m R \cdot \frac{1}{RC} = \frac{g_m}{C}. \tag{64}$$

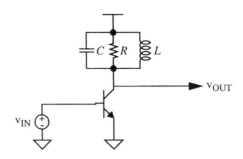

FIGURE 12.35. Amplifier with single tuned load (biasing details omitted)

So for this example, with all of its simplifying assumptions, we obtain a gain–bandwidth product that is *independent* of center frequency, as advertised.

To underscore the profound implications of this last statement, consider two alternative methods for obtaining a gain of 1000 at 10.7 MHz (for the IF section of an FM radio, for example). We could attempt a broadband amplifier design, which would require us to achieve a gain–bandwidth product of over 10 GHz (not a trivially accomplished goal). Or, we could recognize that, for the FM radio example, we need only obtain this gain over a 200-kHz bandwidth,[16] in which case we only need something like a 200-MHz gain–bandwidth product – a considerably easier task.

The fundamental difference between these two approaches, of course, is due to the cancellation of the load capacitance by the inductor in the tuned amplifier. As long as we have direct access to the terminals of any parasitic capacitance (and can make them appear across the tank), we can resonate out this capacitance with an appropriate choice of inductance and obtain a constant gain–bandwidth product at any arbitrary center frequency.

Of course, *real* circuits don't work quite as neatly; we suspect that we probably won't be able to get gain at 100 THz from Jell-O™ transistors, for example, no matter how good our inductor is. But it remains true that, as long as we seek center frequencies that are "reasonable,"[17] tuned loads allow us to obtain roughly constant gain–bandwidth products.

12.5.3 DETAILED ANALYSIS OF THE TUNED AMPLIFIER

The analysis we just performed invokes many simplifying assumptions. In particular, the choice of a zero source resistance and zero base resistance allowed us to absorb the collector–base capacitance into the tank network, permitting the inductance to offset its effects.

---

[16] This value applies to commercial broadcast FM radio; your mileage may vary.
[17] We'll quantify this better a little later, but for now pretend that "reasonable" means "reasonably well below ω_T."

12.5 TUNED AMPLIFIERS

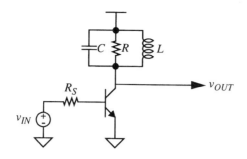

FIGURE 12.36. Amplifier with single tuned load

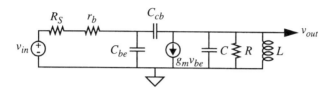

FIGURE 12.37. Incremental model for circuit

Because C_{gd} might have a more serious effect if it were no longer possible to absorb it directly into the tank, let's consider more realistic models for the circuit and examine what happens. Specifically, let's now allow for nonzero source resistance and nonzero series base resistance; see Figure 12.36. The corresponding incremental model is shown in Figure 12.37. Using this model, we can compute two important impedances (actually, admittances, to be precise). First, we'll find the equivalent admittance seen to the left of the RLC tank, then we'll find the admittance seen to the right of the source resistance R_S.

In carrying out this analysis, it is better to apply a test *voltage* source across the tank to find the equivalent admittance seen to its left. Remember, you'll get the same answer whether you use a test voltage or a test current (assuming you make no errors, or at least the same errors), but a test voltage is more convenient here because it most directly fixes the value of v_{be}, the voltage that determines the value of the controlled source.

The precise details are somewhat messy and essentially unrewarding, but the end result is that the admittance seen by the tank consists of an equivalent resistance (which we'll ignore for now) in parallel with an equivalent capacitance. This capacitance is given by:

$$C_{eq} = C_{cb}[1 + g_m R_{eq}] = C_{cb}[1 + g_m(R_S + r_g)]. \tag{65}$$

Notice that C_{eq} can be fairly large. This is actually an alternative manifestation of the Miller effect, this time viewed from the output port. Some fraction of the voltage applied to the collector appears across v_{be}, where it excites the g_m generator. The resulting current adds to that through the capacitors and has to be supplied by the test source, so the source sees a lower impedance. One component of that current is due to a simple capacitive voltage divider and is thus in phase with the applied voltage. It therefore represents a resistive load on the tank, causing a gain reduction. Another component of the current leads the applied voltage and hence represents an additional capacitive load on the tank.

The additional capacitive loading by C_{eq} shifts downward the resonant frequency of the output tank. Although this shift can be compensated by a suitable adjustment of the inductance, it is generally inadvisable to operate in a regime where the resonant frequency depends critically on poorly controlled, poorly characterized, and potentially unstable transistor parasitics. It is therefore desirable to select C relatively large compared with the expected variation in parameters, so that the total tank capacitance remains fairly independent of process and operating point. The unfortunate trade-off is a reduction in the gain–bandwidth product for a given transconductance.

A more serious effect of C_{cb} becomes apparent when we consider the input impedance (or more directly, the input admittance). Since the intermediate details are again of little use outside of deriving the one bit of trivia we're about to state, we'll simply present the result:

$$y_{in} = \frac{y_L y_F}{y_L + y_F} + \frac{g_m y_F}{y_L + y_F}, \qquad (66)$$

where we have set r_b to zero, y_{in} is the admittance seen to the right of the source resistance R_S, y_F is the admittance of C_{cb}, and y_L is the admittance of the RLC tank.[18]

If, as is often the case, the magnitude of the feedback admittance y_F is small compared to that of y_L, then we may write

$$y_{in} \approx y_F + \frac{g_m(j\omega C_{cb})}{y_L}. \qquad (67)$$

The significance of this result becomes apparent when you observe that y_L has a net negative imaginary part at frequencies where the tank looks inductive (i.e., below resonance), so that the second term on the right-hand side of the equation (and, therefore, y_{in}) can have a *negative real* part – that is, the input of the circuit can act as if a negative resistor were connected to it. Having negative resistances around can encourage oscillation (which is just fine if this is your intent, but more typically it isn't). We certainly have all of the necessary ingredients: inductance, capacitance, and negative resistance. If there were no C_{cb}, there would be no such problem.

The difficulty with C_{cb}, then, is that it couples the input and output circuits in potentially deleterious ways. It loads the output tank and decreases gain, detunes the output tank, and can cause instability. This latter problem is particularly severe if one attempts to add a tuned circuit to the input. Furthermore, even before true instability

---

[18] To avoid obscuring the argument any more than it probably already has been, we have neglected the transistor's output admittance in this development; it may be absorbed into y_L if a more exact analysis is desired.

12.6 NEUTRALIZATION AND UNILATERALIZATION

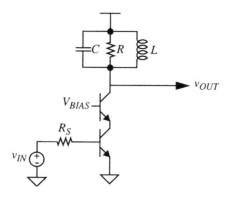

FIGURE 12.38. Cascode amplifier with single tuned load

sets in, the interaction of tuned circuits at both ports may make it extremely challenging to achieve proper tuning.

Unfortunately, C_{cb} will always be nonzero (in fact, it is typically about 30–50% of the main base capacitance, so it is hardly negligible). To mitigate its various undesirable effects therefore requires the use of some topological tricks.

We will shortly explore in greater detail the question of amplifier stability. For now, we will consider a number of techniques that can improve it.

12.6 NEUTRALIZATION AND UNILATERALIZATION

One strategy derives naturally from recognizing that the stability problem stems from coupling the input and output ports. Removing the coupling should therefore be of benefit. This decoupling of output from input should feel familiar – it is precisely what eliminates the Miller effect from common-emitter amplifiers, and what works there works here as well. See Figure 12.38. By providing isolation between input and output ports with the common-base stage, we eliminate (or at least greatly suppress) detuning and the potential for instability, thereby enabling the attainment of larger gain–bandwidth products.

Another topology that achieves these objectives is the emitter-coupled amplifier (which may be viewed as an emitter follower driving a common-base stage); see Figure 12.39. Once again, this structure isolates the output from the input and therefore does not suffer as seriously from the instability and detuning problems of the simple common emitter stage.

Both the cascode and emitter-coupled amplifier behave similarly with regard to isolation. The cascode provides roughly twice the gain for a given total current (because all of this current can be used to set g_m), while the emitter-coupled amplifier requires less total supply voltage (since the two transistors aren't stacked as in the cascode). The choice of which topology to use is usually based on such headroom and gain considerations.

418 CHAPTER 12 AMPLIFIERS

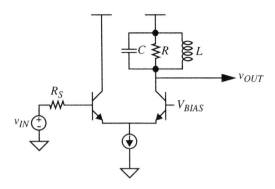

FIGURE 12.39. Emitter-coupled amplifier with single tuned load

The circuits of Figures 12.38 and 12.39 are examples of nearly "unilateral" amplifiers, that is, ones in which signals can flow only one way over large bandwidths. You can well appreciate the value of unilateralization; aside from conferring the circuit benefits we've already discussed, it makes analysis and design much easier by reducing or eliminating unintended and undesired feedback.

If we cannot (or choose not to) eliminate undesired feedback, another approach is to cancel it to the maximum possible extent. Since this cancellation is rarely perfect over large bandwidths, this approach is generally called "neutralization"[19] to distinguish it from more broadband unilateralization techniques that do not depend on cancellations.

The classic neutralized amplifier appears as shown in Figure 12.40. Notice that the inductor has been replaced by something slightly more complex: a tapped inductor, or *autotransformer*. By symmetry, the voltages at the top and bottom of the inductor are exactly 180° out of phase in the connection shown.[20] Therefore, the collector voltage and the voltage at the top of neutralizing capacitor C_N are 180° out of phase. Now, if the undesired coupling from collector to base is due only to C_{gd} then, by symmetry, selection of C_N equal to C_{gd} guarantees that there is no net feedback from collector to base! The current through the neutralizing capacitor is equal in magnitude and opposite in sign to that through C_{gd}; we have removed the coupling from output to input by adding more coupling from output to input (it's just out of phase so that the *net* coupling is zero).

---

[19] Neutralization was developed for AM broadcast radios in the 1920s by Harold Wheeler while working for Louis Hazeltine. His invention allowed the attainment of large, stable gains from tuned RF amplifiers and thus reduced the number of gain stages required (and hence the number of vacuum tubes required) in a typical radio, permitting significant cost reductions over many rival approaches.

[20] Note that autotransformers are not strictly necessary here. They are just a historically common and certainly convenient means of obtaining two voltages that are precise inverses of each other. Clearly, other ways to provide a signal and its inverse exist (consider the example of Figure 12.41).

12.6 NEUTRALIZATION AND UNILATERALIZATION

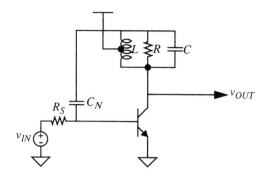

FIGURE 12.40. Neutralized common-emitter amplifier

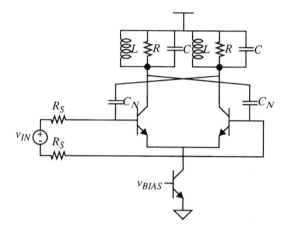

FIGURE 12.41. Neutralized common-emitter amplifier (more practical for ICs)

Neutralization was originally implemented with tapped transformers, but the poor quality of (and large area consumed by) on-chip transformers makes this particular method unattractive for IC implementation. Observe, however, that the tapped transformer is used simply to obtain a signal inversion. Since inversions are easily obtained other ways, practical neutralized IC amplifiers are still realizable. One topology uses a differential pair to obviate the need for a transformer; this is shown in Figure 12.41.

Because perfect neutralization with these techniques depends on feeding back a current that is *precisely* the same as that through C_{gd}, the neutralizing capacitor C_N must match C_{gd} *precisely*. Unfortunately, C_{gd} is somewhat voltage-dependent. Perhaps because of the difficulty of providing precise cancellation in the face of this variability, neutralization has found limited application in semiconductor amplifiers. Vacuum tubes, with their highly linear and relatively constant coupling capacitances, are much better candidates for use of this technique. Nevertheless, with sufficient

FIGURE 12.42. Simple transistor models

diligence, it is possible to obtain usefully large gain–bandwidth improvements in semiconductor-based amplifiers by using neutralization.

12.7 STRANGE IMPEDANCE BEHAVIORS AND STABILITY

One of the many reasons that microwave circuit design is regarded as mysterious by so many is the number of unexpected impedance transformations that arise. Of course, these are unexpected only if you've never encountered them. We now examine a number of these allegedly strange impedance behaviors (SIBs), specifically so they'll cease to be strange.

To illustrate a general analytical approach and at the same time to place many seemingly disconnected phenomena into a unified context, let's consider how the impedances looking into various transistor terminals depend on impedances connected to the other terminals. We will simplify the analyses as appropriate for the RF regime. More comprehensive analyses would lead to more accurate answers, but our aim is to develop design insight, so we will preserve only the minimum complexity consistent with that goal.

First, consider the simple hybrid-π models for a transistor shown in Figure 12.42. The model on the left treats the base–emitter voltage as the independent control variable. This choice is most closely tied to the fundamental physics underlying transistor operation, but it is not the only possible one. As seen on the right of the figure, it is also acceptable to treat the base current as the control variable. The two models are equivalent if we set the current gain as

$$\beta = \frac{g_m v_{be}}{i_b} = \frac{g_m}{sC_{be}} = \frac{\omega_T}{j\omega} = -j\frac{\omega_T}{\omega}. \tag{68}$$

Note the meaning of Eqn. 68 in English: the current gain magnitude is inversely proportional to frequency, dropping to unity at ω_T, as it should. Equation 68 also means that the current gain has a quadrature phase lag associated with it, a result of the quadrature lag in voltage that the input capacitance provides. Finally, Eqn. 68 claims that the current gain grows without bound as the frequency approaches zero. This error tells us that this model is not valid at frequencies where the equation predicts current gains much larger than the DC current gain, β_0, so we must be careful not to apply this model in that frequency regime (the model's lower frequency limits are less constraining for FETs, owing to their vastly superior DC current gain). Similarly, the voltage-controlled model predicts an infinite base–emitter impedance at

12.7 STRANGE IMPEDANCE BEHAVIORS AND STABILITY

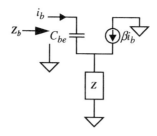

FIGURE 12.43. Circuit model for base impedance calculation

DC, instead of the finite value of a real bipolar device. Again, as long as we only use the model at frequencies where the input current is dominated by the input capacitance, the model will yield useful answers. Finally, note that both of these models apply generally to all transistors, not just bipolars. As a result, the conclusions we will reach are broadly applicable to all transistor circuits.

Within the constraints identified, both models will yield the same answer (they're equivalent), but it may be easier to obtain design insights from one model or the other, depending on circumstances. It's therefore useful to consider both options.

We now use the current-controlled model to derive a simple expression for the impedance looking into the base when the emitter is terminated in an arbitrary impedance, Z. See Figure 12.43. In this case, it's most expedient to drive the base terminal with a test current source. Then, the voltage dropped across the emitter load impedance is just

$$V_Z = Z(\beta + 1)i_b. \tag{69}$$

Thus, from the point of view of the input port, where only i_b flows, the emitter impedance has been multiplied by a factor of $(\beta + 1)$. Adding the input capacitor, we find that the total input impedance is

$$Z_b = \frac{1}{j\omega C_{be}} + Z(\beta + 1) = \frac{1}{j\omega C_{be}} + Z\left(-j\frac{\omega_T}{\omega} + 1\right). \tag{70}$$

Now consider a few special cases. A purely resistive Z turns into a *capacitor* (and a resistor). This capacitance appears in series with the base–emitter capacitance and hence reduces the overall input capacitance. This action may be regarded as the result of bootstrapping the base–emitter capacitance provided by an emitter follower. The larger the value of R, the closer to unity the gain of the follower and the smaller the input capacitance.

Suppose now that the emitter load is a pure inductance. In that case,

$$Z_b = \frac{1}{j\omega C_{be}} + j\omega L\left(-j\frac{\omega_T}{\omega} + 1\right) = \frac{1}{j\omega C_{be}} + \omega_T L + j\omega L. \tag{71}$$

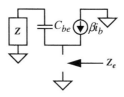

FIGURE 12.44. Circuit model for emitter impedance calculation

That is, the inductive load contributes a pure *resistance* to the input port, aside from an inductance. Just to provide a reference point, a 1-nH inductance in the emitter circuit of a transistor whose ω_T is a modest 10 GHz produces a 63-Ω resistance in series with the input. Obviously, it is critically important to minimize inductive parasitics in the emitter circuit if unwanted resistive input impedances are to be avoided. The faster the transistor, the more challenging this objective. Generating an input resistance by inductive degeneration is not necessarily undesirable, however, as seen in the case of certain low-noise amplifier topologies. In general, an effect large enough to be troublesome in one context may be used to advantage in another.

Now let us examine the effect of a capacitive load. In that case,

$$Z_b = \frac{1}{j\omega C_{be}} + \left(\frac{1}{j\omega C}\right)\left(-j\frac{\omega_T}{\omega} + 1\right) = \frac{1}{j\omega C_{be}} - \frac{\omega_T}{\omega^2 C} + \frac{1}{j\omega C}. \qquad (72)$$

Note that the input impedance is capacitive, in series with a *negative resistance*. If the magnitude of this negative resistance exceeds that of the positive resistance driving the base electrode, instability will result. Even before the onset of truly unstable behavior, the negative resistance can alter the input impedance in ways that produce resonant peaking and otherwise degrade the quality of an input match. Given how common it is to encounter a capacitive load on an emitter follower, it's particularly important to be aware of this mechanism.

Now let us examine how the impedance looking into the emitter varies as a function of the impedance connected to the base terminal. See Figure 12.44. In the emitter-loaded case, the impedance was multiplied by $(\beta + 1)$; here, it is *divided* by the same factor. That is,

$$Z_e = \frac{1}{j\omega C_{be}} + \frac{Z}{(\beta+1)} = \frac{1}{j\omega C_{be}} + \frac{Z}{-j(\omega_T/\omega) + 1}. \qquad (73)$$

To simplify matters even further, assume that we consider frequencies low enough that the magnitude of the current gain is well above unity. In that case,

$$Z_e = \frac{1}{j\omega C_{be}} + \frac{Z}{-j(\omega_T/\omega) + 1} \approx \frac{1}{j\omega C_{be}} + jZ\left(\frac{\omega}{\omega_T}\right). \qquad (74)$$

We see that a resistance in series with the base turns into an inductance when viewed from the emitter. This seeming creation of a reactance out of a resistance is a bit illusory, for it is really the interaction of the base–emitter capacitance with the resistance

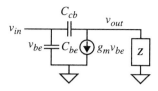

FIGURE 12.45. Circuit model for generalized Miller effect analysis

that produces this result. In fact, the appearance of an inductance is actually the result of *gyration* of the base–emitter capacitance. Again, whether this inductance is beneficial or pernicious depends on the rest of the circuit. Depending on those details, either result is a possibility.

A capacitive base impedance contributes a positive resistance, and an inductance contributes a negative resistance. Just as in the capacitively loaded follower, such a negative resistance can be either an asset or a liability. The engineer must be aware of both cases and either exploit or suppress this mechanism as appropriate. Proceeding as before, we may compute the impedance looking into the emitter as

$$Z_e \approx \frac{1}{j\omega C_{be}} + j\left(\frac{\omega}{\omega_T}\right)(j\omega L) = \frac{1}{j\omega C_{be}} - \left(\frac{\omega^2}{\omega_T}\right)L, \quad (75)$$

so that the real part of the impedance is given by

$$R_{in} \approx -\frac{\omega}{\omega_T}|Z_L|. \quad (76)$$

The ease with which this circuit provides a negative resistance accounts for its popularity in certain types of oscillators. However, it should be obvious that this ease also underscores the importance of minimizing parasitic gate inductance when a negative resistance is *not* desired.

We close by revisiting the Miller effect discussed in Section 12.4, but this time in a slightly different form. To do so properly, we need to restore the collector–base capacitance to our model – see Figure 12.45. The base–emitter capacitance is directly in shunt with the input port, so we may remove it temporarily and then take it into account at the very end. Applying a test voltage to the input port and computing the input impedance (without C_{be}), we obtain

$$Z_{in} = \frac{1 + sC_{cb}Z}{sC_{cb}(1 + g_m Z)}. \quad (77)$$

At frequencies where the admittance of is C_{cb} small compared to that of the load, the input impedance contributes the following component to the input impedance:

$$Z_{in} \approx \frac{1}{sC_{cb}(1 + g_m Z)}. \quad (78)$$

In the special case where the collector load is a pure resistance, we obtain the familiar capacitive multiplication of the classic Miller effect.

If we additionally assume that the transconductance is much larger than the admittance of the load, then

$$Z_{in} \approx \frac{1}{sC_{cb}g_m Z} = \frac{1}{j\omega C_{cb}g_m Z}. \tag{79}$$

From Eqn. 79, we see that a purely inductive load contributes a negative resistance to the input impedance. This result is consistent with the lessons of our previous analysis of the tuned RLC common-emitter amplifier. We also see that a purely capacitive load contributes a pure resistance to the input impedance.

We've seen that impedance transformations of a variety of types can occur in RF circuits. Often these transformations may be exploited, but just as often they are undesired. In both cases, the ability to perform zeroth-order analyses of the types presented here allow you to determine their origins and general dependencies. That information alone is often sufficient to inspire methods for any necessary remediation – without having to resort to more detailed analyses until very near the end of the design process. For example, the appearance of negative resistances at the base terminal (e.g., from inductive collector loads or capacitive emitter loads) can be counteracted by the purposeful addition of suitable positive series resistances. These may be simple resistors or, as is common in discrete RF circuits, ferrite beads. The latter have the advantage of presenting no resistance at DC, thereby minimally disturbing the bias network.

These impedance transformations often force purposeful mismatches at the interfaces to an amplifier. Even though maximizing power transfer is frequently extremely important, it may be impossible to do so and also maintain stable operation. Deliberately mismatching the input or output port (or both) is frequently necessary to assure stability. And, in the case of low-noise amplifier design, such mismatching may be a necessary trade-off in exchange for improved noise performance.

As a final comment, it's again important to adopt a philosophy of moderation. Here, it's important not to design an amplifier to have much more gain or much more bandwidth than necessary. More is definitely not better. For example, it may be that the source and load terminations degrade outside the nominal bandwidth. If the amplifier has excessive bandwidth, these uncontrolled impedances may very well provoke instability.

STABILITY FACTORS, CIRCLES, AND MAXIMUM AVAILABLE GAIN

We've studied various SIBs from a device-level viewpoint in order to develop a fundamental understanding of their sources and cures. In classical microwave design, it is more common to talk about stability in terms of two-port parameters and with plots of stable source and load impedances in Smith-chart form. Although we've already declared our intention not to mention that classical approach, it's worthwhile saying at least a few words about the topic, if for no other reason than to leave the reader at least sufficiently conversant to use such data if it is presented in that form.

12.7 STRANGE IMPEDANCE BEHAVIORS AND STABILITY

First, it is common to encounter various stability factors. One, due to Linvill, provides a simple indication of whether a given two-port can ever go unstable, given open-circuited input and output ports.[21] It is expressed in terms of the admittance two-port parameters of a device as follows:

$$C = \frac{|y_{21} y_{12}|}{2g_{11}g_{22} - \text{Re}[y_{21} y_{12}]}. \tag{80}$$

For unconditional stability, we need $C < 1$. Note that the Linvill stability factor is particularly sensitive to small changes in the reverse transadmittance y_{12}. In terms of model parameters, y_{12} is essentially $j\omega C_{cb}$, so selecting devices with low collector–base (or drain–gate) capacitances is helpful. One must also take great care in layout to avoid inadvertently increasing the capacitance between those two terminals. Alternatively, neutralization or unilateralization techniques may be applied to reduce y_{12} and thereby improve stability.

The equation for the Linvill factor also tells us that resistively loading the input and output ports improves stability. In words, it expresses the intuitively satisfying observation that if we throw away loop gain, stability improves.

Because the two-port parameters for a Linvill test assume open-circuited ports, the C factor does not evaluate amplifier stability under conditions that usually apply to microwave circuits. An alternative stability factor, due to Stern, allows one to assess what combinations of source and load impedances will produce instability.[22] By a somewhat unfortunate arbitrary choice of numerator and denominator, Stern factors *greater* than unity correspond to stable operation. The Stern stability factor is defined as follows:

$$K = \frac{2(g_{11} + G_S)(g_{22} + G_L)}{|y_{21} y_{12}| + \text{Re}[y_{21} y_{12}]}, \tag{81}$$

where G_S and G_L are the conductive parts of the source and load admittances, respectively. As with C, the Stern factor tells us that stability improves as we reduce the feedback capacitance (either through device selection or neutralization) and as we load the input and output ports with increasing conductances.

These two stability factors are not the only ones that have been defined or used, but they are by far the most commonly encountered.

In many microwave transistor data sheets, manufacturers kindly provide information about stability by plotting regions of impedances that correspond to unstable (or stable) operation. This information lets the engineer know whether instability is possible for some combination of source and load impedances and thus, perhaps, suggests how to avoid it. As a simple example, consider the illustration in Figure 12.46. This particular transistor can exhibit instability for the collection of impedances shown in the smaller circle toward the upper right. Some of those impedances are outside of the unit circle, implying negative real parts, for example. Under ordinary conditions

---

[21] J. Linvill and J. Gibbons, *Transistors and Active Circuits*, McGraw-Hill, New York, 1961.
[22] Arthur P. Stern, "Stability and Power Gain of Tuned Transistor Amplifiers," *Proc. IRE*, March 1957.

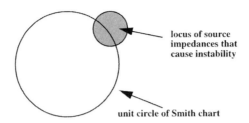

FIGURE 12.46. Illustration of stability circle idea

we would not intentionally drive an amplifier this way, but as we saw in the discussion on SIBs, it is possible to produce such negative resistances unintentionally.

Even if we successfully avoid negative real parts, the particular plot of Figure 12.46 shows that there are some values of source impedance with positive real parts that nonetheless produce amplifier instability. These must be avoided, obviously, but doing so may frustrate the attainment of such other objectives as maximizing gain or minimizing noise figures.

Together with the source stability circle, one also needs to examine the output impedance stability circle in order to assess comprehensively the stability of a proposed design.

Since we've mentioned gain, perhaps it is appropriate here to introduce the figure of merit known as the *maximum available gain* (MAG) of a transistor (or any other two-port). As its name implies, it is the power gain provided when a simultaneous conjugate source and output load match is achieved and when the device has been unilateralized, so that any gain-reducing reverse transadmittance is zero.

We may derive a simple expression for the maximum available gain by considering a two-port consisting of an input and output admittance together with a single transadmittance. The conjugate match condition means that only the conductances survive. The power delivered to the matched output load is thus

$$P_{out} = V_{in}^2 |[\text{voltage gain}]|^2 g_{22} = V_{in}^2 \frac{|y_{21}|^2}{(2g_{22})^2} g_{22} = V_{in}^2 \frac{|y_{21}|^2}{4g_{22}}. \tag{82}$$

The power delivered to the input of the two-port is

$$P_{in} = V_{in}^2 g_{11}, \tag{83}$$

so the power gain is

$$\text{MAG} = \frac{|y_{21}|^2}{4g_{11}g_{22}}. \tag{84}$$

If the transistor is unconditionally stable for all loads, then actually obtaining something close to the MAG may be possible in practice. In the more typical case, however, the impedance corresponding to maximum gain may produce instability. A similar comment applies to the conditions that produce minimum noise. There is no guarantee for a particular device that one can obtain maximum gain, minimum noise figure, and stable operation simultaneously.

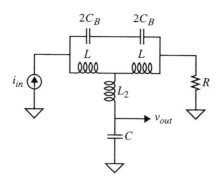

FIGURE 12.47. Bridged T-coil: original circuit

Finally, we reiterate the wisdom of adopting a philosophy of moderation. There is a tendency to believe that more is always better. However, choosing a transistor with huge excesses in high-frequency performance beyond what is strictly required may produce serious problems. The broader the frequency range over which the transistor provides gain, the broader the frequency range over which you are obligated to control the source and load impedances to assure stability. It is too common to encounter cases where a *cheaper* and "less capable" transistor enables a more robust design.

12.8 APPENDIX: DERIVATION OF BRIDGED T-COIL TRANSFER FUNCTION

To the author's knowledge, no one has ever published a derivation of the bridged T-coil's transfer function, although final design equations may be found in the open literature. Regrettably, not all of these sources quite agree (especially with respect to minus signs here and there), and few even bother to state the assumptions underlying the derivations. Because of this situation, we present a detailed derivation in this appendix, preserving many more intermediate steps than we normally would.

Deriving transfer functions in general is straightforward, but tedious. To reduce the pain in this particular case – and thereby reduce the probability of errors – we will use an even- and odd-mode decomposition of the same type we use in analyzing various structures in Chapter 7.

The network we will analyze is shown in Figure 12.47. Note that we've chosen to represent the single bridging capacitance as two series-connected capacitors. The reason for doing so will become clear shortly.

The circuit as drawn in the figure is almost symmetrical. Unfortunately, *almost* is the same as *not,* so you might conclude that there's no option other than to analyze the circuit using textbook brute-force approaches of writing, then solving, a pile of KVL and KCL equations. The network is fortunately of sufficient simplicity that it would yield to persistence with such an approach, although reluctantly. Consequently, it makes a terrific homework problem for students deserving of punishment.

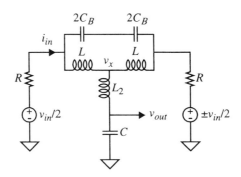

FIGURE 12.48. Bridged T-coil: equivalent circuit with common- and differential-mode sources

We can't hope to avoid altogether the need to solve some equations (it's a fourth-order network, after all), but at least we might be able to decompose the problem into several smaller pieces, each of which is straightforwardly solved (and checked), thereby materially increasing the likelihood of deriving a correct final result. The first step toward that goal is to recast the problem in symmetrical form, as shown in Figure 12.48.

We've now converted the network into a completely symmetrical one, thus making it possible to use the powerful armaments of differential- and common-mode analysis to solve the problem. Although we have added a resistor to the input circuit to enable these decompositions, its presence changes nothing, for we only care about the ratio of output voltage to input current. It's entirely irrelevant how that input current is generated; "amps is amps." That irrelevance also allows us to place a pair of voltage sources in series with the resistors – to be used in evaluating the differential- and common-mode responses separately. Superposition will then yield the overall transfer function we are seeking. Note that this superposition produces a zero net voltage at the bottom of the right-hand resistor, corresponding correctly to the grounded connection of the original network.

12.8.1 DIFFERENTIAL-MODE RESPONSE

For evaluating the odd-mode transfer function, the two voltage sources are of equal magnitude but of opposite sign. Thanks to the antisymmetric excitation of this symmetric circuit, we can ground the midpoint of the bridging capacitor and the main inductors as in Figure 12.49. This grounding of the midpoint means that

$$v_{out,d} = 0, \qquad (85)$$

where the subscript d explicitly identifies the variable as corresponding to a differential analysis.

12.8 APPENDIX: BRIDGED T-COIL TRANSFER FUNCTION

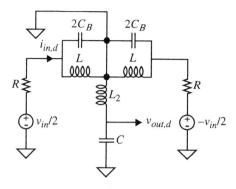

FIGURE 12.49. Equivalent circuit for differential-mode analysis

The input current is readily computed as

$$i_{in,d} = \frac{v_{in}/2}{R + [sL \parallel (1/2sC_B)]} = \frac{v_{in}}{2}\left[\frac{2s^2LC_B + 1}{2s^2RLC_B + sL + R}\right]$$

$$= \frac{v_{in}}{2R}\left[\frac{2s^2LC_B + 1}{2s^2LC_B + s(L/R) + 1}\right]. \tag{86}$$

Note that the differential-mode input impedance has two purely imaginary poles and a pair of complex zeros. The natural frequencies are the same, so this input impedance has the same asymptotic value of $2R$ for DC and infinitely high frequency.

12.8.2 COMMON-MODE RESPONSE

For a common-mode excitation, the bridging capacitance has no voltage across it. Because there is then no current flow through it, we may eliminate it entirely. This is shown in Figure 12.50. The complexity of the remaining network can be reduced further in either of two ways. One is to short together the corresponding mirror-image node pairs (causing L and R to become $L/2$ and $R/2$; the input current we measure is then twice the current we seek). The other is to slice the network down the center (requiring a doubling of L_2 and a halving of C; the input current we measure is then directly the current we're looking for). Both choices will yield the same result, so the choice may be based on other considerations, such as convenience. Here, there's no significant advantage to either one, so we'll randomly choose the former; see Figure 12.51.

The input–output voltage transfer function is simply that of a second-order LC low-pass filter:

$$v_{out,c} = \frac{v_{in}}{2}\left[\frac{1/(sC)}{s\left(\frac{L}{2} + L_2\right) + \frac{R}{2} + \frac{1}{sC}}\right] = \frac{v_{in}}{2}\left[\frac{1}{s^2\left(\frac{L}{2} + L_2\right)C + \frac{sRC}{2} + 1}\right]. \tag{87}$$

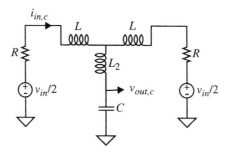

FIGURE 12.50. Equivalent circuit for common-mode analysis

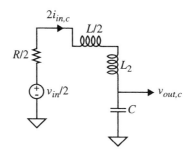

FIGURE 12.51. Simplified equivalent circuit for common-mode analysis

Similarly, the input current (which, again, is twice the current we're looking for) is the output voltage multiplied by the admittance of the output capacitor:

$$2i_{in,c} = (sC)v_{out,c} = \frac{v_{in}}{2}\left[\frac{sC}{s^2\left(\frac{L}{2}+L_2\right)C + \frac{sRC}{2} + 1}\right], \tag{88}$$

or

$$i_{in,c} = \frac{v_{in}}{2}\left[\frac{sC}{s^2(L+2L_2)C + sRC + 2}\right]. \tag{89}$$

12.8.3 THE COMPLETE TRANSFER FUNCTION

Having found the even- and odd-mode output voltages and input currents, we can sum the partial results to find the overall ratio of output voltage to input current. First, the total output voltage is

$$v_{out} = v_{out,d} + v_{out,c} = 0 + \frac{v_{in}}{2}\left[\frac{1}{s^2\left(\frac{L}{2}+L_2\right)C + \frac{sRC}{2} + 1}\right]. \tag{90}$$

The total input current corresponding to that output voltage is

12.8 APPENDIX: BRIDGED T-COIL TRANSFER FUNCTION

$$i_{in} = i_{in,d} + i_{in,c}$$

$$= \frac{v_{in}}{2R}\left[\frac{2s^2LC_B+1}{2s^2LC_B+s\frac{L}{R}+1}\right] + \frac{v_{in}}{2}\left[\frac{sC}{s^2(L+2L_2)C+sRC+2}\right], \quad (91)$$

which may be rewritten as

$$i_{in} = \frac{v_{in}}{2}\left[\left(\frac{1}{R}\right)\frac{2s^2LC_B+1}{2s^2LC_B+s\frac{L}{R}+1} + \frac{sC}{s^2(L+2L_2)C+sRC+2}\right]. \quad (92)$$

The transresistance of the entire bridged-T network is therefore

$$\frac{v_{out}}{i_{in}} = \frac{\dfrac{1}{s^2(\frac{L}{2}+L_2)C+\frac{sRC}{2}+1}}{\left(\dfrac{1}{R}\right)\dfrac{2s^2LC_B+1}{2s^2LC_B+s\frac{L}{R}+1} + \dfrac{sC}{s^2(L+2L_2)C+sRC+2}}. \quad (93)$$

Simplifying in steps, we obtain the following sequence of equations:

$$\frac{v_{out}}{i_{in}} = \left[\left(\frac{1}{R}\right)\frac{(2s^2LC_B+1)\left(s^2(\frac{L}{2}+L_2)C+\frac{sRC}{2}+1\right)}{2s^2LC_B+s\frac{L}{R}+1}\right.$$

$$\left. + \frac{sC\left(s^2(\frac{L}{2}+L_2)C+\frac{sRC}{2}+1\right)}{s^2(L+2L_2)C+sRC+2}\right]^{-1}, \quad (94)$$

and then

$$\frac{v_{out}}{i_{in}} = \frac{2s^2LC_B+s\frac{L}{R}+1}{\left(\frac{1}{R}\right)\left\{(2s^2LC_B+1)\left[s^2(\frac{L}{2}+L_2)C+\frac{sRC}{2}+1\right]\right\}}. \quad (95)$$
$$\phantom{\frac{v_{out}}{i_{in}} = } + \left(\frac{sC}{2}\right)(2s^2LC_B+s\frac{L}{R}+1)$$

Multiplying out and then collecting and ordering the terms in the denominator, we obtain the complete fourth-order transresistance at last:

$$\frac{v_{out}}{i_{in}} = [R]\frac{2s^2LC_B+s\frac{L}{R}+1}{s^4L(L+2L_2)CC_B + 2s^3RLCC_B + s^2(LC+2LC_B+L_2C)+sRC+1}. \quad (96)$$

Observe that there are four poles and two zeros. By inspection, the zeros are complex. We can't say much about the poles by inspection other than to note that, if they are complex, they will appear in conjugate pairs.

Before commencing with an actual design, we need to impose some constraints.

12.8.4 DESIGN EQUATIONS FOR MAXIMALLY FLAT MAGNITUDE RESPONSE

We can simplify considerably our derivation for the maximally flat case by noting that a necessary condition for maximal flatness is that the complex zeros cancel a complex pole pair. It then follows that the remaining pole pair itself must have a maximally flat magnitude response. It's not hard to show that this latter condition is satisfied if the poles are complex, with a damping ratio

$$\zeta = 1/\sqrt{2}. \tag{97}$$

To deduce the conditions that produce the required pole–zero cancellation, we use synthetic division. After executing that operation, the damping-ratio requirement is readily imposed on the remaining second-order polynomial, completing the derivation of the design equations. We begin by performing the following division, starting with the leading term (the bracketed R of Eqn. 96 is just a scaling factor, so we have chosen not to keep it around):

$$\frac{s^4 L(L+2L_2)CC_B + 2s^3 RLCC_B + s^2(LC + 2LC_B + L_2C) + sRC + 1}{2s^2 LC_B + s\frac{L}{R} + 1}. \tag{98}$$

The first step in the division of Eqn. 98 yields the quadratic factor,

$$\frac{s^2}{2}(L+2L_2)C, \tag{99}$$

and a corresponding cubic remainder,

$$s^3 LC\left[2RC_B - \frac{L+2L_2}{2R}\right] + s^2 L\left(2C_B + \frac{C}{2}\right) + sRC + 1. \tag{100}$$

Carrying out the next step in the division generates the linear factor,

$$\frac{s}{2}\frac{C}{C_B}\left[2RC_B - \frac{L+2L_2}{2R}\right], \tag{101}$$

and its corresponding quadratic remainder function,

$$s^2\left[2LC_B - \frac{LC}{2} + \frac{LC(L+2L_2)}{4R^2 C_B}\right] + s\left[\frac{(L+2L_2)C}{4RC_B}\right] + 1. \tag{102}$$

The last step produces the constant factor,

$$1 - \frac{C}{4C_B} + \frac{(L+2L_2)C}{8R^2 C_B^2}, \tag{103}$$

and a final linear remainder function,

$$s\left[\frac{(2L+2L_2)C}{4RC_B} - \frac{L}{R} - \frac{LC(L+2L_2)}{8R^3 C_B^2}\right] + \left[\frac{C}{4C_B} - \frac{(L+2L_2)C}{8R^2 C_B^2}\right]. \tag{104}$$

12.8 APPENDIX: BRIDGED T-COIL TRANSFER FUNCTION

Now, both bracketed terms in the final remainder function must equal zero if the synthetic division has succeeded in yielding the quadratic factor corresponding to the poles not cancelled by the zeros. Thus,

$$\frac{C}{4C_B} - \frac{(L+2L_2)C}{8R^2C_B^2} = 0 \tag{105}$$

and

$$\frac{(2L+2L_2)C}{4RC_B} - \frac{L}{R} - \frac{LC(L+2L_2)}{8R^3C_B^2} = 0. \tag{106}$$

Solving Eqn. 105 gives us one important relationship,

$$R^2 C_B = \frac{L+2L_2}{2}; \tag{107}$$

solving Eqn. 106 yields another,

$$C_B = \frac{C}{4}\left[\frac{L+2L_2}{L}\right]. \tag{108}$$

Combining Eqns. 107 and 108 gives us an equation for an actual element value at last:

$$L = \frac{R^2 C}{2}. \tag{109}$$

Next, note that the terms in Eqn. 103 sum to unity. Thus, synthetic division has discovered the quadratic polynomial corresponding to the uncanceled pole pair:

$$s^2 \frac{(L+2L_2)C}{2} + s\left[RC - \frac{(L+2L_2)C}{4RC_B}\right] + 1. \tag{110}$$

With the aid of Eqn. 107 and Eqn. 108, we may rewrite Eqn. 110 much more simply as

$$s^2 R^2 C_B C + \frac{sRC}{2} + 1. \tag{111}$$

Comparing Eqn. 111 with the standard quadratic form,

$$\frac{s^2}{\omega_n^2} + \frac{2\zeta s}{\omega_n} + 1, \tag{112}$$

we readily find that

$$\omega_n = 1/R\sqrt{C_B C}, \tag{113}$$

$$\zeta = \tfrac{1}{4}\sqrt{C/C_B}. \tag{114}$$

We therefore have another equation that relates the bridging capacitance to the load capacitance. Combining Eqn. 114 with Eqn. 108 gives us

$$C_B = \frac{C}{16\zeta^2} = \frac{C}{4}\left[\frac{L+2L_2}{L}\right]. \tag{115}$$

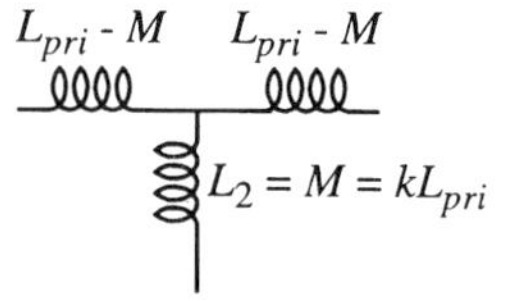

FIGURE 12.52. Equivalent circuit for identical coupled inductors

We can then derive an equation for the remaining component, L_2, by combining Eqn. 109 and Eqn. 115:

$$L_2 = \frac{R^2 C}{4}\left(\frac{1}{4\zeta^2} - 1\right). \tag{116}$$

Note that it is entirely possible for L_2 to take on *negative* values. In fact, we'll see that the bridged T-coil does indeed require a negative inductance because the useful damping ratios exceed 0.5. Although isolated passive inductors cannot have negative values, the mutual inductance within a model for a transformer can possess either sign. Fortunately, L_2 may be realized in precisely this manner in our case.

Observe that, so far, the only condition that we have imposed is pole–zero cancellation. Thus, all of these equations apply more generally than to just the maximally flat magnitude response condition.

To summarize the design procedure, we assume that we are given R and C. We compute the main inductance with Eqn. 109, the bridging capacitance with Eqn. 115, and then the remaining inductance with Eqn. 116 to complete the design. Note that carrying out the design is substantially less difficult than deriving the design equations in the first place!

We now impose a specific condition – maximal magnitude response flatness – to illustrate the design procedure. Following the steps, we first find the main inductance value as

$$L = \frac{R^2 C}{2}. \tag{117}$$

Letting $R = 100\ \Omega$ and $C = 1.5$ pF, we find that $L = 7.5$ nH. Next, the bridging capacitance is readily found from Eqn. 115 to be

$$\left(\zeta = \tfrac{1}{4}\sqrt{C/C_B} = 1/\sqrt{2}\right) \implies C_B = C/8, \tag{118}$$

or 0.1875 pF for our particular component values. Finally,

$$L_2 = \frac{R^2 C}{4}\left(\frac{1}{4\zeta^2} - 1\right) = -\frac{R^2 C}{8}, \tag{119}$$

or -1.875 nH, completing the determination of element values.

We now turn to the problem of how one realizes the required negative inductance. Recall that a transformer may be modeled as an inductive T-network. In the specific case of two identical coupled inductors, the model appears as shown in Figure 12.52.

12.8 APPENDIX: BRIDGED T-COIL TRANSFER FUNCTION

The inductance L_{pri} is the inductance of the primary (or secondary, in this symmetrical case), measured with the secondary open-circuited. It is thus perhaps a more directly useful quantity on which to focus for design, because it is the inductance of the coils you would lay out for the primary and secondary windings. Coupling these windings together affects only the partitioning of that total inductance between the arms; the sum remains constant at L_{pri}.

The sign of the mutual inductance M depends on whether the magnetic fields of the two main inductors aid or oppose one another. If they aid, then the mutual inductance will be negative in sign. Comparing the equivalent circuit of Figure 12.52 with the elements in our bridged-T network, we may write

$$L = L_{pri} - kL_{pri} = L_{pri}(1-k), \tag{120}$$

$$L_2 = kL_{pri} = \frac{kL}{1-k}. \tag{121}$$

Thus, we may express the bridging capacitance equation as

$$C_B = \frac{C}{4}\left[\frac{L+2L_2}{L}\right] = \frac{C}{4}\left[\frac{L_{pri}(1-k)+2(kL_{pri})}{L_{pri}(1-k)}\right] = \frac{C}{4}\left[\frac{1+k}{1-k}\right]. \tag{122}$$

Similarly, the equation for the main inductance may be recast as

$$\left[L = \frac{R^2C}{2} = L_{pri}(1-k)\right] \implies L_{pri} = \frac{R^2C}{2(1-k)}. \tag{123}$$

Now, just to confuse you, the coupling coefficient is most often treated as a positive quantity, so equations generally use the absolute value of k. In that case, Eqn. 122 and Eqn. 123 would appear as

$$C_B = \frac{C}{4}\left[\frac{1-|k|}{1+|k|}\right] \tag{124}$$

and

$$L_{pri} = \frac{R^2C}{2(1+|k|)}, \tag{125}$$

respectively. It is left to the reader to know that the remaining inductor is negative, with value

$$L_2 = -|k|L_{pri}. \tag{126}$$

In most published formulas, the explicit absolute-value bars are absent, so some caution is warranted. In at least one case, the sign confusion has tripped up the author of a paper purportedly offering an alternative design procedure that allegedly obviates the need for coupled windings. Until two-terminal negative inductors become available, that paper's conclusions must be regarded skeptically.

For the (Butterworth) case of maximally flat magnitude response, the coupling coefficient's magnitude is $1/3$.

Finally, let us compute the bandwidth obtained under these conditions to assess the bandwidth boost factor. First, with the bridging capacitance we've computed for maximal flatness,

$$\omega_n = \frac{1}{R\sqrt{C_B C}} = \frac{1}{R\sqrt{(C/8)C}} = \frac{\sqrt{8}}{RC}. \qquad (127)$$

The damping ratio remains

$$\zeta = 1/\sqrt{2}. \qquad (128)$$

As it happens, this particular damping ratio corresponds to the very special case where the −3-dB bandwidth precisely equals ω_n.[23] Thus,

$$\omega_{-3\mathrm{dB}} = \frac{\sqrt{8}}{RC}. \qquad (129)$$

Now, the −3-dB bandwidth with*out* the bridged-T network is simply $1/RC$. Hence, the bandwidth boost factor provided by the network is $\sqrt{8}$, or about 2.83, as stated in the main part of the chapter. Again, the network miraculously provides this near-tripling of bandwidth without increasing power consumption and without requiring better transistors.

12.8.5 DESIGN EQUATIONS FOR MAXIMALLY FLAT DELAY

For this case, we once again require that the zeros cancel a pole pair to leave us with another complex pole pair. The only difference is that the damping ratio of the remaining pole pair is now chosen to maximize delay flatness rather than magnitude response flatness. As a first step, we find the time delay as (minus) the derivative of the phase function:

$$T_D = \frac{d}{dx}\left[\tan^{-1}\left(\frac{2\zeta x}{1-x^2}\right)\right], \qquad (130)$$

where x is the frequency, normalized to ω_n:

$$x \equiv \omega/\omega_n. \qquad (131)$$

Next, it helps to remember that

$$\frac{d}{dx}(\tan^{-1} x) = \frac{1}{x^2+1}. \qquad (132)$$

Then,

$$T_D = \frac{d}{dx}\left[\tan^{-1}\left(\frac{2\zeta x}{1-x^2}\right)\right] = \frac{2\zeta(-x^2+1) + 4\zeta x^2}{4\zeta^2 x^2 + (-x^2+1)^2}. \qquad (133)$$

---

[23] This pole constellation corresponds to that of a second-order Butterworth low-pass filter.

Computing the first derivative and simplifying a little yields

$$\frac{[4\zeta^2 x^2 + (-x^2+1)^2][4(-\zeta x) + 8(\zeta x)]}{[4\zeta^2 x^2 + (-x^2+1)^2]^2} \tag{134}$$

which inspection reveals to be already zero at DC.

Moving on to the second derivative and setting it equal to zero at DC eventually yields

$$4\zeta - (2\zeta)(8\zeta^2 - 4) = 0. \tag{135}$$

Solving for the damping ratio, we find

$$\zeta = \sqrt{3}/2 \tag{136}$$

and a corresponding normalized bandwidth of

$$\frac{\omega_{-3\text{dB}}}{\omega_n} = \left(\frac{\sqrt{5}-1}{2}\right)^{1/2} \approx 0.7862. \tag{137}$$

In turn, the damping ratio implies a capacitance ratio,

$$\frac{C_B}{C} = \frac{1}{12}, \tag{138}$$

which, when combined with the alternative expression for that ratio,

$$\frac{C_B}{C} = \frac{1}{4}\left[\frac{1-|k|}{1+|k|}\right], \tag{139}$$

implies that the coupling coefficient magnitude is $1/2$.

Continuing, we find that the natural frequency is

$$\omega_n = \frac{1}{R\sqrt{C_B C}} = \frac{1}{R\sqrt{(C/12)C}} = \frac{\sqrt{12}}{RC}, \tag{140}$$

allowing us to compute the bandwidth boost factor as

$$\frac{\omega_{-3\text{dB}}}{1/(RC)} = \frac{\omega_{-3\text{dB}}}{\omega_n}\sqrt{12} \approx 2.72. \tag{141}$$

We see that we pay only a small bandwidth penalty to achieve maximally linear phase, relative to maximal gain flatness. Overall, this network is not overly sensitive to variations in element values, so exquisite control over component tolerances is not a requirement.

12.8.6 DESIGN EQUATIONS FOR MAXIMUM BANDWIDTH

So that our derivations will be complete, we now present the design equations for maximum bandwidth – still subject to the constraint of a perfect pole–zero cancellation.

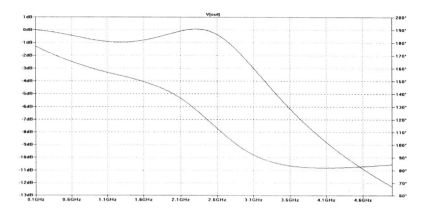

FIGURE 12.53. *LTSpice* simulation of bridged T-coil (maximum bandwidth)

In that case, we first find the damping ratio of a second-order factor that corresponds to maximum bandwidth. The details are unenlightening, so we will simply present the result:

$$\zeta = \left(\frac{5-\sqrt{15}}{2}\right)^{1/2} \approx 0.7507. \tag{142}$$

Then,

$$C_B \approx C/9, \tag{143}$$

$$L_{pri} \approx R^2 C/1.229, \tag{144}$$

$$L_2 \approx 0.38538 L_{pri} \approx 0.31357(R^2 C). \tag{145}$$

The bandwidth boost factor then improves a modest 10% or thereabouts, becoming almost exactly 3, at the cost of a 1-dB passband ripple.

As is evident from Figure 12.53, this choice maximizes bandwidth at the expense of both gain and delay flatness. Given that these objectionable impairments are suffered in exchange for only a small additional bandwidth boost, this choice has not been used in any published implementation, and no references even allude to it. We simply provide this final data point for the sake of completeness.

12.8.7 SUMMARY

As one of two final observations on this network, it's important to recognize the utility of reciprocity. Although all of the derivations assume a particular assignment of input–output ports and variables, we may reverse these assignments without changing the transfer function. Thus, we may drive the "output" capacitor with an input current source, and extract the output voltage across the former input port. This reversal is advantageous if the output capacitance of the transistor that drives this network is greater than the capacitive load the network ultimately drives. Then absorbing the transistor's capacitance into the network will have the maximum beneficial effect overall.

Finally, it is worth regarding this circuit from yet another viewpoint. The bridged-T network may be thought of as one cell of a lumped transmission line model. To the extent that it behaves well, the input impedance of this delay element should appear resistive over a broad band. Thus, the effect of reactive loading that normally limits amplifier bandwidth is suppressed to a large degree.

CHAPTER THIRTEEN

LNA DESIGN

13.1 INTRODUCTION

The first stage of a receiver is typically a low-noise amplifier (LNA), whose main function is to provide enough gain to overcome the noise of subsequent stages (typically a mixer). Aside from providing this gain while adding as little noise as possible, an LNA should accommodate large signals without distortion and frequently must also present a specific impedance, such as 50 Ω, to the input source. This last consideration is particularly important if a filter precedes the LNA, since the transfer characteristics of many filters (both passive and active) are quite sensitive to the quality of the termination.

We will see that one can obtain the minimum noise figure (NF) from a given device by using a particular magic source impedance whose value depends on the characteristics of the device. Unfortunately this source impedance generally differs, perhaps considerably, from that which maximizes power gain. Hence it is possible for poor gain and a bad input match to accompany a good noise figure. One aim of this chapter is to place this trade-off on a quantitative basis to assure a satisfactory design without painful iteration.

We will focus mainly on a single narrowband LNA architecture that it is capable of delivering near-minimum noise figures along with an excellent impedance match and reasonable power gain. The narrowband nature of the amplifier is not necessarily a liability, since many applications require filtering anyway. The LNA we'll study thus exhibits a balance of many desirable characteristics.

Before doing so, however, we need to take a brief detour to study the noise problem in general terms.

13.2 CLASSICAL TWO-PORT NOISE THEORY

In this section we give a macroscopic description of noise in two-ports. Focusing on such system noise models can greatly simplify analysis and lead to the acquisition of useful design insight.

13.2 CLASSICAL TWO-PORT NOISE THEORY

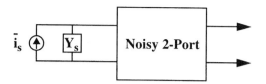

FIGURE 13.1. Noisy two-port driven by noisy source

13.2.1 NOISE FACTOR

A useful measure of the noise performance of a system is the noise factor, usually denoted F. To define it and understand why it is useful, consider a noisy (but linear) two-port (Figure 13.1) driven by a source that has an admittance Y_s and an equivalent shunt noise current $\overline{i_s}$.

If we are concerned only with overall input–output behavior, it is an unnecessary complication to keep track of all of the internal noise sources. Fortunately, the net effect of all of those sources can be represented by just one pair of external sources: a noise voltage and a noise current. This huge simplification allows rapid evaluation of how the source admittance affects the overall noise performance. As a consequence, we can identify the criteria one must satisfy for optimum noise performance.

The noise factor is defined as

$$F \equiv \frac{\text{total output noise power}}{\text{output noise due to input source}}, \qquad (1)$$

where, by convention, the source is at a temperature of 290 kelvins.[1] The noise factor is a measure of the degradation in signal-to-noise ratio that a system introduces. The larger the degradation, the larger the noise factor. If a system adds no noise of its own, then the total output noise is due entirely to the source and the noise factor is thus unity.

In the model of Figure 13.2, all of the noise appears as inputs to the noiseless network, so we may compute the noise figure there. A calculation based directly on Eqn. 1 requires that we compute the total power due to all of the sources and then divide that result by the power due to the input source. An equivalent (and simpler) method is to compute the total short-circuit mean-square noise current and then divide that total by the short-circuit mean-square noise current due to the input source. This alternative method is equivalent because the individual power contributions are proportional to the short-circuit mean-square current, with a proportionality constant (which involves the current division ratio between the source and two-port) that is the same for all of the terms.

---

[1] You might wonder why a relatively cool 290 K is the reference temperature. Then reason is simply that kT is then 4.00×10^{-21} J. Like many practical engineers, Harald T. Friis of Bell Labs preferred round numbers (see his "Noise Figures of Radio Receivers," *Proc. IRE,* July 1944, pp. 419–22). His suggestion of 290 K as the reference temperature had particular appeal in an era of slide-rule computation, and it was adopted rapidly by engineers and ultimately by standards committees.

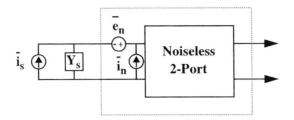

FIGURE 13.2. Equivalent noise model

In carrying out this computation, one generally encounters the problem of combining noise sources that have varying degrees of correlation with one another. In the special case of zero correlation, the individual *powers* superpose. For example, if we assume (as seems reasonable) that the noise powers of the source and of the two-port are uncorrelated, then the expression for noise figure becomes

$$F = \frac{\overline{i_s^2} + \overline{|i_n + Y_s e_n|^2}}{\overline{i_s^2}}. \tag{2}$$

Note that, while we have assumed that the noise of the source is uncorrelated with the two equivalent noise generators of the two-port, Eqn. 2 does *not* assume that the two-port's generators are also uncorrelated with each other.

To accommodate the possibility of correlations between e_n and i_n, express i_n as the sum of two components. One, i_c, is correlated with e_n, and the other, i_u, isn't:

$$i_n = i_c + i_u. \tag{3}$$

Since i_c is correlated with e_n, it may be treated as proportional to it through a constant whose dimensions are those of an admittance:

$$i_c = Y_c e_n. \tag{4}$$

The constant Y_c is known as the *correlation admittance*.

Combining Eqns. 2–4, the noise factor becomes

$$F = \frac{\overline{i_s^2} + \overline{|i_u + (Y_c + Y_s)e_n|^2}}{\overline{i_s^2}} = 1 + \frac{\overline{i_u^2} + |Y_c + Y_s|^2 \overline{e_n^2}}{\overline{i_s^2}}. \tag{5}$$

The expression in Eqn. 5 contains three independent noise sources, each of which may be treated as thermal noise produced by an equivalent resistance or conductance (whether or not such a resistance or conductance actually is the source of the noise):

$$R_n \equiv \frac{\overline{e_n^2}}{4kT\Delta f}, \tag{6}$$

$$G_u \equiv \frac{\overline{i_u^2}}{4kT\Delta f}, \tag{7}$$

$$G_s \equiv \frac{\overline{i_s^2}}{4kT\Delta f}. \tag{8}$$

Using these equivalences, the expression for noise factor can be written purely in terms of impedances and admittances:

$$F = 1 + \frac{G_u + |Y_c + Y_s|^2 R_n}{G_s} = 1 + \frac{G_u + [(G_c + G_s)^2 + (B_c + B_s)^2] R_n}{G_s}, \quad (9)$$

where we have explicitly decomposed each admittance into a sum of a conductance G and a susceptance B.

13.2.2 OPTIMUM SOURCE ADMITTANCE

Once a given two-port's noise has been characterized with its four noise parameters (G_c, B_c, R_n, and G_u), Eqn. 9 allows us to identify the general conditions for minimizing the noise factor. Taking the first derivative with respect to the source admittance and setting it equal to zero yields

$$B_s = -B_c = B_{opt}, \quad (10)$$

$$G_s = \sqrt{G_u/R_n + G_c^2} = G_{opt}. \quad (11)$$

Hence, to minimize the noise factor, the source susceptance should be made equal to the inverse of the correlation susceptance, while the source conductance should be set equal to the value in Eqn. 11.

The noise factor corresponding to this choice is found by direct substitution of Eqns. 10 and 11 into Eqn. 9:

$$F_{min} = 1 + 2R_n[G_{opt} + G_c] = 1 + 2R_n\left[\sqrt{G_u/R_n + G_c^2} + G_c\right]. \quad (12)$$

We may also express the noise factor in terms of F_{min} and the source admittance:

$$F = F_{min} + \frac{R_n}{G_s}[(G_s - G_{opt})^2 + (B_s - B_{opt})^2]. \quad (13)$$

Thus, contours of constant noise factor are non-overlapping circles in the admittance plane;[2] see Figure 13.3.

The ratio R_n/G_s appears as a multiplier in front of the second term of Eqn. 13. For a fixed source conductance, R_n tells us something about the relative sensitivity of the noise figure to departures from the optimum conditions. A large R_n implies a high sensitivity; circuits or devices with high R_n obligate us to work harder to identify, achieve, and maintain optimum conditions. We will shortly see that operation at low bias currents is associated with large R_n, in keeping with the general intuition that achieving high performance only gets more difficult as the power budget tightens.

It is important to recognize that, although minimizing the noise factor has something of the flavor of maximizing power transfer, the source admittances leading to these conditions are generally not the same, as is apparent by inspection of Eqn. 10

---

[2] They are also circles when plotted on a Smith chart because the mapping between the two planes is a bilinear transformation, which preserves circles.

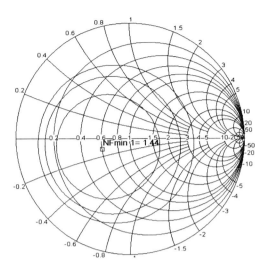

FIGURE 13.3. Example of noise figure circles (for Agilent ATF-551M4 pHEMT, from *AppCAD*)

and Eqn. 11. For example, there is no reason to expect the correlation susceptance to equal the input susceptance (except by coincidence). As a consequence, one must generally accept less than maximum power gain if noise performance is to be optimized, and vice versa.

13.2.3 LIMITATIONS OF CLASSICAL NOISE OPTIMIZATION

The classical theory just presented implicitly assumes that one is given a device with particular, fixed characteristics and then defines the source admittance that will yield the minimum noise figure. This is the usual situation in discrete RF design. However, classical optimization does not directly accommodate the freedom to select device dimensions in integrated circuit design. Thus, additional considerations need to supplement the classical approach in order to proceed rationally toward some optimum. Because of the focus on discrete circuits here, we won't expand on this theme any further. We raise the issue only to point out the incompleteness of the classical approach.[3]

13.2.4 NOISE FIGURE AND NOISE TEMPERATURE

In addition to noise factor, other figures of merit that often crop up in the literature are noise figure and noise temperature. The noise figure is simply the noise factor expressed in decibels.[4]

---

[3] For more on this topic in a CMOS IC context, see T. Lee, *The Design of CMOS Radio-Frequency Integrated Circuits,* 2nd ed., Cambridge University Press, 2004.

[4] Just to complicate matters, the definitions for noise factor and noise figure are switched in some texts.

Table 13.1. *Noise figure, noise factor, and noise temperature*

NF (dB)	F	T_N (kelvins)
0.5	1.122	35.4
0.6	1.148	43.0
0.7	1.175	50.7
0.8	1.202	58.7
0.9	1.230	66.8
1.0	1.259	75.1
1.1	1.288	83.6
1.2	1.318	92.3
1.5	1.413	120
2.0	1.585	170
2.5	1.778	226
3.0	1.995	289
3.5	2.239	359

Noise temperature, T_N, is an alternative way of expressing the effect of an amplifier's noise contribution; it is defined as the increase in temperature required of the source resistance for it to account for all of the output noise at the reference temperature T_{ref} (which is 290 K). It is related to the noise factor as follows:

$$F = 1 + \frac{T_N}{T_{ref}} \implies T_N = T_{ref} \cdot (F - 1). \tag{14}$$

An amplifier that adds no noise of its own has a noise temperature of zero kelvins.

Noise temperature is particularly useful for describing the performance of cascaded amplifiers, as we'll see later in this chapter, and those whose noise factor is quite close to unity (or whose noise figure is very close to 0 dB), since the noise temperature offers a higher-resolution description of noise performance in such cases. This can be seen in Table 13.1. Noise figures in the range of 2–3 dB are generally considered very good, with values around or below 1 dB considered outstanding.

13.3 DERIVATION OF A BIPOLAR NOISE MODEL

Before we can appreciate the attributes (and limitations) of the narrowband LNA topology, it's necessary first to derive an appropriate noise model for a bipolar transistor. To make the analysis tractable and facilitate the acquisition of design insight, we'll need to make a number of simplifying assumptions. These assumptions are not seriously erroneous as long as the device is operated at frequencies well below (say, at least a factor of 5 below) f_T. At still higher frequencies, rapid degradation of other device characteristics (such as gain) militates against the use of the device in the first place and so obviates the need for analysis, accurate or otherwise.

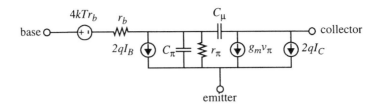

FIGURE 13.4. Noise model for bipolar transistor

Each of the two junctions in a bipolar transistor produces *shot noise,* modeled by a shunt current source whose mean-square spectral density is $2qI_{DC}$, where I_{DC} is the value of the bias current through the junction. The shot noise currents from the two junctions may be treated as uncorrelated for most practical purposes, so we will ignore correlations in all that follows. This neglect will allow us to add noise *powers* directly. That is, a funny (and very useful) kind of superposition is enabled by invoking statistical independence of noise sources.

In addition to the shot noise components (which are in a sense fundamental, because no cleverness in device design can eliminate them), there is also a source of thermal noise: series base resistance, r_b. This noise is represented by a series voltage source whose mean-square density is $4kTr_b$. In modern devices its noise usually dominates (by a good margin) over that due to any series emitter or collector resistance, so we will neglect these. As we'll see, r_b is highly undesirable. Aside from generating noise (and thereby degrading noise figure), its presence often raises to inconvenient values the source resistance that yields minimum noise figure (as we'll soon see).

Although it is tempting to attribute thermal noise to all resistors appearing in a transistor model (e.g., r_π), doing so can amount to double counting. For example, r_π results from linearizing junction behavior, and junction noise is already modeled by shot noise. There is thus a difference between resistances that result from such linearization and those that are simply ordinary resistors. The former do not generate thermal noise, whereas the latter do.

Finally, the collector–emitter output resistance is usually (but not always) large enough to be neglected at high frequencies, so we will omit it in all subsequent analyses.[5]

A small-signal transistor model based on these considerations appears as Figure 13.4. This model, simple as it is, nonetheless captures the most important effects for calculating the noise figure of a bipolar amplifier. It's sufficient for deriving a usefully accurate expression for the noise figure of an amplifier and also for discovering the optimum source resistance.

---

[5] The collector–emitter resistance models the Early effect; it is not thermally noisy because it is the result of linearizing the effect of junction-width variations. Finally, there is a collector–base feedback resistance that also arises from basewidth modulation. Its effect can almost always be completely ignored at RF.

13.3 DERIVATION OF A BIPOLAR NOISE MODEL

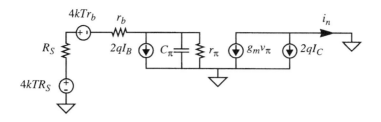

FIGURE 13.5. Model for noise figure calculation

Of the many possible ways to express noise factor, one that is especially useful here is

$$F \equiv \frac{\text{total output noise power}}{\text{output noise power due to source}}, \quad (15)$$

where (as usual) the source temperature is 290 K.

To calculate the noise factor using Eqn. 15, connect a (thermally noisy) source resistance to the circuit of Figure 13.4 and calculate away; see Figure 13.5. Note that the circuit is terminated in a short. In an actual circuit, of course, the output would be loaded with a resistor of some nonzero value – unless the goal is to make a high-tech space heater. However it should be clear from Eqn. 15 that a collector load resistance appears as a multiplier in both the numerator and denominator. As a consequence, it ultimately cancels out and so any value will work for our purposes. We have therefore chosen a zero load resistance, a particularly convenient value.

A considerably sleazier trick is that we have arbitrarily eliminated the collector–base capacitance. Its presence complicates the analysis enough that its removal is necessary simply for clarity. As long as the collector load is a low impedance, this neglect is usually not too serious. In the general case, however, where arbitrary collector loads are to be considered, omitting C_μ can result in significant error. The largest error is in computing the source resistance that leads to the minimum noise figure. Depending on the detailed nature of the load impedance, the optimum source resistance could go up or down. Fortunately, the actual value of that minimum noise figure is usually not greatly affected, so we will proceed to derive the noise figure with a full awareness of the several assumptions that underlie its development.

Given those assumptions and the use of a short-circuit load, the noise factor is simply a ratio of short-circuit currents flowing in the collector branch labeled with i_n. The numerator is the sum of the mean-square short-circuit currents due to all noise sources, and the denominator is the mean-square short-circuit current due only to the source noise. Hence, we have

$$F = \frac{2qI_C + 2qI_B|z_\pi \parallel (r_b + R_S)|^2 g_m^2 + (4kTr_b + 4kTR_S)\left|\frac{z_\pi}{R_S + r_b + z_\pi}\right|^2 g_m^2}{4kTR_S\left|\frac{z_\pi}{R_S + r_b + z_\pi}\right|^2 g_m^2}, \quad (16)$$

where the simple additions of terms in the numerator are a direct consequence of neglecting any correlations among the noise generators.

It's a good idea to study this equation one term at a time to try to make some sense of it. In the denominator, the mean-square voltage spectral density of the source resistor noise, $4kTR_S$, is first multiplied by the square of a voltage divide factor magnitude to find the mean-square voltage across r_π. That squared voltage in turn is multiplied by the square of the transconductance, g_m, to find the squared collector current and thus complete the denominator.

Examining the terms in the numerator from right to left, note that the noise voltage generator of resistor r_b is in series with that of R_S. It therefore undergoes precisely the same transformations, explaining why the last of the three additive terms in the numerator has the form shown.

The base shot noise current sees a total impedance that is a parallel combination of z_π (which, in turn, is r_π in parallel with C_π) and the sum $(r_b + R_S)$. Multiplying the mean-square shot noise current by the squared magnitude of that impedance gives us the mean-square voltage across r_π. Again multiplying that factor by the square of g_m yields the base shot noise contribution to the mean-square collector current.

Finally, the collector shot noise undergoes no scaling or other transformations at all, so it adds directly to all of the other contributions in the numerator.

Equation 16 can be simplified by cancelling some common terms to yield (after a little reordering)

$$F = 1 + \frac{r_b}{R_S} + \frac{2qI_C + 2qI_B|z_\pi \parallel (r_b + R_S)|^2 g_m^2}{4kTR_S|z_\pi/(R_S + r_b + z_\pi)|^2 g_m^2}, \tag{17}$$

which simplifies still further to

$$F = 1 + \frac{r_b}{R_S} + \frac{2qI_C|R_S + r_b + z_\pi|^2}{4kTR_S|z_\pi|^2 g_m^2} + \frac{2qI_B(r_b + R_S)^2}{4kTR_S}. \tag{18}$$

We can continue to cancel common terms to obtain an even simpler form:

$$F = 1 + \frac{r_b}{R_S} + \frac{|R_S + r_b + z_\pi|^2}{2R_S|z_\pi|^2 g_m} + \frac{(r_b + R_S)^2 g_m}{2\beta_F R_S}. \tag{19}$$

In arriving at this last expression, we have made use of the fact that the transconductance of a bipolar transistor is qI_C/kT and that the ratio of collector to base current is β_F.

Note that the second term accounts for noise caused directly by the base resistance, the third term is due to collector shot noise, and the last term is the base current shot noise term. This is the last form of the equation that allows us to make these identifications.

Note also that Eqn. 19 contains three classes of terms (when everything is multiplied out). One is independent of R_S, another is proportional to R_S, and the third is inversely proportional to R_S. At very small source resistance the inversely proportional term dominates, and at very large values the proportional term dominates.

Somewhere between "very small" and "very large" there is an optimum value that minimizes the sum (and hence the noise figure). Before computing the optimum itself, let's understand intuitively why an optimum R_S should exist at all.

At very low source resistances, the contribution by the base resistance is more significant compared to that of the source itself, and noise figure therefore suffers.

At very high source resistances, the contribution to the output noise by the base shot noise is greater (because the impedance it faces is larger, generating a greater voltage across r_π, resulting in a greater current out of the collector). At the same time, the output noise due to the source itself is smaller, because of the harsher voltage divider seen by R_S. The magnitude of the collector shot noise does not change, but its size *relative* to the contribution by R_S is worse, so noise figure degrades further still. The optimum balances the contribution of the base resistance against the effects of base and collector shot noise.

The noise factor equation we will use is a slightly expanded version of Eqn. 19:

$$F = 1 + \frac{r_b}{R_S} + \frac{(R_S + r_b)^2 + |z_\pi|^2 + 2(R_S + r_b)\operatorname{Re}\{z_\pi\}}{2R_S|z_\pi|^2 g_m} + \frac{(r_b + R_S)^2 g_m}{2\beta_F R_S}. \quad (20)$$

Let's now do the math to derive the optimum value for R_S.

OPTIMUM SOURCE RESISTANCE

The procedure for finding this optimum is straightforward enough. Take the first derivative with respect to the source resistance, set it equal to zero, and hope for a minimum:

$$\frac{d}{dR_S}\left(1 + \frac{r_b}{R_S} + \frac{|R_S + r_b + z_\pi|^2}{2R_S|z_\pi|^2 g_m} + \frac{(r_b + R_S)^2 g_m}{2\beta_F R_S}\right) = 0. \quad (21)$$

Grinding inexorably toward the answer generates the sequence

$$\frac{d}{dR_S}\left(\frac{r_b}{R_S} + \frac{(R_S + r_b)^2 + |z_\pi|^2 + 2(R_S + r_b)\operatorname{Re}\{z_\pi\}}{2R_S|z_\pi|^2 g_m} \right.$$
$$\left. + \frac{(r_b^2 + R_S^2 + 2r_b R_S)g_m}{2\beta_F R_S}\right) = 0, \quad (22)$$

$$\frac{d}{dR_S}\left(\frac{r_b}{R_S} + \frac{R_S^2 + r_b^2 + |z_\pi|^2 + 2r_b \operatorname{Re}\{z_\pi\}}{2R_S|z_\pi|^2 g_m} + \frac{(r_b^2 + R_S^2)g_m}{2\beta_F R_S}\right) = 0. \quad (23)$$

In Eqn. 23 we have taken out terms that are independent of R_S and therefore whose derivative is zero (we already took out the unity additive factor in getting to Eqn. 19, in case you were wondering where it went). If you simply want to use the final answer, rather than follow each step of this derivation, feel free to skip ahead!

Separating terms that are proportional to R_S from those that are inversely proportional to it leads us to

$$\frac{d}{dR_S}\left[\frac{1}{R_S}\left(r_b + \frac{r_b^2 + |z_\pi|^2 + 2r_b\,\text{Re}\{z_\pi\}}{2|z_\pi|^2 g_m} + \frac{r_b^2 g_m}{2\beta_F}\right) + R_S\left(\frac{1}{2|z_\pi|^2 g_m} + \frac{g_m}{2\beta_F}\right)\right] = 0. \quad (24)$$

Taking the derivative at last and setting it to zero yields

$$\left(\frac{1}{R_S^2}\right)\left(r_b + \frac{r_b^2 + |z_\pi|^2 + 2r_b\,\text{Re}\{z_\pi\}}{2|z_\pi|^2 g_m} + \frac{r_b^2 g_m}{2\beta_F}\right) = \frac{1}{2|z_\pi|^2 g_m} + \frac{g_m}{2\beta_F}, \quad (25)$$

so that the optimum source resistance (squared) is

$$R_S^2 = \frac{r_b + \dfrac{r_b^2 + |z_\pi|^2 + 2r_b\,\text{Re}\{z_\pi\}}{2|z_\pi|^2 g_m} + \dfrac{r_b^2 g_m}{2\beta_F}}{\dfrac{1}{2|z_\pi|^2 g_m} + \dfrac{g_m}{2\beta_F}}, \quad (26)$$

which reduces a bit to

$$R_S^2 = \frac{2|z_\pi|^2 g_m r_b + r_b^2 + |z_\pi|^2 + 2r_b\,\text{Re}\{z_\pi\} + g_m^2 r_b^2 |z_\pi|^2/\beta_F}{1 + |z_\pi|^2 g_m^2/\beta_F}. \quad (27)$$

Equation 27 is the last form that is traceable directly to our noise model without additional approximations. However, further simplification is possible if we allow one or two very reasonable approximations. One is that the operational frequency is well above $1/r_\pi C_\pi$ ($= \omega_T/\beta$), but not so high that the lumped model is invalid. The other is that the bias current is high enough that C_π is dominated by the diffusion capacitance. With these assumptions, we may write

$$R_S^2 \approx \frac{\left(\dfrac{\omega_T}{\omega}\right)^2\left(\dfrac{2r_b}{g_m} + \dfrac{r_b^2}{\beta_F} + \dfrac{1}{g_m^2}\right) + r_b^2 + \dfrac{2r_b r_\pi}{(\omega/\omega_T)^2 \beta_F^2}}{1 + \dfrac{(\omega_T/\omega)^2}{\beta_F}}. \quad (28)$$

If, as is often the case, the last term in the numerator is small compared to the term preceding it, then

$$R_S^2 \approx \frac{\left(\dfrac{\omega_T}{\omega}\right)^2\left(\dfrac{2r_b}{g_m} + \dfrac{r_b^2}{\beta_F} + \dfrac{1}{g_m^2}\right) + r_b^2}{1 + \dfrac{(\omega_T/\omega)^2}{\beta_F}}. \quad (29)$$

As a specific numerical example, consider using a 2SC3302 microwave transistor at 1 GHz. Assume that the collector bias current is 10 mA, at which the transconductance is 400 mS, $\beta = 80$, and $\omega_T = 10\pi$ Gr/s. The remaining unknown is the value of r_b, which might remain unknown because it is rarely given in data sheets (the 2SC3302 is no exception). Fortunately, however, a plot of input impedance over frequency *is* given, and it shows a resonance at approximately 800 MHz when the bias current is 20 mA (at which we may estimate C_π to be about 23 pF, using other data-sheet information). This resonance is the result of package and lead inductance interacting with C_π. Under that resonant condition, the input resistance is about midway on the Smith chart between the 25-Ω and 50-Ω contours, so we'll estimate the total resistance as 37–38 Ω. This resistance is the sum of r_b and a real term produced by the series emitter inductance associated with the packaging and leads. As is shown in the next section, this induced resistance has a value $\omega_T L_e$. The parasitic inductance is not easily estimated, but one can calculate from the resonance that the total inductance is approximately 1.7 nH. This value is also quite believable from the physical dimensions of the package. Assuming that this total inductance splits evenly between base and emitter (even if it doesn't) allows us to estimate that the contribution by the induced resistance to the total is approximately 30 Ω. Because this value is so close to the total estimated input resistance, our uncertainty in r_b is large. However, we'll press on, and use a value of 7–8 Ω for r_b.

Under these conditions, the optimum source resistance is $\sim$35 Ω (at which the noise figure is 2 dB at 1 GHz), a value close enough to 50 Ω that only a modest NF penalty (of a bit greater than 0.1 dB) is incurred in this case if one performs no impedance transformation. At a bias current of 5 mA, both the noise figure and the penalty for operating at 50 Ω increase (the latter to about 0.2 dB), for an overall minimum noise figure of about 3 dB (again, this value is for operation at 1 GHz).

As a check on our derivations, compare the calculated noise figure of 2 dB to the minimum value of 1.7 dB given in the data sheet for 1-GHz operation. Repeating our calculation for 500-MHz operation yields a 1.6-dB NF (at 5 mA), compared with a data-sheet value of 1.5 dB for that condition. Considering the crude nature of the approximations and parameter extractions, the overall level of agreement is satisfactory.

Finally, remember to keep in mind that an amplifier has to amplify. Achieving a low noise figure is important, but it is only half the battle. For this reason, selection of a suitable bias must take into account gain as well as noise figure. In the specific case of the 2SC3302, somewhat higher gain is obtained at the larger bias current, mainly because f_T is near its maximum value there. Since the minimum achievable noise figure does not change dramatically over the bias current range considered, there is considerable freedom, so other factors (such as implementation issues) may be taken into account as well.

13.4 THE NARROWBAND LNA

The derivations of the previous section show that the source impedance that yields minimum noise factor is generally unrelated to the conditions that maximize power transfer. Furthermore, the high-frequency input impedance of a bipolar transistor is intrinsically capacitive, so providing a good match to a 50-Ω real source without degrading noise performance would appear difficult. Since presenting a known resistive impedance to the external world is almost always a critical requirement of LNAs, we will impose this requirement on our design as well.

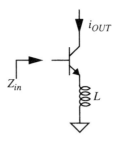

FIGURE 13.6. Inductively degenerated common-emitter amplifier

A particularly good method for producing a real input impedance without degrading noise is to employ inductive emitter degeneration. This method has its origins in analogous vacuum tube amplifiers of the 1930s. It functions equally well for FET-based amplifiers and so, with minor modifications to accommodate biasing differences, the description that follows may be understood to apply to LNAs built with other types of transistors.

With a degenerating inductance, base current undergoes an additional phase shift beyond the ordinary quadrature relationship expected of a capacitor, causing the appearance of a resistive term in the input impedance. An important advantage of this method is that one has control over the value of the real part of the impedance through choice of inductance, as is clear from computing the input resistance of the circuit in Figure 13.6.

To simplify the analysis, consider a device model that includes only a transconductance and a base–emitter capacitance. In that case, it is not hard to show that the input impedance has the following form:

$$Z_{in} = sL + \frac{1}{sC_\pi} + \frac{g_m}{C_\pi}L \approx sL + \frac{1}{sC_\pi} + \omega_T L. \tag{30}$$

Hence, the input impedance is that of a series RLC network, with a resistive term that is directly proportional to the inductance value.

More generally, an arbitrary source degeneration impedance Z is modified by a factor equal to $[\beta(j\omega) + 1]$ when reflected to the gate circuit, where $\beta(j\omega)$ is the current gain:

$$\beta(j\omega) = \frac{\omega_T}{j\omega}. \tag{31}$$

The current gain magnitude goes to unity at ω_T as it should, and has a capacitive phase angle because of C_π. Hence, for the general case,

$$Z_{in}(j\omega) = \frac{1}{j\omega C_\pi} + [\beta(j\omega) + 1]Z = \frac{1}{j\omega C_\pi} + Z + \left[\frac{\omega_T}{j\omega}\right]Z. \tag{32}$$

13.4 THE NARROWBAND LNA

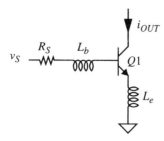

FIGURE 13.7. Narrowband LNA with inductive emitter degeneration (biasing not shown)

Note that a capacitive degeneration would contribute a *negative* resistance to the input impedance.[6] Hence, any parasitic capacitance from emitter to ground offsets the positive resistance from inductive degeneration. It is important to take this effect into account in any actual design (or use it to your advantage).

Whatever the value of this resistive term, it is important to emphasize that it does not bring with it the thermal noise of an ordinary resistor, because a pure reactance is noiseless. We may therefore exploit this property to provide a specified input impedance without degrading the noise performance of the amplifier.

However, the form of Eqn. 30 clearly shows that the input impedance is purely resistive at only one frequency (at resonance), so this method can provide only a narrowband impedance match. Fortunately, there are numerous instances when narrowband operation is not only acceptable but desirable, so inductive degeneration is certainly a valuable technique. The LNA topology we will examine for the rest of this chapter is therefore as shown in Figure 13.7.

The inductance L_e is chosen to provide the desired input resistance (equal to R_s, the source resistance). Since the input impedance is purely resistive only at resonance, an additional degree of freedom, provided by inductance L_b, is needed to guarantee this condition.[7] Now, at resonance, the base-to-emitter voltage is Q times as large as the input voltage. The overall stage transconductance G_m under this condition is therefore

$$G_m = g_{m1} Q_{in} = \frac{g_{m1}}{\omega_0 C_\pi (R_s + \omega_T L_e)} = \frac{\omega_T}{2\omega_0 R_s}, \qquad (33)$$

where we have used the approximation that ω_T is the ratio of g_{m1} to C_π.

The design procedure is thus reasonably straightforward. First select a bias current consistent with the gain and noise figure targets. Then compute the optimum source

---

[6] Capacitively loaded followers are infamous for their poor stability. This negative input resistance is fundamentally responsible, and explains why adding some positive resistance in series with the base or gate circuit helps solve the problem.

[7] It may be that package and other parasitic inductance provides more than this value. In such cases a series *capacitance* may be needed to resonate the input loop at the desired frequency.

resistance to minimize noise figure. Next add enough emitter degeneration inductance to produce an input impedance whose real part is equal to the optimum source resistance, and then add enough of the right kind of impedance (e.g., more inductance) in the base circuit to remove any residual reactive input component, thereby bringing the input loop into resonance. Finally, interpose a lossless matching network (if necessary) between the actual source and the amplifier to transform from 50 Ω (or other source value) to the optimum value of R_S. This matching network often can be merged with whatever inductance (for example) is needed to resonate the input loop.

This particular procedure is attractive because it balances all parameters of interest. An excellent match is guaranteed by the inductive degeneration, and the technique provides nearly the lowest noise figure possible at the given bias conditions. The resonant condition at the input also assures good gain at the same time, since the effective stage transconductance is proportional to ω_T/ω.

The foregoing analysis suffices to highlight the first-order behavior of the circuit. A more detailed analysis accommodates other effects, such as finite transistor r_0. It is straightforward to show that the input impedance of the circuit in Figure 13.7 is

$$Z_{in}(j\omega) = \frac{1}{j\omega C_\pi} + j\omega L + g_m \frac{L}{C_\pi}\left(\frac{r_0}{r_0 + j\omega L + Z_L}\right), \qquad (34)$$

where Z_L is the impedance attached to the collector. Comparing this result with our previous equation, we see that finite output resistance alters the third term in the impedance equation. In particular, we see that the term in parenthesis has a unit magnitude only in the limit. In general, the real part of Z_{in} will be reduced, and the imaginary part may be altered as well, shifting the resonant frequency of the input loop. If, as is common, the load is a parallel resonant tank, then the quantity $|Z_L + j\omega L|$ might be large enough (relative to r_0) at or near its resonance to cause a significant dip in the real part of the input impedance. Depending on the relative resonant frequencies of the input and output loops, it's possible for the dip to appear below, at, or above the desired center frequency for the overall amplifier. Needless to say, the magnitude and location of the dip are important considerations. If the dip occurs far away from the desired operating frequency, its existence may not pose too great a problem. However, it is common for the dip to occur within a couple of percent of the center frequency (because the resonant frequencies of the input and output circuits are usually designed to be close), resulting in poor input match somewhere in the band of interest.

One possible solution is to employ cascoding. However, it's only partially effective because the same r_0 that causes the problem in the first place also limits the effectiveness of cascoding. In stubborn cases, it may be necessary to use several common-base cascoding stages in a stack. Lowering the load resistance of the collector load may also help, but at the cost of reduced gain. Employing some combination of these strategies usually results in a satisfactory design. Simple awareness of the issue usually suffices to avoid unhappy surprises.

13.5 A FEW PRACTICAL DETAILS

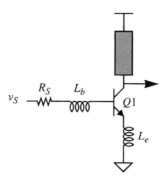

FIGURE 13.8. Narrowband LNA with microstrip load (biasing not shown)

13.5 A FEW PRACTICAL DETAILS

13.5.1 REALIZING THE EMITTER DEGENERATION INDUCTANCE

The narrowband LNA depends on inductive emitter degeneration to produce a real term in the input impedance. Quite often at microwave frequencies, the needed value is so small that it is difficult to produce. For example, continuing with a 2SC3302 biased at 10 mA, we would require ∼2 nH to produce 50 Ω, and perhaps 2–3 times that inductance if the bias were reduced to 5 mA (the increase in the impedance target for minimum NF, plus the reduction in ω_T, causes the needed inductance to increase faster than you might otherwise expect). However, 2 nH is not far from as small as one can expect to achieve without extreme measures, particularly since a fair fraction of this amount is already included in the packaging. Controlling the exact value is therefore challenging. In cases where the packaging and lead inductance already exceed the value you need, the input impedance will actually appear inductive and thus require a capacitance to resonate the input loop. To avoid this necessity, extreme care in layout and construction is essential.

13.5.2 COLLECTOR LOAD

It is generally the case that a resonant collector load is desired. Such a load increases gain by resonating out any output capacitance. Furthermore, the additional filtering of unwanted signals is highly desirable.

There are several practical options for realizing such a load. One is to use a discrete inductor of some appropriate value. A preferable one for our purposes is to implement the inductor out of a suitable length of microstrip, because of its versatility. See Figure 13.8.

The length is adjusted to produce resonance. And, if needed, a downward impedance transformation is readily obtained by merely tapping the output off of some

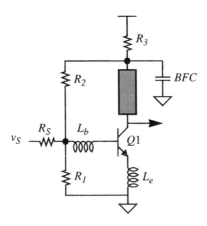

FIGURE 13.9. Narrowband LNA biased by V_{BE} multiplier

intermediate position along the line. Clearly, the impedance is a minimum (zero) at the V_{CC} end of the line and a maximum at the collector end. To a first (and crude) approximation, the impedance varies quadratically along the line.

The sharpness of the resonance can be adjusted by varying the width of the line. The width controls the L/C ratio of the line, and therefore controls Q. The other attractive attribute of the line is that it makes biasing relatively simple, as will be seen in the next section.

13.5.3 BIASING

There are numerous ways to bias a single-ended amplifier at low frequencies. Our options narrow somewhat at microwave frequencies because we cannot always tolerate the impedances that necessarily attend discrete implementations of bias networks. For example, it is very common at lower frequencies to bias the base through a voltage divider and then insert a stabilizing emitter degeneration resistor. To buy back signal gain, a bypass capacitor is placed across this resistor.

In our case, it is probably not practical to use this approach because any "junk" in the emitter circuit only makes our job of implementing tiny inductances tougher. However, since the goal of resistive emitter degeneration is to reduce DC gain through negative feedback, we can seek alternative ways of accomplishing the same net goal. We can apply negative feedback to the base from the collector; as shown in Figure 13.9.

The details of operation are left as an exercise for the reader, but a quick qualitative description is that the DC voltage across R_1 is V_{BE}. If we may neglect base current, then the current through R_1 and R_2 are the same. Consequently the voltage across R_2 is a multiplied-up version of the voltage across R_1 and thus a multiple of V_{BE}. The DC output voltage is therefore

$$V_{OUT} = V_{BE}(1 + R_2/R_1). \tag{35}$$

Since V_{BE} is temperature sensitive, so is the output voltage. However, the variation is small enough for our purposes that it is still a useful circuit.

The collector load resistor, R_3, is bypassed by a capacitor so that the top of the microstrip load will remain a reasonable signal ground. This bypassing need not be perfect, however, because additional inductance here only forces us to shorten the load a little bit. Moving the bias feedback takeoff point from the emitter to the collector thus solves a thorny problem.

As a final note on this bias method, the resistors have to be chosen small enough so that the current flowing through them is large compared with *variations* in transistor base current, if the bias point is to remain roughly insensitive to base current. This requirement is somewhat at odds with the desire to keep the resistors large in order to minimize their contribution to thermal noise. Fortunately, it is usually not difficult to find an acceptable compromise, and net degradations in noise figure can be kept to the level of tenths of a decibel or less.

13.6 LINEARITY AND LARGE-SIGNAL PERFORMANCE

In addition to noise figure, gain, and input match, linearity is also an important consideration because an LNA must do more than simply amplify signals without adding much noise. It must also remain linear even when strong signals are being received. In particular, the LNA must maintain linear operation when receiving a weak signal in the presence of a strong interfering one, for otherwise a variety of pathologies may result. These consequences of intermodulation distortion include desensitization (also known as blocking) and cross-modulation. Blocking occurs when the intermodulation products caused by the strong interferer swamp out the desired weak signal, while cross-modulation results when nonlinear interaction transfers the modulation of one signal to the carrier of another. Both effects are undesirable, of course, so another responsibility of the LNA designer is to mitigate these problems to the maximum practical extent.

The LNA design procedure described in this chapter does not address linearity directly, so we now develop some methods for evaluating the large-signal performance of amplifiers, with a focus on the acquisition of design insight. As we'll see, although the narrowband LNA topology achieves its good noise performance somewhat at the expense of linearity, the trade-off is not serious enough to prevent the realization of LNAs with more than enough dynamic range to satisfy demanding applications.

While there are many measures of linearity, the most commonly used are third-order intercept (IP3) and 1-dB compression point (P_{1dB}).[8] Third-order intercept was first proposed around 1964 as a linearity measure at Avantek. To relate these measures

---

[8] In direct-conversion (homodyne) receivers, the second-order intercept is more important.

to readily calculated circuit and device parameters, suppose that the amplifier's output signal may be represented by a power series.[9] Furthermore, assume that we will evaluate these measures with signals small enough that truncating the series after the cubic term introduces negligible error:

$$i(V_{DC} + v) \approx c_0 + c_1 v + c_2 v^2 + c_3 v^3; \qquad (36)$$

this expression describes the specific case of a transconductance.

Now consider two equal-amplitude sinusoidal input signals of slightly different frequencies:

$$v = A[\cos(\omega_1 t) + \cos(\omega_2 t)]. \qquad (37)$$

Substituting Eqn. 37 into Eqn. 36 allows us, after simplification and collection of terms, to identify the components of the output spectrum.[10] The DC and fundamental components are as follows:

$$[c_0 + c_2 A^2] + [c_1 A + \tfrac{9}{4} c_3 A^3][\cos(\omega_1 t) + \cos(\omega_2 t)]. \qquad (38)$$

Note that the quadratic factor in the expansion contributes a DC term that adds to the output bias. The cubic factor augments the fundamental term, but by a factor proportional to the cube of the amplitude, and thus contributes more than a simple increase in gain. In general, DC shifts come from even powers in the series expansion, while fundamental terms come from odd factors.

There are also second- and third-harmonic terms, which result from the quadratic and cubic factors in the series expansion, respectively:

$$\left[\frac{c_2 A^2}{2}\right][\cos(2\omega_1 t) + \cos(2\omega_2 t)] + \left[\frac{c_3 A^3}{4}\right][\cos(3\omega_1 t) + \cos(3\omega_2 t)]. \qquad (39)$$

In general, nth harmonics come from nth-order factors. Harmonic distortion products, being of much higher frequencies than the fundamental, are usually attenuated enough in tuned amplifiers so that other nonlinear products dominate.

The quadratic term also contributes a second-order intermodulation (IM) product, as in a mixer:

$$\left[\frac{c_2 A^2}{2}\right][\cos(\omega_1 + \omega_2)t + \cos(\omega_1 - \omega_2)t]. \qquad (40)$$

As with the harmonic distortion products, these sum and difference frequency terms are effectively attenuated in narrowband amplifiers if ω_1 and ω_2 are nearly equal, as assumed here.

---

[9] We are also assuming that input and output are related through an anhysteretic (memoryless) process. A more accurate method would employ Volterra series, for example, but the resulting complexity obscures much of the design insight we are seeking.

[10] This derivation makes considerable use of the following trigonometric identity: $(\cos x)(\cos y) = [\cos(x + y) + \cos(x - y)]/2$.

13.6 LINEARITY AND LARGE-SIGNAL PERFORMANCE

Finally, the cubic term gives rise to third-order intermodulation products:

$$\left(\tfrac{3}{4}c_3 A^3\right)[\cos(\omega_1 + 2\omega_2)t + \cos(\omega_1 - 2\omega_2)t \\ + \cos(2\omega_1 + \omega_2)t + \cos(2\omega_1 - \omega_2)t]. \tag{41}$$

Note that these products grow as the cube of the drive amplitude. In general, the amplitude of an nth-order IM product is proportional to the nth power of the drive amplitude.

The sum frequency third-order IM terms are of diminished importance in tuned amplifiers because they typically lie far enough out of band to be significantly attenuated. The difference frequency components, however, can be quite troublesome because their frequencies may lie "in band" if ω_1 and ω_2 differ by only a small amount (as would be the case of a signal and an adjacent channel interferer, for example). It is for this reason that the third-order intercept is an important measure of linearity.

From the previous sequence of equations it is straightforward to compute the input-referred third-order intercept (IIP3) by setting the amplitude of the IM3 products equal to the amplitude of the linear fundamental term:

$$|c_1 A| = \left|\frac{3}{4}c_3 A^3\right| \implies A^2 = \frac{4}{3}\left|\frac{c_1}{c_3}\right|, \tag{42}$$

where we have assumed only a weak departure from linearity in expressing the fundamental output amplitude. It is important to emphasize that the intercept is an extrapolated value because the corresponding amplitudes computed from Eqn. 42 are almost always so large that truncating the series after the third-order term introduces significant error. In both simulations and experiment, the intercept is evaluated by extrapolating trends observed with relatively small amplitude inputs. At such small inputs, the fundamental terms contributed by higher-order nonlinearities will be negligible.

Since Eqn. 42 yields the square of the voltage amplitude, it follows that dividing by twice the input resistance R_s gives us the power at which the extrapolated equality of IM3 and fundamental terms occurs:

$$\text{IIP3} = \frac{2}{3}\left|\frac{c_1}{c_3}\right|\frac{1}{R_s}. \tag{43}$$

We see that IIP3 is proportional to the ratio of the first and third derivatives of the transfer characteristic, evaluated at the bias point. Equivalently, it is proportional to the ratio of the small-signal gain to the second derivative of that gain (again evaluated at the bias point).

Figure 13.10 summarizes the linearity definitions. In this figure, it is customary to plot the output powers as a function of the power of each of the two (equal-amplitude) input tones, rather than their sum.

Since third-order products grow as the cube of the drive amplitude, they have a slope that is three times that of the first-order output when plotted on logarithmic

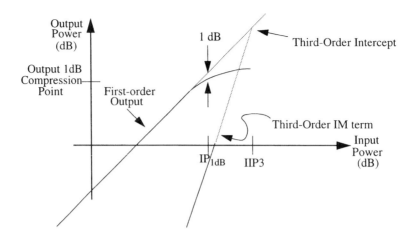

FIGURE 13.10. Illustration of LNA performance parameters

scales, as in the figure. Note that, in Figure 13.10, the 1-dB compression point occurs at a lower input power than IIP3. This general relationship is nearly always the case (by a healthy margin) in practical amplifiers.

Having defined the linearity measures, we now consider ways to estimate IIP3 – with and without the aid of Eqn. 43.

METHODS FOR ESTIMATING IP3

One way to find IP3 is through a transient simulation in which two sinusoidal input signals of equal amplitude and nearly equal frequency drive the amplifier. The third-order intermodulation products of the output spectrum are compared with the fundamental term as the input amplitude varies, and then the intercept is computed.

While simple in principle, there are several significant practical difficulties with the method. First, since the distortion products may be several orders of magnitude smaller than the fundamental terms, numerical noise of the simulator can easily dominate the output unless exceptionally tight tolerances are imposed.[11] A closely related consideration is that the time steps must be *equally spaced* and small enough not to introduce artifacts in the output spectrum.[12] When these conditions are satisfied, the simulations typically execute quite slowly and generate large output files. Pure frequency-domain simulators (e.g., harmonic balance tools) can compute IP3 in much less time, but they are currently less widely available than time-domain simulators such as *Spice*.

---

[11] The tolerances must be *much* tighter, in fact, than the "accurate" default options commonly offered.
[12] This requirement stems from the assumption, made by all FFT algorithms used by practical simulators, that the time samples are uniformly spaced.

Equation 43 offers a simple expression for the third-order intercept in terms of the ratio of two of the power series coefficients, and thus it suggests an alternative method that might be suitable for hand calculations. Though one is rarely given these coefficients directly, it is a straightforward matter to determine them if an analytical expression for the transfer characteristic is available. Even without such an expression, there is an extremely simple procedure, easily implemented in "ordinary" simulators such as *Spice,* that allows rapid estimation of IP3. This technique, which we'll call the three-point method, exploits the fact that knowing the incremental gain at three different input amplitudes is sufficient to determine the three coefficients c_1, c_2, and c_3.[13]

To derive the three-point method, start with the series expansion that relates input and output:

$$i(V_{DC} + v) \approx c_0 + c_1 v + c_2 v^2 + c_3 v^3. \tag{44}$$

The incremental gain (transconductance) is the derivative of Eqn. 36:

$$g(v) \approx c_1 + 2c_2 v + 3c_3 v^2. \tag{45}$$

Although any three different values of v would suffice in principle, particularly convenient ones are 0, V and $-V$, where these voltages are interpreted as deviations from the DC bias value. With those choices, one obtains the following expressions for the corresponding incremental gains:

$$g(0) \approx c_1; \tag{46}$$

$$g(V) \approx c_1 + 2c_2 V + 3c_3 V^2, \tag{47}$$

$$g(-V) \approx c_1 - 2c_2 V + 3c_3 V^2. \tag{48}$$

Solving for the coefficients yields

$$c_1 = g(0), \tag{49}$$

$$c_2 = \frac{g(V) - g(-V)}{4V}, \tag{50}$$

$$c_3 = \frac{g(V) + g(-V) - 2g(0)}{6V^2}. \tag{51}$$

Substituting into Eqn. 43 these last three equations for the coefficients then gives us the desired expression for IIP3 in terms of the three incremental gains:[14]

---

[13] This method is an adaptation of a classic technique from the vacuum tube era that allows estimation of *harmonic* distortion.

[14] Having determined all of the coefficients in terms of readily measured gains, it is easy to derive similar expressions for harmonic and second-order IM distortion. The latter quantity is especially relevant for direct-conversion receivers.

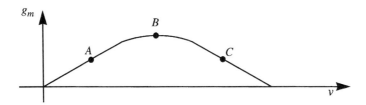

FIGURE 13.11. Hypothetical amplifier g_m

$$\text{IIP3} = \frac{4V^2}{R_s} \cdot \left| \frac{g(0)}{g(V) + g(-V) - 2g(0)} \right|. \quad (52)$$

Finding IIP3 with Eqn. 52 is much faster than via a transient simulation because determining the incremental gains involves so little computation for either a simulator or a human. The three-point method is thus particularly valuable for rapidly estimating IIP3 in the early stages of a design. It is also valuable in guiding the selection of a bias point (or circuit technique) that will maximize IP3. Note from Eqn. 52 that, if we were able to choose a bias point or otherwise arrange the circuit so that the small-signal gain at the bias point equals the *average* of the gains evaluated at V, then IIP3 would actually be infinite. That is, if the second derivative of the gain vanishes at the bias point, IIP3 will go to infinity. If, for example, the small-signal transconductance of a device varies with bias point as in Figure 13.11, then bias points A or C would maximize IIP3. In summary, if maximizing IIP3 is the goal, then one should bias the amplifier to the middle of where the small-signal gain varies linearly with input amplitude.[15]

13.7 SPURIOUS-FREE DYNAMIC RANGE

So far, we have identified two general limits on allowable input signal amplitudes. The noise figure defines a lower bound, while distortion sets an upper bound. Loosely speaking, then, amplifiers can accommodate signals ranging from the noise floor to some linearity limit. Using a dynamic range measure helps designers avoid the pitfall of improving one parameter (e.g., noise figure) while inadvertently degrading another.

This idea has been put on a quantitative basis through a parameter known as the spurious-free dynamic range (SFDR). The term "spurious" means "undesired" and is often shortened to "spur."[16] In the context of LNAs, it usually refers to the third-order products but may occasionally apply to other undesired output spectral components.

---

[15] The corresponding derivation of a two-point method for IP2 estimation reveals that one should bias an amplifier where the first derivative of the small-signal gain is zero – assuming the goal is to maximize IP2. In general, biasing an amplifier for a zero nth derivative of small-signal gain maximizes IP$(n+1)$.

[16] Occasionally (and erroneously), "spurii" is used for the plural of "spurious," even though "spurious" is not a Latin word. "Spurs" is the preferred plural.

13.7 SPURIOUS-FREE DYNAMIC RANGE

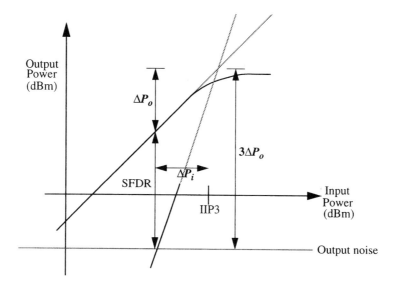

FIGURE 13.12. Spurious-free dynamic range (third-order)

To understand the rationale behind using SFDR as a specific measure of dynamic range, define as a more general measure the lesser of signal-to-noise or signal-to-distortion ratio, and evaluate this measure as one varies the amplitude of the two tones applied to the amplifier. As the input amplitude increases from zero, the first-order output initially has a subunity SNR but eventually emerges from the noise floor. Because third-order distortion depends on the cube of the input amplitude, IM3 products will be well below the noise floor at this point for any practical amplifier. Hence, the dynamic range improves for a while as the input signal continues to increase, since the desired output increases while the undesired output (here, the noise) stays fixed. Eventually, however, the third-order IM terms also emerge from the noise floor. Beyond that input level, the dynamic range decreases, since the IM3 terms grow three times as fast (on a dB basis) as the first-order output.

The SFDR is defined as the signal-to-noise ratio corresponding to the input amplitude at which an undesired product (here, the third-order IM power) just equals the noise power, and is therefore the maximum dynamic range that an amplifier exhibits in the preceding paragraph, as is clear from Figure 13.12.

To incorporate explicitly the noise figure and IIP3 in an expression for SFDR, first define N_{oi} as the input-referred noise power in decibels. Then, since the third-order IM products have a slope of 3 on a decibel scale, the input power below IIP3 at which the input-referred IM3 power equals N_o is given by

$$\Delta P_i = \frac{\text{OIP3} - N_o}{3} \tag{53}$$

(again, all powers are expressed in decibels). The SFDR is just the difference between the output power implied by Eqn. 53 and N_o:

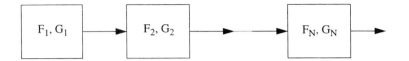

FIGURE 13.13. Cascaded systems for noise figure computation

$$\text{SFDR} = [\text{OIP3} - \Delta P_o] - N_o. \qquad (54)$$

Because the third-order IM products have a slope of 3, we know that

$$3\Delta P_o = \text{OIP3} - N_o. \qquad (55)$$

Hence,

$$\text{SFDR} = \tfrac{2}{3}[\text{OIP3} - N_o] = \tfrac{2}{3}[\text{IIP3} - N_{oi}]. \qquad (56)$$

Observe that the input-referred noise power (in watts this time) is simply the noise factor F multiplied by the noise power $kT\Delta f$. Note also that input-referred quantities may be used in Eqn. 54 because the same gain factor scales both terms.

It is satisfying that SFDR is indeed bounded on one end by IIP3 and on the other by the noise floor, as argued qualitatively at the beginning of this section. The factor 2/3 comes into play because of the particular way in which the limits are defined.

13.8 CASCADED SYSTEMS

The overall noise figure and dynamic range of a cascade of systems depends on the individual noise figures, intercepts, and gains. The dependency on the gain results from the fact that, once the signal has been amplified, the noise of subsequent stages is less important. As a result, system noise figure tends to be dominated by the noise performance of the first couple of stages in a receiver.

How the individual noise figures combine to yield the overall noise figure is complicated by the variety of impedance levels typically found in the system. To develop an equation for the system noise figure, consider the block diagram of Figure 13.13. Here, each F_n is a noise factor and each G_n is a power gain (specifically, the *available gain*, the gain one would obtain with a matched load). Since noise factor depends on source resistance, one must compute the individual noise figures relative to the output impedance of the preceding stage to keep the calculation honest. This issue arises less frequently in discrete designs, where impedance levels are often standardized, but it requires careful attention in IC implementations.

Noise factor may be expressed in several ways, but one form that is particularly useful for the task at hand is

$$F = \frac{R_S + R_e}{R_S} = 1 + N_e, \qquad (57)$$

where R_e is a (possibly) fictitious resistance that accounts for the observed noise in excess of that due to R_S. The quantity N_e is thus an excess noise power ratio, equal to $F - 1$.

Reflecting this power ratio back to the input of the preceding stage simply involves a division by the available power gain of that preceding stage. Reflecting the excess noise contribution of a given stage all the way back to the input thus requires division by the total available gain between that stage and the overall input. The total noise factor is the sum of these individual contributions and is therefore given by the following expression:

$$F = 1 + F_1 - 1 + \frac{F_2 - 1}{G_1} + \frac{F_3 - 1}{G_1 G_2} + \cdots + \frac{F_N - 1}{\prod_{n=1}^{N-1} G_n}, \quad (58)$$

which simplifies to

$$F = F_1 + \frac{F_2 - 1}{G_1} + \frac{F_3 - 1}{G_1 G_2} + \cdots + \frac{F_N - 1}{\prod_{n=1}^{N-1} G_n}. \quad (59)$$

In terms of noise temperature, the cascaded result may be expressed as

$$T_{e12} = T_{e1} + \frac{T_{e2}}{G_{av1}} + \frac{T_{e3}}{G_{av1} G_{av2}} + \cdots. \quad (60)$$

From this last expression, it's easy to see why noise temperature is favored for cascade noise calculations: the overall noise temperature is merely the sum of all the input-referred noise temperatures. The same statement does not apply to noise figure.

It is clear that the system noise figure is in fact dominated by the noise performance of the first few gain stages. Hence, in trying to achieve a good noise figure, most of the design effort will generally focus on the first few stages.

LINEARITY OF CASCADED SYSTEMS

The other figure of merit that bounds system dynamic range is the intercept point. Even though we have discussed only third-order intercepts, it should be mentioned that there are also instances in which the second-order intercept is a relevant linearity measure. A notable example is the degenerate case of a superheterodyne in which the IF is zero. Such *direct-conversion* receivers have become increasingly common in recent years.

A difficulty in developing the desired equation is that the distortion products of one stage combine with those of a later stage in ways that depend on their relative phases. Hence, there is no simple, fixed relationship between the individual and overall intercepts. However, it is possible to derive a conservative (worst-case) estimate by assuming that the amplitudes of the distortion products add directly. This choice in turn makes it most natural to express the gains as voltage ratios, in contrast with the use of power gains in the expression for system noise figure. This is shown in Figure 13.14, where each A_{vn} is a voltage gain and each $IIVM_n$ is an Mth-order input intercept voltage.

To facilitate the derivations, call $V_{dM,n}$ the Mth-order (intermodulation) distortion product at the output of the nth stage due to a voltage V applied to the input of that

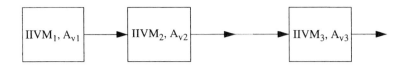

FIGURE 13.14. Cascaded systems for input intercept calculation

stage. Further note from the definition of an input intercept that the input-referred Mth-order IM distortion product may be written as

$$V_{dM} = \frac{V^M}{\text{IIVM}^{M-1}}. \tag{61}$$

Let us carry out the derivation for the specific case of the third-order intercept, and for a cascade of just two stages. The third-order IM at the output of the first stage is

$$V_{d3,1} = \frac{V^3}{\text{IIV3}_1^2}. \tag{62}$$

The third-order IM voltage at the output of the second stage is the sum of two components. One is simply a scaled version of the distortion produced by the first stage, while the other is the distortion produced by the second stage. Adding these directly together yields the following pessimistic estimate:

$$V_{d3,tot} = A_{v2}V_{d3,1} + V_{d3,2}. \tag{63}$$

The input-referred third-order distortion is found by dividing through by the total gain:

$$V_{d3\,in,tot} = \frac{A_{v2}V_{d3,1} + V_{d3,2}}{A_{v1}A_{v2}}. \tag{64}$$

Substituting Eqns. 61 and 62 into Eqn. 64 yields

$$\frac{1}{\text{IIV3}_{tot}^2} = \frac{A_{v1}^2}{\text{IIV3}_1^2} + \frac{(A_{v1}A_{v2})^2}{\text{IIV3}_2^2}. \tag{65}$$

This last equation confirms that the later stages bear a greater burden because of the gain that precedes them. We can also see that the reciprocal IIV3 of a given stage, normalized by the total gain up to the output of that stage, contributes to the overall reciprocal input-referred intercept in root-sum-squared fashion.

Although Eqn. 65 applies strictly to a two-stage cascade, it is readily extended to an arbitrary number of stages as follows:

$$\frac{1}{\text{IIV3}_{tot}^2} = \sum_{j=1}^{n} \left\{ \frac{1}{\text{IIV3}_j^2} \prod_{i=1}^{j} A_{vi}^2 \right\}. \tag{66}$$

One may follow a similar procedure to determine the overall input-referred intercept for distortion products of any order.

Stage Data	Units	Stage 1 Preselector	Stage 2 LNA	Stage 3 Image Filter	Stage 4 Post amp	Stage 5 Mixer	Stage 6 IF Filter	Stage 7 IF amp
Noise Figure	dB	1.5	0.8	3.0	2.0	10	3.0	8
Gain	dB	-1.5	15	-3.0	15	12	-3.0	60
Input IP3	dBm	100	15	100	10	20	100	15
dNF/dTemp	dB/°C	0	0	0	0	0	0	0
dG/dTemp	dB/°C	0	0	0	0	0	0	0
Stage Analysis:								
NF (Temp corr)	dB	1.50	0.80	3.00	2.00	10.00	3.00	8.00
Gain (Temp corr)	dB	-1.50	15.00	-3.00	15.00	12.00	-3.00	60.00
Input Power	dBm	-50.00	-51.50	-36.50	-39.50	-24.50	-12.50	-15.50
Output Power	dBm	-51.50	-36.50	-39.50	-24.50	-12.50	-15.50	44.50
d NF/d NF	dB/dB	0.79	0.94	0.05	0.09	0.02	0.00	0.00
d NF/d Gain	dB/dB	-0.21	-0.06	-0.04	-0.01	0.00	0.00	0.00
d IP3/d IP3	dBm/dBm	0.00	0.00	0.00	0.01	0.03	0.00	0.94

Enter System Parameters:

Input Power	-50	dBm
Analysis Temperature	25	°C
Noise BW	1	MHz
Ref Temperature	25	°C
S/N (for sensitivity)	10	dB
Noise Source (Ref)	290	°K

System Analysis:

Gain =	94.50	dB	Input IP3 =	-19.72	dBm
Noise Figure =	2.61	dB	Output IP3 =	74.78	dBm
Noise Temp =	238.45	°K	Input IM level =	-110.55	dBm
SNR =	61.37	dB	Input IM level =	-60.55	dBC
MDS =	-111.37	dBm	Output IM level =	-16.05	dBm
Sensitivity =	-101.37	dBm	Output IM level =	-60.55	dBC
Noise Floor =	-171.37	dBm/Hz	SFDR =	61.10	dB

FIGURE 13.15. Example of cascade system calculations (*AppCAD*)

Calculations of cascaded noise figure and linearity can get tedious rather quickly, particularly if one is iterating to identify and achieve some optimum. With enough diligent effort, one could always construct a spreadsheet to perform the calculations, and this is precisely what many engineers have done for years. Fortunately, numerous tools that automate the process are now freely available. A popular one is *AppCAD* from Agilent, which can generate examples such as seen in Figure 13.15.

Analyses such as this are invaluable not only for evaluating performance, but also for identifying the performance-limiting blocks in a system. As a convenience, *AppCAD* highlights entries corresponding to the stages that most strongly control the overall system noise figure and linearity. In the example shown, we see that the input preselector filter and LNA have the most influence over noise figure (as expected), while the linearity of the IF amplifier determines the overall system linearity to a large extent (again, as expected). If we were to find the system's performance deficient in either of these dimensions, we would know where our design effort would be most profitably spent.

13.9 SUMMARY

We've seen that an inductively degenerated LNA achieves simultaneously an excellent impedance match, nearly minimum noise figure, and reasonable gain.

The three-point method was also introduced, permitting an approximate, but quantitative, assessment of linearity more quickly than is possible with straightforward time-domain simulators. Even though the method neglects dynamics, measurements on practical amplifiers usually reveal reasonably good agreement with predictions.

Reasonable agreement may generally be expected as long as the device is operated well below ω_T.

If better linearity is required, either power consumption or gain must degrade in exchange for the improved linearity. For example, the bias conditions can be altered to decrease input Q, or negative feedback can be employed. Combining the signal amplitude limitations implied by the noise and distortion figures of merit yields a measure of the maximum dynamic range of an amplifier, the spurious-free dynamic range.

Finally, we examined ways to compute the noise figure and estimate the intercept point of a cascade of systems.

13.10 APPENDIX A: BIPOLAR NOISE FIGURE EQUATIONS

Repeated here are the equations for optimum source resistance (both "exact" and approximate), and the corresponding noise factor:

$$R_S^2 = \frac{2|z_\pi|^2 g_m r_b + r_b^2 + |z_\pi|^2 + 2r_b \operatorname{Re}\{z_\pi\} + g_m^2 r_b^2 |z_\pi|^2/\beta_F}{1 + |z_\pi|^2 g_m^2/\beta_F}, \quad (67)$$

$$R_S^2 \approx \frac{\left(\frac{\omega_T}{\omega}\right)^2 \left(\frac{2r_b}{g_m} + \frac{r_b^2}{\beta_F} + \frac{1}{g_m^2}\right) + r_b^2 + \frac{2r_b r_\pi}{(\omega/\omega_T)^2 \beta_F^2}}{1 + \frac{(\omega_T/\omega)^2}{\beta_F}}; \quad (68)$$

$$F = 1 + \frac{r_b}{R_S} + \frac{(R_S + r_b)^2 + |z_\pi|^2 + 2(R_S + r_b)\operatorname{Re}\{z_\pi\}}{2 R_S |z_\pi|^2 g_m} + \frac{(r_b + R_S)^2 g_m}{2\beta_F R_S}. \quad (69)$$

It might also be helpful to have the following expressions related to the impedance z_π:

$$|z_\pi| = r_\pi/\sqrt{(\omega r_\pi C_\pi)^2 + 1}; \quad (70)$$

$$\operatorname{Re}\{z_\pi\} = r_\pi/(\omega r_\pi C_\pi)^2 + 1. \quad (71)$$

13.11 APPENDIX B: FET NOISE PARAMETERS

13.11.1 THEORY

The basic noise model for a MOSFET, JFET, and MESFET consists of two intrinsic sources. We will initially neglect extrinsic noise sources, such as those associated with the lossiness of gate electrode material. See Figure 13.16.

One intrinsic noise current generator is tied from source to drain and has a mean-square value given by

$$\overline{i_{nd}^2} = 4kT\gamma g_{d0}\Delta f, \quad (72)$$

where γ is theoretically 2/3 and g_{d0} is the drain–source conductance at zero V_{DS}.

13.11 APPENDIX B: FET NOISE PARAMETERS

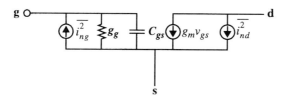

FIGURE 13.16. MOS noise model

Table 13.2. *Summary of FET two-port noise parameters*

Parameter	Expression
G_c	~ 0
B_c	$\omega C_{gs}\left(1 - \alpha|c|\sqrt{\delta/5\gamma}\right)$
R_n	$\dfrac{\gamma g_{d0}}{g_m^2} = \dfrac{\gamma}{\alpha} \cdot \dfrac{1}{g_m}$
G_u	$\dfrac{\delta \omega^2 C_{gs}^2 (1 - |c|^2)}{5 g_{d0}}$

There is also a gate current noise, modeled by a current source tied from gate to source, whose mean-square value is

$$\overline{i_{ng}^2} = 4kT\delta g_g \Delta f, \tag{73}$$

where

$$g_g = \omega^2 C_{gs}^2 / 5 g_{d0} \tag{74}$$

and δ is theoretically $4/3$.

Moreover, the gate noise is correlated with the drain noise, with a correlation coefficient defined formally as

$$c \equiv \overline{i_{ng} \cdot i_{nd}^*} / \sqrt{\overline{i_{ng}^2} \cdot \overline{i_{nd}^2}}. \tag{75}$$

The long-channel value of c is theoretically $j0.395$. We will also neglect C_{gd} in order to simplify the derivation. Wheres the achievable noise figure is little affected by C_{gd}, the input impedance can be a strong function of C_{gd}, and this effect must be taken into account when designing the input matching network.

We will omit a derivation of the noise parameters and simply summarize the results in Table 13.2, where α is the ratio g_m/g_{d0}.

With these parameters, we can determine both the source impedance that minimizes the noise figure as well as the minimum noise figure itself:

$$B_{opt} = -B_c = -\omega C_{gs}\left(1 - \alpha|c|\sqrt{\delta/5\gamma}\right). \tag{76}$$

From Eqn. 76, we see that the optimum source susceptance is essentially inductive in character, except that it has the wrong frequency behavior. Hence, achieving a broadband noise match is fundamentally difficult.

Continuing, the real part of the optimum source admittance is

$$G_{opt} = \sqrt{G_u/R_n + G_c^2} = \alpha\omega C_{gs}\sqrt{(\delta/5\gamma)(1 - |c|^2)}, \tag{77}$$

and the minimum noise figure is given by

$$F_{min} = 1 + 2R_n[G_{opt} + G_c] \approx 1 + \frac{2}{\sqrt{5}}\frac{\omega}{\omega_T}\sqrt{\gamma\delta(1 - |c|^2)}. \tag{78}$$

In this last expression, the approximation is exact if one treats ω_T as simply the ratio of g_m to C_{gs}. Note that if there were no gate current noise (i.e., if δ were zero) then the minimum noise figure would be 0 dB. That unrealistic prediction alone should be enough to suspect that gate noise must indeed exist. Also note that, in principle, increasing the correlation between drain and gate current noise would improve noise figure, although correlation coefficients unrealistically near unity would be required to effect large reductions in noise figure.

The foregoing development ignores several important factors. One is the thermal noise associated with non-superconducting gate electrode material. Although careful device layout by the manufacturer can keep this value small, it often significantly raises the minimum noise figure. We've also neglected the thermal noise associated with the substrate. At microwave frequencies, this component is indeed negligible for typical MOSFETs, but one must verify whether this is the case in any specific instance where it matters.

13.11.2 PRACTICE

The model parameters necessary to compute noise accurately are not always given to us. If design for low noise is to proceed nonetheless, we must develop an alternative approach to obtaining model parameters. A purely empirical method has the appeal of relying on no theoretical assumptions at all, but its practical implementation requires the design of a suitable, finite experimental suite. Fortunately, the very dimensions of the two-port noise parameters themselves suggest a compact method for obtaining data suitable for use over a wide range of conditions – and suitable also for the extraction of relevant noise parameters for other simulation models.

We've already noted that the two-port noise parameters are impedances or admittances. Even though they do not necessarily represent physical quantities that are directly measurable, they nonetheless obey the same scaling laws. For example, suppose we completely characterize the noise parameters for one MOSFET of some unit size. For constant bias voltages (implying constant current densities), the impedance parameters for two such devices in parallel are half that for a single one while the admittance parameters are double. Dimensionless parameters remain unchanged. Generalizing, the noise parameters for a device of width W are

$$G_c = \frac{W}{W_0} G_{c0}, \tag{79}$$

$$B_c = \frac{W}{W_0} B_{c0}, \tag{80}$$

$$G_u = \frac{W}{W_0} G_{u0}, \tag{81}$$

$$R_n = \frac{W_0}{W} R_{n0}. \tag{82}$$

The subscript "0" identifies a parameter corresponding to a unit device of width W_0. The noise factor, F, is a dimensionless quantity and is therefore independent of width (though the source admittance that produces a given F scales with width).

These scaling laws mean that one need only characterize the noise parameters for a single device size (which we'll call a unit cell). Although this characterization must include variation of bias over some range, its elimination of the need to also sweep device width is of self-evident value. The necessary parameters may be extracted from contours of constant noise figure, for example. Extrapolations based on these scaling relationships will be accurate as long as one takes care to select a reference device large enough to make negligible any fixturing and layout parasitics, and as long as one truly uses multiple instances of this unit cell in the final design, which is well-established practice for good matching in analog design.

CHAPTER FOURTEEN

NOISE FIGURE MEASUREMENT

14.1 INTRODUCTION

One of the most important performance metrics for low-level amplifiers is noise figure, NF, or noise factor, F. The two terms are used interchangeably in the literature, but we adopt the following arbitrary convention in this text: We will denote noise figure by NF, and define it as $10 \log F$. We will be somewhat sloppy about using the terms (reflecting common usage), but context should make clear whether or not the decibel version is being discussed.

The definition of noise factor now in use was first formally proposed by Harald Friis[1] of Bell Labs. At its core, the definition involves signal-to-noise ratios (SNRs):

$$F \equiv \frac{\text{SNR}_i}{\text{SNR}_o}. \qquad (1)$$

This definition shows that F is the factor by which an amplifier degrades the signal-to-noise ratio of the input signal. As such, it is never smaller than unity. As simple and straightforward as the definition appears to be, numerous subtleties are buried in it, and it will soon be clear that we have provided an incomplete definition. Accurate measurement of noise figure depends on a full appreciation of all of these subtleties as well as an understanding of how to identify and correct sources of measurement error. As we'll soon see, automated noise figure instruments do not eliminate the need for a knowledgeable operator. As has been noted, "automated equipment merely lets you produce more wrong answers per unit time." The purpose of this chapter is to reduce the rate of erroneous answer generation.

14.2 BASIC DEFINITIONS AND NOISE MEASUREMENT THEORY

One important subtlety concerns the temperature at which the measurement of noise figure is made. Specifically, the temperature of the source has a profound effect on

---

[1] "Noise Figures of Radio Receivers," *Proc. IRE,* July 1944, pp. 419–22.

14.2 BASIC DEFINITIONS AND NOISE MEASUREMENT THEORY

the noise figure. Intuitively, this temperature dependence may be understood as follows. The device under test (DUT) generates its own internal noise, independent of the source temperature. If the latter is very low, then the source noise will be correspondingly low, so the noise added by the DUT will have a comparatively greater effect. The measured noise figure will thus be higher than if the source were hotter (and thus noisier). Because of this sensitivity, a meaningful comparison of noise figures requires that the measurements be made at a standard temperature. Friis proposed a reference temperature, denoted T_0, of 290 kelvins (about 62°F or 17°C), a temperature which is considerably cooler than the interior of most laboratories. An oft-cited reason for this choice is the approximate equality of this temperature with that commonly seen by antennas used in terrestrial wireless communications. However, perhaps a stronger motivation for its selection is simply that kT_0 is then 4.00×10^{-21} J, a round number with undeniable appeal in an era of slide-rule computation, particularly to an eminently practical gentleman like Friis.

The final statement on standard conditions, made by a committee of the Institute of Radio Engineers (IRE, a forerunner of the IEEE), is that the noise figure measurement is to be made with a source whose *available* noise power is the same as that of an input termination whose temperature is 290 K. Recall that available power is defined as the power that could be delivered to a (conjugately) matched load. Hence, even if the source does not in fact happen to drive a matched load, the power remains *available*. Available power is precisely what the words imply: a potential, independent of the actual load. The standards committee accepted Friis's recommendation for basing noise figure on available power, because this parameter can be related directly to the temperature of a thermal noise generator, such as a resistor. Confusion about this definition is all too common and can lead to serious errors, as will be made clear later in this chapter.

A second consideration is that determining input and output signal-to-noise ratios is by no means trivial.[2] Since noise figure is an intrinsic property of the DUT alone (assuming linearity, as we must if noise figure is to be uniquely definable) and thus not of how you drive the DUT, it should be possible to devise a measurement that does not involve the use of an explicit signal. To do so, it is helpful to note that the available noise appearing at the output of the DUT results from two contributions. One is the amplified available source noise power (with the source at $T_0 = 290$ K), which has a value

$$N_{os} = kT_0 B G_{av}, \qquad (2)$$

where B is the noise (brickwall) bandwidth and G_{av} is the available power gain of the DUT.

The other component of output noise is simply the noise added by the DUT itself. We call this noise contribution N_a. The total available output noise power is therefore

---

[2] A third subtlety arises in cases where the system contains frequency-translating elements such as mixers. We defer a discussion of this consideration to Section 14.8.

$$N_1 = kT_0 BG_{av} + N_a. \tag{3}$$

Now let's revisit, and revise, the noise figure definition of Eqn. 1:

$$F \equiv \frac{\text{SNR}_i}{\text{SNR}_o} = \frac{S_i/N_i}{S_o/N_o}. \tag{4}$$

This quantity must be evaluated at 290 K, as stated earlier.

Interpreting all quantities as available powers, the ratio of output signal S_o to input signal S_i is the available gain, G_{av}. The available input noise power is simply $kT_0 B$, and the available output noise power is N_1 as defined in Eqn. 3. So we may write

$$F = \frac{S_i/N_i}{S_o/N_o} = \frac{1}{G_{av}}\left(\frac{N_o}{N_i}\right) = \frac{1}{G_{av}}\left(\frac{N_1}{N_i}\right) = \frac{N_1}{N_{os}} = \frac{kT_0 BG_{av} + N_a}{kT_0 BG_{av}}. \tag{5}$$

The last expression on the right,

$$F = \frac{kT_0 BG_{av} + N_a}{kT_0 BG_{av}}, \tag{6}$$

is the definition officially adopted by the IRE.[3] It initially appears more attractive as a basis for measurement than Eqn. 1 because it contains no terms related to an explicit input or output signal. Using Eqn. 6, measurement of noise figure reduces to the measurement of noise, available gain, and bandwidth. Unfortunately, there are still serious practical difficulties associated with trying to base a measurement directly on this equation. In particular, it is not easy to measure the product of the effective noise bandwidth and available gain, BG_{av}, with high accuracy. The experimental difficulties are perhaps best appreciated after comparing the various noise measurement methods discussed in Section 14.6.

One of these alternative noise figure evaluation methods, which is implemented in commercial instruments such as the HP8970A, cleverly sidesteps the need to make gain–bandwidth measurements by employing a ratio of noise measurements performed at two different source temperatures. As a general philosophy, it is always advantageous to replace absolute measurements with ratiometric ones wherever dimensional considerations permit it. Fortunately noise factor is a dimensionless quantity, so a purely ratiometric measurement is possible. Gain–bandwidth product is not dimensionless, so measuring it should not be fundamentally necessary here.

The basis for the ratiometric technique is that the use of a hot source increases the component of output noise due to the source – without changing the noise added by the DUT. If the ratio of the source temperatures is accurately known, then measuring the output noise powers under the hot and cold conditions permits us to solve for the noise added by the DUT and, hence, compute the noise figure.

Figure 14.1 plots output noise power as a function of source temperature, illustrating how such a ratiometric measurement solves our problem. Comparing features of

---

[3] See *Proc. IRE*, v. 51, no. 3, March 1963, pp. 434–42.

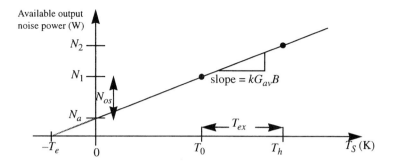

FIGURE 14.1. Output noise power vs. source temperature

this drawing with Eqn. 6, note that the slope and y-intercept tell us everything we need to compute F:

$$F = \frac{kT_0 BG_{av} + N_a}{kT_0 BG_{av}} = 1 + \frac{N_a}{kT_0 BG_{av}} = 1 + \frac{y\text{-intercept}}{(T_0)(\text{slope})}. \quad (7)$$

Clearly, the need to measure gain and bandwidth has disappeared because two points determine a line. In spite of the simplicity seemingly implied by this observation, engineers have devised a surprising number of different ways to use noise data from two points to determine noise figure. Just keep in mind that underlying the apparent complexity in what follows is the extremely simple geometric picture of Figure 14.1.

If we make a noise power measurement at a source temperature, T_h, that is above the reference temperature by an amount T_{ex}, then the available output noise power becomes

$$N_2 = kBG_{av}T_h + N_a = kBG_{av}(T_0 + T_{ex}) + N_a. \quad (8)$$

After combining the hot measurement with the one at T_0 (Eqn. 3), a little algebra allows us to find that the noise factor may be expressed as

$$F = \frac{T_{ex}/T_0}{N_2/N_1 - 1}. \quad (9)$$

The ratio N_2/N_1 is often called the "Y-factor" in the literature (why? because it comes after X...). Figure 14.1 shows a cold temperature equal to the reference temperature, T_0, but it should be clear that any temperature other than T_h could be used to figure out the slope and intercept of the line. More generally, if the cold temperature T_c is not T_0 then the numerator changes, so that the noise factor becomes

$$F = \frac{T_{ex}/T_0 - Y(T_c/T_0 - 1)}{Y - 1}. \quad (10)$$

The ratio T_{ex}/T_0 is a property of the noise source, and it is information that's (almost) supplied by the manufacturer. The qualifier "almost" applies because the manufacturer actually specifies a slightly different quantity called the *excess noise*

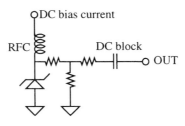

FIGURE 14.2. Typical noise diode

ratio (ENR), which is defined as the ratio of noise powers actually delivered to a 50-Ω load (or occasionally some other standard impedance level). However the ratio T_{ex}/T_0 results from a consideration of *available* powers (as does the Y-factor). The two ratios are equivalent only in the special case where the noise source happens to have an impedance of precisely 50 Ω. Despite the best efforts of manufacturers, this condition is not perfectly satisfied in practice, so substituting ENR for T_{ex}/T_0 is one (generally small) potential source of error. Because it is much easier to determine ENR, however, that's what the NBS (the National Bureau of Standards, now the National Institute for Standards and Technology, NIST) decided to do, and consequently that's what manufacturers measure and report.

In the "old days," actual hot and cold sources were used, commonly with resistors at 77 K (the boiling point of liquid nitrogen) and 373 K (the boiling point of water, although the resistor was electrically heated to this temperature, not immersed in an actual water bath). Clearly, the greater the temperature difference, the more accurately we can compute the slope and intercept for a given magnitude of uncertainty in the power measurement. A limitation on the hot side is the difficulty of accurately determining or controlling the temperature. And the higher the temperature, the more significant the problems of materials properties (e.g., melting).

Nowadays, it is common to use noise diodes[4] (see Chapter 9), which can produce the noise of an exceptionally hot source (e.g., 10,000 K, higher than the melting point of any known metal) while remaining at room temperature. The same diode can provide the cold reference as well, simply by turning it off, causing an internal resistive matching network to provide an available noise power that corresponds to the ambient temperature. See Figure 14.2, where RF choke (RFC) is simply an inductor large enough to be considered an open circuit at all frequencies of interest.

One drawback is that, unlike true hot and cold resistors, such diodes are not fundamental standards; their hot noise cannot be computed from first principles. Since ENR must be known to great accuracy to be useful, it is usually traceable to a primary noise standard (which is a heated or cooled physical resistor) maintained by national laboratories, such as the NIST. This traceability accounts in part for the relatively high cost of noise diodes.[5]

---

[4] See e.g. *HP Journal,* April 1983, p. 26.

[5] The more significant explanation for the high cost, however, is simply that it is what the market will bear.

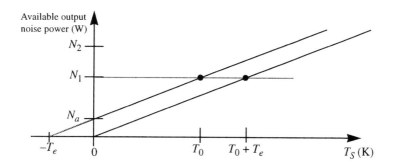

FIGURE 14.3. Noise temperature

14.3 NOISE TEMPERATURE

Noise temperature, T_e, is an alternative figure of merit used in place of noise figure in some cases. As we saw in Figure 14.1, noise temperature is (minus) the extrapolated intercept of the noise power curve with the temperature axis. An intuitively appealing meaning of noise temperature can be extracted by translating the noise power curve to the right by a temperature equal to the noise temperature; see Figure 14.3.

The translated curve is that of a noiseless amplifier (because the noise at zero source temperature is zero) with the same slope (= available gain–bandwidth product, times k) as the original amplifier. As can be seen, this noiseless amplifier produces an available output noise power equal to the available output noise of the original amplifier, if the source is now heated to a temperature $T_0 + T_e$. The increase in available output noise power due to the hotter source is precisely equal to the available noise (N_a) added by the original DUT:

$$N_a = kT_e BG_{av}. \qquad (11)$$

Noise temperature is used in characterizing satellite communications systems for several reasons. One is that objects in the sky generally don't have an effective temperature anywhere near 290 K, so choosing such a reference temperature would have a weaker physical justification. The other is that space communication systems generally have exceptionally low noise figures, and noise temperature is a higher-resolution measure of very low noise figure values. Table 14.1 compares noise figure, noise factor, and noise temperature over a range generally considered to be very low-noise.

It is sometimes helpful to note that, in the very low–noise figure regime (e.g., below about 1 dB), the noise figure in decibels is approximately the noise temperature divided by 70–75. Stated alternatively, each tenth of a decibel corresponds to roughly 7.0–7.5 K.

To relate noise temperature and noise factor, return again to the official IRE noise figure definition:

$$F \equiv \frac{N_1}{N_{os}} = \frac{kT_0 BG_{av} + N_a}{kT_0 BG_{av}}. \qquad (12)$$

Substituting Eqn. 11 for N_a yields

Table 14.1. *Comparison of noise figure, noise factor, and noise temperature*

NF (dB)	F	T_e (kelvins)
0.5	1.122	35.4
0.6	1.148	43.0
0.7	1.175	50.7
0.8	1.202	58.7
0.9	1.230	66.8
1.0	1.259	75.1
1.1	1.288	83.6
1.2	1.318	92.3

$$F = \frac{kT_0 BG_{av} + N_a}{kT_0 BG_{av}} = \frac{kT_0 BG_{av} + kT_e BG_{av}}{kT_0 BG_{av}}, \quad (13)$$

which simplifies to

$$F = 1 + T_e/290. \quad (14)$$

If the noise added by the DUT equals the noise power of the source then the noise figure will be 3 dB, corresponding to a noise temperature of 290 K. Many LNAs with effective noise temperatures well below 100 K (corresponding to noise figures below 1.3 dB) are commercially available.

The noise temperature may be found indirectly by relating Eqn. 14 to Eqn. 9, or directly from the cold and hot noise measurements of Figure 14.1. Pursuing the latter strategy, we may write

$$N_1 = kT_c BG_{av} + N_a = k(T_e + T_c) BG_{av}, \quad (15)$$

$$N_2 = kT_h BG_{av} + N_a = k(T_e + T_h) BG_{av}, \quad (16)$$

so that

$$Y = \frac{N_2}{N_1} = \frac{k(T_e + T_h) BG_{av}}{k(T_e + T_c) BG_{av}}. \quad (17)$$

Solving for T_e yields

$$T_e = \frac{T_h - YT_c}{Y - 1}. \quad (18)$$

The most common reason that noise temperature is used is that the quantity $F - 1$ recurs frequently in calculations of cascaded noise figure, as we shall see in Section 14.4. By rearranging Eqn. 14, it's clear that noise temperature T_e is proportional to $F - 1$, so its use simplifies such calculations considerably.

Because noise figure and noise temperature each fully convey the information of the other (as implied by Eqn. 14, for example), you may use either. The choice of which to use is made largely on the basis of culture and convenience.

14.4 FRIIS'S FORMULA FOR THE NF OF CASCADED SYSTEMS

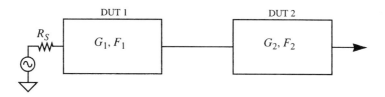

FIGURE 14.4. Cascaded systems

SPOT NOISE FIGURE

In many cases, one is interested in the noise performance of an amplifier as a function of frequency. In those situations, the measurement bandwidth is restricted to some known value (e.g., 4 MHz, as in the HP8970) and then the noise figure for that bandwidth is reported at a specific frequency. Since the parameter is thus a noise figure measured in a narrow band centered around a specific spot, it is known as the spot noise figure. The noise figures most often reported in the literature are spot noise figures.

14.4 FRIIS'S FORMULA FOR THE NOISE FIGURE OF CASCADED SYSTEMS

Computing the noise figure of a cascade of systems is often carried out incorrectly. Once again, the problem is a failure to appreciate certain subtleties. One difficulty is that, unlike gain, individual noise figures do not combine in any simple way to yield the overall cascaded noise figure. Another is that each stage may see a different source impedance, and the noise figure must be computed with respect to that impedance. To understand these and other issues in detail, we now derive the correct equation – called Friis's formula – for the cascaded noise figure.

Consider a noisy system that is driven by yet another noisy system, as shown in Figure 14.4. The first stage has a noise factor F_1 and available power gain G_1 measured with R_S as source resistance. The second stage has an available power gain G_2 and a noise factor F_2 when these quantities are measured *with the output of the previous stage as source resistance*. If there were additional stages, the available gain and noise figure of each one would be determined using the output impedance of the preceding stage as the source resistance. A common error is to use R_S as the source impedance for all stages, but this choice is correct only if the output impedances happen to be R_S. In cascades consisting of discrete modules, that requirement might be satisfied, but it is important not to generalize improperly from that common case.

The easiest way to derive Friis's formula is to use the concept of noise temperature. Because the available output noise power added by each DUT is $kT_e BG_{av}$, the available noise power at the output of the first DUT is

$$N_{o1} = kT_s BG_{av1} + kT_{e1} BG_{av1} = k(T_s + T_{e1}) BG_{av1}. \tag{19}$$

The second stage takes this noise, amplifies it, and adds to it another $kT_e BG_{av}$ of its own:

$$N_{o2} = k(T_s + T_{e1})BG_{av1}G_{av2} + kT_{e2}BG_{av2}. \quad (20)$$

We could just as well regard the overall system as a single amplifier with available gain $G_{av1}G_{av2}$, driven by a source R_s. Hence, we may also write

$$N_{o2} = k(T_s + T_{e12})BG_{av1}G_{av2}, \quad (21)$$

where T_{e12} is the overall noise temperature of the cascade. Combining Eqns. 20 and 21 yields

$$T_{e12} = T_{e1} + \frac{T_{e2}}{G_{av1}}. \quad (22)$$

The overall noise temperature is therefore the noise temperature of the first stage, plus the input-referred noise temperature of the second stage. This formula reflects the understanding that the signal boost provided by the first stage diminishes the effect of noise of subsequent stages. Clearly, Eqn. 22 can be extended to an arbitrary number of stages, yielding one form of Friis's cascade noise figure formula:

$$T_{e12} = T_{e1} + \frac{T_{e2}}{G_{av1}} + \frac{T_{e3}}{G_{av1}G_{av2}} + \cdots. \quad (23)$$

An alternative expression in terms of noise factors is readily derived by using Eqn. 14 to relate noise temperature and noise factor:

$$F_{12} = F_1 + \frac{F_2 - 1}{G_{av1}} + \frac{F_3 - 1}{G_{av1}G_{av2}} + \cdots. \quad (24)$$

From inspection of the last two equations, we see that the expression for cascaded noise temperature is somewhat simpler (none of those pesky -1 terms to clutter up the equation). The noise temperature contributed by the nth stage can be computed simply by dividing through by the product of the available gains of the $(n-1)$ stages preceding it. For this reason, the noise temperature formulation is frequently favored when considering cascaded systems.

14.5 NOISE MEASURE

From Friis's formula we see that, if an amplifier has good noise figure but low gain, suppression of noise from subsequent stages is poor. Unfortunately, classical noise optimization design methods sometimes lead to an "optimum" amplifier design with precisely this combination of characteristics. Because both the noise figure and gain of an amplifier are important in general, another figure of merit known as *noise measure* is sometimes used to guide engineers toward a balanced design. Its formal definition initially seems to combine these two quantities in a puzzling way:

$$M \equiv \frac{F - 1}{1 - 1/G_{av}}. \quad (25)$$

The rationale for this definition becomes clear when we examine Friis's formula for the special case of an infinite cascade of identical amplifiers:

$$F_{tot} = F + \frac{F-1}{G_{av}} + \frac{F-1}{G_{av}^2} + \cdots, \quad (26)$$

which ultimately simplifies to

$$F_{tot} = 1 + \frac{F-1}{1 - 1/G_{av}} = 1 + M. \quad (27)$$

Therefore, this definition of noise measure is actually the normalized noise temperature of the infinite cascade:

$$T_{e,tot} = (F_{tot} - 1)T_0 = MT_0 \implies M = T_{e,tot}/T_0. \quad (28)$$

We see that, for good noise performance, we want the noise measure to be not much greater than the normalized noise temperature of the device itself.

Just to keep you on your toes, noise measure is defined in some references as F_{tot} rather than as $F_{tot} - 1$. Be sure to identify which definition is being used, because the difference can introduce considerable error for low-noise systems. Finally, note that this definition of noise measure has no particular relationship to the definition of noise measure for negative resistance devices, such as Gunn and tunnel diodes (see Chapter 9).

14.6 TYPICAL NOISE FIGURE INSTRUMENTATION

Having derived multiple expressions for noise figure, we're now in a position to examine several different methods for carrying out an actual measurement. As usual, we start with a little history, partly for entertainment but partly because methods that were used long ago tend to be ones that hobbyists can implement economically today.

14.6.1 THE (GOOD?) OLD DAYS

From Figure 14.1 we see that measuring noise figure is equivalent to determining the equation of the noise power-vs.-source temperature line. Measuring two points along the line is sufficient, but so is knowing a single point and the line's slope. The former method is the modern way, but it is worthwhile discussing the latter. Even though it poses nontrivial experimental challenges, the equipment required is within the reach of most RF hobbyists, so a description of this technique merits inclusion here.

Prior to the development of calibrated hot and cold sources, the only noise source available was at room temperature. With that limitation, one can determine the available output noise only at that one (perhaps inaccurately known and poorly controlled) temperature. So immediately, we see an error source: the noise source is probably at a temperature higher than 290 K. Even so, this error source is usually not the dominant one.

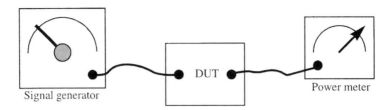

FIGURE 14.5. Signal generator method for noise figure measurement

The tricky part is that of accurately determining the slope of the line, $kG_{av}B$. Boltzmann's constant is pretty solid, but measuring the product of available power gain and noise bandwidth (which generally does not equal the -3-dB bandwidth) is fraught with difficulty. The experimental setup for doing so is straightforward in principle; it's just the practice that's hard.

To measure $G_{av}B$, simply connect a signal generator to the DUT and sweep the frequency to plot the power gain-vs.-frequency curve.[6] See Figure 14.5. In most cases, no provisions are made to ensure a conjugate match (because it is exceedingly tedious to do so at each of many test frequencies), so the power gain that is measured differs from the available gain, leading to more potential errors. The power frequency response curve is integrated (e.g., graphically, or by measuring the -3-dB bandwidth and multiplying by some fudge factor between 1 and 1.57) to find the product GB.

To complete the experiment, the output power N_1 is measured with the noise source (e.g., a simple resistor of value R_s) connected to the input. The noise factor is then

$$F \equiv \frac{N_1}{N_{os}} = \frac{N_1}{kT_0 B G_{av}}. \qquad (29)$$

This measurement method requires simple apparatus: a signal generator, calibrated power meter (or oscilloscope; see Section 14.10 for methods of estimating noise by inspection) and a resistor (which might be provided by simply turning off the generator). Figuring out the gain–bandwidth product from the measured frequency response is rather labor-intensive, but if you'd rather not have to choose between buying a car and buying an automated noise figure meter, the traditional method is the best choice. That said, it is quite difficult to reduce noise figure uncertainties below about 1–2 dB with this method, so characterization of very low-noise amplifiers with this technique is generally out of the question, practically speaking.

Another issue is that the measurement time per frequency point is large, so that it is cumbersome to make real-time evaluations of tweaks made to improve noise figure. It takes patience to use the signal generator method.

There is one case (at least, this is the only one the author can think of) where the signal generator method might actually have an advantage, however. Consider the

---

[6] If the signal generator's output is not constant over the band, it is necessary to measure its output to perform a proper gain calculation. Carrying out this process is tedious (so the temptation to omit it is strong), but failure to do so is a common source of error.

problem of measuring accurately the noise figure of an exceptionally noisy system. In particular, suppose that the DUT is so noisy that the noise temperature greatly exceeds the reference temperature. In this case, it is possible for the output noise powers under the hot and cold conditions to be rather similar, leading to a Y-factor close to unity. Because the formula for noise factor with a hot–cold measurement method contains the term $Y - 1$ in its denominator (see Eqn. 18), the measurement can be quite sensitive to small errors in Y when Y is nearly unity. The signal generator method, on the other hand, does not suffer from this sensitivity because it does not derive slope from measuring noise at two temperatures; there are no subtractions along the way. Thus we can say: for low-noise amplifiers, the hot–cold method is better, but for extremely noisy systems, the signal generator method may be better. The author readily concedes that this last case is somewhat contrived, for a highly accurate measurement of a high noise figure is rarely needed.

14.6.2 ON TO THE MODERN ERA...

The availability of a calibrated hot noise source greatly facilitates accurate noise figure determinations. As mentioned previously, early sources used actual resistors heated or cooled to easily determined or controlled temperatures, such as the boiling points of water and liquid nitrogen. A hot source at the temperature of boiling water is entirely feasible for the home experimenter (just be careful not to burn yourself or start a fire) and also highly accurate – if the water is reasonably pure and corrections are made for boiling-point shifts with altitude. However, this sort of hot source is not nearly as hot as those used in commercial instrumentation, so accuracy is once again degraded.

An alternative to a heated resistor is to exploit the shot noise of a vacuum tube diode. When operated in the temperature-limited regime, such a device exhibits shot noise of a magnitude that is traceable to first principles.[7] Temperature-limited diodes such as the 5722 – which was designed expressly for this purpose – have a mean-square noise current density of $2qI$ and can easily produce ENRs of several (e.g., 5) decibels.

Many commercial cold loads operate at 77 K, but not many hobbyists happen to have a Dewar full of LN_2 about the house.[8] Perhaps a more practical choice for the weekend experimenter is to use a room-temperature cold source, but accurate measurements demand knowledge of the actual room temperature (and still, your measurements won't be that accurate). A modest improvement is possible by using ice in water to provide a 273-K cold temperature. If you have access to (denatured)

---

[7] If operated in the space-charge–limited regime, less than full shot noise is observed, complicating computations. For this reason, such diodes are operated with a combination of abnormally low cathode temperature and abnormally high current density to assure temperature-limited behavior.

[8] Although, if you do, a waveguide aimed at a bucket of liquid nitrogen is allegedly a good way to realize a cold source at 77 K.

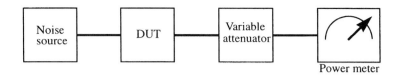

FIGURE 14.6. Y-factor measurement technique (simplified)

ethanol and dry ice, an equilibrium mixture of those two substances will typically have a temperature of about 195 K (−78°C). However, alcohol is flammable, so if you do choose this mixture then be sure to observe all appropriate safety precautions (in particular, keep the alcohol well away from whatever makes the hot source hot).

Once you have both a hot and cold source, there are several measurement options from which to choose. One method, called the "Y-factor method" for reasons that will become clear, avoids the need for a calibrated power meter, replacing it instead with a more easily realized calibrated adjustable attenuator. See Figure 14.6. The attenuator in instruments of this kind is frequently a waveguide operated beyond cutoff, owing to the ease with which the attenuation can be related to mechanical parameters.

This measurement technique relies on the fact that the ratio of output powers with the hot and cold source ($= Y$), plus knowledge of the hot and cold temperatures, is sufficient to compute noise figure. To carry out a measurement with this method, set the attenuation factor to unity, connect the cold load, and note the output power reading on the meter. The absolute value is completely unimportant. Then connect the hot load and adjust the attenuator until you obtain the same power reading as before. Since the attenuation factor is therefore the value that reduces a power N_2 to a value N_1, the attenuation factor is precisely equal to Y. The noise factor is then computed from Eqn. 10:

$$F = \frac{T_{ex}/T_0 - Y(T_c/T_0 - 1)}{Y - 1}. \tag{30}$$

The accuracy achieved depends on the accuracy of the Y-factor determination, as well as on the knowledge of the hot and cold temperatures. With assiduous attention to controlling all error sources, this technique can provide accuracies that are on a par with what can be achieved with commercial instrumentation (assuming equal noise temperatures for the sources). The trade-off is again one of time per measurement.

Actual Y-factor noise determinations are usually carried out with a slightly different configuration to permit measurement of spot noise figure as a function of frequency, rather than a gross noise figure over the entire bandwidth of the amplifier. The typical setup modifies the one shown in Figure 14.6 by adding a mixer, local oscillator, and intermediate-frequency (IF) amplifier, just as in a superheterodyne receiver;[9] see Figure 14.7.

---

[9] An LO, mixer, and IF amplifier can also be added to the setup of Figure 14.5 to improve the signal generator method, readily permitting evaluation of spot noise figure with that system. The noise bandwidth of the IF filter determines the width of the spot, and its value needs to be known to calculate spot noise figure correctly.

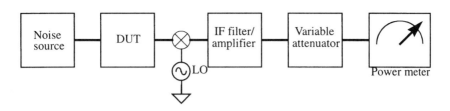

FIGURE 14.7. More typical Y-factor measurement setup

Here, the LO frequency is swept to sample the noise from the DUT at different frequencies. The IF amplifier–filter combination ensures that this noise is measured over some narrow, controlled bandwidth centered about the frequency determined by the LO setting. In many implementations, a filter is additionally interposed between the DUT and the instrumentation to limit bandwidth (perhaps to attenuate image response, for example).

If a spectrum analyzer is used as the power meter in any of these methods, it is necessary to precede it with a high-gain, low-noise preamplifier because spectrum analyzers generally have rather high noise figures (e.g., 30 dB) resulting from design trade-offs made in favor of good large-signal linearity. The gain of the preamp must be large compared with the noise figure of the analyzer in order to effect a substantial reduction in NF. The overall noise figure will then be close to that of the preamp alone. Even so, the noise figure of the preamp–spectrum analyzer combination will generally remain high enough that it cannot be ignored, and Friis's formula for cascaded noise figure should be used to correct the measured values. As a concrete numerical example, assume that the preamp has a noise figure and available power gain of 3 dB and 40 dB, respectively, and that the analyzer has a noise figure of 30 dB. The combination has a noise factor given by Friis's formula,

$$F = 2 + \frac{1000 - 1}{10^4} \approx 2.1, \tag{31}$$

or 3.2 dB, a large improvement over 30 dB and now only slightly greater than the preamp's inherent noise figure. Friis's formula will ultimately be used again, once the DUT is connected to the combination, with 3.2 dB now considered the second stage's noise figure and the DUT's available gain used in the denominator. If the noise figure of the preamp is not known, connect the hot and cold sources directly to its input and then make a measurement of noise figure. Once the combination has been characterized in this manner, it may be used to determine the noise figure of the DUT.

When making spot noise figure measurements with the spectrum analyzer, set the analyzer's resolution bandwidth equal to the desired width of the spot – and as great as possible to minimize the noise of the detected signal. Similarly, choose a video bandwidth (much) narrower than the resolution bandwidth in order to reduce noise in the displayed data (recall that video bandwidth controls the averaging of the *output* signal, after the detector). These considerations derive from the following relationship, which applies for spectrum analyzers that employ a rectify-and-average type of detector:

$$\frac{\text{rms output noise voltage}}{\text{detected DC voltage}} = \sqrt{\frac{b}{2B}}, \quad (32)$$

where b and B are the video and IF noise bandwidths, respectively.

The basic arrangement shown in Figure 14.6 is also quite close to what lies at the core of most modern automatic noise figure instruments. The HP8970, for example, contains all of those components, with the addition of a filter and preamplifier (as in the spectrum analyzer example) and a collection of attenuators at the input and output. With these additional elements, the instrument is able to measure (and correct for) insertion gain (loss) of the DUT and fixturing during the noise figure measurement. Additionally, the noise figure of the meter circuitry must also be known in order to complete an accurate noise figure measurement. What follows is a typical sequence of operations for making a noise figure measurement with a commercial instrument (specifically, the 8970B).

1. Read off the ENR calibration values for the hot/cold noise source, and enter those numbers into the instrument's memory. The 8970B has a list of the common calibration frequencies already in ROM, so the user normally only has to enter the ENR values.

2. Select the start frequency, stop frequency, and frequency increment (step size).

3. Connect the noise source to the instrument in order to permit measurement of the meter's noise figure, set frequency to the desired value, and press the "calibrate" key to initiate the calibration sequence. The meter successively activates and deactivates the noise source to compute the meter's hot and cold noise powers:

$$P_{hm} = k(T_h + T_{em})BG_m; \quad (33)$$

$$P_{cm} = k(T_c + T_{em})BG_m. \quad (34)$$

The ratio of these two powers is completely insensitive to gain–bandwidth product and has only the noise temperature of the meter as an unknown:

$$\frac{P_{hm}}{P_{cm}} = \frac{T_h + T_{em}}{T_c + T_{em}} \implies T_{em} = \frac{P_{cm}T_h - P_{hm}T_c}{P_{hm} - P_{cm}}. \quad (35)$$

The instrument measures the value of T_{em} at three input gain settings.

The 8970 allows the results of several calibration measurements at each frequency to be averaged. The number of measurements is controlled with the "increase" key. Hold it down until the desired number of runs is displayed. This step *precedes* activation of the "calibrate" mode.

4. Insert the DUT between the noise source and the instrument and select "noise figure and gain." To the maximum practical extent, avoid cables. The shorter the fixturing the better – to minimize pre-DUT loss (and thus any errors introduced by uncertainties in its subsequent subtraction) and to reduce RFI pickup. The instrument then measures the hot and cold powers of the cascade (DUT + meter):

$$P_{h,tot} = k(T_h + T_{e,tot})BG_mG_{DUT}; \quad (36)$$

$$P_{c,tot} = k(T_c + T_{e,tot})BG_mG_{DUT}. \quad (37)$$

The ratio of these two powers is also insensitive to gain–bandwidth product and has only the noise temperature of the DUT–meter combination as an unknown:

$$\frac{P_{h,tot}}{P_{c,tot}} = \frac{T_h + T_{e,tot}}{T_c + T_{e,tot}} \implies T_{e,tot} = \frac{P_{c,tot}T_h - P_{h,tot}T_c}{P_{h,tot} - P_{c,tot}}. \tag{38}$$

The ratio of the differences of noise powers enables computation of the DUT's gain:

$$G_{DUT} = \frac{P_{h,tot} - P_{c,tot}}{P_{hm} - P_{cm}}. \tag{39}$$

The gain of the meter has dropped out completely, so its value is theoretically irrelevant. Having computed the gain of the DUT, the noise temperature of the meter, and the noise temperature of the meter and DUT combination, Friis's formula can be used to solve for the noise figure of the DUT alone:

$$T_{e,tot} = T_{DUT} + \frac{T_{em}}{G_{DUT}} \implies T_{DUT} = T_{e,tot} - \frac{T_{em}}{G_{DUT}}. \tag{40}$$

Note that the resulting calculation is correct only if G_{DUT} is equal to the available gain. Mismatches may make these unequal and thereby introduce error.

The 8970 also allows the user to enter the cold temperature. The default is 296.5 K, which is close to typical room temperature.

A separate measurement of fixturing loss (e.g., with a network analyzer) enables correction for any pre-DUT fixturing attenuation. Most instruments allow the user to enter loss values (via the "loss compensation" feature of the 8970, for example) and automatically perform the subtraction of the loss factor. The instrument converts noise temperature into noise figure and displays both NF and G_{DUT}. It takes you much longer to read this description than it does for the instrument to carry out the measurement.

14.7 ERROR SOURCES

There are several ways in which noise figure measurements can go awry. Understanding what these are is a key to making accurate measurements. What follows is a short list of common problems, mistakes, and their fixes.

14.7.1 EXTERNAL NOISE

More than occasionally, external interference couples into the test setup. This interference can be noise radiated by RF sources ranging from TV and radio to digital equipment (particularly computers and their displays). Noise figure measurements are best carried out in a shielded ("screen") room to prevent this interference from injecting into the system. If this option is not available, a poor second choice is to make a spot noise measurement at a frequency removed from the interference, assuming that it is narrowband enough to enable this strategy. Many noise figure measurement systems provide for an oscilloscope connection to monitor the spectrum. If such an output is not available, a normal spectrum analyzer may be used instead. With the aid of a monitor, discrete peaks caused by interference can be identified quite easily, and the measurement frequency moved appropriately away from the interference.

14.7.2 FIXTURING LOSS

Fixturing anomalies are an endless source of errors. For example, a proper measurement of noise figure requires accurate characterization of any loss (e.g., from cable attenuation) that precedes the DUT proper. This loss (in dB) is subtracted from the measured overall NF (if, and only if, the loss is at 290 K) to yield the DUT's true NF. If the loss is large, however, the uncertainty in the final answer can be considerable because the instrument will have subtracted two nearly equal numbers. For example, suppose that the pre-DUT fixturing power loss is 20 dB (a frighteningly large value) and that the DUT itself has a 2-dB noise figure. The noise figure meter will measure a 22-dB noise figure, but inevitably within some error band (say, 0.5 dB). Assume for now that the error results in a composite measured NF of 21.5 dB. A separate measurement of the fixturing loss might have a similar uncertainty of 0.5 dB; suppose we measure 20.5 dB in this case. After subtraction, we compute a DUT NF of 1 dB, instead of the correct value of 2 dB, a huge error. In fact, for amplifiers with very low noise figure and large pre-DUT loss, it is entirely possible to compute negative values! Therefore, be suspicious of noise figure measurements in which a large attenuation has been mathematically removed. As a general rule, it is desirable to limit any such pre-DUT attenuation to values smaller than the anticipated noise figure. The lower this loss, the better.

If the loss is not at 290 K, then the noise temperature of the loss element is

$$T_e = (L-1)T_L, \qquad (41)$$

where L is the loss and T_L is the temperature at which the loss is measured. The resulting value of noise temperature may be used to perform an accurate correction.

14.7.3 SECOND STAGE CONTRIBUTION

Another common error is a failure to take into account the noise of stages that follow the DUT (the "second stage contribution"). A related consideration is that all commercial noise figure meters assume that the measured DUT gain is the same as the available gain. If the DUT has a large output impedance mismatch with that of the noise figure meter's input port, then this assumption will be a poor one and the calculation of the second stage contribution will be in error, as can be seen from Eqn. 40.

Impedance mismatch between the output of the noise source and the input of the DUT is also a concern. Reflections off of the DUT input travel back to the noise source, and any mismatch there causes a re-reflection back toward the DUT. The superposition of the incident power and this reflected power can cause the noise power from the source to differ from what it would be with a matched load. Complicating the situation is that the noise source may have a different impedance in the hot and cold modes, compounding the error, since the noise figure and available gain may consequently change.[10]

Correction for all of these errors requires knowledge of all three of the mismatches, as can be seen from the following formula:[11]

$$K_G = \frac{(1 - |\Gamma_s|^2)|1 - \Gamma_1\Gamma_2|^2}{(1 - |\Gamma_2|^2)|1 - \Gamma_1\Gamma_s|^2}. \tag{42}$$

Here, K_G is the factor by which the measured insertion gain should be multiplied in order to yield the correct value of available gain. The reflection coefficients are referred to various ports as follows: Γ_1 is defined looking into the input of the noise measurement instrumentation, Γ_2 into the output of the DUT, and Γ_s into the output of the noise source. Note that if these reflection coefficients are zero, K_G is unity. Note also that knowledge of both the magnitude and phase of the reflection coefficients is necessary in order to perform the correction. If only the magnitudes of the reflection coefficient are known, the best one can do is bound the error. As a specific example of the latter, assume that the magnitudes of Γ_1, Γ_2, and Γ_s are 0.33, 0.33, and 0.11, respectively. Then the true available gain could be anywhere between about 0.95 and 1.3 times the measured insertion gain.

14.7.4 NOISE SOURCE CALIBRATION UNCERTAINTY

Uncertainty in the ENR of the noise source is an additional error source. As stated previously, noise diodes must be calibrated against a standard. Calibrations are never perfect, and noise diodes may not be perfectly stable (although commercially available solid-state ones are remarkably good). One may typically expect instrument-grade noise diodes (such as the popular HP346B) to possess uncertainties in ENR on the order of 0.1 dB at low frequencies (e.g., 10 MHz), increasing to perhaps 0.2 dB at higher frequencies (e.g., 18 GHz). The percentage error represented by these uncertainties gets progressively more significant as the noise figure of the DUT diminishes.

A noise diode's output is not perfectly constant over the operating frequency range, and neither does it follow any other simple functional law traceable to first principles; hence, noise source calibrations are made at a number of discrete frequencies (10 or 20 is a typical number). In between the calibration points, you (or the noise figure meter) must perform interpolations. The actual noise output may differ from the interpolated value, adding another error term.

---

[10] See N. J. Kuhn, "Curing a Subtle but Significant Cause of Noise Figure Error," *Microwave Journal Magazine,* June 1984.

[11] "Fundamentals of RF and Microwave Noise Figure Measurements," Hewlett-Packard Applications Note 57-1, July 1983.

14.7.5 COLD TEMPERATURE $\neq T_0$

Yet another common problem is that the cold noise source temperature is rarely 290 K. A diode noise source has a cold temperature equal to that of the ambient, and most laboratories are 4–5°C warmer than T_0. As a rough rule of thumb, the measured noise temperature is too low by one degree for each degree the noise source is above T_0. Thus it is typical to underestimate the noise figure of a DUT because of the warm laboratory problem. For more rigorous corrections, an accurate measurement of the cold temperature must be made, with Eqn. 10 then used to compute the adjustment. This correction is most important in the case of very low noise figures. It is also important to consider that noise diodes (and perhaps other sources) heat up during use, so that the source temperature may change as a function of time during the measurement.

14.7.6 FAILURE OF LINEARITY: DIODE DETECTORS

The straight line of Figure 14.1 underlies both the definition and measurement of noise figure. If the device under test is nonlinear, noise figure can't be uniquely defined. A relevant example is that of diodes used as square-law detectors (frequently known as video detectors for historical reasons). In cases such as these, a different figure of merit is used to convey information about noise performance.

One such figure of merit is *tangential signal sensitivity* (TSS). Its original definition is a highly subjective evaluation of noise: An operator observes the noisy output of the detector on an oscilloscope in the absence of any signal and notes the position of the positive noise peaks. Then the signal is turned on, and the operator adjusts the amplitude until the negative-going noise peaks with signal present appear just to touch the positive-going noise peaks noted earlier with the signal absent. Formally, TSS is defined as the level of input signal that produces this condition. The problem with this definition is that noise, being random, has theoretically unbounded peaks. So, the operator has to make an arbitrary judgment when an equality of peaks occurs, and different operators may guess differently (the same operator may also make different determinations at different times). To eliminate this subjectivity, most diode manufacturers now define TSS as the available input signal power that causes the output signal power to exceed the output noise power by 8 dB. These numbers correspond to an output power ratio of about 6.3 and a voltage ratio of 2.5. Note that, if the input power is within the range for which the diode acts as a square-law device, then an output signal-to-noise *voltage* ratio of 2.5 is produced for an input *power* ratio of 2.5. A typical value of TSS for diodes might be −60 dBm.

Another figure of merit is the *nominal detectable signal* (NDS), which is defined as the available input power that results in an output SNR of unity. Both TSS and NDS are generally functions of frequency and bias current, so these must be specified to make TSS and NDS values meaningful.

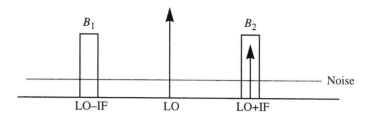

FIGURE 14.8. Spectrum of SSB input to mixer

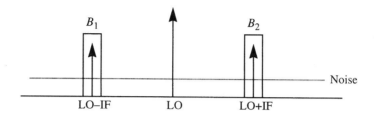

FIGURE 14.9. Spectrum of DSB input to mixer

14.8 SPECIAL CONSIDERATIONS FOR MIXERS

When the DUT is (or contains) a mixer, there is a question of whether one should perform a single-sideband (SSB) or double-sideband (DSB) noise figure measurement. In most cases, the SSB noise figure is the appropriate choice, since few communications systems transmit the same signal in both the main and image bands. The only two exceptions the author is aware of are direct-conversion (homodyne) receivers, in which the main signal occupies the same spectrum as its image, and deep-space radiometry, where noise (that of the universe) *is* the signal. Because DSB noise figure is lower by 3 dB (assuming equal conversion gains for the two sidebands), "specmanship" games are all too frequent, and this figure is often reported instead of SSB.

To place the DSB–SSB issue on a firm foundation, consider that the IRE (now IEEE) noise figure definition has in its numerator all output noise, yet its denominator contains only signal-related noise. If the signal is contained in only one sideband, then the relevant spectra appear roughly as shown in Figure 14.8. Here the signal exists only within bandwidth B_2. From this picture, the correct definition of noise factor is:

$$F_{SSB} \equiv \frac{N_a + kT_0 G_1 B_1 + kT_0 G_2 B_2}{kT_0 G_2 B_2}. \tag{43}$$

Note that the formula allows for the possibility of unequal receiver bandwidths and unequal conversion gains for the two bands.

In the rarer DSB case, the desired signal resides in both bands; see Figure 14.9. The corresponding noise factor is:

$$F_{DSB} \equiv \frac{N_a + kT_0 G_1 B_1 + kT_0 G_2 B_2}{kT_0 G_1 B_1 + kT_0 G_2 B_2}. \tag{44}$$

If the bandwidths are equal and the conversion gains are also equal, then the DSB NF will be 3 dB lower than the SSB value, as stated earlier. More generally, allowing for unequal conversion gains (but still assuming equal bandwidths),

$$F_{SSB} = F_{DSB}(1 + G_1/G_2). \tag{45}$$

In many cases, a mixer is preceded by an image-suppression filter. In this commonly occurring situation, it is appropriate to characterize the combination of the filter and mixer as a unit. Because the job of the filter is to produce unequal conversion gains to the two sidebands, there will no longer be a 3-dB difference between the SSB and DSB NF.

Another important subtlety concerns the nature of terminations on the several mixer ports. Because a mixer has three ports – RF, IF, and LO – misterminations on any of the ports can result in complicated reflections capable of corrupting measurements. A particularly common error is to terminate the IF port of a passive mixer in a load that is matched only at (say) the difference frequency while exhibiting a highly reactive impedance at the sum frequency. Even though we might only care about the difference component, reflections at the sum frequency can cause pathological behavior of both noise and conversion gain.

Additionally, the gain and noise characteristics of mixers typically vary with LO power. For the measurements to be meaningful, then, the LO power must be specified. Preferably, the noise figure (and conversion gain) should be presented as a function of LO power over a range that spans practical values. Finally, mixers are generally sensitive to AM noise on the LO. They may send this noise out the IF port, producing yet another source of error.

14.9 REFERENCES

Various applications notes from Hewlett-Packard (now Agilent Technologies) are excellent sources of information about noise measurement. Some that are of particular interest include "Accurate and Automatic Noise Figure Measurements," (HP Applications Note 64-3, June 1980) and "Fundamentals of RF and Microwave Noise Figure Measurements," (HP Applications Note 57-1, July 1983). Another good source of information is the documentation for the HP8970B noise figure meter, which describes in detail the theory underlying the operation of this instrument.

14.10 APPENDIX: TWO CHEESY EYEBALL METHODS

Making measurements of noise can be rather involved if it is to be done accurately. Typically, a special noise figure instrument (or possibly a spectrum analyzer) is required to determine the noise density as a function of frequency. For very quick assessments of relatively large amounts of noise, a crude measurement is sometimes acceptable. In those cases, an oscilloscope and your eyeball may be the only instruments you need. If we assume that the noise is Gaussian, then the peak-to-peak values very rarely exceed about 5–7 times the rms value. So, the level-zero eyeball measurement is to connect the noisy DUT to the oscilloscope, make some judgment about what the displayed peak-to-peak value seems to be, then divide by about 6 to develop an estimate of the rms value.

This method is *very* crude, of course, and in no small measure because of the difficulty in determining what the "true" peak-to-peak value happens to be. The situation is further complicated by the fact that the oscilloscope brightness setting affects what appear to be the peaks; the brighter the trace, the taller the apparent peaks. The same operator may also make significantly different determinations at different times as a function of sleep deprivation, emotional state, and caffeine levels.

A clever extension of the eyeball technique removes much of this uncertainty by converting the measurement into a differential one.[12] Here, the noisy signal simultaneously drives both channels of a dual-trace oscilloscope, operating in alternating sweep mode rather than chop mode (to avoid introducing a correlation between the two sweeps through the oscilloscope's chopping oscillator). With a sufficiently large initial position difference, there will be a dark band between these two traces. Adjust the position controls until the dark band just disappears, with the two traces merging into a single blurry mess with a monotonically decreasing brightness from the center outward. Note that this description implies an independence of the result on the absolute intensity. Remove the noisy signals, and then measure the distance between the two baselines. The resulting value is twice the rms voltage to a good approximation. Absolute accuracies of about 1 dB are possible with this simple method.

The reason this technique works is that a sum of two identical Gaussian distributions has a maximally flat top when the two distributions are separated by exactly twice the rms value.

Because the eye is an imperfect judge of contrast, it is not possible to establish with infinite precision when the dark band disappears. When following the procedure as outlined, most people will perceive the band to have disappeared a little before it actually does. The error resulting from this uncertainty is on the order of 1 dB for most people. Thus, perhaps 0.5 dB should be subtracted from the measurement if you are very fussy. An alternative is to measure the noise two different ways, one using the procedure given, and another with the two traces initially on top of each other. With the latter initial condition, adjust the spacing until the darker area first seems to *appear*. Average the two readings, and also compute the difference between the two readings as a measure of uncertainty. With care and a little practice, sub–1-dB repeatability is readily achievable.

---

[12] G. Franklin and T. Hatley, "Don't Eyeball Noise," *Electronic Design*, v. 24, 22 November 1973, pp. 184–7.

CHAPTER FIFTEEN

OSCILLATORS

15.1 INTRODUCTION

Given the effort expended in avoiding instability in most feedback systems, it would seem trivial to construct oscillators. Murphy, however, is not so kind; the situation is a lot like bringing an umbrella in order to make it rain. An old joke among RF engineers is that every amplifier oscillates, and every oscillator amplifies.

In this chapter, we consider several aspects of oscillator design. First, we show why purely linear oscillators are a practical impossibility. We then present a linearization technique utilizing *describing functions* that greatly simplify analysis, and help to develop insight into how nonlinearities affect oscillator performance. With describing functions, it is straightforward to predict both the frequency and amplitude of oscillation.

A survey of resonator technologies is included, and we also revisit PLLs, this time in the context of frequency synthesizers. We conclude this chapter with a survey of oscillator architectures. The important issue of phase noise is considered in detail in Chapter 17.

15.2 THE PROBLEM WITH PURELY LINEAR OSCILLATORS

In negative feedback systems, we aim for large positive phase margins to avoid instability. To make an oscillator, then, it might seem that all we have to do is shoot for zero or negative phase margins. We may examine this notion more carefully with the root locus for *positive* feedback sketched in Figure 15.1.[1]

This locus recurs frequently in oscillator design because it applies to a two-pole bandpass resonator with feedback. As seen in the locus, the closed-loop poles lie

---

[1] For those unfamiliar with root loci, these are simply all possible values of a system's closed-loop poles as some parameter is varied. Usually that parameter is the gain around the loop, but it can also be something else, such as a loop-transmission pole frequency. For details, see e.g. T. H. Lee, *The Design of CMOS Radio-Frequency Integrated Circuits,* 2nd ed., Cambridge University Press, 2004.

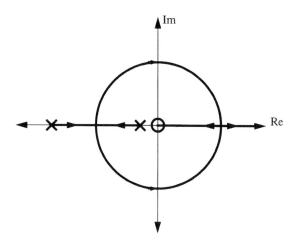

FIGURE 15.1. Root locus for oscillator example

exactly on the imaginary axis for some particular value of loop-transmission magnitude. The corresponding impulse response is therefore a sinusoid that neither decays nor grows with time, and it would seem that we have an oscillator.

There are a couple of practical difficulties with this scenario, however. First, the amplitude of the oscillation depends on the magnitude of the impulse (it is a linear system, after all). This behavior is generally undesirable; in nearly all cases, we want the oscillator to produce a constant-amplitude output that is independent of initial conditions. Another problem is that if the closed-loop poles don't lie *precisely* on the imaginary axis, the oscillations will either grow or decay exponentially with time.

These problems are inherent in any purely linear approach to oscillator design. The solution to these problems therefore lies in a purposeful exploitation of nonlinear effects; *all practical oscillators depend on nonlinearities*. To understand just how nonlinearities can be beneficial in this context, and to develop intuition useful for both analysis and design, we now consider the subject of *describing functions*.

15.3 DESCRIBING FUNCTIONS

We've seen that linear descriptions of systems often suffice, even if those systems are nonlinear. For example, the incremental model of a bipolar transistor arises from a linearization of the device's inherent exponential transfer characteristic. As long as excitations are "sufficiently small," the assumption of linear behavior is well satisfied.

An alternative to linearizing an input–output transfer characteristic is to perform the linearization in the *frequency domain*. Specifically, consider exciting a nonlinear system with a sinusoid of some particular frequency and amplitude. The output will generally consist of a number of sinusoids of various frequencies and amplitudes. A linear description of the system can be obtained by discarding all output components except the one whose frequency matches that of the input. The collection of all possible input–output phase shifts and amplitude ratios for the surviving component comprises the describing function for the nonlinearity. If the output spectrum is dominated by the fundamental component, then results obtained with the describing function approximation will be reasonably accurate.

To validate further our subsequent analyses, we will also impose the following restriction on the nonlinearities: they must generate no subharmonics of the input (DC is a subharmonic). The reason for this restriction will become clear momentarily. For RF systems, this requirement is perhaps not as restrictive as it initially appears, because bandpass filters can often be used to eliminate subharmonic and harmonic components.

As a specific example of generating a describing function, consider an ideal comparator, described by the following equation:

$$V_{out} = B \operatorname{sgn} V_{in}. \tag{1}$$

If we drive such a comparator with a sine wave of some frequency ω and amplitude E, then the output will be a square wave of the same frequency but of a constant amplitude B, independent of the input amplitude. Furthermore, the zero crossings of the input and output will coincide (so there is no phase shift). Hence, the output can be expressed as the following Fourier series:

$$V_{out} = \frac{4B}{\pi} \sum_{1}^{\infty} \frac{\sin \omega n t}{n} \quad (n \text{ odd}). \tag{2}$$

Preserving only the fundamental term ($n = 1$) and taking the ratio of output to input yields the describing function for the comparator:

$$G_D(E) = \frac{4B}{\pi E}. \tag{3}$$

Since there is no phase shift or frequency dependence in this particular case, the describing function depends only on the input amplitude.

Note that the describing function for the comparator shows that the effective gain is *inversely proportional to the drive amplitude,* in contrast with a purely linear system in which the gain is independent of drive amplitude. We shall soon see that this inverse gain behavior can be extremely useful, for it can provide negative feedback to stabilize the amplitude.

15.3.1 A BRIEF CATALOG OF DESCRIBING FUNCTIONS

Having shown how one goes about generating describing functions, in Figures 15.2–15.4 we present (without derivation) describing functions for some commonly encountered nonlinearities.[2] In the example of Figure 15.4, the value of R must be less

---

[2] See e.g. J. K. Roberge's excellent book, *Operational Amplifiers,* Wiley, New York, 1975.

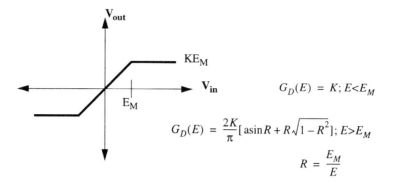

FIGURE 15.2. Transfer characteristic and describing function for saturating amplifier

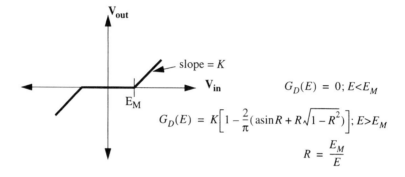

FIGURE 15.3. Describing function for amplifier with crossover distortion

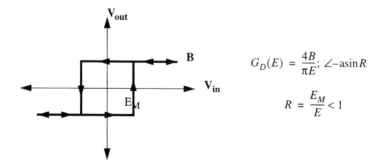

FIGURE 15.4. Transfer characteristic and G_D for Schmitt trigger

than unity; otherwise, the Schmitt never triggers and the output of the comparator will then be only a DC value of either B or $-B$.

It is important to note that describing functions themselves are linear even though the functions that they describe may be nonlinear (got that?). Hence, superposition holds; the describing function for a sum of nonlinearities is equal to the sum of the

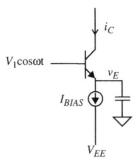

FIGURE 15.5. Large-signal transconductor

individual describing functions. This property is extremely useful for deriving describing functions for nonlinearities not included in the short catalog presented here.

15.3.2 A UNIVERSAL DESCRIBING FUNCTION FOR TRANSISTORS AND VACUUM TUBES

Although the foregoing collection of describing functions is extremely useful, particularly relevant for the RF oscillator design problem are describing functions for one- and two-transistor circuits, since the high frequencies that characterize RF operation are difficult to generate with many transistors in a loop.

To illustrate a general approach, consider the circuit in Figure 15.5. The capacitor is assumed large enough to behave as a short at frequency ω, and the transistor is ideal. We will be using this circuit in tuned oscillators, so the bandpass action provided by the tank guarantees that describing function analysis will yield accurate results.

Before embarking on a detailed derivation of the large-signal (i.e., describing function) transconductance, let's anticipate the qualitative outlines of the result. As the amplitude V_1 increases, the emitter voltage V_E is pulled to higher values, reaching a maximum roughly when the input does. Soon after the base drive heads back downward from the peak, the transistor cuts off as the input voltage falls faster than the current source can discharge the capacitor. Because the current source discharges the capacitor between cycles, the base–emitter junction again forward-biases when the input returns to near its peak value, resulting in a pulse of collector current. The cycle repeats, so the collector current consists of periodic pulses.

Remarkably, we do not need to know any more about the detailed shape of collector current in order to derive quantitatively the large-signal transconductance in the limit of large drive amplitudes. The only relevant fact is that the current pulses consist of relatively narrow slivers in that limit, as in the hypothetical plots[3] of voltage and current in Figure 15.6.

---

[3] The word "hypothetical" is here a euphemism for "wrong." However, even though the detailed waveforms shown are not strictly correct, the results and insights obtained are. In particular, this

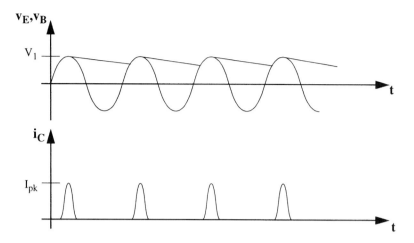

FIGURE 15.6. Hypothetical emitter and base voltage, and collector current for large input voltage

Whatever the current waveform, Kirchhoff's current law demands that its average value equal I_{BIAS}. That is,

$$\langle i_C \rangle = \frac{1}{T} \int_0^T i_C(t)\, dt = I_{BIAS}. \tag{4}$$

Now, the fundamental component of the collector current has an amplitude given by

$$I_1 = \frac{2}{T} \int_0^T i_C(t) \cos \omega t\, dt. \tag{5}$$

We may not know the detailed functional form of $i_C(t)$, but we do know that it consists of narrow pulses in the limit of large drive amplitudes. Furthermore, these current pulses occur roughly when the input is a maximum, so the cosine may be approximated there by unity for the short duration of the pulse. Then,

$$I_1 = \frac{2}{T} \int_0^T i_C(t) \cos \omega t\, dt \approx \frac{2}{T} \int_0^T i_C(t)\, dt = 2 I_{BIAS}. \tag{6}$$

That is, the amplitude of the fundamental component is approximately twice the bias current, again in the limit of large V_1. The magnitude of the describing function is therefore

$$G_m = \frac{I_1}{V_1} \approx \frac{2 I_{BIAS}}{V_1}. \tag{7}$$

It is important to note that this derivation does not depend on detailed transistor characteristics at any step along the way. Because no device-specific assumptions are

---

picture allows us to understand why the describing function transconductance for large drive amplitudes is essentially the same for bipolars and MOSFETs (both long- and short-channel), as well as for JFETs and vacuum tubes.

used, Eqn. 7 is quite general, applying to MOSFETs (both long- and short-channel) as well as to bipolars, JFETs, GaAs MESFETs, and even vacuum tubes.

In deriving Eqn. 7, we have assumed that the drive amplitude, V_1, is "large." To quantify this notion, let us compute the G_m/g_m ratio for long- and short-channel MOSFETs (such as VMOS RF power devices) and bipolar devices.

For long-channel devices, the ratio of g_m to drain current I_{BIAS} may be written as follows:

$$\frac{g_m}{I_{BIAS}} = \frac{2}{V_{GS} - V_t}, \tag{8}$$

so that

$$\frac{G_m}{g_m} = \frac{V_{GS} - V_t}{V_1}. \tag{9}$$

Evidently, "large" V_1 is defined relative to $(V_{GS} - V_t)$ for long-channel MOSFETs. Repeating this exercise for short-channel devices yields[4]

$$\frac{g_m}{I_{BIAS}} = \frac{2}{V_{GS} - V_t} - \frac{1}{E_{sat}L + (V_{GS} - V_t)}, \tag{10}$$

which, in the limit of very short channels, converges to a value precisely half that of the long-channel case. Thus,

$$\frac{V_{GS} - V_t}{V_1} \leq \frac{G_m}{g_m} < \frac{2(V_{GS} - V_t)}{V_1}. \tag{11}$$

Finally, for bipolar devices,

$$\frac{g_m}{I_{BIAS}} = \frac{1}{V_T}, \tag{12}$$

so that

$$\frac{G_m}{g_m} = \frac{2V_T}{V_1}. \tag{13}$$

In bipolar devices, large V_1 is therefore defined relative to the thermal voltage, kT/q.

Although the equation for G_m is valid only for large V_1, practical oscillators usually satisfy this condition, so this restriction is much less constraining than one might think. We will also see in the next chapter that large V_1 is highly desirable for reducing phase noise, so one may argue that all well-designed oscillators automatically satisfy the conditions necessary to validate the approximations used. Nevertheless, it is important to recognize that G_m can never exceed g_m, so one must be careful not to misapply equations such as Eqn. 13. To underscore this point, Figure 15.7 shows, in an approximate way, the actual behavior of G_m/g_m contrasted with the behavior as predicted by Eqn. 13. Although this equation applies strictly to the bipolar case, the overall behavior shown in the figure holds generally.

Having presented numerous describing functions, we now consider an example to illustrate how to use them to analyze oscillators.

---

[4] Here, we have used the approximate, analytic model for short-channel MOSFETs introduced in Chapter 5 of Lee, op. cit. (see footnote 1).

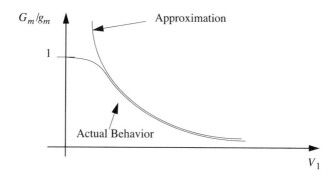

FIGURE 15.7. G_m/g_m vs. V_1

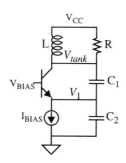

FIGURE 15.8. Colpitts oscillator
(biasing details not shown)

15.3.3 EXAMPLE: COLPITTS OSCILLATOR

Relaxation oscillators – such as the function generator – are rarely used in high-performance transceivers because they generate signals of inadequate spectral purity. Much more common are tuned oscillators, primarily for reasons that we may appreciate only after studying the subject of phase noise. For now, simply accept as an axiom the superiority of tuned oscillators. Our present focus, then, is the use of describing functions to predict the output amplitude of a typical tuned oscillator, such as the Colpitts circuit shown in Figure 15.8.[5] We shall see later in this chapter that a variety of oscillators differing in trivial details are named for their inventors. In keeping with standard practice, we will retain this naming convention, but the reader is advised to focus on operating principles rather than nomenclature.

The basic recipe for these oscillators is simple: Combine a resonator with an active device. The distinguishing feature of a Colpitts oscillator is the capacitively tapped resonator, with positive feedback provided by the active device to make oscillations possible. In Figure 15.8, the resistance R represents the total loading due to finite tank Q, transistor output resistance, and whatever is driven by the oscillator (presumably

---

[5] Edwin Henry Colpitts devised his oscillator in early 1915, while at Western Electric. His colleague Ralph Vinton Lyon Hartley had demonstrated *his* oscillator just a month earlier, on February 10th.

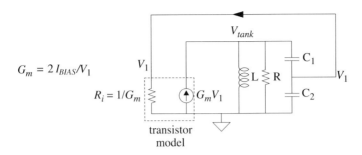

FIGURE 15.9. Describing function model of Colpitts oscillator

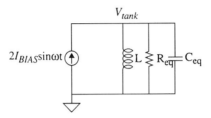

FIGURE 15.10. Simplified model of Colpitts oscillator

the oscillator's output is used somewhere). The current source is frequently replaced by an ordinary resistor in practical implementations, but it is used here to simplify (marginally) the analysis.

From our describing function derivation, we know that the transistor may be characterized by a large-signal transconductance G_m. For the sake of simplicity, we will ignore all dynamic elements of the transistor, as well as all parasitic resistances, although an accurate analysis ought to take these into account. The transistor also has a large-signal source–gate resistance, of course, which must be modeled as well. Taking a cue from describing functions, it seems reasonable to define this resistance as the ratio of the fundamental component of source current to the source–gate voltage. We've actually already found this ratio; it is simply $1/G_m$. Thus, we may model the oscillator with the circuit shown in Figure 15.9.

To simplify the analysis further, first reflect the input resistance R_i across the main tank terminals by treating the capacitive divider as an ideal transformer so that we end up with a simple RLC tank embedded within a positive feedback loop. Note that the resulting circuit has zero phase margin at the resonant frequency of the tank, so that will be the oscillation frequency in this particular case. Note also that the dependent current generator produces an output (sinusoid) whose amplitude is $G_m V_1 = 2I_{BIAS}$ at all times, so it may be replaced with an independent sine generator of this amplitude. Acting on these observations leads to the circuit of Figure 15.10, where C_{eq} is the series combination of the two capacitors,

$$C_{eq} = \frac{C_1 C_2}{C_1 + C_2}, \qquad (14)$$

and
$$\omega = 1/\sqrt{LC_{eq}}. \quad (15)$$

Similarly, R_{eq} is the parallel combination of the original tank resistance R and the reflected large-signal input resistance of the transistor:[6]

$$R_{eq} \approx R \parallel \frac{1}{n^2 G_m}, \quad (16)$$

where n is the capacitive voltage divide factor,

$$n \equiv \frac{C_1}{C_1 + C_2}. \quad (17)$$

The amplitude V_1 is simply the amplitude, V_{tank}, of the tank voltage, multiplied by the capacitive divide factor, so we may write

$$V_{tank} \approx \frac{V_1}{n}. \quad (18)$$

Now we have collected enough equations to get the job done. The amplitude of the tank voltage at resonance is simply the product of the current source amplitude and the net tank resistance:

$$V_{tank} \approx \frac{V_1}{n} \approx 2I_{BIAS} R_{eq} \approx (2I_{BIAS}) \left[R \parallel \frac{1}{n^2 G_m} \right] = (2I_{BIAS}) \cdot \frac{R}{n^2 G_m R + 1}, \quad (19)$$

which ultimately simplifies to

$$V_{tank} \approx 2I_{BIAS} R(1 - n). \quad (20)$$

Thus, the amplitude of oscillation is directly proportional to the bias current and the effective tank resistance. The loading of the tank by the transistor's input resistance is taken into account by the $(1 - n)$ factor, and is therefore controllable by choice of the capacitive divide ratio. Since R also controls Q, it is usually made as large as possible, and adjustment of I_{BIAS} is consequently the main method of defining the amplitude.

As a specific numerical example, consider the ~60-MHz oscillator circuit of Figure 15.11. For the particular element values shown, the capacitive divide factor n is about 0.155.[7] The expected oscillation amplitude (V_{tank}) is therefore about 1.4 V. Measurements made on a bipolar version of this circuit reveal an amplitude of 1.3 V, in good agreement with theoretical predictions. It is important to underscore again that this result is largely independent of the type of active device used to build the

---

[6] In this and related equations, the reason for the "approximately equals" symbol is that we are treating the capacitive divider as an ideal impedance transformer. The approximation is good as long as the in-circuit Q is large.

[7] In practice, it is generally true that best phase performance tends to occur for a C_2/C_1 ratio of about 4, corresponding to $n = 0.2$. This rule of thumb can be put on a more rigorous theoretical basis by making use of the time-varying theory discussed in Chapter 17.

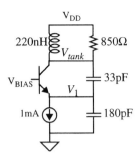

FIGURE 15.11. Colpitts oscillator example

oscillator. The prediction was originally made for a MOSFET design but experimentally verified with a bipolar device.

15.3.4 DETAILED COLPITTS DESIGN EXAMPLE

We now consider a 1-GHz microstrip implementation of a Colpitts oscillator to supplement the lower-frequency example just completed. Also, instead of a single-frequency design, suppose that we elect to make this a voltage-controlled oscillator (VCO).

Many oscillator topologies are almost indistinguishable theoretically, but convergent evolution has led to only a few popular choices. One consideration is that not all are equally amenable to convenient biasing or provide good tuning range. For example, you could choose a topology that grounds the collector of a transistor or that ties it to some positive V_{CC}. Or you could select one that grounds the base. Each of these choices leads to a different collection of bias headaches and sensitivities to parasitic-induced problems.

The Colpitts implementation shown in Figure 15.12 has worked well for many. Here, we have chosen to set the collector's DC potential at ground; *negative* voltages are used to bias the transistor. This choice of polarity is driven by several considerations: Returning the tank to ground is nice because it avoids having to bypass a collector supply (with either chokes or BFCs), and it also eliminates an output coupling capacitor. All these simplifications make it easy to terminate one end of the inductor in an excellent short (locating the inductor close to the edge of a PC board helps to take full advantage of this topology without the inconvenience of vias). Furthermore, the control voltage generated by many PLLs is ground-referenced, so annoying biasing gymnastics are avoided by connecting the varactor to the ground-referenced load structure.

In this configuration, a suitable length of microstrip line acts as an inductor. That inductance resonates with a capacitance formed by the series combination of the varactor and the sum of C_2 and the emitter capacitance. The tuning range is therefore a function of the varactor capacitance versus that other total capacitance. Fortunately, with the particular arrangement shown, it is possible to arrange for the varactor to

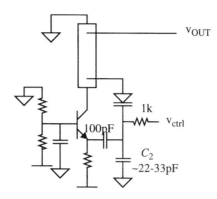

FIGURE 15.12. Slightly simplified schematic of VCO

dominate the total capacitance. This condition is readily produced by choosing a *large* enough C_2, which is what we want in order to maximize tuning range. Unlike some other Colpitts arrangements, then, parasitic capacitances across C_2 are relatively benign. We can't make C_2 arbitrarily large, however, because it is also part of the feedback voltage divider. If it is too large, the feedback may diminish to the point where oscillation ceases. Arbitrarily choosing C_2 equal to the maximum capacitance of the varactor is a reasonable initial choice. The maximum varactor capacitance, in turn, is determined by the minimum value of junction voltage applied across the varactor. Just to attach numbers to the quantities, suppose that the VCO is driven by a PLL that produces a control voltage between 0.5 V and 4.5 V. Then we would compute the varactor capacitance maximum at a junction reverse bias of 0.5 V.

Another important practical consideration is to keep the base bypass path as short as possible. Any inductance in series with the base can result in parasitic oscillations at frequencies you never imagined could exist. Even before such parasitic modes become unstable, they can perturb the oscillator in many undesirable ways.

Load Inductor Design

Assume a net capacitance of 8 pF at the center of the tuning range (calculated by using the geometric mean of what you get when you look at the minimum and maximum varactor capacitances of 8 pF and 18 pF, respectively, in series with 22 pF $C_2 + 4$ pF of emitter cap and strays[8]), you'd want an inductance of about 3 nH. This is a little tricky – but not impossible – to implement by manual cut-and-try means. You must exercise extreme care to minimize the stray inductance of the varactor connection. If you succeed, the calculated tuning range will be about $\pm 15\%$ about the nominal frequency.

---

[8] This is a pessimistic estimate of total "other" capacitance; your mileage may vary, especially if you use a transistor in place of the varactor.

To ease the inductor design problem and simultaneously increase tuning range, if necessary, you can increase the maximum control voltage across the varactor to reduce its minimum capacitance. However, doing so often means that you will have to design and build an amplifier to take the control voltage from the PLL chip and gain it up appropriately.

Bias

From our describing function analysis, we know that the oscillation amplitude is a function of bias current and losses (which include the action of the load resistance). The last item is under a designer's control to a certain extent through choice of output tap location. The shortness of the load inductor limits the ease with which one can select and vary the tap location, so selection of an appropriate bias current will be an important design variable. Depending on the tapping point, a 0–7-dBm output will probably result with bias currents in the range of low to medium numbers of milliamps. Be sure that, in laying out the bias network, you avoid stray capacitance where it would affect the frequency of the tank or degrade the tuning range. As with any other tapped tank, this load structure doesn't guarantee control over both the real and imaginary parts of the output impedance, so additional matching may be required.

Another consideration is that the anode of the varactor may see a zero DC voltage, yet there will also be an AC component, too. The control voltage needs to be sufficiently larger than the peak anode voltage to avoid forward-biasing the varactor. Oscillation might even cease if this problem occurs, causing the loop to get confused. Hence there is a constraint on the tank amplitude.

The simplified schematic shows no values for the biasing components. As a starting point, try dropping a volt or two across the emitter resistor and then selecting the base bias dividers to carry a current of approximately the collector current divided by 10. These rough rules of thumb provide for acceptable bias stability in most cases, but that statement needs to be verified in all situations where it matters. Bias stability in the face of beta variation is improved by increasing the current through the base bias divider, while stability with respect to V_{BE} variation is improved by dropping more voltage across the emitter resistor.

Additional Practical Notes

The tank inductor can be the source of difficulties because of its short length. You might have reasoned that, the narrower the line, the better the tuning range, because a shorter line could then be used for a given impedance, leading the line to act more like a pure inductor than a quasiresonator. After all, resonator loss is only a second-order function of width because most of the loss in FR4 is due to the dielectric, so using narrow lines may not be so unattractive here.

However, shortness is a problem for largely mechanical reasons. To allow a somewhat longer line to be used, consider a *wider* line than you might normally try. Just make sure that all dimensions remain well below a quarter-wavelength, or else strong distributed effects will throw off your calculations (or force you to perform more elaborate calculations). It is preferable in real designs to choose a length that will not produce high impedances at some multiple(s) of the oscillation frequency, because those distortion products will not be filtered out by the tank (remember, the "inductor" you've made is in reality a transmission line that is terminated in a short). In this lab experiment, however, there is no distortion specification, so you don't have to obsess about this issue now. It's just mentioned here so that, when you go out into "the real world," you can't complain that some ivory-tower academic never told you about it.

Another mildly tricky issue is that the tuning range is narrow enough, and the varactor tolerances loose enough, that the center frequency of the VCO may not be quite the value you want. If you're close to the right frequency, you may not have to rip out the inductor and start over. To raise the frequency a little bit, add some copper near the ground end of the line to decrease inductance. To decrease frequency a tiny bit, narrow up the line by slicing a little piece out of it. To decrease frequency further, solder a short length of wider foil onto the line to increase capacitance (again, the closer to the collector, the larger the effect).

If the output power level is too low, the main cure is more bias current. If power varies wildly over the tuning range, the cure is to tap the output from a point closer to the collector end of the line. The loss of the line, and hence its effective resistance, varies with frequency. If this frequency-dependent loss dominates, then the amplitude will vary significantly with frequency. Tapping the output closer to the collector loads down the tank more severely, but at least the load is more constant. An increase in bias current can compensate for the drop in average output power. The trade-off is one of loop gain versus output power flatness versus filtering quality (heavier loading implies degraded Q).

Start-up, Second-Order Effects, and Pathologies

In the foregoing analysis, nothing specific was mentioned about conditions for guaranteeing the start-up of oscillations. From the general root locus of Figure 15.1, however, it should be clear that a necessary condition is a greater-than-unity value of small-signal loop transmission. To evaluate whether start-up might be a problem, one should set the transconductance equal to its small-signal value (an appropriate choice, since the circuit is certainly in the small-signal regime before oscillations have started) and then compute the loop transmission magnitude. If it does not exceed unity, the oscillator will not start up. To fix this problem, adjust some combination of bias current, device size, and tapping ratio.

In the case of the example just considered, let us identify the minimum acceptable transconductance for guaranteeing start-up. That minimum g_m, together with the given value of bias current, defines the width of the device. We use the model of Figure 15.13. The amplitude of the voltage across the tank at resonance is just

$$V_{tank} = \frac{V_1}{n} = g_m V_1 R_{eq} = g_m V_1 \left[R \parallel \frac{1}{n^2 g_m} \right], \tag{21}$$

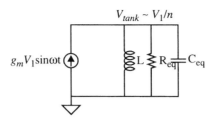

FIGURE 15.13. Start-up model of Colpitts oscillator

which reduces to the following expression for the minimum transconductance:

$$g_m > 1/R[n - n^2]. \tag{22}$$

With $n = 0.155$ and $R = 850\ \Omega$, the absolute minimum acceptable transconductance works out to approximately 9 mS. However, note that merely having enough transconductance to achieve net unity loop gain with no oscillation is not sufficient to make a good oscillator. Additionally, the describing function is accurate only in the limit of large amplitudes and thus only if the small-signal transconductance is substantially larger than the large-signal value. A reasonable choice for a first-cut design is to select g_m to be five times the minimum acceptable value. Hence, we will design for a 45-mS small-signal transconductance.

To estimate the necessary device width, initially assume that the gate overdrive is small enough that the device conforms to square-law behavior. Then we may use Eqn. 8 to estimate the overdrive:

$$\frac{g_m}{I_{BIAS}} = \frac{2}{V_{GS} - V_t} \implies V_{GS} - V_t \approx 44\ \text{mV}. \tag{23}$$

This overdrive is indeed small compared with typical values of $E_{SAT}L$ (e.g., 1–2 V), so we will continue to assume operation in the long-channel regime. Solving for W/L in this regime yields a value of about 6000 for typical values of mobility and C_{ox}. For a 0.5-μm channel length, then, the width should be roughly 3000 μm, which is quite large. This large width is a consequence of the low bias current. A higher bias current would permit the use of a substantially smaller device.

Aside from the neglect of start-up conditions in this development, several other simplifying assumptions were invoked to reduce clutter in the derivations. Transistor parasitics were ignored, for example. We now consider how to modify the analysis to take these into account.

The gate–drain (base–emitter) and drain–bulk (collector–ground) capacitances appear in parallel with the tank, and a first-order correction for their effect simply involves a reduction in the explicit capacitance added externally to keep the oscillation frequency constant. However, these capacitances are nonlinear, so distortion may be unsatisfactorily high if they constitute a significant fraction of the total tank capacitance. Temperature drift properties may also be affected.

The source–gate (emitter–base) and source–bulk (emitter–ground) capacitances appear directly in parallel with C_2, and the same comments apply as for the other device capacitances.

One must also worry about the output resistance of the transistor, for it loads the tank as well. Many high-speed transistors have low Early voltages (e.g., 10–20 volts or less), so this loading can be significant at times. In serious cases, cascoding (or some equivalent remedy) may be necessary to mitigate this problem. In other instances, this loading merely needs to be taken into account to predict the amplitude more accurately.

As a final comment on the issue of amplitude, it must be emphasized that there is always the possibility of amplitude instability, since feedback control of the amplitude is fundamentally involved. That is, instead of staying constant, the amplitude may vary in some manner (e.g., quasisinusoidally). This type of behavior is known as *squegging* and is the bane of oscillator designers. To see how squegging might arise and to develop insights concerning its prevention or cure, we employ the same analytical tools used to evaluate the stability of other feedback systems. That is, we invoke the concepts of loop transmission, crossover frequency, and phase margin. The main subtlety is that we must evaluate these quantities in terms of the *envelope* of the RF signal. Another is that the nonlinearity of amplitude control renders our linearized analyses relevant only near the operating point assumed in the linearization. Proceeding with awareness of those considerations, we would cut the loop at some convenient point (while taking care to preserve all loadings, just as we must in evaluating any loop transmission), and then apply an RF signal to the input of the cut loop. The amplitude of this test signal should be chosen the same as the nominal amplitude that prevails in actual closed-loop operation, to make sure that we evaluate stability under conditions that correspond to normal operation. Given the nonlinear nature of amplitude control, it's also prudent to examine the loop transmission at several amplitudes in order to identify (or preclude) ranges of amplitudes that may result in squegging behavior.

Next, we have a choice of evaluating either the time- or frequency-domain response (or both). In the former, we would examine the loop transmission's response to a step change in *amplitude*. To evaluate the *envelope loop transmission* in the frequency domain, we apply a sinusoidally modulated RF carrier to the input of the cut loop and then sweep the frequency of the modulation, noting the gain and phase of the output modulation relative to the input modulation.

For tuned oscillators with drain-fed tanks, a natural choice is actually to cut the loop at the drain. Inject an RF current into the tank at that point, and then let the RF current's amplitude undergo a step change. The tank by itself provides the equivalent of single-pole filtering of the step, and the capacitive coupling into the source terminal of the transistor contributes additional dynamics.

To illustrate the procedure in detail, consider a Colpitts oscillator. To simplify analysis, we first make an equivalent circuit, shown on the right of Figure 15.14. That

510   CHAPTER 15   OSCILLATORS

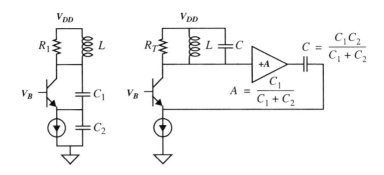

FIGURE 15.14. Colpitts oscillator and equivalent model for evaluation of envelope loop transmission

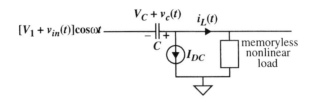

FIGURE 15.15. Capacitively coupled circuit with nonlinear load

the two circuits are in fact equivalent is readily verified by comparing loop transmissions. The drain connection is a particularly convenient point to cut the loops for making this comparison. We see that the two circuits are in fact identical as long as the elements explicitly shown in the schematics include all device parasitics.

Having derived an equivalent circuit for computing the envelope loop transmission, we now decompose the loop transmission into individual pieces. First, we analyze the capacitively coupled circuit shown in Figure 15.15.[9] Here, the load current consists of DC and RF components:[10]

$$i_L(t) = I_{DC} + i_{dc}(t) + [I_O + i_o(t)]\cos\omega t. \qquad (24)$$

The quiescent values of DC current and RF current amplitude are I_{DC} and I_O, respectively, corresponding to an RF drive amplitude of V_1. Perturbing that drive amplitude by an amount $v_{in}(t)$ produces three effects in general. One is a change in the DC value of the voltage across the capacitor by an amount $v_c(t)$, because of rectification by the nonlinear load. If the amplitude of the RF input voltage changes, then that DC capacitor voltage generally changes as well.

---

[9] This analysis is that presented by Kenneth K. Clarke and Donald T. Hess in *Communications Circuits: Analysis and Design*, Krieger, Malabar, FL, 1994.
[10] We use the term "DC" nonrigorously, to distinguish envelope from RF components.

That rectification also changes the DC current through the nonlinear load by an amount $i_{dc}(t)$. Finally, there is a change in the amplitude $i_o(t)$ of the RF current flowing into the nonlinear load. That amplitude change results from direct action by $v_{in}(t)$, compounded by the change in the DC current through the load.

We wish to determine the small-signal admittance $i_o(s)/v_{in}(s)$, but doing so by inspection is nontrivial. Note that our very statement of the problem essentially presumes that small-signal analysis holds. If we validate this assumption by considering only cases where $|v_{in}(t)| \ll V_1$, then we may express each of the currents $i_{dc}(t)$ and $i_o(t)$ as a simple linear combination of the voltages $v_c(t)$ and $v_{in}(t)$. After Laplace transformation, we therefore have

$$i_{dc}(s) = G_{00}v_c(s) + G_{01}v_{in}(s), \tag{25}$$

$$i_o(s) = G_{10}v_c(s) + G_{11}v_{in}(s), \tag{26}$$

where the various constants G_{mn} are conductances to be determined later.

Because it remains true that

$$i_{dc}(s) = -sCv_c(s), \tag{27}$$

we may equate the two expressions for $i_{dc}(s)$ to obtain

$$-sCv_c(s) = G_{00}v_c(s) + G_{01}v_{in}(s) \implies v_c(s) = \frac{-G_{01}}{sC + G_{00}}v_{in}(s). \tag{28}$$

This equation says that the small-signal DC capacitor voltage is simply a low-pass–filtered version of the small-signal RF input amplitude.

Substitution of Eqn. 28 into Eqn. 26 then yields the desired small-signal relationship between input envelope voltage and output envelope current:

$$i_o(s) = G_{10}\frac{-G_{01}}{sC + G_{00}}v_{in}(s) + G_{11}v_{in}(s) \implies \frac{i_o(s)}{v_{in}(s)} = G_{11} - \frac{G_{01}G_{10}}{sC + G_{00}}; \tag{29}$$

after some rearrangement, this becomes

$$\frac{i_o(s)}{v_{in}(s)} = \frac{G_{11}(sC + G_{00}) - G_{01}G_{10}}{sC + G_{00}} = G_{11}\frac{\left(\dfrac{sC}{G_{00}} + 1\right) - \dfrac{G_{01}G_{10}}{G_{00}G_{11}}}{\dfrac{sC}{G_{00}} + 1}. \tag{30}$$

The derivation so far is completely general; Eqn. 30 is not limited to MOSFETs or bipolars, for example. Without knowing anything about the various conductances, we can say that the admittance in question consists of a pole and a zero. This result makes physical sense, for we have one energy storage element (and therefore one pole). Furthermore, the capacitor provides a feedthrough path that emphasizes high-frequency content (of both the carrier and the envelope); that's the action of a zero. It's satisfying that the analysis presented so far passes this macroscopic reasonableness test.

Turning now to the task of figuring out what those conductances are, note that the formal definitions for the various proportionality constants are readily obtained from the time-domain versions of Eqn. 25 and Eqn. 26:

$$G_{00} \equiv \left. \frac{di_{dc}}{dv_c} \right|_{v_{in}=0}, \tag{31}$$

$$G_{01} \equiv \left. \frac{di_{dc}}{dv_{in}} \right|_{v_c=0}, \tag{32}$$

$$G_{10} \equiv \left. \frac{di_o}{dv_c} \right|_{v_{in}=0}, \tag{33}$$

$$G_{11} \equiv \left. \frac{di_o}{dv_{in}} \right|_{v_c=0}. \tag{34}$$

Note that at least two of these conductances should be familiar. From its definition we see that G_{00} is simply the nonlinear load's small-signal ratio of DC current to DC voltage; it is thus the ordinary small-signal conductance, evaluated at the bias point. Similarly, G_{11} is the change in the RF output current amplitude, divided by the change in the RF input voltage amplitude, evaluated at constant capacitor drop. Thus, G_{11} is the describing function conductance of the nonlinear load.

Two conductances we haven't encountered before involve the ratio of a DC term and an RF term. One, G_{01}, is the ratio of the change in rectified DC current, divided by the change in the amplitude of the RF input voltage that produces that rectified current, evaluated at constant capacitor drop. The other, G_{10}, is the change in the amplitude of the RF output current, divided by the change in DC capacitor voltage, for a constant-amplitude RF input voltage.

The last piece we need is a quantitative description of the envelope behavior of the drain tank. Specifically, consider a step change in the envelope of a sinusoidal drive current. The envelope response of the tank voltage will behave as a single-pole low-pass filter's response to a step voltage. Because the single-sided bandwidth of an RC low-pass filter is simply $1/RC$, we anticipate the corresponding time constant for an RLC bandpass filter to be $1/2RC$.[11] At resonance, the drain load thus contributes an *envelope impedance* given by

$$\frac{v_{tank}(s)}{i_o(s)} = \frac{R_T}{2sR_TC + 1}. \tag{35}$$

---

[11] We emphasize *single-sided* to show more clearly the analogy between a low-pass and bandpass filter. It is customary to measure a low-pass filter's bandwidth from DC to the positive-frequency -3-dB corner, rather than between the positive- and negative-frequency -3-dB corners. The pole time constant associated with that single-sided corner controls the risetime and is, of course, simply RC. For a bandpass filter, the single-sided bandwidth is $1/2RC$, meaning that the pole time constant that governs the envelope risetime is $2RC$.

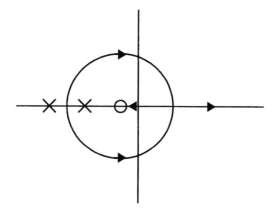

FIGURE 15.16. Possible root locus for envelope feedback loop

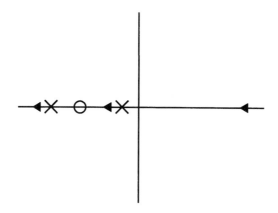

FIGURE 15.17. Another possible root locus for envelope feedback loop

The complete loop transmission is therefore

$$A \frac{i_o(s)}{v_{in}(s)} \frac{v_{tank}(s)}{i_o(s)} = A \left(G_{11} \frac{\left(\frac{sC}{G_{00}} + 1\right) - \frac{G_{01}G_{10}}{G_{00}G_{11}}}{\frac{sC}{G_{00}} + 1} \right) \frac{R_T}{2sR_TC + 1}. \quad (36)$$

Notice that we have two poles and a zero. Note also, somewhat ominously, that the envelope loop transmission is positive in sign. Positive feedback per se does not ensure instability, but we must avoid loop transmission magnitudes that are too large.

One possible root locus corresponding to this loop transmission reveals why squegging can occur; see Figure 15.16. Since we don't know exactly where the zero might be, another possibility is that shown in Figure 15.17. In this case, the poles never become complex, but one of the poles can end up with a positive real part. If any pole enters the right half-plane, the envelope will be unstable. If there is a complex pole pair in that half-plane, the instability will be observed as a (quasi)sinusoidal modulation. If the poles are real, then the modulation will be relaxation-like in character.

Regrettably, we cannot make quantitative statements without considering a specific nonlinear load. Still more regrettably, a rigorous derivation is nigh impossible for real MOSFETs (it's hard enough for a bipolar transistor). Therefore, most practical evaluations of squegging require simulations at some point in the analysis. Discovering G_{01} and G_{10} through simulation is straightforward, though.[12] Thus, even though simple analytical expressions for these parameters may not be readily forthcoming, simulations will yield actual values for them without much trouble.

Even without carrying out such simulations, we can identify general strategies to stop squegging if it occurs. By analogy with the success of dominant pole compensation in ordinary amplifiers, we would consider increasing the tank Q. The attendant narrowing in bandwidth means that the pole it contributes to the amplitude control loop moves to a lower frequency (becomes more dominant). That forces crossover to occur at a lower frequency, where presumably there is greater phase margin.

If the resonator bandwidth cannot be practically narrowed, we may still reduce the amplitude loop's crossover frequency using other methods. For example, we could reduce the loop transmission by varying the capacitive tapping ratio to feed back less signal. And of course we always retain the option of combining strategies.

One possible difficulty is that many of these parameters are interdependent. Depending on how one achieves an increase in tank Q, for example, the envelope impedance of the tank could increase which, in turn, would increase the loop transmission magnitude, frustrating efforts at stabilization. Thus, some deliberation is necessary to identify the best strategies for any given circuit.

In very stubborn cases, it may be necessary to impose external amplitude control (e.g., by explicitly measuring the amplitude, comparing it with a reference voltage, and then appropriately adjusting the bias current; see Figure 15.18). This decoupling of amplitude control from fundamental oscillator operation not only permits the exercise of additional degrees of freedom to solve the stability problem, it also allows one to design the oscillator without having to make compromises for factors such as start-up reliability (or speed) and amplitude stability.

Two additional terms that describe unwanted oscillator behaviors are *frequency pulling* and *supply pushing*. Both of these terms refer to shifts in oscillator frequency that occur due to parasitic effects. Pulling may occur from numerous sources, ranging from load changes to parasitic coupling from other periodic signals. Buffering and other isolation strategies help reduce pulling.

Supply pushing reflects the unfortunate fact that oscillator frequency is not entirely independent of supply voltage. For example, device capacitances may change as a

---

[12] The reader may reasonably ask why we should not simply simulate the entire oscillator. The answer is that squegging frequencies are usually quite a bit lower than the main oscillation frequency, so simulation times can quickly get out of hand. Using the results of a more rigorous analysis allows you to identify which simulations ought to be run.

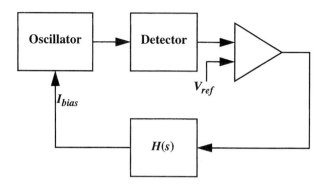

FIGURE 15.18. Oscillator with separate amplitude control

function of bias voltage, causing shifts in the oscillation frequency. Supply pushing is suppressed by selecting tank element values that swamp out such parasitic elements and by employing regulated supplies. It is important to filter the latter very well – and also to watch out for $1/f$ fluctuations in the supply voltage, since these can cause (close-in) phase modulations of the oscillator.

15.4 RESONATORS

The previous describing function example analyzed a tuned oscillator. Since tuned circuits inherently perform a bandpass filtering function, distortion products and noise are attenuated relative to the fundamental component. Not surprising, then, is that the performance of these circuits is intimately linked to the quality of available resonators. Before proceeding onward to a detailed discussion of oscillator circuitry, then, we first survey a number of resonator technologies.

RESONATOR TECHNOLOGIES

Quarter-Wave Resonators

Aside from the familiar and venerable RLC tank circuit, there are many ways to make resonators. At high frequencies, it becomes increasingly difficult to obtain adequate Q from lumped resonators because required component values are often impractical to realize.

One alternative is to use a resonator made out of a $\lambda/4$ piece of transmission line terminated in a short. For small displacements about the resonant condition, the line appears very much like a parallel RLC network. The required physical dimensions generally favor practical realization in discrete form in the UHF band and above. As an example, the free-space wavelength at 300 MHz is one meter, so that a $\lambda/4$ resonator would be about 10 inches (25 cm). On FR4, it becomes roughly 6 inches (15 cm) long or thereabouts.

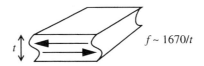

FIGURE 15.19. Illustration of bulk shear mode

Dimensions become compatible with IC realizations at mid-gigahertz frequencies. At 3 GHz, for example, the free-space $\lambda/4$ is about 1 inch (2.5 cm). With dielectric materials that are commonly available, $\lambda/4$ IC resonators smaller than half an inch (∼1 cm) or so are possible.

There is an important difference, however, between a shorted $\lambda/4$ line and a lumped RLC resonator: the line appears as an infinite (or at least large) impedance at all odd multiples of the fundamental resonance. Sometimes this periodic behavior is desired, but it can also result in oscillation simultaneously on multiple frequencies or in a chaotic hopping from one mode to another. Additional tuned elements may be required to suppress oscillation on unwanted modes.

Quartz Crystals

The most common non-RLC resonator is made of quartz. The remarkable properties and potential of quartz for use in the radio art were first seriously appreciated around 1920 by Walter G. Cady of Bell Laboratories.[13] Quartz is a piezoelectric material and thus exhibits a reciprocal transduction between mechanical strain and electric charge. When a voltage is applied across a slab of quartz, the crystal physically deforms. When a mechanical strain is applied, charges appear across the crystal.[14]

Most practical quartz crystals used at radio frequencies[15] employ a bulk shear vibrational mode; see Figure 15.19. In this mode, the resonant frequency is inversely proportional to the thickness of the slab, according to the rough formula in the figure (SI units assumed).[16]

Even though quartz does not exhibit a particularly large piezoelectric effect, it has other properties that make it extremely valuable for use in RF circuitry. Chief among

---

[13] W. G. Cady, "The Piezo-Electric Resonator," *Proc. IRE*, v. 10, April 1922, pp. 83–114. His first oscillator was somewhat complex, based as it was on a two-port piezoelectric filter.

[14] Piezoelectricity's mechanical-to-electrical transduction was discovered by Jacques and Pierre Curie (before the latter met and married Marie Sklodowska). See "Développement, par pression, de l'électricité polaire dans les cristaux hémièdres à faces inclinées" [Development, by Pressure, of Electrical Polarization in Hemihedral Crystals with Inclined Faces], *Comptes Rendus des Séances de l'Académie des Sciences*, v. 91, 1880, pp. 294–5. Their friend, physicist Gabriel Lippman, then predicted the existence of the inverse effect on thermodynamic grounds, with verification by the Curies shortly afterward.

[15] The crystals used in digital watches employ a torsional mode of vibration to allow resonance at a low frequency (32.768 kHz) in a small size.

[16] The formula given neglects the influence of the other dimensions.

15.4 RESONATORS

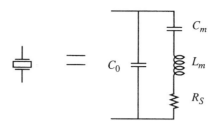

FIGURE 15.20. Symbol and model for crystal

them is the exceptional stability (both electrical and mechanical) of the material. Furthermore, it is possible to obtain crystals with very low temperature coefficients by cutting the quartz at certain angles.[17] Additionally, the transduction is virtually lossless, and Q-values are in the range of 10^4 to 10^6.[18]

An electrical model for a quartz resonator is shown in Figure 15.20. The capacitance C_0 represents the parallel-plate capacitance associated with the contacts and the lead wires, while C_m and L_m represent the mechanical energy storage. Resistance R_S accounts for the nonzero lossiness that all real systems must exhibit.

To a very crude approximation, the resistance of well-made crystals is inversely proportional to the resonant frequency and generally follows a relationship like this:

$$R_S \approx \frac{5 \times 10^8}{f_0}. \tag{37}$$

This formula is a quasi-empirical one and should be used only if measurements aren't available.[19]

The values of C_m and L_m can be computed if R_S, Q, and the resonant frequency are given. In general, because of the extraordinarily high Q-values that quartz crystals possess, the effective inductance value will be surprisingly high while the series capacitance value is vanishingly small. For example, a 1-MHz crystal with a Q of 10^5 has an effective inductance of about 8 henries (no typo here, that's really 8 *henries*) and a C_m of about 3.2 fF (again, no typo). It is apparent that crystals offer significant advantages over lumped LC realizations, where such element values are unattainable for all practical purposes.

Above about 20–30 MHz, the required slab thickness becomes impractically small. For example, a 100-MHz fundamental-mode crystal would be only about 17 μm thick. However, crystals of reasonable thickness can still be used if higher vibrational modes

---

[17] It is also possible to obtain controlled, nonzero temperature coefficients. This property has been exploited to make temperature-to-frequency transducers that function at temperatures too extreme for ordinary electronic circuits.

[18] At lower frequencies, damping by air lowers Q significantly. The higher Q-values correspond to crystals mounted inside a vacuum.

[19] This formula strictly applies only to "AT-cut" crystals operating in the fundamental mode.

are used. The boundary conditions are such that only odd overtones are allowed. Because of a variety of effects, the overtones are not exactly integer multiples of the fundamental (but they're close, off by 0.1% or so in the high direction). Third- and fifth-overtone crystals are fairly common, and seventh- or even ninth-overtone oscillators are occasionally encountered. However, as the overtone order increases, so does the difficulty of guaranteeing oscillation on only the desired mode.

As another extremely crude rule of thumb, the effective series resistance grows as the square of the overtone mode. Hence,

$$R_S \approx \frac{5 \times 10^8}{f_0} N^2, \tag{38}$$

where f_0 is here interpreted as the frequency of the Nth overtone.

Because the overtones are not at exact integer multiples of the fundamental mode, the crystal must be cut to the correct frequency at the desired overtone. Well-cut overtone crystals possess Q-values similar to those of fundamental-mode crystals.

Quartz crystal fabrication technology is an extremely advanced art. Crystals with resonant frequencies guaranteed within 50 ppm are routinely available, and substantially better performance can be obtained, although at higher cost. The general chemical inertness of quartz guarantees excellent stability over time, and a judicious choice of cut in conjunction with passive or active temperature compensation and/or control can lead to temperature coefficients of well under 1 ppm/°C. For these reasons, quartz oscillators are nearly ubiquitous in communications equipment and instrumentation (not to mention the lowly wristwatch, where a one-minute drift in a month corresponds to an error of only about 20 ppm).

Surface Acoustic Wave (SAW) Devices

Because quartz crystals operate in bulk vibrational modes, high-frequency operation requires exceedingly thin slabs. A 1-GHz fundamental-mode quartz crystal would have a thickness of only about 1.7 μm, for example. Aside from obvious fabrication difficulties, thin slabs break easily if the electrical excitation is too great. Because of their high Q, it is easy to develop large-amplitude vibrations with very modest electrical drive. Even before outright fracture occurs, the extreme bending results in a host of generally undesired nonlinear behavior.

One way to evade these limitations is to employ surface, rather than bulk, acoustic waves. If the material supports such surface modes, then the effective thickness can be much smaller than the physical thickness, allowing high resonant frequencies to be obtained with crystals of practical dimensions.

Lithium niobate ($LiNbO_3$) is a piezoelectric material that supports surface acoustic waves with little loss, and it has been used extensively to make resonators and filters at frequencies practically untouchable by quartz. Control of frequency to quartz crystal accuracy is not yet obtainable at low cost, unfortunately, but performance is adequate to satisfy high-volume, low-cost applications such as automatic garage-door

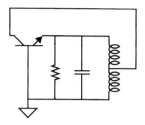

FIGURE 15.21. Hartley oscillator
(biasing not shown)

openers, which work around 250–300 MHz, typically, as well as front-end filters for cellular telephones.

Sadly, neither quartz nor lithium niobate is compatible with ordinary IC fabrication processes. Also disappointing is the lack of any piezoelectric activity in silicon. Hence, no inherently high-Q resonator can be made with layers normally found in ICs.

15.5 A CATALOG OF TUNED OSCILLATORS

There seems to be no limit to the number of ways to combine a resonator with a transistor or two to make an oscillator, as will become evident shortly. In the examples that follow, only the most minimal explanations are usually provided, perhaps leaving the reader in doubt as to which topology is "best." It is generally true that, with sufficient diligence and care, just about any of these topologies can be made to perform well enough for a given application. When we consider the issue of phase noise, more rational selection criteria will become evident.

15.5.1 BASIC *LC* FEEDBACK OSCILLATORS

The basic ingredients in these oscillators are simple: one transistor plus a resonator. Many of the oscillators are named after the fellows who first came up with the topologies but, as we'll see, a more or less unified description of these designs is possible.

We've already met one version of the Colpitts oscillator. In alternative versions, the feedback is from emitter back to the base, rather than from collector to emitter. That is, the transistor may be connected either as an emitter follower or as a common-emitter amplifier. Either way, there is net positive feedback.

The Hartley oscillator is essentially identical to the Colpitts, but it uses a tapped inductor for feedback instead of a tapped capacitor; see Figure 15.21. The Hartley oscillator has its origins in the very early days of radio, when tapped inductors were readily available. It is much less common today.

One could also use a tapped resistor, in principle, but that particular configuration doesn't seem to have a name attached to it.

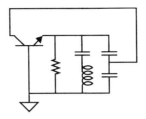

FIGURE 15.22. Clapp oscillator (biasing *still* not shown)

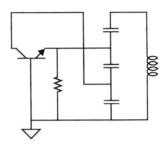

FIGURE 15.23. Redrawn Clapp oscillator

The Clapp oscillator (Figure 15.22) is a modified Colpitts oscillator, with a series LC replacing the lone inductor.[20] The Clapp oscillator is actually just a Colpitts oscillator with an additional tap on the capacitive divider chain, as is evident in the redrawn schematic of Figure 15.23. The extra tap allows the voltage swing across the inductor (and capacitive divider) to exceed considerably that of either the collector or emitter – and therefore to exceed the supply and even device breakdown voltages. The larger signal energy helps overcome the effect of various noise processes to improve spectral purity (specifically, phase noise, as discussed in Chapter 17).

Of these topologies, the Colpitts is almost certainly the most commonly encountered. Its use of tapped capacitors is most compatible with IC and microstrip implementations, although the inductor is generally not. One other important reason for the popularity of the Colpitts configuration is that it is capable of excellent phase noise performance, as we'll see.

Another oscillator idiom actually owes its existence to the instability of some tuned amplifiers. Recall that it is possible for a common-source amplifier to have a negative input admittance if it operates with a tuned load below the resonant frequency of the load (so that it looks inductive).[21] This negative resistance can be used to overcome the loss in another resonant circuit to produce oscillations. The TITO oscillator

---

[20] See James K. Clapp, "An Inductive-Capacitive Oscillator of Unusual Frequency Stability," *Proc. IRE,* v. 36, 1948, pp. 356–8, 1261. Clapp invented his modification of the Colpitts oscillator while working for the General Radio Corporation.

[21] Satisfying this condition is not sufficient, however.

15.5 A CATALOG OF TUNED OSCILLATORS

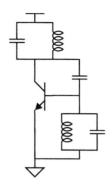

FIGURE 15.24. Tuned input–tuned output (TITO) oscillator (bias details incomplete)

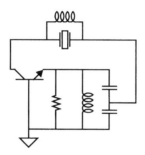

FIGURE 15.25. Colpitts crystal oscillator

(Figure 15.24) uses a Miller-effect coupling capacitor. In many designs (particularly at very high frequencies), an explicit coupling capacitor is unnecessary; the device's inherent feedback capacitance is sufficient to provide the desired negative resistance. This observation underscores the difficulty of using tuned circuits in both the input and output circuits of nonunilateral amplifiers at high frequencies.

Because of its pair of tuned circuits, the TITO oscillator is theoretically capable of producing signals with good spectral purity. However, its need of two inductors makes this topology unattractive for IC implementation. An additional strike against it is the need for careful tuning of two resonators if proper operation is to be obtained.

15.5.2 CRYSTAL OSCILLATOR POTPOURRI

Many crystal oscillators are recognizably derived from LC counterparts. In Figure 15.25, for example, the crystal is used in its series resonant mode (where it appears as a low resistance) to close the feedback loop only at the desired frequency.

The inductance across the crystal is frequently (but not always) needed in practical designs to prevent unwanted off-frequency oscillations due to feedback provided by the crystal's parallel capacitance (C_0). The inductance resonates out this capacitor so that only the series RLC arm of the crystal controls the feedback.

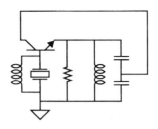

FIGURE 15.26. Modified Colpitts crystal oscillator

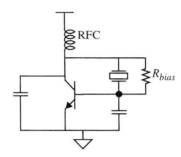

FIGURE 15.27. Pierce crystal oscillator

A variation on this theme is shown in Figure 15.26. In this particular configuration, the capacitive divider off of the tank provides the feedback as in a classic *LC* Colpitts. However, the crystal grounds the base only at the series resonant frequency of the crystal, permitting the loop to have sufficient gain to sustain oscillations at that frequency only. This topology is useful if one terminal of the crystal must be grounded.

Yet another topology is the Pierce oscillator,[22] depicted in Figure 15.27. In this oscillator, assume that the capacitors model transistor and stray parasitics, so that the transistor itself is ideal. Given this assumption, the only way to satisfy the zero–phase margin criterion is for the oscillation frequency to occur a bit above the series resonance of the crystal. That is, the crystal must look inductive at the oscillation frequency. This property confers the advantage that no external inductance is therefore

---

[22] Radio pioneer, entrepreneur, and Harvard professor George Washington Pierce made many contributions to the wireless art, of which the crystal oscillator bearing his name is but one example. See G. W. Pierce, "Piezoelectric Crystal Resonators and Crystal Oscillators Applied to the Precision Calibration of Wavemeters," *Proc. Amer. Acad. of Arts and Sci.*, v. 59, October 1923, pp. 81–106. Also see his U.S. Patent #2,133,642, filed 25 February 1924, granted 18 October 1938. He made these developments soon after Cady demonstrated an early piezoelectric oscillator to him. In addition to his work in oscillators, Pierce gave us the name "crystal rectifier" for point-contact diodes and painstakingly disproved a thermally based explanation of their operation. His 1909 textbook, *Principles of Wireless Telegraphy*, was the first runaway best-seller of the technology's early days.

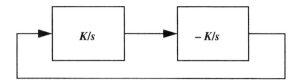

FIGURE 15.28. Quadrature oscillator block diagram

required for this oscillator to function (the RF choke may be replaced by a large-valued resistor or current source). Hence, it is more amenable to integration than a Colpitts, for example, particularly at low frequencies.

That the crystal must look inductive can be argued as follows. If we are to have a phase margin of zero and if the transistor's transconductance already provides a 180° phase shift, then the passive elements must supply the other 180°. There's no way for a two-pole *RC* network to provide 180° (close, but close doesn't count here), so the crystal must look inductive.

Because the output frequency of a Pierce oscillator thus does not coincide with the series resonance of the crystal, one must use a crystal that has been cut to oscillate at the desired frequency *with a specified load capacitance* (in this case, the value of the two capacitors in series).

As a final note on the Pierce oscillator, it happens to form the basis of many "digital" oscillators. An ordinary CMOS inverter, for example, can act as the gain element if biased to its linear region (e.g., with a large-valued feedback resistor R_{bias}, just as with the single-transistor implementation). Just add the appropriate amount of input and output capacitance, toss in the crystal from input to output, and chances are very good that you'll have an oscillator. Generally one or two stages of buffering (with more inverters of course) are necessary to obtain full CMOS swings and also to isolate load changes from the oscillator core.

15.5.3 OTHER OSCILLATOR CONFIGURATIONS

In some applications, it is desirable to have two outputs in quadrature. One oscillator architecture that naturally provides quadrature outputs (at least in principle) uses a pair of integrators in a feedback loop; this is shown in Figure 15.28. From the magnitude condition, we can deduce that the frequency of oscillation is

$$\omega_{osc} = K. \qquad (39)$$

Thus tuning may be effected by varying the integrator gain. Furthermore, the desired quadrature relationship is obtained across any of the integrators.

In practice, unmodeled dynamics cause a departure from ideal behavior. Consider, for example, the effect of additional poles on the root locus. Rather than consisting of a purely imaginary pair, the locus with additional poles breaks away from the imaginary axis. Furthermore, these unmodeled parasitics tend to be rather unreliable, so

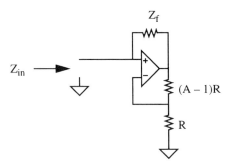

FIGURE 15.29. Generalized impedance converter

that allowing the oscillation frequency to depend on them is undesirable. Despite these obstacles, however, 1-GHz quadrature oscillators with reasonable quadrature phase (error under 0.5 degrees) have been reported.[23]

15.6 NEGATIVE RESISTANCE OSCILLATORS

A perfectly lossless resonant circuit is very nearly an oscillator, but lossless elements are difficult to realize. Overcoming the energy loss implied by the finite Q of practical resonators with the energy-supplying action of active elements is one potentially attractive way to build practical oscillators, as in the TITO example.

The foregoing description is quite general, covering both feedback and open-loop topologies. Among the former is a classic textbook circuit, the negative impedance converter (NIC). The NIC can be realized with a simple op-amp circuit that employs both positive and negative feedback. Specifically, consider the configuration shown in Figure 15.29.

If ideal op-amp behavior is assumed, it is easy to show that the input impedance is related to the feedback impedance as follows:

$$Z_{in} = Z_f/(1-A). \qquad (40)$$

If the closed-loop gain A is set equal to precisely 2, then the input impedance will be the algebraic inverse of the feedback impedance. If the feedback impedance is in turn chosen to be a pure positive resistance, then the input impedance will be a purely negative resistance. This negative resistance may be used to offset the positive resistance of all practical resonators to produce an oscillator.

As usual, the inherent nonlinearities of all real active devices will limit amplitudes, and describing functions can be used to estimate the oscillation amplitude, if desired. Describing functions may also be used to verify that the oscillator will, in fact, oscillate.

---

[23] R. Duncan et al., "A 1 GHz Quadrature Sinusoidal Oscillator," *IEEE CICC Digest,* 1995, pp. 91–4.

15.6 NEGATIVE RESISTANCE OSCILLATORS

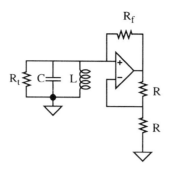

FIGURE 15.30. Negative resistance oscillator

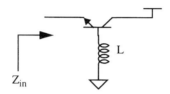

FIGURE 15.31. Canonical RF negative resistance (biasing not shown)

As a specific example, consider the oscillator of Figure 15.30. To guarantee oscillation, we require the net resistance across the tank to be negative. Thus, we must satisfy the inequality

$$R_t > R_f. \qquad (41)$$

The nonlinearity that most typically limits the amplitude at low frequencies is the finite output swing of all real amplifiers. Since there is a gain of 2 from the tank to the op-amp output, the signal across the tank will generally limit to a value somewhat greater than half the supply, corresponding to periodic saturation of the amplifier output.

At higher frequencies, it is possible for the finite slew rate of the amplifier to control the amplitude (partially, if not totally). In general, this situation is undesirable because the phase lag associated with slew limiting can cause a shift in oscillation frequency. In extreme cases, the amplitude control provided by slew limiting (or almost any other kind of amplitude limiting) can be unstable, and squegging can occur.

Finally, the various oscillator configurations presented earlier (e.g., Colpitts, Pierce, etc.) may themselves be viewed as negative resistance oscillators.

A more practical negative resistance is easily obtained by exploiting yet another "parasitic" effect: Inductance in the base circuit of a common-base amplifier can cause a negative resistance to appear at the emitter terminal, as seen in Figure 15.31. A straightforward analysis reveals that Z_{in} has a negative real part for frequencies greater than the resonant frequency of the inductor and C_{be} (if C_{cb} is neglected). For frequencies much larger than that resonant frequency but much smaller than ω_T, the real part of Z_{in} is approximately

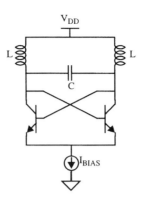

FIGURE 15.32. Simple differential negative resistance oscillator

$$R_{in} \approx -\frac{\omega^2 L}{\omega_T} = -\frac{\omega}{\omega_T}|Z_L|. \tag{42}$$

The ease with which this circuit provides a negative resistance accounts for its popularity. However, it should be obvious that this ease also underscores the importance of minimizing parasitic base inductance when a negative resistance is *not* desired.

A circuit that has become a frequently recurring idiom in recent years uses a cross-coupled differential pair to synthesize the negative resistance. "It is left as an exercise for the reader" to analyze the circuit of Figure 15.32.

As will be shown in Chapter 17, spectral purity improves if the signal amplitudes are maximized (because this increases the signal-to-noise ratio). In many oscillators, such as the circuit of Figure 15.32, the allowable signal amplitudes are constrained by the available supply voltage or breakdown voltage considerations. Since it is the energy in the tank that constitutes the "signal," one could take a cue from the Clapp oscillator and employ tapped resonators to allow peak tank voltages that exceed the device breakdown limits or supply voltage, as in the negative resistance oscillator[24] shown in Figure 15.33.[25]

The differential connection might make it a bit difficult to see that this circuit indeed employs a tapped resonator, so consider the simplified half-circuit of Figure 15.34. In the simplified half-circuit, the transistors are replaced by a negative resistor and the positive resistors are not shown at all. Furthermore, the two capacitors are replaced by their series equivalent, while the junction of the two inductors corresponds to the drain connection of the original circuit.

---

[24] J. Craninckx and M. Steyaert, "A CMOS 1.8GHz Low-Phase-Noise Voltage-Controlled Oscillator with Prescaler," *ISSCC Digest of Technical Papers,* February 1995, pp. 266–7. The inductors are bondwires stitched across the die.

[25] It is best to have a tail current source to constrain the swing, but we omit this detail in the interest of simplicity.

15.6 NEGATIVE RESISTANCE OSCILLATORS

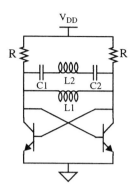

FIGURE 15.33. Negative resistance oscillator with modified tank (simplified)

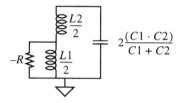

FIGURE 15.34. Simplified half-circuit of negative resistance oscillator

It should be clear that the swing across the equivalent capacitance (or across $L2$) can exceed the supply voltage (and even the transistor breakdown voltage) because of the tapped configuration, so that this oscillator is the philosophical cousin of the Clapp configuration. Useful output may be obtained either through a buffer interposed between the oscillator core and load or through a capacitive voltage divider to avoid spoiling resonator Q. As a consequence of the large energy stored in the tank with either a single- or double-tapped resonator, this topology is capable of excellent phase noise performance, as will be appreciated in Chapter 17.

Tuning of this (and all other LC) oscillators may be accomplished by realizing all or part of $C1$ or $C2$ as a variable capacitor (such as the junction capacitor formed with a p+ diffusion in an n-well) and tuning its effective capacitance with an appropriate bias control voltage. Since many junction capacitors have relatively poor Q, it is advisable to use only as much junction capacitance as necessary to achieve the desired tuning range. In practice, tuning ranges are frequently limited to below 5–10% if excessive degradation of phase noise is to be avoided. A simple (but illustrative) example of a voltage-controlled oscillator using this method is shown in Figure 15.35.

As a final comment on negative resistance oscillators, it should be clear that many (if not all) oscillators may be considered as negative resistance oscillators because, from the point of view of the tank, the active elements cancel the loss due to finite Q

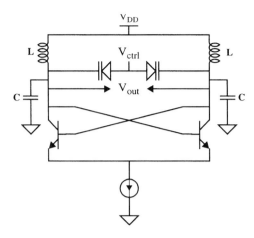

FIGURE 15.35. Voltage-controlled negative resistance oscillator (simplified)

of the resonators. Hence, whether to call an oscillator a "negative resistance" type is actually more a philosophical decision than anything fundamental.

15.7 SUMMARY

In this chapter we examined how the amplitude of oscillation can be stabilized through nonlinear means and also extended feedback concepts to include a particular type of linearized nonlinearity: describing functions. Armed with describing functions and knowledge of the rest of the elements in a loop transmission, both oscillation frequency and amplitude can be determined.

We looked at a variety of oscillators, of both open-loop and feedback topologies. The Colpitts and Hartley oscillators use tapped tanks to provide positive feedback, while the TITO oscillator employs the negative resistance that a tuned amplifier with Miller feedback can provide. The Clapp oscillator uses an extra tap to allow resonator swings that exceed the supply voltage, permitting signal energy to dominate noise.

Crystal oscillator versions of LC oscillators were also presented. Since a quartz crystal behaves much like an LC resonator with extraordinarily high Q, it permits the realization of oscillators with excellent spectral purity and low power consumption. The Colpitts configuration oscillates at the series-resonant frequency of the crystal and thus requires an LC tank. The Pierce oscillator operates at a frequency where the crystal looks inductive, and therefore it requires no external inductance. The off-resonant operation, however, forces the use of crystals that have been cut specifically for a particular load capacitance.

A random sampling of other oscillators was also provided, including a quadrature oscillator using two integrators in a feedback loop as well as several negative resistance oscillators. Again, tapped resonators were seen to be beneficial for improving phase noise.

CHAPTER SIXTEEN

SYNTHESIZERS

16.1 INTRODUCTION

Phase-locked loops (PLLs) have become ubiquitous in modern communications systems because of their remarkable versatility. As one important example, a PLL may be used to generate an output signal whose frequency is a programmable and rational multiple of a fixed input frequency. Such *frequency synthesizers* are often used to provide the local oscillator signal in superheterodyne transceivers. PLLs may also be used to perform frequency modulation and demodulation as well as to regenerate the carrier from an input signal in which the carrier has been suppressed. Their versatility also extends to purely digital systems, where PLLs are indispensable in skew compensation, clock recovery, and the generation of clock signals.

To understand in detail how PLLs may perform such a vast array of functions, we will need to develop linearized models of these feedback systems. But first, of course, we begin with a little history to put this subject in its proper context.

16.2 A SHORT HISTORY OF PLLs

The earliest description of what is now known as a PLL was provided by de Bellescize in 1932.[1] This early work offered an alternative architecture for receiving and demodulating AM signals, using the degenerate case of a superheterodyne receiver in which the intermediate frequency is zero. With this choice, there is no image to reject, and all processing downstream of the frequency conversion takes place in the audio range.

To function correctly, however, the *homodyne* or *direct-conversion* receiver requires a local oscillator (LO) whose frequency is *precisely* the same as that of the incoming carrier. Furthermore, the local oscillator must be in phase with the incoming carrier for maximum output. If the phase relationship is uncontrolled, the gain could be as small as zero (as in the case where the LO happens to be in quadrature with

---

[1] "La réception synchrone," *L'Onde Électrique,* v. 11, June 1932, pp. 230–40.

the carrier) or vary in some irritating manner. De Bellescize described a way to solve this problem by providing a local oscillator whose phase is locked to that of the carrier.

For various reasons, the homodyne receiver did not displace the ordinary superheterodyne receiver, which had come to dominate the radio market by about 1930. However, there has recently been a renewal of interest in the homodyne architecture because its relaxed filtering requirements possibly improve amenability to integration.[2]

The next PLL-like circuit to appear was used in televisions for over three decades. In standard broadcast television, two sawtooth generators provide the vertical and horizontal deflection ("sweep") signals. To allow the receiver to synchronize the sweep signals with those at the studio, timing pulses are transmitted along with the audio and video signals.

To perform synchronization in older sets, the TV's sweep oscillators were adjusted to free-run at a somewhat lower frequency than the actual transmitted sweep rate. In a technique known as *injection locking*,[3] the timing pulses caused the sawtooth oscillators to terminate each cycle prematurely, thereby effecting synchronization. As long as the received signal had relatively little noise, the synchronization worked well. However, as signal-to-noise ratio degraded, synchronization suffered either as timing pulses disappeared or as noise was misinterpreted as timing pulses. In the days when such circuits were the norm, every TV set had to have vertical and horizontal "hold" controls to allow the consumer to fiddle with the free-running frequency and, therefore, the quality of the lock achieved. Improper adjustment caused vertical rolling or horizontal "tearing" of the picture. In modern TVs, true PLLs are used to extract the synchronizing information robustly even when the signal-to-noise ratio has degraded severely. As a result, vertical and horizontal hold adjustments thankfully have all but disappeared.

The next wide application of a PLL-like circuit was also in televisions. When various color television systems were being considered in the late 1940s and early 1950s, the Federal Communications Commission (FCC) imposed a requirement of compatibility with the existing black-and-white standard, and it further decreed that the color television signal could not require any additional bandwidth. Since monochrome

---

[2] However, the homodyne requires exceptional front-end linearity and is intolerant of DC offsets. Furthermore, since the RF and LO frequencies are the same, LO leakage back out of the antenna is a problem. Additionally, this LO leakage can sneak back into the front end, where it mixes with the LO with some random phase, resulting in a varying DC offset that can be several orders of magnitude larger than the RF signal. These problems are perhaps as difficult to solve as the filtering problem, and are considered in greater detail in Chapter 19 of T. H. Lee, *The Design of CMOS Radio-Frequency Integrated Circuits,* 2nd ed., Cambridge University Press, 2004.

[3] See Balth. van der Pol, "Forced Oscillations in a Circuit with Nonlinear Resistance (Reception with Reactive Triode)," *Philosophical Magazine,* v. 3, January 1927, pp. 65–80, as well as R. B. Adler, "A Study of Locking Phenomena in Oscillators," *Proc. IRE,* v. 34, June 1946, pp. 351–7. The circadian rhythms of humans provide another example of injection locking. In the absence of a synchronizing signal from the sun, a "day" for most people exceeds 24 hours. Note that the free-running frequency is again somewhat lower than the locked frequency.

television had been developed without looking forward to a colorful future, it was decidedly nontrivial to satisfy these constraining requirements. In particular, it seemed impossible to squeeze a color TV signal into the same spectrum as a monochrome signal without degrading something. The breakthrough was in recognizing that the 30-Hz frame rate of television results in a comblike (rather than continuous) spectrum, with peaks spaced 30 Hz apart. Color information could thus be shoehorned in between these peaks without requiring additional bandwidth. To accomplish this remarkable feat, the added color information is modulated on a *subcarrier* of approximately 3.58 MHz.[4] The subcarrier frequency is carefully chosen so that the sidebands of the chroma signal fall precisely midway between the spectral peaks of the monochrome signal. The combined monochrome (also known as the luminance or brightness signal) and chroma signals subsequently modulate the final carrier that is ultimately transmitted. The U.S. version of this scheme is known as NTSC (for National Television Systems Committee).

Color information is encoded as a vector whose phase with respect to the subcarrier determines the hue and whose magnitude determines the amplitude ("saturation") of the color. The receiver must therefore extract or regenerate the subcarrier quite accurately to preserve the $0°$ phase reference; otherwise, the reproduced colors will not match those transmitted.

To enable this phase locking, the video signal includes a "burst" of a number of cycles (NTSC specifications dictate a minimum of 8) of a 3.58-MHz reference oscillation transmitted during the retrace of the CRT's electron beam as it returns to the left side of the screen. This burst signal feeds a circuit inside the receiver whose job is to regenerate a continuous 3.58-MHz subcarrier that is phase-locked to this burst. Since the burst is not applied continuously, the receiver's oscillator must free-run during the scan across a line. To prevent color shifts, the phase of this regenerated subcarrier must not drift. Early implementations did not always accomplish this goal successfully, leading some wags to dub NTSC "never twice the same color."[5]

Europe (with the exception of France[6]) chose to adopt a similar chroma scheme, but there the phase drift problem was addressed by alternating the polarity of the reference every line. This way, phase drifts tend to average out to zero over two successive lines, reducing or eliminating perceived color shifts. Thus was born the phase-alternating line (PAL) system.

---

[4] If you really want to know, the exact frequency is 3.579545 MHz, derived from the 4.5-MHz spacing between the video and audio carrier frequencies, multiplied by 455/572.

[5] In fact, the very earliest such circuits dispensed with an oscillator altogether. Instead, the burst signal merely excited a high-Q resonator (a quartz crystal), and the resulting ringing was used as the regenerated subcarrier. The ringing had to persist for over 200 cycles without excessive decay. Cheesy!

[6] The French color television system is known as SECAM, for Séquentiel Couleur avec Mémoire. In this system, luminance and chrominance information are sent serially in time and reconstructed in the receiver.

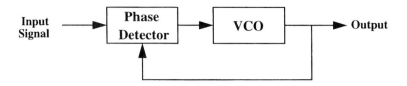

FIGURE 16.1. Phase-locked loop architecture

Another early application of PLL-like circuits was in stereo FM radio. Again, to preserve backward compatibility, the stereo information is encoded on a subcarrier, this time at 38 kHz. Treating the monaural signal as the sum of a left and right channel (and bandlimited to 15 kHz), stereo broadcast is enabled by modulating the subcarrier with the *difference* between the left and right channels. This L − R difference signal is encoded as a double-sideband, suppressed-carrier (DSB-SC) signal. The receiver then regenerates the 38-kHz subcarrier and recovers the individual left and right signals through simple addition and subtraction of the L + R monaural and L − R difference signals. To simplify receiver design, the transmitted signal includes a low-amplitude *pilot* signal at precisely half the subcarrier frequency, which is doubled at the receiver and used to demodulate the L − R signal. As we'll see shortly, a PLL can easily perform this frequency-doubling function even without a pilot, but for the circuits of 1960, it was a tremendous help.

Early PLLs were mainly of the injection-locked variety because the cost of a complete, textbook PLL was too great for most consumer applications. Except for a few exotic situations, such as satellite communications and scientific instrumentation, such "pure" PLLs didn't exist in significant numbers until the 1970s, when IC technology had advanced enough to provide a true PLL for stereo FM demodulation. Since then, the PLL has become commonplace, found in systems ranging from the mundane to the highly specialized.

From the foregoing, it should be clear that phase locking enables a rich variety of applications. With that background as motivation, we now turn to the task of modeling "textbook" PLLs.

16.3 LINEARIZED PLL MODEL

The basic PLL architecture, shown in Figure 16.1, consists of a phase detector and a voltage-controlled oscillator (VCO).[7] The phase detector compares the phase of an incoming reference signal with that of the VCO and then produces an output that is some function of the phase difference. The VCO simply generates a signal whose frequency is some function of the control voltage.

---

[7] Some oscillators control frequency through an adjustment of current. Nevertheless, it is common practice to refer to both current- and voltage-controlled oscillators as VCOs, unless there is some overriding need to make a distinction.

16.3 LINEARIZED PLL MODEL

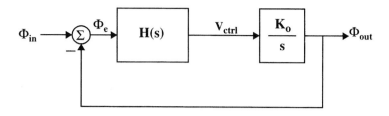

FIGURE 16.2. Linearized PLL model

The general idea is that the output of the phase detector drives the VCO frequency in a direction that reduces the phase difference; that is, it's a negative feedback system. Once the loop achieves lock, the phase of the input reference and VCO output signals ideally have a fixed phase relationship (most commonly 0° or 90°, depending on the nature of the phase detector).

Although both the phase detector and VCO may be highly nonlinear in practice, it is customary to assume linearity when analyzing loops that have achieved lock. We will eventually consider a more general case (including the acquisition process), but we have to begin somewhere, and it's best to start simple and add complexity as we go along.

Let us begin with a linearized PLL model, as seen in Figure 16.2. Because we are generally interested in the phase relationship between the input and output signals, the input and output variables are phases in this model, rather than the time waveforms of the actual inputs and outputs. Hence, if you are accustomed to thinking of signals as voltages in a block diagram, the input and output voltages are now proportional to phases.

Another consequence of choosing phase as the input–output variable is that the VCO, whose output frequency depends on a control voltage, is modeled as an integrator, since phase is the integral of frequency. The VCO gain constant K_o has units of radians per second per volt; it merely describes what change in output frequency results from a specified change in control voltage. Also note that, unlike ordinary amplifiers whose outputs are bounded, the VCO is a true integrator. The longer we wait, the more phase we accumulate (unless someone turns off the oscillator).

The phase detector is modeled as a simple subtractor that generates a phase error output Φ_e that is the difference between the input and output phases. To accommodate gain scaling factors and the option of additional filtering in the loop, a block with transfer function $H(s)$ is included in the model as well.

16.3.1 FIRST-ORDER PLL

The simplest PLL is one in which the function $H(s)$ is simply a scalar gain (call it K_D, with units of volts per radian). Because the loop transmission then possesses just a single pole, this type of loop is known as a first-order PLL. Aside from simplicity, its main attribute is the ease with which large phase margins are obtained.

Offsetting those positive attributes is an important shortcoming, however: bandwidth and steady-state phase error are strongly coupled in this type of loop. One generally wants the steady-state phase error to be zero, independent of bandwidth, so first-order loops are infrequently used.

We may use our linear PLL model to evaluate quantitatively the limitations of a first-order loop. Specifically, the input–output phase transfer function is readily derived:

$$\frac{\Phi_{out}(s)}{\Phi_{in}(s)} = \frac{K_0 K_D}{s + K_0 K_D}. \tag{1}$$

The closed-loop bandwidth is therefore

$$\omega_h = K_0 K_D. \tag{2}$$

To verify that the bandwidth and phase error are linked, let's now derive the input-to-error transfer function:

$$\frac{\Phi_e(s)}{\Phi_{in}(s)} = \frac{s}{s + K_0 K_D}. \tag{3}$$

If we assume that the input signal is a constant-frequency sinusoid of frequency ω_i, then the phase ramps linearly with time at a rate of ω_i radians per second. Thus, the Laplace-domain representation of the input signal is

$$\Phi_{in}(s) = \frac{\omega_i}{s^2}, \tag{4}$$

so that

$$\Phi_e(s) = \frac{\omega_i}{s(s + K_0 K_D)}. \tag{5}$$

The steady-state error with a constant frequency input is therefore

$$\lim_{s \to 0} s \Phi_e(s) = \frac{\omega_i}{K_0 K_D} = \frac{\omega_i}{\omega_h}. \tag{6}$$

The steady-state phase error is thus simply the ratio of the input frequency to the loop bandwidth; a one-radian phase error results when the loop bandwidth equals the input frequency. A small steady-state phase error therefore requires a large loop bandwidth; the two parameters are tightly linked, as asserted earlier.

An intuitive way to arrive qualitatively at this result is to recognize that a nonzero voltage is required in general to drive the VCO to the correct frequency. Since the control voltage derives from the output of the phase detector, there must be a nonzero phase error. To produce a given control voltage with a smaller phase error requires an increase in the gain that relates the control voltage to the phase detector output. Because an increase in gain raises the loop transmission uniformly at all frequencies, a bandwidth increase necessarily accompanies a reduction in phase error.

To produce zero phase error, we require an element that can generate an arbitrary VCO control voltage from a zero phase detector output, implying the need for an infinite gain. Yet to decouple the steady-state error from the bandwidth, this element needs to have infinite gain only at DC, rather than at all frequencies. An integrator has the prescribed characteristics, and its use leads to a second-order loop.

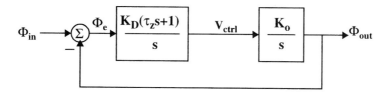

FIGURE 16.3. Model of second-order PLL

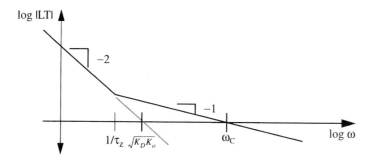

FIGURE 16.4. Loop transmission of second-order PLL

16.3.2 SECOND-ORDER PLL

The model for a second-order PLL is shown in Figure 16.3. The 90° negative phase shift contributed by the added integrator has to be offset by the positive phase shift of a loop-stabilizing zero. As with any other feedback system compensated in this manner, the zero should be placed well below the crossover frequency to obtain acceptable phase margin.

In this model, the constant K_D has the units of volts per second because of the extra integration. Also thanks to the added integration, the loop bandwidth may be adjusted independently of the steady-state phase error (which is zero here), as is clear from studying the loop transmission magnitude behavior graphed in Figure 16.4. The stability of this loop can be explored with the root-locus diagram of Figure 16.5.

As the loop transmission magnitude increases (by increasing $K_D K_0$), the loop become progressively better damped because an increase in crossover frequency allows more of the zero's positive phase shift to offset the negative phase shift of the poles. For very large loop transmissions, one closed-loop pole ends up at nearly the frequency of the zero, while the other pole heads for infinitely large frequency.

In this PLL implementation, the loop-stabilizing zero comes from the forward path. Hence, this zero also shows up in the closed-loop transfer function.

It is straightforward to show that the phase transfer function is

$$\frac{\Phi_{out}}{\Phi_{in}} = \frac{\tau_z s + 1}{s^2/K_D K_0 + \tau_z s + 1}, \qquad (7)$$

from which we determine that

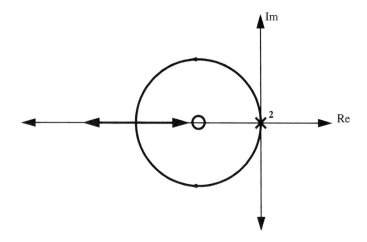

FIGURE 16.5. Root locus of second-order PLL

$$\omega_n = \sqrt{K_D K_0} \tag{8}$$

and

$$\zeta = \frac{\omega_n \tau_z}{2} = \frac{\tau_z \sqrt{K_D K_0}}{2}. \tag{9}$$

Furthermore, the crossover frequency for the loop may be expressed as

$$\omega_c = \left[\frac{\omega_n^4}{2\omega_z^2} + \omega_n^2 \sqrt{\frac{1}{4}\left(\frac{\omega_n}{\omega_z}\right)^4 + 1} \right]^{1/2}, \tag{10}$$

which simplifies considerably if the crossover frequency is well above the zero frequency (as it is frequently is):

$$\omega_c \approx \frac{\omega_n^2}{\omega_z}. \tag{11}$$

Figure 16.4 and Eqn. 10 both show that the crossover frequency always exceeds ω_n, which – from Figure 16.4 and Eqn. 8 – is the extrapolated crossover frequency of the loop with no zero. Finally, it should be clear increasing the zero's time constant improves the damping, given a fixed ω_n. Thus, the bandwidth and stability of a second-order loop may be adjusted as desired while preserving a zero steady-state phase error.

16.4 PLL REJECTION OF NOISE ON INPUT

It can be shown that maximizing the bandwidth of the PLL helps to minimize the influence of disturbances that alter the frequency of, say, a voltage-controlled oscillator. This insight is not too deep – making a system faster means that it recovers more quickly from errors, whatever the source.

However, there is a potential drawback to maximizing the bandwidth, above and beyond the stability issue. As the loop bandwidth increases, the loop gets better at

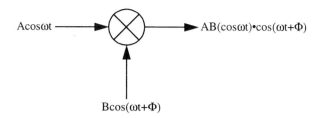

FIGURE 16.6. Multiplier as phase detector

tracking the input. If the input is noise-free (or at least less noisy than the PLL's own VCO), then there is a net improvement overall. However, if the input signal is *noisier* than the PLL's VCO, then the high-bandwidth loop will faithfully reproduce this input noise at the output. Hence, there is a trade-off between sensitivity to noise on the input to the loop (a consideration that favors smaller loop bandwidths) and sensitivity to noise that disturbs the VCO frequency (which favors larger loop bandwidths).

In general, tuned oscillators (e.g., *LC* or crystal-based) are inherently less (often much less) noisy, at a given power level, than relaxation oscillators (such as ring or *RC* phase-shift oscillators). Hence, if the reference input to the PLL is supplied from a tuned oscillator while the VCO is based on a relaxation oscillator topology, larger bandwidths are favored. If, instead, the situation is the reverse (a rarer occurrence) and a relaxation oscillator supplies the reference to a crystal oscillator–based PLL, then smaller loop bandwidths will generally be favored.

16.5 PHASE DETECTORS

We've taken a look at the classical phase-locked loop at the block diagram level, with a particular focus on the linear behavior of a second-order loop in lock. We now consider a few implementation details to see how real PLLs are built and how they behave. In this section, we'll examine several representative phase detectors.

16.5.1 THE ANALOG MULTIPLIER AS A PHASE DETECTOR

In PLLs that have sine-wave inputs and sine-wave VCOs, the most common phase detector by far is the multiplier, often implemented with a Gilbert-type topology. For an ideal multiplier, it isn't too difficult to derive the input–output relationship. See Figure 16.6.

Using some trigonometric identities, we find that the output of the multiplier may be expressed as:

$$AB \cos \omega t \cos(\omega t + \Phi) = \frac{AB}{2}[\cos(\Phi) - \cos(2\omega t + \Phi)]. \qquad (12)$$

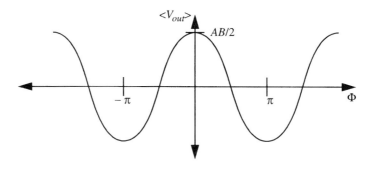

FIGURE 16.7. Multiplier phase detector output vs. phase difference

Note that the output of the multiplier consists of a DC term and a double-frequency term. For phase detector operation, we are interested only in the DC term. Hence, the average output of the phase detector is

$$\langle AB \cos \omega t \cos(\omega t + \Phi) \rangle = \frac{AB}{2}[\cos \Phi]. \quad (13)$$

We see that the phase detector gain "constant" is a function of the phase angle and is given by

$$K_D = \frac{d}{d\Phi} \langle V_{out} \rangle = -\frac{AB}{2}[\sin(\Phi)]. \quad (14)$$

If we plot the average output as a function of phase angle, we get something that looks roughly as shown in Figure 16.7. Notice that the output is periodic. Further note that the phase detector gain constant is zero when the phase difference is zero and is greatest when the input phase difference is 90°. Hence, to maximize the useful phase detection range, the loop should be arranged to lock to a phase difference of 90°. For this reason, a multiplier is often called a *quadrature* phase detector.

When the loop is locked in quadrature, the phase detector has an incremental gain constant given by

$$K_D|_{\Phi=\pi/2} = \frac{d}{d\Phi} \langle V_{out} \rangle \bigg|_{\Phi=\pi/2} = -\frac{AB}{2}. \quad (15)$$

In what follows, we will glibly ignore minus signs. The reason for this neglect is that a loop may servo to either a 90° or −90° phase difference (but not to both), depending on the net number of inversions provided by the rest of the loop elements.

Because there are two phase angles (within any given 2π interval) that result in a zero output from the phase detector, there would seem to be two equilibrium points to which the loop could lock. However, one of these points is a stable equilibrium while the other is a *metastable* point from which the loop must eventually diverge. That is, only one of these lock points corresponds to negative feedback.

When speaking of phase errors for a quadrature loop, we calculate the departure from the equilibrium condition of a 90° phase difference. Thus, when the phase difference is 90° in an ideal quadrature loop, the phase *error* is considered to be zero.

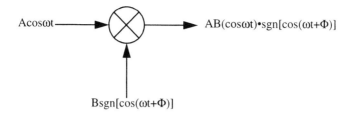

FIGURE 16.8. Multiplier with one square-wave input

16.5.2 THE COMMUTATING MULTIPLIER AS A PHASE DETECTOR

In the previous section, we assumed that both inputs to the loop were sinusoidal. However, one or both of these inputs may be well approximated by a square wave in many cases of practical interest, so let us now modify our results to accommodate a single square-wave input. In this case, we have the situation depicted in Figure 16.8, where "sgn" is the signum function, defined as:

$$\text{sgn}(x) = \begin{cases} 1 & \text{if } x > 0, \quad (16) \\ -1 & \text{if } x < 0, \quad (17) \end{cases}$$

Now recall that a square wave of amplitude B has a fundamental component whose amplitude is $4B/\pi$. If we assume that we care only about the fundamental component of the square wave, then the average output of the multiplier is

$$\langle V_{out} \rangle = \frac{4}{\pi} \frac{AB}{2} [\cos(\Phi)] = \frac{2}{\pi} AB [\cos(\Phi)]. \quad (18)$$

The corresponding phase detector gain is similarly just $4/\pi$ times as large as in the purely sinusoidal case:

$$K_D|_{\Phi=\pi/2} = \frac{d}{d\Phi} \langle V_{out} \rangle \bigg|_{\Phi=\pi/2} = -\frac{2AB}{\pi}. \quad (19)$$

Although the expressions for the phase detector output and gain are quite similar to those for the purely sinusoidal case, there is an important qualitative difference between these two detectors. Because the square wave consists of more than just the fundamental component, the loop can actually lock onto harmonics or subharmonics of the input frequency. Consider, for example, the case where the B square-wave input is at precisely one third the frequency of the sinusoidal input frequency. Now, square waves[8] consist of odd harmonics, and the third harmonic will then be at the

---

[8] We are implicitly assuming that the square waves are of 50% duty cycle. Asymmetrical square waves will also contain even as well as odd harmonic components, providing an "opportunity" to lock to even multiples of the incoming reference in addition to odd multiples.

same frequency as the input sine wave. Those two signals will provide a DC output from the multiplier.

Because the spectrum of a square wave drops off as $1/f$,[9] the average output gets progressively smaller as we attempt to lock to higher and higher harmonics. The attendant reduction in phase detector gain constant thus makes it more difficult to achieve or maintain lock at the higher harmonics, but this issue must be addressed in all practical loops that use this type of detector. Sometimes harmonic locking is desirable, and sometimes it isn't. If it isn't, then the VCO frequency range usually has to be restricted (or acquisition carefully managed) to prevent the occurrence of harmonic locking.

Another observation worth making is that multiplication of a signal by a periodic signum function is equivalent to inverting the phase of the signal periodically. Hence, a multiplier used this way can be replaced by switches (also known as commutators, by analogy with a component of rotating machines). The passive diode ring mixers function as commutating mixers. In some IC technologies (such as CMOS), commutating mixers supplement Gilbert-type multipliers.

16.5.3 THE EXCLUSIVE-OR GATE AS A PHASE DETECTOR

If we now drive an analog multiplier with square waves on *both* inputs, we could analyze the situation by using the Fourier series for each of the inputs, multiplying them together, and so forth. However, it turns out that analyzing this particular situation in the *time* domain is much easier, so that's what we'll do. The reader is welcome (indeed, encouraged) to explore the alternative method and perform the analysis in the frequency domain as a recreational exercise.

In this case, the two square-wave inputs produce the output shown in Figure 16.9. As we change the input phase difference, the output takes the form of a square wave of varying duty cycle, with a 50% duty cycle corresponding to a quadrature relationship between the inputs. Since the duty cycle is in fact proportional to the input phase difference, we can readily produce a plot (Figure 16.10) of the average output as a function of the input phase difference.

The phase detector constant *is* a constant in this instance; it is equal to

$$K_D = 2AB/\pi. \tag{20}$$

We see that, within a scale factor, this phase detector has the same essential behavior as an analog multiplier with sinusoidal inputs, again interpreting phase errors relative to quadrature.

---

[9] Here's another fun piece of trivia with which to amaze party guests: In general, the spectrum of a signal will decay as $1/f^n$, where n is the number of derivatives of the signal required to yield an impulse. Hence, the spectrum of an ideal sine wave has an infinitely fast rolloff (since no number of derivatives ever yields an impulse), that of an impulse doesn't roll off (since $n = 0$), that of a square wave rolls off as $1/f$, that of a triangle wave as $1/f^2$, and so on.

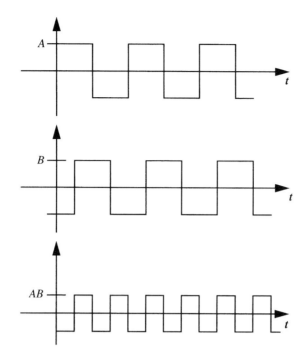

FIGURE 16.9. Multiplier inputs and output

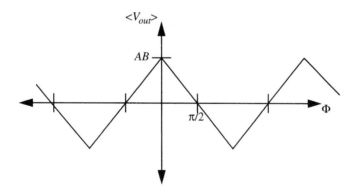

FIGURE 16.10. Multiplier characteristic with two square-wave inputs

As in the case with one square-wave input, this phase detector allows the loop to lock to various harmonics of the input. Again, depending on the application, this property may or may not be desirable.

If we examine the waveforms for this detector more closely, we see that they have precisely the same shape as would be obtained from using a digital exclusive-OR gate, the only difference being DC offsets on the inputs and outputs as well as an inversion here or there. Hence, an XOR may be considered an overdriven analog multiplier. For the special case where the inputs and output are logic levels that swing between

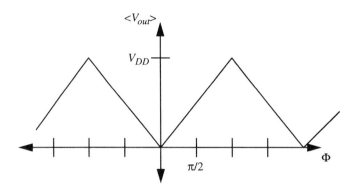

FIGURE 16.11. Characteristic of XOR as quadrature phase detector

ground and some supply voltage V_{DD} (as in CMOS), the phase detector output has an average value that behaves as graphed in Figure 16.11.

The corresponding phase detector gain is then

$$K_D = V_{DD}/\pi. \tag{21}$$

Because of the ease with which they are implemented, and because of their compatibility with other digital circuitry, XOR phase detectors are frequently found in simple IC PLLs.

16.6 SEQUENTIAL PHASE DETECTORS

Loops that use multiplier-based phase detectors lock to a quadrature phase relationship between the inputs to the phase detector. However, there are many practical instances (de Bellescize's homodyne AM detector is one example) where a *zero* phase difference is the desired condition in lock. Additionally, the phase detector constants at the metastable and desired equilibrium points have the same magnitude, resulting in potentially long residence times in the metastable state, perhaps delaying the acquisition of lock.

Sequential phase detectors can provide a zero (or perhaps 180°) phase difference in lock, and they also have vastly different gain constants for the metastable and stable equilibrium points. In addition, some sequential phase detectors have an output that is proportional to the phase error over a span that exceeds 2π radians.

Sequential detectors do have some disadvantages. Since they operate only on transitions, they tend to be quite sensitive to missing edges (although there are modifications that can reduce this sensitivity), in contrast with multipliers that look at the whole waveform. Another consequence of their edge-triggered nature is that they introduce a sampling operation into the loop. As we will see later, sampling inherently adds something similar to a time delay into the loop transmission. The associated increasing negative phase shift with increasing frequency imposes an upper bound on the allowable crossover frequencies that is often substantially more restrictive than if a different phase detector were used.

16.6 SEQUENTIAL PHASE DETECTORS

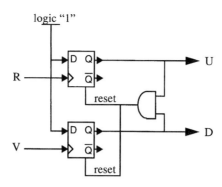

FIGURE 16.12. Phase detector with extended range

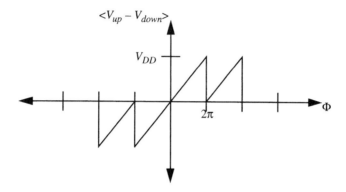

FIGURE 16.13. Characteristic of extended-range phase detector

16.6.1 SEQUENTIAL DETECTORS WITH EXTENDED RANGE

A widely used circuit that provides both extended lock range and an in-phase lock condition consists of two D flip-flops and a reset gate; see Figure 16.12. The designations R and V stand for "reference" and "VCO," while U and D stand for "up" and "down" – terms that will mean something shortly.

For this circuit, the up and down outputs have an average *difference* that behaves as shown in Figure 16.13. Note that the linear input range now spans 4π radians, with a constant phase detector gain of

$$K_D = V_{DD}/2\pi. \tag{22}$$

One characteristic that occasionally causes trouble is the potential for the generation of runt pulses. If the reset path in Figure 16.12 acts too fast, then the minimum pulsewidth generated at the U and D outputs may be too narrow for the next stage to function reliably. This problem occurs when the R and V inputs are very close to each other, and thus it degrades behavior near the locking point. This degradation typically takes the form of an inability to resolve phase errors reliably near lock. This "dead zone" problem is readily solved by simply slowing the reset path. The insertion of some appropriate number of inverters after the AND gate will guarantee that the U and D outputs will be of a width consistent with proper operation of subsequent stages. In lock, both U and D outputs are asserted simultaneously for identical amounts of time.

16.6.2 PHASE DETECTORS VERSUS FREQUENCY DETECTORS

In many applications, it is important (or at least useful) to have some information about the magnitude of any *frequency* difference between the two inputs. Such information could be used to aid acquisition, for example.

Multiplier-based phase detectors cannot provide such information, but sequential phase detectors can. Consider the extended range phase detector of the previous section. If the frequency of the VCO exceeds that of the reference then the U output will have a high duty cycle, because it is set by a rising edge of the higher-frequency VCO but isn't cleared until there is another rising edge on the lower-frequency reference. Hence, this type of phase detector provides not only a large and linear phase detection range but also a signal that is indicative of the sign and magnitude of the frequency error. These attributes account for this detector's enormous popularity. Detectors with this frequency discrimination property are known collectively as *phase-frequency* detectors.

It should be mentioned that this detector does have some problems, however. Being a sequential detector, it is sensitive to missing edges. Here, it would misinterpret a missing edge as a frequency error and the loop would be driven to "correct" this error. Additionally, the slope of the phase detector characteristic near zero phase error may actually be somewhat different from what is shown in Figure 16.13 because both the U and D outputs are narrow slivers in the vicinity of the lock point. Because all real circuits have limited speed, the nonzero risetimes will cause a departure from the ideal linear shape shown, since the areas of the slivers will no longer have a linear relationship to the input time (phase) differences.

In some systems, this problem is solved by intentionally introducing a DC offset into the loop so that the phase detector output must be nonzero to achieve lock. By biasing the balanced condition away from the detector's center, the nonlinearities can be greatly suppressed. Unfortunately, this remedy is inappropriate for applications that require small error, since the added offset translates into a static phase error.

16.7 LOOP FILTERS AND CHARGE PUMPS

So far, we've examined the behavior of PLLs using a linear model, as well as a number of ways to implement phase detectors. We now consider how to implement the rest of the loop. We'll take a look at various types of loop filters and work through an actual example to illustrate a typical design procedure.

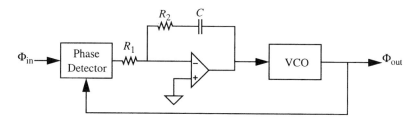

FIGURE 16.14. PLL with typical loop filter

The VCO requires some control voltage to produce an output of the desired frequency. To provide this control voltage with a zero output from the phase detector (and hence zero phase error), the loop filter must provide an integration. Then, to ensure loop stability, the loop filter must also provide a zero. A classic PLL architecture that satisfies these requirements appears as Figure 16.14.

It should be easy to deduce the general properties of the loop filter without resorting to equations. At very low frequencies, the capacitor's impedance dominates the op-amp's feedback, so the loop filter behaves as an integrator. As the frequency increases, though, the capacitive reactance decreases and eventually equals the series resistance R_2. Beyond that frequency, the capacitive reactance becomes increasingly negligible compared with R_2, and the gain ultimately flattens out to simply $-R_2/R_1$.

Stating these observations another way, we have a pole at the origin and a zero whose time constant is R_2C. Furthermore, the value of R_1 can be adjusted to provide whatever loop transmission magnitude we'd like, so the op-amp circuit provides us with the desired loop filter transfer function.

It must be mentioned that PLLs need not include an active loop filter of the type shown. In the simplest case, a passive RC network could be used to connect the phase detector with the VCO. However, the static phase error will then not be zero, and the loop bandwidth will be coupled (inversely) with the static phase error. Because of these limitations, such a simple loop filter is used only in noncritical applications.

The circuit of Figure 16.14 is commonly used in discrete implementations, but a different (although functionally equivalent) approach is used everywhere else. The reason is that it is wasteful to build an entire op-amp simply to obtain the desired loop filter transfer function. A considerable reduction in complexity and area (not to mention power consumption) can be obtained by using an element that is specially designed for this single purpose: a *charge pump,* working in tandem with an RC network. Here, the phase detector controls one or more current sources, and the RC network provides the necessary loop dynamics.

Figure 16.15 shows how a charge pump provides the necessary loop filter action. Here, the phase detector is assumed to provide a digital "pump up" or "pump down" signal. If the phase detector determines that the VCO output is lagging the input reference, it activates the top current source, depositing charge onto the capacitor (pumping up). If the VCO is ahead, the bottom current source is activated, withdrawing charge from the capacitor (pumping down).

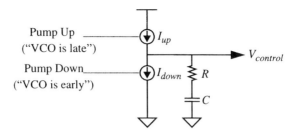

FIGURE 16.15. Basic charge pump with loop filter

If there were no resistor, then we would have a pure integration. As usual, the series resistor provides the necessary loop-stabilizing zero by forcing the high-frequency asymptotic impedance to a nonzero value.

Since switched current sources are easily implemented with a very small number of transistors, the charge pump approach allows the synthesis of the desired loop filter without the complexity, area, and power consumption of a textbook op-amp. The nature of the control also meshes nicely with the many digital phase detectors (e.g., sequential phase detectors) that exist, such as the one shown in Figure 16.12. When that detector is used with the charge pump of Figure 16.15, the net pump current is given by

$$I = I_{pump} \frac{\Delta \Phi}{2\pi}, \qquad (23)$$

where $I_{pump} = I_{up} = I_{down}$. This current, multiplied by the impedance of the filter network connected to the current sources, gives the VCO control voltage.

Control-Line Ripple and Higher-Order Poles

Even when the charge pump is well designed, we must assume the existence of some ripple on the control voltage. The loop-stabilizing zero improves stability at the expense of degraded high-frequency filtering. Consequently, there can be significant high-frequency content on the control line that drives the VCO. This "hash" can come from having to compensate for charge pump leakage, from the higher-order mixing products in a multiplier-type detector (i.e., essentially the double frequency term), or from asymmetries in the charge pump or phase detector. Many of these components are periodic and so produce stationary sidebands (spurs). One obsession of synthesizer designers is the systematic eradication of spurs. Unfortunately, spurs arise very easily from noise injected into the control line, including noise from the supply, substrate, or even from external fields coupling into the chip. A typical RF VCO may possess tuning sensitivities of tens or hundreds of megahertz per volt, so even a few millivolts of noise can generate noticeable spectral artifacts. The resultant modulation of the VCO frequency may be unacceptable in many applications.

The design of a practical charge pump is somewhat more difficult than might be supposed from examining Figure 16.15. The subtleties involved are perhaps best appreciated by studying a representative design; see Figure 16.16. Analysis of this

16.7 LOOP FILTERS AND CHARGE PUMPS

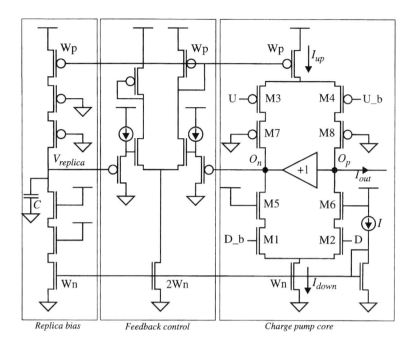

FIGURE 16.16. Example of PLL charge pump

CMOS circuit highlights some of the more important design considerations associated with charge pump design. Transistors M1 through M4 are differential switches operated by the up and down commands from the phase detector. Depending on the state of those commands, either source current I_{up} or sink current I_{down} is steered to the output node O_p. Thus, I_{out} equals I_{up} or I_{down}, depending on the phase detector state.

The switches are cascoded by transistors M5 to M8 for high output impedance because any leakage increases spur power. To understand why, consider the locked condition. With low leakage, very little net charge needs to be delivered by the charge pump per cycle. There is thus very little ripple on the control line and hence very little modulation of the VCO. As leakage increases, however, the charge pump must make up for an increasing amount of lost charge, implying the necessity for an increasing static phase error. For example, if the leakage is such that the control voltage droops between phase measurements, then the phase error must increase until the net charge deposited as a result of up pulses is just enough greater than that deposited from the down pulses to compensate for the leakage. Cascoding helps reduce control line ripple by reducing leakage, and therefore reduces the spur energy (and static phase error). Because the voltage droops between corrections, which occur with a frequency equal to that of the reference input, the control-line ripple also has a fundamental periodicity equal to that of the reference. The spurs are therefore displaced from the carrier by an amount equal to the reference frequency. The existence of large reference frequency spurs is usually a sign of poor charge pump design; see

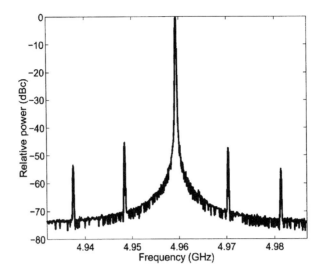

FIGURE 16.17. Output spectrum of synthesizer with somewhat leaky charge pump

Figure 16.17. Clearly visible are the reference spurs spaced 11 MHz away from the 4.96-GHz carrier. There are additional spurs (spaced integer multiples of 11 MHz away) corresponding to the Fourier components of the ripple on the control line.

For similar reasons it is also important to have equal up and down currents. If one is stronger than the other, a compensating static phase error must again appear, with its attendant consequences for control-line ripple. To mitigate this problem, the charge pump design here uses relatively large devices (to reduce threshold mismatch) and operates them at moderately large overdrive. In addition, a simple unity-gain buffer forces the unused charge pump output to have the same common-mode voltage as the main output, thus removing systematic mismatch that would arise from operation with unequal drain–source voltages. Supplementing that strategy is a replica bias loop, whose output voltage is compared with the voltage at the unused output of the charge pump. A simple op-amp drives these two voltages to equality (compensation capacitor C is for loop stability) and thus ensures that all conducting devices in the main core have the same bias voltages as in the replica. The resulting up and down tail currents are then equal, within the limits of random mismatch.

Attention to these sorts of details enables the suppression of the reference spurs by large factors, as is apparent from Figure 16.18. Spurs are invisible in this plot and are thus below the noise floor of -70 dBc. The >25-dB reduction in reference spur power represents an improvement by a factor of more than 300 (on a power basis).

We see the critical role played by the loop filter in removing the "teeth" produced by the phase detection process (which, if you recall, is fundamentally a sampled system in digital implementations) in addition to other noise that may couple there. We need to consider how to design the best possible loop filters. For a given loop bandwidth, a higher-order filter provides more attenuation of out-of-band components. However, the higher the order, the harder it is to make the loop stable. For this reason,

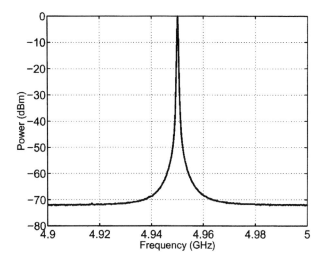

FIGURE 16.18. Spectrum of improved synthesizer

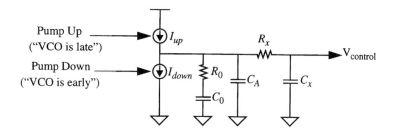

FIGURE 16.19. Idealized PLL charge pump with third-order loop filter

many simple synthesizer loops are second order, but these rarely provide competitive performance. Designing higher-order loops requires ever increasing vigilance to guard against instability – and also taxes our mathematical skills. The law of diminishing returns leads us to discover that a fourth-order loop is about optimum, so we will focus on the design of a third-order loop filter. See Figure 16.19.

Remembering that the VCO adds another pole (at the origin), we see that choosing a three-pole loop filter results in the creation of a fourth-order loop. Elements C_A, R_x, and C_x supply additional filtering beyond that provided by a simple second-order loop. In the past, no simple closed-form design method existed, so designing such a filter involved staring at lots of plots before giving up and going back to a second- or third-order loop. Luckily this situation has changed quite recently, and we can offer a simple cookbook recipe that is close enough to optimum for most purposes.[10] Alternatively, many widely available CAD tools (e.g., National Semiconductor's

---

[10] H. R. Rategh and T. H. Lee, *Multi-GHz Frequency Synthesis and Division*, Kluwer, Dordrecht, 2001.

PLL_LpFltr) automate the design, leading to somewhat better loop filters than produced by our cookbook procedure.

Step 1. *Specify a phase margin.* Once this value is chosen, it sets a constraint on capacitor values. Specifically,

$$(PM) \approx \tan^{-1}(\sqrt{b+1}) - \tan^{-1}(1/\sqrt{b+1}), \tag{24}$$

where

$$b = \frac{C_0}{C_A + C_X}. \tag{25}$$

It's probably prudent to choose a phase margin a few degrees above the target value in order to absorb the inevitable negative phase contributions by the sampled nature of the loop, unmodeled poles, and other destabilizing sources. For example, suppose the specified phase margin target is 45°. If we therefore design for 50°, we find (through iteration, for example) that b should be about 6.5. This would be a typical design value.

Step 2. *Select loop crossover frequency,* based on specifications on tracking bandwidth, for example. Combined with the results of step 1, we find the location of the loop stabilizing zero as follows.

We know that maximizing the loop bandwidth maximizes the frequency range over which the presumably superior phase noise characteristics of the reference oscillator are conferred upon the output. Unfortunately, the loop is a sampled data system, and we can push up the crossover frequency to only about a tenth of the phase comparison frequency before the phase lag inherent in a discrete-time phase detector starts to degrade phase margin seriously. As a specific example, assume that the reference frequency (and hence the phase comparison frequency) is 2 MHz. Choosing a crossover frequency of 200 kHz is safe, since it is a decade below the reference frequency. It would be imprudent to target a crossover frequency much higher than this value.

For the crossover frequency we have

$$\omega_c \approx \frac{\sqrt{b+1}}{\tau_z} = \frac{\sqrt{b+1}}{R_0 C_0}. \tag{26}$$

Step 3. *Calculate C_0, the value of the zero-making capacitor:*

$$C_0 = \frac{I_P}{2\pi} \frac{K_0}{N} \frac{b}{\sqrt{b+1}} \frac{1}{\omega_c^2}, \tag{27}$$

where I_P is the charge pump current, N is the divide modulus, and K_0 is the VCO gain constant in radians per second per volt.

Step 4. *Calculate $R_0 = \tau_z/C_0$.* This completes the design of the main part of the loop filter.

Step 5. *Select $\tau_x = R_X C_X$ within the following range:*

$$0.01 < \tau_x/\tau_z < 0.1. \tag{28}$$

Within these wide limits is considerable freedom of choice. You can choose to design for the arithmetic mean, or the geometric mean, or some other kind of mean. Typically, one selects τ_x to be 1/30 to 1/20 of τ_z. A bigger time constant results in somewhat better filtering action but tends to be associated with lower stability. Since loop constants aren't constant, it is prudent to design for some margin.

Step 6. *Complete the remaining calculations.* Back in step 1, we developed a constraint on the capacitance ratios. Having found one of the capacitances, we now know the sum of C_A and C_X. You are free to select the individual values over a quite wide range, as long as they sum to the correct value. Arbitrarily setting them equal is a common choice.[11] Having done so then allows us to determine their absolute values, which subsequently allows us to determine the value of R_X.

This completes the design of the loop filter.

16.8 FREQUENCY SYNTHESIS

Oscillators built with high-Q resonators exhibit the best spectral purity but cannot be tuned over a range of more than several hundred parts per million or so. Since most transceivers must operate at a number of different frequencies that span a considerably larger range than that, one simple way to accommodate this lack of tuning capability is to use a separate resonator for each frequency. Clearly, this straightforward approach is practical only if the number of required frequencies is small.

Instead, virtually all modern gear uses some form of frequency synthesis, in which a single quartz-controlled oscillator is combined with a PLL and some digital elements to provide a multitude of output frequencies that are traceable to that highly stable reference. In the ideal case, then, one can obtain a wide operating frequency range and high stability from one oscillator.

Before undertaking a detailed investigation of various synthesizers, however, we need to digress briefly to examine an issue that strongly influences architectural choices. A frequency divider is used in all of the synthesizers we shall study, and it is important to model properly its effect on loop stability.

16.8.1 DIVIDER "DELAY"

Occasionally, one encounters the term "divider delay" in the literature on PLL synthesizers in the context of evaluating loop stability. We'll see momentarily that the phenomenon is somewhat inaccurately named, but there is indeed a stability issue associated with the presence of dividers in the loop transmission.

---

[11] The noise generated by the resistors in the filter will produce broadband modulation of the VCO, resulting in phase noise. Minimizing the phase noise would impose additional constraints on the loop filter design, complicating the situation enough that the cookbook procedure offered here is all we'll consider. It is a good idea to select values that make the overall realization less dependent on parasitics. Generally speaking, using the largest capacitors consistent with the required time constants will help reduce broadband noise modulation of the control voltage.

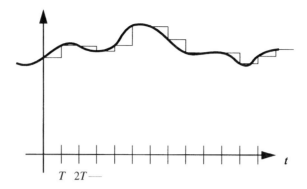

FIGURE 16.20. Action of sample-and-hold

The use of a frequency divider generally implies that the phase detector is digital in nature.[12] As a consequence, knowledge about phase error is available to the loop only at discrete instants. That is, the loop is a sampled-data system. If a divider is present, the loop samples the phase error less frequently than might be implied by the VCO frequency. To model the PLL correctly, then, we need to account properly for this sampled nature.

To develop the necessary insight, consider a process in which a continuous-time function is sampled and held periodically; see Figure 16.20. The sample-and-hold (S/H) operation shown introduces a phase lag into the process. Mathematics need not be invoked to conclude that this is so; simply "eyeball" the sampled-and-held waveform and consider its time relationship with the original continuous-time waveform. You should be able to convince yourself that the best fit occurs when you slide the original waveform to the right by about a half of a sample time.

More formally, the "hold" part of the S/H operation can be modeled by an element whose impulse response is a rectangle of unit area and T-second duration, as seen in Figure 16.21. This element, known formally as a zero-order hold (ZOH), has a transfer function given by[13]

$$H(s) = \frac{1 - e^{-sT}}{sT}. \tag{29}$$

The magnitude of the transfer function is

$$|H(j\omega)| = \frac{\sin \omega(T/2)}{\omega(T/2)}, \tag{30}$$

while the phase is simply

$$\angle[H(j\omega)] = -\omega(T/2). \tag{31}$$

---

[12] Although there are exceptions (e.g., subharmonic injection-locked oscillators), we will limit the present discussion to the more common implementations.

[13] If you would like a quick derivation of the transfer function, note that the impulse response of the zero-order hold is the same as that of the difference of two integrations (one delayed in time). That should be enough of a hint.

16.8 FREQUENCY SYNTHESIS

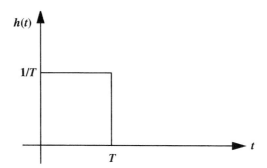

FIGURE 16.21. Impulse response of zero-order hold

The time delay is thus $T/2$ seconds. The same result is obtained by recognizing that a rectangular impulse response centered about zero seconds has zero phase by symmetry. Shifting that response to the right by $T/2$ seconds gives us the graph shown in Figure 16.21. Independently of how we compute it, the fact that there is a delay is the reason for the term "divider delay." However, since the magnitude term is not constant with frequency, the term "delay" is not exactly correct.[14]

Now let's apply this information to the specific example of a PLL with a frequency divider in the loop transmission. From the expression for phase shift, we can see the deleterious effect on loop stability that dividers can introduce. As the divide modulus increases, the sampling period T increases (assuming a fixed VCO output frequency). The added negative phase shift thus becomes increasingly worse, degrading phase margin. As a consequence, loop crossover must be forced to a frequency that is low compared with $1/T$ in order to avoid these effects. Since the sampling rate is determined by the output of the *dividers* and hence of the frequency at which phase comparisons are made (rather than the output of the VCO), a high division factor can result in a severe constraint on loop bandwidth, with all of the attendant negative implications for settling speed and noise performance. It is therefore common to choose loop crossover frequencies that are about a tenth of the phase comparison rate.

Finally, practical dividers are not jitter-free. That is, a noise-free input does not produce a noise-free output, so one must expect some degradation in the synthesized output. A common rule of thumb is to accommodate this reality by designing for a synthesizer output noise power that is 2–3 dB lower than necessary. That rule of thumb should only be used in the absence of detailed information about the magnitude of this noise. That said, this extra margin suffices for a great many practical designs.

16.8.2 SYNTHESIZERS WITH STATIC MODULI

Having developed an understanding of the constraints imposed by the presence of a frequency divider in the loop transmission, we now turn to an examination of various synthesizer topologies.

---

[14] The magnitude is close to unity, however, for $\omega T < 1$.

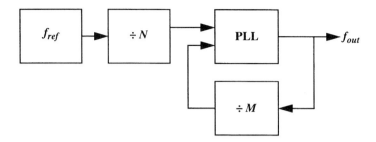

FIGURE 16.22. Classic PLL frequency synthesizer

A simple PLL frequency synthesizer uses one reference oscillator and two frequency dividers, as shown in Figure 16.22. The loop forces the VCO to a frequency that makes the inputs to the PLL equal in frequency. Hence, we may write:

$$\frac{f_{ref}}{N} = \frac{f_{out}}{M}, \quad (32)$$

so that

$$f_{out} = \frac{M}{N} f_{ref}. \quad (33)$$

Thus, by varying the divide moduli M and N, any rational multiple of the input reference frequency can be generated. The long-term stability of the output (i.e., the average frequency) is every bit as good as that of the reference, but stability in the shortest term (phase noise) depends on the net divide modulus as well as on the properties of the PLL's VCO and loop dynamics. Within the PLL's loop bandwidth, the output phase noise will be M/N times that of the reference oscillator, since a phase multiplication necessarily accompanies a frequency multiplication. Outside of the PLL loop bandwidth, feedback is ineffective, and the output phase noise will therefore be that of the PLL's own VCO. In practice, additional sources of noise (e.g., divider and phase detector) will cause the synthesized phase noise to be larger than the theoretical minimum. For this reason, the design target should be set at a level that is, say, 2 dB more stringent than required.

Note that the output frequency can be incremented in steps of f_{ref}/N and that this frequency represents the rate at which phase detection is performed in the PLL. Stability considerations as well as the need to suppress control-voltage ripple force the use of loop bandwidths that are small compared with f_{ref}/N. To obtain the maximum benefit of the (presumed) low noise reference, however, we would like the PLL to track that low noise reference over as wide a bandwidth as possible. Additionally, a high loop bandwidth speeds settling after a change in modulus. These conflicting requirements have led to the development of alternative architectures.

One simple modification that is occasionally used is shown in Figure 16.23. For this synthesizer, we may write:

$$f_{out} = \frac{M}{NP} f_{ref}. \quad (34)$$

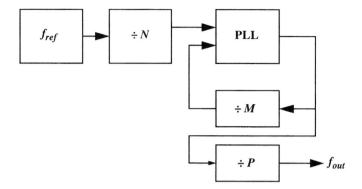

FIGURE 16.23. Modified PLL frequency synthesizer

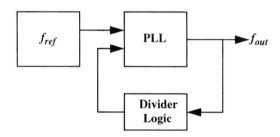

FIGURE 16.24. Integer-N frequency synthesizer

The minimum output frequency increment is evidently f_{ref}/NP, but the loop compares phases at f_{ref}/N, or P times as fast as the previous architecture. This modification therefore improves the loop bandwidth constraint by a factor of P, at the cost of requiring the PLL to oscillate P times faster and requiring the $\div M$ counter to run that much faster as well.

Yet another modification is the integer-N synthesizer of Figure 16.24. In this widely used synthesizer, the divider logic consists of two counters and a dual-modulus prescaler (divider). One counter, called the channel spacing (or "swallow") counter, is made programmable to enable channel selection. The other counter, which we'll call the frame counter (also known as the program counter), is usually fixed and determines the total number of prescaler cycles that comprise the following operation: The prescaler initially divides by $N+1$ until the channel spacing counter overflows, then divides by N until the frame counter overflows; the prescaler modulus is reset to $N+1$, and the cycle repeats.

Let S be the maximum value of the channel spacing counter and F the maximum value of the frame counter. Then the prescaler divides the VCO output by $N+1$ for S cycles and by N for $F-S$ cycles before repeating. The effective overall divide modulus M is therefore

$$M = (N+1)S + (F-S)N = NF + S. \tag{35}$$

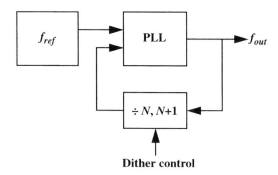

FIGURE 16.25. Block diagram for frequency synthesizer with dithered modulus

The output frequency increment is thus equal to the reference frequency. This architecture is the most popular way of implementing the basic block diagram of Figure 16.22, and it gets its name from the fact that the output frequency is an integer multiple of the reference frequency.

16.8.3 SYNTHESIZERS WITH DITHERING MODULI

In the synthesizers studied so far, the desired channel spacing directly constrains the loop bandwidth. An approach that eases this problem is to dither between two divide moduli to generate channel spacings that are smaller than the reference frequency; this is shown in Figure 16.25. As an illustration of the basic idea, consider that dividing alternately by (say) 4 then 5 with a 50% duty ratio is equivalent to dividing by 4.5 on average. Changing the percentage of time spent on any one modulus thus changes the effective (average) modulus, so that the synthesized output can be incremented by frequency steps smaller than the input reference frequency.

There are many strategies for switching between two moduli that yield the same average modulus, of course, yet not all of them are equally desirable because the *instantaneous* frequency is also of importance. The most common strategy is used by the fractional-N synthesizer, where we divide the VCO output by one modulus (call it $N+1$) every K VCO cycles and by the other modulus (N) for the rest of the time. The average divide factor is thus

$$N_{eff} = (N+1)\left(\tfrac{1}{K}\right) + N\left(1 - \tfrac{1}{K}\right) = N + \tfrac{1}{K}, \tag{36}$$

so that

$$f_{out} = N_{eff} f_{ref} = \left(N + \tfrac{1}{K}\right) f_{ref}. \tag{37}$$

We see that the resolution is determined by K, so that the minimum frequency increment can be much *smaller* than the reference frequency. However, unlike the other synthesizers studied so far, the phase detector operates with inputs whose frequency is *much higher* than the minimum increment (in fact, the phase detector is driven with signals of frequency f_{ref}), thus providing a much-desired decoupling of synthesizer frequency resolution from the PLL sampling frequency.

To illustrate the operation of this architecture in greater detail, consider the problem of generating a frequency of 1.57542 GHz with a reference input of 10 MHz. The integral modulus N therefore equals 157, while the fractional part ($1/K$) equals 0.542. Thus, we wish to divide by 158 ($= N + 1$) for 542 out of every 1000 VCO cycles (for example) and by 157 ($= N$) the other 458 cycles.

Of the many possible strategies for implementing this desired behavior, the most common (but not necessarily optimum) one is to increment an accumulator by the fractional part of the modulus (here 0.542) every cycle. Each time the accumulator overflows (here defined as equalling or exceeding unity), the divide modulus is set to $N + 1$.

The residue after overflow is preserved, and the loop continues to operate as before. It should be apparent that the resolution is set by the size of the accumulator and is equal to the reference frequency divided by the total accumulator size. In our example of a 10-MHz reference, a five-digit BCD accumulator would allow us to synthesize output steps as small as 100 Hz.

There is one other property of fractional-N synthesizers that needs to be mentioned. Since the loop operates by periodically switching between two divide moduli, there is necessarily a periodic modulation of the control voltage and, hence, of the VCO output frequency. Therefore, even though the output frequency is correct on *average*, it may not be on an *instantaneous* basis, and the output spectrum therefore contains sidebands. Furthermore, the size and location of the sidebands depend on the particular moduli as well as on loop parameters.

In practical loops of this kind, compensation for this modulation is generally necessary. Many forms of compensation are enabled by the deterministic nature of the modulation – we know in advance what the control-line ripple will be. Hence, a compensating control voltage variation may be injected to offset the undesired modulation. In practice, this technique (sometimes called API, for *analog phase interpolation*) is capable of providing between 20 and 40 dB suppression of the sidebands. Achieving the higher levels of suppression (and beyond) requires intimate knowledge of the control characteristics of the VCO, including temperature and supply voltage effects, so details vary considerably from design to design.[15]

An alternative to this type of cancellation is to employ two identical loops. To the extent that the two match, both synthesizers will have the same ripple. One may extract the ripple component (e.g., through a DC blocking capacitor), invert it, and then inject it into the first loop. On a steady-state basis, this feedforward correction cancels the ripple component. The drawback is the need to build two loops (thus doubling complexity, area, and power), assure that they're identical, and prevent the two synthesizers from interacting in some undesired way.

One may also eliminate the periodic control voltage ripple altogether by employing a more sophisticated strategy for switching between the two moduli. For example, one might randomize this switching to decrease the amplitudes of spurious spectral components at the expense of increasing the noise floor. A powerful improvement on

---

[15] See e.g. V. Mannassewitsch, *Frequency Synthesizers,* 3rd ed., Wiley, New York, 1987.

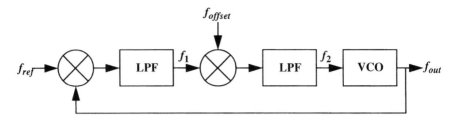

FIGURE 16.26. Offset synthesizer loop

that strategy is to use *delta-sigma* techniques to distribute the noise *nonuniformly*.[16] If the spectrum is shaped to push the noise out to frequencies far from the carrier, subsequent filtering can readily remove the noise. The loop itself takes care of noise near the carrier, so the overall output can possess exceptional spectral purity.

16.8.4 COMBINATION SYNTHESIZERS

Another approach is to combine the outputs of two or more synthesizers. The additional degree of freedom so provided can ease some of the performance trade-offs, but at the expense of increased complexity and power consumption.

The most common expression of this idea is to mix the output of a fixed frequency source with that of a variable one. The *offset synthesizer* (Figure 16.26) is one architecture that implements that particular choice. With this architecture, the loop does not servo to an equality of output and reference frequencies because the additional intermediate mixing offsets the equilibrium point. Without an intermediate mixing, note that the balance point would correspond to a zero frequency output from the low-pass filter that follows the (first and only) mixer. In the offset loop, then, the balance point corresponds to a zero frequency output from the final low-pass filter. Armed with this observation, it is a straightforward matter to determine the relationship between f_{out} and f_{ref}.

The low-pass filters selectively eliminate the sum-frequency components arising from the mixing operations. Hence, we may write

$$f_1 = f_{out} - f_{ref}, \tag{38}$$

$$f_2 = f_1 - f_{offset}$$
$$= f_{out} - f_{ref} - f_{offset}. \tag{39}$$

Setting f_2 equal to zero and solving for the output frequency yields

---

[16] The classic paper on this architecture is by Riley et al., "Sigma-Delta Modulation in Fractional-*N* Frequency Synthesis," *IEEE J. Solid-State Circuits*, v. 28, May 1993, pp. 553–9. The terms "delta-sigma" and "sigma-delta" are frequently used interchangeably, but the former nomenclature was used by the inventors of the concept.

$$f_{out} = f_{ref} + f_{offset}. \tag{40}$$

Thus, the output frequency is the sum of the two input frequencies.

An important advantage of this approach is that the output frequency is not a multiplied version of a reference. Hence, the phase noise similarly undergoes no multiplication, making it substantially easier to produce a low–phase noise output signal. A related result is that any phase or frequency modulation on either of the two input signals is directly transferred to the output without scaling by a multiplicative factor. As a consequence of these attributes, the offset synthesizer has found wide use in transmitters for FM/PM systems, particularly for GSM.

There are other techniques for combining two frequencies to produce a third. For example, one might use two complete PLLs and combine the outputs with a mixer. To select out the sum rather than the difference (or vice versa), one would conventionally use a filter. One may also ease the filter's burden by using a single-sideband mixer (also known as a complex mixer) to reduce the magnitude of the undesired component. However, such loops are rarely used in IC implementations owing to the difficulty of preventing two PLLs from interacting with each other. A common problem is for the two loops to (attempt to) lock to each other parasitically through substrate coupling or incomplete reverse isolation through amplifiers and other circuitry. These problems are sufficiently difficult to solve that such dual-loop synthesizers are rarely used at present.

16.8.5 DIRECT DIGITAL SYNTHESIS

There are some applications that require the ability to change frequencies at a relatively high rate. Examples include certain frequency-hopped spread-spectrum systems, in which the carrier frequency changes in a pseudorandom pattern.[17] Conventional synthesizers may be hard-pressed to provide the fast settling required, so alternative means have been developed. The fastest-settling synthesizers are open-loop systems, which can evade the constraints imposed by the stability considerations of feedback systems (such as PLLs).

One extremely agile type of synthesizer employs direct digital synthesis (DDS). The basic block diagram of such a synthesizer is shown in Figure 16.27. This synthesizer consists of an accumulator (ACC), a read-only memory (ROM) lookup table (with integral output register), and a digital-to-analog converter (DAC). The accumulator accepts a frequency command signal (f_{inc}) as an input and then increments its output by this amount every clock cycle. The output therefore increases linearly until an overflow occurs and the cycle repeats. The output Φ thus follows a sawtooth pattern. A useful insight is that phase is the integral of frequency, so the output of the accumulator is analogous to the integral of the frequency input command. The

---

[17] This strategy is particularly useful in avoiding detection and jamming in military scenarios, for which it was first developed, because the resulting spectrum looks very much like white noise.

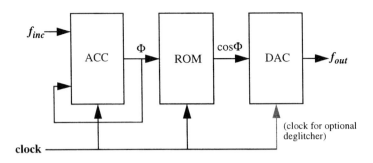

FIGURE 16.27. Direct digital frequency synthesizer

frequency of the resulting sawtooth pattern is then a function of the clock frequency, accumulator word length, and input command.

The phase output of the accumulator then drives the address lines of a ROM cosine lookup table that converts the digital phase values into digital amplitude values.[18] Finally, a DAC converts those values into analog outputs. Generally, a filter follows the DAC to improve spectral purity to acceptable levels.

The frequency can be changed rapidly (with a latency of only a couple clock cycles), and in a phase-continuous manner, simply by changing the value of f_{inc}. Furthermore, modulation of both frequency and phase is trivially obtained by adding the modulation directly in the digital domain to f_{inc} and Φ, respectively. Finally, even amplitude modulation can be added by using a multiplying DAC (MDAC), in which the analog output is the product of an analog input (here, the amplitude modulation) and the digital input from the ROM.[19]

The chief problem with this type of synthesizer is that the spectral purity is markedly inferior to that of the PLL-based approaches considered earlier. The number of bits in the DAC set one bound on the spectral purity (*very* loosely speaking, the carrier-to-spurious ratio is about 6 dB per bit), while the number of ROM points per cycle determines the location of the harmonic components (with a judicious choice of the n points, the first significant harmonic can be made to occur at $n - 1$ times the fundamental). Since the clock will necessarily run much faster than the output frequency ultimately generated, these types of synthesizers produce signals whose frequencies are a considerably smaller fraction of a given technology's ultimate speed than VCO/PLL-based synthesizers. Frequently, the output of a DDS is upconverted through mixing with the output of a PLL-based synthesizer (or used as one of the inputs in an offset synthesizer) to effect a compromise between the two.

---

[18] With a little additional logic, one can easily reduce the amount of ROM required by 75%, since one quadrant's worth of values is readily reused to reconstruct an entire period.

[19] One may also perform the amplitude modulation in the digital domain simply by multiplying the ROM output with a digital representation of the desired amplitude modulation before driving the DAC.

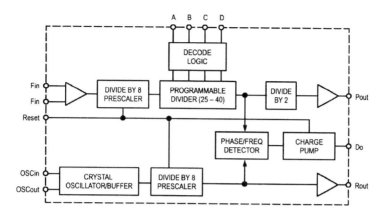

FIGURE 16.28. Simplified block diagram of MC12181 frequency synthesizer (from the data sheet)

16.9 A DESIGN EXAMPLE

At this point, working through a practical design example is useful for putting all of the foregoing into context. Suppose that we wish to design a synthesizer whose output frequency is to be 1 GHz, starting with a reference oscillator of 25 MHz. To make things easier, we will use an off-the-shelf IC (the MC12181 from Motorola; see Figure 16.28) that contains a phase detector, charge pump, reference oscillator, and divider logic. All we have to do is connect a crystal (for the reference oscillator), a handful of auxiliary passive components, and a VCO – and then design a loop filter to go with it.

The frequency multiplication factor can be set to any one of sixteen integer values, from 25 to 40, using four configuration bits. The reference frequency that is multiplied upward by that programmable factor is controlled by an external crystal connected to two pins of the chip. Here we encounter the first subtlety: the oscillator inside the 12181 is a traditional Pierce topology that acts much like a common-emitter bipolar amplifier, although the actual circuit is more complicated. An external resistor of around 50 kΩ must be connected across the crystal to establish the bias for the oscillator. Furthermore, two capacitors are needed to provide additional phase shift (beyond what the crystal provides) to satisfy the conditions for oscillation. As mentioned in the previous chapter, the crystal in a Pierce thus operates at a frequency somewhat above its series resonance, so that it presents a net inductive impedance under normal operation. The exact oscillation frequency depends on both the crystal *and* the two capacitors, although it is much more sensitive to the crystal's resonant frequency (because of the vastly steeper reactance-vs.-frequency curve of crystals, relative to that of capacitors). Each crystal intended for use in a Pierce is specially cut to oscillate on frequency only with a specified capacitive load. If absolute frequency accuracy is important (as it almost always is) and if that capacitive load is uncertain, then the capacitors should be variable ones that allow tuning to the correct frequency.

High-frequency crystals are very thin, making them fragile and difficult to manufacture. The designers of the 12181 know this and consequently specify a 25-MHz upper frequency, which corresponds to about the maximum frequency at which manufacturers will provide an inexpensive fundamental-mode crystal. Unfortunately, 25 MHz is still high enough that some manufacturers prefer to make an 8.33-MHz crystal to save cost, expecting the user to operate it on the third overtone. An ordinary oscillator circuit, unfortunately, may satisfy conditions for oscillation at both the fundamental and overtone frequencies, so a modification must be made to poison the loop conditions at the fundamental mode frequency lest some very weird and undesirable effects result. The easiest way to accomplish this feat is to add an inductance in series with one of the capacitors to produce a series resonance at the fundamental. This resonance produces a short to ground that reduces loop gain to zero, preventing oscillation at that frequency. To restore proper operation at the third overtone, a capacitor in shunt with the added inductor would have to be provided. Here, let us assume that we have procured a 25-MHz fundamental-mode crystal, allowing us to sidestep that complexity.

The MC12181 immediately divides the crystal oscillator frequency by 8, then feeds that divided-down signal to the phase detector. The loop thus performs phase comparisons at a rate $f_{ref}/8$, or 3.125 MHz in this case. To avoid having to account for divider delay, we would like to set the loop's crossover frequency to a small fraction of the comparison frequency.

Now let us suppose that we use the third-order loop filter of Figure 16.19 to create a fourth-order PLL. We need to specify the phase margin and the loop crossover frequency. We also need to know the VCO gain constant for the particular oscillator we will be connecting to the synthesizer chip. For purposes of this exercise, suppose the specified phase margin target is 45° and that the loop is to cross over at 100 kHz (628 krps), well below the 3.125-MHz comparison frequency. Finally, assume that the VCO happens to have a gain constant of 100 Mrps/V. With that data in hand, we now follow the procedure presented previously (in Section 16.7).

Step 1. Our specified phase margin target is 45°, so we conservatively design for 50°, for which b is about 6.5, where

$$b = \frac{C_0}{C_A + C_X}. \tag{41}$$

Step 2. Having chosen a crossover frequency of 628 krps, we can readily find the time constant corresponding to the loop-stabilizing zero:

$$\omega_c \approx \frac{\sqrt{b+1}}{\tau_z} = \frac{\sqrt{b+1}}{R_0 C_0}. \tag{42}$$

For our numbers, $\tau_z = R_0 C_0$ works out to about 4.4 μs.

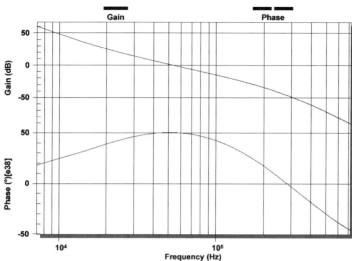

FIGURE 16.29. *PLL_LpFltr* design values and Bode plot of loop transmission

Step 3. We next calculate C_0 from

$$C_0 = \frac{I_P}{2\pi} \frac{K_0}{N} \frac{b}{\sqrt{b+1}} \frac{1}{\omega_c^2}. \tag{43}$$

For the MC12181, the nominal charge pump current I_P is 2 mA. For our design, N is 320 (remember, there is a built-in divide-by-8 prescaler ahead of the 4-bit programmable divider) and K_0 is 100 Mrps/V, so C_0 is about 600 pF.

Step 4. We have $R_0 = \tau_z/C_0 = 7.3$ kΩ, completing the design of the main part of the loop filter.

Step 5. We now turn to the design of the additional ripple filter. We will arbitrarily set $\tau_x = R_X C_X$ to 1/20 of τ_z, or 220 ns.

Step 6. Having set C_0, we now know the sum of C_A and C_X; here, it is 92 pF. Arbitrarily setting the capacitances equal yields 46 pF each. The nearest standard value is 47 pF, which is close enough to the calculated value not to matter. Having determined those capacitances, we can now compute the value of R_X as about 4.7 kΩ, which just happens to be a standard 10% value as well.

Design of a third-order loop filter is also facilitated by a number of CAD tools. For example, National Semiconductor's *PLL_LpFltr* program will carry out designs for second- and third-order loop filters and also plot the gain and phase of the loop transmission for the PLL it's just designed. Because it uses a more sophisticated algorithm than the simplified procedure outlined here, the values it yields for the loop filter components differ a bit from the ones we've computed. See Figure 16.29.

Once we have carried out all of the loop filter calculations, the design of the synthesizer is complete.

PRACTICAL CONSIDERATIONS

In the design example just described, we assumed that the VCO gain constant *is* truly constant. Furthermore, we implicitly assumed that the VCO's control voltage range matches the output voltage swing provided by the MC12181. Finally, we completely neglected any loading of the charge pump by the control port of the VCO. Not all of these assumptions are well satisfied in practical designs, so we need to consider what happens in actual PLLs.

First, a nonconstant VCO gain implies that the loop dynamics will vary as the control voltage varies. To avoid endangering loop stability at some value(s) of control voltage, the loop filter must be designed conservatively to accommodate the worst-case VCO gain. Doing so is quite tedious. Rather than carry out those tedious analytical solutions, many designers use macromodels of the synthesizer and then perform Monte Carlo analyses to evaluate the distribution of phase margins that may result. Fortunately, suitable simulation tools are readily available. For example, *Spice* works fine for this purpose as long as a good macromodel is used. A transistor-level model is generally much too detailed for a practical simulation.

If the output voltage range of the synthesizer chip is not adequate to drive the VCO, then level shifting and amplification of the charge pump output may be required. In addition, if the VCO's control port is not a high impedance then it will load the output of the charge pump. Because such loading may degrade stability and cause spur generation, buffering may be needed as well. This auxiliary circuitry must have a high enough bandwidth not to add any appreciable phase lag to the loop transmission. A good rule of thumb is to make sure that the added circuitry has its first pole at least a decade above the highest crossover frequency possessed by the loop over the control voltage range. Also, the noise of the level shifter must be kept low because any control voltage noise can modulate the VCO output. Within the PLL bandwidth, this noise can be tracked out. Outside of the loop bandwidth, however, this control voltage noise will cause the noise floor of the VCO output spectrum to rise.

Finally, the capacitance of the VCO control port needs to be small relative to C_X, for otherwise stability may be endangered once again.

16.10 SUMMARY

We examined a number of frequency synthesizers in this chapter. Stability considerations force loop crossover frequencies well below the phase comparison frequency, while phase noise considerations favor large loop bandwidths. Because the output frequency increment is tightly coupled to the phase comparison frequency in simple architectures, it is difficult to synthesize frequencies with fine increments while additionally conferring to the output the good phase noise of the reference. The fractional-N synthesizer decouples the frequency increment from the phase comparison rate, allowing the use of greater loop bandwidths. However, while phase noise is thereby improved, various spurious components can be generated owing to ripple on the control voltage. Suppression of these spurious tones is possible either by cancellation of the ripple (since it is deterministic in the case of the classical fractional-N architecture) or via use of randomization or noise shaping of the spectrum.

16.11 APPENDIX: INEXPENSIVE PLL DESIGN LAB TUTORIAL

Designing a microwave-frequency PLL synthesizer can involve rather expensive hardware, both for construction as well as testing. It's possible to convey the essence of PLL operation to students without incurring such expense by operating at much lower frequencies.

The particular examples we'll study use a commercially available PLL chip, the 4046. It is a very inexpensive (∼\$0.25–\$1) CMOS device that contains two phase detectors (one XOR and one sequential phase detector) and a VCO. We will consider the design of a PLL with each of the phase detectors and a couple of loop filters.

Although the 4046 is a relatively slow device (with a maximum oscillation frequency of only about 1 MHz or so), the design procedure we'll follow is generally applicable to PLLs whose output frequency is much higher, so what follows isn't a purely academic exercise. In any event, the device remains useful for many applications even today, and it is certainly an exceptionally inexpensive PLL tutorial lab-on-a-chip.

16.11.1 CHARACTERISTICS OF THE 4046 CMOS PLL

Phase Comparator I

The chip contains two phase detectors. One, called "Phase Comparator I" by its manufacturer, is a simple XOR gate. Recall from the section on phase detectors that an XOR has a gain constant given by

$$K_D = V_{DD}/\pi \text{ V/rad}. \tag{44}$$

Throughout these design examples, we will use a power supply voltage of 5 V, so the specific numerical value for our designs will be

$$K_D = V_{DD}/\pi \approx 1.59 \text{ V/rad}. \tag{45}$$

Phase Comparator II

The chip's other phase comparator is a sequential phase detector that operates only on the positive edges of the input signals. It has two distinct regions of behavior depending on which input is ahead.

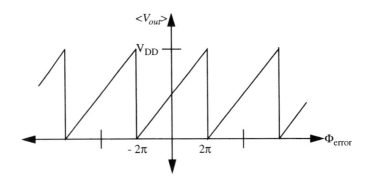

FIGURE 16.30. Characteristic of Phase Comparator II

If the signal input edge precedes the VCO feedback edge by up to one period, the output of the phase detector is set high (that is, to V_{DD}) by the signal edge and sent into a *high-impedance* state by the feedback edge. (We'll see momentarily why it can be advantageous to have this high-impedance state.) If, however, the signal input edge *lags* the VCO output by up to one period, then the output is set low (to ground) by the VCO edge and sent into a high-impedance state by the signal input edge. And that's all there is to this phase detector.

The high-impedance state allows one to reduce the amount of ripple on the control line when in the locked state. Hence, the amount of unintended phase and frequency modulation of the VCO during lock can be much smaller than when other detectors are used. It should also be clear that a PLL that uses this sequential phase detector forces a zero phase difference in lock, in contrast with the quadrature condition that results with an XOR detector.

The other bit of information we need in order to carry out a design is the phase detector gain constant. Unfortunately, this particular detector does not have a particularly well-defined K_D because the output voltage in the high-impedance state depends on external elements, rather than on the phase error alone. A good solution to this problem is to remove the uncertainty by forcing the output voltage to $V_{DD}/2$ during the high-impedance condition (e.g., with a simple resistive divider). With this modification, K_D can be determined.

For phase errors of less than one period (signal input leading), the average output voltage will be linearly proportional to the phase error. The minimum output is $V_{DD}/2$ for zero phase error and is a maximum value of V_{DD} with a 2π phase error. The minimum output is determined by the added resistive divider, while the maximum output is simply controlled by the supply voltage.

Similarly, in the case of a lagging input signal, the average output voltage will be $V_{DD}/2$ for zero phase error and zero volts for a 2π phase error. Hence, the phase detector characteristic looks as shown in Figure 16.30. After solving Schrödinger's equation with appropriate boundary conditions, we find that the slope of the line is:

$$K_D = V_{DD}/4\pi \text{ V/rad}. \tag{46}$$

For our assumed V_{DD} of 5 V, the phase detector gain is approximately 0.40 V/rad.

VCO Characteristics

The VCO used in the 4046 is reminiscent of the emitter-coupled multivibrator used in many bipolar VCOs. Here, an external capacitor is alternately charged one way, then the other, by a current source. A simple differential comparator switches the polarity of the current source when the capacitor voltage exceeds some trip point. The feedback polarity is chosen to keep the circuit in oscillation.

The main VCO output is a square wave, derived from one output of the differential comparator. An approximation to a triangle wave is also available across the capacitor terminals. The triangle wave signal is useful if a sine-wave output is desired, since either a filtering network or nonlinear waveshaper can be used to convert the triangle wave into some semblance of a sine wave.

Frequency control is provided through adjustment of the capacitive charging current. Both the center frequency and VCO gain can be adjusted independently by choosing two external resistors. One resistor, R_2, sets the charging current (and hence the VCO frequency) in the absence of an input, thus biasing the output frequency-vs.-control voltage curve. The other resistor, R_1, sets the transconductance of a common-source stage and therefore adjusts the VCO gain.

Conspicuously absent from the data sheets, however, is an explicit formula for relating the VCO frequency to the various external component values. A quasi-empirical (and highly approximate) formula that provides this crucial bit of information is as follows:[20]

$$\omega_{osc} \approx \frac{2\left(\dfrac{V_C - 1}{R_1} + \dfrac{4}{R_2}\right)}{C}. \tag{47}$$

From this formula, the VCO gain constant is easily determined by taking the derivative with respect to control voltage:

$$K_0 \approx \frac{2}{R_1 C} \text{ rad/s/V}. \tag{48}$$

Miscellany

Notice that the phase detector gains are functions of the supply voltage. Additionally, the VCO frequency is a function of V_{DD} as well. Hence, if the supply voltage varies then so will the loop dynamics, for example. If power supply variations (including noise) are not to influence loop behavior, it is necessary to provide a well-regulated

---

[20] This formula is the result of measurements on only one particular device with a 5-V power supply. Your mileage may vary, especially if you use resistance values below about 50–100 kΩ (the VCO control function gets quite nonlinear at higher currents). *Caveat nerdus.*

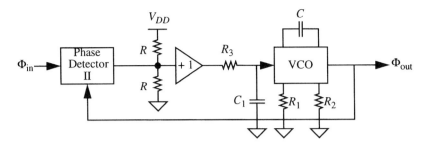

FIGURE 16.31. PLL with Phase Comparator II

and well-filtered supply. As a convenience, the 4046 includes a 5.2-V (plus or minus about 15%) zener diode that may be used for this purpose.

The 4046 also includes a simple source follower to buffer the control voltage. This feature is handy for those applications in which the PLL is being used as an FM demodulator, for example. The demodulated signal is equal to the VCO control voltage, so a buffered version of this control signal is convenient for driving external circuitry.

Finally, the chip includes an "inhibit" control signal that shuts off the oscillator and source follower to reduce chip dissipation to the 100-μW range (even less if the signal input is a constant logic level).

16.11.2 DESIGN EXAMPLES

Second-Order PLL with Passive RC Loop Filter and PD II

We know that active filters can provide superior performance, particularly with regard to steady-state error. However, there are some applications for which fully passive filters are adequate and thus for which active filters are simply devices that consume additional area and power.

Suppose we use Phase Comparator II and a simple RC low-pass loop filter (without a loop-stabilizing zero). Design a circuit to meet the following specifications.

crossover frequency: 1 krad/s
phase margin: 45°
center frequency: 20 kHz

Solution. First, we recognize that the high-impedance characteristic of this phase detector requires the use of the resistive divider, as mentioned earlier. Then, to provide the ability to drive an arbitrary RC network, it is advisable to add a buffer. The resulting PLL appears as shown in Figure 16.31.

The value of R is not particularly critical, but it should be large enough to avoid excessive loading of the phase detector's weak outputs. Values on the order of tens of kilohms are acceptable. Note that the loop transmission is

$$-L(s) = K_D H_f(s) \frac{K_0}{s} = \frac{V_{DD}}{4\pi} \cdot \frac{1}{sR_3C_1 + 1} \cdot \frac{K_0}{s}. \tag{49}$$

16.11 APPENDIX: INEXPENSIVE PLL DESIGN LAB TUTORIAL

The phase margin specification *requires* us to choose the pole frequency of the loop filter equal to the desired crossover frequency, since we do not have a loop-stabilizing zero. Having made that choice, we adjust the VCO gain through selection of R_1C. Finally, we choose R_2 to satisfy the center frequency specification.

Carrying out these steps – and being mindful that resistance values should be no lower than about 50 kΩ to validate the quasi-empirical VCO equation – yields the following sequence of computations, half-truths, and outright lies.

1. As already stated, the phase margin specification requires a loop filter time constant of 1 ms. Somewhat arbitrarily choose $R_3 = 100$ kΩ, so that $C_1 = 0.01$ μF. Both values happen to correspond to standard component values.

2. Because the crossover frequency must be 1 krps while R_3C_1 and the phase detector gain constant are both known, K_0 must be chosen to yield the desired crossover frequency:

$$|L(j\omega_c)| = K_D \cdot \frac{1}{\sqrt{2}} \cdot \frac{K_0}{10^3 \text{ rps}} = 1 \implies R_1C = 0.582 \text{ ms}. \qquad (50)$$

Arbitrarily choose the capacitor equal to a standard value, 0.001 μF, so that the required resistance is 582 kΩ (not quite a standard value, but close to 560 kΩ, which is). Just for reference, the corresponding VCO gain constant is about 3.56 krps/V.

3. Now select R_2 to yield the desired center frequency (here defined as the VCO frequency that results with a control voltage of $V_{DD}/2$) with the VCO capacitor chosen in step 2. From the quasi-empirical VCO formula, we find that R_2 should be approximately 67.3 kΩ (the closest standard 10% value is 68 kΩ). Because of variability from device to device, it is advisable to make R_2 variable over some range if the VCO center frequency must be accurately set.

That completes the design.

With the parameters as chosen, let us compute the VCO tuning range, the steady-state phase error throughout this range, and the lock range (something we haven't explicitly discussed before). The lock range is defined here as the range over which we may vary the input frequency before the loop loses lock.

For the frequency tuning range, we again use the VCO formula. With the chosen values, the VCO can tune about 1 kHz above and below the center frequency. This range sets an upper bound on the overall PLL frequency range.

Because of the passive loop filter, the static phase error will not be zero in general since a nonzero phase detector output is required to provide a nonzero VCO control voltage.[21] If we now assume that the VCO gain constant is, well, *constant*, then we can compute precisely how much control voltage change is required to adjust the frequency over the range computed in step 1. If the corresponding phase error exceeds the $\pm 2\pi$ span of the phase detector, the loop will be unable to maintain lock over the entire ± 1-kHz frequency range.

---

[21] Here, zero control voltage is interpreted as a deviation from the center value of $V_{DD}/2$.

The voltage necessary to move the output frequency is found from K_0 and is related to the phase detector gain constant and the phase error as follows:

$$\Delta V_{ctrl} = \frac{\Delta \omega}{K_0} = K_D \Phi_{error}. \tag{51}$$

Using our component values, the phase error is predicted to be about 4.4 rad at 1 kHz off the center frequency. Actual measurements typically reveal that, at the lower frequency limit (1 kHz below center), the phase error is about 4.3 rad.

At 1 kHz *above* center, though, the typical measured phase error is actually about 5.9 rad. The reason for this rather significant discrepancy is that the VCO frequency isn't quite linearly related to the control voltage at higher control voltages. It turns out that a larger-than-expected control voltage is required to reach the upper frequency limit. Hence, a larger phase detector output is required and so a larger corresponding phase error results. Since angles of both 4.3 rad and 5.9 rad are still within the phase detector's linear range, however, it is the VCO's limited tuning range – rather than the phase detector's characteristics – that determines the overall PLL's lock range in this particular case.

Second-Order PLL with Passive *RC* Loop Filter and PD I

It is instructive to re-do the previous design with the XOR phase detector replacing the sequential phase detector. Because the XOR has four times the gain of PD II, the value of K_0 must be adjusted downward by this factor to maintain the crossover frequency. We may adjust K_0 by increasing R_1 to four times its previous value. Maintaining a 20-kHz center frequency requires that R_2 be adjusted as well (downward). Because the XOR does not have a high-impedance output state, the resistive divider and buffer may be eliminated.

Once these changes have been made, the locked loop displays dynamics that are similar to those observed with the previous design. However, the VCO modifications alter the VCO tuning range and, therefore, the corresponding phase error:

$$\Delta V_{ctrl} = \frac{\Delta \omega}{K_0} = K_D \Phi_{error} \implies \Phi_{error} = \frac{\Delta \omega}{K_0 K_D}. \tag{52}$$

Because R_1 has been changed upward, the VCO tuning range has decreased to a fourth of its previous value while the product of phase detector gain and VCO gain remains unchanged. Now, the XOR is linear over only a fourth of the phase error span of the sequential phase detector. Hence, for a given crossover frequency and damping, use of the XOR phase detector can cause the loop to possess a narrower lock range.

It is left as an exercise for the reader to carry out actual numerical calculations to verify these assertions. (In this case, it turns out that the VCO tuning range is still the limiting factor, but just barely.)

As a few final notes on the use of the XOR, it should be mentioned that this type of detector is sensitive to the duty cycle of the input signals. The ideal triangular

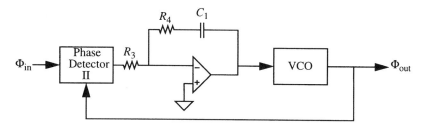

FIGURE 16.32. PLL with active loop filter (defective)

characteristic of the XOR phase detector is obtained only when both inputs possess a 50% duty cycle. If there are any asymmetries, the average output will no longer reach both supply rails at the extremes of phase error. The sequential phase detector is an edge-triggered device and so does not suffer this duty-cycle sensitivity.

Another important note is to reiterate that the XOR phase detector allows locking onto harmonics of the input, since the action of the XOR is equivalent to multiplying two sine waves together. The rich harmonic content of square waves provides many opportunities for a coincidence in frequency between components of the input and VCO output, permitting lock to occur. If harmonic locking is undesirable, then use of an XOR phase detector may cause some problems.

Second-Order PLL with Active RC Loop Filter and PD II

Now let's consider replacing the simple passive RC loop filter with an active filter. Let this filter provide a pole at the origin to drive the steady-state phase error to zero. Additionally, assume that we want to achieve precisely the same crossover frequency and phase margin as in the earlier design, but with the additional requirement that the loop maintain lock at least ± 10 kHz away from the center frequency.

To satisfy the phase margin requirement, we need to provide a loop-stabilizing zero to offset the negative phase contribution of our loop filter's integrator. Our first-pass PLL then should look something like Figure 16.32 (VCO components not shown).

Why "first-pass"? The circuit has a small embarrassment: if the input is ahead of the VCO, the phase detector provides a positive output. The inverting loop filter then drives the VCO toward a lower frequency, exacerbating the phase error; we have a positive feedback loop. To fix this problem, we must provide an additional inversion in the control line.

There is another problem with the circuit: The op-amp's noninverting terminal is grounded. The implication is that the output of the loop filter can never integrate up, since the minimum output of the phase detector is ground. To fix this last (known) problem, we need to connect the noninverting terminal to $V_{DD}/2$, as shown in Figure 16.33. Now we can set about determining the various component values.

First, note that our loop transmission is as follows:

$$-L(s) = K_D H_f(s) \frac{K_0}{s} = \frac{V_{DD}}{4\pi} \cdot \frac{sR_4C_1+1}{sR_3C_1} \cdot \frac{K_0}{s}. \tag{53}$$

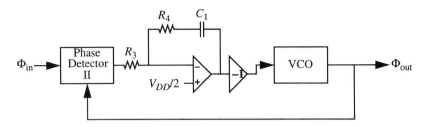

FIGURE 16.33. PLL with active loop filter (fixed)

In order to achieve a 45° phase margin, the zero must be placed at crossover, since the two poles at the origin contribute a total phase shift of −180°. Hence, $R_4 C_1$ must equal 1 ms. Choosing values with the same moderately constrained arbitrariness as in the passive filter case, we let $R_4 = 100$ kΩ, so that the value of C_1 is 0.01 μF.

Next, note that the loop transmission magnitude is controlled by both R_3 and K_0, so we would have an underconstrained problem if achieving a specified crossover frequency were the only consideration. Since there is a requirement on the lock range of the loop, however, there is an additional constraint that allows us to fix both R_3 and K_0. Specifically, the control voltage has an effect on VCO frequency only from about 1.2 to 5 volts, according to the empirical formula.[22] The center of this voltage range is 3.1 V, not the 2.5 V implicitly assumed. If we continue to use 2.5 V as our definition of center, though, the lock range will not be symmetrical about 20 kHz. But since there is no specification about a symmetrical lock range, we will remain consistent in our use of 2.5 V as the control voltage that corresponds to the center frequency of the VCO.

With that choice, the lower frequency limit is smaller than the higher one. To satisfy our 10-kHz specification, we must be able to change the VCO frequency by 10 kHz (or more) with the control voltage at its minimum value of 1.2 V, corresponding to a deviation of 1.3 V from the center. Hence, we require

$$K_0 > \frac{2\pi \cdot 10 \text{ kHz}}{1.3 \text{ V}} \approx 4.8 \times 10^4 \text{ rps/V}. \tag{54}$$

Maintaining a center frequency of 20 kHz with this VCO gain constant leads to the following choices for the three VCO components:

$$C = 0.001 \ \mu\text{F}, \quad R_1 = 42 \text{ k}\Omega, \quad R_2 = 130 \text{ k}\Omega.$$

Here, the closest standard (10% tolerance) resistors for R_1 and R_2 are 39 kΩ and 120 kΩ, respectively.

Finally, having determined everything else, the crossover frequency requirement fixes the value of the op-amp input resistor:

---

[22] The control voltage term is not allowed to take on a negative value in the formula.

$$R_3 C_1 = \frac{K_D K_0}{\omega_c^2} \cdot \sqrt{2} \approx 27.7 \text{ ms}. \tag{55}$$

Therefore, R_3 equals 2.8 megohms (2.7 meg is the closest standard value), and the design is complete.

Note that, for this design, it is definitely the VCO tuning range (and not the phase detector characteristics) that determines the lock range. With a loop filter that provides an integration, any steady-state VCO control voltage can be obtained with zero phase error. As a result, the phase detector characteristics are irrelevant with respect to the steady-state lock range.

16.11.3 SUMMARY

The design examples presented in this lab exercise are representative of typical practice, although they are a tiny subclass of the vast universe of possible PLL applications. The frequencies involved here are below 1 MHz, yet the basic principles remain valid at microwave frequencies. The frequency reduction (by a factor of 3–4) enables students to gain experience with PLLs without the expense and fixturing headaches that afflict gigahertz-frequency PLL design.

CHAPTER SEVENTEEN

OSCILLATOR PHASE NOISE

17.1 INTRODUCTION

We asserted in Chapter 15 that tuned oscillators produce outputs with higher spectral purity than relaxation oscillators. One straightforward reason is simply that a high-Q resonator attenuates spectral components removed from the center frequency. As a consequence, distortion is suppressed, and the waveform of a well-designed tuned oscillator is typically sinusoidal to an excellent approximation.

In addition to suppressing distortion products, a resonator also attenuates spectral components contributed by sources such as the thermal noise associated with finite resonator Q, or by the active element(s) present in all oscillators. Because amplitude fluctuations are usually greatly attenuated as a result of the amplitude stabilization mechanisms present in every practical oscillator, phase noise generally dominates – at least at frequencies not far removed from the carrier. Thus, even though it is possible to design oscillators in which amplitude noise is significant, we focus primarily on phase noise here. We show later that a simple modification of the theory allows for accommodation of amplitude noise as well, permitting the accurate computation of output spectrum at frequencies well removed from the carrier.

Aside from aesthetics, the reason we care about phase noise is to minimize the problem of *reciprocal mixing.* If a superheterodyne receiver's local oscillator is completely noise-free, then two closely spaced RF signals will simply translate downward in frequency together. However, the local oscillator spectrum is not an impulse and so, to be realistic, we must evaluate the consequences of an impure LO spectrum.

In Figure 17.1, two RF signals heterodyne with the LO to produce a pair of IF signals. The desired RF signal is considerably weaker than the signal at an adjacent channel. Assuming (as is typical) that the front-end filter does not have sufficient resolution to perform channel filtering, downconversion preserves the relative amplitudes of the two RF signals in translation to the IF. Because the LO spectrum is of nonzero width, the downconverted RF signals also have width. The tails of the LO spectrum act as parasitic LOs over a continuum of frequencies. Reciprocal mixing is the heterodyning of RF signals with those unwanted components. As is evident

17.1 INTRODUCTION

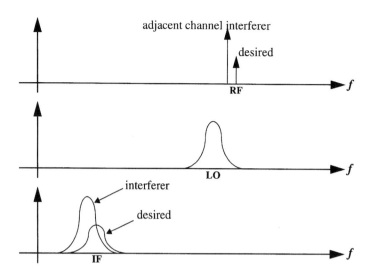

FIGURE 17.1. Illustration of reciprocal mixing due to LO phase noise

from the figure, reciprocal mixing causes the undesired signal to overwhelm the desired signal in this particular case. Reduction of LO phase noise is essential in order to minimize the occurrence and severity of reciprocal mixing.

The theoretical and practical importance of oscillators has motivated the development of numerous treatments of phase noise. The sheer number of publications on this topic underscores the importance attached to it. At the same time, many of these disagree on rather fundamental points, and it may be argued that the abundance of such conflicting research quietly testifies to the inadequacies of many of those treatments. Complicating the search for a suitable theory is that noise in a circuit may undergo frequency translations before ultimately becoming oscillator phase noise. These translations are often attributed to the presence of obvious nonlinearities in practical oscillators. The simplest theories nevertheless simply ignore the nonlinearities altogether and frequently ignore the possibility of time variation as well. Such linear, time-invariant (LTI) theories manage to provide important qualitative design insights, but these theories are understandably limited in their predictive power. Chief among the deficiencies of an LTI theory is that frequency translations are necessarily disallowed, begging the question of how the (nearly) symmetrical sidebands observed in practical oscillators can arise.

Despite this complication, and despite the obvious presence of nonlinearities necessary for amplitude stabilization, the noise-to-phase transfer function of oscillators nonetheless may be treated as linear. However, a quantitative understanding of the frequency translation process requires abandonment of the principle of time invariance implicitly assumed in most theories of phase noise. In addition to providing a quantitative reconciliation between theory and measurement, the time-*varying* phase noise model presented in this chapter identifies an important symmetry principle, which may be exploited to suppress the upconversion of $1/f$ noise into close-in phase

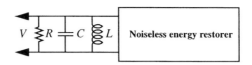

FIGURE 17.2. "Perfectly efficient" RLC oscillator

noise. At the same time, it provides an explicit accommodation of cyclostationary effects – which are significant in many practical oscillators – and of amplitude-to-phase (AM–PM) conversion as well. These insights allow a reinterpretation of why certain topologies, such as the venerable Colpitts oscillator, exhibit good performance. Perhaps more important, the theory informs design, suggesting novel optimizations of well-known oscillators and even the invention of new circuit topologies. We examine some tuned LC oscillator circuit examples to reinforce the theoretical considerations developed, concluding with a brief consideration of practical simulation issues.

We first need to revisit how one evaluates whether a system is linear or time-invariant. This question rarely arises in the analysis of most systems, and perhaps more than a few engineers have forgotten how to tell the difference. Indeed, we find that we must even take care to define explicitly what is meant by the word *system*. We then identify some very general trade-offs among key parameters, such as power dissipation, oscillation frequency, resonator Q, and circuit noise power. Then, we study these trade-offs qualitatively in a hypothetical ideal oscillator in which linearity of the noise-to-phase transfer function is assumed, allowing characterization by an impulse response.

Although the assumption of linearity is defensible, we shall see that time invariance fails to hold even in this simple case. That is, oscillators are linear, time-varying (LTV) systems, where *system* is defined by the noise-to-phase transfer characteristic. Fortunately, complete characterization by an impulse response depends only on linearity, not time invariance. By studying the impulse response, we discover that periodic time variation leads to frequency translation of device noise to produce the phase noise spectra exhibited by real oscillators. In particular, the upconversion of $1/f$ noise into close-in phase noise is seen to depend on symmetry properties that are potentially controllable by the designer. Additionally, the same treatment easily subsumes the cyclostationarity of noise generators. As we'll see, that accommodation explains why class-C operation of active elements within an oscillator can be beneficial. Illustrative circuit examples reinforce key insights of the LTV model.

17.2 GENERAL CONSIDERATIONS

Perhaps the simplest abstraction of an oscillator that still retains some connection to the real world is a combination of a lossy resonator and an energy restoration element. The latter precisely compensates for the tank loss to enable a constant-amplitude oscillation. To simplify matters, assume that the energy restorer is noiseless (see Figure 17.2). The tank resistance is therefore the only noisy element in this model.

In order to gain some useful design insight, first compute the signal energy stored in the tank:

$$E_{sig} = \tfrac{1}{2}CV_{pk}^2, \qquad (1)$$

so that the mean-square signal (carrier) voltage is

$$\overline{V_{sig}^2} = \frac{E_{sig}}{C}, \qquad (2)$$

where we have assumed a sinusoidal waveform.

The total mean-square noise voltage is found by integrating the resistor's thermal noise density over the noise bandwidth of the RLC resonator:

$$\overline{V_n^2} = 4kTR \int_0^\infty \left|\frac{Z(f)}{R}\right|^2 df = 4kTR \cdot \frac{1}{4RC} = \frac{kT}{C}. \qquad (3)$$

Combining Eqn. 2 and Eqn. 3, we obtain a noise-to-carrier ratio (the reason for this "upside-down" ratio is simply one of convention):

$$\frac{N}{S} = \frac{\overline{V_n^2}}{\overline{V_{sig}^2}} = \frac{kT}{E_{sig}}. \qquad (4)$$

Sensibly enough, one therefore needs to maximize the signal levels to minimize the noise-to-carrier ratio.

We may bring power consumption and resonator Q explicitly into consideration by noting that Q can be defined generally as proportional to the energy stored divided by the energy dissipated:

$$Q = \frac{\omega_0 E_{sig}}{P_{diss}}. \qquad (5)$$

Hence, we may write

$$\frac{N}{S} = \frac{\omega_0 kT}{Q P_{diss}}. \qquad (6)$$

The power consumed by this model oscillator is simply equal to P_{diss}, the amount dissipated by the tank loss. The noise-to-carrier ratio is here inversely proportional to the product of resonator Q and the power consumed, and it is directly proportional to the oscillation frequency. This set of relationships still holds approximately for many real oscillators, and it explains the traditional obsession of engineers with maximizing resonator Q, for example.

Other important design criteria become evident by coupling these considerations with additional knowledge of practical oscillators. One is that oscillators generally operate in one of two regimes that are distinguished by their differing dependence of output amplitude on bias current (see Figure 17.3), so that one may write

$$V_{sig} = I_{BIAS} R, \qquad (7)$$

where R is a constant of proportionality with the dimensions of resistance. This constant, in turn, is proportional to the equivalent parallel tank resistance, so that

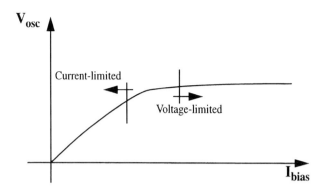

FIGURE 17.3. Oscillator operating regimes

$$V_{sig} \propto I_{BIAS} R_{tank}, \tag{8}$$

implying that the carrier power may be expressed as

$$P_{sig} \propto I_{BIAS}^2 R_{tank}. \tag{9}$$

The mean-square noise voltage has already been computed in terms of the tank capacitance as

$$\overline{V_n^2} = \frac{kT}{C}, \tag{10}$$

but it may also be expressed in terms of the tank inductance:

$$\overline{V_n^2} = \frac{kT}{C} = \frac{kT}{1/\omega_0^2 L} = kT\omega_0^2 L. \tag{11}$$

An alternative expression for the noise-to-carrier ratio in the current-limited regime is therefore

$$\frac{N}{C} \propto \frac{kT\omega_0^2 L}{I_{BIAS}^2 R_{tank}}. \tag{12}$$

Assuming operation at a fixed supply voltage, a constraint on power consumption implies an upper bound on the bias current. Of the remaining free parameters, then, only the tank inductance and resistance may be practically varied to minimize the N/C ratio. That is, optimization of such an oscillator corresponds to minimizing L/R_{tank}. In many treatments, maximizing tank inductance is offered as a prescription for optimization. However, we see that a more valid objective is to minimize L/R_{tank}.[1] Since generally the resistance is itself a function of inductance, it follows that identifying (and then achieving) this minimum is not always trivial. An additional consideration is that, below a certain minimum inductance, oscillation may

---

[1] D. Ham and A. Hajimiri, "Concepts and Methods in Optimization of Integrated LC VCOs," *IEEE J. Solid-State Circuits,* June 2001.

cease. Hence, the optimization prescription here presumes oscillation – and in a regime where the output amplitude is proportional to the bias current.

17.3 DETAILED CONSIDERATIONS: PHASE NOISE

To augment the qualitative insights of the foregoing analysis, let us now determine the actual output spectrum of the ideal oscillator.

Assume that the output in Figure 17.2 is the voltage across the tank, as shown. By postulate, the only source of noise is the white thermal noise of the tank conductance, which we represent as a current source across the tank with a mean-square spectral density of

$$\frac{\overline{i_n^2}}{\Delta f} = 4kTG. \tag{13}$$

This current noise becomes voltage noise when multiplied by the effective impedance facing the current source. In computing this impedance, however, it is important to recognize that the energy restoration element must contribute an average effective negative resistance that precisely cancels the positive resistance of the tank. Hence, the net result is that the effective impedance seen by the noise current source is simply that of a perfectly lossless LC network.

For a relatively small offset frequency $\Delta\omega$ from the center frequency ω_0, the impedance of an LC tank may be approximated by

$$Z(\omega_0 + \Delta\omega) \approx -j \cdot \frac{\omega_0 L}{2(\Delta\omega/\omega_0)}. \tag{14}$$

We may write the impedance in a more useful form by incorporating an expression for the unloaded tank Q:

$$Q = \frac{R}{\omega_0 L} = \frac{1}{\omega_0 G L}. \tag{15}$$

Solving Eqn. 15 for L and substituting into Eqn. 14 yields

$$|Z(\omega_0 + \Delta\omega)| \approx \frac{1}{G} \cdot \frac{\omega_0}{2Q\Delta\omega}. \tag{16}$$

Thus, we have traded an explicit dependence on inductance for a dependence on Q and G.

Next, multiply the spectral density of the mean-square noise current by the squared magnitude of the tank impedance to obtain the spectral density of the mean-square noise voltage:

$$\frac{\overline{v_n^2}}{\Delta f} = \frac{\overline{i_n^2}}{\Delta f} \cdot |Z|^2 = 4kTR\left(\frac{\omega_0}{2Q\Delta\omega}\right)^2. \tag{17}$$

The power spectral density of the output noise is frequency-dependent because of the filtering action of the tank, falling as the inverse square of the offset frequency. This $1/f^2$ behavior simply reflects the facts that (a) the voltage frequency response of an

RLC tank rolls off as $1/f$ to either side of the center frequency and (b) power is proportional to the square of voltage. Note also that an increase in tank Q reduces the noise density (all other parameters held constant), underscoring once again the value of increasing resonator Q.

In our idealized *LC* model, thermal noise causes fluctuations in both amplitude and phase, and Eqn. 17 accounts for both. The *equipartition theorem* of thermodynamics tells us that, if there were no amplitude limiting, noise energy would split equally into amplitude and phase noise domains. The amplitude-limiting mechanisms present in all practical oscillators remove most of the amplitude noise, leaving us with about half the noise given by Eqn. 17.

Additionally, we are often more interested in how large this noise is relative to the carrier, rather than its absolute value. It is consequently traditional to normalize the mean-square noise voltage density to the mean-square carrier voltage and then report the ratio in decibels, thereby explaining the "upside down" ratios presented previously. Performing this normalization yields the following equation for phase noise:

$$L\{\Delta\omega\} = 10\log\left[\frac{2kT}{P_{sig}} \cdot \left(\frac{\omega_0}{2Q\Delta\omega}\right)^2\right]. \tag{18}$$

The units of phase noise are thus proportional to the log of a density. Specifically, they are commonly expressed as "decibels below the carrier per hertz" (dBc/Hz), specified at a particular offset frequency $\Delta\omega$ from the carrier frequency ω_0. For example, one might speak of a 2-GHz oscillator's phase noise as "-110 dBc/Hz at a 100-kHz offset." Purists may complain that the "per hertz" actually applies to the argument of the log, not to the log itself; doubling the measurement bandwidth does not double the decibel quantity. Nevertheless, as lacking in rigor as "dBc/Hz" is, it is a unit in common usage.

Equation 18 tells us that phase noise (at a given offset) improves as both the carrier power and Q increase, as predicted earlier. These dependencies make sense. Increasing the signal power improves the ratio simply because the thermal noise is fixed, while increasing Q improves the ratio quadratically because the tank's impedance falls off as $1/Q\Delta\omega$.

Because many simplifying assumptions have led us to this point, it should not be surprising that there are some significant differences between the spectrum predicted by Eqn. 18 and what one typically measures in practice. For example, although real spectra do possess a region where the observed density is proportional to $1/(\Delta\omega)^2$, the magnitudes are typically quite a bit larger than predicted by Eqn. 18 because there are additional important noise sources besides tank loss. For example, any physical implementation of an energy restorer will be noisy. Furthermore, measured spectra eventually flatten out for large frequency offsets, rather than continuing to drop quadratically. Such a floor may be due to the noise associated with any active elements (such as buffers) placed between the tank and the outside world, or it can even reflect limitations in the measurement instrumentation itself. Even if the output were taken directly from the tank, any resistance in series with either the inductor

17.3 DETAILED CONSIDERATIONS: PHASE NOISE

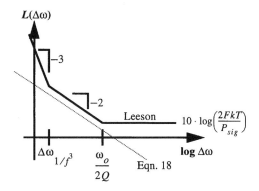

FIGURE 17.4. Phase noise: Leeson versus Eqn. 18

or capacitor would impose a bound on the amount of filtering provided by the tank at large frequency offsets and thus ultimately produce a noise floor. Finally, there is almost always a $1/(\Delta\omega)^3$ region at small offsets.

A modification to Eqn. 18 provides a means to account for these discrepancies:

$$L\{\Delta\omega\} = 10\log\left[\frac{2FkT}{P_{sig}}\left\{1 + \left(\frac{\omega_0}{2Q\Delta\omega}\right)^2\right\}\left(1 + \frac{\Delta\omega_{1/f^3}}{|\Delta\omega|}\right)\right]. \quad (19)$$

These modifications, due to Leeson, consist of a factor F to account for the increased noise in the $1/(\Delta\omega)^2$ region, an additive factor of unity (inside the braces) to account for the noise floor, and a multiplicative factor (the term in the second set of parentheses) to provide a $1/|\Delta\omega|^3$ behavior at sufficiently small offset frequencies.[2] With these modifications, the phase noise spectrum appears as in Figure 17.4.

The Leeson model is extremely valuable for the intuitive insights it may provide about oscillators. However, it is important to note that the factor F is an empirical fitting parameter and therefore must be determined from measurements, diminishing the predictive power of the phase noise equation. Furthermore, the model asserts that $\Delta\omega_{1/f^3}$, the boundary between the $1/(\Delta\omega)^2$ and $1/|\Delta\omega|^3$ regions, is precisely equal to the $1/f$ corner of device noise. However, measurements frequently show no such equality, and thus one must generally treat $\Delta\omega_{1/f^3}$ as an empirical fitting parameter as well. Also it is not clear what the corner frequency will be in the presence of more than one noise source, each with an individual $1/f$ noise contribution (and generally differing $1/f$ corner frequencies). Finally, the frequency at which the noise flattens out is not always equal to half the resonator bandwidth, $\omega_0/2Q$.

Both the ideal oscillator model and the Leeson model suggest that increasing resonator Q and signal power are ways to reduce phase noise. The Leeson model additionally introduces the factor F, but without knowing precisely what it depends on,

---

[2] D. B. Leeson, "A Simple Model of Feedback Oscillator Noise Spectrum," *Proc. IEEE*, v. 54, February 1966, pp. 329–30.

it is difficult to identify specific ways to reduce it. The same problem exists with $\Delta\omega_{1/f^3}$ as well. Finally, blind application of these models has periodically led to earnest but misguided attempts by some designers to use active circuits to boost Q. Sadly, increases in Q through such means are necessarily accompanied by increases in F as well because active devices contribute noise of their own, and the anticipated improvements in phase noise fail to materialize. Again, the lack of analytical expressions for F can obscure this conclusion, and one continues to encounter various doomed oscillator designs based on the notion of active Q boosting.

That neither Eqn. 18 nor Eqn. 19 can make quantitative predictions about phase noise is an indication that at least some of the assumptions used in the derivations are invalid, despite their apparent reasonableness. To develop a theory that does not possess the enumerated deficiencies, we need to revisit, and perhaps revise, these assumptions.

17.4 THE ROLES OF LINEARITY AND TIME VARIATION IN PHASE NOISE

The preceding derivations have all assumed linearity and time invariance. Let's reconsider each of these assumptions in turn.

Nonlinearity is clearly a fundamental property of all real oscillators, as its presence is necessary for amplitude limiting. It seems entirely reasonable, then, to try to explain certain observations as a consequence of nonlinear behavior. One of these observations is that a single-frequency sinusoidal disturbance injected into an oscillator gives rise to two equal-amplitude sidebands, symmetrically disposed about the carrier.[3] Since LTI systems cannot perform frequency translation and nonlinear systems can, nonlinear mixing has often been proposed to explain phase noise. As we shall see momentarily, amplitude-control nonlinearities certainly do affect phase noise – but only *indirectly*, by controlling the detailed shape of the output waveform.

An important insight is that disturbances are just that: perturbations superimposed on the main oscillation. They will always be much smaller in magnitude than the carrier in any oscillator worth using or analyzing. Thus, if a certain amount of injected noise produces a certain phase disturbance, we ought to expect that doubling the injected noise will double the disturbance. Linearity would therefore appear to be a reasonable (and experimentally testable) assumption *as far as the noise-to-phase transfer function is concerned*. It is therefore particularly important to keep in mind that, when assessing linearity, it is essential to identify explicitly the input–output variables. It is also important to recognize that this assumption of linearity is not equivalent to a neglect of the nonlinear behavior of the active devices. Because it is a linearization around the steady-state solution, it therefore already takes the effect of device nonlinearity into account. It is precisely analogous to amplifier analysis,

---

[3] B. Razavi, "A Study of Phase Noise in CMOS Oscillators," *IEEE J. Solid-State Circuits,* v. 31, no. 3, March 1996.

17.4 THE ROLES OF LINEARITY AND TIME VARIATION IN PHASE NOISE

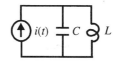

FIGURE 17.5. LC oscillator excited by current pulse

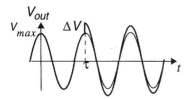
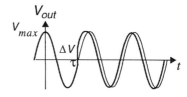

FIGURE 17.6. Impulse responses of LC tank

where small-signal gains are defined around a bias solution found using large-signal (nonlinear) equations. There is thus no contradiction here with the prior acknowledgment of nonlinear amplitude control. Any seeming contradiction is due to the fact that the word *system* is actually ill-defined. Most take it to refer to an assemblage of components and their interconnections, but a more useful definition is based on the particular input–output variables chosen. With this definition, a given circuit may possess nonlinear relationships among certain variables and linear ones among others. Time invariance is also not an inherent property of the entire circuit; it is similarly dependent on the variables chosen.

We are left only with the assumption of time invariance to re-examine. In the previous derivations we extended time invariance to the noise sources themselves, meaning that the measures that characterize noise (e.g., spectral density) are time-invariant (stationary). In contrast with linearity, the assumption of time invariance is less obviously defensible. In fact, it is surprisingly simple to demonstrate that oscillators are fundamentally time-varying systems. Recognizing this truth is the main key to developing a more accurate theory of phase noise.[4]

To test whether time invariance holds, consider explicitly how an impulse of current affects the waveform of the simplest resonant system, a lossless LC tank (Figure 17.5). Assume that the system has been oscillating forever with some constant amplitude; then consider how the system responds to an impulse injected at two different times, as shown in Figure 17.6.

If the impulse happens to coincide with a voltage maximum (as in the left plot of the figure), the amplitude increases abruptly by an amount $\Delta V = \Delta Q/C$, but because the response to the impulse superposes exactly in phase with the pre-existing

---

[4] A. Hajimiri and T. Lee, "A General Theory of Phase Noise in Electrical Oscillators," *IEEE J. Solid-State Circuits*, v. 33, no. 2, February 1998, pp. 179–94.

oscillation, the timing of the zero crossings does not change. Thus, even though we have clearly changed the energy in the system, the amplitude change is not accompanied by a change in phase. On the other hand, an impulse injected at some other time generally affects both the amplitude of oscillation and the timing of the zero crossings, as in the right-hand plot.

Interpreting the zero-crossing timings as a measure of phase, we see that the amount of phase disturbance for a given injected impulse depends on when the injection occurs; time invariance thus fails to hold. An oscillator is therefore a linear yet (periodically) time-varying (LTV) system. It is especially important to note that it is theoretically possible to leave unchanged the energy of the system (as reflected in the constant tank amplitude of the right-hand response) if the impulse injects at a moment near the zero crossing when the net work performed by the impulse is zero. For example: a small positive impulse injected when the tank voltage is negative extracts energy from the oscillator, whereas the same impulse injected when the tank voltage is positive delivers energy to the oscillator. Just before the zero crossing, an instant may be found where such an impulse performs no net work at all. Hence the amplitude of oscillation cannot change, but the zero crossings will be displaced.

Because linearity (of noise-to-phase conversion) remains a good assumption, the impulse response still completely characterizes that system – even with time variation present. The only difference relative to an LTI impulse response is that the impulse response here is a function of *two* arguments, the observation time t and the excitation time τ. Because an impulsive input produces a step change in phase, the impulse response may be written as

$$h_\phi(t, \tau) = \frac{\Gamma(\omega_0 \tau)}{q_{max}} u(t - \tau), \tag{20}$$

where $u(t)$ is the unit step function. Dividing by q_{max}, the maximum charge displacement across the capacitor, makes the function $\Gamma(x)$ independent of signal amplitude. This normalization is a convenience that allows us to compare different oscillators fairly. Note that $\Gamma(x)$, called the impulse sensitivity function (ISF), is a dimensionless, frequency- and amplitude-independent function periodic in 2π. As its name suggests, the ISF encodes information about the sensitivity of the oscillator to an impulse injected at phase $\omega_0 t$. In the *LC* oscillator example, $\Gamma(x)$ has its maximum value near the zero crossings of the oscillation and a zero value at maxima of the oscillation waveform. In general, it is most practical (and most accurate) to determine $\Gamma(x)$ through simulation, but there are also analytical methods (some approximate) that apply in special cases.[5] In any event, to develop a feel for typical shapes of ISFs, consider two representative examples: the *LC* and ring oscillators of Figure 17.7.

---

[5] F. X. Kaertner, "Determination of the Correlation Spectrum of Oscillators with Low Noise," *IEEE Trans. Microwave Theory and Tech.*, v. 37, no. 1, January 1989. Also see A. Hajimiri and T. Lee, *The Design of Low-Noise Oscillators*, Kluwer, Dordrecht, 1999.

17.4 THE ROLES OF LINEARITY AND TIME VARIATION IN PHASE NOISE

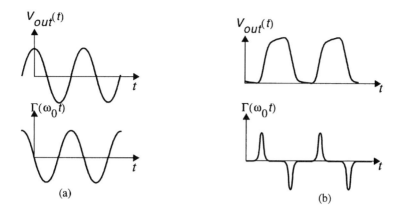

FIGURE 17.7. Example ISF for (a) *LC* oscillator and (b) ring oscillator

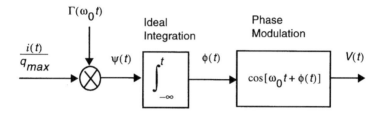

FIGURE 17.8. Equivalent block diagram of the process described (in part) by Eqn. 21

Once the impulse response has been determined (by whatever means), we may compute the excess phase due to an *arbitrary* noise signal through use of the superposition integral. This computation is valid here because superposition is linked to linearity, not to time invariance:

$$\phi(t) = \int_{-\infty}^{\infty} h_\phi(t,\tau) i(\tau)\, d\tau = \frac{1}{q_{max}} \int_{-\infty}^{t} \Gamma(\omega_0 \tau) i(\tau)\, d\tau. \qquad (21)$$

The equivalent block diagram shown in Figure 17.8 helps us visualize this computation in ways that are familiar to telecommunications engineers, who will recognize a structure reminiscent of a superheterodyne system (more on this viewpoint shortly).

To cast this superposition integral into a more practically useful form, note that the ISF is periodic and therefore expressible as a Fourier series:

$$\Gamma(\omega_0 \tau) = \frac{c_0}{2} + \sum_{n=1}^{\infty} c_n \cos(n\omega_0 \tau + \theta_n), \qquad (22)$$

where the coefficients c_n are real and where θ_n is the phase of the nth harmonic of the ISF. (We will ignore θ_n in all that follows because we assume that noise components are uncorrelated and so their relative phase is irrelevant.) The value of this

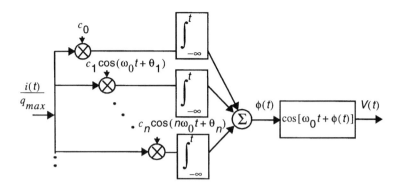

FIGURE 17.9. Equivalent system for ISF decomposition

decomposition is that – like many functions associated with physical phenomena – the series typically converges rapidly, so that it is often well approximated by just the first few terms of the series.

Substituting the Fourier expansion into Eqn. 21 and then exchanging summation and integration, one obtains

$$\phi(t) = \frac{1}{q_{max}} \left[\frac{c_0}{2} \int_{-\infty}^{t} i(\tau)\, d\tau + \sum_{n=1}^{\infty} c_n \int_{-\infty}^{t} i(\tau) \cos(n\omega_0 \tau)\, d\tau \right]. \quad (23)$$

The corresponding sequence of mathematical operations is shown graphically in the left half of Figure 17.9. Note that the block diagram again contains elements that are analogous to those of a superheterodyne receiver. The normalized noise current is a broadband "RF" signal, whose Fourier components undergo simultaneous downconversions (multiplications) by a "local oscillator" signal that is the ISF, whose harmonics are multiples of the oscillation frequency. It is important to keep in mind that multiplication is a linear operation if one argument is held constant, as it is here. The relative contributions of these multiplications are determined by the Fourier coefficients of the ISF. Equation 23 thus allows us to compute the excess phase caused by an arbitrary noise current injected into the system, once the Fourier coefficients of the ISF have been determined (typically through simulation).

We have already noted the common observation that signals (noise) injected into a nonlinear system at some frequency may produce spectral components at a different frequency. We now show that a linear but time-varying system can exhibit qualitatively similar behavior, as implied by the superheterodyne imagery invoked earlier. To demonstrate this property explicitly, consider injecting a sinusoidal current whose frequency is near an integer multiple m of the oscillation frequency, so that

$$i(t) = I_m \cos[(m\omega_0 + \Delta\omega)t], \quad (24)$$

where $\Delta\omega \ll \omega_0$. Substituting Eqn. 24 into Eqn. 23 and noting that there is a negligible net contribution to the integral by terms other than when $n = m$, we derive the following approximation:

17.4 THE ROLES OF LINEARITY AND TIME VARIATION IN PHASE NOISE

$$\phi(t) \approx \frac{I_m c_m \sin(\Delta \omega t)}{2 q_{max} \Delta \omega}. \qquad (25)$$

The spectrum of $\phi(t)$ therefore consists of two equal sidebands at $\pm \Delta \omega$, even though the injection occurs near some integer multiple of ω_0. This observation is fundamental to understanding the evolution of noise in an oscillator.

Unfortunately, we're not quite done: Eqn. 25 allows us to figure out the spectrum of $\phi(t)$, but we ultimately want to find the spectrum of the output voltage of the oscillator, which is not quite the same thing. However, the two quantities are linked through the actual output waveform. To illustrate what we mean by this linkage, consider a specific case where the output may be approximated as a sinusoid, so that $v_{out}(t) = \cos[\omega_0 t + \phi(t)]$. This equation may be considered a phase-to-voltage converter; it takes phase as an input, producing from it the output voltage. This conversion is fundamentally nonlinear because it involves the phase modulation of a sinusoid.

Performing this phase-to-voltage conversion and assuming "small" amplitude disturbances, we find that the single-tone injection leading to Eqn. 25 results in two equal-power sidebands symmetrically disposed about the carrier:

$$P_{SBC}(\Delta\omega) \approx 10 \cdot \log\left(\frac{I_m c_m}{4 q_{max} \Delta\omega}\right)^2. \qquad (26)$$

Note that the amplitude dependence is linear (the squaring operation simply reflects the fact that we are dealing with a power quantity here). This relationship has been verified experimentally for an exceptionally wide range of practical oscillators.

This result may be extended to the general case of a white noise source:

$$P_{SBC}(\Delta\omega) \approx 10 \cdot \log\left[\frac{(\overline{i_n^2}/\Delta f) \sum_{m=0}^{\infty} c_m^2}{4 q_{max}^2 \Delta\omega^2}\right]. \qquad (27)$$

Together, Eqns. 26 and 27 imply both upward and downward frequency translations of noise into the noise near the carrier, as illustrated in Figure 17.10. This figure summarizes what the preceding equations tell us: Components of noise near integer multiples of the carrier frequency all fold into noise near the carrier itself.

Noise near DC is upconverted, with relative weight given by coefficient c_0, so $1/f$ device noise ultimately becomes $1/f^3$ noise near the carrier; noise near the carrier stays there, weighted by c_1; and white noise near higher integer multiples of the carrier undergoes downconversion, turning into noise in the $1/f^2$ region. Note that the $1/f^2$ shape results from the integration implied by the step change in phase caused by an impulsive noise input. Since an integration (even a time-varying one) gives a white voltage or current spectrum a $1/f$ character, the power spectral density will have a $1/f^2$ shape.

It is clear from Figure 17.10 that minimizing the various coefficients c_n (by minimizing the ISF) will minimize the phase noise. To underscore this point quantitatively, we may use Parseval's theorem to write

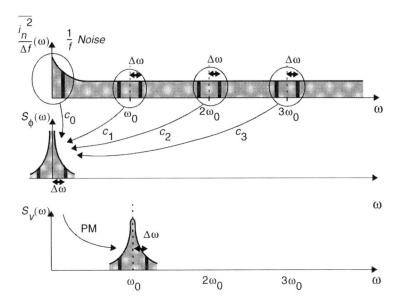

FIGURE 17.10. Evolution of circuit noise into phase noise

$$\sum_{n=0}^{\infty} c_m^2 = \frac{1}{\pi} \int_0^{2\pi} |\Gamma(x)|^2 \, dx = 2\Gamma_{rms}^2, \tag{28}$$

so that the spectrum in the $1/f^2$ region may be expressed as

$$L(\Delta\omega) = 10 \cdot \log\left[\frac{(\overline{i_n^2}/\Delta f)\Gamma_{rms}^2}{2q_{max}^2 \Delta\omega^2}\right], \tag{29}$$

where Γ_{rms} is the rms value of the ISF. All other factors held equal, reducing Γ_{rms} will reduce the phase noise at all frequencies. Equation 29 is the rigorous equation for the $1/f^2$ region, and it is one key result of this phase noise model. Note that no empirical curve-fitting parameters are present in this equation.

Among other attributes, Eqn. 29 allows us to study quantitatively the upconversion of $1/f$ noise into close-in phase noise. Noise near the carrier is particularly important in communication systems with narrow channel spacings. In fact, the allowable channel spacings are frequently constrained by the achievable phase noise. Unfortunately, it is not possible to predict close-in phase noise correctly with LTI models.

This problem disappears if the new model is used. Specifically, assume that the current noise behaves as follows in the $1/f$ region:

$$\overline{i_{n,1/f}^2} = \overline{i_n^2} \cdot \frac{\omega_{1/f}}{\Delta\omega}, \tag{30}$$

where $\omega_{1/f}$ is the $1/f$ corner frequency. Using Eqn. 27 yields the following for the noise in the $1/f^3$ region:

17.4 THE ROLES OF LINEARITY AND TIME VARIATION IN PHASE NOISE

$$L(\Delta\omega) = 10 \cdot \log\left[\frac{(\overline{i_n^2}/\Delta f)c_0^2}{8q_{max}^2\Delta\omega^2} \cdot \frac{\omega_{1/f}}{\Delta\omega}\right]. \tag{31}$$

The $1/f^3$ corner frequency is then

$$\Delta\omega_{1/f^3} = \omega_{1/f} \cdot \frac{c_0^2}{4\Gamma_{rms}^2} = \omega_{1/f} \cdot \left(\frac{\Gamma_{dc}}{\Gamma_{rms}}\right)^2, \tag{32}$$

from which we see that the $1/f^3$ phase noise corner is not necessarily the same as the $1/f$ device/circuit noise corner; it will generally be lower. In fact, since Γ_{dc} is the DC value of the ISF, there is a possibility of reducing by large factors the $1/f^3$ phase noise corner. The ISF is a function of the waveform and hence is potentially under the control of the designer, usually through adjustment of the rise and fall time symmetry. This result is not anticipated by LTI approaches, and it is one of the most powerful insights conferred by this LTV model. This result has particular significance for technologies with notoriously poor $1/f$ noise performance, such as GaAs MESFETs and CMOS. Specific circuit examples of how one may exploit this observation are presented in Section 17.5.

One more extremely powerful insight concerns the influence of cyclostationary noise sources. As alluded to previously, the noise sources in many oscillators cannot be well modeled as stationary. A typical example is the nominally white drain noise current in a FET, or the shot noise in a bipolar transistor. Noise currents are a function of bias currents, and the latter vary periodically and significantly with the oscillating waveform. The LTV model is able to accommodate a cyclostationary white noise source with ease, because such a source may be treated as the product of a stationary white noise source and a periodic function:[6]

$$i_n(t) = i_{n0}(t) \cdot \alpha(\omega_0 t). \tag{33}$$

In this equation, i_{n0} is a stationary white noise source whose peak value is equal to that of the cyclostationary source, and the *noise modulation function* (NMF) $\alpha(x)$ is a periodic dimensionless function with a peak value of unity. See Figure 17.11. Substituting the expression for noise current into Eqn. 21 allows us to treat cyclostationary noise as a stationary noise source, provided we define an effective ISF as follows:

$$\Gamma_{eff}(x) = \Gamma(x) \cdot \alpha(x). \tag{34}$$

Figure 17.12 shows $\Gamma(x)$, $\alpha(x)$, and $\Gamma_{eff}(x)$ for a Colpitts oscillator, all plotted over one cycle. The quasisinusoidal shape of $\Gamma(x)$ is perhaps to be anticipated on the basis of the ideal LC oscillator ISF examined earlier, where the output voltage and ISF were approximately the same in shape but in quadrature. The NMF is near zero most of the time, which is consistent with the Class-C operation of the transistor in

---

[6] W. A. Gardner, *Introduction to Random Processes,* McGraw-Hill, New York, 1990.

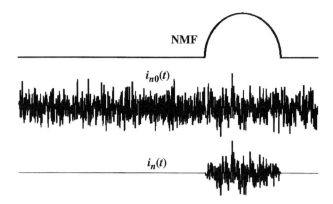

FIGURE 17.11. Cyclostationary noise as product of stationary noise and NMF

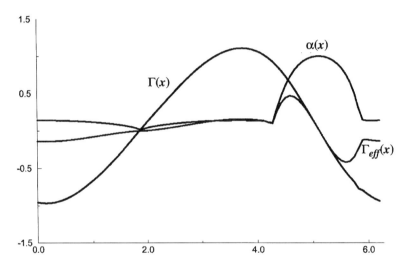

FIGURE 17.12. Accommodation of cyclostationarity

a Colpitts circuit; the transistor replenishes the lost tank energy over a relatively narrow window of time, as suggested by the shape of $\alpha(x)$. The product of these two functions, $\Gamma_{eff}(x)$, has a much smaller rms value than $\Gamma(x)$, explicitly showing the exploitation of cyclostationarity by this oscillator.

This example underscores that cyclostationarity is therefore easily accommodated within the framework we have already established. None of the foregoing conclusions changes as long as Γ_{eff} is used in all of the equations.[7]

---

[7] This formulation might not apply if *external* cyclostationary noise sources are introduced into an oscillator, such as might be the case in injection-locked oscillators. For a detailed discussion, see P. Vanassche et al., "On the Difference between Two Widely Publicized Methods for Analyzing Oscillator Phase Behavior," *Proc. IEEE/ACM/ICCAD,* Session 4A, 2002.

17.4 THE ROLES OF LINEARITY AND TIME VARIATION IN PHASE NOISE

Having identified the factors that influence oscillator noise, we're now in a position to articulate the requirements that must be satisfied in order to make a good oscillator. First, in common with the revelations of LTI models, both the signal power and resonator Q should be maximized, all other factors held constant. In addition, note that an active device is always necessary to compensate for tank loss, and that active devices always contribute noise. Note also that the ISFs tell us there are sensitive and insensitive moments in an oscillation cycle. Of the infinitely many ways that an active element could return energy to the tank, the best strategy is to deliver all of the energy at once, where the ISF has its minimum value. Thus in an ideal LC oscillator, the transistor would remain off almost all of the time, waking up periodically to deliver an impulse of current at the signal peak(s) of each cycle. The extent to which real oscillators approximate this behavior determines in large part the quality of their phase noise properties. Since an LTI theory treats all instants as equally significant, such theories are unable to anticipate this important result.

The prescription for impulsive energy restoration has actually been practiced for centuries, but in a different domain. In mechanical clocks, a structure known as an *escapement* regulates the transfer of energy from a spring to a pendulum. The escapement forces this transfer to occur impulsively and only at precisely timed moments (coincident with the point of maximum pendulum velocity), which are chosen to minimize the disturbance of the oscillation period. Although this historically important analogue is hundreds of years old, having been designed by intuition and trial and error, it was not analyzed mathematically until 1826 by Astronomer Royal George Airy.[8] Certainly its connection to the broader field of electronic oscillators has only recently been recognized.

Finally, the best oscillators will possess the symmetry properties that lead to small Γ_{dc} for minimum upconversion of $1/f$ noise. After examining some additional features of close-in phase noise, we consider in the following section several circuit examples of how to accomplish these ends in practice.

CLOSE-IN PHASE NOISE

From the development so far, one expects the spectrum $S_\phi(\omega)$ to have a close-in behavior that is proportional to the inverse cube of frequency. That is, the spectral density grows without bound as the carrier frequency is approached. However, most measurements fail to show this behavior, and this failure is often misinterpreted as the result of some new phenomenon or as evidence of a flaw in the LTV theory. It is therefore worthwhile to spend some time considering this issue in detail.

The LTV theory asserts only that $S_\phi(\omega)$ grows without bound. Most "phase" noise measurements, however, actually measure the spectrum of the oscillator's output

---

[8] G. B. Airy, "On the Disturbances of Pendulums and Balances, and on the Theory of Escapements," *Trans. Cambridge Philos. Soc.*, v. 3, pt. I, 1830, pp. 105–28. The author is grateful to Mr. Byron Blanchard for bring this reference to his attention.

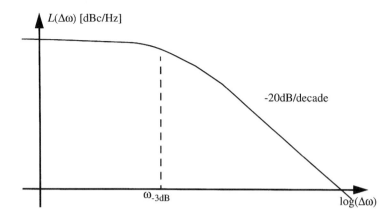

FIGURE 17.13. Lorentzian spectrum

voltage. That is, what is often measured is actually $S_V(\omega)$. In such a case, the output spectrum will not show a boundless growth as the offset frequency approaches zero, reflecting the simple fact that a cosine function is bounded even for unbounded arguments. This bound causes the measured spectrum to flatten as the carrier is approached; the resulting shape is *Lorentzian,*[9] as shown in Figure 17.13.

Depending on the details of how the measurement is performed, the −3-dB corner may or may not be observed. If a spectrum analyzer is used, the corner typically *will* be observed. If an ideal phase detector and a phase-locked loop were available to downconvert the spectrum of $\phi(t)$ and measure it directly, no flattening would be observed at all. A −3-dB corner will generally be observed with real phase detectors (which necessarily possess finite phase detection range), but the precise value of the corner will now be a function of the instrumentation; the measurement will no longer reflect the inherent spectral properties of the oscillator. The lack of consistency in measurement techniques has been a source of great confusion in the past.

17.5 CIRCUIT EXAMPLES – *LC* OSCILLATORS

Having derived expressions for phase noise at low and moderate offset frequencies, it is instructive to apply to practical oscillators the insights gained. We examine first the popular Colpitts oscillator and its relevant waveforms (see Figure 17.14 and Figure 17.15). An important feature is that the drain current flows only during a short interval coincident with the most benign moments (the peaks of the tank voltage). Its corresponding excellent phase noise properties account for the popularity of this configuration. It has long been known that the best phase noise occurs for a certain

---

[9] W. A. Edson, "Noise in Oscillators," *Proc. IRE,* August 1960, pp. 1454–66. Also see J. A. Mullen, "Background Noise in Nonlinear Oscillators," *Proc. IRE,* August 1960, pp. 1467–73. A Lorentzian shape is the same as a single-pole low-pass filter's power response. It just sounds more impressive if you say "Lorentzian."

17.5 CIRCUIT EXAMPLES – LC OSCILLATORS

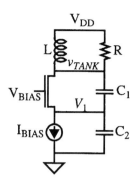

FIGURE 17.14. Colpitts oscillator (simplified)

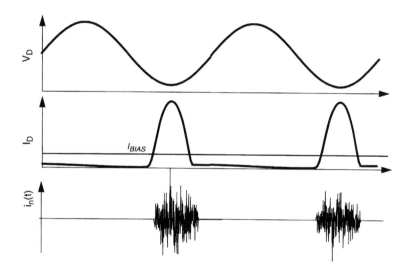

FIGURE 17.15. Approximate incremental tank voltage and drain current for Colpitts oscillator

narrow range of tapping ratios (e.g., a 3 : 1 or 4 : 1 C_2/C_1 capacitance ratio), but before LTV theory there was no theoretical basis for explaining a particular optimum.

The cyclostationary nature of the drain noise is evident in the graphs of Figure 17.15. Because the noise is largest when the ISF is relatively small, the effective ISF (the product of the ISF and the noise modulating function) is much smaller than the ISF.

Both LTI and LTV models point out the value of maximizing signal amplitude. In order to evade supply voltage or breakdown constraints, one may employ a tapped resonator to decouple resonator swings from device voltage limitations. A common configuration that does so is Clapp's modification to the Colpitts oscillator (reprised in Figure 17.16). Differential implementations of oscillators with tapped resonators have also appeared in the literature.[10] These types of oscillators are either of Clapp

---

[10] J. Craninckx and M. Steyaert, "A 1.8GHz CMOS Low-Phase-Noise Voltage-Controlled Oscillator with Prescaler," *IEEE J. Solid-State Circuits*, v. 30, no. 12, December 1995, pp. 1474–82. Also

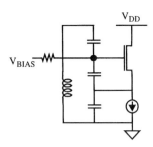

FIGURE 17.16. Clapp oscillator

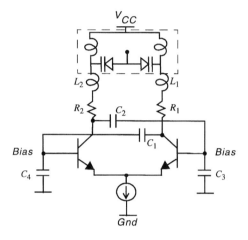

FIGURE 17.17. Simplified schematic of the VCO in Margarit et al. (1999)

configurations or the dual (with tapped inductor). The Clapp configuration becomes increasingly attractive as supply voltages scale downward, where conventional resonator connections lead to V_{DD}-constrained signal swings. Use of tapping allows signal energy to remain high even with low supply voltages.

Phase noise predictions using the LTV model are frequently more accurate for bipolar oscillators owing to the availability of better device noise models. In Margarit et al. (see footnote 10), impulse response modeling (see Section 17.8) is used to determine the ISFs for the various noise sources within the oscillator, and this knowledge is used to optimize the noise performance of a differential bipolar VCO. A simplified schematic of this oscillator is shown in Figure 17.17.

A tapped resonator is used to increase the tank signal power, P_{sig}. The optimum capacitive tapping ratio is calculated to be around 4.5 (corresponding to a capacitance

see M. A. Margarit, J. I. Tham, R. G. Meyer, and M. J. Deen, "A Low-Noise, Low-Power VCO with Automatic Amplitude Control for Wireless Applications," *IEEE J. Solid-State Circuits,* v. 34, no. 6, June 1999, pp. 761–71.

17.5 CIRCUIT EXAMPLES – LC OSCILLATORS

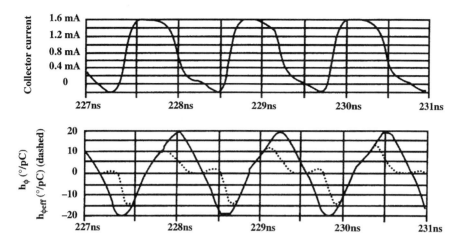

FIGURE 17.18. ISF for shot noise of each core transistor (after Margarit et al. 1999)

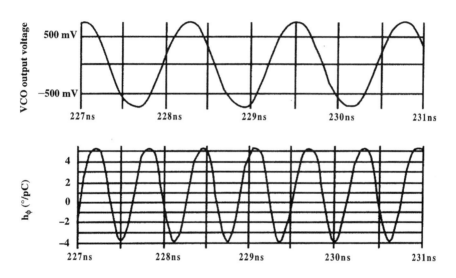

FIGURE 17.19. ISF for shot noise of tail current (after Margarit et al. 1999)

ratio of 3.5), based on simulations that take into account the cyclostationarity of the noise sources. Specifically, the simulation accounts for noise contributions by the base spreading resistance and collector shot noise of each transistor and also by the resistive losses of the tank elements. The ISFs (taken from Margarit et al., in which these are computed through direct evaluation in the time domain as described in Section 17.8) for the shot noise of the core oscillator transistors and for the bias source are shown in Figures 17.18 and 17.19, respectively. As can be seen, the tail current noise has an ISF with double the periodicity of the oscillation frequency, owing to the differential topology of the circuit (the tail voltage waveform contains a component at twice the oscillator frequency). Noteworthy is the observation that tail noise thus

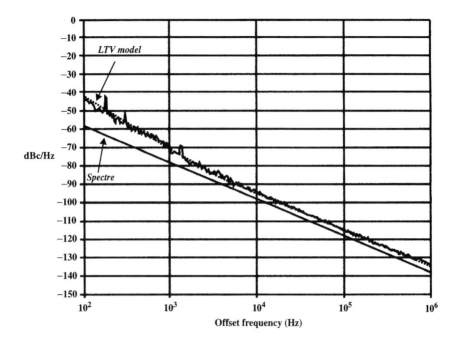

FIGURE 17.20. Measured and predicted phase noise of VCO in Margarit et al. (1999)

contributes to phase noise only at even multiples of the oscillation frequency. If the tail current is filtered through a low-pass (or bandstop) filter before feeding the oscillator core, then the noise contributed by the tail source can be reduced substantially; decreases of 10 dB or more have been noted.[11] Only the tail current's $1/f$ noise would remain as a noise contributor. The individual ISFs are used to compute the contribution of each corresponding noise source, and the contributions are then summed.

The reduction of $1/f$ noise upconversion in this topology is clearly seen in Figure 17.20, which shows a predicted and measured $1/f^3$ corner of 3 kHz – in comparison with an individual device $1/f$ noise corner of 200 kHz. Note that the then-current version of one commercial simulation tool, *Spectre*, fails in this case to identify a $1/f^3$ corner within the offset frequency range shown, resulting in a 15-dB underestimate at a 100-Hz offset. The measured phase noise in the $1/f^2$ region is also in excellent agreement with the LTV model's predictions. For example, the predicted value of -106.2 dBc/Hz at 100-kHz offset is negligibly different from the measured value of -106 dBc/Hz. As a final comment, this particular VCO design is also noteworthy for its use of a separate automatic amplitude control loop; this allows for independent

---

[11] A. Hajimiri and T. Lee, in *The Design of Low-Noise Oscillators* (Kluwer, Dordrecht, 1999), describe a simple shunt capacitor across the tail node to ground. E. Hegazi et al., in "A Filtering Technique to Lower Oscillator Phase Noise" (*ISSCC Digest of Technical Papers*, February 2001), use a parallel tank between the tail source and the common-source node to achieve a 10-dB phase noise reduction.

17.6 AMPLITUDE RESPONSE

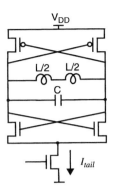

FIGURE 17.21. Simple symmetrical negative resistance oscillator

optimization of the steady-state and start-up conditions, with favorable implications for phase noise performance.

As mentioned, a key insight of the LTV theory concerns the importance of symmetry, the effects of which are partially evident in the preceding example. A configuration that exploits this knowledge more fully is the symmetrical negative resistance oscillator shown in Figure 17.21.[12] This configuration is not new by any means, but an appreciation of its symmetry properties is. Here, it is the half-circuit symmetry that is important, because noise in the two half-circuits is only partially correlated at best. By selecting the relative widths of the PMOS and NMOS devices appropriately to minimize the DC value of the ISF (Γ_{dc}) for each half-circuit, one may minimize the upconversion of $1/f$ noise. Through exploitation of symmetry in this manner, the $1/f^3$ corner can be dropped to exceptionally low values, even when device $1/f$ noise corners are high (as is typically the case for CMOS). Furthermore, the bridgelike arrangement of the transistor quad allows for greater signal swings, compounding the improvements in phase noise. As a result of all of these factors, a phase noise of -121 dBc/Hz at an offset of 600 kHz at 1.8 GHz has been obtained with low-Q (estimated to be 3–4) on-chip spiral inductors consuming 6 mW of power in a 0.25-μm CMOS technology (see footnote 12). This result rivals what one may achieve with bipolar technologies, as seen by comparison with the bipolar example of Margarit et al. (1999). With a modest increase in power, the same oscillator's phase noise becomes compliant with specifications for GSM1800.

17.6 AMPLITUDE RESPONSE

While the close-in sidebands are dominated by phase noise, the far-out sidebands are greatly affected by amplitude noise. Unlike the induced excess phase, the excess

---

[12] A. Hajimiri and T. Lee, "Design Issues in CMOS Differential LC Oscillators," *IEEE J. Solid-State Circuits*, May 1999.

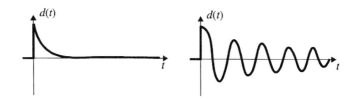

FIGURE 17.22. Overdamped and underdamped amplitude responses

amplitude $A(t)$ due to a current impulse decays with time. This decay is the direct result of the amplitude-restoring mechanisms always present in practical oscillators. The excess amplitude may decay very slowly (e.g., in a harmonic oscillator with a high-quality resonant circuit) or very quickly (e.g., in a ring oscillator). Some circuits may even demonstrate an underdamped second-order amplitude response. The detailed dynamics of the amplitude-control mechanism have a direct effect on the shape of the noise spectrum.

In the context of the ideal LC oscillator of Figure 17.5, a current impulse with an area Δq will induce an instantaneous change in the capacitor voltage, which in turn will result in a change in the oscillator amplitude that depends on the instant of injection (as shown in Figure 17.6). The amplitude change is proportional to the instantaneous normalized voltage change, $\Delta V/V_{max}$, for small injected charge $\Delta q \ll q_{max}$:

$$\Delta A = \Lambda(\omega_0 t)\frac{\Delta V}{V_{max}} = \Lambda(\omega_0 t)\frac{\Delta q}{q_{max}}, \quad \Delta q \ll q_{swing}, \tag{35}$$

where the amplitude impulse sensitivity function $\Lambda(\omega_0 t)$ is a periodic function that determines the sensitivity of each point on the waveform to an impulse; it is the amplitude counterpart of the phase impulse sensitivity function $\Gamma(\omega_0 t)$. From a development similar to that for phase response, the amplitude impulse response can be written as

$$h_A(t,\tau) = \frac{\Lambda(\omega_0 t)}{q_{max}} d(t-\tau), \tag{36}$$

where $d(t-\tau)$ is a function that defines how the excess amplitude decays. Figure 17.22 shows two hypothetical examples: $d(t)$ for a low-Q oscillator with overdamped response and for a high-Q oscillator with underdamped amplitude response.

As with our evaluation of the phase response, we invoke a small-signal linear approximation here. Again, we are not neglecting the fundamentally nonlinear nature of amplitude control; we are simply taking advantage of the fact that amplitude noise will certainly be small enough to validate a small-signal linear approximation for any oscillator worth the analysis effort. We will assume without loss of generality that the amplitude-limiting system of most oscillators can be approximated as first or second order, again for small disturbances. The function $d(t-\tau)$ will thus typically be either a dying exponential or a damped sinusoid.

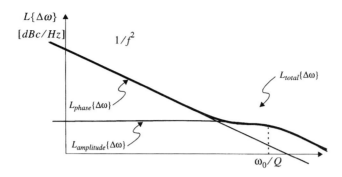

FIGURE 17.23. Phase, amplitude, and total sideband powers for the overdamped amplitude response

For a first-order system,

$$d(t - \tau) = e^{-\omega_0(t-\tau)/Q} \cdot u(t - \tau). \tag{37}$$

Therefore, the excess amplitude response to an arbitrary input current $i(t)$ is given by the superposition integral,

$$A(t) = \int_{-\infty}^{t} \frac{i(\tau)}{q_{max}} \Lambda(\omega_0 \tau) e^{-\omega_0(t-\tau)/Q} \, d\tau. \tag{38}$$

If $i(t)$ is a white noise source with power spectral density, then the output power spectrum of the amplitude noise, $A(t)$, can be shown to be

$$L_{amplitude}\{\Delta\omega\} = \frac{\Lambda_{rms}^2}{q_{max}^2} \cdot \frac{\overline{i_n^2}/\Delta f}{2 \cdot [\omega_0^2/Q^2 + (\Delta\omega)^2]}, \tag{39}$$

where Λ_{rms} is the rms value of $\Lambda(\omega_0 t)$. If L_{total} is measured then the sum of both $L_{amplitude}$ and L_{phase} will be observed and hence there will be a pedestal in the phase noise spectrum at ω_0/Q, as shown in Figure 17.23. Also note that the significance of the amplitude response depends greatly on Λ_{rms}, which in turn depends on the topology.

As a final comment on the effect of amplitude-control dynamics, an underdamped response would result in a spectrum with some peaking in the vicinity of ω_0/Q.

17.7 SUMMARY

The insights gained from LTI phase noise models are simple and intuitively satisfying: One should maximize signal amplitude and resonator Q. An additional, implicit insight is that the phase shifts around the loop generally must be arranged so that oscillation occurs at or very near the center frequency of the resonator. This way, there is a maximum attenuation by the resonator of off-center spectral components.

Deeper insights provided by the LTV model are that the resonator energy should be restored impulsively at the ISF minimum instead of evenly throughout a cycle, and that the DC value of the effective ISF should be made as close to zero as possible in order to suppress the upconversion of $1/f$ noise into close-in phase noise. The theory also shows that the inferior broadband noise performance of ring oscillators may be offset by their potentially superior ability to reject common-mode substrate and supply noise.

17.8 APPENDIX: NOTES ON SIMULATION

Exact analytical derivations of the ISF are usually not obtainable for any but the simplest oscillators. Various approximate methods are outlined in the reference of footnote 4, but the only generally accurate method is a direct evaluation of the time-varying impulse response. In this direct method, an impulsive excitation perturbs the oscillator and the steady-state phase perturbation is measured. The timing of the impulse with respect to the unperturbed oscillator's zero crossing is then incremented, and the simulation is repeated until the impulse has been "walked" through an entire cycle.

The impulse must have a small enough value to ensure that the assumption of linearity holds. Just as an amplifier's step response cannot be evaluated properly with steps of arbitrary size, one must judiciously select the area of the impulse rather than blindly employing some fixed value (e.g., 1 C). If one is unsure whether the impulse chosen has been sized properly, linearity may always be tested explicitly by scaling the size of impulse by some amount and then verifying that the response scales by the same factor.

Finally, some confusion persists about whether the LTV theory properly accommodates the phenomenon of amplitude-to-phase conversion exhibited by some oscillators. As long as linearity holds, the LTV theory does accommodate AM-to-PM conversion – provided that an exact ISF has been obtained. This is because changes in the phase of an oscillator arising from an amplitude change appear in the impulse response of the oscillator. A slight subtlety arises from the phase relationships among sidebands generated by these two mechanisms, however. Summed contributions from these two sources may result in sidebands with unequal amplitudes, in contrast with the purely symmetrical sidebands that are characteristic of AM and PM individually.

CHAPTER EIGHTEEN

MEASUREMENT OF PHASE NOISE

18.1 INTRODUCTION

The importance and origin of oscillator phase noise are discussed in the chapters on oscillators and phase noise. We now take up the problem of how to measure it. Even more than is the case with amplifier noise figure, the measurement of phase noise is easily corrupted by numerous sources of error, so an awareness of what these errors are (and of how to reduce them) is essential. Complicating the task is that many references present several approximations without explicitly stating what these approximations are. In all that follows, we'll endeavor to identify any approximations, with a particular focus on their domain of validity.

In keeping with the guiding philosophy of this book, phase noise measurement methods suitable for the weekend experimenter are presented in addition to techniques more commonly used in professional laboratories.

18.2 DEFINITIONS AND BASIC MEASUREMENT METHODS

As discussed in Chapter 15, all real oscillators exhibit some variation in phase and amplitude:

$$V_{out}(t) = V_m[1 + \varepsilon(t)]\cos[\omega t + \phi(t)]. \tag{1}$$

Because of the undesired amplitude and phase modulations, represented by $\varepsilon(t)$ and $\phi(t)$, the output spectrum has broadly distributed energy at frequencies other than the nominal oscillation frequency. A typical spectrum of V_{out} might thus appear as in the plot of Figure 18.1, which shows clearly that the noise power varies with frequency. In general, the displayed spectrum is a combination of both amplitude and phase noise, but we are often interested primarily in that portion that is attributable to phase variations. The reason is that some form of amplitude limiting is inherent in all real oscillators. No analogous corrective mechanism exists to limit phase variations (there is no way for the oscillator to establish where the "true" time origin is), so they tend to dominate. However, it should not be inferred from this statement that amplitude noise is inconsequential: Amplitude noise can cause all sorts of odd behavior,

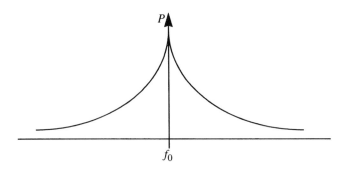

FIGURE 18.1. Idealized double-sided output spectrum of a real oscillator

particularly at frequencies well removed from the carrier.[1] That said, phase noise is the main preoccupation of most oscillator designers, so it will be ours as well.

The following definition of phase noise is often used to characterize oscillators:

$$\mathcal{L}(\Delta f) \equiv \frac{P_{SSB}(1\ \text{Hz})}{P_{sig}}, \tag{2}$$

where $P_{SSB}(1\ \text{Hz})$ is the output *phase noise* power *density* measured at an offset frequency, Δf, away from the carrier and where P_{sig} is the power of the carrier itself. Thus $\mathcal{L}(\Delta f)$, known as "script-L," is a density normalized to the power of the carrier. It is more conventionally reported in decibel form (relative to the carrier) as some number of dBc/Hz, even though the "per hertz" part applies to the argument of the log, not to the log itself. Despite this lack of rigor, it is the convention. Just don't be misled: doubling the bandwidth does not double the decibel level of noise.

If (and this can be a big "if") amplitude noise is known to be negligible, then phase noise can be directly read off of a spectrum analyzer display. To do so, first normalize the displayed noise power (P_{noise}) at the desired offset (Δf) to a 1-Hz bandwidth by dividing it by the resolution noise bandwidth setting. In many spectrum analyzers, the internal IF filters that set the resolution bandwidth are synchronously tuned (i.e., a cascade of identical filter stages tuned to the same center frequency) and therefore produce a Gaussian response shape. Their noise bandwidth is thus approximately 1.2 times the −3-dB resolution bandwidth. A final normalization by the signal power then yields $\mathcal{L}(\Delta f)$, as shown in Figure 18.2.

As a specific numerical example, suppose the carrier power P_{sig} is 50 mW (17 dBm) and that P_{noise} is 2 nW (−57 dBm) when measured with a resolution bandwidth of 100 Hz (20 dBHz) at an offset of 600 kHz. Then the phase noise is approximately

$$\mathcal{L}(\Delta f) = \frac{P_{SSB}(1\ \text{Hz})}{P_{sig}} = \frac{2\ \text{nW}}{50\ \text{mW}} \cdot \frac{1}{(1.2)(100\ \text{Hz})} = 3.33 \times 10^{-10}\ \text{Hz}^{-1}, \tag{3}$$

---

[1] For those who wish to see integrals galore, *The Design of Low-Noise Oscillators* (A. Hajimiri and T. Lee, Kluwer, Dordrecht, 1999) discusses in great detail precisely how device noise becomes oscillator noise.

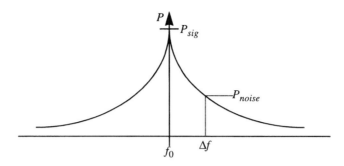

FIGURE 18.2. Estimating $\mathcal{L}(\Delta f)$ from spectrum analyzer display

or about -95 dBc/Hz. Again, the assumptions are that amplitude noise is negligible at this frequency offset and that a Gaussian filter is used to set the resolution bandwidth. Many spectrum analyzers include a mode for performing the set of normalizations automatically: just move the cursor to the desired offset, and the instrument calculates and displays its approximation to the phase noise $\mathcal{L}(\Delta f)$ (again, the "approximate" part is our assuming that any amplitude noise is negligible).

Note that the sensitivity of a spectrum analyzer–based measurement is limited by the dynamic-range problems associated with the presence of the carrier. Increasing the gain in an effort to improve sensitivity can easily result in carrier powers that would overload the analyzer (a typical spurious-free dynamic range for a spectrum analyzer might be around 70 dB). Interposing a notch filter between the device under test (DUT) and the analyzer to remove the carrier can ease the dynamic range problem at the expense of distorting the spectrum near the carrier. If this distortion can be tolerated (or at least characterized for later removal), then the notch filter is a simple way to increase the useful range of the spectrum analyzer technique.

Another way to characterize phase noise is to report the actual power spectral density of the phase noise itself:

$$S_\phi(\Delta f) \equiv \overline{\phi^2}/B, \tag{4}$$

where B is the noise bandwidth over which the measurement is performed. In the case of small maximum phase deviations ($|\phi| \ll 1$), we may invoke the approximation that half the phase "modulation" power goes into one sideband and the rest into the other. In that special case, we may make the following approximation:

$$\mathcal{L}(\Delta f) \approx S_\phi(f)/2. \tag{5}$$

The factor of 2 arises simply because $\mathcal{L}(\Delta f)$ is defined as a single-sideband quantity, and phase modulation of the carrier results in two sidebands.

Another measure that is occasionally used is the power spectral density of the frequency noise. Now frequency is the time derivative of phase, and multiplication by frequency is the spectral analogue of taking a time derivative. Hence,

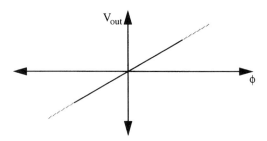

FIGURE 18.3. Ideal phase detector characteristic

$$s_f(\Delta f) \equiv \overline{\Delta f^2}/B = f^2 S_\phi(f) \approx 2\mathcal{L}(\Delta f), \tag{6}$$

where (once again) the last approximation is based on assuming sufficiently small phase deviations.

18.3 MEASUREMENT TECHNIQUES

Having provided the fundamental definitions of phase noise and having described a simple direct measurement method using a spectrum analyzer, we now consider alternative techniques for measuring phase noise.

The spectrum analyzer example underscores the dynamic-range problems attendant upon any method that seeks to measure noise in the presence of a carrier signal. As noted previously, using a notch filter reduces these problems, permitting an increase in sensitivity. An alternative is to employ heterodyne architectures to downconvert toward DC signals near the carrier, permitting a measurement of noise in the absence of an interfering carrier. The next three methods describe different strategies for eliminating the carrier from the measurement.

18.3.1 PLL-BASED PHASE DETECTOR TECHNIQUE

Since we wish to characterize phase fluctuations, it shouldn't surprise you that the most sensitive phase noise measurements are made with phase detectors at the core of the instrumentation. Because of the phase detector's central role, understanding its characteristics is important.

An ideal phase detector would produce an output that is completely insensitive to amplitude and exactly proportional to the phase difference between its two input signals (e.g., one from the DUT and one generated from within the instrument); see Figure 18.3. Because this characteristic is precisely what we want, Murphy guarantees that it's not what we get. Typically, an analog multiplier (Figure 18.4) is pressed into service as an approximation to a phase detector. Using some trigonometric identities, we find that the output of the multiplier may be expressed as

$$AB \cos \omega t \cos(\omega t + \phi) = k_d[\cos \phi - \cos(2\omega t + \phi)], \tag{7}$$

where k_d is a constant.

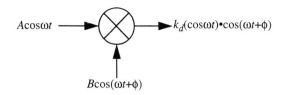

FIGURE 18.4. Multiplier as phase detector

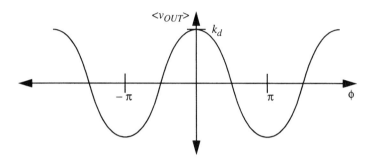

FIGURE 18.5. Multiplier phase detector output vs. phase difference

Note that the output of the multiplier consists of a DC term and a double-frequency term. For phase detector operation, we are interested only in the DC term because it is the only one that is a pure function of phase difference. This term is just the time-averaged output of the phase detector:

$$\langle AB \cos \omega t \cos(\omega t + \phi) \rangle = k_d [\cos \phi]. \tag{8}$$

If we plot the average output as a function of phase angle, the result will look something like Figure 18.5. Notice that the output is periodic and decidedly nonlinear, even within a period. As a result of the periodicity, phase errors that are very large are not distinguishable from those that are very small. Further note that the phase detector's sensitivity (slope) is zero when the phase difference is zero and at its maximum when the input phase difference is 90°. Hence, for the phase detector to be useful, there must be a nominal quadrature phase relationship between the output of the DUT and that of the reference oscillator.

From Figure 18.5 the peak output of the phase detector is seen to be k_d volts, and from Eqn. 8 k_d is also the slope of the phase detector characteristic (volts/radian) around the quadrature point. Thus, for small phase errors the output is linearly proportional to phase error to a good approximation,

$$\langle v_{OUT} \rangle \approx k_d \phi. \tag{9}$$

Not only do we want a linear proportionality to phase, we also desire insensitivity to amplitude noise. Fortunately, this sensitivity is automatically minimized when a quadrature relationship is satisfied, as can be seen from Eqn. 8. Maintenance

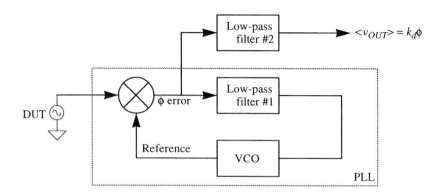

FIGURE 18.6. Phase noise measurement with PLL

of quadrature is therefore important for several reasons in this type of phase noise measurement.

The block diagram of Figure 18.6 shows how a phase-locked loop (PLL) guarantees the desired quadrature relationship. The PLL, shown within the dotted boundary, is designed to have a very low bandwidth, since its only job is to establish a nominal 90° phase relationship between the two oscillators. Because the PLL causes the voltage-controlled oscillator (VCO) to track the DUT's output within the PLL's bandwidth, phase noise at offset frequencies smaller than that bandwidth will be artificially attenuated. Hence, only phase noise measurements made at offsets beyond the PLL's bandwidth will be trustworthy. This limitation must be taken into account when designing the PLL's low-pass filter (#1). Low-pass filter #2 averages the phase error output from the multiplier over its bandwidth. It therefore sets the resolution bandwidth of the phase noise measurement.

To calibrate this setup, disable the PLL and then adjust the VCO frequency to be close to that of the DUT. The second filter passes the difference-frequency component but rejects the sum-frequency component. Monitor the output of that filter with a spectrum analyzer or oscilloscope. The spectrum analyzer measures rms values, so multiply them by $\sqrt{2}$ to obtain k_d. After determining this constant, enable the PLL; the setup is now ready to perform calibrated phase noise measurements.

In practice, two isolation amplifiers (not shown) at both input ports of the mixer are almost always needed to prevent unwanted phase locking of one oscillator to the other. These oscillators should have independent power supplies in order to prevent injection locking through supply noise. If a parasitic PLL were allowed to control the system, then the two oscillators would track each other over the bandwidth of the parasitic loop, improperly reducing the measured phase noise over some uncontrolled bandwidth.

The phase detector method generally requires that the VCO possess much lower phase noise than the oscillator under test, since the VCO is used as the reference oscillator that defines the "correct" zero-crossing instants against which the DUT's output is compared. It is possible to perform a calibration in which the reference

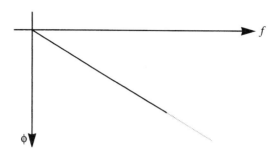

FIGURE 18.7. Phase shift of pure time delay

oscillator's phase noise is characterized, allowing it to be subtracted out from subsequent measurements. The only requirement is that reference noise be small enough to permit a reliable subtraction. It is generally accepted that the VCO's phase noise should be at least as good as that of the DUT if accurate subtractions are to be made.

18.3.2 DELAY LINE DISCRIMINATOR TECHNIQUE

Although the PLL-based measurement is capable of exquisite sensitivity, it suffers from the need for two oscillators and the complexity that always attends the building of any PLL. A method that requires only one oscillator is the delay line technique. Although its sensitivity is not as good as that of the two-oscillator PLL technique, its simplicity makes the delay line method attractive for some applications.

The basis for this measurement method is simply the phase-vs.-frequency behavior of a delay line. Since a pure time delay exhibits a linearly increasing negative phase shift with frequency, a time-delay element may be viewed as a frequency-to-phase converter; see Figure 18.7. The specific relationship between frequency (in hertz) and phase (in radians) is:

$$\phi = -2\pi f \tau. \qquad (10)$$

If a phase detector is used to compare a signal with its time-delayed version, the output will correspond to fluctuations in the relative frequency between the two signals. If the noise is truly random, then the noise of the two inputs will be uncorrelated and an accurate measurement will result. If there are correlations, however, the measurements will be in error. For example, $1/f$ noise causes a slow frequency modulation. If the line delay is short compared with the period of these $1/f$ phenomena, both inputs to the phase detector will move together, leading to a false indication of low phase noise.

Since frequency fluctuations are related to phase fluctuations through Eqn. 6, this technique yields an indirect measurement of phase noise. The output is fundamentally proportional to the frequency noise, as seen from

$$\Delta v_{OUT} = k_d \Delta \phi = k_d 2\pi \Delta f \tau \implies \Delta f = \frac{\Delta v_{OUT}}{k_d 2\pi \tau}, \qquad (11)$$

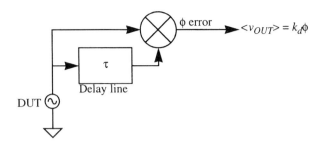

FIGURE 18.8. Phase noise measurement using delay line frequency discriminator (simplified)

so that the power spectral density (PSD) of the frequency noise is

$$S_f(\Delta f) \equiv \frac{\overline{\Delta f^2}}{B} = \frac{\overline{\left(\frac{\Delta v_{OUT}}{k_d 2\pi \tau}\right)^2}}{B}. \tag{12}$$

In turn, Eqn. 6 relates the PSDs of phase and frequency noise to each other:

$$S_\phi(\Delta f) \equiv \frac{1}{f^2}\left(\frac{\overline{\Delta f^2}}{B}\right) = \frac{1}{\Delta f^2} \frac{\overline{\left(\frac{\Delta v_{OUT}}{k_d 2\pi \tau}\right)^2}}{B}. \tag{13}$$

It also allows us to approximate script-L for small phase errors as

$$\mathcal{L}(\Delta f) \approx \frac{S_\phi(\Delta f)}{2} = \frac{1}{2\Delta f^2} \frac{\overline{\left(\frac{\Delta v_{OUT}}{k_d 2\pi \tau}\right)^2}}{B} = \frac{1}{2\Delta f^2} \frac{\overline{\Delta v_{OUT}^2}}{(k_d 2\pi \tau)^2 B}. \tag{14}$$

The measurement setup thus appears (in simplified form) as seen in Figure 18.8. To make a measurement, use a spectrum analyzer to observe the output noise as a function of frequency. Because spectrum analyzers typically have poor noise figures (e.g., 30 dB), sensitivity can be improved by preceding the analyzer with a high-gain, low-noise amplifier (LNA). Because the carrier has been effectively removed, using a preamplifier in this case causes no overload.

From Eqn. 14, the measured noise power must be scaled by several constant factors as well as by the square of the frequency, Δf^2, because the display is actually of the frequency noise. As a numerical example, suppose B is 100 Hz (20 dBHz) and that $k_d 2\pi \tau$ is 10^{-5} s. Suppose further that the power measured over the resolution bandwidth of the spectrum analyzer at 100 kHz is 2 nW (-57 dBm). Then the phase noise at that 100-kHz offset from the carrier is -110 dBc/Hz.

A subtle issue is that the delay time must be chosen to produce a quadrature relationship between the two mixer input signals. For a given oscillator frequency, this requirement implies that the allowable time-delay values are quantized to odd multiples of a quarter-period of oscillation. In practice, either a variable phase shifter is added in series with the delay line or the oscillator frequency is made adjustable, thereby producing the necessary quadrature relationship. If the time delay or oscillator frequency drifts during the measurement, errors will result because of departures from quadrature. This unfortunate sensitivity is a distinct disadvantage relative to the PLL method.

As noted earlier, a larger time delay allows a more accurate measurement of phase noise close to the carrier. From Eqn. 14, we see that it also increases the output voltage for a given level of phase noise, thereby improving the sensitivity of the measurement at all offset frequencies. The main limitation on the length of the delay is imposed by the need to keep phase differences within the phase detector's linear range. Arbitrarily calling the latter ~1 rad, the delay must be chosen to satisfy

$$\Delta\phi = 2\pi f \tau \leq 1 \implies \tau \leq 1/2\pi f, \tag{15}$$

where f is the lowest offset frequency at which the phase noise is to be measured with reasonable accuracy.

As a specific numerical example, suppose we wish to characterize phase noise at an offset of 1 kHz. Equation 15 implies that the time delay must be of the order of 160 μs for an accurate measurement. About 30 kilometers (!) of typical coaxial cable would be needed to make this possible. Obviously, this is an impractical length. Even if one were willing to pay for that much cable, the attenuation at the carrier frequency would eliminate any residual hope of making a measurement. If we arbitrarily take 10 meters of cable as an upper bound (and this is still a lot of cable), it would appear that offset frequencies of a few megahertz represent a practical lower range. Below such values of offset frequency, sensitivity degrades rapidly.

For far-out noise, the discriminator can approach the PLL method in sensitivity, but the limitation on allowable time delays prevents measurements of close-in phase noise with high accuracy if physical lines are used. If one is willing to accept some degradation in sensitivity, phase noise measurements at offsets somewhat below 1 MHz (perhaps approaching 100 kHz) are possible, but accurate measurements closer to the carrier than that are achievable only through the PLL method.

18.3.3 RESONANT DISCRIMINATOR TECHNIQUE

To overcome the practical difficulties of implementing sufficient time delays with long cables, engineers have devised alternative methods. To understand how these alternatives work, first recall that delay is proportional to the slope of a system's phase shift with frequency:

$$t_D = -\frac{d\phi}{d\omega}. \tag{16}$$

Instead of a length of cable, then, one might substitute another element whose phase shift decreases linearly with frequency to a good approximation. As suggested by its name, an ordinary resonator's phase-vs.-frequency characteristic is exploited in the resonant discriminator method. See Figure 18.9.

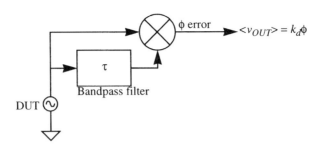

FIGURE 18.9. Phase noise measurement using resonator frequency discriminator (simplified)

With a resonator, there is a relationship between Q and phase shift, and hence between Q and time delay. To determine these relationships, first recognize that the phase shift of a bandpass filter (such as a simple RLC tank) may be expressed as

$$\phi = \frac{\pi}{4} - \operatorname{atan}\frac{\omega/\omega_n}{Q[1-(\omega/\omega_n)^2]}, \quad (17)$$

where "atan" is short for "arctangent" and ω_n is the center frequency. Evaluating the time delay at ω_n using Eqn. 16, we ultimately find that

$$t_D|_{\omega=\omega_n} = \frac{2Q}{\omega_n} = \frac{2}{B}, \quad (18)$$

where B is the -3-dB bandwidth in radians per second. The effect of a long cable therefore can be mimicked to a certain extent simply by using a resonator with a suitably low bandwidth. The main limitation is that, like all networks with a finite number of poles, the total amount of phase shift is bounded, so the resonator can only approximate a time delay over a finite bandwidth. Equation 18 shows us that the greater the delay, the narrower this bandwidth. Toward the passband edges, the delay diminishes, ultimately heading toward zero. The range over which the delay remains usefully close to the value given by Eqn. 18 is conventionally (and arbitrarily) taken to be half the bandwidth.

It is important to consider the behavior of the system outside of the nominal delay range. For spectral components further away from the carrier than the resonator bandwidth, the filter's attenuation will be significant. In such a case, we may assume that it has cleaned up the spectrum, causing one input to the phase detector to be effectively noise-free. The measurement conditions thus converge to those in the PLL-based setup, and the output will be the phase noise:

$$\Delta v_{OUT} = k_d \Delta\phi. \quad (19)$$

In summary, then, the resonator technique causes us to measure frequency noise within the half-bandwidth of the resonator. This frequency noise must be scaled by the offset in order to compute phase noise. Well beyond the resonator bandwidth, phase noise is measured directly; no scaling is required. In between these two regimes, the displayed values are some mixture of phase and frequency noise, and meaningful measurements are not to be expected.

It is also important to keep in mind that the resonator method shares with the delay line technique a requirement of nominal quadrature. A variable phase shifter (or the ability to vary the oscillator frequency) is thus also required to guarantee satisfaction of this requirement.

Both discriminator techniques, limited though they may be, are good methods for implementation by the hobbyist because of their simplicity. Unless there is a pressing need to characterize an oscillator's phase noise very close to the carrier, these methods will yield excellent results with modest equipment.

18.4 ERROR SOURCES

To make accurate phase noise measurements with any of these methods, it's necessary to identify and quantify the various sources of error. As might be expected, not all methods are equally sensitive to all sources of error. Nonetheless, a unified noise model is still possible. In all that follows, the noise may be assumed to appear as additive sources just prior to the spectrum analyzer. What does differ from technique to technique is the actual measurement consequence of this noise.

18.4.1 SPECTRUM ANALYZER METHOD

To see what factors affect the spectrum analyzer technique, recall that

$$\mathcal{L}(\Delta f) \approx \frac{S_\phi(f)}{2}, \tag{20}$$

which may also be expressed as

$$\mathcal{L}(\Delta f) \approx \frac{S_\phi(f)}{2} = \frac{\overline{\Delta v_{OUT}^2}}{4v_{peak}^2}, \tag{21}$$

where v_{peak} is the peak output voltage of the mixer. The noise floor of this measurement is therefore set by the minimum value of Δv_{out}, which in turn is determined by the noise of the mixer and the effective input noise of the spectrum analyzer:

$$\mathcal{L}(\Delta f) \approx \frac{S_\phi(f)}{2} = \frac{\overline{\Delta v_{OUT}^2}}{4v_{peak}^2} = \frac{1}{4} \frac{\overline{\Delta v_{n,mixer}^2} + \overline{\Delta v_{n,SA}^2}}{v_{peak}^2}. \tag{22}$$

Mixer output noise and spectrum analyzer input noise are assumed to be uncorrelated, so their powers are added directly together in Eqn. 22. Sensitivity is clearly improved by using mixers capable of producing large outputs while exhibiting low noise figures. Diode ring mixers in which each branch is a series combination of several diodes (i.e., high-level mixers) are commonly used. Preceding the analyzer with an LNA is also extremely helpful, since an LNA reduces the effective input noise, $\Delta v_{n,SA}$. Commercially available instruments are capable of achieving noise floors as low as -180 dBc/Hz.

18.4.2 DELAY LINE DISCRIMINATOR

In both discriminator methods, the spectrum analyzer displays the frequency noise directly; the user must normalize the measurement by the square of the frequency offset in order to convert the measurement into phase noise. Hence, any noise that appears at the input to the spectrum analyzer appears as an effective frequency noise:

$$S_f(\Delta f) \equiv \frac{\overline{\Delta f^2}}{B} = \frac{\overline{\left(\frac{\Delta v_{OUT}}{k_d 2\pi\tau}\right)^2}}{B} = \frac{\overline{\Delta v_{n,mixer}^2} + \overline{\Delta v_{n,SA}^2}}{B(k_d 2\pi\tau)^2}. \qquad (23)$$

In many practical implementations, a buffer amplifier is inserted between the delay line output and the mixer input. If such an amplifier is used, its noise must also be considered. Since the output noise of such an amplifier appears as an input to one port of the mixer, it too acts as a frequency noise term – although scaled by a factor equal to the phase detector's gain. Hence, the total frequency noise power spectral density is given by

$$S_f(\Delta f) \equiv \frac{\overline{\Delta f^2}}{B} = \frac{\overline{\left(\frac{\Delta v_{OUT}}{k_d 2\pi\tau}\right)^2}}{B} = \frac{\overline{\Delta v_{n,mixer}^2} + \overline{\Delta v_{n,SA}^2}}{B(k_d 2\pi\tau)^2} + \frac{\overline{\Delta v_{n,buffer}^2}}{B(2\pi\tau)^2}. \qquad (24)$$

The phase noise measurement floor is thus

$$\mathcal{L}(\Delta f) \approx \frac{S_f(\Delta f)}{2(\Delta f)^2} = \frac{1}{2}\frac{1}{B[2\pi(\Delta f)\tau]^2}\left(\frac{\overline{\Delta v_{n,mixer}^2} + \overline{\Delta v_{n,SA}^2}}{k_d} + \overline{\Delta v_{n,buffer}^2}\right). \qquad (25)$$

Note that the floor rises rapidly at low frequencies, consistent with our earlier observations on the weaknesses of this method.

18.5 REFERENCES

The following notes from Hewlett-Packard (now Agilent Technologies) contain much useful information about phase noise measurements: "Phase Noise Characterization of Microwave Oscillators" (Product Note 11729B-1, August 1983); and "The Art of Phase Noise Measurement" (Dieter Scherer, RF & Microwave Measurement Symposium and Exhibition, October 1984).

CHAPTER NINETEEN

SAMPLING OSCILLOSCOPES, SPECTRUM ANALYZERS, AND PROBES

19.1 INTRODUCTION

Oscilloscopes and spectrum analyzers are ubiquitous pieces of test equipment in any RF laboratory. The reason, of course, is that it is useful to study signals in both time and frequency domains, despite the fact that both presentations theoretically provide equivalent information.

Most electrical engineers are familiar with basic operational principles of lower-frequency oscilloscopes. However, an incomplete understanding of how probes behave (particularly with respect to grounding technique) is still remarkably widespread. The consequences of this ignorance only become worse as the frequency increases and so, after a brief review of a conventional low-frequency scope, our primary focus will be the additional considerations one must accommodate when using scopes at gigahertz frequencies. Also, because the sampling oscilloscopes commonly used at high frequencies have subtle ways of encouraging "pilot error," we'll spend some time studying how they work and how to avoid being fooled by them. High-speed sampling circuits are interesting in their own right, so these types of scopes give us a nice excuse to spend a little bit of time examining how samplers function.

Another amazing instrument is the modern spectrum analyzer (with cost approximately proportional to the square of amazement), which is capable of making measurements over a wide dynamic range (e.g., 80–100 dB SFDR) and over a large frequency span (e.g., near DC to 20 GHz in a single instrument).

To maximize the utility of this equipment and to avoid common measurement errors, it's important to understand their internal architecture as well as the characteristics of the probes or other fixturing that connect the instruments to the device under test. We begin with a brief overview of ordinary continuous time oscilloscopes.

19.2 OSCILLOSCOPES

It used to be safe to assume that most engineers were familiar with a typical analog oscilloscope, but modern laboratories and classrooms are increasingly populated with digital instruments. Students and engineers consequently encounter ordinary scopes with decreasing frequency. More typically nowadays, a layer of software insulates the user from nature, inhibiting the acquisition of important insights. So, it's worth spending a little time with a quick review of the classic analog scope.

19.2.1 "PURE" ANALOG SCOPES

A typical continuous time (i.e., nonsampling) oscilloscope is portrayed in Figure 19.1. The precision broadband attenuator accepts the input signal and scales it by a calibrated amount to prevent overload of the vertical amplifier. The latter block provides enough gain to drive the deflection plates of the CRT. A delay line following the vertical amplifier gives the trigger and sweep circuitry enough time to begin working before the input signal arrives at the CRT. This delay thus allows the oscilloscope to display parts of the waveform that actually precede the triggering event. The delay line may be implemented as a specially designed low-dispersion cable or as a lumped approximation to a transmission line. If the former, it is often several meters in length. Whether the former or the latter, the attenuation and dispersion characteristics of the delay element are critical to the operation of the scope. Equalization circuitry is universally used to compensate for the delay line's distortion.

All classic analog scopes use electrostatically deflected CRTs because the inductance of magnetic deflection coils is too large to allow practical operation at frequencies beyond a few megahertz. Televisions and computer monitors use magnetic deflection because large deflection sensitivity is easily obtained through the use of many turns, significantly reducing the depth (and thus the weight) of the CRT for a given screen size. The modest deflection bandwidth requirements enable the use of magnetic deflection in those applications. Electrostatic deflection is the only practical option for high-frequency oscilloscopes, at the cost of rather long (deep) tubes.

The sweep generator generates a sawtooth voltage to drive the horizontal deflection plates. The sawtooth linearly deflects the CRT's electron beam to produce a constant-velocity horizontal sweep. A trigger circuit synchronizes this sawtooth with the input signal by initiating the sweep only when the input exhibits some predetermined characteristic, such as exceeding a particular voltage threshold. To produce a comforting baseline even in the absence of signal, scopes have an optional trigger mode ("auto") in which a sawtooth is self-initiated if no triggering events have occurred after some time.

This basic oscilloscope architecture has been around for quite some time. Ferdinand Braun (of crystal rectifier fame) invented a primitive form of oscilloscope

19.2 OSCILLOSCOPES

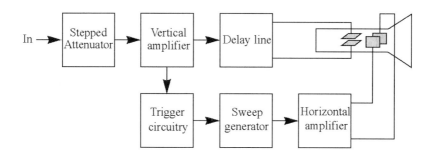

FIGURE 19.1. Typical analog oscilloscope block diagram

around 1897.[1] Consequently, the CRT is still sometimes known as a Braunsche Röhre ("Braun tube") in Germany. Refinements during subsequent decades include greatly increased bandwidth and sensitivity. By the 1930s, electrostatically deflected CRTs had advanced to the point where tens of volts would produce a beam deflection of one centimeter. These types of CRTs are still used and are compatible with bandwidths up to about 100 MHz.

As frequency increases, it becomes progressively more difficult to obtain large voltage swings from practical amplifiers. To underscore this difficulty, note that the maximum slope of a 100-V–amplitude, 1-GHz sine wave is over 600 V per *nano-second*, or 600 kilovolts per microsecond. Building amplifiers that can supply peak voltages of 100 V and produce such high slew rates at the same time is a decidedly nontrivial task!

In order to understand the developments that have enabled improved deflection sensitivity, it's useful to note that ordinary plates have only a short time during which they have any opportunity to deflect the electron beam. To solve this problem, *distributed* vertical deflection structures may be used instead of simple parallel plates to increase the span of time over which the electron beam interacts with the deflecting voltage. In such a structure, each plate is broken up into segments that are connected together through a small inductance. The distributed deflection "plate" is thus actually a transmission line, as depicted in Figure 19.2. In practice, the entire distributed structure is implemented as a box made out of spiral-wound flat stock. This arrangement properly distributes the capacitance and inductance continuously throughout to maximize the line's useful bandwidth.

The electron velocity is matched to the delay of the line to maximize the interaction time between the deflection voltage and the electron beam. An order-of-magnitude

---

[1] "Ueber ein Verfahren zur Demonstration und zum Studium des zeitlichen Verlaufes variabler Ströme" [On a Method of Demonstrating and Studying the Time Dependence of Variable Currents], *Annalen der Physik und Chemie*, February 1897. This first oscilloscope was magnetically deflected and was used to study the output of electric utilities. Electronic amplifiers did not exist then, so this early instrument was quite insensitive.

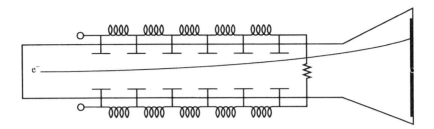

FIGURE 19.2. Distributed vertical deflection plates

improvement in vertical deflection sensitivity to about 3 V/cm results, allowing scopes to function beyond 500 MHz with 10-mV/cm sensitivity after amplification.[2] The slew rate requirement drops significantly to 30 kV/μs (still a most impressive number).

Extension of this general architecture into the gigahertz realm was achieved just as the 1970s drew to a close. This bandwidth improvement was enabled primarily by advances in CRT design.[3] The use of distributed structures for both horizontal and vertical deflection and of *microchannel plate* CRTs improves deflection sensitivities by another factor of 3, greatly relaxing the demands on amplifier design.

The microchannel plate CRT exploits a simple observation: It is easier to deflect an electron beam before it has acquired the high energies necessary to produce a bright trace on the screen. So, do that. To obtain sufficient beam energy after deflection, the low-energy electron stream passes through a thin semiconducting plate that is full of minute cylindrical channels (see Figure 19.3) and across which is imposed a large enough voltage to produce a significant accelerating field. As electrons strike the walls of these microscopic channels, they release secondary electrons that multiply the beam current by large factors, producing a readily visible trace on the screen. The tremendous sensitivity of this CRT allows the Tektronix 7104 to produce displays of *single-shot* 350-ps events that are visible to the naked eye in ordinary room light. At the same time, it is capable of supporting sweep velocities greater than the speed of light!

---

[2] The CRT of the first commercial unit to do so, the Tektronix 7904, has distributed vertical deflection plates only; the horizontal deflection electrodes are of the ordinary parallel-plate variety. As with other "classical" scopes, it is available on the surplus market for quite modest sums and should be considered by the serious home experimenter. Its nominal 500-MHz bandwidth is not a hard limit, for one may drive the CRT's vertical deflection plates directly. The resulting sensitivity is about 3–4 V/cm, with a bandwidth well in excess of 1 GHz.

[3] Actually, GHz scopes first became commercially available around 1960, but these connected an input signal directly to the CRT plates. Sensitivities were correspondingly poor (quite a few volts were required to obtain any appreciable deflection), and the instruments were of limited usefulness as a consequence. These units occasionally appear on the surplus market; avoid them (unless you want to warp the local gravitational field – they are huge!). The 7104 has a nominal 1-GHz bandwidth at 10-mV/cm sensitivity when equipped with the model-7A29 plug-in amplifier, and it can function up to about 2–3 GHz if the CRT plates are driven directly (resulting in a still useful sensitivity of about 1 V/cm).

19.2 OSCILLOSCOPES

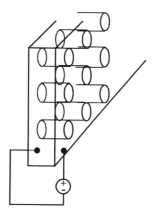

FIGURE 19.3. Microchannel plate

The advantage of this kind of architecture is that – within the limits of noise, distortion, and fixturing quality – what you see is what's really there. Unfortunately, this fidelity comes at the cost of extremely sophisticated CRT technology. Worse, it is unclear how one may extend the technology to achieve bandwidths significantly beyond 1 GHz. The scaling relationships are daunting: A faster sweep means that the beam spends less time on the screen, leading to a dimmer trace. Increasing the acceleration voltage to increase brightness makes the beam harder to deflect, even with a microchannel plate. Yet transistors capable of high-frequency operation have lower breakdown voltages as a consequence of the smaller feature sizes necessary to enable high bandwidth, reducing swing capability. And on it goes.

The rules of that game being what they are, engineers have understandably chosen to play a different game.

19.2.2 SAMPLING SCOPES

That different game is *sampling*. Pursuing a sampling architecture shifts the design problem to a purely electronic domain, enabling the use of CRTs that are only marginally more sophisticated than those used in a typical television. The first publication on a modern sampling scope architecture appeared in 1950.[4]

The key idea underlying a sampling oscilloscope is best illustrated with the graph shown in Figure 19.4. Here, the higher-frequency sine wave is sampled at instants indicated by the dots. The sampled waveform is then filtered to yield the lower-frequency sine wave in the figure. The lower-frequency wave is then easily displayed on a conventional low-frequency oscilloscope. From this description, it's clear that a sampling scope is analogous to a flashing strobe light that appears to slow down

---

[4] J. M. L. Janssen, "An Experimental 'Stroboscopic' Oscilloscope for Frequencies up to About 50Mc/s – I: Fundamentals," *Philips Tech. Rev.*, v. 12, no. 3, August 1950, pp. 52–9. The actual hardware used a pentode vacuum tube as the sampling element and functioned up to about 30 MHz.

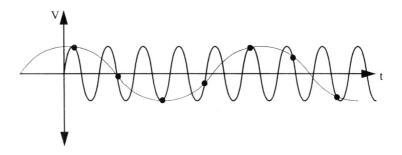

FIGURE 19.4. Illustration of sampling

motion. Thinking about the sampling scope in this manner is very useful in understanding its properties, including the ways in which it can mislead.

Clearly, the main magic in this architecture is in the sampling circuitry itself. Aside from subtleties in the trigger and sweep subcircuits, the rest is relatively mundane. The high-frequency limit is controlled by the time resolution of the sampling circuitry. The narrower the sampling window (aperture), the higher this bandwidth. To a reasonable approximation, the product of window width and bandwidth (in Hz) is about 0.35. Sampling apertures of several picoseconds represent the state of the art for commercial instruments and correspond to bandwidths in excess of approximately 50 GHz. Samplers based on superconducting technology are able to function at still higher frequency. Laboratory demonstrations have shown the feasibility of operation to several hundred gigahertz.

Since the sampling gate lies at the heart of this type of scope, it is worthwhile examining a few methods by which such narrow sampling apertures are achieved. The block diagram of Figure 19.5 shows the structure of a typical sampler from the early 1960s to the 1980s.[5] Here, the transistor acts as an avalanche pulser, much like the one described in Chapter 8. The collector supply voltage is adjusted a bit below the value that would result in spontaneous avalanching. When a positive-going trigger pulse is applied to the base, the transistor goes into full avalanche breakdown.[6] The collector voltage then plummets while the emitter voltage simultaneously spikes up. A typical transition time here is in the neighborhood of a couple of hundred picoseconds. The signals from the avalancher are thus impressively fast in their own right and can be used directly to operate the sampling bridge for bandwidths up to, say, approximately 1 GHz.

For still faster operation, the output of the avalancher needs to be conditioned to produce even shorter rise and fall times. A broadband transformer, acting as a

---

[5] This schematic is adapted from that for the Tektronix model S-6 14-GHz sampling head.

[6] The first published use of avalanching for sampling scopes is evidently due to G. B. B. Chaplin et al., "A Sensitive Transistor Oscillograph with D.C. to 300 Mc/s Bandwidth," *International Transistors and Association of Semiconductor Developers Convention,* May 1959, pp. 815–23. Also see U.S. Patent #3,069,559, granted 19 December 1962. The first commercial sampling scope was introduced by Lumatron in 1959, although the 1960 debut of the HP 185A – with its 500-MHz sampling plug-in unit – had more influence on subsequent developments in the field.

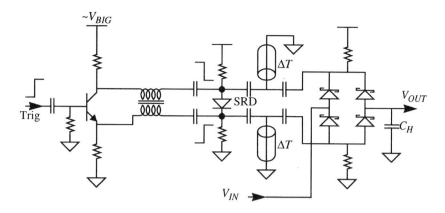

FIGURE 19.5. Representative sampling gate circuitry (simplified)

common-mode choke, first removes any small common-mode component arising from slightly different delays between the collector and emitter.[7] The near-perfect differential voltage at the output of the transformer reverse-biases the ordinarily forward-biased snap diode SRD. Eventually the diode snaps off, generating an exceptionally fast step (as described in Chapter 9). The characteristics of the SRD are such that the risetime speeds up by an order of magnitude, to values in the 10-ps range.[8]

The fast step is converted into a narrow pulse through the use of shorted transmission line segments. If each line has a one-way time-of-flight delay ΔT, then the total pulse width is of the order of $2\Delta T$. The typical risetimes of about 10 ps thus correspond to pulse widths on the order of 20–30 ps. A delay of $\sim$50 ps/cm is typical for commonly realized transmission line segments, so fast samplers employ rather short lines (typically a few millimeters in length). Low dispersion is important in this application, so the lines are often implemented with air as the dielectric.

The narrow pulses drive, say, a four-diode switch whose diodes are normally biased off. The pulses forward-bias the diodes, creating a conductive path from input to output and thereby charging up the output hold capacitor C_H to the value of the input. This sampled-and-held voltage changes only at the sampling repetition frequency (which might be as low as a few kHz), so subsequent stages need not have particularly high bandwidth. In fact, the circuit of Figure 19.5 can be used in front of a standard oscilloscope to extend its bandwidth well beyond a gigahertz. (Generating an appropriate trigger remains a challenge, however.)

Moving from avalanching transistors to SRDs provides greater than an order-of-magnitude speedup, permitting sampling scopes to function up to $\sim$20 GHz. Another similar improvement factor is provided by a *shockwave* transmission line.[9] Imagine

---

[7] In less demanding applications, this transformer can be eliminated, with compensation for relative phase then provided simply by adjusting the interconnect lengths to the collector and emitter.
[8] The HP 186A, introduced in 1962, was the first instrument to use SRDs.
[9] The basic underlying idea had been described as early as 1960, but it remained a largely academic curiosity until Mark Rodwell published his work on these types of lines in the late 1980s.

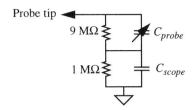

FIGURE 19.6. Simplified model of general-purpose probe loaded by oscilloscope

an artificial line in which the distributed capacitance is provided by the voltage-dependent capacitance of semiconductor junctions. If we assume that the bias is such that falling voltages produce increases in reverse junction bias, then edges will experience reduced capacitance as they fall; they thus speed up. A pulse propagating down such a line will experience continual improvements in falling-edge speed until limited by some other mechanism (e.g., dispersion or finite junction bandwidth). Step risetimes below ∼700 fs have been reported, corresponding to bandwidths in excess of ∼700 GHz, using Schottky diodes connected along a coplanar line.[10]

19.2.3 PROBES, COMPENSATION, NOISE, AND GROUNDING

It is remarkable how many engineers pay for a sensitive spectrum analyzer or high-speed oscilloscope – only to mate it with a probe used in a fashion that guarantees high noise, low bandwidth, unflat passband, and otherwise erroneous measurements. To understand this problem, we must first consider the detailed nature of a scope probe.

Contrary to what one might think, a probe is most emphatically *not* just a glorified piece of wire with a tip on one end and a connector on the other. Think about this fact: Most "10 : 1" probes (so-called because they provide a factor-of-10 attenuation) present a 10 MΩ impedance to the circuit under test yet may provide a bandwidth of hundreds of megahertz. But intuition would suggest that the maximum allowable capacitance consistent with this bandwidth is under about 100 *atto*farads (0.1 femtofarads)! So, how can probes provide such a large bandwidth while presenting a 10-MΩ impedance? The answer is that the combination of a probe and oscilloscope isn't an "ordinary" *RC* network.

A simplified model of the scope–probe combination is shown in Figure 19.6. The 1-MΩ resistor represents the oscilloscope input resistance, while C_{scope} represents the scope's input capacitance.

---

[10] U. Bhattacharya, S. T. Allen, and M. J. W. Rodwell, "DC-725 GHz Sampling Circuits and Subpicosecond Nonlinear Transmission Lines Using Elevated Coplanar Waveguide," *IEEE Microwave and Guided Wave Lett.*, v. 5, no. 2, February 1995, pp. 50–2. Elevating the lines reduces the coupling of energy into the semiconductor substrate, reducing loss and dispersion.

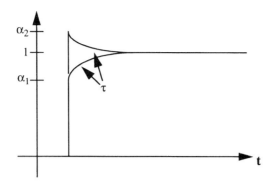

FIGURE 19.7. Possible step responses of pole–zero doublet

Inside the probe, there is a 9-MΩ resistor to provide the necessary 10:1 attenuation at low frequencies.[11] To avoid the tremendous bandwidth degradation that would result from use of a simple 9-MΩ resistor, the probe also has a capacitor in parallel with that resistor. At high frequencies, the 10:1 attenuation is actually provided by the capacitive voltage divider. It may be shown (and it isn't hard to show it) that when the top *RC* equals the bottom *RC*, the attenuation is exactly a factor of 10, *independent of frequency*. There is a zero that cancels precisely the slow pole, leading to a transfer function that has no bandwidth limitation.

Because it is impossible to guarantee perfect compensation with fixed elements, all 10:1 probes have an adjustable capacitor. To appreciate more completely the necessity for such an adjustment, let's recall the effect of imperfect pole–zero cancellation with the following transfer function (see Chapter 12):

$$H(s) = \frac{\alpha \tau s + 1}{\tau s + 1}. \tag{1}$$

The initial value is α and the final value is unity. We can sketch (see Figure 19.7) one step response with $\alpha_1 < 1$ and one with $\alpha_2 > 1$. In both cases the response jumps immediately to α but then settles to the final value, with a time constant of the pole. If $\alpha = 1$ then the response would reach final value in zero time, though the additional poles in all practical circuits limit the risetime to a nonzero value.

From Figure 19.7, it is easy to see the importance of adjusting the capacitor to avoid gross measurement errors. This compensation is most easily performed by examining the response to a square wave of sufficiently low frequency to make visible the slow settling due to the doublet's pole and then adjusting the capacitor for the flattest time response. Many scopes provide a square-wave output specifically for the purpose of adjusting the probe. This adjustment needs to be checked before making a measurement, or else large errors are possible.

Another common error is to use the generous ground leads frequently supplied with most probes. Though convenient, long ground leads virtually guarantee erroneous measurements. The reasons are simple: the inductance of the ground wire

---

[11] Probes for use at the highest frequencies often employ special cable with distributed resistance.

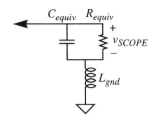

FIGURE 19.8. Circuit model for scope and probe with ground clip

FIGURE 19.9. Grounding arrangement for higher-frequency probing

(about 0.5 nH/mm, typically) adds an impedance that creates a low-pass filter with the impedance of the scope. Near the cutoff frequency, resonant peaking is possible.

As an example (see Figure 19.8), suppose the ground lead is 20 cm long (not an atypical value, unfortunately). The parasitic inductance therefore might be of the order of 50–100 nH (the exact value depends on the path shape). Given a typical scope input capacitance of 15 pF, the low-pass filter formed has a cutoff frequency below about 200 MHz. Clearly, it makes no sense to use this sort of arrangement with a 1-GHz (or faster) scope; the bandwidth is limited to low values well before the signals actually get to the scope! Aside from the bandlimiting, the frequency response may be far from flat, resulting in erroneous determinations of amplitude.

In addition to acting as an inductance, the ground lead can also form a fairly effective loop antenna with the probe, allowing it to pick up noise from far and wide. The commonly observed fuzziness of traces is not, as many believe, an inherent property. Reduction of the ground connection to the absolute minimum goes a long way toward cleaning up such anomalies.

To avoid these problems, never use the long ground lead. Period. Immediately pack it up and store it in your toolbox and let it gather dust. Instead, get into the practice of using the small spring-like attachment that fits around the ground collar of the probe tip. If these springs have been lost (as many frequently are), it's easy enough to make replacements out of a suitable length of bus wire; see Figure 19.9. The author is fond of soldering a number of these onto prototype boards at key test points to save the trouble of hunting down these ground springs at a later time. The reduction of inductance by roughly two orders of magnitude permits probing at the maximum frequencies supported by the probe.

Commercially available passive 10:1 probes are generally capable of satisfactory operation up to a couple of hundred megahertz. At gigahertz frequencies, most (but not all) probes are 50 Ω to maintain compatibility with other microwave gear. Unfortunately, this low value limits what nodes of a circuit can be studied; high-impedance nodes will be loaded down by the probe, rendering the measurement useless. To solve this problem, one may use an active FET probe, which contains a source follower near the probe tip to minimize resistive and capacitive loading. This reduction in capacitance (to ~100 fF to 1-pF values; even less with compensation) permits a higher impedance level (e.g., 1–100 kΩ) – at the expense of degraded DC stability, reduced absolute accuracy, and diminished bandwidth. Such probes can function beyond a few gigahertz, however, and this range is adequate for a great many situations. An example is the Agilent Model 1158A 10:1 active probe. The integrated buffer in the probe body enables a bandwidth in excess of 4 GHz while presenting an input resistance and capacitance of 100 kΩ and 800 fF, respectively.

The probing of nodes on a microwave IC (or very small discrete modules) presents special problems of its own, both because of the mechanical challenges involved and because it is difficult to avoid a poor transition from the IC to the probe, and then ultimately to whatever instrument is connected to the probe. The difficulties are indeed substantial, but fortunately not insurmountable. The GGB Industries Model 35 active probe, for example, operates from DC to 26 GHz, presenting an input resistance and capacitance of 1.25 MΩ and 50 fF. A passive 50-Ω probe, the Model 110H, operates up to 110 GHz, the upper limit compatible with a 1.0-mm W coax connector. By using a WR-5 waveguide connection, the Model 220 extends operation to 220 GHz.[12] The waveguide connection implies, of course, an inability to function down to DC. Indeed, if we wish to maintain operation at one mode, the bandwidth is constrained not to exceed a single octave.

When using such probes for impedance measurement with a vector network analyzer, it's necessary to perform calibrations first, just as with any other VNA impedance measurement. To facilitate carrying out TRL or LRM calibrations (see Section 8.4.2), several companies offer special substrates on which various calibration structures have been fabricated. The geometries are almost always coplanar and usually feature ground–signal or ground–signal–ground configurations.

Probes for microwave work, whether active or passive, tend to be rather expensive and easily damaged. For students and hobbyists, an economical intermediate option is to build your own passive probe based on a minor modification of the basic 10:1 scope probe circuit, shown in Figure 19.10. The resistor R is chosen to provide the basic attenuation desired, and the compensation capacitor C is selected to maximize the bandwidth over which the response is flat to within some tolerance. Table 19.1 gives values for R and C for different attenuation factors.

---

[12] The "X" in a WR-X waveguide designation refers to the critical (wide) dimension in hundredths of an *inch*. Thus WR-5 waveguide has a 50-mil (not millimeter) critical dimension, or about 1.27 mm. The wavelength at cutoff is about 2.54 mm, which corresponds roughly to 120 GHz. We thus expect WR-5 to operate over a frequency range somewhat smaller than 120–240 GHz, and indeed the specified operating range spans 140–220 GHz.

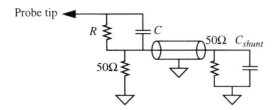

FIGURE 19.10. Basic schematic for N : 1 homemade passive probe

Table 19.1. *Probe component values*

Attenuation	R (Ω)	C_{shunt}/C
10 : 1	225	9
20 : 1	475	19
50 : 1	1.225k	49
100 : 1	2.475k	99

The capacitance C_{shunt} models the input and stray capacitance of the instrument connected to the probe and also the parasitic capacitance across the termination resistor in the probe proper. Given that systems intended to operate over large bandwidths are typically designed to approximate a purely real 50-Ω input impedance, a typical value for C_{shunt} is of the order of 1 pF (if that much). A brief look at the table of component values should give us pause, for the required compensation capacitance quickly plummets as the attenuation factor increases. For example, if we try to build a 100 : 1 probe to produce a 2.5-kΩ input resistance, the required capacitance would be of the order of 10 fF. It is essentially impossible to produce values this small reliably using manual methods. In any case, the compensation capacitance has to be adjustable. The capacitance for a 10 : 1 probe would be easier to implement, but a 250-Ω input resistance is probably too low for many applications. Better attenuation choices would be 20 : 1 or 50 : 1, but these still require that we produce sub-picofarad capacitances.

One way of avoiding these problems is to let the tip capacitance C_{comp} be whatever it is and then effect compensation by *increasing* the capacitance C_{shunt}. To construct a probe using these ideas, mount all components in microstrip fashion on a suitable piece of board (e.g., FR4). A small needle may be soldered down as the probe tip in series with a surface-mount resistor R. Depending on the attenuation you have chosen, the necessary C_{shunt} may be provided by a commercial trimmer capacitor (such as a piston-type, which has very low series resistance and inductance) or by taking advantage of the 2.5-pF/cm^2 capacitance of 1.6-mm FR4. Make the capacitor out of copper foil tape to facilitate its adjustment by the addition or trimming of tape. The probe then might appear somewhat as in Figure 19.11.

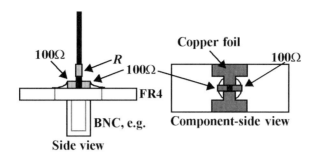

FIGURE 19.11. Homemade probe

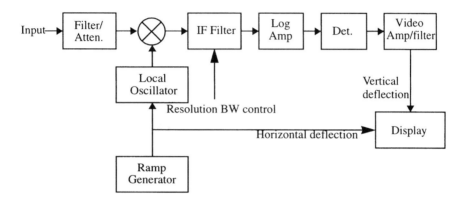

FIGURE 19.12. Swept frequency spectrum analyzer block diagram

It is entirely practical to build 20 : 1 and 50 : 1 probes using this method. By carefully trimming the copper foil, bandwidths in the low-gigahertz range are readily achieved. Mechanical stability can be enhanced somewhat by putting a piece of heat-shrink tubing around the probe.

19.3 SPECTRUM ANALYZERS

Several architectures have been used for building spectrum analyzers, but commercial RF analyzers are all based on the superheterodyne. See Figure 19.12. The input signal, after some optional attenuation, is mixed to the intermediate frequency (IF) using a local oscillator whose frequency is controlled by a ramp (or its digital equivalent). This ramp also feeds the horizontal deflection circuitry of the display to establish a frequency axis. The local oscillator has the difficult task of operating over a very wide frequency range (e.g., several decades). The LO is commonly implemented with a YIG sphere (pronounced to rhyme with "fig"; YIG stands for yttrium-iron-garnet) at its core.

This device (see Figure 19.13) is the dual of a piezoelectric crystal, and thus it behaves as a parallel LC resonator with a parasitic series inductance (as opposed to

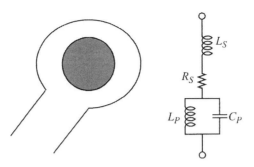

FIGURE 19.13. YIG sphere, coupling loop, and one possible equivalent circuit

 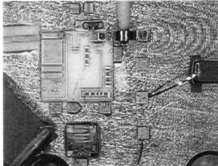

FIGURE 19.14. YIG sphere and coupling loop (left), and in context [*courtesy of David Straight, M/A-Com*]

the series resonator with a parasitic shunt capacitance that models a quartz crystal). Typical unloaded Q-values are in excess of 10^4, resulting in excellent spectral purity. Of greater relevance for spectrum analyzers is that the resonant frequency is linearly proportional to the strength of an applied magnetic field, permitting tuning over the exceptional range demanded of laboratory instruments.

As seen in Figure 19.14, the sphere is typically mounted on the end of a dielectric support rod. A single-turn coupling loop made of a wide conductor ribbon (to reduce the parasitic series inductance) surrounds the sphere and connects to the rest of the oscillator. Not shown is the frequency-control magnet and associated electronics. The complete assembly would be enclosed in a structure that is well shielded electrostatically and magnetically to prevent ambient fields from perturbing the oscillator.

The IF filter's noise bandwidth is known as the *resolution bandwidth,* and it can usually be chosen from a set of several discrete values. This bandwidth determines the frequency range over which the spectral power is measured. The IF filter is usually implemented as a cascade of synchronously tuned bandpass stages, resulting in a Gaussian response shape. Such a filter has good phase linearity (or, alternatively, approximately constant group delay) and a noise bandwidth that is about 1.2 times the −3-dB bandwidth. The difference in noise bandwidth (about 0.8 dB on a ratio basis) needs to be taken into account when integrating the noise density to find the total power.

To provide an amplitude measurement calibrated in decibels, a logarithmic amplifier compresses the output of the IF filter. Because noise peaks are therefore diminished in magnitude relative to the average, the log amp introduces some measurement distortion. In this case, the measurement of power is low by about 1.45 dB. Failure to take logarithmic warping into account is a common mistake. Just remember, for example, that you cannot simply average an ensemble of measurements expressed in decibels. You must average the raw values first and then subsequently convert to decibel values.

After filtering and logarithmic amplification, a "video" detector drives the vertical deflection circuitry of the display.[13] Another filter (known, appropriately enough, as the video filter) smooths the output of the detector. The characteristics of the detector in measuring random noise introduce a calibration issue of their own. Most detectors are calibrated to read the proper rms values under the assumption of a sinusoidal input, multiplying the raw detector output by $1/\sqrt{2}$. However, random noise is not sinusoidal; the envelope of band-limited noise turns out to be Rayleigh distributed, resulting in an average value that is about a factor of 1.25 times the rms value. Together these considerations suggest we should add about 1.05 dB to the measured values in order to correct for the detector's characteristics. Overall, to compensate for both the log amp and envelope detector's bias, one must add about 2.5 dB to the raw measured power data.

19.3.1 RESOLUTION BANDWIDTH VERSUS VIDEO BANDWIDTH

The presence of filters both ahead of and past the detector confuses many spectrum analyzer users. The IF filter sets the bandwidth over which noise power is integrated before detection. As a consequence, its value determines the displayed value of noise. A factor-of-10 reduction in resolution bandwidth diminishes the displayed noise power by 10 dB.

The video bandwidth, on the other hand, determines the total amount of noise superimposed on the detector output. Since the desired signal from the detector is all at DC, it follows that the narrower the video bandwidth, the more noise-free the displayed levels. The trade-off is that the video filter's settling time increases, forcing slower sweep rates as the video bandwidth is reduced. Thus, the resolution bandwidth determines the measured value noise, and the video bandwidth reduces the noisiness of the displayed noise.

---

[13] The nomenclature traces its origins to the early days of radar, where the detector output similarly drove a video display.

19.3.2 TRACKING GENERATOR

Some spectrum analyzers allow an external oscillator to share the control voltage used by the analyzers's internal VCO. This separate oscillator, known as a tracking generator because its frequency tracks that of the analyzer's VCO, can be used as a signal source to a device under test, allowing the measurement of the DUT's frequency response. As a consequence, the addition of a tracking generator converts a spectrum analyzer into a scalar network analyzer. (It is scalar because phase is ignored, precluding a full vector measurement.) Measuring the frequency response magnitude of filters and amplifiers is rapidly and conveniently carried out with such a setup. A tracking generator is thus a highly desirable accessory, and most commercially available spectrum oscillators accommodate an optional tracking generator.

19.3.3 CAVEATS

It is always important to be aware of the performance limits of any instrument. This advice is especially important in the digital age, where instruments happily provide data with many digits regardless of the underlying integrity of the data. Just because you paid for all those digits does not necessarily mean they're all trustworthy.

One basic requirement is to observe the dynamic range limits of the analyzer. If the input signal is too large then something in the analyzer will overload, generating distortion within the instrument. Such spurious responses may be improperly attributed to the DUT, causing the engineer to waste a great deal of time chasing a phantom problem. Fortunately, spectrum analyzer overload is readily detected and avoided. If increasing the attenuation (decreasing the sensitivity) causes spurious tones to change in relative power (or to disappear altogether), then the spectrum analyzer is the source of the problem. As a routine procedure, one should vary the input sensitivity in order to verify that the measurement is truly attributable to the DUT and not to the analyzer.

It's also important to acknowledge the existence of a maximum allowed input power. Exceeding this absolute upper limit can result in actual damage to the front end. This problem almost never arises when characterizing low-level amplifiers. However, when working with power amplifiers, it may be necessary to use external attenuators with appropriate power ratings. On the other hand, if the input signal is too small, the signal from the DUT will be corrupted (or even buried) by the noise floor of the analyzer. Be wary of measurements made within a few decibels of the inherent noise floor of the analyzer.

Another consideration is that the useful resolution bandwidth is limited by the phase noise of the analyzer's internal oscillator. Well-designed instruments are "self-aware" in the sense that the IF filters are chosen to have a minimum bandwidth consistent with the VCO characteristics. Alas, not all instruments are well designed, so it is important to study the specifications carefully to understand the true limits of the equipment.

In reading noise levels from the display, it is important to correct for the noise bandwidth not only of the IF filters but also of the log amp and envelope detector. As mentioned previously, some instruments include a cursor option for automatic correction of these effects, but not all do. For the latter, one must perform the corrections manually.

Finally, note that harmonic distortion in the analyzer's LO can result in multiple responses to a single frequency input. This distortion is kept very small through careful design, but it's never zero. For large input powers, the distortion can result in multiple responses that are noticeably above the noise floor, possibly causing one to misinterpret the display as indicative of some pathology in the DUT. Again, a careful understanding of a particular instrument's characteristics and limitations will allow the user to avoid being fooled by such artifacts.

19.4 REFERENCES

The following notes provide additional information of value: "Spectrum Analyzer Measurement Seminar" (Hewlett-Packard, February 1988); and "Spectrum Analysis ... Spectrum Analyzer Basics" (Hewlett-Packard Applications Note 150, April 1974).

A wonderful paper by Mark Kahrs, "50 Years of RF and Microwave Sampling" (*IEEE Trans. Microwave Theory and Tech.*, v. 51, no. 6, June 2003, pp. 1787–1805), provides an exceptional history of the subject of sampling. With an extensive reference list of over 250 papers, this review covers the subject with unusual thoroughness.

CHAPTER TWENTY

RF POWER AMPLIFIERS

20.1 INTRODUCTION

In this chapter, we consider the problems of efficiently and linearly delivering RF power to a load. Simple, scaled-up versions of small-signal amplifiers are fundamentally incapable of high efficiency, so we have to consider other approaches. As usual, there are trade-offs – here, between spectral purity (distortion) and efficiency.

In a continuing quest for increased channel capacity, more and more communications systems employ amplitude and phase modulation together. This trend brings with it an increased demand for much higher linearity (possibly in both amplitude and phase domains). At the same time, the trend toward portability has brought with it increased demands for efficiency. The variety of power amplifier topologies reflects the inability of any single circuit to satisfy all requirements.

SMALL- VERSUS LARGE-SIGNAL OPERATING REGIMES

Recall that an important compromise is made in analyzing circuits containing nonlinear devices (such as transistors). In exchange for the ability to represent, say, an inherently exponential device with a linear network, we must accept that the model is valid only for "small" signals. It is instructive to review what is meant by "small" and to define quantitatively a boundary between "small" and "large."

In what follows, we will decompose signals into their DC and signal components. To keep track of which is which, we will use the following notational convention: DC variables are in upper case (with upper-case subscripts); small-signal components are in lower case (with lower-case subscripts); and the combination of DC and small-signal components is denoted by a combination of a lower-case variable and an upper-case subscript.

Let's start with the familiar exponential v_{BE} law:

$$i_C = I_S e^{v_{BE}/V_T}, \qquad (1)$$

where we have neglected the -1 term, as is traditional.

Now express the base–emitter voltage as the sum of a DC and "small-signal" component:

$$v_{BE} = V_{BE} + v_{be}. \tag{2}$$

Next, substitute into the exponential law and use a series expansion:

$$i_C = I_S e^{v_{BE}/V_T} = I_S e^{V_{BE}/V_T} \left[1 + \frac{v_{be}}{V_T} + \frac{1}{2}\left(\frac{v_{be}}{V_T}\right)^2 + \cdots \right]. \tag{3}$$

In traditional incremental analysis, we preserve only the DC and first-order terms. To get a feel for when the neglect of higher-order terms is justified, let's find the value of v_{be} (the incremental base–emitter voltage) that gives us a second-order term that is no larger than a tenth as large as the first-order term:

$$\frac{1}{2}\left(\frac{v_{be}}{V_T}\right)^2 \leq \left|\frac{v_{be}}{10V_T}\right|. \tag{4}$$

Solving for v_{be} yields

$$|v_{be}| \leq \tfrac{1}{5}V_T. \tag{5}$$

At room temperature, V_T is about 25 mV, so the maximum allowable excursion in base–emitter voltage is a measly ±5 mV, which corresponds to a collector current change of approximately ±20% about the quiescent value. Even smaller excursions are permitted if the second-order term is to be smaller than 10% of the first-order term. Thus, you can well appreciate that "small signal" *means* small signal.

An analogous derivation for a FET reveals that *small* is defined there as relative to the gate overdrive, $V_{GS} - V_T$, where V_T is the threshold voltage. In terms of absolute volts, the linear range of an FET is greater than that of a bipolar. However, this apparent superiority comes at the expense of reduced transconductance.

Transistors in a typical power amplifier may traverse all regions of operation. Consequently the base–emitter (or gate–source) voltage swings will generally exceed the small-signal limits we've just identified. As a result, the amplifiers will exhibit significant distortion that must be accommodated in both analysis and design. In narrowband amplifiers, this distortion can be reduced most conveniently by using filters with sufficiently high Q to pass only the carrier and modulation sidebands. In broadband PAs, one must use other means, such as negative feedback or feedforward – and generally at the cost of less efficiency or more complexity.

20.2 CLASSICAL POWER AMPLIFIER TOPOLOGIES

There are four types of power amplifiers, distinguished primarily by bias conditions, that may be termed "classic" because of their historical precedence. These are labeled Class A, AB, B and C, and all four may be understood by studying the single model sketched in Figure 20.1.[1] The model isn't unique, but it is representative.

---

[1] Many variations on this theme exist, but the operating features of all of them may still be understood with this model.

632 CHAPTER 20 RF POWER AMPLIFIERS

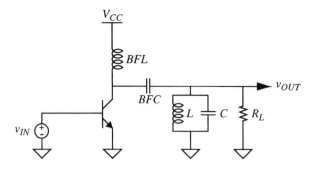

FIGURE 20.1. General power amplifier model for classic topologies

In this general model, the resistor R_L represents the load into which we are to deliver the output power. A "big, fat" inductance, BFL, feeds DC power to the collector and is assumed large enough that the current through it is substantially constant. The collector is connected to a tank circuit through capacitor BFC to prevent any DC dissipation in the load. One advantage of this particular configuration is that the transistor's output capacitance can be absorbed into the tank, as in a conventional small-signal amplifier. Another is that the filtering provided by the tank cuts down on out-of-band emissions caused by the ever-present nonlinearities that accompany large-signal operation. To simplify analysis, we assume that the tank has a high enough Q that the voltage across the tank is reasonably well approximated by a sinusoid, even if it is fed by nonsinusoidal currents.

20.2.1 CLASS A AMPLIFIERS

The defining characteristic of a Class A PA is that bias levels are chosen to ensure that the transistor conducts all the time. The primary distinction between Class A power amplifiers and small-signal amplifiers is that the signal currents in a PA are a substantial fraction of the bias level, and one would therefore expect potentially serious distortion. The Class A amplifier moderates this distortion at the expense of efficiency because there is always dissipation due to the bias current, even when there is no input signal. To understand quantitatively why the efficiency is poor, assume that the collector current is reasonably well approximated by

$$i_C = I_{DC} + i_{rf} \sin \omega_0 t, \tag{6}$$

where I_{DC} is the bias current, i_{rf} is the amplitude of the signal component of the collector current, and ω_0 is the signal frequency (and also the resonant frequency of the tank). Although we have glibly ignored distortion, the errors introduced are not serious enough to invalidate what follows.

The output voltage is simply the product of a signal current and the load resistance. Since the big, fat inductor BFL forces a substantially constant current through

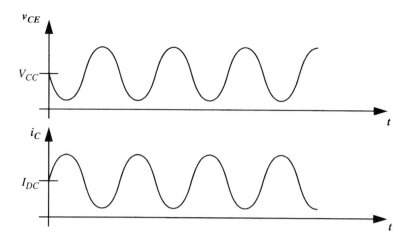

FIGURE 20.2. Collector voltage and current for ideal Class A amplifier

it, KCL tells us that the signal current is none other than the signal component of the collector current. Therefore,

$$v_o = -i_{rf} R \sin \omega_0 t. \tag{7}$$

Finally, the collector voltage is the sum of the DC collector voltage and the signal voltage. The big, fat inductor BFL presents a DC short, so the collector voltage swings symmetrically about V_{CC}.[2] The collector voltage and current are therefore offset sinusoids that are 180° out of phase with each other, as shown in Figure 20.2.

If it isn't clear from the equations, it should be clear from the figure that the transistor always dissipates power because the product of collector current and collector voltage is always positive. To evaluate this dissipation quantitatively, compute the efficiency by first calculating the signal power delivered to the resistor R:

$$P_{rf} = i_{rf}^2 R/2. \tag{8}$$

Next, compute the DC power supplied to the amplifier. Let us assume that the quiescent collector current, I_{DC}, is made just large enough to guarantee that the transistor does not ever cut off. That is,

$$I_{DC} = i_{rf}, \tag{9}$$

so that the input DC power is

$$P_{DC} = I_{DC} V_{CC} = i_{rf} V_{CC}. \tag{10}$$

---

[2] This is not a typographical error. The collector voltage actually swings *above* the positive supply. One way to argue that this must be the case is to recognize that an ideal inductor cannot have any DC voltage across it (otherwise infinite currents would eventually flow). Therefore, if the collector voltage swings below the supply, it must also swing above it. This kind of thinking is particularly helpful in deducing the characteristics of various types of switched-mode power converters.

The ratio of RF output power to DC input power is a measure of efficiency (usually called the collector efficiency), and is given by

$$\eta \equiv \frac{P_{rf}}{P_{DC}} = \frac{i_{rf}^2(R/2)}{i_{rf}V_{CC}} = \frac{i_{rf}R}{2V_{CC}}. \quad (11)$$

Now, the absolute maximum that the product $i_{rf}R$ can have is V_{CC}, for otherwise the transistor would saturate on the (negative) peaks. Hence, the maximum theoretical collector efficiency is just 50%. If one makes due allowance for nonzero saturation voltage, variation in bias conditions, nonideal drive amplitude, and inevitable losses in the tank and interconnect, values substantially smaller than 50% often result – particularly at lower supply voltages, where $V_{CE,sat}$ represents a larger fraction of VCC. Consequently, collector efficiencies of 30–35% are not at all unusual for practical Class A amplifiers.[3]

Aside from efficiency, another important consideration is the stress on the output transistor. In a Class A amplifier, the maximum collector-to-emitter voltage is $2V_{CC}$, while the peak collector current has a value of $2V_{CC}/R$. Hence, the device must be rated to withstand peak voltages and currents of these magnitudes, even though both maxima do not occur simultaneously.

One common way to quantify these requirements is to define another figure of merit, the "normalized power-handling capability," which is simply the ratio of the actual output power to the product of the maximum device voltage and current. For this type of amplifier, the maximum value of this dimensionless figure of merit is

$$P_N \equiv \frac{P_{rf}}{v_{CE,max} i_{C,max}} = \frac{V_{CC}^2/(2R)}{(2V_{CC})(2V_{CC}/R)} = \frac{1}{8}. \quad (12)$$

This figure of merit is also known as the *utilization factor*.

The Class A amplifier thus provides reasonable linearity at the cost of low efficiency and relatively large device stresses. For this reason, pure Class A amplifiers have been rare in RF power applications[4] and relatively rare in audio power applications (particularly so at the higher power levels).[5]

It is important to underscore once again that the 50% efficiency value represents an upper limit. If the collector swing is less than the maximum assumed in our discussion or if there are additional losses anywhere else, the efficiency drops. As the swing approaches zero, the collector efficiency also approaches zero because the signal power delivered to the load goes to zero while the transistor continues to burn DC power.

---

[3] Another factor is that relative distortion drops with the output power. In applications where low distortion is important, efficiency is often traded off for linearity, resulting in quite low efficiency.

[4] Except, perhaps, in low-level applications, or in the early stages of a cascade.

[5] An exception is the high-end audio crowd, of course, for whom power consumption is often not a constraint. More recently, the linearity demands of third-generation (3G) wireless have resulted in a resurgence of Class A amplifiers.

20.2 CLASSICAL POWER AMPLIFIER TOPOLOGIES

As a final comment on this topology, note that even though the amplifier distorts because of large-signal operation, the amplifier does have linear *modulation* characteristics since the output fundamental is in fact proportional to the fundamental component of the drive current. As long as the drive current's fundamental is itself proportional to the desired modulation, the voltage developed across the load will also be proportional to it, and linear modulation results. Thus, the amplifier is still linear in the describing function sense (see Chapter 15).

20.2.2 CLASS B AMPLIFIERS

A clue to how one might achieve higher efficiency than a Class A amplifier is actually implicit in the waveforms of Figure 20.2. It should be clear that if the bias were arranged to reduce the fraction of a cycle over which collector current and collector voltage are simultaneously nonzero, then transistor dissipation would diminish.

In the Class B amplifier, the bias is arranged to shut off the output device half of every cycle. An exact 50% conduction duty cycle is a mathematical point, of course, so true Class B amplifiers do not actually exist. Nevertheless, the concept is useful in constructing a taxonomy. In any case, with intermittent conduction, we expect a gross departure from linear operation. However, we must distinguish between distortion in the output (an earmark of nonlinearity), and proportionality (or lack thereof) between input and output powers (evaluated at the fundamental). A single-ended Class B amplifier may produce a nonsinusoidal output yet still act linearly in this sense of input–output power proportionality. We still care about out-of-band spectral components, of course, and a high-Q resonator (or other filter) is absolutely mandatory in order to obtain an acceptable approximation to a sinusoidal output. However, despite this distortion, the Class B amplifier can possess linear modulation characteristics, again in the describing function sense, and may therefore be used in applications requiring linear amplification.

Although the single-transistor version of a Class B amplifier is what we'll analyze here, it should be mentioned that most practical Class B amplifiers are push–pull configurations of two transistors (more on this topic later).

For this amplifier, then, we assume that the collector current is sinusoidal for one half-cycle and zero for the other half-cycle;

$$i_C = i_{rf} \sin \omega_0 t \quad \text{for } i_C > 0. \tag{13}$$

The output tank filters out the harmonics of this current, leaving a sinusoidal collector voltage as in the Class A amplifier. The collector current and collector voltage therefore appear approximately as shown in Figure 20.3.

To compute the output voltage, we first find the fundamental component of the collector current and then multiply this current by the load resistance:

$$i_{fund} = \frac{2}{T} \int_0^{T/2} i_{pk}(\sin \omega_0 t)(\sin \omega_0 t)\, dt = \frac{i_{pk}}{2}, \tag{14}$$

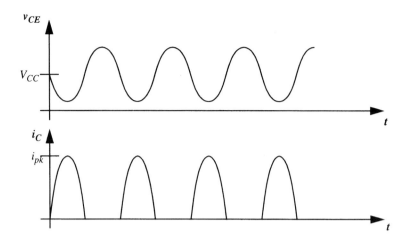

FIGURE 20.3. Collector voltage and current for ideal Class B amplifier

a result we expect after cutting out half the output waveform. Thus,

$$v_o \approx \frac{i_{pk}}{2} R \sin \omega_0 t. \tag{15}$$

Since the maximum possible value of v_o is V_{CC}, it is clear from Eqn. 15 that the maximum value of i_{pk} is

$$i_{pk,max} = \frac{2V_{CC}}{R}. \tag{16}$$

The peak collector current and maximum output voltage are therefore the same as for the Class A amplifier.[6]

Computing the collector efficiency as before, we first calculate the output power as

$$P_o = \frac{v_{rf}^2}{2R}, \tag{17}$$

where v_{rf} is the amplitude of the swing across the load resistor. The maximum value of the swing amplitude remains V_{CC}, so that the maximum output power is

$$P_{o,max} = \frac{V_{CC}^2}{2R}. \tag{18}$$

Computing the DC input power is a little more complicated, but straightforward nonetheless. The average collector current is

$$\overline{i_C} = \frac{1}{2\pi} \int_0^\pi \frac{2V_{CC}}{R} \sin \Theta \, d\Theta = \frac{2V_{CC}}{\pi R}, \tag{19}$$

---

[6] The assumption of half-sinusoidal current pulses is, necessarily, an approximation. The collector current in practical circuits differs mainly in that the transition to and from zero current is not abrupt. Hence, the true device dissipation is somewhat greater, and the efficiency somewhat lower, than predicted by ideal theory.

so that the DC power supplied is

$$P_{DC} = V_{CC}\overline{i_C} = \frac{2V_{CC}^2}{\pi R}. \tag{20}$$

Finally, the maximum collector efficiency for a Class B amplifier is

$$\eta = \frac{P_{o,max}}{P_{DC}} = \frac{\pi}{4} \approx 0.785. \tag{21}$$

The theoretical maximum collector efficiency is thus considerably higher than that for the Class A PA. Again, however, the actual efficiency of any practical implementation will be somewhat lower than given by the analysis shown here as a result of effects that we have neglected. Nonetheless, it remains true that, all other things held equal, the Class B amplifier offers substantially higher efficiency than its Class A cousin.

The normalized power capability of this amplifier is 1/8, the same as for the Class A, since the output power, maximum collector voltage, and maximum collector current are the same.[7]

With the Class B amplifier, we have accepted some distortion (but retained modulation linearity) and reduced gain in exchange for a significant improvement in efficiency. Since this trade-off is effected by reducing the fraction of a period that the transistor conducts current, it is natural to ask whether further improvements might be possible by reducing the conduction angle even more. Exploration of this idea leads to the Class C amplifier.

20.2.3 CLASS C AMPLIFIER

In a Class C PA, the bias is arranged to cause the transistor to conduct less than half the time. Consequently, the collector current consists of a periodic train of pulses. It is traditional to approximate these pulses by the top pieces of sinusoids to facilitate a direct analysis.[8] Specifically, one assumes that the collector current is of the following form:

$$i_C = I_{DC} + i_{rf} \sin \omega_0 t \quad \text{for } i_C > 0, \tag{22}$$

where the offset I_{DC} (which is analogous to the bias current in a linear amplifier) is actually negative for a Class C amplifier. Of course, the *overall* collector current i_C is always positive or zero. That is, the collector current is a piece of a sine wave when the transistor is active and zero when the transistor is in cutoff. We continue

---

[7] A two-transistor push–pull Class B amplifier has a normalized power capability that is twice as large.
[8] See e.g. Krauss, Bostian, and Raab, *Solid-State Radio Engineering,* Wiley, New York, 1981. They provide a more complete analysis than presented here.

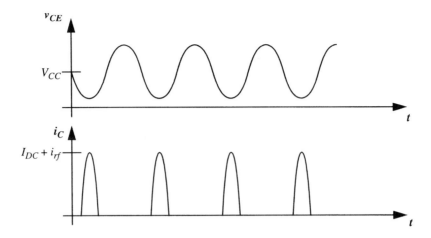

FIGURE 20.4. Collector voltage and current for ideal Class C amplifier

to assume that the transistor behaves at all times as a current source (high output impedance).[9]

We also continue to assume that the output tank has a high enough Q that the voltage across the load remains substantially sinusoidal. The collector voltage and collector current therefore appear as shown in Figure 20.4. Because we're interested in what distinguishes a Class C from a Class B PA, let's focus our attention on the behavior as the collector current pulses get relatively narrow. In that case, we can simplify matters considerably by idealizing the pulses as triangles with the same peak value and width; see Figure 20.5.

The average value of collector current is simply the area of the triangular pulse divided by the period:

$$\overline{i_C} = \frac{1}{T}\int_0^T i_C(t)\,dt \approx \frac{\frac{1}{2}(2\Phi)(i_{PK})}{2\pi} = \frac{(\Phi)(i_{PK})}{2\pi}. \qquad (23)$$

The average power supplied by V_{CC} is therefore

$$P_{DC} = V_{CC}\overline{i_C} \approx V_{CC}\frac{(\Phi)(i_{PK})}{2\pi}. \qquad (24)$$

Now, the fundamental component of the collector current has an amplitude given by

$$i_{fund} = \frac{2}{T}\int_0^T i_C(t)\cos\omega t\,dt \approx 2\overline{i_C} = \frac{(\Phi)(i_{PK})}{\pi}, \qquad (25)$$

---

[9] Violation of this assumption leads to an exceedingly complex situation. Maximum efficiency is typically obtained when the output power is nearly saturated. Under such conditions, Class C amplifiers force bipolar transistors into saturation (and MOSFETs into triode) for some fraction of a period, making exact analysis difficult.

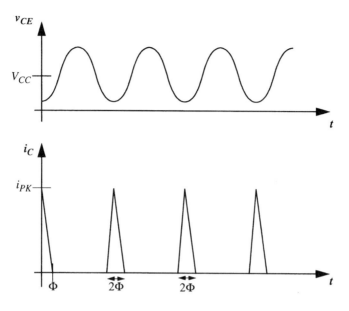

FIGURE 20.5. Idealized collector current waveform

where we have assumed that the pulses are so narrow that the cosine function is well approximated by unity for the duration of the pulse. Note that Eqn. 25 says that, independent of pulse shape, the fundamental component of current is always about double the DC current – provided the collector current flows in narrow pulses. Thus, idealization of the collector current pulses as triangular introduces no serious errors in this regime, while simplifying the analysis enough to facilitate acquisition of useful design insight.

One important insight is the nonlinearity implied by Eqn. 25. The peak output current increases with the input voltage, and so does the conduction angle. There's no reason to expect the product of these two increases to be linear, and it isn't (in fact, the output amplitude is more closely approximated as depending on the square of the input amplitude). Thus the output power depends nonlinearly on the input power, unlike the Class A and B amplifiers.

The signal power delivered to the load is

$$P_{rf} = \frac{i_{fund}^2 R}{2} \approx \frac{\left[i_{PK}\frac{\Phi}{\pi}\right]^2}{2}. \tag{26}$$

The collector efficiency is readily calculated as

$$\eta = \frac{P_{rf}}{P_{DC}} \approx \frac{\frac{\left[i_{PK}\frac{\Phi}{\pi}\right]^2 R}{2}}{V_{CC}\frac{(\Phi)(i_{PK})}{2\pi}} = \frac{\left[i_{PK}\frac{\Phi}{\pi}\right]R}{V_{CC}} = \frac{i_{fund}R}{V_{CC}}. \tag{27}$$

Now the maximum output power is produced for

$$i_{fund,max} = \frac{V_{CC}}{R}, \qquad (28)$$

so the collector efficiency approaches 100% for a sufficiently large product of peak current and (narrow) conduction angle. We see from Eqns. 23 and 25 that adjusting the base drive to produce narrower collector current pulses requires proportional increases in the peak collector current if the output power is to remain constant. It should thus be clear that the normalized power-handling capability of the Class C amplifier approaches zero as the conduction angle approaches zero, because then the peak collector current approaches infinity. Practical conduction angles are not near zero for this reason.

Although the foregoing development treats the conduction angle as an independent variable, it is more commonly the result of choosing a conveniently realized input bias (such as zero volts) in combination with whatever input drive is available. Accommodating that truth – plus balancing gain, efficiency, and power-handling capability – typically results in conduction angles between 135° and 150°.[10] The primary virtue in carrying out the exercise is to develop some general intuition useful for design: efficiency can be high, but at the cost of reduced power handling capability, gain, and linearity.

20.2.4 CLASS AB AMPLIFIERS

We've seen that Class A amplifiers conduct 100% of the time, Class B amplifiers 50% of the time, and Class C PAs somewhere between none and 50% of the time. The Class AB amplifier, as its name suggests, conducts somewhere between 50% and 100% of a cycle, depending on the bias levels chosen. As a result, its efficiency and linearity are intermediate between those of a Class A and Class B amplifier. This compromise is frequently satisfactory, as one may deduce from the popularity of this PA. All real Class B amplifiers are Class AB or C amplifiers, depending on the conduction angle.

20.2.5 CLASS D AMPLIFIERS

The classic PAs use the active device as a controlled current source. An alternative is to use transistors as switches. This alternative is attractive because a switch ideally dissipates no power. At any given instant, there is either zero voltage across it or zero current through it. Since the switch's *V–I* product is therefore always zero, the transistor dissipates no power and the efficiency must be 100%.

---

[10] The astute reader will note that this conduction angle range is not small in the sense of the approximations used in the derivations presented. Nevertheless, the overall qualitative conclusions remain valid.

20.2 CLASSICAL POWER AMPLIFIER TOPOLOGIES

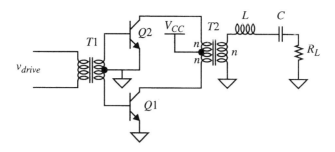

FIGURE 20.6. Class D amplifier

One type of amplifier that exploits this observation is the Class D amplifier. At first glance, it looks the same as a push–pull, transformer-coupled version of a Class B amplifier. See Figure 20.6. In contrast with the parallel tanks we've typically seen, a series RLC network is used in the output of this amplifier, since switch-mode amplifiers are the duals of the current-mode amplifiers studied previously. As a consequence, the output filters are also duals of each other.

The input connection guarantees that only one transistor is driven on at a given time, with one transistor handling the positive half-cycles and the other the negative half-cycles, just as in a push–pull Class B. The difference here is that the transistors are driven hard enough to make them act like switches, rather than as linear (or quasilinear) amplifiers.

Because of the switching action, each primary terminal of the output transformer $T2$ is alternately driven to ground, yielding a square-wave voltage across the primary (and hence across the secondary). When one collector goes to zero volts, transformer action forces the other collector to a voltage of $2V_{CC}$. The output filter allows only the fundamental component of this square wave to flow into the load.

Since only fundamental currents flow in the secondary circuit, the primary current is sinusoidal as well. As a consequence, each switch sees a sinusoid for the half-cycle that it is on, and the transformer current and voltage therefore appear as in Figures 20.7 and 20.8. Because the transistors act like switches (in principle, anyway), the theoretical efficiency of the Class D amplifier is 100%.

The normalized power efficiency for this amplifier happens to be[11]

$$\frac{P_o}{v_{CE,on} \cdot i_{C,pk}} = \frac{1}{\pi} \approx 0.32, \tag{29}$$

which is considerably better than a Class A amplifier and somewhat better than a push–pull Class B. Of course, the Class D amplifier cannot normally provide linear modulation, but it does provide potentially high efficiency and does not stress the devices very much.

---

[11] It may help to keep in mind that the amplitude of the fundamental component of a square wave is $4/\pi$ times the amplitude of the square wave.

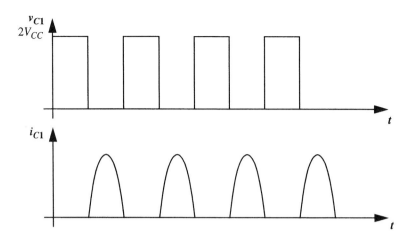

FIGURE 20.7. Q1 collector voltage and current for ideal Class D amplifier

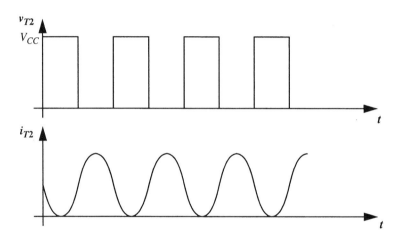

FIGURE 20.8. T2 secondary voltage and current for ideal Class D amplifier

One practical problem with this (or any other switching) PA is that there is no such thing as a perfect switch. Nonzero saturation voltage guarantees static dissipation in the switches, while finite switching speeds imply that the switch V–I product is nonzero during the transitions. Hence, switch-mode PAs function well only at frequencies well below f_T. Furthermore, a particularly serious reduction in efficiency can result in bipolar implementations if, due to charge storage in saturation, one transistor fails to turn completely off before the other turns on. Transformer action then attempts to apply the full supply voltage across the device that is not yet off, and the V–I product can be quite large.

20.2.6 CLASS E AMPLIFIERS

As we've seen, using transistors as switches has the potential for providing greatly improved efficiency, but it's not always trivial to realize that potential in practice

20.2 CLASSICAL POWER AMPLIFIER TOPOLOGIES

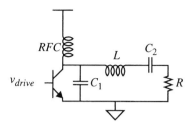

FIGURE 20.9. Class E amplifier

owing to imperfections in real switches (transistors). The associated dissipation degrades efficiency. To prevent gross losses, the switches must be quite fast relative to the frequency of operation. At high carrier frequencies, it becomes increasingly difficult to satisfy this requirement.

If there were a way to modify the circuit so that the switch voltage were zero for a nonzero interval of time about the instant of switching, then the dissipation would decrease. The Class E amplifier uses a high-order reactive network that provides enough degrees of freedom to shape the switch voltage to have both zero value *and* zero slope at switch turn-on, thus reducing switch losses. It does nothing for the turn-off transition, which is often the more troublesome edge, at least in bipolar designs. Another issue, as we'll see later, is that the Class E amplifier has rather poor normalized power-handling capability (worse, in fact, than a Class A amplifier), requiring the use of rather oversized devices to deliver a given amount of power to a load – despite the high potential efficiency (theoretically 100% with ideal switches) of this topology. If these constraints are not a bother, then the Class E topology is capable of excellent performance.

Another virtue of the Class E amplifier is that it is straightforward to design. Unlike typical Class C amplifiers, practical implementations require little post-design tweaking to obtain satisfactory operation.

With that preamble out of the way, let's take a look at the Class E topology drawn in Figure 20.9. The RF choke (*RFC*, the equivalent of a *BFL*) simply provides a DC path to the supply and approximates an open circuit at RF. Note additionally that the capacitor C_1 is conveniently positioned, for any device output capacitance can be absorbed into it.

Derivation of the design equations is sufficiently involved (and the payoff in terms of design insight sufficiently small) that we'll omit the details. Interested readers are directed to the relevant literature for derivation of the following equations:[12]

---

[12] N. O. Sokal and A. D. Sokal, "Class E, a New Class of High-Efficiency Tuned Single-Ended Power Amplifiers," *IEEE J. Solid-State Circuits,* v. 10, June 1975, pp. 168–76. Invention of the Class E amplifier is usually traced to that paper, but Gerald D. Ewing's doctoral thesis, "High-Efficiency Radio-Frequency Power Amplifiers" (Oregon State University, Corvallis, 1964), is the earliest exposition of the concept. I am grateful to Prof. David Rutledge of Caltech for providing this historical note.

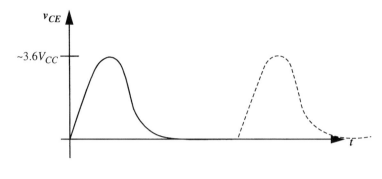

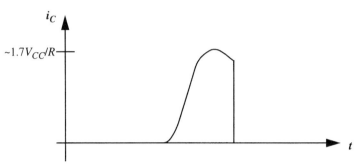

FIGURE 20.10. Idealized waveforms for Class E amplifier

$$L = \frac{QR}{\omega}; \quad (30)$$

$$C_1 \approx \frac{1}{\omega(R \cdot 5.447)} \approx \frac{0.184}{\omega R}, \quad (31)$$

$$C_2 \approx C_1\left(\frac{5.447}{Q}\right)\left(1 + \frac{1.42}{Q - 2.08}\right). \quad (32)$$

Although we provide equations for the output series LC network elements, they are based on a simple, parasitic-free model for the transistor. In practice, one will have to search for element values that maximize the efficiency while providing the desired output power.

Once everything is tuned up, the collector current and voltage waveforms resemble those shown in Figure 20.10. Note that the collector voltage has zero slope at turn-on, although the current is nearly a maximum when the switch turns off. Hence, switch dissipation can be significant during that transition if the switch isn't infinitely fast (as is the case with most switches you're likely to encounter). This dissipation can offset much of the improvement obtained by reducing the dissipation during the transition to the "on" state.

Additionally, note that each of the waveforms has a rather dramatic peak-to-average ratio. In fact, a detailed analysis shows that the peak collector voltage is approximately $3.6V_{CC}$, while the peak collector current is roughly $1.7V_{CC}/R$.

20.2 CLASSICAL POWER AMPLIFIER TOPOLOGIES

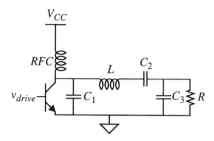

FIGURE 20.11. Class E amplifier with output filter and impedance transformer

The maximum output power delivered to the load is

$$P_o = \frac{2}{1+\pi^2/4} \cdot \frac{V_{CC}^2}{R} \approx 0.577 \cdot \frac{V_{CC}^2}{R}, \tag{33}$$

and the device utilization factor is therefore

$$\frac{P_o}{v_{CE,pk} \cdot i_{C,pk}} \approx 0.098. \tag{34}$$

As you can see, the Class E is more demanding of its switch specifications than even a Class A amplifier. As long as this property can be accommodated, the Class E amplifier is capable of excellent performance.

Finally, note that the drain current waveforms are distinctly nonsinusoidal. Additional filtering is therefore almost certainly necessary if statutory limits on out-of-band emissions are to be satisfied. Quite often, the required filtering may be combined with an impedance transformation (for correct output power) and/or partially absorbed into the Class E's reactive network elements.

In Figure 20.11, C_3 forms part of a downward-transforming L-match network. Choosing the low-pass version provides some filtering for free. The series inductance L serves double duty, completing the L-match and functioning as part of the Class E series LC output network. If any additional filtering is required, it can be implemented with series LC traps shunting the output load or with additional elements interposed between the Class E stage proper and the output load.

There is a tendency to focus on the output side of a PA, but we shouldn't overlook drive requirements on the input side. At higher power levels in particular, the power device is often so large that its input impedance ends up being quite low (independent of device technology). It is not uncommon for the input impedance to be as low as a single ohm (resistive plus reactive), so coupling power efficiently into the input of this amplifier can be challenging on occasion.

As a final note on the Class E amplifier, bipolar implementations typically don't perform as well as ones built with FETs. The reason is that bipolar transistors exhibit

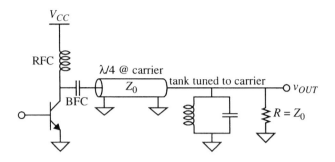

FIGURE 20.12. Class F amplifier

a turn-off delay phenomenon due to minority carrier storage.[13] The reactive network of a Class E eases *V–I* overlap for the turn-on transition only; as mentioned earlier, it does nothing to help the more problematic turn-off transition of a bipolar implementation.

20.2.7 CLASS F AMPLIFIERS

Implicit in the design of Class E amplifiers is the concept of exploiting the properties of reactive terminations in order to shape the switch voltage and current waveforms to your advantage. Perhaps the most elegant expression of this concept is found in the Class F amplifier. As shown in Figure 20.12, the output tank is tuned to resonance at the carrier frequency and is assumed to have a high enough Q to act as a short circuit at all frequencies outside of the desired bandwidth.

The length of the transmission line is chosen to be precisely a quarter-wavelength at the carrier frequency. Recall that the input impedance of such a line is proportional to the reciprocal of the termination impedance:

$$Z_{in} = Z_0^2/Z_L. \tag{35}$$

We may deduce from this equation that a *half*-wavelength piece of line presents an input impedance equal to the load impedance, since two quarter-wave sections give us two reciprocations that undo each other.

With that quick review out of the way, we can figure out the nature of the impedance seen by the collector. At the carrier frequency, the collector sees a pure resistance of $R = Z_0$, since the tank is an open circuit there, and the transmission is therefore terminated in its characteristic impedance.

At the second harmonic of the carrier, the collector sees a short, because the tank is a short at all frequencies away from the carrier (and its modulation sidebands), and the transmission line now appears as a half-wavelength piece of line. Clearly,

---

[13] For a more detailed discussion of this phenomenon, see the supplementary material included on the CD-ROM that accompanies this book.

20.2 CLASSICAL POWER AMPLIFIER TOPOLOGIES

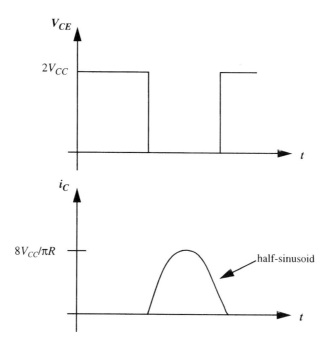

FIGURE 20.13. Collector voltage and current waveforms for Class F PA

at *all* even harmonics of the carrier, the collector sees a short, since the transmission line appears as some integer multiple of a half-wavelength at all even harmonics. Conversely, the collector sees an open circuit at all *odd* harmonics of the carrier, because the tank still appears as a short circuit; the transmission line appears as an odd multiple of a quarter-wavelength and thus provides a net reciprocation of the load impedance.

Now, if the transistor is assumed to act as a switch, then the reactive terminations guarantee that all of the odd harmonics of the collector voltage will see no load (other than that associated with the transistor's own output impedance) and hence a square-wave voltage ideally results at the collector (recall that a square wave with 50% duty ratio has only odd harmonics).

Because of the open-circuit condition imposed by the transmission line on all odd harmonics above the fundamental, the only current that flows into the line is at the fundamental frequency. Hence, the collector current is a sinusoid when the transistor is on. And of course, the tank guarantees that the output voltage is a sinusoid even though the transistor is on for only a half cycle (as in a Class B amplifier).

By cleverly arranging for the square-wave voltage to see no load at all frequencies above the fundamental, the switch current is ideally zero at both switch turn-on and turn-off times. The high efficiencies possible are suggested by the waveforms depicted in Figure 20.13. The total peak-to-peak collector voltage is seen to be twice the supply voltage. Therefore, the peak-to-peak component of V_{CE} at the carrier frequency is

$$(4/\pi)2V_{CC}. \qquad (36)$$

Note that the fundamental has a peak-to-peak value that actually exceeds the total V_{CE} swing, thanks to the magic of Fourier transforms.

Now, since only the fundamental component survives to drive the load, the output power delivered is

$$P_o = \frac{[(4/\pi)V_{CC}]^2}{2R}. \qquad (37)$$

Since the switch dissipates no power, we can conclude that the Class F amplifier is capable of 100% efficiency in principle. In practice, one can obtain efficiency superior to that of Class E amplifiers. Additionally, the Class F PA has substantially better normalized power-handling capability, since the maximum voltage is just twice the supply while the peak collector current is

$$i_{C,pk} = \frac{2V_{CC}}{R} \cdot \frac{4}{\pi} = \frac{8}{\pi} \cdot \frac{V_{CC}}{R}. \qquad (38)$$

The normalized power handling capability is therefore

$$\frac{P_o}{v_{CE,pk} \cdot i_{C,pk}} = \frac{\frac{[(4/\pi)V_{CC}]^2}{2R}}{2V_{CC} \cdot \left(\frac{8}{\pi} \cdot \frac{V_{CC}}{R}\right)} = \frac{1}{2\pi} \approx 0.16, \qquad (39)$$

or exactly half that of the Class D amplifier. In some respects, the Class F amplifier may be considered equivalent to a single-ended Class D amplifier.

It should be emphasized that Class C, D, E, and F amplifiers are all essentially *constant-envelope* amplifiers. That is, they do not normally provide an output that is proportional to the input and thus tend to perform best when all we ask of them is a constant-amplitude output (as would be suitable for FM, for example). Nonetheless, we will see later that it is still possible to use these amplifiers in applications requiring linear operation. This capability is important because there is presently a shift toward more spectrally efficient modulation methods (e.g., QAM) that involve amplitude modulation and for which linear operation is thus necessary. At present, this requirement has frequently forced the use of Class AB amplifiers, with a corresponding reduction in efficiency relative to constant-envelope PA topologies. A general method for providing linear operation at constant-envelope efficiencies remains elusive. In Section 20.3 we will consider in more detail the problem of modulating power amplifiers.

Inverse Class F (F^{-1})

The dual of the Class F is itself a power amplifier with the same theoretical bounds on efficiency as its cousin.[14] Whereas the Class F amplifier's termination appears

---

[14] See e.g. S. Kee et al., "The Class E/F Family of ZVS Switching Amplifiers," *IEEE Trans. Microwave Theory and Tech.*, v. 51, May 2003.

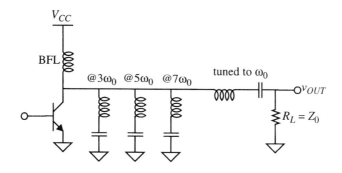

FIGURE 20.14. Inverse Class F amplifier (three-resonator lumped element example shown)

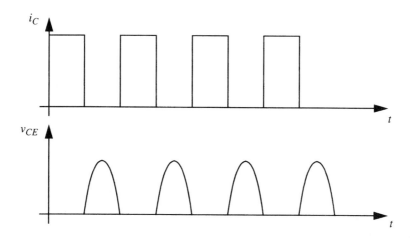

FIGURE 20.15. Collector voltage and current for ideal inverse Class F amplifier

as an open circuit at odd harmonics of the carrier beyond the fundamental and as a short circuit at even harmonics, the *inverse Class F* (often denoted by the shorthand F^{-1}) employs a termination that appears as an open circuit at even harmonics and as a short circuit at the odd harmonics. See Figure 20.14.

Again, a transmission line may replace the lumped resonators when it is advantageous or otherwise practical to do so. Here, a piece of line whose length is $\lambda/2$ at the fundamental frequency replaces the paralleled series resonators and is interposed between the drain and the output series LC tank.

For an infinite number of series resonators, the drain voltage waveform thus appears ideally as a (half) sinusoid and the current waveform as a square wave; see Figure 20.15. Once again, the lack of *V–I* overlap at the switching transitions accounts for the high theoretical efficiency of this architecture, just as with the Class E and standard Class F amplifiers.

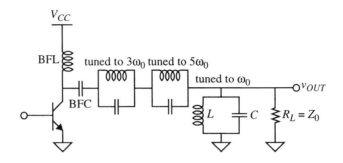

FIGURE 20.16. Alternative Class F amplifier

Alternative Class F Topology

The topology shown in Figure 20.12 is elegant, but the transmission line may be inconveniently long in many applications. Furthermore, the benefits of an infinite (or nearly infinite) impedance at odd harmonics other than the fundamental are somewhat undermined in practice by the transistor's own output capacitance. Hence, a lumped approximation frequently performs nearly as well as the transmission line version.

To create such a lumped approximation, replace the transmission line with a number of parallel-resonant filters connected in series. Each of these resonators is tuned to a different odd harmonic of the carrier frequency. Quite often, simply one tank tuned to $3\omega_0$ is sufficient. Significant improvement in efficiency is rarely noted beyond the use of the two tanks shown in Figure 20.16. For example, use of one tank tuned to the third harmonic boosts the drain efficiency maximum to about 88%, compared to Class B's maximum of about 78%. Addition of tanks tuned to the fifth and seventh harmonics increases the Class F efficiency limit to 92% and 94%, respectively.[15] Given that practical tank elements are not lossless, the law of diminishing returns rapidly makes the use of additional resonators worse than futile.

20.3 MODULATION OF POWER AMPLIFIERS

20.3.1 CLASS A, AB, B, C, E, F

Modulating a Class A or B amplifier is straightforward in principle because the output voltage is ideally proportional to the amplitude of the signal component of the collector current, i_{rf}. Hence, if i_{rf} is itself proportional to the input drive, then linear modulation results. An approximation to this proportionality is achievable with short-channel MOS devices and other FETs, which possess constant transconductance with sufficient gate voltage. Bipolar devices can provide reasonable linearity

---

[15] F. H. Raab, "Class-F Power Amplifiers with Maximally Flat Waveforms," *IEEE Trans. Microwave Theory and Tech.*, v. 45, no. 11, November 1997, pp. 2007–12.

as a result of series base resistance, either externally provided or simply that of the device itself. Nevertheless, linearity requirements have become increasingly severe as wireless systems have evolved. A crude (but almost universally used) linearization "method" is power *backoff,* meaning that we only ask for, say, 1 W out of an amplifier capable of 10 W.[16]

The rationale for backoff is readily understood by considering the same sort of weakly nonlinear amplifier model we invoked in Chapter 13 to define IP3. Because third-order IM terms drop 3 dB for every 1-dB drop in input power, the ratio between the fundamental and third-order components improves 2 dB for each 1-dB reduction in input power (similarly, the corresponding ratios for fifth-order and seventh-order IM terms theoretically improve by 4 dB and 6 dB, respectively, per 1-dB input power drop). If this trend holds, then there is some input power level below which the output IM3 (and other) distortion products are acceptably low in power relative to the carrier. Because Class A efficiency and output power both diminish as the amount of backoff increases, one should use the minimum value consistent with achieving the distortion objectives. Typical backoff values once were generally below 6–8 dB (and sometimes even as low as 1–3 dB) relative to the 1-dB compression point. These days, it is not unusual to find that backoff values must be as high as 10–20 dB in order to satisfy the stringent linearity requirements of some systems. Because it's sometimes easy to lose track of perspective when expressing quantities in decibels, let's re-examine that last interval of values: It says that after you beat your brains out to design a ten-watt RF amplifier, you might find that it meets the specifications for spectral purity only if output powers are kept below a few hundred milliwatts.

Compared to Class A amplifiers, the output IM products of a Class AB amplifier exhibit weaker dependencies on input power (e.g., 2-dB drop in IM3 power per dB reduction of input power), owing to the latter topology's greater inherent nonlinearity. Worse, it is unfortunately not unusual to encounter cases in which *no* amount of backoff will result in acceptable distortion. Finally, as with the Class A amplifier, backoff often degrades efficiency to unacceptably low levels (e.g., to below 5–10% in some cases). We will shortly discuss appropriate linearization alternatives that one might use to relax some of these trade-offs.

The Class C amplifier poses an even more significant challenge, as may be appreciated by studying the equation for the output current derived earlier:

$$i_{fund} = \Phi i_{pk}/\pi. \tag{40}$$

Despite appearances, the fundamental component of the current through the resistive load is generally *not* linearly proportional to i_{pk} because the conduction angle is also a function of i_{pk}.[17] Hence, Class C amplifiers do not normally provide linear

---

[16] If someone else uses a cheesy technique, it's a hack. If *you* use it, it's a method.
[17] That is, with the exception of Class A or B operation. For the other cases, proportionality between drive and response does not occur and so linear modulation is not an inherent property.

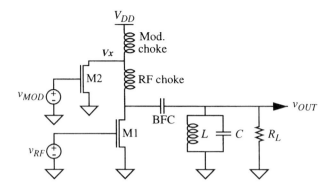

FIGURE 20.17. MOSFET Heising modulator with Class C RF stage (simplified)

modulation capability and are therefore generally unsuitable for amplitude modulation, at least when a modulated carrier drives the base circuit.

To obtain linear amplitude modulation from a nonlinear amplifier (e.g., Class C, D, E, or F), it is often advantageous to consider the *power supply* terminal (drain circuit) as an input port. The general idea is simple: Varying the supply voltage varies the output power. The control there can actually be more linear than at the standard input (e.g. the gate). The first to act on this insight was apparently Raymond Heising of AT&T, around 1919, for vacuum tube amplifiers (of course).[18] The Heising modulator (also known as the *constant current* modulator because of its use of a choke) is shown as Figure 20.17 in its simplest MOSFET incarnation.

The modulation amplifier M2 is loaded by a choke ("Mod. choke"), chosen large enough to have a high reactance at the lowest modulation frequencies. The voltage V_x is the sum of V_{DD} and the modulation voltage developed across that choke. That sum in turn is the effective supply voltage for M1 (biased to operate as a Class C amplifier), fed to the drain through the RF choke as in our standard model for all of the classic PA topologies. Since the two transistors share a common DC supply and since the voltage V_x only approaches ground, transistor M1's output can never quite go to zero. Consequently, the basic Heising modulator is inherently incapable of modulation depths of 100% (this property is a virtue in some instances, because overmodulation – and its attendant gross distortion – becomes inherently impossible). Typical maximum modulation percentages of 60–80% are not uncommon among commercial examples. In applications where the full modulation range is required, a quick fix is to place a capacitively bypassed resistor in series with the RF choke. The DC drop across the resistor causes M1 to operate with a lower supply voltage than does M2. The disadvantage is that the improvement in modulation depth comes at the cost of degraded efficiency, owing to the dissipation in this added resistor.

---

[18] See E. B. Craft and E. H. Colpitts, "Radio Telephony," *AIEE Trans.*, v. 38, 1919, p. 328. Also see R. A. Heising, "Modulation in Radio Telephony," *Proc. IRE*, v. 9, August 1921, pp. 305–22, and also *Radio Review*, February 1922, p. 110.

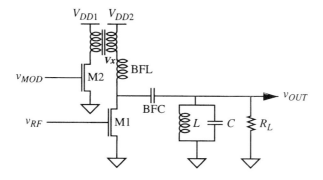

FIGURE 20.18. Alternative drain modulation example

There are alternative forms of drain modulation that do not suffer this painful trade in efficiency for improved modulation depth. A popular example is shown in Figure 20.18. Here, one can find many combinations of supply voltage, transformer turns ratio, and modulation amplitude that will force the drain voltage of M1 to go to zero (not merely approach it) at an extremum of the modulation. For example, assume that we choose a 1:1 transformer. Because we want the secondary voltage to be able to go to zero, the swing at the primary must have an amplitude V_{DD2}, too. In turn, that requirement forces us to choose V_{DD1} somewhat larger than V_{DD2} (to accommodate nonzero drops across M2) as well as an appropriate gate drive level to produce the desired modulation swing.

More commonly, a single drain supply voltage is used, leading to the need for other than a 1:1 transformer turns ratio. By choosing a suitable voltage step-up ratio, 100% modulation can be achieved.

It is important to recognize here that the modulator is itself a power amplifier. As such, these high-level modulators suffer from essentially the same trade-offs between efficiency and linearity as do the RF stages they modulate. Without care, the power dissipated by the modulator could exceed that of the main RF power amplifier. In a popular variation intended to address this problem, M2 is replaced by a push–pull Class B stage for improved efficiency. In that case, the output drains of the Class B stage connect to a transformer's primary, whose center tap provides the connection point for the DC supply. An even higher-efficiency alternative is to generate the voltage Vx with a stage operating as a switch-mode (e.g., Class D) amplifier. A challenge is to filter the switching noise sufficiently to meet stringent spectral purity requirements, but the high potential efficiency often justifies the engineering effort. Delta-sigma modulation is occasionally used in switching modulators to shape the noise spectrum in a way that eases filtering requirements.

These few examples show that there are many ways to effect drain (high-level) modulation.[19] However, even though drain modulation permits the nominally linear modulation of nonlinear amplifiers, the degree of linearity may still be insufficient

---

[19] Although much of the literature makes no distinction between drain and Heising modulation, we point out that the latter is a subset of the former.

to satisfy stringent requirements on spectral purity. That shortcoming motivates consideration of a variety of enhancements and alternatives.

20.3.2 LINEARIZATION TECHNIQUES

Envelope Feedback

It's probably a bit generous to refer to power backoff and drain modulation as linearization techniques. Backoff pays for linearity with efficiency, and efficiency is too precious a currency to squander. Drain modulation, although superior to gate modulation for certain topologies, still ultimately relies on open-loop characteristics and so the distortion is not under the direct control of the designer. In this subsection, we consider a number of ways to improve the linearity of RF power amplifiers at a minimal cost in efficiency.

When faced with the general problem of linearizing an amplifier, negative feedback naturally comes to mind. Closing a classic feedback loop around an RF power amplifier is fraught with peril, however. If you use resistive feedback, the dissipation in the feedback network can actually be large enough in high-power amplifiers to present a thermal problem, to say nothing of the drop in efficiency that always attends dissipation. Reactive feedback doesn't have this problem, but then one must take care to avoid spurious resonances that such reactances may produce. Next, there is the matter of loop transmission magnitude sufficiency, an issue that applies to all amplifiers. It can be shown that nonlinearities are suppressed by a factor equal to the magnitude of the loop transmission (actually, the return difference, but for large loop transmission magnitudes these quantities are approximately equal), at the cost of an equal reduction in closed-loop gain. A factor-of-10 reduction in closed-loop gain will accompany a factor-of-10 improvement in third-order IM distortion (normalized to the fundamental). One therefore needs an ample supply of excess gain to enable large improvements in linearity. At radio frequencies, it is regrettably often true that available gain is difficult enough to come by on an open-loop basis. Consequently it may be hard to obtain significant improvements in linearity without reducing the closed-loop gain to the point that one wins only by losing.

Attempts at other than a Pyrrhic victory by simply cascading a number of gain stages cause the appearance of the classic stability problem. This problem increases in severity as we seek larger bandwidths owing to the greater likelihood that parasitic poles will fall in band and degrade stability margins.

The astute reader will note that the linearization need only be effective over a bandwidth equal to that of the modulation, and that this bandwidth need not be centered about the carrier. As a specific exploitation of this observation, suppose that we feed back a signal corresponding to the envelope of the output signal (with a demodulator, for example, which can be as crude as a diode envelope detector in noncritical applications) and then use this demodulated signal to close the loop.[20]

---

[20] See e.g. F. E. Terman and R. R. Buss, "Some Notes on Linear and Grid-Modulated Radio Frequency Amplifiers," *Proc. IRE,* v. 29, 1941, pp. 104–7.

Closing the loop at baseband frequencies is potentially advantageous because it is then considerably easier to obtain the requisite excess loop gain over the bandwidth of interest. Still, meeting all of the relevant requirements is not necessarily trivial, particularly if one seeks large improvements in linearity over a large bandwidth. A brief numerical example should suffice to highlight the relevant issues.

Suppose that we want to reduce distortion by 40 dB over a bandwidth of 1 MHz. Then we must have 40 dB of excess gain at 1 MHz. If the feedback loop is well modeled as single pole, then the corresponding loop crossover frequency will be 100 MHz, implying the need for a stable closed-loop bandwidth of 100 MHz as well. Assuring that the loop indeed behaves as a single-pole system over this bandwidth is not impossible, but neither is it trivial.[21] From the numbers, it's readily apparent that the difficulty increases rapidly if one seeks greater linearity improvements over a broader bandwidth.

Even if one needs only relatively modest improvements in amplitude linearity, constraints on the phase performance (referenced to baseband) could still present design difficulties. Recall that the phase lag of a single-pole system is 45° at the −3-dB frequency. If there is a tight specification on the permissible phase shift over the passband (e.g. to constrain group delay variation), then the only recourse for a single-pole system is to increase the bandwidth. If the allowable phase error is 5.7°, then the bandwidth must be chosen a decade above the baseband bandwidth. If that error budget shrinks to 0.57°, then the required bandwidth increases another order of magnitude – to one hundred times the baseband bandwidth.[22]

These calculations all presume optimistically that the only error source is in the forward path; the feedback is assumed perfect in all respects. This requirement translates to the need for an exceptionally linear demodulator over a wide dynamic range, because a negative feedback system is desensitized only to imperfections in the *forward* path. The overall system's performance is limited by the quality of the feedback, so any nonlinearities and phase shifts in the demodulator bound the effectiveness of the loop.

These difficulties are sufficiently daunting that a collection of other techniques have evolved as alternatives or supplements to classical negative feedback. Some of these are purely open-loop techniques and thus are not constrained by stability concerns. Furthermore, the linearization techniques we'll present may be used singly or in combination with other techniques, depending on the particular design objectives.

Feedforward

One open-loop linearization technique, feedforward, was devised by Harold Black (before he invented the negative feedback amplifier); see Figure 20.19.

---

[21] Relaxing the single-pole restriction can help moderate the excess bandwidth requirement but at the risk of potentially creating a conditionally stable feedback system.

[22] One may reduce the demand for excess bandwidths by employing suitable phase-compensating all-pass filters.

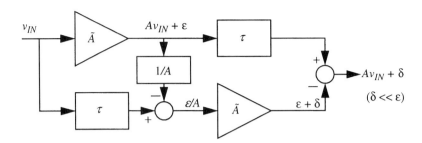

FIGURE 20.19. Feedforward amplifier

We note that the bandwidth over which feedforward provides significant linearity improvements depends in part on the bandwidth over which the group delay of the individual amplifiers may be tracked accurately by realizable time-delay elements.[23] This tracking must remain accurate over time and in the presence of variations in temperature and supply voltage. In many commercial examples, such as some GSM base-station power amplifiers, the delay elements are largely realized with suitable lengths of low-loss coaxial cable. As with most techniques that rely on matching, one might expect improvements of perhaps 30 dB in practice (maybe above 40 dB with great care). In certain cases, it may be possible to implement automated trimming techniques, some of which rely on pilot signals sent through the amplifier. Automatic calibration of this nature can enable feedforward to provide excellent linearity with great consistency. If linearity must be improved further still, one retains the option of combining this technique with others.

Despite the relatively high bandwidth achievable with feedforward, the low efficiency that results from consuming power in two identical amplifiers is certainly a drawback. Although the partial redundancy provided by having two separate gain paths is sometimes a compelling and compensating asset, the efficiency is low enough (typically below 10%) that the general trend is away from feedforward RF power amplifiers for most applications.

Pre- and Postdistortion

Another approach to open-loop linearization exploits the fact that cascading a nonlinear element with its mathematical inverse results in an overall transfer characteristic that is linear. Logically enough, the compensating element is called a *predistorter* if it precedes the nonlinear amplifier and a *postdistorter* if it follows it. Predistortion is by far the more common of the two (because the power levels are lower at the input to the PA proper) and may be applied either at baseband or at RF. Baseband predistortion is extremely popular because the frequencies are lower and because

---

[23] Because real amplifiers generally do not exhibit constant group delay, designing the compensating delay elements to track this real behavior is quite difficult.

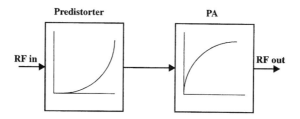

FIGURE 20.20. Illustration of RF predistortion

practical options include both analog and digital techniques (with the latter enjoying increasing popularity owing to digital's characteristic flexibility). Another attribute of baseband predistortion is that it may also correct for nonlinearities suffered during upconversion to RF; see Figure 20.20.

Because the principal nonlinearity in an amplifier is associated with gain compression, predistorters succeed only to the extent to which they are able to undo the compression accurately by providing an increasing gain as the input increases. It's important, however, to keep in mind that a predistorter cannot increase the saturated output power of an amplifier, and consequently we should expect little or no improvement in the 1-dB compression point. Because IP3 for "well-behaved" nonlinearities is at least somewhat related to compression point, it shouldn't surprise you that predistortion rarely succeeds in reducing IM3 products by much more than about a dozen decibels. If much greater reductions are needed, predistortion alone is unlikely to succeed.[24]

Corrections for phase errors (including those that may be the result of AM-to-PM conversion) may be provided by a phase shifter placed in series with the input. Most amplifiers tend to exhibit larger phase lag for small amplitude inputs, so the control for the phase shifter must be arranged to provide a compensating phase shift. Constraints here are usually less severe than for amplitude correction, but devising an analog control circuit is complex enough that digital control has become popular. Then, once you go to the trouble of building a digital controller, you might as well use it to control both the gain and phase correctors. It's just a short hop from there to closing a true feedback loop around both the amplitude and phase paths, at which point you've implemented *polar feedback*, a topic about which we'll have more to say shortly.

To achieve even the modest dozen-decibel improvement just alluded to – and regardless of whether the predistortion is implemented as a purely analog circuit or as a digitally controlled element – one must solve the problem of accurately producing the desired inverse transfer characteristic and subsequently assuring that this inverse

---

[24] Occasionally, claims of much large values are encountered in some of the literature. Upon close examination, however, it turns out that many of these claims are for systems that combine backoff (or some other method) with predistortion, whether or not backoff is explicitly acknowledged.

remains correct over time (and with potentially varying load) in the face of the usual variations in process, voltage, and temperature.[25]

Because a fixed predistorter may prove inadequate to accommodate such drifts, it is natural to consider adaptive predistortion as an alternative. Such a predistorter uses real-time measurements of voltage and temperature, for example, in computing and updating the inverse function periodically. Successful implementation therefore requires a model for the system as well as sensors to measure the relevant input variables. Sadly, system modeling is quite a difficult task, particularly if some of the important variables (such as output load, which can vary wildly in portable applications) cannot be measured conveniently (if at all). Compounding the difficulty is that the nonlinearity may be hysteretic (have memory) owing to energy storage. In such cases, the present value of the output is a function not solely of the input but also of the past values of the input. These limitations do not imply that predistortion is valueless (quite the contrary, in fact, as many broadcast television transmitters rely on this technique) but they do explain why it's difficult for predistortion to provide large linearity improvements on a sustained basis. As with the other techniques, predistortion may be applied in combination with other methods to achieve overall linearity objectives.

Envelope Elimination and Restoration

Originally developed by Leonard Kahn to improve single-sideband (SSB) transmission systems, envelope elimination and restoration (EER) is not a linearization technique per se but rather a system for enabling linear amplification from nonlinear (constant-envelope) amplifiers through drain modulation.[26] In EER, a modulated RF signal to be linearly amplified is split into two paths; see Figure 20.21. One feeds a *limiting amplifier* (essentially a comparator) to produce a constant-envelope RF signal that is subsequently amplified at high efficiency by a constant-envelope (e.g., Class C) amplifier; the other path feeds an envelope detector (demodulator). The extracted modulation is then reapplied to the constant-envelope amplifier using drain modulation. Because EER is not itself a linearization method (it's better to regard it as an efficiency-boosting technique), achievement of acceptable spectral purity may require the supplementing of EER with true linearization techniques.[27]

Now, it is difficult to build an ideal element of any kind, particularly at RF. Consequently it is worthwhile examining what we actually require of the limiter.

---

[25] Predistortion as a method for providing overall linear operation should be familiar to you. A current mirror, for example, actually relies on a pair of inverse, nonlinear conversions (first from current to voltage, then back to current again) to provide truly linear behavior in the current domain. A more sophisticated example that relies on the same basic cascade of nonlinear transductions (I to V, then back again) is a true Gilbert gain cell.

[26] L. R. Kahn, "Single Sideband Transmissions by Envelope Elimination and Restoration," *Proc. IRE,* v. 40, 1952, pp. 803–6. It really is *Kahn* and not *Khan,* by the way; the latter was Capt. Kirk's nemesis.

[27] D. Su and W. McFarland, "An IC for Linearizing RF Power Amplifiers Using Envelope Elimination and Restoration," *IEEE J. Solid-State Circuits,* v. 33, December 1998, pp. 2252–8.

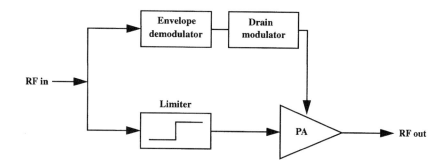

FIGURE 20.21. Kahn EER system

The role of the limiter in Kahn's EER system is simply to provide adequate drive to the PA stage to assure efficient operation. It so happens, however, that it may actually be advantageous for the input to the PA to follow the envelope of the RF input (at least coarsely), rather than remaining fixed, in order to avoid unnecessarily large (and wasteful) PA drives when the envelope is small. Design of a practical limiter may be considerably easier as a result, because the problem essentially reduces to one of building an amplifier instead of a hard limiter – but without much concern for amplitude linearity.[28] Depending on the characteristics of a particular PA, the limiter may have to provide relatively high gain when the input amplitude is low in order to guarantee that the PA stage is always driven hard enough to assure high efficiency and keep the noise floor low.[29] It may also be necessary to insert a compensating delay (generally in the RF path) to assure that the drain modulation is properly time-aligned with the PA drive. Failure of alignment may affect the ability of EER to function well with low power inputs, thereby resulting in a reduction in the usable dynamic range of output power. Still, it is challenging in practice to achieve dynamic range values much larger than about 30 dB.

Chireix Outphasing (RCA *Ampliphase*) and LINC

A general term for techniques that may obtain linear modulation by combining the outputs of nonlinear amplifiers has come to be called LINC (*li*near amplification with *n*onlinear *c*omponents).[30] The first expression of a LINC idea in the literature is *outphasing modulation,* developed by Henri Chireix (pronounced a bit like "she wrecks") around 1935. Outphasing produces amplitude modulation through the vector addition

---

[28] We must still be conscious of careful design to avoid AM-to-PM conversion in those communications systems where such conversion may be objectionable. This comment applies to the entire amplifier and thus holds regardless of whether or not one seeks to implement a classic filter.

[29] F. Raab, "Drive Modulation in Kahn-Technique Transmitters," *IEEE MTT-S Digest,* v. 2, June 1999, pp. 811–14.

[30] D. C. Cox, "Linear Amplification with Nonlinear Components," *IEEE Trans. Commun.,* December 1974, pp. 1942–5.

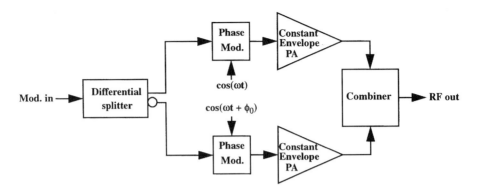

FIGURE 20.22. Block diagram of outphasing modulator

of two constant-amplitude signals of differing phase.[31] The constant-amplitude characteristic allows the use of highly efficient constant-envelope RF amplifiers, while vector addition obviates the need for drain modulation and thus avoids its associated dissipation.

Outphasing modulation enjoyed intermittent and modest commercial success in the two decades following its invention, but its popularity soared when RCA chose this technology for use in their famous line of broadcast AM transmitters, starting with the 50-kW BTA-50G in 1956.[32] The *Ampliphase*, as RCA's marketing literature called it, would dominate broadcast AM radio transmitter technology for the next fifteen years.

To implement the outphasing method, first perform a single-ended to differential conversion of a baseband signal (if it isn't already available in differential form), and then use the outputs to phase-modulate a pair of isochronous – but not synchronous – RF carriers. Amplify the phase-modulated RF signals with highly efficient constant-envelope amplifiers; then sum the two amplified phase-modulated signals together using a simple passive network. The amplitude-modulated RF signal appears across the output of the combiner, as seen in Figure 20.22.

In typical implementations, the quiescent phase shift ϕ_0 between the two amplifier outputs is chosen equal to 135°. The two signal paths are designed to produce maximum phase deviations of 45° and −45° each, so that the total phase difference between the two inputs to the combiner swings between 90° and 180°. When the phase difference is the former, the two signals add to produce the maximum output. When the phase difference is 180°, the two signals cancel, producing zero output. These two extremes correspond to the peaks of a modulated signal with 100% depth.

---

[31] H. Chireix, "High-Power Outphasing Modulation," *Proc. IRE*, v. 23, 1935, pp. 1370–92.

[32] D. R. Musson, "Ampliphase ... for Economical Super-Power AM Transmitters," *Broadcast News*, v. 119, February 1964, pp. 24–9. Also see *Broadcast News*, v. 111, 1961, pp. 36–9. Outphasing boosts efficiency enough to enable the construction of practical transmitters with at least 250 kW of output power.

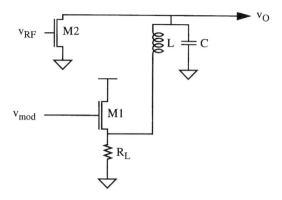

FIGURE 20.23. Ampliphase phase modulator (simplified CMOS version; bias details omitted)

Much of the design effort in an outphasing system concerns (a) obtaining linear phase modulation and (b) realizing a combiner that has low loss yet prevents the pulling of one amplifier by the other from degrading hard-won efficiency, linearity, and stability. In the *Ampliphase* system, the phase modulator exploits the change in phase obtained from varying the Q of a tank, whose center frequency is offset from the carrier frequency by some amount; see Figure 20.23.

The output resistance of transistor M1 acts as a variable resistor, whose value varies with the modulation. As the modulation voltage goes up, M1's output resistance goes down and so increases the Q of the output tank. Transistor M2 is simply a transconductor, converting the RF voltage into a current. The phase angle of the output voltage relative to that of the RF current in the drain of M1 is thus the same (within a sign here or there) as the phase angle of the tank impedance. That angle in turn is a function of the tank's Q and is therefore a function of the modulation. Note that a linear dependence of phase shift on modulation voltage is hardly guaranteed with this circuit. Predistortion is used to obtain nominally linear modulation.

Another source of design difficulty is the output power combining network. In the *Ampliphase* transmitter, the combiner is basically a pair of CLC π-networks (necessary for impedance transformation anyway) whose outputs are tied together (each π-network acts as a lumped approximation to a quarter-wave line). This is shown in Figure 20.24.

This combiner appears simple, but looks are deceiving.[33] Although shown to be workable in a highly successful commercial design, it also illustrates a basic problem with outphasing. The effective impedance seen by the drain of each transistor depends not only on the load connected to the final output but also on the relative phase of the signal at the other transistor's drain. To avoid having to accommodate

---

[33] The appearance of simplicity is enhanced if we replace the π-networks with suitable pieces of transmission line, although the variable compensating capacitances would still be required.

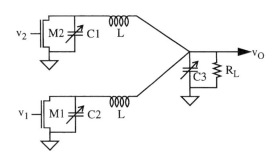

FIGURE 20.24. *Ampliphase* output combiner (simplified CMOS version; bias details omitted)

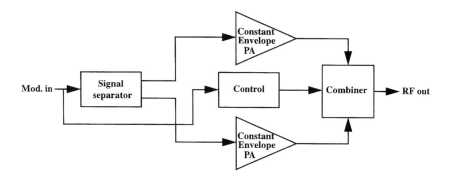

FIGURE 20.25. Block diagram of generalized LINC system

wild variations in drain load impedance, capacitors C1 and C2 must vary (in opposite directions) as a function of instantaneous phase angle of the drive. Needless to say, this requirement for a linear, controllable capacitance only serves to increase the level of design difficulty. In fact, the quest for a practical, low-loss, linear combiner that also provides a high degree of isolation remains unfulfilled. It is principally for this reason that LINC does not dominate today, despite its architectural appeal.

The difficulties of obtaining linear phase modulation and arranging for the correct quiescent phase – to say nothing of maintaining proper operation over time, temperature, and supply voltage – are great enough that broadcast engineers, with a mixture of affection and derision, occasionally took to referring to these transmitters as *Amplifuzz*. By the mid-1970s, the *Ampliphase* line had been, well, phased out.[34]

Since that time, engineers have hardly given up on LINC. The availability of sophisticated digital signal processing capability has inspired many to apply that computational power to overcome some of LINC's impairments. A very general block diagram for the resulting LINC appears as Figure 20.25. Regrettably, signal processing gives us only some of what is needed. Design of the combiner in particular remains, for the most part, an ongoing exercise in futility.

---

[34] Sorry.

20.3 MODULATION OF POWER AMPLIFIERS

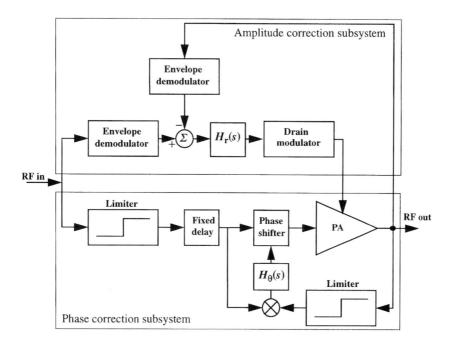

FIGURE 20.26. Example of power amplifier linearized with polar feedback

Polar Feedback

Because of the general need to correct both phase and amplitude nonlinearities in any signal path, it seems logical to employ a feedback loop around each separately, as we hinted in the discussion of predistortion. The *polar feedback* loop directly implements this idea.[35] Polar feedback is often used in tandem with EER (for the amplitude component), supplemented by a phase detector and phase shifter. See Figure 20.26.

The two control loops are readily identifiable in the figure. The amplitude control loop compares the envelope of the output with that of the input, and the difference drives a drain modulator, just as in EER. The gain function $H_r(s)$ shapes control over the dynamics of the amplitude feedback loop. For example, if it contains an integration, the steady-state amplitude error can be driven to zero. Because the bulk of the loop transmission gain may be obtained at baseband through gain block $H_r(s)$ (rather than at RF), it is possible in principle to suppress nonlinearities by large factors.

The phase control loop examines the phase difference between amplitude-limited versions of the input and output signals. A limiter is a practical necessity because most phase detectors are sensitive to both amplitude and phase (or otherwise require some minimum amplitude in order to function properly). Their presence is convenient as well, considering that the basic architecture for EER requires one anyway. The phase error signal drives a phase shifter placed in series with the input to the PA

---

[35] V. Petrovic and W. Goslin, "Polar Loop Transmitter," *Electronics Letters,* v. 15, 1979, pp. 1706–12.

stage and adjusts it accordingly, with dynamics again controlled by a gain block, this time of transfer function $H_\theta(s)$.

One important consideration is to assure that the amplitude and phase corrections line up properly in time. Because the phase and amplitude control subsystems are generally realized with rather different elements, however, there is no guarantee that their delays (or any other relevant characteristics) will match. For example, the bandwidth of the amplitude control loop is a function of the drain modulator's bandwidth. As mentioned before, a switch-mode modulator is often used to keep efficiency high. High bandwidth in turn demands exceptionally high switching frequencies in the modulation amplifier. As a result, bandwidths much in excess of 1 MHz are difficult with current technology.

The phase shift loop typically suffers from far fewer constraints and is consequently much higher in bandwidth, with a correspondingly smaller delay. It's therefore usually necessary to insert a fixed compensating delay to assure that the delays through the two control paths match. (In principle the phase shifter could bear the burden of delay compensation, but making it do so complicates unnecessarily the design of the shifter in most cases.)

Because proper operation of polar feedback requires matching the delays of two very different types of control loops, it is decidedly challenging to obtain high performance consistently. Another complication arises from ever-present AM-to-PM conversion, which couples the two loops in ways that degrade stability. The stability challenge is compounded by the amplitude dependency of AM-to-PM conversion. Polar feedback remains a topic of active research, but thus far the difficulties of achieving the necessary levels of matching (to say nothing of maintaining same over time in the face of variations in supply, process, and temperature), as well as assuring stability over the full bandwidth and dynamic range of inputs, have proven large enough to prevent large-scale commercialization at the frequencies and bandwidths of greatest interest for mobile communications.

Observe that one important idea behind polar feedback is that of decomposing an RF signal into two orthogonal components and then closing a feedback loop around each separately. Given that the polar variables of magnitude and phase represent only one possible choice, perhaps it is a worthwhile exercise to consider another.

Cartesian Feedback

Polar- and rectangular-coordinate representations of a signal are equivalent, so instead of a decomposition into magnitude and phase, we can decompose a signal into in-phase *I* and quadrature *Q* components, for example. This rectangular (*Cartesian*) representation has favorable practical implications, so Cartesian feedback has received considerable attention.[36] The block diagram of Figure 20.27 shows that,

---

[36] V. Petrovic and C. N. Smith, "The Design of VHF Polar Loop Transmitters," *IEE Comms. 82 Conference,* 1982, pp. 148–55. Also see D. Cox, "Linear Amplification by Sampling Techniques: A New Application for Delta Coders," *IEEE Trans. Commun.,* August 1975, pp. 793–8.

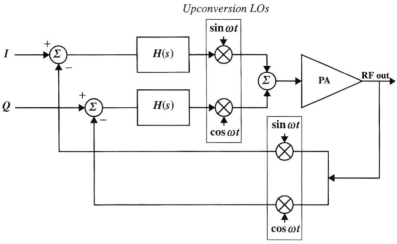

FIGURE 20.27. Transmitter linearized with Cartesian feedback

in contrast with a polar loop, a Cartesian feedback loop consists of two electrically identical paths.

Here, the output undergoes a pair of orthogonal downconversions. Baseband symbols I and Q are compared with their corresponding demodulated counterparts.[37] The baseband error signals are computed separately, amplified, upconverted back to RF, and finally summed at the input to the PA stage. Most of the loop gain is obtained at baseband from $H(s)$, rather than at RF, greatly easing loop design.

The fact that both feedback paths are identical in architecture means that Cartesian feedback is free of the matching problem that vexes polar feedback. However, there remain difficult design problems that, once again, have inhibited widespread use of the architecture.

The most significant problem arises from a lack of strict orthogonality between the two loops. Only if orthogonality holds will the two loops act independently and allow design to proceed with relatively few concerns. If the two loops are coupled, the dynamics may change in complex (Murphy-degraded) ways. Worse, the amount by which the loops are not orthogonal may change with time, temperature, and voltage – and also as the RF carrier is tuned over some range. This problem is hardly unique to Cartesian feedback; it's potentially a concern in any system that possesses multiple feedback paths (such as the polar feedback loop).

To evaluate the system-level consequences of this problem, consider a phase misalignment of ϕ between the upconversion and downconversion LOs (we still assume that each pair of LOs consists of orthogonal signals). As with any feedback loop, we

---

[37] We assume that we have either performed a quadrature downconversion in order to obtain the I and Q signals, or that we are in fact performing an upconversion of baseband symbols that we have generated digitally. Thus the figure describes a transmitter or an amplifier.

may evaluate the loop transmission by breaking the loop, injecting a test signal, and observing what comes back. Doing so, we obtain

$$L_{eff}(s,\phi) = L_{one}(s)\cos\phi + \frac{[L_{one}(s)\sin\phi]^2}{1 + L_{one}(s)\cos\phi}, \qquad (41)$$

where $L_{one}(s)$ is the transmission around each individual loop.[38]

The effective loop transmission expression helps us understand why Cartesian feedback loops can exhibit "odd behaviors." Depending on the amount of phase misalignment, the overall loop transmission can range from that of a single loop (when the misalignment is zero) all the way to a *cascade of two* single loops (when the misalignment is $\pi/2$). As is true for many control loops, $H(s)$ would be designed to contain an integration (to drive steady-state error to zero, for example). If the misalignment remains zero, that choice presents no problem. However, as the misalignment grows, $H(s)$ now contributes two integration poles to the loop transmission, leading to zero phase margin at best. Any negative phase shift from other sources drives phase margin to negative values.

Identification of this mechanism as one significant source of problems with Cartesian feedback is relatively recent. Solutions include automatic phase alignment (to eliminate the source of the stability problem) and carefully crafting $H(s)$ to tolerate the wide variation in loop dynamics as the misalignment varies.[39] Slow rolloff compensation is one possible choice for $H(s)$ if the latter strategy is pursued, which may also be used in tandem with the former. Implementation of these corrective measures enables Cartesian feedback to provide exceptionally large improvements in linearity over extremely wide bandwidths.

20.3.3 EFFICIENCY-BOOSTING TECHNIQUES

Having presented a number of linearization methods, we now focus attention on efficiency-boosting techniques and architectures.

Adaptive Bias

The efficiency of any amplifier with nonzero DC bias current degrades as the RF input power decreases (the Class A is worse than others in this regard). There are many RF PAs, such as those in cell phones, that operate at less than maximum output power a considerable fraction of the time and thus for which the average efficiency is terrible. To improve efficiency at lower power levels, a time-honored technique is to employ *adaptive bias* strategies.[40] Varying the bias current and supply voltage dynamically in accordance with the instantaneous demands on the amplifier can moderate considerably the degradation in efficiency (at least in principle). At high modulation values,

---

[38] J. Dawson and T. Lee, "Automatic Phase Alignment for a Fully Integrated CMOS Cartesian Feedback PA System," *ISSCC Digest of Technical Papers,* February 2003.
[39] Ibid.
[40] F. E. Terman and F. A. Everest, "Dynamic Grid Bias Modulation," *Radio,* July 1936, p. 22.

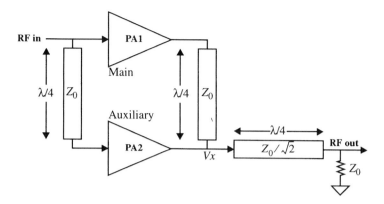

FIGURE 20.28. Doherty amplifier

the PA stage would operate with a relatively high supply voltage (with correspondingly higher gate bias, for example). At low modulations, the drain supply and gate bias voltages would drop as well. This strategy makes the efficiency a much weaker function of signal amplitude, as desired. Thanks to the advent of flexible, inexpensive digital control circuitry, it is now considerably easier to implement adaptive bias than it once was.

The controllable drain supply is essentially identical to the drain modulation amplifier in an EER system, with all of the same design challenges. An additional challenge of adaptive biasing overall is that varying so many significant parameters more or less simultaneously is hardly a prescription for linear behavior. Nevertheless, adaptive bias presents an additional degree of freedom to be exercised in the never-ending series of trade-offs between efficiency and linearity.

The Doherty and Terman–Woodyard Composite Amplifiers

Another efficiency-boosting strategy is to use multiple amplifiers, each responsible for amplification over some subset of the overall power range. By using only the minimum number of amplifiers necessary to provide the desired output power, it's possible to reduce unnecessary dissipation. In effect, we implement the electronic equivalent of a turbocharger. The earliest realization of this idea uses two amplifiers and is due to Doherty;[41] see Figure 20.28.

Amplifiers PA1 and PA2 are the main and auxiliary amplifiers, respectively. The auxiliary amplifier is arranged to be cut off for low amplitude inputs. Assuming that PA2's output is an open circuit in this mode, it is straightforward to deduce that the impedance seen by the output of PA1 is then $2Z_0$.[42]

---

[41] W. H. Doherty, "A New High-Efficiency Power Amplifier for Modulated Waves," *Proc. IRE*, v. 24, 1936, pp. 1163–82.

[42] The reader is reminded that the quarter-wave lines may be approximated by CLC π-networks when appropriate or convenient to do so. See Section 2.9.2.

At some predetermined threshold, the auxiliary amplifier turns on and begins to contribute its power to the output. The $\lambda/4$ delay in feeding PA2 matches the $\lambda/4$ delay coupling the output of PA1 to the output of PA2. The contribution of PA2 to Vx is thus in phase with that of PA1. The fact that Vx is larger when PA2 is active implies an increase in the impedance seen by the main amplifier's output delay line; PA2 bootstraps PA1. When reflected back through the $\lambda/4$ line to PA2's output, this increased impedance at Vx is seen as a reduction in impedance. In turn, this reduction in PA2's load resistance increases the power supplied by that amplifier. When both amplifiers are contributing their maximum power, each amplifier sees a load impedance of Z_0 and contributes equally to the output.

After some reflection, one recognizes that this composite amplifier shares a key characteristic with the push–pull Class B amplifier: half of the power is handled by half of the circuit. In fact, the limiting peak efficiencies are identical, at about 78%, when the two amplifiers are partitioned as in Doherty's original design. Average efficiency is theoretically somewhat less than half this value.

The question of how best to select the threshold at which the auxiliary amplifier begins to contribute to the output is answered in large part by examining the envelope probability density function (PDF) of the signals to be processed by the amplifier. This consideration is particularly important in view of the trend toward complex modulation methods. For example, a 16-QAM signal theoretically exhibits a 17-dB peak-to-average ratio. A 16-QAM signal with a 16-dBm (40-mW) average output power thus may have occasional peaks as high as 33 dBm (2 W). Designing a conventional 2-W amplifier and operating it with 40 mW of average power virtually assures terrible average efficiency. A Doherty-like technique would seem well suited to accommodate modulations with such characteristics, since a highly efficient, low-power main amplifier would be bearing the burden most of the time – with the auxiliary amplifier activated only intermittently to handle the relatively rare high-power peaks. We see that a PDF heavily weighted toward lower powers implies that we should lower the threshold. If we were using some other modulation whose PDF were heavily weighted toward higher power, then we would wish to raise the threshold. Implementing arbitrary power division ratios may be accomplished a number of ways, including adjustment of the coupling impedances and operating the two amplifiers with different supply voltages.[43]

Further improvements in efficiency are possible by subdivision into more than two power ranges. Although the complexity of the load structure that effects the power combining rapidly increases in magnitude, the theoretical boosts in efficiency can be substantial. A doubling in average efficiency is not out of the question, for example.[44]

The similarity with Class B doesn't end with efficiency calculations, unfortunately. A problem akin to crossover distortion afflicts the Doherty amplifier as well. It is

---

[43] M. Iwamoto et al., "An Extended Doherty Amplifier with High Efficiency over a Wide Power Range," *IEEE MTT-S Digest*, May 2001, pp. 931–4.

[44] F. H. Raab, "Efficiency of Doherty RF Power Amplifier Systems," *IEEE Trans. Broadcast.*, v. 33, September 1987, pp. 77–83.

20.3 MODULATION OF POWER AMPLIFIERS

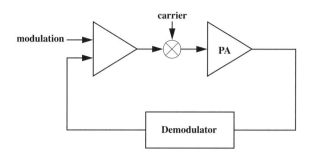

FIGURE 20.29. Negative feedback for improving modulation linearity

perhaps not surprising that efficiency is obtained at the expense of distortion, and one must expend a great deal of engineering effort to suppress nonlinearities (e.g., by embedding a Doherty amplifier within a feedback loop) in both amplitude and phase domains. And as with the outphasing system, the impedance seen at the output of one amplifier is a function of the other amplifier's output. Hence there is ample opportunity for a host of misbehaviors arising from unanticipated interactions.

An extension of the Doherty amplifier is the modulator–amplifier combination of Terman and Woodyard.[45] It is similar to the Doherty amplifier in its use of two amplifiers (driven by $\lambda/4$-delayed RF carrier signals) and of the same output combiner. The difference lies in the modulation capability, which is provided by injecting modulation in phase to the gate circuit of both amplifiers simultaneously. Because modulation is thus the result of nonlinearities inherent in the device transfer characteristic, less than faithful modulation results. However, by wrapping a feedback loop around the envelope signal, for example, large improvements in linearity may be obtained at low cost in efficiency.

Finally, with all of these methods, negative feedback may be employed to reduce distortion. In order to relax the requirements on gain–bandwidth product for the feedback loop, one may sample the output signal, demodulate it, and then use the demodulated signal to close the loop. See Figure 20.29. Such architectures are often distinguished by the demodulation method used. For example, if the demodulator consists of a pair of mixers driven with carriers in quadrature, then the PA is sometimes said to be linearized through *Cartesian feedback*.

20.3.4 PULSEWIDTH MODULATION

Another technique for obtaining nearly linear modulation is through the use of pulsewidth modulation (PWM). Amplifiers using this technique are occasionally known as Class S amplifiers, although this terminology is by no means universal.

---

[45] F. E. Terman and J. R. Woodyard, "A High-Efficiency Grid-Modulated Amplifier," *Proc. IRE,* v. 26, 1938, pp. 929–45.

Such amplifiers do not perform modulation through variation of drive amplitude. Rather, it is accomplished by controlling the duty cycle of constant-amplitude drive pulses. The pulses are filtered so that the output power is proportional to the input duty cycle, and the goal of linear operation at high efficiency is achieved in principle.

While PWM works well at relatively low frequencies (e.g., for switching power converters up to the low-MHz range), it is fairly useless at the gigahertz carrier frequencies of cellular telephones. The reason is not terribly profound. Consider, for example, the problem of achieving modulation over a 10:1 range at a carrier of 1 GHz. With a half-period of 500 ps, modulation to 10% of the maximum value requires the generation of 50-ps pulses. Even if we were able to generate such narrow pulses (*very* difficult), it is unlikely that the switch would actually turn on completely, leading to large dissipation. Therefore, operation of PWM amplifiers over a large dynamic range of output power is essentially hopeless at high frequencies. Stated another way, the switch (and its drive circuitry) has to be n times faster than in a non-PWM amplifier, where n is the desired dynamic range. As a result, it becomes increasingly difficult to use pulsewidth modulation once carrier frequencies exceed roughly 10 MHz.

20.3.5 OTHER TECHNIQUES

Gain or Power Boost by Cascading

Cascading is so obvious a method for increasing gain and power levels that even mentioning it invites scorn. However, there are enough subtleties in cascading PA stages that it's worth the risk of providing a brief discussion.

Power levels generally scale upward as we proceed from stage to stage in a cascade of amplifiers. If the power consumed by the early stages is low enough, it may be prudent to design those stages with a focus on linearity, deferring to a later stage (or stages) the problem of obtaining high efficiency. In practice, then, the earliest stages may be implemented as Class A amplifiers, with a transition to (say) Class B or C for the last stage, for example.

When using drain modulation with a cascade of stages (e.g., Class C), the level of drain modulation should scale with the power level so that the drive for each stage in the cascade also scales. Without this scaling, there is a risk of overdriving one or more stages, leading to overload-related effects such as slow recovery from peaks, excessive AM-to-PM conversion, and poor linearity.

Finally, cascading always involves a risk of instability, particularly when the stages are tuned. The risk is greatest with Class A amplifiers and is mitigated by using the same general collection of techniques that are effective for low-level amplifiers. In stubborn cases, the *losser* method offers relief when all others fail. As its name implies, the method (if one could so dignify it) employs a resistor placed somewhere in the circuit to throw away gain and Q. Resistors in series with the gate, or across the gate–source terminals, are particularly effective.

Gain Boost by Injection Locking

We alluded to injection locking in Section 16.2 and offer only a brief description here. You should simply accept that it's often possible to lock an oscillator's phase to that of a signal injected into an appropriate point in an oscillator circuit, provided certain conditions are met (e.g. frequency of injected signal close enough to the free-running oscillator frequency, amplitude of injected signal within a certain window, etc.). In fact, unwanted injection locking (perhaps caused by signals coupled through the substrate) is a very real problem in RF integrated circuits. As with other parasitic phenomena, virtually every effect that is unwanted in one context can be turned into an advantage in another.

Whereas cascading is an obvious means to increase gain, the relatively low inherent gain of CMOS devices generally implies the necessity for more stages than would be the case in other technologies, such as bipolar. Aside from the increased complexity, it's quite likely that the power consumed would be greater as well. To evade these limits, it's sometimes advantageous to consider building an *oscillator* and somehow influencing its phase or frequency. After all, an oscillator provides an RF output signal with *no* RF input; the "gain" is infinite. *It's easier to influence a signal than to produce it.* Because the power required to effect locking can be small, the apparent gain can be quite large.

Injection locking as an amplification technique is normally limited to constant-envelope modulations because, as its very name implies, the input signal primarily affects the phase. In principle, amplitude modulation could be provided as well (e.g., by varying the bias that controls the oscillation amplitude), but AM-to-PM conversion is usually so serious that this combination is practical only for shallow amplitude modulation depths.

An alternative possibility would be to combine injection locking with outphasing to produce amplitude modulation. Such a combination would theoretically exhibit high gain as well as high efficiency and, as a free bonus, present many subtle design problems. Verifying this statement experimentally is left as an exercise for the reader.

Power Boost by Combining

Power combiners may work instead of – or in concert with – impedance transformers to allow the attainment of higher output power than would otherwise be practical. One popular combiner is the Wilkinson combiner introduced in Chapter 7. The Wilkinson is attractive because it theoretically allows lossless power combining when operating into a matched load;[46] see Figure 20.30. Hence, the Wilkinson combiner enables an ensemble of low-power amplifiers to provide high output power.

If larger than 2:1 boosts in output power are needed, several stages of combining may be used in a structure known as a *corporate combiner*. This is shown in the

---

[46] E. J. Wilkinson, "*N*-Way Hybrid Power Combiner," *IRE Transactions MTT*, 1960. Unequal power splitting factors are also possible with an asymmetrical structure.

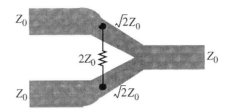

FIGURE 20.30. Wilkinson power combiner

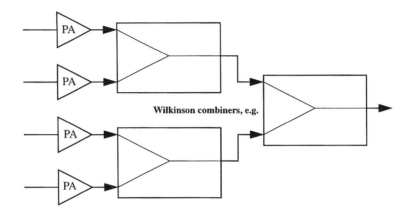

FIGURE 20.31. Corporate power combiner

schematic of Figure 20.31. These techniques routinely succeed in discrete form, but integration presents numerous challenges. First, the transmission line segments of the combiner are a quarter-wavelength long. As a result, the die area consumed by combiners for power amplifiers in the low-gigahertz frequency range would be impractically large. As a specific example, consider that an on-chip quarter-wave line would be about 4 cm long at 1 GHz. Another factor is the loss of on-chip transmission lines. Again, a 1-dB attenuation is a 21% power loss, and it is difficult to keep losses even this low. Furthermore, the Wilkinson combiner won't be lossless if it is imperfectly terminated. Reflections resulting from any mistermination are absorbed by the bridging resistor, with a consequent loss in efficiency.

Another use of multiple stages is to supply inputs in parallel and then take the outputs in series; the voltage boost reduces the need for absurd impedance transformation ratios. One very simple implementation of this idea is to build a differential power amplifier. In the ideal case, the differential output voltage swing is twice that of a single-ended stage, permitting a quadrupling of output power for a given supply voltage as well as a moderation of any additional impedance transformations that might be required. A balun can be used to convert the balanced differential output to a single-ended ground-referenced one.

A particularly elegant structure that extends this idea is the *distributed active transformer* (DAT).[47] The name is perhaps a bit of a misnomer in that this amplifier is not a distributed system in the same way that a transmission line is, for example; no distributed parameters are needed to describe it. Rather, the architecture gets its name from distributing the total power burden among a number of devices, a concrete expression of "divide-and-conquer."

Suppose that we require more than a doubling of voltage swing in order to achieve our power goal (or to moderate any additional impedance transformation ratio). We could quadruple the swing (and boost power by a factor of 16) by summing the contributions of two differential amplifiers. The summation may be accomplished conveniently by driving the primaries of transformers with the differential stages and then simply connecting the secondaries in series. Figure 20.32 (p. 674) illustrates this principle, extended to four differential stages, for a theoretical eightfold boost in voltage swing (relative to that of a single device), corresponding to a power boost of a factor of 64. (For simplicity, capacitances for producing a resonant load are not shown in schematic or layout.)

In the simplified layout shown, each center-tapped drain load is the primary of an on-chip transformer, realized out of coupled lines. The secondary here is a one-turn square inductor, each arm of which is coupled to its corresponding center-tapped primary. Because the four arms are connected in series, the voltage contributions add as desired and so produce a boosted output voltage across the load R_L. The attendant impedance transformation implies that the current flowing in the secondary is smaller than in the primary by a factor of N, permitting the use of narrower lines than in the primary, as suggested by the relative line widths (shaded areas) in the figure.

Generalizing to N differential pairs with N output transformers, we see that the maximum boost in voltage is $2N$ (again, relative to that of a single transistor) with a corresponding power boost of $4N^2$. Losses in real circuits certainly diminish performance below those maximum limits, but the DAT remains a practical alternative nonetheless. As a benchmark, an early realization of this concept delivered 2.2 W of saturated output power with 35% drain efficiency (31% power-added efficiency) at 2.4 GHz, using a 0.35-μm CMOS process (see footnote 47). Gain is around 8.5 dB for input and output impedances of 50 Ω.

20.3.6 PERFORMANCE METRICS

Prior to the advent of complex modulation schemes, it was largely sufficient to frame the design of transmit chains in terms of the specifications we've already presented,

---

[47] I. Aoki et al., "Distributed Active Transformer – A New Power-Combining and Impedance-Transformation Technique," *IEEE Trans. Microwave Theory and Tech.*, v. 50, January 2002, pp. 316–31.

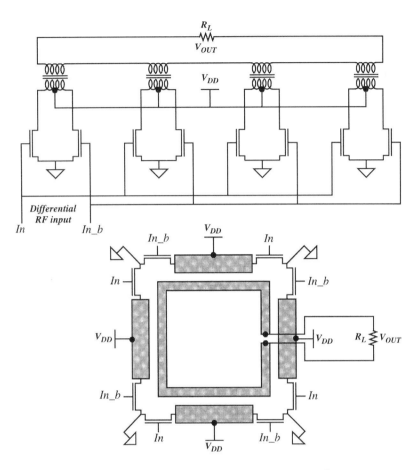

FIGURE 20.32. Illustration of distributed active transformer with simplified layout (after Aoki et al. 2002)

such as saturated output power, third-order intercept, 1-dB compression point, power-added efficiency, and the like. Engineers could rapidly estimate the level of backoff necessary to achieve a certain level of distortion, for example, and that was often enough to construct systems that functioned well. The relatively loose statutory constraints on out-of-band emissions reflected a concession to the crude state of the art. That situation has changed as a result of continuing efforts to reduce waste of precious spectrum through the use of more sophisticated modulation methods. That increased sophistication brings with it an increased sensitivity to certain impairments and thus an obligation to specify and control performance more tightly.

Adjacent channel power ratio (ACPR) is one example of such a performance metric. Developed in response to the difficulty of using a conventional two-tone test to predict adequately the interference generated by a transmitter using complex digital modulations (e.g., CDMA systems), ACPR characterizes interference potential by using representative modulations and then directly measuring the out-of-band power

20.3 MODULATION OF POWER AMPLIFIERS

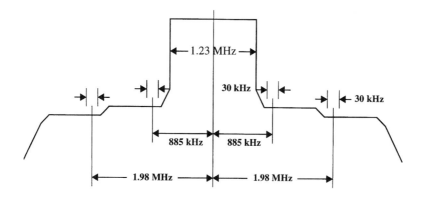

FIGURE 20.33. Example of ACPR specification

at frequency offsets corresponding to the location of adjacent channels.[48] See Figure 20.33.

Even though quantitatively relating ACPR to IP3 may not be possible in all cases, it remains typically true that ACPR improves with backoff in much the same way IP3 does. That is, for every 1-dB power backoff you can expect perhaps 2-dB ACPR improvement, with the out-of-band skirts moving as an ensemble. This statement holds if third-order nonlinearity dominates, as is often (but not always) the case.

Methods for measuring ACPR are not quite standardized, so in order to interpret reported values properly it's important to know the measurement methods to which the numbers correspond. As an example, handset ACPR for IS-95 CDMA needs to be better than −42 dBc when measured at an 885-kHz offset frequency.[49] A subtlety is that some techniques measure the ratio of integrated power densities and others the ratio of the densities themselves. Further differences involve the choice of integration bandwidths. For example, we could integrate the power density at an offset of 885 kHz over the 30-kHz bandwidth indicated in the figure, then divide by the integral of the power density over the 1.23-MHz bandwidth of the central lobe; strictly speaking, it's *that* ratio that needs to be −42 dBc or better.

In other (much more common) measurement methods, the power density is integrated over a 30-kHz bandwidth centered about both measurement frequencies; then a correction factor is applied to extrapolate the measured ratio to correspond to measurements made using the previous method. That is, given certain assumptions, the two raw ratios will differ by a correction factor

---

[48] Analytical approaches relating two-tone measurements to ACPR can still yield useful insights, however. For a comprehensive discussion of this approach, see Q. Wu et al., "Linear and RF Power Amplifier Design for CDMA Signals: A Spectrum Analysis Approach," *Microwave Journal*, December 1998, pp. 22–40.

[49] Strictly speaking, IS-95 defines the air interface, IS-97 specifies performance of the base station, and IS-98 that of the mobile units.

$$\Delta \text{ACPR} = 10 \log \frac{1.23 \text{ MHz}}{30 \text{ KHz}} \approx 16.13 \text{ dB}. \tag{42}$$

Thus, about 16.1 dB needs to be subtracted from the ACPR measured by the second method in order to correspond to the first.[50] This second method assumes that the average power density in the 30-kHz window about the carrier is the same as in the rest of the 1.23-MHz band. It is important in this context to note that an IS-95 signal typically exhibits 2 dB or so of ripple over that band.

Another measurement subtlety concerns the nature of signals used in ACPR evaluations. For CDMA systems, the modulations are "noiselike" in nature. As a result, it's tempting to use suitably band-limited noise as a signal for ACPR tests. However, it's important to understand that *noise* and *noiselike* are two different things, much as it might be important to keep track of the difference between *food* and *foodlike*. ACPR is a measure of distortion and, as such, is sensitive to average power level and envelope details such as peak-to-average ratio. Those, in turn, are functions of the code set used in generating the modulations. Different noiselike waveforms with identical average power can cause a given amplifier to exhibit rather different ACPR. It is not unusual to see values vary over a 15-dB range as a function of stimulus.

Along with new ways to characterize the effects of distortion comes new terminology. The organic-sounding term *spectral regrowth* refers to the broadening in spectrum that results from distortion. Because distortion increases with power level and also as signals progress through the various stages in a transmitter, it is important to accommodate spectral regrowth in allocating a distortion budget for the various elements in the chain. Thus, to meet (say) a -42-dBc ACPR specification for the entire transmitter, it would be prudent to design the PA stage proper to have a worst-case ACPR at least a few (e.g., 3) decibels better than strictly needed to meet the ACPR specifications.

The philosophical underpinnings of ACPR can be summed up as "be a good neighbor." Specifications are chosen with the hope that compliance will assure that one transmitter minimally interferes with unintended receivers within reception range. In general, however, specifying the out-of-band power at a few discrete frequencies may be insufficient. An ACPR specification by itself, for example, does not preclude relatively strong narrowband emissions. In such cases, it may be necessary to specify a *spectral mask* instead. As its name implies, a spectral mask defines a continuum of limits on emission. Three representative examples are shown in Figure 20.34. One is for GSM (from version 05.05 of the standard), another is for indoor ultrawideband (UWB) systems (as defined in the *FCC Report and Order* of February 2002), and the third is for 802.11b wireless LAN.

The UWB mask in particular is notable for its complexity. The notch between 0.96 GHz and 1.61 GHz exists primarily in order to prevent interference with the

---

[50] For more detailed information on measurement methods and interpretation of data, see "Testing CDMA Base Station Amplifiers," Agilent Applications Note AN 1307.

20.3 MODULATION OF POWER AMPLIFIERS

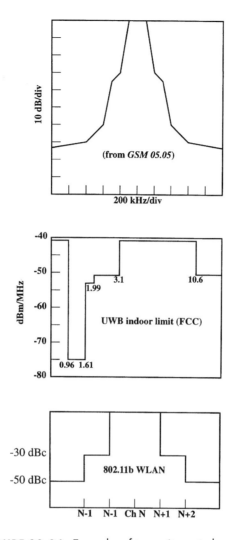

FIGURE 20.34. Examples of transmit spectral masks

global positioning system (GPS). The mask is least constraining from 3.1 GHz to 10.6 GHz and so this slice of spectrum is frequently cited as "the" UWB spectrum allocation. Also notice that the mask specifications are in terms of power spectral density. The absence of a carrier in UWB systems makes the familiar "dBc" and "dBc/Hz" units inapplicable.

Satisfying the "good neighbor" dictum by conforming to such masks is necessary, but not sufficient. We must also ensure that our transmitter produces modulations that *intended* receivers can demodulate successfully. The *error vector magnitude* (EVM) is particularly well suited for quantifying impairments in many digitally modulated transmitters. The error vector concept applies naturally to systems employing vector modulations of some type (e.g., QAM).

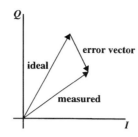

FIGURE 20.35. Illustration of error vector

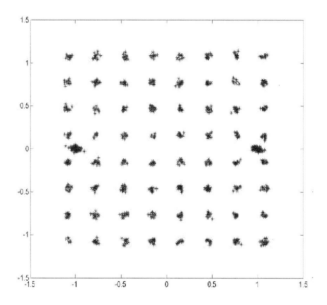

FIGURE 20.36. Measurement of EVM (802.11a example; 64-QAM constellation shown)

The error vector magnitude is simply the length of the error vector, as seen in Figure 20.35. Every symbol or chip has its own error vector. For 802.11b WLAN at 11 Mb/s, EVM is defined as the rms value over 1000 chips and must be less than a very generous 35%. For 802.11a operating at 54 Mb/s (the maximum specified in the standard), the allowable EVM is 5.6% – a considerably tighter specification than for 802.11b.

When measuring EVM, instrumentation produces plots that look like the one shown in Figure 20.36. In the ideal case, the constellation would appear as a perfect 8 × 8 square array of dots; the smudges would be points. The normalized rms smearing out is the EVM. In the particular case shown, the EVM is approximately 2% – well within specifications.

20.4 ADDITIONAL DESIGN CONSIDERATIONS

20.4.1 POWER-ADDED EFFICIENCY

In the foregoing development, collector efficiency is used to characterize the PAs. However, the definition of collector efficiency involves only the RF output power and the DC input power, so it can assign a high efficiency to a PA that has no power gain. Another measure of efficiency has therefore been developed to yield a figure of merit that takes power gain into account. Power-added efficiency (PAE) simply replaces RF output power with the difference between output and input power in the collector efficiency equation:

$$\text{PAE} \equiv \frac{P_{out} - P_{in}}{P_{DC}}. \qquad (43)$$

Clearly, power-added efficiency will always be less than the collector efficiency.

20.4.2 PA INSTABILITY

Amplifiers of any kind can be unstable with certain combinations of load and source impedances, and power amplifiers are no exception. One extremely important problem results from collector-to-base coupling (or drain-to-gate coupling). As noted in Chapter 12, this coupling can cause the input impedance to have a negative real part. In small-signal amplifiers, this problem can be reduced or eliminated entirely by using the various unilateralization techniques described earlier. Unfortunately, these tricks are generally inappropriate for power amplifiers because the requirement for high efficiency precludes the use of any technique (such as cascoding) that diminishes supply headroom. In general, the problem is usually solved through the brute-force means of degrading the input impedance (e.g., through the use of a simple resistor across the input terminals) to make the feedback less significant. Unfortunately, this action has the side effect of reducing gain. In general, MOSFETs – with their larger inherent input impedances – exhibit this stability problem to a greater degree than bipolar devices. In any case, there is usually a significant stability–gain trade-off due to the feedback capacitance. And of course, thoughtful layout is required to avoid augmenting the inherent device feedback capacitance from an unfortunate juxtaposition of input and output wires.

Instability can even occur in switching-type amplifiers. Because all practical drive waveforms have finite slopes, it is possible for the switching devices to dwell in the linear region long enough for oscillations to occur. Even before onset of actual oscillation, the input impedance (for example) may undergo rather dramatic shifts. Design of the previous stage must accommodate this possibility in order to avoid destabilization there.

20.4.3 BREAKDOWN PHENOMENA

MOS Devices

Downward impedance transformations are required to deliver the desired amount of power into the output load in many cases. Clearly, the transformation ratio could be reduced if a higher power supply voltage were permitted, and the reader may reasonably ask why one could not simply demand a higher voltage be made available. The reason is that devices have finite breakdown voltages. Furthermore, as semiconductor technology scales to ever-smaller dimensions to provide ever-faster devices, breakdown voltages tend to diminish as well. Thus, increasing transformation ratios are required as devices scale if one wishes to deliver a certain fixed amount of power to the load.

In MOS devices, one may identify four primary limits to allowable applied voltages in PAs. These are drain (or source) diode zener breakdown, drain–source punchthrough, time-dependent dielectric breakdown (TDDB), and gate oxide rupture.

The drain and source regions are quite heavily doped to reduce their resistivity. As a consequence, the diodes they form with the substrate have a relative low breakdown voltage, with typical values of the order of 10–12 V for 0.5-μm technologies. Drain–source punchthrough is analogous to base punchthrough in bipolar devices and occurs when the drain voltage is high enough to cause the depletion zone around the drain to extend all the way to the source, effectively eliminating the channel. Current flow then ceases to be controlled by the gate voltage.

Time-dependent dielectric breakdown is a consequence of gate oxide damage by energetic carriers. With the high fields typical of modern short-channel devices, it is possible to accelerate carriers (primarily electrons) to energies sufficient for them to cause the formation of traps in the oxide. Any charge that gets trapped there then shifts the device threshold. In NMOS transistors, the threshold increases so that the current obtained for a given gate voltage decreases; in PMOS devices, the opposite happens. TDDB is cumulative, so it places a limitation on device lifetime. Typically, TDDB rules are designed with a goal of no more than 10% degradation in drive current after 10 years.

Bipolar Devices

Bipolar transistors have no gate oxide to rupture, but junction breakdown and base punchthrough impose important limits on allowable supply voltages. The collector–base junction can undergo avalanche breakdown in which fields are sufficiently high to cause significant hole–electron pair generation and multiplication. In well-designed devices, this mechanism imposes the more serious constraint, although the extremely thin bases that are characteristic of high-f_T devices can often cause base punchthrough to be important as well.

Another, more subtle problem that can plague bipolar devices is associated with irreducible terminal inductances that act in concert with large di/dt values. When turning off the device, significant base current can flow in the reverse direction until the base charge is pumped out (see the CD-ROM chapter on the charge control model). When the base charge is gone, the base current abruptly ceases to flow, and the large di/dt can cause large reverse voltage spikes across base to emitter. Recall that the base–emitter junction has a relatively low reverse breakdown voltage (e.g., 6–7 V, although some power devices exhibit significantly larger values) and that the damage from breakdown depends on the energy and is cumulative. Specifically, β degrades (and the device also gets noisier). Hence gain decreases, possibly causing incorrect bias, and the spectrum of the output can show an increase in distortion products as well as a steadily worsening noise floor. In performing simulations of power amplifiers, it is therefore important to look specifically for this effect and to take corrective action if necessary.[51] Options include clamping diodes connected across the device (perhaps integral with the device itself to reduce inductances between the clamp diode and the output transistor) or simply reducing $L\,di/dt$ through improved layout or better drive control.

It is possible (but rare) for a similar phenomenon to occur in MOS implementations. As the gate drive diminishes during turn-off, the gate capacitance drops abruptly once the gate voltage goes below the threshold. Again, the $L\,di/dt$ spike may be large enough to harm the device.

20.4.4 THERMAL RUNAWAY

Another problem concerns thermal effects. In order to achieve high-power operation, it is common to use paralleled devices. In bipolars, the base–emitter voltage for a constant collector current has a temperature coefficient of about -2 mV/°C. Therefore, as a device gets hotter, it requires less drive to maintain a specified collector current. Thus, for a fixed drive, the collector current increases dramatically as temperature increases.

Now consider what happens in a parallel connection of bipolars if one device happens to get a little hotter than the others. As its temperature increases, the collector current increases. The device gets hotter still, steals more current, and so on. This thermoelectric positive feedback loop can run out of control if the loop transmission exceeds unity, resulting in rapid device destruction. To solve the problem, some small resistive degeneration in each transistor's emitter leg is extremely helpful. This way, as the collector current tries to increase in any one device, its base–emitter voltage decreases – offsetting the negative temperature coefficient – and thermal runaway is avoided. Many manufacturers integrate such degeneration (often known as *ballasting*) into the device structure so that none has to be added externally. Even so, it is not uncommon to observe temperature differences of 10°C or more in high-power amplifiers because of this positive feedback mechanism.

Thermal runaway is normally not a problem in MOS implementations because mobility degradation with increasing temperature causes drain current to diminish,

---

[51] Again, it is important to have a good model that is based on the actual physical structure.

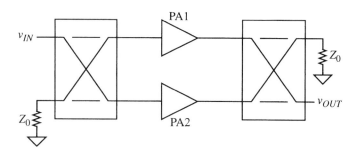

FIGURE 20.37. Quadrature hybrids for producing input and output match

rather than increase, with a fixed gate–source drive. A subtle exception can occur if feedback control is used to force the gate–source voltage to increase with temperature in order to maintain a constant drive current. In that case, device losses increase with temperature, reviving the possibility of a thermal runaway scenario.

For either bipolar or MOS PAs, it is often prudent to include some form of thermal protection to guard against overload. Fortunately, it is trivial in IC implementations to measure temperature with an on-chip thermometer and arrange to reduce device drive accordingly.

20.4.5 LARGE-SIGNAL IMPEDANCE MATCHING

Driving transistors at the large signal levels found in PAs presents a serious challenge if a decent impedance match to the driving source is to be obtained. This statement applies to bipolars and FETs alike. For example, the base–emitter junction of a bipolar transistor is a diode, so the input impedance is inherently highly nonlinear. Recognizing this difficulty, manufacturers of bipolar transistors often specify the input impedance at a specified power level and frequency. However, since there is generally no reliable guide as to how this impedance might change with power level or other operating conditions and since is not even guaranteed at any set of conditions, the designer is left with limited design choices.[52] The traditional way of solving this problem for low-to-moderate power PAs is to swamp out the nonlinearity with a small-valued resistor connected from base to emitter. If the resistance is low enough, it dominates the input impedance. In high-power designs, the required resistance can easily drop well below 1 Ω. Again, this statement applies to both FETs and bipolars.

In cases where a crude shunt resistance is unacceptable and increased circuit complexity is allowed, excellent input and output matching can be obtained through the use of quadrature hybrids. Both 3-dB Lange and branchline couplers may be used in the configuration of Figure 20.37.

---

[52] This despite helpful applications notes from device manufacturers with optimistic titles such as "Systematic methods make Class C amplifier design a snap."

Reflections caused by mismatches at the inputs or outputs of the two PAs are absorbed by the termination resistors shown, so that the input and output ports see theoretically perfect matches. Each individual PA can operate with purposeful mismatches, if desired, to maximize output power or stability, for example. At the same time, the quadrature couplers are also power combiners, so that each PA stage only needs to provide half of the total output power. However, it is important to keep in mind that the worse the nominal mismatch, the greater the required dissipation rating of the resistors.

Finally, statutory requirements on spectral purity cannot always be satisfied with simple output structures such as the single tanks used in these examples. Additional filter sections usually must be cascaded in order to guarantee acceptably low distortion. Unfortunately, every filter inevitably attenuates to some degree. In this context, it is important to keep in mind that just 1 dB of attenuation represents a whopping 21% loss. Assiduous attention to managing all sources of loss is therefore required to keep efficiency high.

20.4.6 LOAD-PULL CHARACTERIZATION OF PAs

All of the examples we've considered so far assumed a purely resistive load. Unfortunately, real loads are rarely purely resistive, except perhaps by accident. Antennas in particular hardly ever present their nominal load to a power amplifier, because their impedance is influenced by such uncontrolled variables as proximity to other objects (e.g., a human head in cell-phone applications).

To explore the effect of a variable load impedance on a power amplifier, one may systematically vary the real and imaginary parts of the load impedance, plotting contours of constant output power (or gain) in the impedance plane (or, equivalently, on a Smith chart). The resulting contours are collectively referred to as a *load-pull* diagram.

The approximate shape of an output power load-pull diagram may be derived by continuing to assume that the output transistor behaves as a perfect controlled-current source throughout its swing. The derivation that follows is adapted from the classic paper by S. L. Cripps,[53] who first applied it to GaAs PAs.

Assume that the amplifier operates in Class A mode. Then, the load resistance is related to the supply voltage and peak collector current as follows:

$$R_{opt} \equiv \frac{2V_{CC}}{I_{C,pk}}, \tag{44}$$

with an associated output power of

$$P_{opt} \equiv \left[\tfrac{1}{2} I_{C,pk}\right]^2 R_{opt}. \tag{45}$$

---

[53] "A Theory for the Prediction of GaAs FET Load-Pull Power Contours," *IEEE MTT-S Digest*, 1983, pp. 221–3.

Now, if the magnitude of the load impedance is less than this value of resistance, then the output power is limited by the current $I_{C,pk}$. The power delivered to a load in this current-limited regime is therefore simply

$$P_L = \left[\tfrac{1}{2} I_{C,pk}\right]^2 R_L, \quad (46)$$

where R_L is the resistive component of the load impedance.

The peak collector voltage is the product of the peak current and the magnitude of the load impedance:

$$V_{pk} = I_{C,pk} \cdot \sqrt{R_L^2 + X_L^2}. \quad (47)$$

Substituting for the peak collector current from Eqn. 44 yields

$$V_{pk} = \frac{2V_{CC}}{R_{opt}} \cdot \sqrt{R_L^2 + X_L^2}. \quad (48)$$

To maintain linear operation, the value of V_{pk} must not exceed $2V_{CC}$. This requirement constrains the magnitude of the reactive load component:

$$|X_L|^2 \leq (R_{opt}^2 - R_L^2). \quad (49)$$

We may interpret this sequence of equations as follows: For load impedance magnitudes smaller than R_{opt}, the peak output current limits the power; contours of constant output power are lines of constant resistance R_L in the impedance plane, up to the reactance limit in Eqn. 49.

If the load impedance magnitude exceeds R_{opt} then the power delivered is constrained by the supply voltage. In this voltage-swing–limited regime, it is more convenient to consider a load admittance, rather than load impedance, so that the power delivered is

$$P_L = [V_{CC}/2]^2 G_L, \quad (50)$$

where G_L is the conductance term of the output load admittance.

Following a method analogous to the previous case, we compute the collector current as

$$i_C = 2V_{CC}\sqrt{G_L^2 + B_L^2}, \quad (51)$$

where B_L is the susceptance term of the output load admittance. The maximum value that the collector current in Eqn. 51 may have is

$$i_{C,pk} = 2V_{CC} G_{opt}. \quad (52)$$

Substituting Eqn. 52 into Eqn. 51 and solving the inequality then yields

$$|B_L|^2 \leq (G_{opt}^2 - G_L^2). \quad (53)$$

Our interpretation of these equations is that, for load impedance magnitudes *larger* than R_{opt}, contours of constant power are lines of constant conductance G_L, up to the susceptance value given by Eqn. 49. The contours for the two impedance regimes together comprise the load-pull diagram.

Load-pull contours are valuable for assessing the sensitivity of a PA to load variations, identifying optimum operating conditions, and possibly revealing hidden vulnerabilities. Experienced PA designers often acquire the ability to diagnose a number of pathologies by inspection of a load-pull contour.

20.4.7 LOAD-PULL CONTOUR EXAMPLE

To illustrate the procedure, let's construct the output power load-pull diagram for a hypothetical Class A amplifier for which the peak voltage is 6.6 V and the peak current is 1.65 A, leading to a 4-Ω R_{opt}. To find the locus of all load admittances (impedances) that allow us to deliver power within, say, 1 dB of the optimum design value, we first compute that a 1-dB deviation from 4 Ω corresponds to about 3.2 Ω and 5.0 Ω. The former value is used in the current-limited regime, the latter in the voltage-swing–limited regime.

In the current-limited regime, we follow the 3.2-Ω constant-resistance line up to the maximum allowable reactance magnitude of about 2.6 Ω, whereas in the swing-limited regime we follow the constant-conductance line of 0.2 S up to the maximum allowable susceptance magnitude of 0.15 S.

Rather than plotting the contours in the impedance and admittance planes, it is customary to plot the diagram in Smith-chart form. Since circles in the impedance or admittance plane remain circles in the Smith chart (and lines are considered to be circles of infinite radius), the finite-length lines of these contours become circular arcs in the Smith chart, and the corresponding diagram appears as Figure 20.38, here normalized to 5 Ω and 0.2 S (instead of 50 Ω and 0.02 S) to make the contour big enough to see clearly.

The power delivered to a load will therefore be within 1 dB of the maximum value for all load impedances lying inside the intersection of two circles: one of constant resistance (whose value is 1 dB less than the optimum load resistance) and the other of constant conductance (whose value is 1 dB less than the optimum load conductance). Note from this description that one need not compute at all the reactance or susceptance magnitude limits; the intersection of the two circles automatically takes care of this computation graphically. Hence, construction of theoretical load-pull diagrams is considerably easier than the detailed derivations might imply.

It should be emphasized that the foregoing development assumes that the transistor behaves as an ideal, parasitic-free, controlled current source. Device and packaging parasitics, combined with an external load impedance, comprise the total effective load for the diagram. Plus, real amplifiers are nonlinear to some degree. In constructing practical load-pull diagrams, one has knowledge only of the externally imposed impedance. Because of all these factors, real load-pull contours will generally be translated, rotated, and distorted relative to the parasitic-free case.

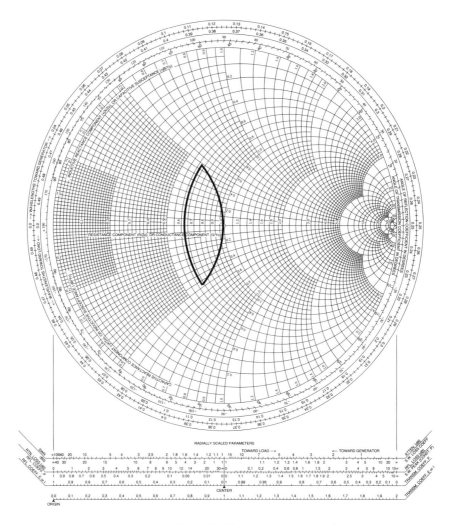

FIGURE 20.38. 1-dB load-pull contour (normalized to 5 Ω) for Class A amplifier example

Practical Notes

Load-pull characterization of power amplifiers depends on the availability of adjustable, calibrated tuners. The real and imaginary parts are successively set to desired values and then the gain and output power are measured at each step along the way. As one might imagine, carrying out a complete set of measurements can be an exceedingly tedious affair. Automated instruments are available for reducing the manual labor, but they are certainly expensive.

There are two general types of commercial load-pull instruments. One is based on stepper motor–driven mechanical tuners. These are somewhat large and slow but are capable of excellent accuracy and repeatability. The other class of instrument uses electronically switched elements (e.g., PIN diodes or other semiconductor switch) to effect the tuning. These instruments are more compact and faster than the classical motor-driven units. However, the parasitics associated with the tuners are much worse for the all-electronic version, requiring the user to pay much more attention to calibration and compensation.

20.5 SUMMARY

We've seen that Class A amplifiers offer good linearity but poor power-handling capability (0.125 on a normalized basis) and low efficiency (50% at absolute maximum). Class B amplifiers improve on the efficiency ($\sim$78.5% at best) by reducing the fraction of a period during which the transistor is active while maintaining the potential for linear modulation.

Class C amplifiers offer efficiencies approaching 100%, but normalized power-handling capability and power gain both approach zero at the same time. Also, they sacrifice linear operation to obtain their improvements in efficiency. Additionally, bipolar Class C amplifiers actually do not satisfy many of the assumptions used in our derivations and hence are difficult to design and construct; MOS and vacuum tube implementations tend to be much less troublesome in this regard.

Amplifiers based on switching concepts do not readily provide linear modulation capability either, but they (theoretically) offer 100% efficiency at nonzero power-handling capability. Although such perfection is unrealizable, at least the limitation is not an inherent property of the topology.

Class D amplifiers offer a normalized power-handling capability of approximately 0.16, but they suffer from potentially large "crowbar" dissipation due to noninfinite switching speeds. Class E PAs solve the dissipation problem for the turn-on transition but impose rather large stresses on the switch. Both Class E and Class F employ reactive loading to shape the voltage and current waveforms in ways that reduce switch dissipation, highlighting a powerful tool for improving efficiency.

We also examined numerous methods for enhancing linearity and efficiency, and concluded by considering experimental characterization methods.

CHAPTER TWENTY-ONE

ANTENNAS

21.1 INTRODUCTION

As we noted early in this book, it is important to remember that conventional lumped circuit theory results from approximating the way the universe behaves (in particular, from setting to zero some terms in Maxwell's equations, effectively treating the speed of light as infinite). The much vaunted "laws" of Kirchhoff[1] are not really laws at all; they are consequences of making simplifying approximations, and so they ultimately break down.[2] The lumped descriptions of circuit theory – in which it is possible to identify elements as individual resistances, capacitances, and inductances – are allowable only when the elements are small relative to a wavelength. Although a rigorous proof of this length criterion is somewhat outside of the spirit of a volume allegedly devoted to practical matters, perhaps a brief plausibility argument might be permissible and sufficient.

If you are willing to accept as an axiom that the finiteness of the speed of light is not noticeable when the propagation delay T_D along a circuit element of length Δl is a small fraction of the shortest period of interest T_{min}, then we would require

$$\left(T_D = \frac{\Delta l}{v}\right) \ll T_{min} = \frac{1}{f_{max}}, \tag{1}$$

where v is the propagation velocity and f_{max} is the maximum frequency of interest. When rewritten, this inequality may be expressed as

$$\Delta l \ll \frac{v}{f_{max}} = \lambda_{min}. \tag{2}$$

The wavelength λ_{min} is that of the highest frequency of interest.

---

[1] Please, two hs and two fs, and pronounced "keerk off" rather than "kirtch off."
[2] Failure to acknowledge this fact is the source of an infinite variety of false conundrums, many of which are debated ad nauseam on various internet chat sites ("proof that physics is broken" and that sort of thing, written by folks who are often wrong but never in doubt).

21.1 INTRODUCTION

Conventional circuit analysis is thus valid as long as circuit elements are "very small" compared to the shortest relevant wavelength. You might be tempted to argue that the restriction to "small" elements is not a serious practical constraint, because we may always subdivide a large structure into suitably small elements, each of which might be described accurately by a lumped approximation. However, the problem with such an approach is that it implicitly assumes that all the energy in the network is confined to the space occupied by the circuit elements themselves. In this chapter, we remove that assumption by allowing for the possibility of radiation of electromagnetic energy. In so doing, we identify the conditions that must be satisfied for significant radiation to occur. We shall see that radiation is theoretically possible from conductors of any length yet is facilitated by structures whose dimensions are at least a significant fraction of a wavelength. Understanding this length dependency explains why we may almost always neglect radiation at low frequencies and why most classical antennas are as big as they are.

As with filters, the subject of antennas is much too vast for comprehensive treatment in just one chapter, of course.[3] The main goals here are (a) to develop intuitive insights that are infrequently provided by (or perhaps difficult to extract from) rigorous mathematical treatments found in some texts and (b) to supplement the brief explanations commonly offered by many "how-to" books. Because this chapter is thus intended to complement, rather than replace, existing descriptions, be forewarned that we will sometimes (actually, often) sacrifice some rigor in favor of the development of design insight. In fact, there may be so much handwaving that you will occasionally need a windbreaker. You will not see a single integral involving the magnetic vector potential, for example.

Aside from a refreshing breeze, the most important tangible products of such an approach are the development of simple analytical circuit models for antennas and an appreciation of why there are so many different antenna configurations.

Although the book's focus is on planar circuits, we will begin with a study of the (electric) dipole antenna, not only because it is so widely used but also because its analysis elucidates many issues common to all classical antennas. A clear understanding of a dipole's limitations explains why certain modifications, such as the addition of "capacity hats" or loading coils, can greatly improve the radiation properties of short dipoles. As will be shown, this same understanding reveals a relationship among normalized length, efficiency, and achievable bandwidth that is reminiscent of the gain–bandwidth trade-offs found in many amplifiers.

Equations describing the dipole also lead directly to a description of the magnetic loop antenna because they are duals; the loop antenna is a magnetic dipole antenna.

---

[3] Three excellent texts on this topic are *Antenna Theory and Design* by Stutzman and Thiele (Wiley, New York, 1998), the classic *Antennas* by Kraus (McGraw-Hill, New York, 1988), and *Antenna Theory* (2nd ed.) by Constantine A. Balanis (Wiley, New York, 1996). Much practical information on antenna construction for amateur radio work may be found in *The ARRL Antenna Handbook* and numerous other books by the American Radio Relay League.

In keeping with our planar viewpoint, the chapter spends a fair amount of time examining the microstrip patch antenna, which has become extremely popular in recent years because it is easily made with the same low-cost mass-production techniques that are used to make printed circuit boards. As will be seen, the intuitive foundations established during a study of the dipole serve well in understanding the patch antenna.

Because our preoccupation will be with equivalent circuit models for antennas, other important characteristics (such as radiation patterns, directivity, and gain) are sadly omitted here. The interested reader is directed to the references cited in footnote 3.

21.2 POYNTING'S THEOREM, ENERGY, AND WIRES

To develop a unified viewpoint that explains when a wire is a wire and when it's an antenna, it is critically important to discard the mental imagery – of electricity-as-a-fluid traveling down wires-as-pipes – that is consciously implanted in students before and during their undergraduate education. Instead understand that ideal wires, strictly speaking, *do not carry electromagnetic energy at all.*[4] Many (perhaps most) students and engineers, to say nothing of lay people, find this statement somewhat controversial. Nevertheless, the statement that wires do not carry energy is correct, and it is easy to show.

To do so, start with the formula for power from ordinary low-frequency circuit theory:

$$P = \tfrac{1}{2} \text{Re}\{VI^*\}. \tag{3}$$

In simple words, delivery of real power requires voltage, current, *and* the right phase relationship between them (the asterisk in Eqn. 3 denotes complex conjugation). If either V or I is zero, then no power can be delivered to a load. Furthermore, even if both are nonzero, a pure quadrature (90° phase) relationship still results in an inability to deliver real power.

The corresponding field-theoretical expression of the same ideas is Poynting's theorem, which states that the (real) power associated with an electromagnetic wave is proportional to the vector cross-product of the electric and magnetic fields:[5]

$$P = \tfrac{1}{2} \text{Re}\{\mathbf{E} \times \mathbf{H}^*\}. \tag{4}$$

To deliver real power, one must have $\mathbf{E}$, $\mathbf{H}$, and the right phase between them. If either $\mathbf{E}$ or $\mathbf{H}$ is zero or if they are in precise quadrature, no power can be delivered. Now the electric field inside a perfect conductor is zero. So, by Poynting's theorem,

---

[4] Heaviside was perhaps the first to express this idea explicitly.

[5] As implausible as it may seem, the theorem is named for a real person, John Henry Poynting, who presented it in 1884 before Hertz had begun his experiments. It's just one of those remarkable coincidences that the Poynting vector points in the direction of energy flow.

no (real) energy flows inside such a wire; if there is to be any energy flow, it must take place entirely in the space *outside* of the wire.[6] Many students who comfortably and correctly manipulate Poynting's theorem to solve advanced graduate problems in field theory nonetheless have a tough time when this particular necessary consequence is expressed in words, for it seems to defy common sense and ordinary experience ("I get a shock only when I *touch* the wire").

The resolution to this seeming paradox is that conductors *guide* the flow of electromagnetic energy. This answer may seem like semantic hair-splitting, but it is actually a profound insight that will help us to develop a unified understanding of wires, antennas, cables, waveguides, and even optical fibers. So for the balance of this text (and of your professional careers), retain this idea of conductors as guides, rather than conduits, for the electromagnetic energy that otherwise pervades space. Then many apparently different ways to deliver electromagnetic energy will be properly understood simply as variations on a single theme.

21.3 THE NATURE OF RADIATION

More than a few students have caught on to the fact that electrodynamic equations – rife with gradient, divergence, and curl – are a devious invention calculated to torment hapless undergraduates. And from a professor's perspective, that is unquestionably the most valuable attribute of E&M (S&M?) theory.

But, perhaps understandably, the cerebral hemorrhaging associated with this trauma frequently causes students to overlook important questions: What is radiation, exactly? How does a piece of wire know when and when not to behave as an antenna? What are the terminal electrical characteristics of an antenna? How are these affected by proximity to objects? Who invented liquid soap, and why?[7]

Let's begin with a familiar example from lumped circuit theory. Without loss of generality, consider driving a pure reactance (e.g., a lossless capacitor or inductor) with a sinusoidal source. If we examine the relationship between voltage and current, we find that they are precisely in time quadrature ("ELI the ICE man" and all that[8]). The *average* power delivered by the source to any pure reactance is zero because energy simply flows back and forth between the source and the reactance. In one quarter-cycle, say, some amount of energy flows to the reactance, and in the next, that entire amount returns to the source. To deliver nonzero average power requires that there be an in-phase component of voltage and current. Adding a resistance across (or in series with) a reactance produces a shift in phase from a pure

---

[6] This argument changes little when real conductors are considered. In that case, all that happens is the appearance of a small tangential component of electric field, which is just large enough to account for ohmic loss.

[7] John Cusack in *The Sure Thing* (Embassy Films Associates, 1985, D: Rob Reiner).

[8] Just in case this mnemonic is unfamiliar to you, it is a way of keeping track of the impedance phase relations in inductors and capacitors. "ELI" tells us that E leads (comes before) I in inductors, and "ICE" tells us that I comes before E in capacitors.

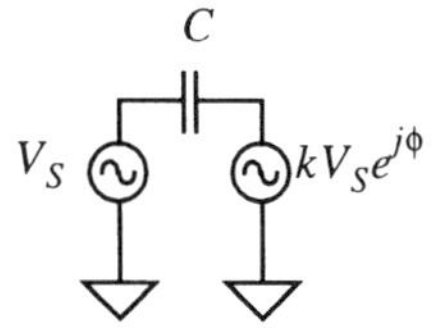

FIGURE 21.1. Capacitance driven by two isochronous sources

90° relationship, enabling one to produce just such an in-phase component and an associated power dissipation.

The question of power flow in electromagnetic fields involves the exact same considerations. Whenever the electric and magnetic fields are in precise time quadrature, there can be no real power flow. If we define radiation as the conveyance of power to a remotely located load, lack of real power flow therefore implies a lack of radiation. We already know (from lumped circuit theory) that quadrature relationships prevail in nominal reactances at frequencies where all circuit dimensions are very short compared with the shortest wavelength of interest. For example, we treat as an inductance a short length of wire connected to ground, and as a capacitance a conductor suspended above a ground plane. These treatments are possible because the fields surrounding the conductors are changing "slowly."

Just as with those lumped reactance examples, real power delivery requires other than a pure 90° phase relationship between the electric and magnetic fields. Classical antennas produce such a departure with the assistance of the finite speed of light to add extra delay.[9]

To understand concretely how the finite speed of light helps produce (actually, enables) radiation, consider a finite length of conductor driven at one end, say, by a sinusoidal voltage source. Near the source, the magnetic and electric fields may be well approximated as being in quadrature. However, because it takes nonzero time for the signal to propagate along the conductor, the voltage (and hence the electric field) at the tip of the conductor is somewhat delayed relative to the voltage and electric field at the driven end. The currents (and their associated magnetic fields) at the two ends are similarly shifted in time. Thus the electric field at the far end is no longer precisely 90° out of phase with the magnetic field at the source end, and nonzero average power is consequently delivered by the driving source. A lumped circuit analogy that exhibits qualitatively similar features is the network in Figure 21.1, where a capacitance is driven by two sinusoidal generators of equal frequency.

From a casual inspection of this particular network, one might be tempted to assert that there can be no power dissipation because a capacitor is a pure reactance. In fact, this is the most common answer given by prospective Ph.D. candidates during

---

[9] Of course, this departure from quadrature also occurs when a real resistance is in the circuit. Our focus here is on radiation, so we will not consider dissipative mechanisms any further.

qualifying examinations. Let's directly evaluate the correctness of this assertion by computing the impedance seen by, say, the left source. The current through the capacitor is simply the voltage across it, divided by the capacitive impedance. So,

$$Z_{eq} = \frac{V_S}{V_S(1 - ke^{j\phi})sC} = \frac{1}{sC(1 - ke^{j\phi})} = \frac{1}{j\omega C[1 - k(\cos\phi + j\sin\phi)]}. \quad (5)$$

The constant k is any real value (it is not meant to represent Boltzmann's constant). Notice that the factor in brackets has a purely real value whenever the phase angle ϕ is either zero or 180°. Under those conditions, the phase angle of the impedance is ±90°, implying zero dissipation. Energy is simply stored in the capacitor in one quarter-cycle, then returned to the sources in the next. *Any* other phase angle produces the equivalent of a real component of impedance as seen by the sources. Despite the presence of a pure reactance, dissipation is nonetheless possible. The capacitor certainly continues to dissipate zero power, but there are still the two sources to consider. A nonzero average power transfer between these isochronous (i.e., equal-frequency) sources is possible. That is, one source can perform work on the other.

Analogous ideas apply to the radiation problem. Because of the finiteness of propagation velocity, the electric and magnetic field components that normally simply store energy in the space around the conductors suddenly become capable of delivering real power to some remotely located load; this is radiation. As a consequence, the signal source that drives the conductor must see the equivalent of a resistance in addition to any reactance that might be present. One way to think about it is that this resistance, and radiation, *may result from work performed by moving charges in one part of the antenna on charges in other parts of the antenna.*[10] The fields associated with radiation are actually present all the time (energy isn't in the conductors, it's in space), but radiation results only when the proper phase relationships exist.

From the foregoing description of radiation, it is also not difficult to understand why the length of an antenna is important. If the conductor (antenna) is very short then the time delay will be very short, leading to negligible departure from quadrature. More precisely, when the length of the conductor is very small *compared to a wavelength,* the resistive component of the antenna impedance will be correspondingly small. Normalization by the wavelength makes sense because a given length produces a fixed amount of time delay, and this time delay in turn represents a linearly increasing phase shift as frequency increases (wavelength decreases).

Now that we have deduced that radiation is a necessary consequence of a lack of pure quadrature, let us see if we can deduce the distance dependency of the radiation. Recalling that the electric field of an isolated, stationary charge in free space falls

---

[10] Richard Feynman, the late Caltech physics Nobelist, described the process most succinctly of all: "If you shake an electron, light comes out." That is, radiation results not only from the fields of accelerated charges acting on the fields of other charges (either in the antenna or in surrounding media) but also from the action of an accelerated charge interacting with its own field. That is, the nonquadrature **E** and **H** fields that give rise to radiation need not arise from different parts of a structure.

as the inverse square of distance, we might be tempted to argue that radiation must also exhibit an inverse square law.[11] To test this idea (again with a minimum of field theory), suppose we have a source of electromagnetic energy (it is completely unnecessary at this point to be more specific). Let's follow the outward flow of energy from the source through two successive (and concentric) spheres. If there is to be radiation then the total energy passing through the two spherical shells must be equal; otherwise the total energy would increase or decrease with distance, implying destruction or creation of energy.[12] We may therefore write

$$\text{energy} = \bar{P}_1 A_1 = \bar{P}_2 A_2, \tag{6}$$

where $\bar{P}$ is the areal power density and A is the area. Now, because the surface area of a sphere is proportional to the square of the radius, constancy of total energy implies that the power density must decrease with the inverse distance squared. In free space, the electric and magnetic fields are proportional to each other. Coupling this fact with Poynting's theorem, we know that the power density is proportional to the square of the field strength:

$$\bar{P} \propto |E|^2 \propto |H|^2 \propto 1/r^2. \tag{7}$$

Hence, we see that there must exist a component of electric or magnetic field whose amplitude falls as the *first* power of distance in order for radiation to be possible.[13] This development is remarkable, for if we had to depend solely on fields with an inverse-square spatial dependence (such as that of an isolated stationary charge) then long-distance communications would be very difficult indeed (a $1/r^4$ power rolloff would be a catastrophe). Fortunately, a miracle of electrodynamics produces components of time-varying electric and magnetic fields that roll off much less dramatically (again, in free space). These radiation components are what make wireless communications practical. Although we certainly have not derived the precise form of the fields, we have nonetheless deduced important facts about them from very elementary arguments.

In addition to allowing us to associate radiation with the existence of inverse-distance fields, the foregoing tells us that the radiation of energy must be indistinguishable from energy dissipated in a resistor, from the point of view of the source. Correspondingly, we shall see that radiation contributes a resistive component to an antenna's driving-point impedance, as asserted earlier.

---

[11] Because there's no such thing as absolute velocity, we may anticipate (from elementary relativity considerations) that radiation cannot result from a uniform motion of charge; acceleration is required.

[12] Or a monotonically increasing storage of energy in free space, which is incompatible with the assumption of a steady state.

[13] It is important to keep in mind that this conclusion depends on the assumption of free-space propagation. If this assumption is violated (e.g., by the presence of lossy media) then other conclusions may result.

Note also that the development here actually makes a rather strong statement: *no distance dependency other than inverse distance can be associated with free-space radiation*. For example, if the fields were to fall off more rapidly, energy would have to accumulate in the space between two successive concentric spheres. If the fields decayed more slowly, energy would have to be supplied by that space. Since neither of these two conditions is compatible with the steady state, we conclude that such field components cannot support radiation. Instead, those other components must represent, at best, stored (reactive) energy, which flows back and forth between the source and the surrounding volume. Thus, their effect is accounted for with either inductive or capacitive elements in an antenna's circuit model (depending on whether the energy is primarily stored in the magnetic or electric field), to whose development we will turn shortly.

Having extracted about as much insight as is possible without resorting to any higher mathematics, we now turn to a description of antenna performance characteristics before considering the practical problem of constructing and modeling real antennas.

21.4 ANTENNA CHARACTERISTICS

Antennas are often characterized by their performance relative to a hypothetical isotropic radiator. As its name implies, an isotropic antenna radiates its energy uniformly in all directions. Even though a true isotropic antenna does not exist, it serves as a useful normalizing reference with which real antennas may be compared.

Practical antennas, of course, do not radiate energy in all directions with equal strength. Measures of this anisotropy include *directive gain* and *directivity*. To understand these measures, first suppose that an antenna radiates a total power P_{tot}. The antenna's equivalent isotropic power density (EIPD) at a distance r is simply that total radiated power divided by the surface area of a sphere of radius r:[14]

$$\text{EIPD} = \frac{P_{tot}}{4\pi r^2}. \tag{8}$$

The directive gain is defined as the actual radiated power density in a particular direction, divided by the EIPD. Directivity is simply the maximum value of the directive gain.

The foregoing definitions all concern the radiated power. Because there are always losses in any real antenna, the total radiated power will be less than the power delivered to the antenna terminals. To account for losses, one may speak of an electrical efficiency

$$\eta = \frac{P_{rad}}{P_{tot}}, \tag{9}$$

where P_{tot} is the total power supplied to the antenna's electrical terminals.

---

[14] The normalizations are usually performed in connection with a unit radius sphere, so the r^2 term in Eqn. 8 is often absent.

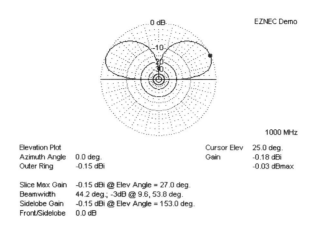

FIGURE 21.2. Representative antenna radiation plot
(*EZNEC* 3.0 simulation; elevation plot)

The catch-all term *antenna gain* accommodates the combined effects of directivity and subunity electrical efficiency. One often encounters the unit "dBi" in connection with these measures. The reference unit "i" refers to the isotropic radiator's performance. Hence, an antenna gain of 3 dBi means that the antenna radiates 3 dB more power in that direction than a hypothetical isotropic radiator. Thus, the term *gain* when used in connection with antennas does not imply actual power gain (an absurdity, given an antenna's passive nature) but gain *relative to a standard antenna*. That standard antenna is typically an isotropic radiator but is sometimes a dipole antenna.

To confuse matters, the term *omnidirectional antenna* often appears in the literature. Contrary to what one might think, it is not just another term for isotropic antenna. It's commonly meant to describe an antenna whose directive gain is independent of angle within some plane, not necessarily all planes. Thus, all isotropic antennas are omnidirectional, but not all omnidirectional antennas are isotropic.

Given the anisotropic radiation of real antennas, a single number such as directivity or gain provides only an incomplete picture of an antenna's performance. The *beamwidth* parameter is thus used to supplement those measures by describing the angle over which the power density stays within 3 dB (or some other standardized value) of the peak density.

Finally, many antennas are designed to focus their energy primarily in one direction but do so imperfectly. Their radiation patterns may reveal the existence of other directions in which energy may be transmitted. Control of such *sidelobe* radiation is the aim of many antenna designs.

The plot in Figure 21.2 shows a representative radiation pattern (in this case, for a conventional dipole antenna, to be discussed in more detail shortly). The view is from the side (an elevation plot). The antenna is a straight rod extending vertically from the center of the circular plot, mounted over a conducting earth that occupies the lower half of the circle. We see that there is no radiation in the direction of the

21.5 THE DIPOLE ANTENNA

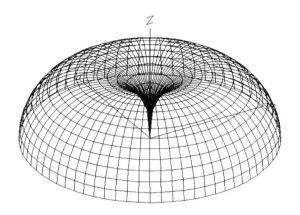

FIGURE 21.3. 3-D radiation plot for antenna of Figure 21.2

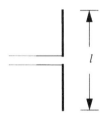

FIGURE 21.4. Short center-fed dipole antenna

antenna element itself, nor any along the ground. The radiation is a maximum for an angle intermediate between those two extremes. In this case, the maximum occurs at an elevation around 27° (near the position of the cursor), and the antenna's beamwidth (measured at the −3-dB points) is about 44°.

It's also instructive to view the 3-D radiation pattern, because it shows clearly the omnidirectional nature of the radiation (again, omnidirectional in a given plane); see Figure 21.3.

21.5 THE DIPOLE ANTENNA

The most common antenna is without question the dipole (both in electric and magnetic versions). The electric version of Figure 21.4 is probably quite familiar to you. Countless millions of dipoles in the form of "rabbit ears" have sat on top of television sets for decades (and still do), and countless millions more are presently found in cell phones and on automobiles (over 400 million cell phones were sold worldwide in 2002, for example). The lowly, ubiquitous AM radio also uses a dipole in the form of its magnetic dual. As we'll see, the dipole operates on principles that follow directly from the description of radiation we've given.

21.5.1 RADIATION RESISTANCE

We've argued that radiation must give rise to a resistive component of the input impedance. This *radiation resistance* is extremely important because it determines, for example, how effectively energy from a source can be coupled into radiated energy. At the same time, by the principle of reciprocity we know that the antenna's circuit model is the same for transmitting as for receiving.

Deriving the radiation resistance of a dipole from first principles is difficult enough that such antennas were used for a long time before such a derivation was actually carried out. An extremely useful engineering approximation is readily derived, however, by simply *assuming* a current distribution along the antenna. We can be guided toward a reasonable assumption by thinking about the dipole as approximately a two-wire transmission line (yes, it's a bent one, but if you insist on more rigor, you will regret it) that is terminated in an open circuit. Then an approximately sinusoidal current distribution results, with a boundary condition of nearly zero current at the open end of the wire.[15]

Using this assumed current distribution and assuming that the antenna is made of infinitesimally thin superconductors, one can derive the following approximation of the radiation resistance of a short dipole:

$$R_r \approx 20\pi^2 (l/\lambda)^2. \tag{10}$$

The formula agrees well with numerical simulations (based on the same assumptions) for l/λ up to about 0.3. At a half-wavelength, the approximate formula predicts a radiation resistance of about 50 Ω. In practice, the impedance of real antennas typically exhibits a real part at resonance closer to 73 Ω. However, at frequencies a little bit above resonance, the impedance is closer to 60 Ω, providing a good match to 50-Ω lines (the minimal improvements that might be provided by a matching network are almost always more than offset by the losses inherent in any practical implementation). The approximate formula of Eqn. 10 is therefore practically useful over a wider range than is usually appreciated.

The accuracy of the formula can be improved a little bit if one treats the antenna as slightly longer than its physical length. This effective extension results from the fact that the current along the antenna doesn't quite go to zero at the tip because of fringing field. We will later see that one may purposefully enhance this effect to improve (increase) the radiation resistance of short dipoles.

The length correction factor is a somewhat complicated (but rather weak) function of the radius-to-wavelength ratio and is commonly taken as approximately 5% for typical dipole antennas. That is, the physical length should be multiplied by roughly 1.05, and that product is inserted into Eqn. 10.

---

[15] The current doesn't quite go to zero at the end because there is some nonzero fringing capacitance, but assuming that it does go to zero incurs a small enough error that the subsequent derivation is usefully accurate.

21.5 THE DIPOLE ANTENNA

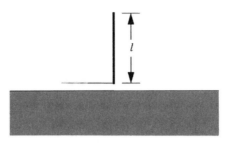

FIGURE 21.5. Short monopole antenna over ground plane

One of Marconi's key inventions is a valuable variation on the dipole antenna in which image charges induced in the earth (or other conducting plane) effectively double the length of the antenna. For such a vertical *monopole* antenna over an ever-elusive perfect ground plane (Figure 21.5), the radiation resistance will be precisely double the value given by Eqn. 10.[16]

The doubling of effective length contributes a quadrupling of the radiation resistance. However, only the physical part of the vertical monopole (not the image) actually radiates, halving that quadrupling (got that?). The radiation resistance of a short monopole antenna is thus

$$R_r \approx 40\pi^2 (l/\lambda)^2. \tag{11}$$

This equation is reasonably accurate for (l/λ)-values up to about 0.15–0.20. An infinitesimally thin superconducting quarter-wavelength monopole will have an impedance of approximately 36 Ω, compared with the formula's prediction of 25 Ω. Again, real monopoles are likely to exhibit a minimum standing-wave ratio (SWR) somewhere above resonance, and a typical length correction remains on the order of ∼5%.

In the simulated antenna SWR plot of Figure 21.6, we see that the SWR stays below 2:1 over a 15% total bandwidth. The cursor is positioned at the frequency where the imaginary part of the antenna impedance goes to zero. At that frequency, the resistive part is 36 Ω in this example. The minimum SWR occurs at a somewhat higher frequency, where the antenna impedance's reactive part is slightly inductive. Although the SWR never gets to 1:1, it is nonetheless low enough over a usefully large frequency range that little improvement in system performance would typically be gained by adding a matching network (especially given that lossless matching networks do not exist).

As does an open-circuited transmission line, the monopole and dipole each exhibit periodic resonances. The functional form of the radiation resistance varies somewhat

---

[16] A monopole is often also called a dipole antenna because there is no fundamental difference between their operating principles and current distribution. In this text we will use both terms, with the precise meaning to be inferred from the context.

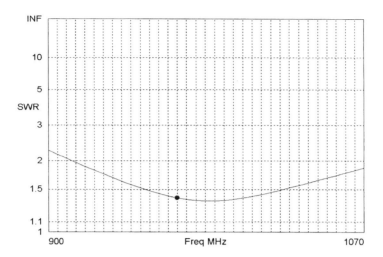

FIGURE 21.6. Plot of ~1-GHz monopole antenna voltage SWR vs. frequency (*EZNEC* 3.0)

Table 21.1. *Approximate radiation resistance for short and medium-length center-fed dipoles*

Normalized conductor length, l/λ	R_{rad}
0–0.25	$20[\pi(l/\lambda)]^2$
0.25–0.5	$24.7[\pi(l/\lambda)]^{2.4}$
0.5–0.64	$11.1[\pi(l/\lambda)]^{4.17}$

as a function of resonant mode. For the center-fed dipole, very approximate equations for the radiation resistance are presented in Table 21.1.[17]

The formulas in the table can be modified for use with monopoles as well. Just remember that the image doubles the effective length of the antenna; plug this doubled length into the formulas in the table (with ~5% length corrections). Then, remember that only the true antenna (and not its image) actually radiates, causing the radiation resistance to drop by a factor of 2. Finally, the formulas will then apply for a range of normalized lengths that are half the values given in the first column. Again, the length is that of the actual conductor, not including the image.

---

[17] Stutzman and Thiele, op. cit. (see footnote 3), p. 171. These formulas continue to assume infinitesimally thin conductors and thus yield values that may not accurately track values exhibited by real antennas. Also, the astute reader will note that the formulas are not continuous across the length boundaries. At exactly a quarter-wavelength, for example, the two formulas straddling that length do not agree, leaving one puzzled as to how to proceed. Such is the nature of approximations!

Finally, it's important to understand that we rarely encounter perfect ground planes. If we are trying to produce one with a conductive sheet, then it should be at least a quarter-wavelength in radius. If a sheet is impractical (as at lower frequencies), several rod conductors of that length will often produce a satisfactory approximation to a ground plane.

21.5.2 REACTIVE COMPONENTS OF ANTENNA IMPEDANCE

Because radiation carries energy away, its effect is modeled with a resistance, as is loss. In general, however, some energy also remains in the vicinity of the antenna, flowing back and forth between the source and the surrounding volume. This near-field nonradiative component thus represents stored energy, so it contributes an imaginary component to the terminal impedance.

To derive *highly* approximate expressions for the effective reactance (inductance and capacitance) of short antennas, we again use the idea that a dipole antenna behaves much like an open-circuited transmission line. If we assume TEM propagation and unit values of relative permittivity and permeability, then the speed of light is expressed as

$$c = 1/\sqrt{LC}, \tag{12}$$

where L and C are here the inductance and capacitance *per length*. Now, we already have the following equation for the approximate inductance per length of a wire with circular cross-section (see Chapter 6):

$$L \approx \frac{\mu_0}{2\pi}\left[\ln\left(\frac{2l}{r}\right) - 0.75\right]. \tag{13}$$

The capacitance per unit length (in farads per meter) is thus very approximately

$$C \approx \left\{c^2 \frac{\mu_0}{2\pi}\left[\ln\left(\frac{2l}{r}\right) - 0.75\right]\right\}^{-1} \approx \frac{2\pi\varepsilon_0}{\ln(2l/r) - 0.75} \approx \frac{5.56 \times 10^{-11}}{\ln(2l/r) - 0.75}. \tag{14}$$

For typical dimensions, the capacitance per length is *very* roughly of the order of 10 pF/m. For example, a 10-cm length of 18-gauge conductor (about 1 mm in radius) has a capacitance of almost exactly 1 pF, according to the formula. Note that the inductance grows somewhat faster – and the capacitance somewhat more slowly – than linearly with length. Thus the capacitance per length is not a constant, but the 10-pF/m (or 10-fF/mm) estimate serves well for many back-of-the-envelope calculations. As with everything else, if you need better accuracy then you will have to do more work. Fortunately, electromagnetic field solvers are readily available to handle these more challenging situations.

Again treating the dipole antenna as an open-circuited transmission line, we expect short dipoles to exhibit a primarily capacitive reactance, changing to a pure resistance as we lengthen the line toward resonance (at half-wavelength), then to an inductance as we pass resonance. This general trend is periodic, repeating every wavelength,

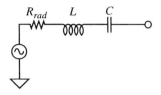

FIGURE 21.7. Circuit model for dipole antenna (one mode only)

but the peak-to-peak variation in impedance diminishes because of the increasing cumulative loss.[18] We may also infer that the driving-point impedance of a monopole changes periodically every *half*-wavelength.[19]

As a final comment, it must be reiterated that all of these equations assume that the antenna is in free space, without any other objects nearby (except for a ground plane in the case of the monopole). Measurements on real antennas often show significant deviations from the predictions of these simple equations, partly because of the simplemindedness underlying their derivation but mainly because one is rarely able in practical circumstances to arrange for all objects to be very far removed from the antenna. Objects less than a few wavelengths away from the antenna can have an important influence on both the reactance and the radiation resistance. Loosely and unreliably speaking, antenna reactance is primarily sensitive to the proximity of dielectric substances (if the antenna is primarily dependent on an electric field) or of magnetic substances (if the antenna is primarily dependent on the magnetic field). The real term is generally most sensitive to nearby lossy substances.

Summarizing the results of this section, simple lossless dipoles may be modeled by the simple circuit shown in Figure 21.7. In this model, the generator represents the voltage induced by a received signal. For short dipoles, this voltage is simply the product of electric field strength and the length of the antenna. When the antenna is used as a transmitter, the generator is set to zero value (a short), and the power radiated may be computed as the power delivered to R_{rad}. Any loss (arising, say, from skin effect) would be modeled by an additional resistance in series with the radiation resistance.

21.5.3 CAPACITIVELY LOADED DIPOLE

We've seen that the radiation resistance of a short dipole varies as the square of normalized length. Hence, good radiation requires a dipole to be a reasonable fraction

---

[18] This observation dovetails nicely with an observation expressed in the chapter on power amplifiers: One expedient method for providing a nonradiative load for high-power transmitter testing is simply to use a great length of coaxial cable. The cumulative loss is large enough to prevent significant reflection, thus assuring a good match. At the same time, the distributed nature of the loss implies that the power is dissipated over a large volume, thereby reducing thermal problems. This trick permits testing to proceed in the absence of specially designed high-power resistive loads.

[19] Again, loss due to radiation and any other dissipative mechanism causes the variation in impedance to diminish.

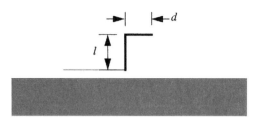

FIGURE 21.8. L-antenna

of a wavelength, or else the radiation resistance will be too low to permit coupling energy into (or out of) it with high efficiency. Unfortunately, it is not always practical to lengthen an antenna arbitrarily to satisfy this requirement, particularly at low frequencies (remember, the free-space wavelength at 1 MHz is 300 meters). Sometimes an important constraint is imposed by a mechanical engineering problem: that of supporting a tall, skinny thing.

One way to finesse the problem is to bend the antenna (it's easier to support a long horizontal thing than a tall vertical thing). To understand why this is potentially beneficial, recall the observation that the fringing field of a straight dipole causes the antenna to act somewhat longer than its physical length. The capacitance associated with the fringing field prevents the current from going all the way to zero at the end, increasing the average current along the antenna, thus raising the radiation resistance. Although the effect is normally small, resulting in a length correction of only ~5% for ordinary dipole antennas, fringing can be purposefully enhanced to make short dipoles act significantly longer.[20] In applications where longer dipoles are not permitted because of space limitations in the vertical dimension, one can employ capacitive loading – using what are known as *capacity* (or *capacitive*) *hats* to increase the current at the end as well as the average current over the length of the dipole. Various conductor arrangements may be used, including flat disks, spheres, and horizontal wires (the latter is used in the L- and T-antenna). Alas, accurate equations for these different cases are not easily derived.

Nonetheless, in the special case of an L-antenna (Figure 21.8), we can derive an approximate formula by making the following assumption (windbreaker required here): Pretend that the current distribution along a straight conductor is only moderately perturbed when the antenna is bent into an L-shape. If this cheesy assumption holds, then we have already derived the relevant formula:

$$R_r \approx 40\pi^2 \left(\frac{l+d}{\lambda}\right)^2, \qquad (15)$$

where l and d are as defined in the figure and the total length is assumed to be short compared with a wavelength. This equation is so approximate that one should expect

---

[20] It may be shown that the absolute theoretical maximum radiation resistance boost factor is 4, corresponding to a constant-amplitude current all along the dipole. In practice, the boost factors achieved are considerably smaller than allowed by theory.

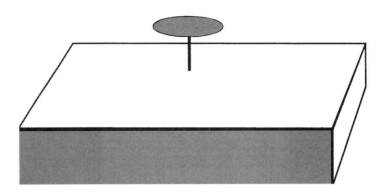

FIGURE 21.9. Antenna with capacity hat

the need to trim the antenna to the proper length. However, it is a reasonable guide to establish rough dimensions for an initial design. Certainly, the use of antenna analysis programs such as *EZNEC* reduces the number of physical iterations required.

If the primary value of the horizontal segment is in boosting capacitance, then further improvements might be enabled by using more segments. Commonly two (for a T-antenna), three, and four horizontal conductors are used, symmetrically arranged about the vertical portion. The capacity hat may be considered the limit of using an infinite number of radial conductors; see Figure 21.9. Other capacitive structures, such as spheres and spheroids, have also been used in place of the flat disk shown in the figure.

21.5.4 INDUCTIVELY LOADED DIPOLE

After all our discussion about how radiation is generally insignificant until conductor dimensions are some reasonable fraction of a wavelength, it may be somewhat surprising that the signal power available from a dipole antenna at any single frequency is actually independent of its length. This invariance can be understood by observing first that shorter dipoles deliver lower *voltages*. To first order, one may take the open-circuit voltage as equal to the product of antenna length and received electric field strength, so voltage scales linearly with length for short dipoles (up to a point). At the same time, we've seen that the radiation resistance varies as the square of length. Hence the ratio of voltage squared to resistance is independent of length. As the dipole length diminishes, the lower voltage is delivered from lower Thévenin resistances (the radiation resistances) and so the available power remains constant (neglecting losses). For a lossless monopole, for example, the available power is

$$P_{av} = \frac{(E_{pk}l)^2/8}{40\pi^2(l/\lambda)^2} = \frac{(E_{pk}\lambda)^2}{320\pi^2}. \tag{16}$$

Clearly, the available power is independent of length, depending instead on the field strength and wavelength. Thus, for the lossless dipole assumed, theory says that we could make antennas out of infinitesimally short segments. This conclusion is seemingly at odds with ordinary experience, where radiation from ordinary wires is routinely ignored with impunity in low-frequency circuit analysis and where AM radio stations use antennas of such size that they must be supported by very tall towers. The resolution to this apparent paradox is that the radiation resistance forms a voltage divider with the Thévenin equivalent resistance of the driving source, augmented by the ever-present resistive losses in any circuit. At frequencies where the antenna is a tiny fraction of a wavelength, the radiation resistance is so small relative to the resistance of the circuit connected to it that negligible power is delivered to the radiation resistance. It is this gross impedance mismatch that commonly allows us to ignore so glibly the possibly of radiation from short wires used as interconnect. Any wire is theoretically capable of radiating at any time, but if it's electrically short then the impedance mismatch prevents significant radiation. *That's* how a wire knows when and when not to act as an antenna in ordinary circuits.

Suppose, though, that we were able to avoid this impedance mismatch. After all, impedance transformers are readily designed. Could we then make antennas arbitrarily short? The answer is a qualified Yes. One qualification can be appreciated after recognizing that bandwidth and normalized antenna length are actually coupled. Because short dipoles have a capacitive reactive component, addition of a suitable inductance will permit the antenna circuit to be brought into resonance at a given desired frequency of operation. Electrically speaking, the antenna acts longer insofar as the disappearance of a reactive term is concerned. These loading inductances are usually placed either at the base of the dipole (i.e., at the feedpoint) or near the center of the dipole. However, as the antenna shrinks, so does its capacitance. To maintain resonance, the compensating (loading) inductance must increase. Recalling that the Q of a series resonant circuit is

$$Q = \sqrt{L/C}/R, \tag{17}$$

it should be clear that Q increases as a dipole antenna shortens (assuming no losses), since inductance and capacitance are both roughly proportional to length and since the radiation resistance is proportional to the square of the length. Therefore,

$$Q = \sqrt{L/C}/R \propto l^{-2} \implies B \propto l^2. \tag{18}$$

The bandwidth is therefore proportional to the square of length. As a result, allowable reductions in antenna length are limited by the desired communication bandwidth. Furthermore, the narrower the bandwidth, the more sensitive the antenna's center frequency to the proximity of objects. Purposeful addition of series resistance to mitigate this sensitivity and improve bandwidth is accompanied by an unfortunate increase in loss. Even if no additional resistance is provided intentionally, there is always some loss. If efficiency is to remain high, the additional series resistance representing this loss must be small compared with the radiation resistance. To underscore the practical difficulties involved, consider a monopole antenna that is 1%

FIGURE 21.10. Loop antenna
(air core version shown)

of a wavelength. The radiation resistance is then about 0.04 Ω. Needless to say, it is exceedingly difficult to arrange for all RF losses to be small compared to resistances of forty milliohms! The fundamental trade-off between efficiency and bandwidth thus tightly constrains the practical extent to which a dipole may be shortened. The coupling among bandwidth, normalized length, and efficiency drives most antenna designs to at least as long as about 10% of wavelength. Practical dipole antennas are rarely much shorter than this value, except for applications where the available signal power is so large that inefficient antennas are acceptable.

Occasionally, one will encounter antennas that employ both capacitive and inductive loading (e.g., capacity hat plus loading coil). The resulting additional degree of freedom can permit one to relax the trade-off to a certain extent but can never fully eliminate it.

21.5.5 MAGNETIC LOOP ANTENNA

The dual of an electric dipole is the magnetic dipole formed by a loop of current. Just as the dipole antenna is sensitive primarily to the electric field, the loop antenna (Figure 21.10) is sensitive mainly to the magnetic-field component of an incoming wave. We'll see momentarily that this duality makes the loop antenna attractive in many situations where the electric dipole antenna suffers from serious problems. In particular, at low frequencies, a loop antenna design is often more practical than its electric dipole counterpart, explaining why loop antennas are almost universally used in portable AM radios and in many pagers, for example.

The following equation for the effective radiation resistance of a loop antenna assumes that the diameter is very short compared to a wavelength and that no magnetic materials are used:[21]

$$R_{rad} \approx 320\pi^4 \left(\frac{nA}{\lambda^2}\right)^2 \approx 31{,}200 \left(\frac{nA}{\lambda^2}\right)^2, \quad (19)$$

where n, λ, and A (respectively) are the number of turns, the wavelength, and the loop area.

---

[21] A. Alford and A. G. Kandoian, "Ultrahigh-Frequency Loop Antennas," *Trans. AIEE,* v. 59, 1940, pp. 843–8; also see Kraus, op. cit. (footnote 3). The formula also assumes zero conductor loss.

Just as the short electric dipole antenna produces a net capacitive reactance, the magnetic loop antenna has a net inductive reactance. Wheeler's famous formula can be used to estimate the inductance of an air-core loop:

$$L \approx \frac{10\pi\mu_0 n^2 r^2}{9r + 10l}, \quad (20)$$

which assumes dimensions and inductance in SI units – unlike Wheeler's original formulation, which uses dimensions in inches to yield inductances in microhenries.

It is important to take note of a new degree of freedom not present in the dipole case: one can add more turns in order to increase radiation resistance. This improvement comes about in the same way as does the impedance transformation of a conventional transformer. The changing magnetic field of the incoming wave induces the same voltage in each turn, so we get n times the per-turn voltage at the antenna terminals. Since energy must be conserved, the current must drop by this same factor n, and so the resistance (the ratio of voltage to current, says Professor Ohm) increases by n^2. Thus, even if the area of the loop is much smaller than λ^2, the radiation resistance may still be boosted to usable values.

We may now appreciate how the loop antenna can solve the thorny problem of AM radio reception. Signals at the lower end of the AM band possess a wavelength of almost 600 meters. The maximum allowable dimensions of any practical portable device will necessarily be an absurdly small fraction of this wavelength. A standard dipole antenna of any human-sized dimensions would thus have an infinitesimal radiation resistance, making efficient operation practically impossible. The loop antenna offers a welcome alternative. It is chiefly for this reason that loop antennas are the only type of antenna used in portable AM radios. Further improvements are provided by winding the antenna around a ferrite rod, whose large permeability concentrates the magnetic field. These "loopstick" antennas dominate portable applications up to frequencies where the lossiness of ferrites negates their usefulness (perhaps as high as the VHF range). Loop antennas are also the choice in pagers, where the desire for a very small form factor makes it difficult to realize an efficient dipole. The loop is conveniently shaped as a rectangle and mounted inside the case of the pager.

21.6 THE MICROSTRIP PATCH ANTENNA

We've seen that radiation becomes practical whenever a conductor is an appreciable fraction of a wavelength. This effect is not always wanted; for example, radiation losses increase the attenuation of microstrip lines. Although undesirable in that context, such radiation is of course precisely what is required to make antennas. When built out of microstrip, these radiators are known as *patch* antennas, so named because of their shape. They have become extremely popular because of their planar nature, making them amenable to inexpensive batch fabrication, just as any other printed circuit. Despite a number of important limitations (*excessive Q* or, equivalently, excessively narrow bandwidth), their convenience and compactness more than compensate in many applications.

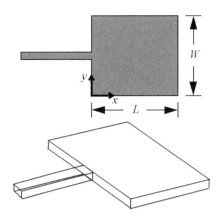

FIGURE 21.11. Half-wave patch antenna (conductor pattern and perspective view)

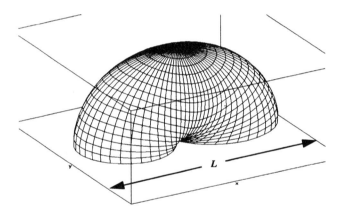

FIGURE 21.12. Typical radiation pattern for patch antenna of Figure 21.11

As seen in Figure 21.11, the patch is typically implemented in the form of a rectangular piece of conductor (over a ground plane). To first order, the patch antenna can be considered the limiting case of connecting a planar array of thin dipoles in parallel so that they form a sheet. As such, the primary radiation is normal to the surface of the patch; see Figure 21.12, where the patch lies in the xy-plane.

The precise nature of the radiation pattern can be adjusted within fairly wide limits by controlling how one feeds the antenna. In the most basic configuration, patches are fed at one end, at the center of an edge (as in Figure 21.11). However, one may also use off-center feeds (offset feeds) to excite other than linear polarizations. This ability is highly valuable, for many microwave communications systems employ polarizations to provide a measure of multipath mitigation.[22] One may also capacitively

---

[22] Reflection off of an object reverses the sense of polarization, changing a counterclockwise polarization into a clockwise one, for example. Using an antenna that selectively rejects one of these components thus reduces a communications link's susceptibility to troublesome reflections.

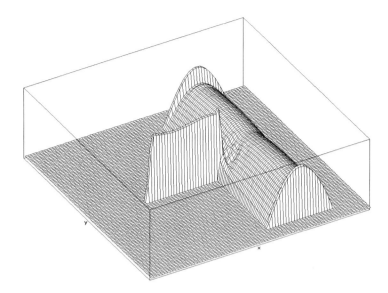

FIGURE 21.13. Current density for a 1-GHz rectangular patch antenna at center frequency (*Mstrip40*)

couple the feedline to the patch, either laterally or with an additional layer of conductor beneath the patch proper.

For the antenna in Figure 21.11, assume that the length L is chosen equal to a half-wavelength. In that case, the current is zero at $x = 0$ and $x = L$, with a maximum at $L/2$, as in a classic dipole antenna. At the same time, the voltage is a minimum at $L/2$ and a maximum at the source and far end, again just as in a classic dipole.

The current density plot of Figure 21.13, generated by the program *Mstrip40*, reveals that the current density does in fact vary sinusoidally, as expected, at least to within "eyeball" resolution. Note also that the microstrip feedline's current has a fairly constant amplitude along the line, indicating the achievement of a good match. The plot corresponds to the same patch orientation as in Figure 21.11.

One important characteristic of a patch antenna is its relatively narrow bandwidth (typically on the order of 1%). To understand the origin of this property, it is useful to think of the patch as more than a flat array of dipoles. Specifically, consider the structure as a cavity resonator (box) that is missing four out of six sides. Viewed in this way, the radiation from the antenna is the result of energy leaking out of the resonant cavity (specifically, radiation is due primarily to energy leaking from the two gaps of width W). Because the thickness of typical dielectric layers is extremely small relative to the other dimensions of a patch, the energy leaking out of the box is much smaller than the energy stored within it. The resulting high Q-factor endows the patch with a powerful ability to filter off-frequency signals, but it also demands much more accuracy in design and manufacturing as well as exceptional stability of material properties. The variability of FR4 virtually guarantees numerous cut-and-try iterations, limiting most use of this material to prototyping and hobbyist applications if ordinary patch geometries are to be used.

Because allowing more energy to leak out reduces Q, one simple way to increase the bandwidth of a patch antenna is simply to use thicker substrates. The difficulty of obtaining inexpensive, low-loss PC boards of suitable thickness occasionally motivates antenna designers to use air as a substantial portion (or even all) of the dielectric. Bandwidths of tens of percent are possible with such arrangements. To zeroth order, the fractional bandwidth (neglecting losses) for a 2:1 VSWR is given *very* approximately by the following empirical formula:[23]

$$B \approx \frac{\varepsilon_r - 1}{2\varepsilon_r} \frac{WH}{L\lambda}. \tag{21}$$

Keeping in mind that radiation takes place primarily along the two edges of width W, we would expect energy leakage, and thus the bandwidth, to be proportional to the area (WH) of the associated apertures, explaining the appearance of that factor in the numerator. On the other hand, increasing L increases the energy stored (the cavity increases in size); hence it belongs in the denominator, as shown.

To highlight the difficulties involved, let's take a look at a specific numerical example. Suppose we have a 1-GHz patch made with 1.6-mm–thick lossless dielectric, a 1.8:1 width/length ratio, and a relative dielectric constant of 4.6. In that case, the estimated fractional bandwidth is about 0.7%. This small value underscores one of the challenges of using a patch antenna, for tiny imprecisions in manufacture lead to antennas that do not radiate efficiently at the desired frequency.

We have deferred presentation of an equation for the radiation resistance of patch antennas for a reason: Numerous design equations have appeared in the literature, spanning a broad range of complexity. Unfortunately, many of these disagree with each other to the first order, so the reader is cautioned not to place too much faith in them. In the interest of preserving the maximum level of intuitive value consistent with usefulness, the equations presented here are simple (but *highly* approximate). Furthermore, published equations generally neglect dielectric loss. Regrettably, FR4 is hardly a zero-loss material, so we'll have to modify the equations to reduce the approximation error in that case.

As noted previously, the classic patch antenna is designed as a half-wave radiator, so its electrical length is chosen equal to a half-wavelength:

$$L_{eff} = \lambda/2. \tag{22}$$

In relating electrical and physical lengths, it's important to consider both fringing field and the effective dielectric constant:

---

[23] This equation is a modification of one originally presented by D. R. Jackson and N. G. Alexopoulos, "Simple Approximate Formulas for Input Resistance, Bandwidth, and Efficiency of a Resonant Rectangular Patch," *IEEE Trans. Antennas and Propagation,* v. 3, March 1991, pp. 407–10. In the original, instead of a factor of 2 in the denominator, there is a factor of 3.77 in the numerator. The modification in Eqn. 21 makes the formula's predictions match much better the field-solver and experimental values that we have encountered.

$$L_{eff} \approx \sqrt{\varepsilon_{r,eff}}\,[L + 2(H/2)] = \sqrt{\varepsilon_{r,eff}}\,(L + H), \quad (23)$$

where H is the thickness of the dielectric and where the approximate length correction per edge, $H/2$, is the same as derived in Section 7.9. More elaborate length corrections abound, but $H/2$ is good enough for most practical purposes.

The effective dielectric constant is given by

$$\varepsilon_{r,eff} \approx 1 + 0.63 \cdot (\varepsilon_r - 1) \cdot (W/H)^{0.1255}, \quad W/H > 0.6, \quad (24)$$

which is the same formula as used for ordinary microstrip lines (as is the correction for fringing in Eqn. 23). Marginally easier to remember is the following alternative approximation:

$$\varepsilon_{r,eff} \approx 1 + \tfrac{5}{8} \cdot (\varepsilon_r - 1) \cdot (W/H)^{1/8}, \quad W/H > 0.6. \quad (25)$$

Since the width W of a typical patch antenna is so much greater than the dielectric thickness H, the effective dielectric constant is usually quite close to the dielectric constant of the material (say, only 5–10% below it). For that reason, design formulas presented in many references rarely make a distinction between these two dielectric constants.

We need to perform a similar accommodation of fringing effects on the effective width:

$$W_{eff} \approx \sqrt{\varepsilon_{r,eff}}\,[W + 2(H/2)] = \sqrt{\varepsilon_{r,eff}}\,(W + H). \quad (26)$$

Continuing with our design equations, we obtain

$$f_0 \approx \frac{1.5 \times 10^8}{L_{eff}}, \quad (27)$$

at which resonant frequency the driving-point impedance is theoretically[24]

$$Z_0 \approx \frac{90 \left[\varepsilon_{r,eff} \frac{L+H}{W+H} \right]^2}{\varepsilon_{r,eff} - 1}. \quad (28)$$

Notice that the width-to-length ratio has the greatest influence on the edge impedance of a patch antenna. For patches made on 1.6-mm–thick FR4 with an effective relative dielectric constant of 4.2, this formula says that a $\sim$50-Ω feedpoint impedance results from a W/L ratio of a bit more than 3 and that a square patch should present an edge impedance of about 500 Ω. A more rigorous calculation, based on the general-purpose electromagnetic field solver *Mstrip40,* says that a W/L ratio of about 4.7 is needed to produce an $\sim$50-Ω edge impedance. An antenna design–specific program (*Pcaad* 2.1) based on a leaky cavity model yields a larger value still, in excess of 5. So, you can see that there is imperfect agreement among the authorities.[25] At

---

[24] Jackson and Alexopoulos, ibid.
[25] That said, it is important to note that the predictions of *Mstrip40* are in close accord with the author's own lab experience.

least they all agree in the basic prediction that the width needs to be rather large to achieve 50 Ω. It is desirable for the width and lengths to be similar to avoid excitation of modes along the width, because polarization might be affected as well as the input impedance. In practice, the width should not be chosen larger than twice the length and usually should be set nearly equal to it.

The situation becomes even murkier when we seek to take into account the nonnegligible dielectric loss of materials such as FR4. Because dielectric loss is equivalent to a shunt conductance, real patch antennas present a driving-point impedance that is lower than predicted by lossless models. To accommodate the effect of dielectric loss, recall that the effective shunt conductance across the terminals of a capacitor built with a lossy dielectric is simply

$$G = \omega\varepsilon(WL/h)\tan\delta. \quad (29)$$

This conductance is what one would measure across the capacitor terminals when the plate dimensions are so small compared to a wavelength that each plate behaves as a unipotential surface. Antennas necessarily violate this criterion, so we have to do a little extra work to get the right answer. In the case of a half-wave patch, the voltage varies roughly cosinusoidally along the length, with extrema at the driven and far edges of the patch. We neglect any voltage variation along the width of the antenna. Consequently, the power dissipated in the dielectric is smaller near $L/2$ than at the ends, and we therefore expect the effective conductance (as viewed from the driving source) to be smaller than G. Now, for a voltage of amplitude V varying sinusoidally in *time*, we know that the power dissipated in a resistor R is simply $V^2/2R$. By analogy, the total dissipation produced by a (co)sinusoidal *spatial* voltage variation is also $V^2/2R$, where V is the voltage at the driven end.

Because the spatial variation introduces a factor-of-2 reduction in dissipation (relative to the case where the voltage is constant), we may represent the total dielectric loss by a single shunt resistor of value $2/G$, connected between the driven end and ground. The total driving-point impedance of the half-wave patch antenna is thus

$$Z_{ideal} \parallel \frac{2}{G} = Z_{ideal} \parallel \frac{2}{\omega\varepsilon(WL/h)\tan\delta} = Z_{ideal} \parallel \frac{2h}{\omega\varepsilon WL \tan\delta}. \quad (30)$$

Experience with FR4, with its typical loss tangent of 0.022, shows that a W/L ratio closer to 1.8 produces an ~50-Ω feedpoint impedance for a 1-GHz half-wave patch ($W = 12.6$ cm, $L = 7.0$ cm), meaning that the width must be reduced by a factor of 2.6 over the lossless case in order to provide a match.[26] Aside from producing a more acceptable form factor (at the cost of reduced efficiency), the loss also widens the bandwidth (here arbitrarily defined once again as where SWR < 2) to about 2%, from about 0.7% in the lossless case. As noted earlier, the narrow bandwidth means that – more so than with other microstrip structures – extensive cut-and-try is likely to be required to produce satisfactory patch antennas.

---

[26] Again, the predictions of *Mstrip40* are in good agreement with this experience.

21.6 THE MICROSTRIP PATCH ANTENNA

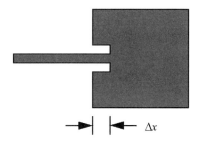

FIGURE 21.14. Half-wave patch antenna with impedance-transforming inset feed (top view)

We've already noted that using large widths in an effort to provide a lower driving-point impedance carries a risk of difficult-to-predict behaviors. A patch is therefore often made square in many applications, necessitating impedance transformations. A classical method is to interpose a quarter-wavelength segment of line between the source and the antenna. If the transforming line's characteristic impedance is made equal to the geometric mean of the source and load impedances, a match results.

As a specific numerical example, suppose we need to design a 50-Ω patch antenna for use in a portable application in the 2.5-GHz ISM (industrial, scientific, and medical) frequency band. Using a hypothetical lossless counterpart to FR4, a rectangular patch would have dimensions of about 27 mm by 80 mm. The length is very reasonable, but the width isn't quite compatible with the form factors of many portable devices. In any case, a 3:1 width-to-length ratio is risky owing to the possible excitation of transverse modes. Suppose that we choose a square patch instead, whose impedance is theoretically as high as 500 Ω. A 160-Ω quarter-wavelength line would perform the necessary transformation. Realizing such an impedance requires a narrow line, and manufacturing tolerances are consequently important in such a case. Easing the burden is that a real patch of those dimensions would more likely present an impedance nearly a factor of 2 lower, permitting the use of a matching line whose impedance is closer to a quite comfortable 100 Ω. Notice that we are unconcerned about the bandwidth of the impedance transformers, because the patch is the bandwidth-constraining element.

Yet another impedance transformer option is available with patches (indeed, with any resonant antenna). Because of the standing-wave setup in the antenna, voltages and currents vary along the patch. In the half-wave case we've been studying, the boundary conditions force the current to be a minimum at the feedpoint and at the far end of the patch, with a maximum in the middle. At the same time, the voltage is a minimum in the middle and a maximum at the source and far end. The impedance, being the ratio of voltage to current, therefore varies along the antenna – from a maximum at the normal feedpoint to a minimum at the middle of the patch – much as it would for an open-circuited half-wave transmission line. One may exploit this impedance variation by using an inset feed, as shown in Figure 21.14.

To a first approximation, we again treat the structure as a transmission line. Thus we again assume that the standing waves (voltage and current) vary (co)sinusoidally along the length of the patch (recall that these assumptions are the same as those used in deriving the radiation resistance of an ordinary dipole). The current is nearly zero at both ends of the patch, increasing roughly sinusoidally as one moves toward the center. At the same time, the voltage is a peak at the ends, sinusoidally decaying toward zero in the center. Therefore, as one moves the feedpoint toward the center, the ratio of voltage to current varies approximately quadratically, because voltage decreases sinusoidally at the same rate that the current increases. The impedance is thus multiplied by the following factor:

$$Z \approx (Z_{edge})\left(\cos \pi \frac{\Delta x}{L}\right)^2, \tag{31}$$

where Z_{edge} is the driving-point impedance in the absence of an inset feed.

For our example of the ever-elusive lossless FR4 patch antenna, we need to transform downward by the comparatively large factor of about 10.8, implying that

$$\Delta x \approx \frac{L}{\pi} \cos^{-1}\left(\frac{1}{\sqrt{10.8}}\right) \approx 0.4L. \tag{32}$$

Note that the calculated inset position is nearly all the way to the middle. Because the change in voltage with distance is large near the center,[27] the precise value of the impedance is an extremely sensitive function of distance in the vicinity of this inset's location. That, plus the uncertainty inherent in our approximations, means again that considerable empirical adjustment will probably be necessary to obtain the correct impedance. Another consideration is that such a deep inset represents a first-order perturbation, and one ought to expect degradation in polarization and other parameters of interest. For real FR4, the dielectric loss lowers the edge impedance of a square patch, implying that a more moderate impedance transformation ratio will suffice.

One convenient method for trimming such an inset patch for prototyping purposes is first to use a deeper-than-nominal inset. Then, upward impedance adjustments are easily effected by placing a shorting strip across some portion of the inset. See Figure 21.15. This particular method avoids the need for precise cutting and also facilitates multiple iterations. Soldering (and unsoldering) a piece of copper foil tape is much easier than gouging out segments of copper cladding.

A disadvantage of the inset feed is that it perturbs the field distributions by an amount that increases with the depth of the inset.[28] The three impedance-matching methods (controlling W/L, quarter-wave transformer, and inset feed) provide important degrees of freedom for trading off parameters of interest. For example, if a

---

[27] The voltage is a minimum in the center, but its spatial derivative is a *maximum* there.
[28] This sensitivity is frequently exploited. A properly positioned off-center feed excites longitudinal and transverse modes simultaneously, whose superposition produces circular polarizations.

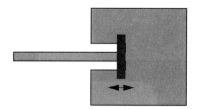

FIGURE 21.15. Adjustment method for inset patch (top view)

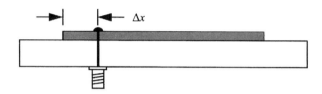

FIGURE 21.16. Patch antenna with coaxial feed (side view)

rectangular patch possesses dimensions that produce a 200-Ω impedance, one could use an inset feed to drop the impedance to 100 Ω and then complete the match with a 71-Ω quarter-wave line to get to 50 Ω. Since the impedance transformations at each step along the way involve relatively small ratios, a more practical and robust design results.[29] Again, the lossiness of FR4 somewhat reduces the need for dramatic impedance transformations (and broadens the bandwidth, albeit at the cost of reduced efficiency), but it remains true that the basic patch antenna is a narrowband structure.

An alternative to the inset feed is to leave the patch proper untouched and to feed the patch at the appropriate tap point directly opposite a connector. This is shown in Figure 21.16. Although this method requires a via (or, more commonly, a hole to accommodate a short length of wire), the inconvenience is modest enough that many commercial patch antennas are built in this manner. However, it's important to be aware of two potential problems with a coaxial feed. If one uses relatively thick substrates in an effort to increase bandwidth, then the inductance of the feed may be large enough to alter the tuning and impedance. At the same time, the feedline may itself radiate to a certain extent, causing distortion in the overall radiation pattern.

One obvious solution to the matching problem produced by feedline inductance is simply to use a conventional impedance-matching network. However, an interesting alternative is to modify the geometry of the patch itself to provide the necessary compensating capacitive reactance. For example, suppose that we cut slots around

---

[29] Of course, one could also use a sequence of quarter-wave transformers or some other variations. There are many ways to accomplish the needed transformation, and the reader is invited to explore alternatives independently.

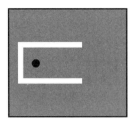

FIGURE 21.17. Patch antenna with U-slot (top view)

the feedpoint to produce an approximation to a parallel-plate capacitor in the vicinity of the feed; see Figure 21.17.

When the U-slot is properly positioned and dimensioned, something quite unexpected occurs: the 2:1 VSWR bandwidth jumps to as much as 40%.[30] This result is completely contrary to the ordinary workings of Murphy and is due to the creation of multiple resonant modes. Not only is there the basic resonance of the patch (we could call that the common-mode resonance), there is a resonance associated with the feedline inductance interacting with the U-capacitance. There is also a differential-mode resonance in which the parts of the patch on opposite sides of the U-boundary operate in antiphase. If these different modes are spaced apart by the right amounts (as in a Chebyshev filter, for example), then the overall bandwidth can be considerably larger than that of a basic patch.

From the qualitative description, you might get the impression that designing for this miraculous condition is difficult, and you would be correct. There are currently no simple analytical formulas to guide the design of these antennas, so many iterations with a field solver are necessary to converge on an acceptable configuration. And as with just about everything else in engineering, there are many variations on this basic theme (e.g., as choice of feed, shape of boundary), further broadening the search space. The variety of possible behaviors has stimulated active ongoing research in this field. For example, instead of a relatively constant gain over a broad frequency band, it is sometimes desirable to have multiple peaks centered about several discrete frequencies that correspond to separate communications bands. This principle underlies the patch antennas used in many multiband cellular phones, although nonobvious (non-U) slot shapes are the norm.

BROADBAND PATCHES

The U-slot modification broadens the bandwidth of a patch antenna by an order of magnitude. Impressive as that is, there remain applications for which even a 30–40%

---

[30] T. Huynh and K. F. Lee, "Single-Layer Single-Patch Wideband Microstrip Antenna," *Electronics Letters*, v. 31, no. 16, 1995, pp. 1310–12.

bandwidth is insufficient. For example, recent interest in ultrawideband (UWB) communications has created a demand for antennas capable of operation over a frequency range exceeding 3 : 1 (or even exceeding 10 : 1). Obtaining such large bandwidths unfortunately requires exploitation of antenna concepts beyond simple modifications of standard antennas.

Before presenting some ultrawideband antennas, it's useful to make a few distinctions. The word *bandwidth* is a little ambiguous, for example, because it could refer to the impedance bandwidth, the bandwidth over which the directivity stays above a certain value, the bandwidth over which the polarization remains within some error band, or the bandwidth over which the main radiation lobe has a certain minimum width. Depending on the context, one or more of these definitions may be relevant. An ideal UWB antenna would possess a constant radiation pattern, gain, and impedance over the entire bandwidth. At the same time, some UWB systems (e.g., pulse-modulated systems) additionally require low dispersion over that same wide bandwidth. Simultaneous achievement of all goals is decidedly challenging, to say the least.

A valuable guiding principle in the design of UWB antennas concerns the relationship between the impedance of an antenna made with a particular conductor pattern and that of an antenna whose conductor pattern is the precise complement of the first. By applying *Babinet's principle*[31] to these complementary structures, we find that the two impedances are related to each other through the impedance of free space in the following manner:

$$Z_1 \bar{Z}_1 = \eta^2/4, \tag{33}$$

where the overbar identifies the impedance corresponding to the complementary conductor pattern.[32]

Now suppose we are able to design an antenna *whose conductor pattern is its own complement*. In that case $Z_1 = \bar{Z}_1$. To the extent that we succeed, we expect such an antenna to have an impedance (in free space) of approximately

$$Z = \sqrt{Z_1 \bar{Z}_1} = \eta/2 \approx 188.5 \; \Omega. \tag{34}$$

That is, the antenna will have a constant impedance over a theoretically infinite bandwidth. Now, it is impractical to satisfy this condition exactly (because, taken literally, we'd have to permit conductors of infinite extent, if you think about it for a moment), but seeking to approximate it leads to practically useful antennas. If the antenna is built in or on a dielectric material, then the impedance will scale from

---

[31] Articulated by the 19th-century French physicist Jacques Babinet.

[32] Babinet's principle might be more familiar to you from an optics lecture in physics class as a way to simplify the analysis of certain classic diffraction problems, but its adaptation to the antenna *design* problem is arguably more important. Briefly, if one generates an E' from ηH and an H' from E/η, the resulting transformed fields (identified as primed variables) still satisfy Maxwell's equations, thanks to the symmetry properties of the latter. As a consequence, replacement of a conductor pattern with its complement results in the impedance relationship given.

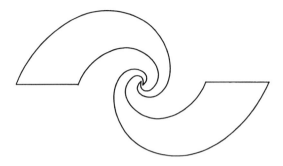

FIGURE 21.18. Two-arm, 1.5-turn truncated equiangular (logarithmic) spiral antenna

the free-space value by the square root of the effective relative dielectric constant. Even accounting for this factor, many UWB antennas present an impedance below the expected "magic" value (largely because practical implementations aren't perfectly self-complementary). Typically, one may expect the impedance of practical antennas to be 60–80% of the theoretical value calculated on the basis of perfect self-complementarity.

Our focus on Babinet's principle should not be taken to mean that self-complementary antennas are the only ultrawideband antennas. It's just that the UWB implications of this principle allow us to focus design effort in a conceptually simple way to generate a large number of broadband antennas. A specific example is a classic broadband antenna known as the *equiangular* (or *logarithmic*) *spiral*, which is based on the exponential relationship

$$r = r_0 e^{\phi/\phi_0}, \tag{35}$$

which may also be expressed as

$$\phi = \phi_0 \ln(r/r_0), \tag{36}$$

where r_0 is the radius at zero-angle ϕ, and ϕ_0 controls the rate at which the radius grows with the phase angle.[33] A typical value for ϕ_0 is 4–5 radians, implying that the radius increases by a factor of about 4 each turn. That factor in turn is called the *expansion ratio,* while ϕ_0 is sometimes known as the *flare angle* (its reciprocal is the *flare rate*).

The foregoing equations define the general shape but do not fully define the conductor pattern. Each arm of the spiral possesses two boundaries, both of the general form given by Eqn. 35 but offset from each other by some angle. Furthermore, practical considerations preclude extension of the arms to infinite radius, so one must make a somewhat arbitrary decision about how and when to stop. Generally, the best results (as measured by constancy of impedance over the passband) are obtained by tapering the ends, but truncation is more easily implemented (as in Figure 21.18).

---

[33] J. D. Dyson, "The Equiangular Spiral Antenna," *IRE Trans. Antennas and Propagation*, v. 7, October 1959, pp. 329–34.

21.6 THE MICROSTRIP PATCH ANTENNA

FIGURE 21.19. Example of two-branch, 1.75-turn Archimedean spiral conductor pattern

The maximum radius determines the lower frequency limit. To a crude approximation, the low-frequency wavelength limit is approximately four times the maximum radius. The high-frequency limit is determined by the radius (r_0) of the spiral at the feedpoint, with the upper frequency wavelength limit roughly equal to four times this minimum radius.[34]

Experimentally, it has been found that 1.5 turns typically leads to a spiral antenna whose characteristics are reasonably insensitive to fabrication tolerances. Hence the layout in Figure 21.18 features this number of turns. A particular implementation of this idea in FR4 exhibits a return loss of better than 10 dB from 400 MHz to 3.8 GHz, as well as an 80-Ω impedance over that band.[35] This decade frequency coverage, as remarkable as it is, by no means represents an absolute limit. From our design guidelines, we see that one could reduce the lower limit (by extending the spiral to larger radii) and increase the upper limit (by reducing the minimum radius). To ensure that the full bandwidth potential is realized, one must exercise great care in connecting to the driving terminals. Reactive discontinuities and other parasitic effects can easily reduce the bandwidth to small fractions of the theoretical maxima.

Another type of self-complementary antenna is based on linear, rather than exponential, spirals. In these *Archimedean spirals,* each arm's boundary is based on equations of the basic form[36]

$$r = r_0 \phi. \tag{37}$$

See Figure 21.19. As with the logarithmic spiral antenna, design degrees of freedom here include the number of arms, the minimum and maximum radii, and the flare rate. The same general design criteria apply: The lower (upper) frequency limit is set by the total (minimum) radius, both determined by the quarter-wavelength condition.

---

[34] Stutzman and Thiele, op. cit. (footnote 3).
[35] J. Thaysen et al., "A Logarithmic Spiral Antenna for 0.4 to 3.8 GHz," *Applied Microwave and Wireless,* February 2001, pp. 32–45.
[36] Stutzman and Thiele, op. cit. (footnote 3).

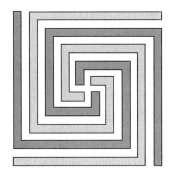

FIGURE 21.20. Example of four-arm square spiral conductor pattern

Yet another option dispenses with these curved geometries to allow even simpler Manhattan layouts. One possible result is the square or rectangular spiral, as shown in Figure 21.20. As the reader will appreciate, it is considerably easier to lay out a square spiral than it is to generate layouts for the spirals. Degrees of freedom here are the number of arms, as well as the minimum and maximum radii. One may also select a linear or exponential flare. Finally, although a square is convenient, other geometries are clearly options as well.

Self-similarity, in addition to self-complementarity, is a useful guiding principle in broadband antenna design. Qualitatively speaking, it seems reasonable that self-similarity at different length scales should permit an antenna to have broadband characteristics, for then the antenna will have the same normalized *electrical* length at different frequencies. Because fractals are geometric objects with precisely this property of self-similarity, fractal antennas have received a fair amount of attention in recent years. In this context, however, it must be observed that the venerable *log-periodic* dipole array is a self-similar structure, so fractal antennas are not quite as novel as one might think. That is, the antenna type preceded its classification.[37]

21.7 MISCELLANEOUS PLANAR ANTENNAS

Not all antennas fall neatly into the types we've presented. Often, constraints on packaging, cost, and frequency range preclude the use of textbook antennas, forcing considerable improvisation. For example, the design of cell phones and other consumer products must often make concessions to aesthetic considerations. More often these days, antennas for such devices must conform to the packaging, not the other way around.

One broadband antenna is based on the *biconical* antenna invented by Sir Oliver Lodge in the 19th century to accommodate the broadband spectrum of the spark

---

[37] For a discussion of many of these classic antennas, see J. D. Kraus as well as *The ARRL Antenna Handbook* (both cited in footnote 3).

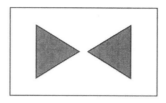

FIGURE 21.21. Microstrip bow-tie patch antenna (top view)

signals then in use (see *Who Really Invented Radio?* in Chapter 1). Translated into planar form, it becomes the *bow-tie* antenna, popular for decades as a television antenna for covering the octave-wide UHF band. It works equally well in microstrip form, where it is popular for its simplicity; see Figure 21.21.

The triangular patches are fed at their apices – either through a pair of microstrip lines on the surface or by lines originating on different conductor layers – just as with any other patchlike structure. Some experimental UWB systems use the bow-tie antenna, where it is sometimes called the *bi-fin* antenna, perhaps to impart a cachet that *bow-tie* lacks.

We should also mention that any of the antenna configurations we've presented may be used as elements within an array. Depending on how the individual antennas are designed and driven, one may perform beamforming or broadbanding (or both).

As another example, consider a typical wireless LAN card. The first generation of those products generally used short dipole antennas that jutted out vertically from the plane of the card. The performance of such cards was excellent, but as the product category evolved, engineers responded to a demand for more compact form factors by moving to planar antennas. An example is shown in Figure 21.22.

As is evident, the two hook-shaped antennas are not readily categorized as patches or dipoles, although they're closer to the latter than the former. In fact, they're just lousy. The manufacturers quietly acknowledge this truth by providing for an external antenna! As can be seen in the photo, the PC board has holes drilled to accommodate an SMA connector, to which a good antenna may be attached. The author has disassembled a number of WLAN cards, and found that a majority of them have this hidden provision. Addition of an SMA (plus one or two surface-mount components) typically enables the use of an external antenna, with which one may extend the useful communication range of these cards by 50–100%.

21.8 SUMMARY

We've seen that radiation is fundamentally the result of the finite propagation speed of light. The need for sufficient phase shift to produce a reasonably high radiation resistance explains why real antennas are a reasonable fraction of a wavelength in extent, at minimum.

FIGURE 21.22. Representative planar antennas for a commercial 802.11b WLAN card

Not only are elementary dipole antennas (both balanced and grounded) quite commonly used, they serve as an important basis for understanding more complex antennas. Short dipoles have low radiation resistances and are primarily capacitive. Capacity hats can be used to increase radiation resistance, and inductances can be used to tune out any capacitance. Such measures are effective up to a limit imposed by the need for providing a given minimum bandwidth or efficiency, and for producing an antenna whose characteristics are not overly sensitive to small changes in dimensions or environmental conditions. The trade-offs are such that antennas much shorter than about a tenth of a wavelength are frequently regarded as unsatisfactory.

The magnetic loop antenna may be viewed as the dual of the electric dipole. Unlike the dipole, the radiation resistance depends on the number of turns, endowing it with an additional degree of freedom to enable compact realizations.

The patch antenna can be viewed as a continuous parallel connection of an infinite number of infinitesimally thin dipoles. Its excessive Q is a definite disadvantage in many situations, however. The addition of a U-shaped slot potentially increases bandwidth by an order of magnitude, permitting a relaxation in manufacturing tolerances.

Finally, we examined a number of ultrawideband antennas, whose design is guided by the self-complementarity inspired by Babinet's principle. Practical antennas with bandwidths exceeding a decade are thus made possible.

CHAPTER TWENTY-TWO

LUMPED FILTERS

22.1 INTRODUCTION

The subject of filter design is so vast that we have to abandon all hope of doing justice to it in any subset of a textbook. Indeed, even though we have chosen to distribute this material over two chapters, the limited aim here is to focus on important qualitative ideas and practical information about filters – rather than attempting a comprehensive review of all possible filter types or supplying complete mathematical details of their underlying theory. For those interested in the rigor that we will tragically neglect, we will be sure to provide pointers to the relevant literature. And for those who would rather ignore the modest amount of rigor that we do provide, the reader is invited to skip directly to the end of this chapter for the tables that summarize the design of several common filter types in "cookbook" form.

Although our planar focus would normally imply a discussion limited to microstrip implementations, many such filters derive directly from lower-frequency lumped prototypes. Because so many key concepts may be understood by studying those prototypes, we will follow a roughly historical path and begin with a discussion of lumped filter design. It is definitely the case that certain fundamental insights are universal, and it is these that we will endeavor to emphasize in this chapter, despite differences in implementation details between lumped and distributed realizations.

We consider only passive filters here, partly to limit the length of the chapter to something manageable. Another reason is that, compared to passive filters, active filters generally suffer from higher noise and nonlinearity, limited operational frequency range, higher power consumption, and relatively high sensitivity to parameter variations – particularly at the gigahertz frequencies with which we are primarily concerned.

22.2 BACKGROUND – A QUICK HISTORY

The use of frequency-selective circuits certainly dates back at least to the earliest research on electromagnetic waves. In his classic experiments Hertz himself used dipole and loop antennas (ring resonators) to clean up the spectrum generated by his spark-gap apparatus and thereby impart a small measure of selectivity to his primitive receivers. Wireless pioneer Sir Oliver Lodge coined the term *syntony* to describe the action of tuned circuits, showing a conscious appreciation of the value of such tuning despite the hopelessly broadband nature of the spark signals then in use.[1] At nearly the same time, Nikola Tesla and Guglielmo Marconi developed tuned circuits of their own (Marconi's British Patent #7777 was so valuable that it became the subject of bitter and protracted litigation[2]) for the specific purpose of rejecting unwanted signals, anticipating the advent of sinusoidal carrier-based communications.

Despite that foundation, however, modern filter theory does not trace directly back to those early efforts in wireless. Rather, the roots go back even further in time: it is research into the properties of transmission lines for telegraphy (then telephony) that primarily informs early filter theory. In 1854 William Thomson (later to become Lord Kelvin) carried out the first analysis of a transmission line, considering only the line's distributed resistance and capacitance. His work, carried out as a consultant for what was to be the 4000-kilometer Atlantic Cable Project, established a relationship between practical transmission rates and line parameters.[3] A bit over twenty years later, Oliver Heaviside and others augmented Kelvin's analysis by including distributed inductance, thereby extending greatly the frequency range over which transmission line behavior could be described accurately.[4] Following up on one particular implication of Heaviside's work, about 1900 both George Ashley Campbell of the American Bell Company and Michael Idvorsky Pupin of Columbia University suggested inserting lumped inductances at regularly spaced intervals along telephone transmission lines to reduce dispersion (the smearing out of pulses).[5] This suggestion is relevant to the filter story because Heaviside recognized that a lumped line differs from a continuous one in that the former possesses a definite cutoff frequency. Campbell and Pupin provided design guidelines for guaranteeing a certain minimum bandwidth.[6]

---

[1] See Hugh Aitken's excellent book, *Syntony and Spark* (Princeton University Press, Princeton, NJ, 1985), for a technically detailed and fascinating account of early work in wireless.

[2] As mentioned in Chapter 1, the U.S. Supreme Court cited prior work by Lodge, Tesla, and Stone in invalidating the U.S. version of Marconi's "four sevens" patent in 1943.

[3] He was knighted for his key contributions to the success of this remarkable endeavor, which joined the Old and New World for the first time. Only one ship, *The Great Eastern*, had been big enough to carry and lay the eight million–kilogram cable. After four frustrating and costly failures starting in 1857, a cable finally connected Valentia, Ireland, to Heart's Content, Newfoundland, on 27 July 1866. The cable carried telegraph traffic continuously until 1965.

[4] For additional background on this story, see Paul J. Nahin's excellent book, *Oliver Heaviside: Sage in Solitude*, IEEE Press, New York, 1987.

[5] As with many key ideas of great commercial import, a legal battle erupted over this one. It is a matter of record that the Bell System was already experimenting with loading coils developed by Campbell well before publication of Pupin's 1900 paper. Nahin (ibid.) observes that Pupin's self-promotional abilities were superior and so he was able to obtain a patent nonetheless. He eventually earned royalties of over $400,000 from Campbell's employer (at a time when there was no U.S. income tax) for his "invention." To add to the insult, Pupin's Pulitzer Prize–winning autobiography of 1924 fails to acknowledge Campbell and Heaviside.

[6] A. T. Starr, *Electric Circuits and Wave Filters*, 2nd ed., Pitman & Sons, London, 1948.

22.2 BACKGROUND – A QUICK HISTORY

In true engineering fashion, the apparent liability of a lumped line's limited bandwidth was quickly turned into an asset, thus establishing the main evolutionary branch of filter design. The first published formalism is Campbell's, whose classic 1922 paper describes in fuller detail ideas he had developed and patented during WWI.[7] Karl Willy Wagner also developed these ideas at about the same time, but German military authorities delayed publication, giving Campbell priority.[8] It is now acknowledged that these two pioneers should share credit for having independently and nearly simultaneously hit upon the same great idea.

Campbell's colleague, Otto J. Zobel, published a much-referenced extension of Campbell's work, but one that still derives from transmission line ideas.[9] In the developments of subsequent decades one sees an evolving understanding of how closely one may approach in practice the theoretical ideal of a perfectly flat passband, constant group delay, and an infinitely steep transition to an infinitely attenuating stopband. Conscious acknowledgment that this theoretical ideal is unattainable leads to the important idea that one must settle for approximations. Some of the more important, practical, and well-defined of these approximations are the Butterworth, Chebyshev, and Cauer (elliptic) filter types we'll study in this chapter. By the 1930s, methods for the direct synthesis of filters to meet particular passband and stopband objectives had been developed by folks like Sidney Darlington.

Throughout and after the Second World War, the subject of filter design advanced at an accelerated pace. Investigation into methods for accommodating finite-Q elements in the design lumped filters offered hope for improved predictability and accuracy. In the microwave domain, filter topologies based directly on lumped prototypes came to be supplemented by ones that exploit, rather than ignore, distributed effects. Many of these are readily implemented in microstrip form and are the focus of the next chapter.

The advent of transistors assured that the size of active devices no longer dominated that of a circuit. Numerous active filter topologies evolved to respond to a growing demand for miniaturization, replacing bulky passive inductor–capacitor circuits in many instances. Aside from enabling dramatic size reductions, some active filters are also electronically tunable. However, these attributes do come at a price: active filters consume power, suffer from nonlinearity and noise, and possess diminished upper operational frequencies because of the need to realize gain elements with well-controlled characteristics at high frequencies. These trade-offs become increasingly serious as microwave frequencies are approached. This statement should not be taken to mean that microwave active filters can never be made to work well enough for some applications (because successful examples certainly abound), but it remains true that the best performance at such frequencies continues to be obtained from passive implementations. It is for this reason that we consider passive filters exclusively.

---

[7] G. A. Campbell, "Physical Theory of the Electric Wave-Filter," *Bell System Tech. J.*, v. 1, no. 2, November 1922, pp. 1–32. See also his U.S. Patent #1,227,113, dated 22 May 1917.

[8] "Spulen- und Kondensatorleitungen" [Inductor and Capacitor Lines], *Archiv für Electrotechnik*, v. 8, July 1919.

[9] O. J. Zobel, "Theory and Design of Uniform and Composite Electric Wave-Filters," *Bell System Tech. J.*, v. 2, no. 1, January 1923, pp. 1–46.

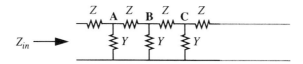

FIGURE 22.1. Infinite ladder network as artificial line

The arrival of transistors also coincided with (and helped drive) a rapidly decreasing cost of computation.[10] No longer limited to considering only straightforward analytical solutions, theorists became free to pose the filter approximation problem much more generally; for example, "Place the poles and zeros of a network to minimize the mean-square error (or maximum error, or some other performance metric) in a particular frequency interval, relative to an ideal response template, while accommodating component tolerances, with less than a 1% yield fallout." Direct accommodation of so many factors is difficult or impossible with classical analytical approaches. Using these modern synthesis methods, the resulting filters are optimum in the sense that one cannot do better (as evaluated by whatever design criteria were imposed in the first place) for a given filter order. The trade-off is that the design cannot be carried out by hand and thus may be less understandable. The main purpose here is therefore to provide an intuitive explanation for how these filters work, leaving detailed execution to machine computation. Because these intuitive explanations are limited in complexity, the equations we will present are suitable mainly for providing a starting point from which a satisfactory final design may emerge after iteration.

22.3 FILTERS FROM TRANSMISSION LINES

We start with the "electric wave filters" of Campbell, Wagner, and Zobel. As mentioned, these derive from lumped approximations to transmission lines, so we begin by examining such "artificial" lines to see how a limited bandwidth arises.

CONSTANT-k ("IMAGE PARAMETER") FILTERS

For convenience, we repeat here some of the calculations from Chapter 2. Recall that we first consider the driving-point impedance, Z_{in}, of the infinite ladder network shown in Figure 22.1. Solving for Z_{in} yields

$$Z_{in} = \frac{Z \pm \sqrt{Z^2 + 4(Z/Y)}}{2} = \frac{Z}{2}\left[1 \pm \sqrt{1 + \frac{4}{ZY}}\right], \qquad (1)$$

where passivity considerations lead us to choose only the solution with a nonnegative real part.

---

[10] Regrettably, space limitations force us to neglect the fascinating story of Teledeltos paper, electrolytic tanks, and other analog computers used to design filters based on potential theory.

As a specific (but typical) case, consider a low-pass filter in which $Y = j\omega C$ and $Z = j\omega L$. Then, the input impedance of the infinite artificial line is

$$Z_{in} = \frac{j\omega L}{2}\left[1 - \sqrt{1 - \frac{4}{\omega^2 LC}}\right]. \qquad (2)$$

At very low frequencies, the factor under the radical is negative and large in magnitude, making the term within the brackets almost purely imaginary. The overall Z_{in} in that frequency range is therefore largely real, with

$$Z_{in} \approx \sqrt{Z/Y} = \sqrt{L/C} = k. \qquad (3)$$

Because the ratio Z/Y is constant, such filters are often known as *constant-k* filters;[11] the literature also refers to them as *image parameter* filters. (For filters made of iterated sections such as those we consider here, a quantity known as the image impedance is the same as the characteristic impedance k.)

As long as the input impedance has a real component, nonzero average power will couple into the line from the source. Above some particular frequency, however, the input impedance becomes purely imaginary, as can be seen from inspection of Eqn. 2. Under this condition, no real power can be delivered to the network, and the filter thus attenuates heavily.[12] For self-evident reasons, the frequency at which the input impedance becomes purely imaginary is called the cutoff frequency, which for this low-pass filter example is given by

$$\omega_h = 2/\sqrt{LC}. \qquad (4)$$

Any practical filter must employ a finite number of sections, of course, leading one to question the relevance of any analysis that assumes an infinite number of sections. Intuitively, it seems reasonable that a "sufficiently large" number of sections would lead to acceptable agreement. Based on lumped network theory, we also expect the filter order to control the ultimate rate of rolloff. Hence, the desired rolloff behavior determines the number of sections to which the network is truncated (we'll have more to say on this subject later). The greater the number of sections, the greater the rate of rolloff. As we'll see, there is also some (but practically limited) flexibility in the choice of Z and Y, permitting a certain level of trade-off among passband, transition band, and stopband characteristics. However, it remains true that one limitation of filters based on concepts of an artificial line is the difficulty of directly incorporating specifications on those characteristics in the design process. Note, for example, the

---

[11] Campbell used the symbol k in precisely this context, but it was Zobel (op. cit.; see footnote 9) who apparently first used the actual term "constant-k" in print.

[12] Attenuation without dissipative elements might initially seem intuitively unpalatable. However, consider that a filter might also operate by *reflecting* energy, rather than dissipating it. That is, a filter can function by producing a purposeful impedance mismatch over some band of frequencies. Modern filter synthesis methods are based directly on manipulation of the reflection coefficient as a function of frequency.

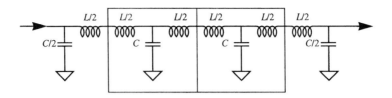

FIGURE 22.2. Low-pass constant-k filter example using two cascaded T-sections

conspicuous absence of any discussion about how the filter behaves near cutoff. We don't know if the transition from passband to stopband is gradual or abrupt, monotonic or oscillatory. We also don't know the precise shape of the passband. Finally, we don't have any guide for how to modify the transition shape should we find it unsatisfactory. As we'll see, these shortcomings lead us to consider other filter design approaches.

Once the filter order is chosen (by whatever means), the next problem is one of termination. Note that the infinite artificial line analysis assumes that the filter is terminated in an impedance that behaves as described by Eqn. 2. That is, our putative finite filter must be terminated in the impedance produced by the prototype *infinite* ladder network: It must have a real impedance at low frequencies, then become purely imaginary above the cutoff frequency. Stated another way, rigorous satisfaction of the criteria implied by Eqn. 2 absurdly requires that we terminate the filter with an element that is itself the filter we desire! We should therefore not be too surprised to discover that a practical realization involves compromises, all intimately related to the hopeless task of using a finite structure to mimic the impedance behavior of an infinite one. For example, the near-universal choice is to terminate the circuit of Figure 22.2 with a simple resistance R that is equal to k.

A source with a Thévenin resistance also of value k is assumed to drive the network. Note that this example uses two complete T-sections (shown in the boundaries), with a *half*-section placed on each end. (Alternatively, one may also regard this structure as consisting of three complete cascaded π-sections.) Termination in half-sections is the traditional way to construct such filters. The series-connected inductors, shown individually to identify clearly the separate contributions of the unit T-sections, are combined into a single inductance in practice. Alternatively, one may implement the filter with π-sections (again, we may consider this filter to consist of three complete cascaded T-sections); see Figure 22.3.

The choice of configuration is sometimes based on inductor count (three vs. four, in these two examples), or on the basis of which topology most gracefully accommodates parasitics at the input and output interfaces. If these parasitics are primarily capacitive in nature, then the T-section implementation is favored because such capacitances may be absorbed into the capacitances at the ends of the filter. Similarly, inductive parasitics are most readily accommodated by a filter using internal π-sections.

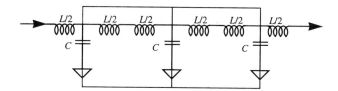

FIGURE 22.3. Low-pass constant-k filter example using two cascaded π-sections

Table 22.1. *Characteristics of ideal constant-k filters*

n_s	Attenuation at cutoff frequency (dB)	Normalized -3-dB bandwidth	Normalized -6-dB bandwidth	Normalized -60-dB bandwidth	Normalized -10-dB S_{11} bandwidth
0	3.0	1.000	1.201	10.000	0.693
1	7.0	0.911	0.980	3.050	0.810
2	10.0	0.934	0.963	1.887	0.695
3	12.3	0.954	0.969	1.486	0.773
4	14.2	0.967	0.976	1.302	0.696
5	15.7	0.976	0.981	1.203	0.756

The design equations for both filter topologies are readily derived from combining Eqn. 3 and Eqn. 4:

$$C = \left(\frac{2}{\omega_h}\right)\frac{1}{R}; \qquad (5)$$

$$L = \left(\frac{2}{\omega_h}\right)R. \qquad (6)$$

Thus, once one specifies R, the characteristic impedance, the desired cutoff frequency, and the total number of sections, the filter design is complete.

Regrettably, deducing the number of sections required can be a bit of a cut-and-try affair in practice. There are equations that can provide guidance, but they are either cumbersome or inaccurate enough that one often simply increases the number of sections until simulations reveal that the filter behaves as desired.[13] Furthermore, the unsophisticated termination of a simple resistance leads to degradation of important filter characteristics, often resulting in a hard-to-predict insertion loss and passband flatness as well as in reduced stopband attenuation (relative to predictions based on true, infinite-length lines). These difficulties are apparent from an inspection of Table 22.1, which shows the attenuation at the cutoff frequency – as well as

---

[13] For example, one frequently cited formula is based on the attenuation characteristics of an infinite ladder. Clearly, such a formula, simple as it is, cannot be expected to yield accurate predictions of a finite, resistively terminated structure.

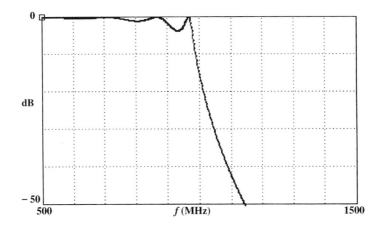

FIGURE 22.4. Frequency response of 1-GHz low-pass constant-k filter ($n_s = 5$)

the -3-dB and -6-dB bandwidths (expressed as a fraction of the cutoff frequency) – of constant-k filters (both T- and π-implementations) as a function of order. In the table, n_s is the number of complete T- (or π-) sections in the central core of the filter. The overall filter order n is therefore $2n_s + 3$.

Note that both the bandwidth and the attenuation at the nominal cutoff frequency are dependent on the number of filter sections. Further note that the cutoff frequency (as computed by Eqn. 4) equals the -3-dB bandwidth only for $n_s = 0$ and is as much as 10% beyond the -3-dB bandwidth in the worst case. In critical applications, the cutoff frequency target may have to be altered accordingly to achieve a specified bandwidth.

Figure 22.4 is a frequency response plot (from *Puff*) for a constant-k filter that consists of five full sections and a terminating half-section on each end. Note that the frequency axis is linear, not logarithmic. Aside from the large ripple evident in the figure, it is also unfortunate that the bandwidth over which the return loss exceeds 10 dB turns out to be only $\sim$70–80% of the cutoff frequency. A considerable improvement in the impedance match bandwidth is possible by using filter sections whose impedance behavior better approximates a constant resistance over a broader normalized frequency range. One example, developed by Zobel, uses "m-derived" networks either as terminating structures or as filter sections (or both). In Figure 22.5, one may regard the structure as three cascaded π-sections or two cascaded T-sections (the latter with terminating half-sections).

As with the prototype constant-k filter, this structure is both driven and terminated with a resistance of value k ohms. The m-derived filter, which itself is a constant-k structure, is best understood by noting that the prototype constant-k filter previously analyzed has a response that generally attenuates more strongly as the cutoff frequency ω_1 is approached. At small fractions of the cutoff frequency, the response is fairly flat, so it should seem reasonable that increasing the cutoff frequency to some value ω_2 should produce a more constant response within the original bandwidth ω_1.

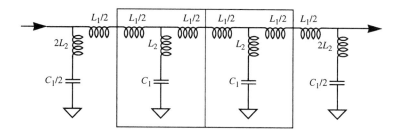

FIGURE 22.5. Low-pass m-derived filter using two cascaded T-sections

The first step in designing an m-derived filter, then, consists simply of increasing the cutoff frequency of a prototype constant-k filter. In the absence of inductor L_2, we see that scaling the values of L_1 and C_1 each by, say, a factor m (with m ranging from 0 to 1) increases the cutoff frequency by a factor of $1/m$, from a value ω_1 to $\omega_2 = \omega_1/m$. The characteristic impedance remains unchanged at k because the *ratio* of L_1 to C_1 is unaffected by this scaling.

Now, to restore the original cutoff frequency, add an inductance L_2 to produce a series resonance with C. At the resonant frequency, this series arm presents a short circuit, creating a notch in the filter's transmission. If this notch is placed at the correct frequency (just a bit above the desired cutoff frequency), the filter's cutoff frequency can be brought back down to ω_1. However, be aware that the filter response does pop back up above the notch frequency (where the resonant branch then looks much like a simple inductance). This characteristic needs to be taken into account when using the m-derived filter and its cousins, the inverse Chebyshev and elliptic filters (which we'll study shortly).

Following a procedure exactly analogous to that used in determining the cutoff frequency of ordinary constant-k filters, we find that the cutoff frequency of an m-derived filter may be expressed as

$$\omega_1 = \frac{2(R/L_1)}{\sqrt{4(L_2/L_1)+1}}. \tag{7}$$

To remove L_1 from the equation, note that the cutoff frequency may also be expressed as

$$\omega_1 = 2m/\sqrt{L_1 C_1}, \tag{8}$$

while the characteristic impedance is given by

$$R = \sqrt{L_1/C_1}. \tag{9}$$

Combining these last three equations allows us to solve for L_2:

$$L_2 = \frac{(1-m^2)R}{2m\omega_1}. \tag{10}$$

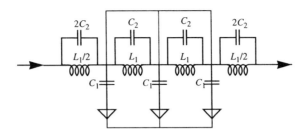

FIGURE 22.6. Low-pass constant-k filter example using two cascaded π-sections with parallel resonators

Solving Eqns. 8 and 9 for L_1 and C_1 yields

$$C_1 = \left(\frac{2m}{\omega_1}\right)\frac{1}{R}, \tag{11}$$

$$L_1 = \left(\frac{2m}{\omega_1}\right)R. \tag{12}$$

An alternative to a series resonator in the shunt arm of each filter section is a parallel resonator in each series arm. This is shown in Figure 22.6. The equations for L_1 and C_1 are as before:

$$C_1 = \left(\frac{2m}{\omega_1}\right)\frac{1}{R}; \tag{13}$$

$$L_1 = \left(\frac{2m}{\omega_1}\right)R. \tag{14}$$

The equation for C_2 is

$$C_2 = \frac{1 - m^2}{2m\omega_1 R}. \tag{15}$$

Both types of m-derived filters behave the same. As with the prototype constant-k filters, the choice of topology in practice is often determined by which implementation uses more easily or conveniently realized components, or which better accommodates the dominant parasitic elements.

Use of the foregoing equations requires that the designer have an idea of what value of m is desirable. As m approaches unity, the circuit converges to an ordinary constant-k filter (and therefore exhibits an increasing passband error), whereas passband peaking increases as m approaches zero. A compromise resides somewhere between these two behaviors, and practical values of m are typically within 25–30% of 0.5 and are most commonly chosen equal to 0.6. This latter value yields a reasonably broad frequency range over which the transmission magnitude remains roughly constant. Table 22.2 enumerates (to more digits than are practically significant) some of the more relevant characteristics of m-derived filters for the specific value of $m = 0.6$. As in Table 22.1, the parameter n_s is the number of complete T- (or π-) sections

Table 22.2. *Characteristics of ideal m-derived filters* ($m = 0.6$)

n_s	Attenuation at cutoff frequency (dB)	Normalized −3-dB bandwidth	Normalized −6-dB bandwidth	Normalized −10-dB S_{11} bandwidth	Minimum stopband attenuation (dB)
0	1.34	1.031	1.063	0.965	8.21
1	3.87	0.993	1.013	0.956	21.24
2	6.27	0.988	0.999	0.969	34.25
3	8.30	0.989	0.996	0.979	47.09
4	10.00	0.991	0.995	0.954	59.81
5	11.44	0.993	0.996	0.961	72.43

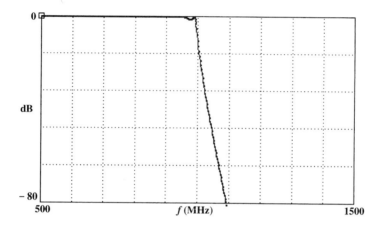

FIGURE 22.7. Frequency response of 1-GHz *m*-derived low-pass filter ($m = 0.6$, $n_s = 5$)

used in the filter. The column labeled "Minimum stopband attenuation" gives the worst-case value of attenuation above the transmission notch frequency where the filter response pops back up.

Note in the table that the cutoff frequency and −3-dB bandwidth are much more nearly equal than for the prototype constant-*k* case (the worst-case difference here is about 3%). The bandwidth over which the return loss exceeds 10 dB is also a much greater fraction of the cutoff frequency (above 95%, in fact). Note also from the table entries that the minimum stopband attenuation (in dB) for this range of values is approximately $(8 + 13n_s)$, so one may readily estimate the number of sections required to provide a specified stopband attenuation.

Figure 22.7 illustrates how the use of *m*-derived sections improves the magnitude response (note that the vertical axis now spans 80 dB, rather than 50 dB). Compared with Figure 22.4, this response shows significantly less passband ripple as well as a much steeper transition to stopband (owing to the stopband notch).

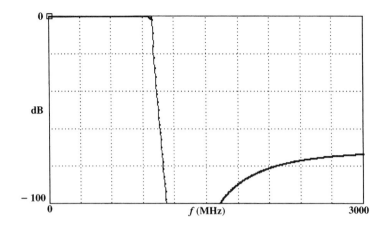

FIGURE 22.8. Response of 1-GHz m-derived low-pass filter, plotted over wider range ($m = 0.6$, $n_s = 5$)

On the frequency scale of Figure 22.7, the characteristic notch is invisible, as is the popping up of the response at higher frequencies. The plot in Figure 22.8 shows these features more clearly. Aside from the potential for improved flatness over the passband, the notches that are inherent in m-derived filters can be used to null out interfering signals at a specific frequency (or frequencies, if sections with differing values of m are used). Later, we will see that judiciously distributed notches are used by both the inverse Chebyshev and elliptic filters, providing dramatic transitions from passband to stopband.

If the precise location of a notch is of importance, it is helpful to know that the frequency of the null ω_∞ is related to m as follows:

$$\frac{\omega_\infty}{\omega_1} = \frac{1}{\sqrt{1-m^2}}, \tag{16}$$

so that the value of m needed to produce a notch at a specified frequency ω_∞ is

$$m = \sqrt{1 - \left(\frac{\omega_1}{\omega_\infty}\right)^2}. \tag{17}$$

A value of 0.6 for m corresponds to a notch frequency that is a factor of 1.25 times the cutoff frequency.

Table 22.3 summarizes the design of constant-k and m-derived lowpass filters. Component values (again, to more digits than you will ever need) are for the specific case of a termination (and source) resistance of 50 Ω and a cutoff frequency of 1 GHz. The left two columns are for the simple constant-k case, and the last four columns give values for the two m-derived configurations (for a specific $m = 0.6$). For filters with a cutoff frequency other than 1 GHz, simply multiply all component values by the ratio of 1 GHz to the desired cutoff frequency. For a different characteristic

22.3 FILTERS FROM TRANSMISSION LINES

Table 22.3. *Component values for 1-GHz constant-k and m-derived filters* ($Z = 50\ \Omega$, $m = 0.6$)

constant-k		m-derived			
L	C	L_1	C_1	L_2	C_2
15.9155 nH	6.3662 pF	9.5493 nH	3.8197 pF	4.2441 nH	1.6976 pF

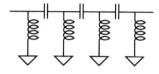

FIGURE 22.9. High-pass filter?

impedance, multiply all component *impedances* by the ratio of the desired impedance to 50 Ω.

One may also combine ordinary constant-k and m-derived sections because the individual sections for both are constant-k in nature. Such a *composite* filter may be desirable, for example, to effect a compromise between flatness and the production of notches at specific frequencies. Unfortunately, the design of such a filter is very much an ad hoc affair. One simply mixes and matches sections as seems sensible, then simulates to verify whether the design indeed functions satisfactorily.

High-Pass, Bandpass, and Bandstop Shapes

At least in principle, a high-pass constant-k filter is readily constructed from the low-pass constant-k prototype simply by swapping the positions of the inductors and capacitors; the values remain the same. Thus one may design (say) a 1-GHz constant-k low-pass filter using the values of Table 22.3 and then interchange the Ls and Cs to synthesize a 1-GHz high-pass filter.

The reason for the qualifier "at least in principle" is that high-pass filters often exhibit serious deviations from desired behavior. These deviations often motivate microwave filter designers to avoid high-pass filters that are based on lower-frequency prototypes. Although there are many ways – too numerous to mention, in fact – in which a practical filter of any kind can fall short of expectations, perhaps the following lumped high-pass filter example will suffice to illustrate the general nature of the problem. Specifically, consider Figure 22.9.

Every practical inductor is shunted by some parasitic capacitance and thus exhibits a resonance of its own. Above the resonant frequency, the "inductor" actually appears as a capacitance. Similarly, every practical capacitance has in series with it some parasitic inductance. Above the corresponding series resonance, the capacitor actually appears inductive. Hence, at sufficiently high frequencies, a high-pass filter

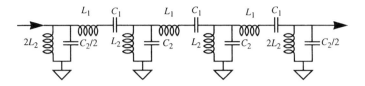

FIGURE 22.10. Bandpass constant-k filter example using two cascaded T-sections

may actually act as a low-pass filter. A complementary effect may afflict low-pass filters where, at high frequencies, it is possible for the response to pop back up.

In cases where a workaround is needed, it is sometimes useful to employ a bandpass filter with a sufficiently wide passband to approximate the desired filter shape. Of course, that solution presupposes knowledge of how to construct bandpass filters. Fortunately, the constant-k structure works here, too (we'll later examine alternative bandpass implementations). As a general strategy for deriving a bandpass filter from a low-pass prototype, replace the inductance of a low-pass prototype with a series LC combination and the capacitance with a parallel LC combination. The added elements are chosen to resonate with their respective mates at the center frequency; see Figure 22.10. Unlike our previous figures, the individual T-sections are not shown (in order to simplify the schematic).

Note that this structure continues to exhibit the correct qualitative behavior even if inductors ultimately become capacitors and vice versa. This property is fundamental to the potentially reduced sensitivity of this topology to parasitic effects.

The formula for the inductance L_1 of the series resonator is the same as that for the inductance in the prototype low-pass filter, except that the *bandwidth* (defined as the difference between the upper and lower cutoff frequencies) replaces the cutoff frequency. The capacitance C_1 is then chosen to produce a series resonance at the center frequency (defined here as the *geometric mean* of the two cutoff frequencies[14]). Hence:

$$L_1 = \frac{2}{\omega_2 - \omega_1} R; \tag{18}$$

$$C_1 = \frac{\omega_2 - \omega_1}{2\omega_0^2} \frac{1}{R}. \tag{19}$$

Similarly, the equation for the capacitance of the low-pass prototype is modified for the bandpass case by replacing the cutoff frequency with the bandwidth. The resonating inductance is again chosen to produce a resonance at the center frequency:

---

[14] In some of the literature, it is unfortunately left unclear as to what sort of mean should be used. For the common case of small fractional bandwidths, this ambiguity is acceptable, for there is then little difference between an arithmetic and geometric mean. Practical component tolerances make insignificant such minor differences. However, the discrepancy grows with the fractional bandwidth, and the error can become quite noticeable at large fractional bandwidths if the arithmetic mean is used.

22.3 FILTERS FROM TRANSMISSION LINES

Table 22.4. *Component values for a 100-MHz–bandwidth, constant-k, 1-GHz bandpass filter* ($Z = 50\ \Omega$)

L_1	C_1	L_2	C_2
159.15 nH	0.15955 pF	0.39888 nH	63.662 pF

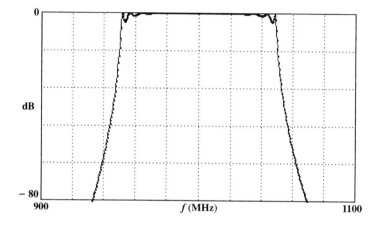

FIGURE 22.11. Frequency response for bandpass filter derived from constant-k prototype ($n_s = 5$)

$$C_2 = \frac{2}{\omega_2 - \omega_1} \frac{1}{R}; \quad (20)$$

$$L_2 = \frac{\omega_2 - \omega_1}{2\omega_0^2} R. \quad (21)$$

Values for a constant-k bandpass filter with cutoff frequencies of 950 MHz and 1.05 GHz (corresponding to a center frequency of approximately 998.75 MHz) are given in Table 22.4. As is the case for its low-pass counterpart, half-sections terminate the bandpass filter. Each half-section consists of components of value $L_1/2$, $2C_1$, $2L_2$, and $C_2/2$. The resulting filters have the same characteristics enumerated in Table 22.1 if the comparisons are performed on the basis of bandwidth rather than center frequency.

As a specific example, consider the frequency response of a bandpass filter derived from a low-pass constant-k filter with $n_s = 5$ (Figure 22.11). The design bandwidth is 100 MHz, centered at 1 GHz. As expected, the behavior at the passband edges resembles that of the low-pass prototype.

For a different bandwidth, multiply C_1 and L_2 by the ratio of the new bandwidth to 100 MHz, and reduce L_1 and C_2 each by the same factor. For a different center frequency, reduce C_1 and L_2 each by the square of the ratio of the new center frequency to 1 GHz. Finally, for a different characteristic impedance, increase the impedance of all four components by the ratio of the new impedance to 50 Ω.

The bandpass filter can be converted into a bandstop (also known as a band-reject) filter simply by swapping the positions of the series and parallel resonators. As in the conversion from low pass to high pass, the values remain unchanged.

From the tables and examples given, it is clear that the constant-k and m-derived filters are extremely simple to design because they consist of identical iterated sections. This simplicity is precisely their greatest attribute. However, as stated before, this ease of design comes at the cost of not being permitted to specify certain details (such as passband ripple), because the design method does not incorporate any specific constraints on response shape. It is clear from the tables, for example, that the cutoff frequency doesn't correspond to a certain fixed attenuation value (such as −6 dB), and monotonicity is far from guaranteed. Stopband behavior is similarly uncontrolled. Shortcomings such as these are what motivated the development of modern filter design methods. Those methods allow one to manipulate the response in far more detailed ways to meet a greater array of design specifications. In turn, that power obligates us to spend a little time identifying and defining the key filter performance metrics that we will now be able to specify. We therefore consider a brief sidebar and introduce these parameters.

22.4 FILTER CLASSIFICATIONS AND SPECIFICATIONS

Filters may be classified broadly by their general response shapes – for example, low-pass, bandpass, band-reject, and high-pass – and may be further subdivided according to bandwidth, *shape factor* (or skirt selectivity), and amount of ripple (in either the phase or magnitude response, and in either the passband or stopband, or both). This subdivision is an acknowledgment that ideal, brickwall filter shapes are simply unrealizable (not merely impractical). Different approaches to approximating ideal characteristics result in different trade-offs, and the consequences of these compromises require characterization.

Bandwidth is perhaps the most basic descriptive parameter and is conventionally defined using −3-dB points in the response. However, it is important to recognize that 3 dB is quite an arbitrary choice (there is nothing fundamental about the half-power point, after all), and we will occasionally use other bandwidth definitions that may be more appropriate, depending on the situation. It is certainly an incomplete specification, because there are infinitely many filter shapes that share a common −3-dB bandwidth. *Shape factor* is an attempt to convey some information about the filter's response at frequencies well removed from the −3-dB point; it is defined as the ratio of bandwidths measured at two different attenuation values (i.e., values at two different points on the *skirt*). As an arbitrary example, a "6/60" shape factor specification is defined as the bandwidth at −60-dB attenuation divided by the bandwidth at −6-dB attenuation; see Figure 22.12.

From the definition of shape factor, values approaching unity clearly imply response shapes that approach infinitely steep transitions from passband to stopband.

22.4 FILTER CLASSIFICATIONS AND SPECIFICATIONS

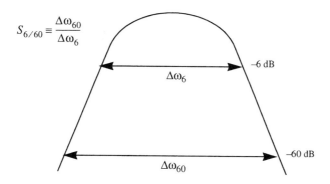

FIGURE 22.12. Illustration of 6/60 shape factor

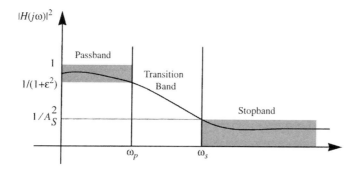

FIGURE 22.13. General filter response template (shown for the low-pass case)

A single-pole low-pass filter (or a standard single-LC bandpass resonator) has a 6/60 shape factor of roughly 600, a value generally regarded as pathetically large.[15] This trio of numbers is easily remembered because of the decimal progression. Because the relevance of a given shape factor depends very much on context, there cannot be a single, universally relevant definition. Thus, although 6/60 happens to be a common one, other specifications are often encountered.

As stated earlier, the inability of practical filters to provide perfectly flat passbands and infinitely steep transitions to infinitely attenuating stopbands implies that we must always accept approximations to the ideal. In the best case we have the opportunity to quantify and specify bounds on the approximation error. The traditional way of doing so is to specify the parameters displayed in Figure 22.13. Note that the square of the magnitude is plotted in the figure, rather than the magnitude itself, because the former is proportional to power gain. This convention isn't universally followed, but it is quite common owing to the RF engineer's typical preoccupation with power gain.

---

[15] The actual number is closer to 577, but it has less mnemonic value than 600.

Note also the pervasiveness of reciprocal quantities on the vertical axis. This annoying feature is avoided by plotting attenuation (rather than gain) as a function of frequency, explaining why many treatments present data in precisely that manner.

Observe further that the filter response template accommodates some amount of variation within the passband (whose upper limit is denoted ω_p), with a maximum permitted deviation of $1/(1 + \varepsilon^2)$. Additionally, a finite transition between the passband and stopband (whose lower frequency limit is denoted ω_s) is also permitted, with a minimum allowed power attenuation of A_s^2 in the stopband. Specification of these parameters thus allows the design of real filters. We now consider several important classes of approximations that make use of these parameters.

22.5 COMMON FILTER APPROXIMATIONS

The constant-k filter's limitations ultimately derive from a synthesis procedure that isn't a synthesis procedure. It ignores the control over filter response afforded by direct manipulation of the pole (and zero) locations to meet specific performance objectives. This limitation is a natural consequence of the transmission line theoretical basis for constant-k filters; because transmission lines are infinite-order systems, consideration of pole locations there is unnatural and in any case leads to numerous analytical difficulties.

However, if one no longer insists on treating filters from a transmission line viewpoint, these difficulties disappear (but are replaced by new ones). Additional, and powerful, techniques then may be brought to bear on the filter analysis and synthesis problem. In this section, we underscore this point by following a procedure not possible with the constant-k filter: Starting from a specification of the desired frequency response, compute a corresponding pole–zero constellation and then synthesize a lumped network that possesses poles and zeros at those locations.

The class of filters we study in this section all have a magnitude characteristic that is expressible in the following general way:

$$|H(j\omega)|^2 = \frac{1}{1 + \varepsilon^2 F^2(j\omega)}. \qquad (22)$$

The various filter types are distinguished by the particular form of the function F. In the simplest cases, F is a polynomial, implying that the overall filter transfer function contains only poles. In more sophisticated filters, the function is a ratio of polynomials (i.e., a rational function), allowing the filter transfer function to have finite zeros as well. As we'll see, a rational function provides additional degrees of freedom beyond the order n, allowing us to reduce the filter order needed to satisfy a given set of specifications.

22.5.1 BUTTERWORTH FILTERS

Some applications are entirely intolerant of ripple, limiting the number of options for response shape. As do all practical filters, the Butterworth seeks to approximate the ideal rectangular brickwall shape. The Butterworth filter's monotonic magnitude response minimizes the approximation error in the vicinity of zero frequency by maximizing the number of derivatives whose value is zero there. As the filter order approaches infinity, the filter shape progressively better approximates the ideal brickwall shape.

A natural (but potentially undesirable) consequence of a design philosophy that places greater importance on the approximation error at low frequencies is that the error grows as the cutoff frequency is approached. If this characteristic is indeed undesirable, then one must seek shapes other than the Butterworth. Subsequent sections will examine some of these alternatives.[16]

The Butterworth's response magnitude (squared) as a function of frequency is given for the low-pass case by the following expression:

$$|H(j\omega)|^2 = \frac{1}{1 + \varepsilon^2(\omega/\omega_p)^{2n}} \qquad (23)$$

where the passband edge ω_p is the frequency at which the power attenuation is $(1+\varepsilon^2)$.[17] The parameter n is the order (or *degree*) of the filter, and it is equal to the number of independent energy storage elements as well as to the power of ω with which the response magnitude ultimately rolls off. One may readily verify that the number of derivatives of Eqn. 23 that we may set equal to zero at DC is $2n-1$. From the equation, it is straightforward to conclude that the response is indeed monotonic.

In designing a Butterworth filter, one often specifies the 3-dB attenuation frequency ω_c, which (depending on the allowable passband ripple) may or may not equal the passband edge.[18] To maintain consistency with the template of Figure 22.13, we first express the power gain at the 3-dB frequency as

$$\frac{1}{1 + \varepsilon^2(\omega_c/\omega_p)^{2n}} = \frac{1}{2}, \qquad (24)$$

from which we readily determine that

$$\omega_p = \varepsilon \omega_c. \qquad (25)$$

Next, we compute the required filter order, using the equation for the attenuation at the stopband edge:

$$\frac{1}{A_s^2} = \frac{1}{1 + \varepsilon^2(\omega_s/\omega_p)^{2n}}. \qquad (26)$$

---

[16] As will be discussed later, one of these alternatives – the inverse or Type II Chebyshev – actually achieves better passband flatness than the Butterworth (making it flatter than maximally flat) by permitting stopband ripple (while preserving passband monotonicity).

[17] Although not rigorously correct (because of the possibility of unequal input and output impedances), we will frequently use the term "power gain" interchangeably with the more cumbersome "response magnitude squared."

[18] Many discussions of Butterworth filters consider only the particular case where $\varepsilon = 1$, corresponding to $\omega_c = \omega_p$. We are considering the general case here, so do not get confused when comparing our results with those published elsewhere.

Solving for the required filter order, n, yields

$$n = \frac{\ln\left(\sqrt{A_s^2 - 1}/\varepsilon\right)}{\ln(\omega_s/\omega_p)} \approx \frac{\ln(A_s/\varepsilon)}{\ln(\omega_s/\omega_p)}. \qquad (27)$$

The approximation holds well if the square of the stopband attenuation is large compared to unity (as it usually is).

Once the maximum passband attenuation, minimum stopband attenuation, and normalized stopband frequency are specified, the required filter order is immediately determined. Because Eqn. 27 generally yields noninteger values, one chooses the next higher integer as the filter order. In that case, the resulting filter will exhibit characteristics that are superior to those originally sought. One way to use the "surplus" performance is to retain the original ω_p, in which case the filter will exhibit greater attenuation at ω_s than required. Alternatively, one may instead retain the original ω_s, in which case the filter exhibits smaller attenuation (i.e. smaller error) at the passband edge than originally targeted. Or, one may elect a balanced strategy that is intermediate between these two choices.

Because its approximation error is very small near DC, the Butterworth shape is also described as maximally flat.[19] However, it is important to recognize that *maximally* flat does not imply *perfectly* flat.[20] Rather, it implies the flattest passband that can be achieved subject to the constraint of monotonicity (later, we will see that it is possible to have an even flatter passband response if we are willing to permit ripple in the stopband).

As a design example, let us continue the exercise that we began with the constant-k topology. We now have the ability to specify more filter parameters than in that case, so we will. Here, arbitrarily allow a 1-dB loss (gain of 0.794) at the passband edge of 1 GHz, and require a 30-dB minimum attenuation at a 3-GHz stopband edge.

From the passband specification, we compute ε as

$$\varepsilon = \sqrt{10^{(1\text{dB})/10} - 1} \approx 0.5088. \qquad (28)$$

From the stopband specification, we see that A^2 is 1000. As a result, the minimum filter order required to meet the specifications is

$$n = \frac{\ln\left(\sqrt{999}/0.5088\right)}{\ln(3)} \approx 3.76, \qquad (29)$$

---

[19] This term was evidently introduced by V. D. Landon in "Cascade Amplifiers with Maximal Flatness," *RCA Review*, v. 5, January 1941, pp. 347–62. Coining of the term thus follows by more than a decade Butterworth's own exposition of the subject in "On the Theory of Filter Amplifiers," *Wireless Engineer*, v. 7, October 1930, pp. 536–41. Although others published similar results earlier, *Butterworth* and *maximal flatness* are now seemingly linked forever.

[20] In this way, "maximally flat" is a bit like "creme filling" in describing the ingredients of an Oreo™ cookie; it means something a little different from how it sounds.

22.5 COMMON FILTER APPROXIMATIONS

Table 22.5. *Component values for 1-GHz, 1-dB, fourth-order low-pass Butterworth filter* ($Z = 50\ \Omega$)

L first	C first
$L_1 = 5.1441$ nH	$C_1 = 2.0576$ pF
$C_2 = 4.9675$ pF	$L_2 = 12.419$ nH
$L_3 = 12.419$ nH	$C_3 = 4.9675$ pF
$C_4 = 2.0576$ pF	$L_4 = 5.1441$ nH

which we round upward to 4. In the computations that follow, we will assume this value of n and select a passband edge of precisely 1 GHz. Again, the excess performance could be distributed some other way, but in the absence of other constraint information, we make this arbitrary choice in order to proceed.

The next step is to compute the element values. Using methods based on those outlined in Section 22.6, one may derive the following equation for the element values of an nth-order Butterworth low-pass filter, normalized to a 1-Ω impedance level and to a 1-rad/s passband edge:

$$b_k = 2(\varepsilon^{1/n}) \sin\left[\frac{(2k-1)\pi}{2n}\right], \tag{30}$$

where k ranges from 1 to n.

As in the constant-k examples, Butterworth filters can start with a shunt capacitor or series inductor. The foregoing equation generates the normalized values for both equivalent configurations. Elements with odd k are capacitors if the filter begins with a shunt capacitor and are inductors if the filter begins with a series inductor.

The following equations denormalize those computed values to yield actual component values:

$$L_k = \frac{R}{\omega_p} b_k; \tag{31}$$

$$C_k = \frac{1}{\omega_p R} b_k. \tag{32}$$

Use of these equations yields the component values shown in Table 22.5, where values for two equivalent realizations are shown side by side in separate columns. Notice that the two designs are practically the same, with their only difference being the assignment of input and output ports. In general, even-order Butterworth filters share this characteristic whereas odd-order filters are symmetrical, beginning and ending with the same type of component (and of the same value).

It is always wise to simulate any design to make sure that no computational errors have crept in. Beware even if you are using tabulated values – some published tables have typographical (or worse) errors! In demanding applications, simulation is also

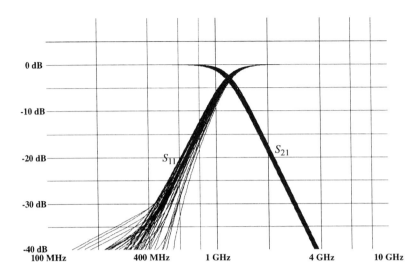

FIGURE 22.14. Monte Carlo analysis of Butterworth filter magnitude response

valuable for assessing the sensitivities of the filter to practical variations in component values or to other imperfections (such as finite element Q or the presence of parasitics). The program *RFSim99* is noteworthy for the ease with which one may rapidly assess the effect of component tolerances on the filter response. Although Monte Carlo analysis is offered by many other circuit simulators, *RFSim99* invokes it with the fewest keystrokes or mouse clicks. The plot of Figure 22.14 shows that this Butterworth filter is not overly sensitive to component tolerance (in this case, the capacitors and inductors both have 5% tolerances, considered tight for ordinary components). Even so, the simulation still shows that a fairly large percentage of the designs do not meet specifications. Either tighter component tolerances or a higher-order filter design would be required to guarantee high yield with loose tolerances.

As a final note, Butterworth high-pass, bandpass, and bandstop filters are readily realized with the same transformations used in the constant-k case.

22.5.2 CHEBYSHEV (EQUIRIPPLE OR MINIMAX) FILTERS

Although passband and stopband monotonicity certainly have an esthetic appeal, insisting on them constrains other valuable filter shape properties. These other properties include the steepness of transitions from passband to stopband as well as the stopband attenuation for a given filter order. Alternative filters, based on nonmonotonic frequency response, are named after the folks who invented them or who developed the underlying mathematics. The Chebyshev filter, an example of the latter, allows a reduction in filter order precisely by relaxing the constraint of monotonicity.[21] In

---

[21] Contrary to common belief, Pafnuti L'vovich Chebyshev never worked on filters at all (he was born in 1821 and died in 1894, certainly well before the establishment of filter theory). In fact, he

contrast with the Butterworth approximation, which is preoccupied with minimizing error at low frequencies, the Chebyshev minimizes the maximum approximation error (relative to the ideal brickwall shape) throughout the entire passband. The resulting *minimax* response shape thus exhibits some ripple, the amount of which may be specified by the designer. For a given order, the Chebyshev filter shape offers a more dramatic transition from passband to stopband than a Butterworth offers. The steepness of the transition is also a function of the passband ripple one allows; the greater the permissible ripple, the steeper the transition.

A consequence of minimizing the maximum error is that the ripples of a Chebyshev response are all of equal amplitude. A rigorous proof of the minimax optimality of an equiripple shape is surprisingly involved, so we won't attempt one here. However, it should seem intuitively reasonable that equiripple behavior would be optimal in the minimax sense, for if any one error peak were larger than any other then a better approximation could probably be produced by reducing it – at the cost of increasing the size of one or more of the others. Such tradings-off would proceed until nothing would then be left to trade for anything else; all error peaks would be equal.

Similar advantages also accrue if the stopband, rather than the passband, is allowed to exhibit ripple. The inverse Chebyshev filter (also known as a Type II Chebyshev filter) is based on this idea; it actually combines a flatter-than-Butterworth passband with an equiripple stopband.

To understand how the simple act of allowing either stopband or passband ripple provides these advantages, we need to review the properties of a complex pole pair. First recall that one standard (and perfectly general) form for the transfer function of such a pair is:

$$H(s) = \left[\frac{s^2}{\omega_n^2} + \frac{2\zeta s}{\omega_n} + 1 \right]^{-1}, \tag{33}$$

where ω_n is the distance to the poles from the origin and ζ (zeta) is the damping ratio; see Figure 22.15.

Damping ratio is a particularly significant parameter when examining time-domain behavior. A zero damping ratio corresponds to purely imaginary poles, and a damping ratio of unity corresponds to a pair of poles coincident on the real axis. The former condition applies to an oscillator, and the latter defines critical damping. Above a damping ratio of unity the two poles split – with one moving toward the origin and the other toward negative infinity – all the while remaining on the real axis. Whatever the value of damping, the frequency ω_n always equals the geometric mean of the pole frequencies.

---

developed his equations during a study of mechanical linkages used in steam engines, as described in "Théorie des mécanismes connus sous le nom de parallélogrammes" [Theory of Mechanisms Known under the Name of Parallelograms], *Oeuvres,* vol. I, St. Petersburg, 1899. "Parallelograms" translate rotary motion into an approximation of rectilinear motion. By the way, the spelling of his name here is just one of many possible transliterations of Пафнутий Львович Чебышев. German-language journals generally render it as *Tchebyscheff.* We've selected the transliteration most likely to lead an English speaker to a close approximation of the correct pronunciation. The commonly encountered *Chebychev* lacks this property.

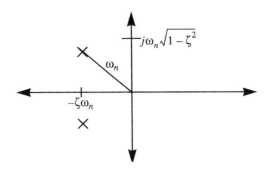

FIGURE 22.15. Two-pole constellation

In filter design, we are generally most concerned with frequency response, and it is then helpful to relate damping ratio to more directly relevant parameters. Specifically, damping ratio is directly related to Q as follows:

$$Q = 1/2\zeta. \tag{34}$$

Although it is true that all two-pole systems with no finite zeros have a frequency response that ultimately rolls off as ω^{-2}, both the frequency response magnitude and the slope *in the vicinity of the peak* are very much functions of Q, with each parameter increasing as Q increases (see the lower panel of Figure 22.16). For Q below $1/\sqrt{2}$, the frequency response exhibits no peaking. Above that value of Q, peaking increases without bound as Q approaches infinity. For large values of Q, the peak gain is proportional to Q. Stated alternatively, higher Q-values lead to greater ultimate attenuation, relative to the peak gain, and to slopes that are normally associated with systems of higher (and perhaps much higher) order.

Now consider ways a filter might exploit this Q-dependent behavior. Specifically, suppose we use a second-order section to improve the magnitude characteristics of a single-pole filter. If we arrange for the peak of the second-order response to compensate (boost) the response of the first-order section beyond where the latter has begun a significant rolloff, then the frequency range over which the magnitude of the cascade remains roughly constant can be increased. At the same time, the rolloff beyond the compensation point can exhibit a rather high initial slope, providing an improved transition from passband to stopband. Clearly, additional sections may be used to effect even larger improvements, with increases in the maximum Q. This latter requirement stems from the need to provide larger boosts to compensate for ever larger attenuations.

Having developed this understanding, we may revisit the Butterworth and Chebyshev approximations. The Butterworth condition results when the poles of the transfer characteristic are arranged so that the modest amount of frequency response peaking of a complex pole pair offsets, to a certain extent, the rolloff of any poles of lower frequency. The resulting combination extends the frequency range over which there is a roughly flat transmission magnitude. Formally, one may deduce from Eqn. 23 that

22.5 COMMON FILTER APPROXIMATIONS

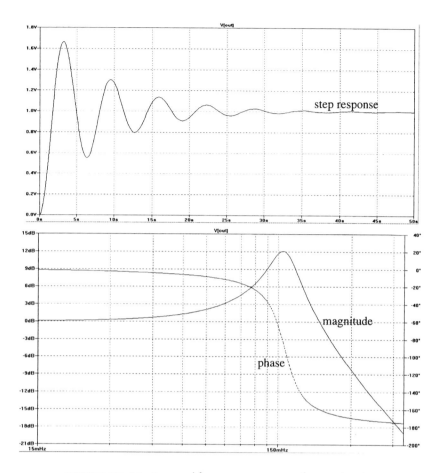

FIGURE 22.16. Step and frequency response of second-order low-pass section ($\omega_n = 1$ rps, $\zeta = 0.125$)

all of the poles lie on a semicircle in the s-plane, distributed as if there were twice as many poles disposed at equal angles along the circumference, the right half-plane poles being ignored.[22] For example, a third-order Butterworth (Figure 22.17) has a single pole on the real axis as well as a complex conjugate pair at 60° angles with the real axis. The distance from the origin to the poles is the 3-dB cutoff frequency.

The Chebyshev filter goes further by allowing passband (or stopband) ripple. Continuing with our third-order example, the response of the real pole is allowed to drop below the low-frequency value by some specified amount (the permissible ripple) before the complex pair's peaking is permitted to bring the response back up. The damping ratio of the complex pair must be lower than that in the Butterworth case to produce enough additional peaking to compensate for greater amounts of attenuation.

---

[22] Okay, perhaps it isn't quite "intuitively obvious," but finding the roots of Eqn. 23 to discover the factoid about Butterworth poles lying on a circle isn't all that bad.

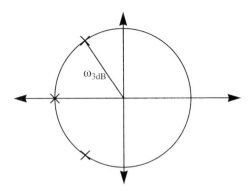

FIGURE 22.17. Pole constellation for third-order Butterworth low-pass filter

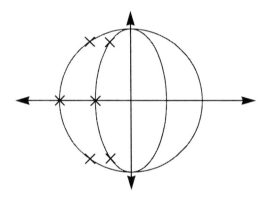

FIGURE 22.18. Third-order Butterworth and Chebyshev low-pass filter pole constellations

A side effect of this lower damping is that there is a more dramatic rolloff beyond the cutoff frequency. In this manner the Chebyshev filter permits the designer to trade passband flatness for better stopband attenuation.

Although it is even less intuitively obvious, the poles of a Chebyshev low-pass filter are located along a (semi)ellipse, remarkably with imaginary parts that are equal to those of a corresponding Butterworth low-pass filter.[23] Increasing the eccentricity of the ellipse increases the ripple; see Figure 22.18. Mathematically, the Chebyshev response is of the general form

$$|H(j\omega)|^2 = \frac{1}{1 + \varepsilon^2 C_n^2(\omega/\omega_p)}, \tag{35}$$

---

[23] There are many references that provide excellent derivations of the Butterworth and Chebyshev conditions. A particularly enlightening presentation may be found in chapters 12 and 13 of R. W. Hamming's *Digital Filters* (2nd ed., Prentice-Hall, Englewood Cliffs, NJ, 1983).

where ω_p once again is the frequency at which the response magnitude squared has dropped to a value of

$$\frac{1}{1+\varepsilon^2}. \tag{36}$$

For self-evident reasons, ε is known as the ripple parameter, and it is specified by the designer. The function $C_n(x)$ is known as a Chebyshev polynomial of order n. The most relevant property of such polynomials is that they oscillate between -1 and $+1$ as the argument x varies over the same interval. This property fairly distributes the approximation burden by allowing the filter's (power) response to oscillate between 1 and $1/(1+\varepsilon^2)$ within the passband. Recall that a Butterworth filter uses all of its approximation power at DC, allowing the error to grow monotonically as the passband edge is approached. The Chebyshev filter achieves its better performance by spreading its approximation error over the entire passband. Outside of this interval, a Chebyshev polynomial's magnitude grows rapidly (as x^n in fact), corresponding to monotonically increasing filter attenuation.

There are a couple of ways of generating Chebyshev polynomials algorithmically. One is through the recursion formula

$$C_n(x) = 2xC_{n-1}(x) - C_{n-2}(x), \tag{37}$$

where knowing that $C_0 = 1$ and $C_1 = x$ will get you started. As can be seen from the formula, the leading coefficient of Chebyshev polynomials is 2^{n-1}, a fact we shall use later in comparing Chebyshev and Butterworth polynomials.

Another method for generating the Chebyshev polynomials is in terms of some trigonometric functions from which the oscillation between -1 and $+1$ (for $|x| < 1$) is directly deduced:

$$C_n(x) = \cos(n \cos^{-1} x) \quad \text{for } |x| < 1. \tag{38}$$

For arguments larger than unity, the formula changes a little bit:

$$C_n(x) = \cosh(n \cosh^{-1} x) \quad \text{for } |x| > 1. \tag{39}$$

Although it is probably far from obvious at this point, these functions are likely familiar to you as Lissajous figures, formed and displayed when sine waves drive both the vertical and horizontal deflection plates of an oscilloscope. That is, suppose that the horizontal deflection plates are driven by a signal

$$x = \cos t, \tag{40}$$

so that

$$t = \cos^{-1} x. \tag{41}$$

Further suppose that the vertical plates are simultaneously driven by a signal

$$y = \cos nt. \tag{42}$$

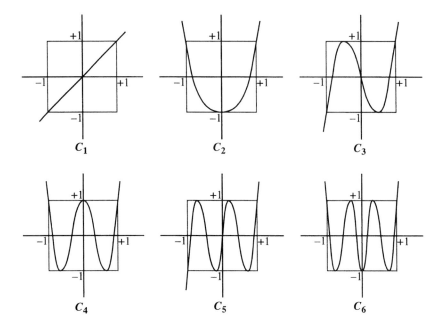

FIGURE 22.19. Rough sketches of some Chebyshev polynomials

Table 22.6. *First ten Chebyshev polynomials*

Order, n	Polynomial
0	1
1	x
2	$2x^2 - 1$
3	$4x^3 - 3x$
4	$8x^4 - 8x^2 + 1$
5	$16x^5 - 20x^3 + 5x$
6	$32x^6 - 48x^4 + 18x^2 - 1$
7	$64x^7 - 112x^5 + 56x^3 - 7x$
8	$128x^8 - 256x^6 + 160x^4 - 32x^2 + 1$
9	$256x^9 - 576x^7 + 432x^5 - 120x^3 + 9x$

Substituting Eqn. 41 into Eqn. 42 to remove the time parameter then yields

$$y = \cos(n \cos^{-1} x), \tag{43}$$

which is seen to be the same as Eqn. 38. That is, what's displayed on an oscilloscope driven in this fashion is actually the Chebyshev polynomial for that order n, for values of $|x|$ up to unity. Over that interval, the function displayed looks very much like a sinusoid sketched on a piece of paper, wrapped around a cylinder, and then viewed from a distance.

A few Chebyshev polynomials are sketched crudely in Figure 22.19, and expressions for the first ten Chebyshev polynomials are listed in Table 22.6. From the foregoing equations, we may derive an expression for the filter order required to satisfy the specified constraints:

$$n = \frac{\cosh^{-1}(\sqrt{A_s^2 - 1}/\varepsilon)}{\cosh^{-1}(\omega_s/\omega_p)} \approx \frac{\cosh^{-1}(A_s/\varepsilon)}{\cosh^{-1}(\omega_s/\omega_p)}. \qquad (44)$$

This set of equations is similar in form to that for the Butterworth case, with the only difference being the replacement of the natural logarithm by $\cosh^{-1}$. Again, the approximation holds well if the square of the stopband attenuation is much greater than unity. And as with the Butterworth, the order as computed by Eqn. 44 should be rounded upward to the next integer value. Again, the resulting "excess" performance can be used to improve some combination of passband and stopband characteristics.

One way in which the Chebyshev is superior to a Butterworth is in the ultimate stopband attenuation provided. At high frequencies, a Butterworth with $\varepsilon = 1$ provides an attenuation that is approximately

$$\left| A\left(j\frac{\omega}{\omega_p}\right) \right|^2 \approx \left(\frac{\omega}{\omega_p}\right)^{2n}. \qquad (45)$$

Compare that asymptotic behavior with that of a Chebyshev (again with $\varepsilon = 1$):[24]

$$\left| A\left(j\frac{\omega}{\omega_p}\right) \right|^2 \approx 2^{2n-2}\left(\frac{\omega}{\omega_p}\right)^{2n}. \qquad (46)$$

Clearly the Chebyshev filter offers higher ultimate attenuation by an amount that is equal to $3(2n-2)$ dB for a given order. As a specific example, a seventh-order Chebyshev ultimately provides 36 dB more stopband attenuation than does a seventh-order Butterworth.

As another comparison, the relationship between the poles of a Butterworth and those of a Chebyshev of the same order can be put on a quantitative basis by normalizing the two filters to have precisely the same -3-dB bandwidth. It also may be shown (but not by us) that the -3-dB bandwidth of a Chebyshev may be reasonably well approximated by[25]

$$\cosh\left[\frac{1}{n}\sinh^{-1}\left(\frac{1}{\varepsilon}\right)\right]. \qquad (47)$$

Since the diameter of a Butterworth's circular pole constellation is the -3-dB bandwidth, we normalize the Chebyshev's ellipse to have a major axis defined by Eqn. 47. The imaginary parts of the poles of a Chebyshev filter are the same as for the Butterworth, while the real parts of the Butterworth prototype are merely scaled by the factor

$$\tanh\left[\frac{1}{n}\sinh^{-1}\left(\frac{1}{\varepsilon}\right)\right] \qquad (48)$$

---

[24] This comparison should not mislead you into thinking that such large ripple values are commonly used. In fact, such filters are typically designed with ripple values below 1 dB.
[25] See e.g. M. E. Van Valkenburg, *Introduction to Modern Network Synthesis* (Wiley, New York, 1960), pp. 380–1. The original method is due to E. A. Guillemin.

to yield the real parts of the poles of a Chebyshev filter. Thus the design of a Chebyshev filter may be based on a prototype Butterworth, and it's trivial to design the latter.

There is one subtlety that requires discussion, however, and this concerns the source and termination impedances of a passive Chebyshev filter. From both the sketches and equations, it's clear that only odd-order Chebyshev polynomials have a zero value for zero arguments. Hence, the DC value of the filter transfer function will be unity for such polynomials (that is, the passband's first dip below unity occurs at some frequency above DC). For even-order Chebyshev filters, however, the filter's transfer function starts off at a dip, with a DC power transmission value of $1/(1 + \varepsilon^2)$, implying a termination resistance that is less than the source resistance. If – as is usually the case – such an impedance transformation is undesired, choose an odd-order Chebyshev filter or add an impedance transformer to an even-order Chebyshev filter. As the former is less complex, odd-order Chebyshev realizations are the near-universal choice in practice.

Finally, recognize that the elliptical pole distribution implies that the ratio of the imaginary to real parts of the poles, and hence the Qs of the poles, are higher for Chebyshevs than for Butterworths of the same order. As a result, Chebyshev filters are more strongly affected by the finite Q of practical components. The problem increases rapidly in severity as the order of the filter increases. This important practical issue must be kept in mind when choosing a filter type.

Given that the Chebyshev's pole locations are closely related to those of the Butterworth, it should not be surprising that the element values for the two filters are related as well. In fact, element values for a Chebyshev filter may be derived from those of a Butterworth with the aid of the following sequence of equations.[26] First we compute a parameter, β, to simplify the expressions that follow:

$$\beta = \sinh\left(\frac{\tanh^{-1}\left(1/\sqrt{1+\varepsilon^2}\right)}{n}\right). \tag{49}$$

The element values (again, normalized to 1 rps and 1 Ω) are then

$$c_1 = b_1/\beta \tag{50}$$

and

$$c_k = \frac{b_k b_{k-1}}{c_{k-1}\left(\beta^2 + \left\{\sin\left[\frac{(k-1)\pi}{n}\right]\right\}^2\right)}, \tag{51}$$

where the various b_k are again the normalized Butterworth element values (for $\varepsilon = 1$):

$$b_k = 2\sin\left[\frac{(2n-1)\pi}{2n}\right]. \tag{52}$$

---

[26] One may derive these equations (or their equivalents) using the methods outlined in Section 22.6, but the particular ones presented here (with minor changes in variables) are those found in David B. Rutledge's excellent text, *The Electronics of Radio* (Cambridge University Press, 1999).

Table 22.7. *Element values for third-order 1-dB–1-GHz, 30-dB–3-GHz Chebyshev filter*

L first	C first
$L_1 = 16.104$ nH	$C_1 = 6.442$ pF
$C_2 = 3.164$ pF	$L_2 = 7.911$ nH
$L_3 = 16.104$ nH	$C_3 = 6.442$ pF

It's important to note that these equations apply only for odd values of n, because they assume equal source and load terminations. As mentioned, even-order Chebyshev filters require unequal source and load terminations (unless transformers are allowed) and are therefore used less frequently than those of odd order.[27] The equation set presented here thus suffices for the vast majority of applications.

Carrying out the computations for our ongoing filter design example, we first determine the minimum order required from

$$n = \frac{\cosh^{-1}\left(\sqrt{A_s^2 - 1}/\varepsilon\right)}{\cosh^{-1}(\omega_s/\omega_p)} \approx \frac{\cosh^{-1}(31.6/0.5088)}{\cosh^{-1}([3 \text{ GHz}]/[1 \text{ GHz}])} \approx \frac{4.8222}{1.7267} = 2.73. \quad (53)$$

So, for these specifications, a third-order Chebyshev suffices. As with the Butterworth example, we will arbitrarily choose to meet the specification exactly at the passband, leaving the excess performance for the stopband. A more practical choice would be to distribute the excess performance between the two, but we will continue nonetheless. The resulting third-order filter has the nominal component values listed in Table 22.7.

The frequency response for the nominal design appears as shown in Figure 22.20; with 5% component tolerances, the spread in filter transfer characteristics appears as shown in Figure 22.21.

22.5.3 TYPE II (INVERSE) CHEBYSHEV FILTERS

We have alluded several times to the possibility of realizing a flatter-than-maximally flat transfer characteristic. The Type II (also known as an inverse or reciprocal) Chebyshev filter achieves such flatness by permitting ripple in the stopband while continuing to provide passband monotonicity.

The Type II filter derives from the Type I (ordinary) Chebyshev through a pair of simple transformations. In the first step, the Type I Chebyshev response is simply subtracted from unity, leading to the conversion of a low-pass filter into a high-pass

---

[27] However, note that an impedance transformation is sometimes desired. In such cases, even-order Chebyshev filters might be quite useful.

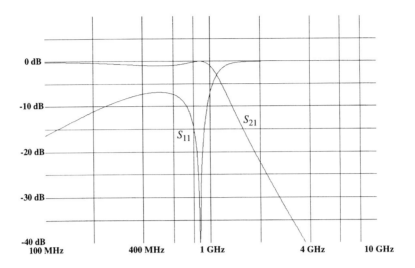

FIGURE 22.20. Frequency response of nominal design

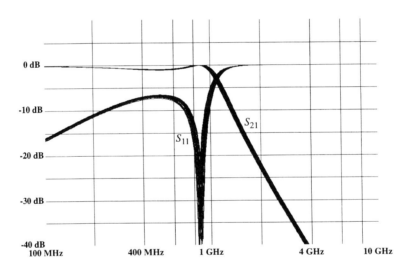

FIGURE 22.21. Monte Carlo simulation of 1-GHz, 1-dB, third-order Chebyshev filter (5% tolerance)

one. The resulting response is monotonic in the new passband, because the Type I response is monotonic in its stopband. All we have to do next is figure out a way to convert this high-pass filter back into a low-pass filter while preserving this monotonicity. The key to this second step is to replace ω by $1/\omega$. Since high frequencies are thus mapped into low ones and vice versa, this second transformation indeed converts the filter shape back into a low-pass response, but in a way that exchanges the ripple at low frequencies with ripple at high frequencies. This transformation thus restores a monotonic passband, and it also happens to map the Type I passband edge into the new stopband edge. The trade-offs are such that a flatter passband response may be obtained at the expense of a lack of stopband monotonicity.

22.5 COMMON FILTER APPROXIMATIONS

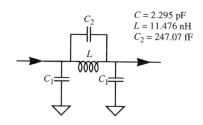

FIGURE 22.22. Low-pass 1-GHz inverse Chebyshev example

Mathematically, the transformations described result in the following power response for a Type II filter:

$$|H(j\omega)|^2 = 1 - \frac{1}{1+\varepsilon^2 C_n^2(\omega_p/\omega)} = \frac{\varepsilon^2 C_n^2(\omega_p/\omega)}{1+\varepsilon^2 C_n^2(\omega_p/\omega)}. \tag{54}$$

Normalized inverse Chebyshev filters thus have poles located at the reciprocals of the "normal" (Type I) Chebyshev in addition to purely imaginary zeros distributed in some complicated fashion. Just as a complex pole pair provides peaking, a complex zero pair provides nulling. We've seen this behavior already, where the purely imaginary zeros of m-derived filters provide notches of infinite depth. The inverse Chebyshev filter exploits these nulls to provide a flat passband without degrading the transition from passband to stopband. The resulting pole–zero constellation roughly resembles the Greek letter Ω rotated counterclockwise by 90°.

Although the Type II filter is not encountered as often as the Butterworth, its relative rarity should not be taken to imply a corresponding lack of utility. Despite the superior passband flatness provided by the inverse Chebyshev, the lack of simple equations for the component values has allowed the Butterworth filter to dominate in those applications where passband monotonicity is allegedly prized. Fortunately, the ready availability of filter design software is changing this situation. For example, the program *LADDER* synthesizes a wide variety of passive filters, among which is the inverse Chebyshev. Using that program, we find the network of Figure 22.22 for a low-pass filter (1-dB error at 1 GHz, at least 30-dB attenuation at 3 GHz).

We have omitted synthesis of the dual network with series inductances at the two ports, as well as a series *LC* trap between their common point and ground. Figure 22.23 shows that the inverse Chebyshev does indeed provide a more rapid transition from passband to stopband than a comparable Butterworth, while maintaining monotonic passband response. At the same time, the average return loss in the passband is larger than for the ordinary Chebyshev.

22.5.4 ELLIPTIC (CAUER) FILTERS

We have seen that allowing ripple in the passband or stopband confers desirable attributes, so perhaps it is not surprising that further improvements in transition steepness may be provided by allowing ripple in both the passband and stopband

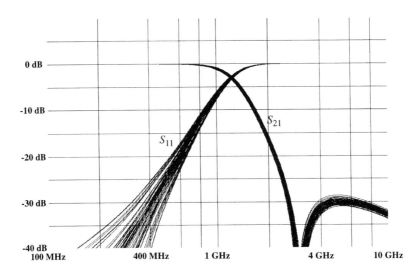

FIGURE 22.23. Monte Carlo simulation of inverse Chebyshev (5% tolerance)

simultaneously.[28] As in the inverse Chebyshev and m-derived filters, Cauer (also known as elliptic) filters exploit the nulls provided by finite zeros to create a dramatic transition from passband to stopband – at the expense of a stopband response that later bounces back some amount beyond the null frequency (again, just as in m-derived and inverse Chebyshev filters, and for the same reasons). As discussed further in Section 22.7, the name *elliptic* comes from the appearance of elliptic functions in the mathematics and should not be confused with the elliptic pole distribution of a Chebyshev filter.

Wilhelm Cauer is the inventor whose deep physical insights (and intimate familiarity both with the nulls of Zobel's m-derived filters and with elliptic functions in general) allowed him first to recognize that these additional degrees of freedom existed and then to exploit them, even though his first public disclosure of the elliptic filter offered no formal mathematical proof of the correctness of his ideas.[29] At a time when minimizing component count was an obsession, Cauer was able to use fewer

---

[28] These are also sometimes known as Darlington, Cauer–Chebyshev, generalized Chebyshev, or Zolotarev filters. Igor Ivanovich Zolotarev (1847–1878), who had studied with Chebyshev, evidently derived Chebyshev functions a decade or so before Chebyshev did. His significant contributions include numerical approximations to elliptic functions and the use of rational functions to generate minimax approximations. In 1878 he also demonstrated, by direct experiment, the tragic results of train–human momentum transfer.

[29] Cauer (1900–1945) became familiar with elliptic functions while studying at the University of Göttingen with the brilliant and infamously absentminded David Hilbert. Once Hilbert suddenly asked a close friend, physicist James Franck, "Is your wife as mean as mine?" Franck managed to respond, "Why, what has she done?" Hilbert answered, "I discovered today that my wife does not give me an egg for breakfast. Heaven only knows how long this has been going on."

Regrettably, stories about Cauer are not as lighthearted. He was shot to death during the Soviet occupation of Berlin in the closing days of WWII in a manner sadly reminiscent of the end of Archimedes (see ⟨http://www-ft.ee.tu-berlin.de/geschichte/th_nachr.htm⟩).

inductors than the best filters that were then in use. According to lore, publication of his patent reportedly sent Bell Labs engineers and mathematicians scurrying off to the New York City Public Library to bone up for weeks on the then- (and still-) obscure literature on elliptic functions.[30] Given that it took the brains at Bell Labs that amount of time, we have no hope of doing much more than present an outline. So that's what we'll do.

Elliptic filters have the following power transmission behavior:

$$|H(j\omega)|^2 = \frac{1}{1 + \varepsilon^2 F_n^2(\omega/\omega_p)}, \tag{55}$$

where $F_n(x)$ is a ratio of two polynomials rather than a simple polynomial alone (as in the Butterworth or Chebyshev case, for example). A key observation is that, for a given order, such a rational function has additional degrees of freedom (more coefficients we can manipulate to advantage) relative to a simple polynomial. These additional degrees of freedom permit the satisfaction of specifications with a lower-order filter. So the basic motivating idea isn't at all hard to grasp, but the devil's definitely in the details.

Just as with Chebyshev polynomials, these rational functions $F_n(x)$ (known as Chebyshev rational functions) have a magnitude which oscillates between 0 and +1 for arguments $|x|$ smaller than unity (corresponding to the passband) and which grows rapidly for arguments outside of that range and as the order n increases.[31] However, unlike Chebyshev polynomials, whose magnitudes grow monotonically outside of that range, $F_n(x)$ oscillates in some fashion between infinity and a specified finite value. Hence the filter response exhibits stopband ripples, with a finite number of frequencies at which the filter transmission is zero. The attenuation poles correspond to transmission zeros (notches) in whose proximity the filter response changes rapidly. Thus, perhaps you can see how permitting such ripples in the stopband enables a much more dramatic transition from passband to stopband, thus allowing one to combine the passband attributes of ordinary Chebyshev filters with the stopband attributes of inverse Chebyshev filters.

We can deduce some important facts if we perform the same operations on the elliptic filter's transfer function as those that convert a Chebyshev into an inverse Chebyshev. First express the rational Chebyshev function explicitly as the ratio of two polynomials:

$$|H(j\omega)|^2 = \frac{1}{1 + \varepsilon^2 F_n^2(\omega/\omega_p)} = \frac{1}{1 + \varepsilon^2 \dfrac{N(\omega/\omega_p)}{D(\omega/\omega_p)}}$$

$$= \frac{D(\omega/\omega_p)}{D(\omega/\omega_p) + \varepsilon^2 N(\omega/\omega_p)}, \tag{56}$$

where N and D are the numerator and denominator polynomials of F_n^2, respectively.

---

[30] M. E. Van Valkenburg, *Analog Filter Design*, Harcourt Brace Jovanovich, New York, 1982, p. 379.
[31] Strictly speaking, Chebyshev polynomials are a special case of Chebyshev rational functions in which the denominator is unity.

Continuing, we subtract the power transfer function from unity and then reciprocate arguments:

$$|H_2(j\omega)|^2 = 1 - |H(j\omega)|^2 = \frac{\varepsilon^2 N(\omega/\omega_p)}{D(\omega/\omega_p) + \varepsilon^2 N(\omega/\omega_p)}; \qquad (57)$$

$$|H_3(j\omega)|^2 = \frac{\varepsilon^2 N(\omega_p/\omega)}{D(\omega_p/\omega) + \varepsilon^2 N(\omega_p/\omega)}. \qquad (58)$$

We see that H_3 is also an elliptic filter and of the same order as the original, H. Given that an elliptic filter is minimax optimal in both passband and stopband, the uniqueness of the optimality itself implies that H and H_3 have the same poles and zeros. From comparing the numerators, we deduce that the roots of N are just the reciprocals of the roots of D. Thus, computation of the roots of D (which are the zero locations of the overall filter) directly allows computation of the roots of N. Once we combine that knowledge with the value of the specified passband ripple, we have all of the information necessary to complete the transfer function.

As with nearly everything else related to elliptic filters, derivation of the equations for the required filter order is difficult and would take us too deep into arcane areas of mathematical trivia (the reader may feel that this has already occurred). As long as you are willing to set aside your natural curiosity about where the equation comes from, however, you can perform the necessary computations nonetheless:

$$n = \frac{F(m)F(1-m')}{F(m')F(1-m)}, \qquad (59)$$

where $F(m)$ is the complete elliptic integral of the first kind (see Section 22.7 for its definition and some numerical methods for computing it), m is a function of the normalized stopband frequency,

$$m = 1/\omega_s^2, \qquad (60)$$

and m' is a function of the stopband and passband ripple parameters,

$$m' = (\varepsilon_s/\varepsilon_p)^2. \qquad (61)$$

Although an exact closed-form expression for $F(m)$ doesn't exist, a highly accurate approximation is[32]

$$F(m) \approx (a_0 + a_1 m + a_2 m^2 + a_3 m^3 + a_4 m^4) \\ + (b_0 + b_1 m + b_2 m^2 + b_3 m^3 + b_4 m^4) \ln(1/m), \qquad (62)$$

where the various coefficients are as listed in Table 22.8. For $0 < m < 1$, the approximation error is no greater than 2×10^{-8}. Note that the parameter m is simply the square of the elliptic modulus k described in Section 22.7:

$$m = k^2. \qquad (63)$$

---

[32] M. Abramowitz and I. Stegun, *Handbook of Mathematical Functions*, Dover, New York, 1965.

Table 22.8. *Coefficients for approximation to complete elliptic integral of the first kind*

a_n	b_n
$a_0 = 1.38629436112$	$b_0 = 0.5$
$a_1 = 0.09666344259$	$b_1 = 0.12498593597$
$a_2 = 0.03590092383$	$b_2 = 0.06880248576$
$a_3 = 0.03742563713$	$b_3 = 0.03328355346$
$a_4 = 0.01451196212$	$b_4 = 0.00441787012$

Once the required order n is determined, the next task is to find the normalized zero locations. Here again, as long as we don't ask for a derivation first, the calculations themselves are not too bad. There are numerous, highly accurate numerical approximations for computing the elliptic functions that give the locations of the zeros directly (and solely as functions of the ratio ω_s/ω_p). Formally, these are given by

$$\pm j \frac{1}{k\{\text{cd}[(2i-1)\frac{K}{n}, k]\}}, \quad i = 1, 2, \ldots, \text{Int}\left(\frac{n}{2}\right). \tag{64}$$

Again, various methods abound for obtaining actual numerical values from that expression, but a simple (yet highly accurate) closed-form equation for the zero locations is:

$$\pm j \frac{1}{\sqrt{k}} \frac{F_4(i)}{F_3(i)}, \quad i = 1, 2, \ldots, \text{Int}\left(\frac{n}{2}\right). \tag{65}$$

The functions F_3 and F_4 are given by

$$F_3(i) = F_{34e}(i) - F_{34o}(i), \tag{66}$$

$$F_4(i) = F_{34e}(i) + F_{34o}(i), \tag{67}$$

where

$$F_{34e}(i) \approx \alpha^i + \alpha^{(8n-3i)} + \alpha^{(8n+5i)} + \alpha^{(32n-7i)} + \alpha^{(32n+9i)}, \tag{68}$$

$$F_{34o}(i) \approx \alpha^{(2n-i)} + \alpha^{(2n+3i)} + \alpha^{(18n-5i)} + \alpha^{(18n+7i)}$$
$$+ \alpha^{(50n-9i)} + \alpha^{(50n+11i)}. \tag{69}$$

In turn,

$$\alpha^{2n} = \lambda + 2\lambda^5 + 15\lambda^9 + 150\lambda^{13} + 1707\lambda^{17} + \cdots, \tag{70}$$

where

$$\lambda = \frac{1}{2}\left[\frac{1-\sqrt{k}}{1+\sqrt{k}}\right]. \tag{71}$$

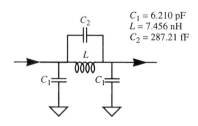

FIGURE 22.24. Low-pass 1-GHz elliptic filter example

The foregoing sequence of equations follows closely (but not uniformly) the notational conventions used by Orchard.[33] The approximations are carried out to more terms than you will ever need for any practical filter design but are still simple enough to use in spreadsheets, for example.

Finding the zero locations enables the subsequent synthesis of an actual network, using one of any number of methods that have been developed over the decades.[34]

As with the inverse Chebyshev filter, one may choose how to distribute the frequencies of the resonant traps along the filter. When carrying out a synthesis, it is not uncommon to discover that some choices produce difficulties, such as requiring negative inductors. Experience shows that assigning the lowest frequency null to the resonator in the central position – and then progressively working outward in alternation – greatly reduces (but does not eliminate) the probability of such difficulties. For example, if there are five null frequencies numbered 1–5 in order of increasing frequency, then the best implementation would likely involve resonator tunings ordered as 4-2-1-3-5. The program *LADDER* uses a synthesis method that avoids problems with realizability.

Continuing with our low-pass example, *LADDER*'s synthesis results are as shown in Figure 22.24. Running a Monte Carlo simulation on this filter produces the plots in Figure 22.25. We see that the elliptic filter has the most dramatic transition to stopband of all the implementations we've examined. In this particular case, the stopband response pops back up to a maximum of about -46 dB (not visible on the plot).

---

[33] H. J. Orchard, "Computation of Elliptic Functions of Rational Fractions of a Quarterperiod," *IRE Trans. Circuit Theory,* December 1958, pp. 352–5. Although in subsequent publications Orchard renounces the use of these closed-form numerical techniques (based on *theta functions*) in favor of iterative algorithms (in no small measure because they have the pedagogical advantage of explicitly showing the relationship between the poles of Chebyshev and elliptic filters), it's still more convenient to use the former in constructing spreadsheets, for example.

[34] A particularly ingenious method is presented by Pierre Amstutz in "Algorithms for Elliptic Filter Design on Small Computers," *IEEE Trans. Circuits and Systems,* December 1978. Unfortunately, Amstutz chooses an idiosyncratic normalization of elliptic functions. The unique notational conventions he uses make it *much* harder than necessary for the uninitiated to follow the mathematical details in the paper. He compensates for this transgression by providing complete source code (in Fortran) for his filter design algorithm. His algorithm nonetheless occasionally produces designs that are unrealizable.

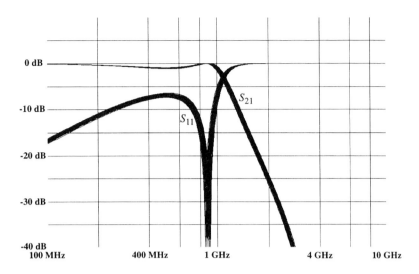

FIGURE 22.25. Monte Carlo simulation of elliptic filter of Figure 22.24 (5% tolerance)

22.5.5 BESSEL–THOMSON MAXIMALLY FLAT DELAY FILTERS

The filters we've examined so far are designed to meet specifications on the magnitude response. Because the design methods thus neglect phase behavior, many of these filters can exhibit significant delay variation over their passband. That is, the filter may not delay all Fourier components of an input signal by a uniform amount. As these misaligned Fourier components superpose to form the output, there may be serious distortion of the waveform's shape. This dispersive behavior is a particularly serious concern in digital systems, where pulse shapes need to be preserved to avoid intersymbol interference.

Because a constant time delay implies a phase shift that is linearly proportional to frequency, dispersion is minimized in a filter whose phase approximates this linear behavior as closely as possible. Just as conventional filter design neglects phase behavior in its focus on magnitude response, so will we neglect magnitude behavior in crafting the phase response of maximally flat time-delay filters.

The most elegant and efficient derivation of the filter transfer functions is due to Storch.[35] We begin by noting that the transfer function of a unit time delay is simply

$$H(s) = e^{-s}, \qquad (72)$$

and then note that e^s may be expressed as the sum of two hyperbolic functions as follows:

$$e^s = \sinh s + \cosh s. \qquad (73)$$

---

[35] L. Storch, "Synthesis of Constant Time Delay Ladder Networks Using Bessel Polynomials," *Proc. IRE*, v. 42, 1954, pp. 1666–75.

Next, express sinh s and cosh s each as a simple series, and divide one by the other to yield a continued fraction expansion of coth s:

$$\sinh s = s + \frac{s^3}{3!} + \frac{s^5}{5!} + \cdots \tag{74}$$

and

$$\cosh s = 1 + \frac{s^2}{2!} + \frac{s^4}{4!} + \cdots, \tag{75}$$

so that

$$\coth s = \frac{1}{s} + \cfrac{1}{\cfrac{3}{s} + \cfrac{1}{\cfrac{5}{s} + \cfrac{1}{\cfrac{7}{s} + \cdots}}}. \tag{76}$$

Continued fraction expansions are attractive for easily generating a sequence of progressively better approximations. Simply truncate the expansion after the number of terms corresponding to the desired filter order. Summing together the numerator and denominator polynomials therefore provides an approximation to the sum of sinh and cosh, and thus to e^s, according to Eqn. 73. Storch's remarkable observation (with accompanying proof) is that the approximations formed this way have maximally flat time delay, whereas those formed directly from the conventional series expansion for e^s do not. Important too is his identification of these polynomials with a known class of polynomials previously studied by Bessel. Despite all of this marvelous work, however, the corresponding class of filters is nonetheless known as Bessel, Thomson, or Bessel–Thomson, because one W. E. Thomson had published a paper about them about five years earlier – although with less elegant methods and without recognizing the relationship with Bessel polynomials.[36] This last consideration is significant because Bessel polynomials are readily generated by well-known (to mathematicians, anyway) recurrence formulas. For the particular class of Bessel polynomials relevant to the optimal delay case, one may use the following simple recurrence relation to bypass the need for evaluating continued fractions:

$$P_n(s) = (2n-1)P_{n-1}(s) + s^2 P_{n-2}(s), \tag{77}$$

where the first two polynomials are

$$P_1(s) = s + 1 \tag{78}$$

and

$$P_2(s) = s^2 + 3s + 3 \tag{79}$$

to help get you started.

Coefficients for the first six Bessel polynomials are given in Table 22.9 (we omit the leading coefficient, since it is unity in all cases).

---

[36] W. E. Thompson, "Delay Networks Having Maximally Flat Frequency Characteristics," *Proc. IRE*, pt. 3, v. 96, 1949, pp. 487–90.

22.5 COMMON FILTER APPROXIMATIONS

Table 22.9. *Bessel polynomial coefficients*

n	a_5	a_4	a_3	a_2	a_1	a_0
1						1
2					3	3
3				6	15	15
4			10	45	105	105
5		15	105	420	945	945
6	21	210	1260	4725	10395	10395

One subtlety is that the polynomials generated by either of these two methods do not have unit DC value in general, so one must provide a suitable normalization constant; thus,

$$H(s) = \frac{P_n(0)}{P_n(s)}. \qquad (80)$$

A second subtlety is the need for a normalization in frequency: The various $H(s)$ generated by the polynomial recursion formula arise from approximating a unit time delay, so bandwidth is not controlled to a uniform value. To normalize to a constant bandwidth, compute the 3-dB corner frequency for each $H(s)$ and then scale s (and component values) accordingly to produce a 1-rps corner frequency. These normalizations presume that the goal is to produce a filter with a specified corner frequency while providing a close approximation to a uniform time delay. If one is instead interested in simply providing the best approximation to a time delay over the widest possible bandwidth, there is no need to perform such normalizations.[37]

Note that the methods described directly yield the transfer function rather than the magnitude squared, reducing the work expended in searching for a "Hurwitz polynomial" in the network synthesis recipe described in Section 22.6. Regrettably, and unlike the Butterworth and Chebyshev cases, there do not seem to be simple formulas for Bessel filters that give the component values directly. We therefore resort to the use of our standard synthesis recipe to generate the entries in a table of normalized component values (Table 22.10; more entries may be found in Section 22.8).

There also seem to be no simple design formulas for selecting the order based, say, on a specification of the acceptable deviation from a prescribed nominal delay. However, inspection of delay-frequency curves for Bessel filters of various orders reveals that the product of the delay bandwidth BW_D (where the delay has dropped

---

[37] However, far better alternatives are available if all one desires is a time-delay approximation. The Bessel–Thomson filter is a minimum phase filter (all zeros are in the left half-plane) and is thus limited in its approximating power. By allowing the use of right half-plane zeros, one can double the number of degrees of freedom for a given filter order. A class of approximations known as *Padé approximants* exploits this observation by providing poles and zeros in mirror-image pairs. The corresponding networks are all-pass in nature and thus provide no filtering action.

Table 22.10. *Normalized element values for 1-rps Bessel–Thomson low-pass filters*

$C_1\ (L_1)$	$L_2\ (C_2)$	$C_3\ (L_3)$	$L_4\ (C_4)$	$C_5\ (L_5)$	$L_6\ (C_6)$
2.0000					
2.1478	0.5755				
2.2034	0.9705	0.3374			
2.2404	1.0815	0.6725	0.2334		
2.2582	1.1110	0.8040	0.5072	0.1743	
2.2645	1.1126	0.8538	0.6392	0.4002	0.1365

Table 22.11. *Denormalized element values for fifth-order, 3-dB, 1-GHz Bessel–Thomson low-pass filter*

C_1	L_2	C_3	L_4	C_5
7.1881 pF	8.8411 nH	2.5592 pF	4.0362 nH	0.5548 pF

Table 22.12. *Final element values for fifth-order, 1-dB, 1-GHz Bessel–Thomson low-pass filter*

C_1	L_2	C_3	L_4	C_5
4.2194 pF	5.1897 nH	1.5022 pF	2.3692 nH	0.3256 pF

3 dB from its low-frequency value) and the nominal time delay is given by a simple (and very crude) approximation:

$$(BW_D)(T_D) \approx n - 0.4. \tag{81}$$

A tighter specification on allowable delay deviation requires an increase in filter order, and Eqn. 81 at least provides some guidance. In general, as the delay bandwidth increases well beyond the ordinary (magnitude) bandwidth, nonlinear phase becomes progressively less of a concern, simply because improperly delayed Fourier components are not a significant problem as long as they are strongly attenuated. In any event, filter design software such as *LADDER* generates Bessel filters with ease.

Let's work out an example, just to round out our collection of designs. Arbitrarily choose a fifth-order, 1-GHz filter with shunt capacitances at the ends. Denormalizing the values from Table 22.10 for a 50-Ω system and this frequency yields the values listed in Table 22.11.

Simulation of this filter reveals that the −1-dB point occurs at about 587 MHz, so we need to perform a second renormalization. In this case, we need to multiply all element values by 0.587 to move the −1-dB frequency to 1 GHz; see Table 22.12.

The Monte Carlo simulations for this filter (Figure 22.26) reveal a fairly low sensitivity to parameter variation. However, the filter does not provide a 30-dB attenuation

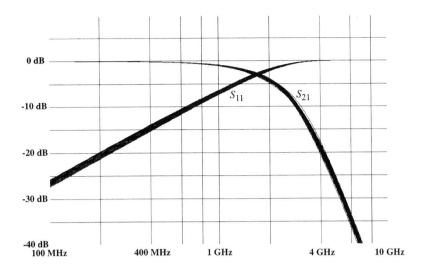

FIGURE 22.26. Monte Carlo simulation of fifth-order, 1-dB, 1-GHz, Bessel–Thomson low-pass filter (5%)

until around 5 GHz, instead of the design target of 3 GHz. To meet stopband specifications, we would need to increase the order.

In comparing the various filter types, we observe a general tendency: The more ideal a filter's magnitude characteristic, the worse its time-delay flatness. For example, a Bessel filter has a relatively poor magnitude flatness but good time-delay uniformity. An elliptic filter meeting similar magnitude characteristics will generally have worse (sometimes much worse) time-delay behavior. One intuitive reason is that elliptic filters meet magnitude specifications with fewer poles. A lower order therefore implies less overall potential phase shift. A smaller total possible phase shift in turn implies a diminished ability to provide a linearly increasing phase shift over a given frequency range.

Space constraints prevent us from doing more than simply alluding to other possibilities for approximating linear phase behavior. The Bessel–Thomson filter provides maximally flat delay and is therefore the delay counterpart of the Butterworth magnitude filter. We may extend this idea of delay-magnitude counterparts and imagine the delay counterpart of a Chebyshev filter, for example, which would provide an equiripple approximation to constant delay. Such a filter would provide a larger delay bandwidth for a given order.

Another option is to compensate for a filter's poor phase response with another filter known as a phase equalizer. Such an equalizer would have a constant gain at all frequencies (and therefore would be an all-pass network), just like a piece of wire, yet possess a controllable nonzero phase shift (unlike a piece of wire). Such filters have zeros in the right half-plane and mirror-image poles in the left half-plane. Zeros in the right half-plane have the same magnitude behavior as those in the left half-plane but the *phase* behavior of left half-plane *poles*. As such, these are nonminimum phase networks.

As a final comment, it's important to note that the various transformations that generate other filter shapes from low-pass prototypes preserve only the magnitude characteristics, without exercising any explicit control over the phase behavior. Thus the low-pass–bandpass transformation regrettably does *not* preserve the uniform time-delay characteristic of Bessel–Thomson filters. A direct synthesis of a linear-phase bandpass filter is required to obtain a maximally flat bandpass delay characteristic.

In the next chapter we'll examine how to transform these lumped filters into forms that are amenable to realization with microstrip components.

22.6 APPENDIX A: NETWORK SYNTHESIS

The path from filter response specifications to a realizable network is perhaps not as straightforward as one would like. Indeed, there are many possible paths in general, each with its own particular trade-offs among intuitive appeal, computational complexity, and robustness. Unfortunately, we can't hope to provide a comprehensive examination of so sophisticated a subject (an ongoing apology throughout this chapter), but we offer this brief synopsis to make the overall process of synthesis perhaps a little less mysterious and to orient the interested reader toward the relevant literature for proceeding further. As an adjunct, we also provide some background material on elliptic functions, because that subject in particular is ignored in most electrical engineering curricula.

The birth of modern network synthesis is often dated to the publication of Otto Brune's doctoral thesis in 1931.[38] Sidney Darlington (inventor of the Darlington pair, among many other achievements) extended Brune's work on one-ports to the synthesis of two-ports, publishing his own much-referenced doctoral thesis in 1939.[39] Additional important contributions by Foster, Cauer, and others helped place the subject on a firm footing. Emphasis gradually shifted away from a preoccupation with proofs of realizability and toward the development of practical methods for synthesis. As these methods evolved, so did their ability to accommodate imperfect components and more sophisticated constraints on filter response. Thanks to these developments, abetted by the increasing availability of machine computation, modern filter synthesis finally began to dominate around the late 1950s.

We now consider an infinitesimal subset of that work, examining one possible way to synthesize a filter that consists only of purely lossless elements.

A RECIPE FOR LOSSLESS LADDER NETWORK SYNTHESIS

A classical method for synthesizing lossless ladder networks relies on the relationship between the transfer function of a lossless two-port and its input impedance. This

---

[38] "Synthesis of a Finite Two-Terminal Network whose Driving Point Impedance Is a Prescribed Function of Frequency," *J. Math. Phys.*, v. 10, 1931, pp. 191–236.

[39] "Synthesis of Reactance 4-Poles which Produce Prescribed Insertion Loss Characteristics," *J. Math. Phys.*, v. 18, September 1939, pp. 257–353.

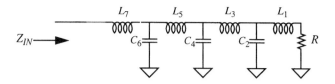

FIGURE 22.27. Terminated *LC* ladder network

approach makes sense because methods for synthesizing one-port networks with a prescribed impedance were developed first. Thus, we determine first the input impedance function implied by the desired filter transfer function (which, in turn, is determined from the design constraints applied to the chosen filter type) and then expand the input impedance function in continued fraction form. In many cases of practical relevance, the component values are easily read off by inspection of the terms in the expansion (this ease motivated Cauer's enthusiastic advocacy of continued fraction expansions). Fortunately, the Butterworth, Chebyshev, and Bessel–Thomson filters are among those for which relatively simple synthesis using these methods is possible, provided we assume the use of ideal, lossless inductors and capacitors. Synthesis of filters with lossy elements and/or with imaginary zeros (such as the inverse Chebyshev and Cauer types) regrettably requires the use of somewhat more sophisticated procedures than we may describe here.[40]

To proceed, let's first derive an expression for the input impedance of a ladder network in terms of a continued fraction expansion in order to explain Cauer's enthusiasm. Specifically, consider the ladder network of Figure 22.27, which is terminated by a resistance R (an LC ladder is shown as a particular example, but the derivation that follows applies more generally).[41]

The input impedance for such a network may be found a number of ways. The natural choice for most people is to start at the input side and proceed toward the load end, but there is an advantage to starting at the load and working backward to the input, as we'll see. So, we start with the impedance seen to the right of C_2:

$$Z_1 = sL_1 + R. \tag{82}$$

In turn, the parallel combination of Z_1 and the impedance of C_2 has an admittance

$$Y_2 = sC_2 + \frac{1}{Z_1} = sC_2 + \frac{1}{sL_1 + R}. \tag{83}$$

Working back one more step to the impedance seen to the right of C_4, we have

---

[40] For details on more general network synthesis methods, see M. E. Van Valkenburg, *Introduction to Modern Network Synthesis,* Wiley, New York, 1960. Also see the extensive design data presented in A. Zverev, *Handbook of Filter Synthesis,* Wiley, New York, 1967.

[41] The basic method outlined here certainly can be generalized to accommodate unequal source and load terminations (particularly useful for synthesizing Chebyshev filters of even order), but for the sake of simplicity we will consider only the case of equal resistances.

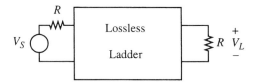

FIGURE 22.28. Terminated lossless ladder

$$Z_3 = sL_3 + \frac{1}{Y_2} = sL_3 + \frac{1}{sC_2 + \frac{1}{sL_1 + R}}. \quad (84)$$

Clearly, we may continue working backward until we finally obtain an expression for the input impedance. From inspection of the form of the expressions, you can see why the term *continued fraction expansion* is appropriately descriptive. More important, notice that the first new term of each successive step in the expansion magically yields the individual ladder element values. If we are able to express an input impedance in this form, we can readily read off the component values, essentially by inspection.

To use this observation in network synthesis, we need to find a way to link input impedance explicitly with the transfer function to be synthesized. Fortunately, it's conceptually straightforward to do so. The power transfer function tells us how much input power makes it to the load; we can then easily compute how much is reflected back to the source. Using the same relationship between load impedance and reflection coefficient used in developing the Smith chart permits the derivation of the filter input impedance.

Darlington's method for implementing this set of operations begins with a simple application of energy conservation to the system depicted in Figure 22.28. Because the ladder itself is lossless, the average power delivered to the load must equal that supplied to the input of the ladder. Recognizing that power is transferred only to the real part, R_{in}, of the input impedance Z_{in}, the power balance criterion implies

$$|I_{in}|^2 R_{in} = \frac{|V_L|^2}{R}. \quad (85)$$

Noting in turn that the input current is simply

$$I_{in} = \frac{V_S}{R + Z_{in}} \quad (86)$$

allows us to write

$$\left(\left|\frac{V_S}{R + Z_{in}}\right|^2 R_{in} = \frac{|V_L|^2}{R}\right) \implies \left|\frac{V_L}{V_S}\right|^2 = \frac{RR_{in}}{|R + Z_{in}|^2}. \quad (87)$$

Having expressed the desired filter's transfer function in terms of the known termination and a network input impedance, we next determine that network impedance indirectly through an intermediate step (due to Darlington) employing an *auxiliary function* defined as follows:[42]

$$|A(j\omega)|^2 = 1 - 4\left|\frac{V_L}{V_S}\right|^2 = 1 - 4\frac{RR_{in}}{|R+Z_{in}|^2} = \frac{|R+Z_{in}|^2 - 4RR_{in}}{|R+Z_{in}|^2}. \tag{88}$$

We next expand the numerator of the final equation by expressing the input impedance explicitly in terms of its real and imaginary parts, leading to the following sequence of equations:

$$\frac{|R+Z_{in}|^2 - 4RR_{in}}{|R+Z_{in}|^2} = \frac{(R+R_{in})^2 + X_{in}^2 - 4RR_{in}}{|R+Z_{in}|^2} = \frac{R^2 + R_{in}^2 + X_{in}^2 - 2RR_{in}}{|R+Z_{in}|^2}, \tag{89}$$

which simplifies to

$$|A(j\omega)|^2 = \frac{|R-Z_{in}|^2}{|R+Z_{in}|^2}. \tag{90}$$

Letting $s = j\omega$ and solving for Z_{in} finally yields two expressions for the input impedance in terms of the auxiliary function and the termination resistance:

$$Z_{in} = R\frac{[1-A(s)]}{[1+A(s)]}; \tag{91}$$

$$Z_{in} = R\frac{[1+A(s)]}{[1-A(s)]}. \tag{92}$$

Note that these two solutions correspond to two distinct but equivalent networks. Furthermore, observe that the two expressions for Z_{in} are in fact reciprocals of each other if the termination resistance is of unit value. Using normalized values throughout the synthesis procedure thus simplifies the computation of the element values for the two solutions. One need only denormalize values as a final step after the hard part of the synthesis has been completed.

An illustrative example will help to elucidate details of the method and simultaneously expose some subtleties that require careful handling. Specifically, consider the Butterworth filter example of Section 22.5.1. The normalized squared-magnitude response of the desired filter is

$$|H(j\omega)|^2 = \frac{1}{1+\varepsilon^2(\omega/\omega_p)^{2n}}. \tag{93}$$

To maintain consistency with the development in this appendix, we note that

---

[42] The auxiliary function may seem somewhat mysterious initially, but note from its form that it is precisely the reflected power as a function of frequency. As stated in the main body of the text, there is a bi-unique relationship between reflection and load impedance (where the latter is here the input impedance of the terminated ladder), so knowing the reflection coefficient is equivalent to knowing the input impedance of the ladder. You may also recall that this bi-unique relationship between reflection coefficient and impedance is the basis for the Smith chart.

$$4\left|\frac{V_L}{V_S}\right|^2 = |H(j\omega)|^2 = \frac{1}{1+\varepsilon^2(\omega/\omega_p)^{2n}}, \qquad (94)$$

where the factor of 4 accounts for the fact that $|V_L/V_S|$ has a value of 0.5 at DC, according to the block diagram description of our filter. Computing the auxiliary function is then a straightforward exercise that, after normalizing to a unit passband frequency, yields

$$|A(j\omega)|^2 = 1 - 4\left|\frac{V_L}{V_S}\right|^2 = 1 - |H(j\omega)|^2 = \frac{\varepsilon^2\omega^{2n}}{1+\varepsilon^2\omega^{2n}}. \qquad (95)$$

We see that the factor of 4 inherent in the definition of Darlington's auxiliary function cancels out when a filter's transfer function template, $|H(j\omega)|$, has been normalized to unit DC gain, simplifying the math somewhat.

Now let $s = j\omega$ (so that $\omega^2 = -s^2$) and recognize that $|A(j\omega)|^2 = A(s)A(-s)$ with this substitution. That is, the squared magnitude of the auxiliary function is the product of two terms whose poles are algebraic inverses of each other. Here we encounter the first subtlety: We seek the function $A(s)$, but we are given only its magnitude squared. We actually want only that part of $|A(j\omega)|^2$ that has no poles or zeros in the right half-plane; in the language of network theorists, we require a *Hurwitz* (minimum phase) polynomial. We therefore need to find the roots of the numerator and denominator polynomials of Eqn. 95, discard the right half-plane roots, and then reconstruct a polynomial from the surviving roots in order to discover the $A(s)$ we're looking for.[43] If the polynomial cannot be factored readily by inspection or other convenient means, one must use a numerical root finder. This step is surprisingly fraught with peril for the unwary, because the numerical accuracy required in computing the roots can sometimes be ludicrously greater than the accuracy required of the final component values.[44] The numerator in the example of Eqn. 95 requires no special handling, so this root-finding business is limited to the denominator here.

In this particular example, we are fortunate because we can find the roots without excessive agony. We start with:

$$\left(|A(j\omega)|^2 = \frac{\varepsilon^2\omega^{2n}}{1+\varepsilon^2\omega^{2n}}\right) \Longrightarrow A(s)A(-s) = \frac{\varepsilon^2 s^{2n}}{1+\varepsilon^2 s^{2n}}, \qquad (96)$$

---

[43] If, as in this example, there are multiple roots at the origin, preserve half of them (they will always occur in pairs).

[44] Of the great many root-finding algorithms in existence, filter designers overwhelmingly favor Laguerre's method. Despite the lack of any formal proof of its general convergence properties, extensive experience has shown that this venerable method is surprisingly robust in practice. In particular, it handles the especially challenging case of repeated roots with remarkable grace. For detailed information on this algorithm, see W. H. Press et al., *Numerical Recipes*, Cambridge University Press (any edition). Also valuable is H. J. Orchard, "The Laguerre Method for Finding the Zeros of Polynomials," *IEEE Trans. Circuits and Systems*, v. 36, no. 11, November 1989, pp. 1377–81.

22.6 APPENDIX A: NETWORK SYNTHESIS

from which we see that we wish to solve for the poles of $A(s)$ by finding the values of s for which

$$1 + \varepsilon^2 s^{2n} = 0. \tag{97}$$

Equivalently, we wish to solve

$$s^{2n} = -\frac{1}{\varepsilon^2}. \tag{98}$$

We therefore seek values of s that produce a purely negative number. This observation is the key to the solution, for expressing the number -1 as an appropriate imaginary power of e allows us to write

$$s^{2n} = -\frac{1}{\varepsilon^2} = \frac{1}{\varepsilon^2} e^{j(2k+1)\pi}, \tag{99}$$

where the index k takes on values from 0 to $2n - 1$ (actually, any sequence of $2n$ consecutive integers will do).

The solutions to this equation are therefore

$$s_k = \frac{1}{\varepsilon^{(1/n)}} e^{[j(2k+1)\pi]/(2n)}. \tag{100}$$

For our fourth-order example:

$$s_0 = \frac{1}{\varepsilon^{(1/n)}} e^{j(\pi/8)}; \tag{101}$$

$$s_1 = \frac{1}{\varepsilon^{(1/n)}} e^{j(3\pi/8)}. \tag{102}$$

Those first two roots are in the right half-plane and so are ignored.

Continuing, we have

$$\left(s_2 = \frac{1}{\varepsilon^{(1/n)}} e^{j(5\pi/8)} \right)$$

$$\implies \tau_2 = \frac{1}{s_2} = \varepsilon^{(1/n)} e^{j(-5\pi/8)} \approx \varepsilon^{(1/n)} [-0.38268 - j0.92388]; \tag{103}$$

$$\tau_3 = \varepsilon^{(1/n)} e^{j(-7\pi/8)} \approx \varepsilon^{(1/n)} [-0.92388 - j0.38268], \tag{104}$$

$$\tau_4 = \varepsilon^{(1/n)} e^{j(-9\pi/8)} \approx \varepsilon^{(1/n)} [-0.92388 + j0.38268], \tag{105}$$

$$\tau_5 = \varepsilon^{(1/n)} e^{j(-11\pi/8)} \approx \varepsilon^{(1/n)} [-0.38268 + j0.92388], \tag{106}$$

which completes the discovery of the four left half-plane roots.

Note that we have chosen to express the roots as time constants. This maneuver allows reconstruction of the auxiliary function by multiplying together terms of the form

$$(\tau s + 1), \tag{107}$$

making it easier to express $A(s)$ with the correct coefficients to give it the required unit magnitude at high frequencies (see Eqn. 96).

We discard the remaining two roots,

$$s_6 = \frac{1}{\varepsilon^{(1/n)}} e^{j(13\pi/8)}, \tag{108}$$

$$s_7 = \frac{1}{\varepsilon^{(1/n)}} e^{j(15\pi/8)}, \tag{109}$$

because they have positive real parts.

Now that we have the four relevant roots, we can complete the derivation of the auxiliary function:

$$A(s) = \frac{\varepsilon s^4}{(\tau_2 s - 1)(\tau_3 s - 1)(\tau_4 s - 1)(\tau_5 s - 1)}. \tag{110}$$

Normalizing the termination to unit resistance gives us the desired input impedance functions:

$$Z_{in} = \frac{1 - \dfrac{\varepsilon s^4}{(\tau_2 s - 1)(\tau_3 s - 1)(\tau_4 s - 1)(\tau_5 s - 1)}}{1 + \dfrac{\varepsilon s^4}{(\tau_2 s - 1)(\tau_3 s - 1)(\tau_4 s - 1)(\tau_5 s - 1)}}; \tag{111}$$

$$Z_{in} = \frac{1 + \dfrac{\varepsilon s^4}{(\tau_2 s - 1)(\tau_3 s - 1)(\tau_4 s - 1)(\tau_5 s - 1)}}{1 - \dfrac{\varepsilon s^4}{(\tau_2 s - 1)(\tau_3 s - 1)(\tau_4 s - 1)(\tau_5 s - 1)}}. \tag{112}$$

These expressions simplify a little to

$$Z_{in} = \frac{(\tau_2 s - 1)(\tau_3 s - 1)(\tau_4 s - 1)(\tau_5 s - 1) - \varepsilon s^4}{(\tau_2 s - 1)(\tau_3 s - 1)(\tau_4 s - 1)(\tau_5 s - 1) + \varepsilon s^4}, \tag{113}$$

$$Z_{in} = \frac{(\tau_2 s - 1)(\tau_3 s - 1)(\tau_4 s - 1)(\tau_5 s - 1) + \varepsilon s^4}{(\tau_2 s - 1)(\tau_3 s - 1)(\tau_4 s - 1)(\tau_5 s - 1) - \varepsilon s^4}. \tag{114}$$

To save labor from this point on, we will ignore the synthesis option implicit in Eqn. 114. It will be clear from what follows how one may base a network on that equation, so neglect of this second option represents no conceptual loss.

Substituting the appropriate roots into Eqn. 113 yields, after some simplification,

$$Z_{in} = \frac{[k^2 s^2 + s(2ka) + 1][k^2 s^2 + s(2kb) + 1] - \varepsilon s^4}{[k^2 s^2 + s(2ka) + 1][k^2 s^2 + s(2kb) + 1] + \varepsilon s^4}, \tag{115}$$

where

$$k = \varepsilon^{(1/4)}; \quad a \approx 0.38268, \quad b \approx 0.92388. \tag{116}$$

Further simplifications yield

$$Z_{in} = \frac{\varepsilon s^4 + s^3[2k^3(a+b)] + s^2[2k^2(1+2ab)] + s[2k(a+b)] + 1 - \varepsilon s^4}{\varepsilon s^4 + s^3[2k^3(a+b)] + s^2[2k^2(1+2ab)] + s[2k(a+b)] + 1 + \varepsilon s^4} \tag{117}$$

22.6 APPENDIX A: NETWORK SYNTHESIS

Table 22.13. *Component values for 1-GHz, 1-dB, fourth-order low-pass Butterworth filter* $(Z = 50\ \Omega)$

L first
$L_1 = 5.1441$ nH
$C_2 = 4.9675$ pF
$L_3 = 12.419$ nH
$C_4 = 2.0576$ pF

and so, at last, we have

$$Z_{in} = \frac{s^3[2k^3(a+b)] + s^2[2k^2(1+2ab)] + s[2k(a+b)] + 1}{2\varepsilon s^4 + s^3[2k^3(a+b)] + s^2[2k^2(1+2ab)] + s[2k(a+b)] + 1}. \quad (118)$$

Next, generate a continued fractions expression for Z_{in}. At every step of the synthetic division, divide the polynomial of higher order by the one of lower order, reversing operands as needed all along the way (you will notice that terms drop out in pairwise fashion as the division proceeds). For example, since the numerator of Eqn. 118 is of lower order, we divide it into the denominator polynomial as the first step of the expansion.

Executing the synthetic division yields the following sequence of terms, each of which yields an inductive reactance and capacitive admittance in alternation (it doesn't matter which choice you begin with):

$$s\frac{k}{a+b} = sL_1; \quad (119)$$

$$s\frac{k(a+b)}{2ab} = sC_2; \quad (120)$$

$$s\frac{8k(ab)^2}{(a+b)(4ab-1)} = sL_3; \quad (121)$$

and, finally,

$$s\frac{k(a+b)(4ab-1)}{2ab} = sC_4. \quad (122)$$

Completing the design by denormalizing the component values, we obtain the "L first" version of the filter presented in the section on Butterworth filters; see Table 22.13. Thanks to reciprocity combined with equal terminations, you may reverse the roles of input and output with no change in the filter transfer function. Doing so would yield the "C first" filter option. This flexibility is sometimes of value in allowing you to absorb parasitics that might be different (either in magnitude or character) at the two ports.

You can see from this example that manual synthesis can involve considerable labor and also presents many opportunities for error (imagine carrying out these procedures for, say, an eleventh-order filter, and you'll get the idea). The straightforward nature of the process, however, lends itself nicely to automation. Numerous filter design tools are readily available to do the hard work for you. Particularly attractive is the program *LADDER*, in part because it is free but also because its education-focused feature set includes options to display the results of each synthesis step as the design proceeds.[45]

22.7 APPENDIX B: ELLIPTIC INTEGRALS, FUNCTIONS, AND FILTERS

As mentioned in the main body of this chapter, the design of Cauer filters depends on the mathematics of elliptic functions. But if the word *elliptic* has nothing to do with the Chebyshev filter's distribution of poles along (half) an ellipse, where does the term come from? And how can it possibly relate to filter design? This appendix, a brief tutorial on the subject of elliptic functions, is intended to bridge the gap between the expositions of some elementary filter handbooks and more advanced treatments of the subject. Readers who are uninterested in the mathematical details are invited (indeed, urged) to skip over this material, and feed the pages to a goat.

22.7.1 WHY ARE THEY "ELLIPTIC"?

The need to solve a physical problem often stimulates initial work on a particular topic in mathematics, but it is also often the case that the field subsequently moves far away from those origins. This process is frequently abetted by those mathematicians who don't want to taint their work with any obvious connection to physical reality. And so it is with elliptic functions and integrals.

If the reader has taken a course in integral calculus, the following problem (and the corresponding method of solution) may be familiar: Calculate the length of a circular arc from, say, 15° to 38° for a circle of radius a. A calculus-based solution is to integrate the equation for differential arc length over the indicated interval. It's not considered a particularly difficult problem, and it shows up routinely as a homework exercise or illustrative classroom example in introductory integral calculus courses (partly because the answer can be verified with a simple, non–calculus based derivation).

If you attempt to answer the analogous question for the arc length or circumference of an *ellipse*, however, the situation changes from almost trivial to nigh impossible. You obtain an integral that stubbornly defies evaluation by methods taught in standard calculus courses. In fact, it has been proven rigorously that the general evaluation

---

[45] R. D. Koller and B. Wilamowski, "LADDER – A Microcomputer Tool for Passive Filter Design and Simulation," *IEEE Trans. Education*, v. 39, no. 4, November 1996, pp. 478–87.

22.7 APPENDIX B: ELLIPTIC INTEGRALS, FUNCTIONS, AND FILTERS

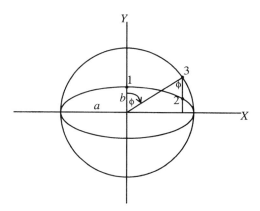

FIGURE 22.29. Ellipse inscribed within a circle

of these elliptical integrals requires the use of numerical methods. Prior to the ready availability of computational tools, vast tables (perhaps computed with the use of series expansions) were the only way for most engineers to obtain actual answers in other than symbolic form.

Computational impediments are hardly the only barriers to comprehension. For example, the authors of *Numerical Recipes* note, with dry understatement, that "one of the harder things about using elliptic integrals is the notational thicket that surrounds the subject in the literature."[46] Differing notational conventions abound. Furthermore, what today are called elliptical integrals were once called elliptical functions. The latter term now refers to a certain class of functions defined in reference to elliptical integrals, as we'll see. Just be aware of this semantic difference when reading through the literature on the subject.

Here, we follow the notational conventions of E. Jahnke and F. Emde.[47] The rationale for this choice can be understood with reference to Figure 22.29, where we have inscribed an ellipse inside a circle of radius a. Additionally, we have drawn a ray from the origin to point 3 on this circle and have also dropped a perpendicular down to the x-axis from point 3. Finally, note that we define positive angles clockwise, relative to the *vertical* axis.

The equation for the ellipse is

$$\frac{x^2}{a^2} + \frac{y^2}{b^2} = 1, \tag{123}$$

where a and b are the lengths of the semimajor and semiminor axes, respectively.

---

[46] W. H. Press et al. (Cambridge University Press, 1986).

[47] An excellent introduction to elliptic integrals may be found in H. W. Reddick and F. H. Miller, *Advanced Mathematics for Engineers,* 3rd ed., Wiley, New York, 1955. That text follows the same notational conventions as do Jahnke and Emde in their *Tables of Functions with Formulae and Curves* (Dover, New York, 1945).

The arc length along the ellipse from, say, point 1 to point 2 is readily found by first computing the coordinates of the point 2. By inspection, the x-coordinate is simply

$$x = a \cdot \sin \phi. \tag{124}$$

Finding the y-coordinate of point 2 requires only a little more work. Substituting Eqn. 124 into Eqn. 123 and solving for y yields

$$y = \frac{b}{a}\sqrt{a^2 - x^2} = \frac{b}{a}\sqrt{a^2 - (a \cdot \sin \phi)^2} = b \cos \phi. \tag{125}$$

There's probably an elegant geometric construction that would yield this result more directly, but we're happy to find any way at all.

Now, arc length in general is expressed as

$$s = \int \sqrt{dx^2 + dy^2}. \tag{126}$$

Substituting appropriately for the quantities in the integrand yields

$$s = \int \sqrt{(a \cos \phi)^2 + (b \sin \phi)^2}\, d\phi = \int \sqrt{a^2[1 - (\sin \phi)^2] + (b \sin \phi)^2}\, d\phi. \tag{127}$$

After simplifying and cleaning up the notation a bit (using ψ to denote the dummy variable of integration and reserving ϕ for the upper limit of integration), we obtain

$$s = a \int_0^\phi \sqrt{1 - k^2 (\sin \psi)^2}\, d\psi, \tag{128}$$

where k is the eccentricity of the ellipse:

$$k = \sqrt{1 - \frac{b^2}{a^2}} = \frac{\sqrt{a^2 - b^2}}{a}. \tag{129}$$

From the figure it should be clear that k, which is zero in the special case of a circle, cannot exceed unity.[48]

When the semimajor axis a is of unit length, the integral is known as the elliptic integral of the second kind:

$$E(k, \phi) = \int_0^\phi \sqrt{1 - k^2 (\sin \psi)^2}\, d\psi. \tag{130}$$

The choice of E may be construed mnemonically as directly representing *elliptical* arc length in the ordinary geometric sense of the word.

Because the foregoing is relatively clear and certainly physically relevant, we must now begin some artful obfuscation. In the language of the subject, the eccentricity k

---

[48] Nonetheless, extensions of this branch of mathematics can accommodate values of k in excess of unity (and complex values, as well), but it is difficult there to articulate a simple connection to the original picture.

is renamed the *modulus* and the upper limit ϕ the *amplitude*. Just to make a confusing subject even more so, many texts and tables instead employ the *modular angle* (simply the arcsine of the modulus k) rather than k itself:

$$k = \sin\theta. \tag{131}$$

In addition to the modulus, one often encounters the *complementary* modulus:

$$k' = \cos\theta, \tag{132}$$

so that

$$k' = \sqrt{1 - k^2}. \tag{133}$$

The reason that the integral under consideration is of the *second* kind is simply due to the order in which mathematicians (principally Abel and Jacobi, after pioneering work by Legendre) chose to classify them.

Returning now to our ellipse, it is evident that setting the amplitude equal to $\pi/2$ and evaluating the integral yields the arc length over one quadrant. Because of the ellipse's symmetry, knowing this arc length is sufficient to compute the entire circumference. The corresponding integral is thus known as the *complete* elliptic integral of the second kind. It is denoted by $E(k)$, or often simply E:

$$E(k) = \int_0^{\pi/2} \sqrt{1 - k^2(\sin\psi)^2}\, d\psi. \tag{134}$$

The total circumference of an ellipse is thus $4aE(k)$.

Elliptic integrals of the *first* kind are similar, except that the integrand is the reciprocal of that for integrals of the second kind:

$$F(k, \phi) = \int_0^{\phi} \frac{1}{\sqrt{1 - k^2(\sin\psi)^2}}\, d\psi. \tag{135}$$

The designation F reminds us that the corresponding integral is of the first kind. These are the integrals we encounter in designing elliptic filters, and also in solving rigorously for the oscillation period of a swinging pendulum.[49]

22.7.2 ELLIPTIC FUNCTIONS

Earlier, we mentioned that elliptic integrals were once called elliptic functions, and that the latter term is now reserved for a collection of functions defined in terms of elliptic integrals. We now define those elliptic functions, beginning by recalling that the phase angle ϕ is also known as the amplitude of the elliptic integral of the first

---

[49] Since you're no doubt dying to know, the oscillation period of a pendulum in a vacuum is given rigorously by $4(L/g)^{1/2}[F(k, \pi/2)]$, where L is the length of the pendulum, g is the acceleration due to gravity, and k is the sine of half the pendulum's peak angular displacement with respect to the vertical axis.

kind. If the integral itself is regarded as a function of this angle, then the inverse of this function is the amplitude. By convention, the integral in this context is usually known not as F but rather by u (again, don't ask why):

$$\phi = \operatorname{am} u. \tag{136}$$

If one desires to remind the reader explicitly that u depends on the modulus k as well, one may also write:

$$\phi = \operatorname{am}(u, \operatorname{mod} k). \tag{137}$$

Since the amplitude is itself a phase angle, one may use it as an argument of standard trigonometric functions, defining at last three elliptic functions:

$$\sin \phi \equiv \sin \operatorname{am} u \equiv \operatorname{sn} u, \tag{138}$$

$$\cos \phi \equiv \sqrt{1 - (\operatorname{sn} u)^2} \equiv \cos \operatorname{am} u \equiv \operatorname{cn} u, \tag{139}$$

$$\sqrt{1 - k^2 (\sin \phi)^2} \equiv \Delta \phi \equiv \operatorname{dn} u. \tag{140}$$

When $k = 0$, the first two elliptic functions converge to ordinary $\sin \phi$ and $\cos \phi$, respectively. When $k = 1$, the former function converges to the conventional hyperbolic tangent. In a real sense, then, elliptic functions may be viewed as generalized trigonometric functions that continuously change from ordinary to hyperbolic functions as the modulus varies from zero to unity. The continuity of this behavior is the basis of many methods for accurately computing the values of elliptic functions.

Various ratios of these functions define still other elliptic functions. The first and second letter of the function name come from the first letters of the numerator and denominator functions comprising it. For example,

$$\operatorname{cd} u = \frac{\operatorname{cn} u}{\operatorname{dn} u} \tag{141}$$

is one such function. It also just happens to arise directly in the computation of the zero locations of elliptic filters.

22.7.3 NUMERICAL EVALUATION OF ELLIPTIC FUNCTIONS

As stated earlier, elliptic integrals can't be evaluated using elementary functions (i.e., algebraic, trigonometric, and exponential functions – and their inverses). A straightforward (but quite inefficient) numerical method is to expand the integrand in a power series (using, e.g., the binomial theorem) and then integrate the result term by term. For those in need of a refresher, here's the binomial series:

$$(1 + x)^n = 1 + nx + \frac{n(n-1)x^2}{2!} + \frac{n(n-1)(n-2)x^3}{3!} + \cdots . \tag{142}$$

It is with similar approaches that the first tables of elliptic integrals were originally assembled. In this age of readily available computation, it is often easier to implement an algorithm than to locate such tables.

It is easiest to provide series expansions for the complete elliptic integrals, so that's all we'll present here. Even though several intermediate steps are not shown, the reader should be able to fill in the blanks if determined to do so. The procedure outlined in the development so far ultimately results in relatively compact expressions, as seen in the following series:

$$F(k) = \frac{\pi}{2}\left[1 + \left(\frac{1}{2}\right)^2 k^2 + \left(\frac{1\cdot 3}{2\cdot 4}\right)^2 k^4 + \left(\frac{1\cdot 3\cdot 5}{2\cdot 4\cdot 6}\right)^2 k^6 + \cdots\right]; \quad (143)$$

$$E(k) = \frac{\pi}{2}\left[1 - \left(\frac{1}{2}\right)^2 k^2 - \left(\frac{1\cdot 3}{2\cdot 4}\right)^2 \frac{k^4}{3} - \left(\frac{1\cdot 3\cdot 5}{2\cdot 4\cdot 6}\right)^2 \frac{k^6}{5} - \cdots\right]. \quad (144)$$

We'll present significantly more computationally efficient methods a little later on, but the straightforward series converge fast enough for values of k not too close to unity that using them is not out of the question. A perfectly respectable alternative is to use the approximating series presented in the main part of the chapter – if a 20-ppb worst-case error is acceptable (as it almost always is).

For amplitudes other than $\pi/2$, following our straightforward recipe generally requires the repeated use of the following recursion formula for the integral of $\sin^n \psi$:

$$\int (\sin\psi)^n \, d\psi = -\frac{(\sin\psi)^{n-1}\cos\psi}{n} + \frac{n-1}{n}\int (\sin\psi)^{n-2} \, d\psi. \quad (145)$$

Computation of incomplete elliptic integrals in this manner thus can be a somewhat messy affair. Worse, the resulting series may converge rather slowly (particular for moduli close to unity). Consequently it's not surprising that clever folks have labored hard to develop efficient, rapidly converging methods. However, explaining their derivation is harder than describing their operation, so we'll simply present the following algorithms and point the reader to the relevant literature for the details.[50]

In order to compute the incomplete integral of the first kind, $F(k,\phi)$, first define a sequence (F_n, k_n, ϕ_n) of three numbers as follows:

$$F_{n+1} = \frac{F_n}{1+\sqrt{1-k_n^2}}, \quad (146)$$

$$k_{n+1} = \frac{2}{1+\sqrt{1-k_n^2}} - 1; \quad (147)$$

$$\tan(\phi_{n+1} - \phi_n) = (\tan\phi_n)\left(\sqrt{1-k_n^2}\right), \quad (148)$$

with initial values $\phi_0 = \phi$, $k_0 = k$, and $F_0 = 1$.

Eqn. 148 is probably more intuitively appealing (but less computationally efficient) when expressed as

$$\phi_{n+1} = \phi_n + \operatorname{atan}\left[(\tan\phi_n)\sqrt{1-k_n^2}\right], \quad (149)$$

where "atan" denotes arctangent.

---

[50] The algorithms presented here are adapted from those in J. A. Ball's *Algorithms for RPN Calculators* (Wiley, New York, 1978). Ball, in turn, is implementing an extremely useful method originally due to Landen.

Iterate until $k_n = 0$ within some tolerance (e.g., 10^{-10}). Then,

$$F(k,\phi) = F_N \phi_N, \qquad (150)$$

where N is the value of n during the final iteration. Convergence is usually so rapid that fewer than a dozen iterations are normally required to achieve quite high accuracies.

There are also analogous procedures for evaluating incomplete elliptic integrals of the second kind, but since we don't need them in filter design, we omit their consideration.

Computation of the complete integrals (both of them) is also easily performed with rapidly converging methods. Here, first define a pair of numbers (a_n, g_n) and then successively compute their arithmetic and geometric means:

$$a_{n+1} = \frac{a_n + g_n}{2}; \qquad (151)$$

$$g_{n+1} = \sqrt{a_n g_n}, \qquad (152)$$

with initial values $a_{-1} = (1+k)$ and $g_{-1} = (1-k)$.

Iterate, while keeping track of the intermediate number pairs. You will notice that the arithmetic and geometric mean values converge. Continue iterating until $a_n = g_n$ within the desired tolerance (e.g., 10^{-10}). Then,

$$F(k) = \frac{\pi}{2a_N}, \qquad (153)$$

where N is again the value of n at the final iteration, and

$$E(k) = \frac{F(k)[2 - (a_0^2 - g_0^2) - 2(a_1^2 - g_1^2) - 4(a_2^2 - g_2^2) - \cdots]}{2}, \qquad (154)$$

where the coefficients of the difference-of-squares terms are simple powers of 2.

Finally, anyone wishing to understand elliptic functions from an electrical engineer's point of view is absolutely required to read Harry Orchard's superbly written papers on the subject. The most recent of these summarizes his wisdom accumulated over a half-century of experience with elliptic filter design and is an invaluable resource for the uninitiated in particular.[51]

22.7.4 CHEESY APPROXIMATE FORMULAS

Although this chapter is allegedly about filter design, we introduced elliptic functions by considering the problem of computing elliptical arc length. We now close by presenting a supplement to the more formal methods already presented. Here is a simple closed-form approximation for the circumference of an ellipse, one of several from the remarkable Ramanujan:

---

[51] H. J. Orchard and A. N. Willson, "Elliptic Functions for Filter Design," *IEEE Trans. Circuits and Systems I*, v. 44, no. 4, April 1997, pp. 273–87.

$$\text{circumference} \approx \pi\left[3(a+b) - \sqrt{(3a+b)(3b+a)}\,\right]. \tag{155}$$

His formula allows us to deduce that

$$E(k) \approx \frac{\pi}{4}\left[3\left(1+\sqrt{1-k^2}\right) - \sqrt{\left(3+\sqrt{1-k^2}\right)\left(3\sqrt{1-k^2}+1\right)}\,\right]. \tag{156}$$

This formula yields values that agree with those in published tables to the fourth decimal place for values of k from zero up to a bit more than 0.93. The error grows as k approaches unity from there, but it remains below 0.5%.

22.8 APPENDIX C: DESIGN TABLES FOR COMMON LOW-PASS FILTERS

Here we provide tables of component values for several of the low-pass filter types discussed in the main part of this chapter. Normalized element values for the constant-k, Butterworth, Chebyshev, and Bessel low-pass filters are given up to the ninth order. In all the tables that follow, the normalizations are to 1 Ω and 1-rps passband frequency (or cutoff frequency in the case of constant-k and m-derived types). For a different impedance level, scale the element impedances proportionately. For a 50-Ω system, for example, multiply all inductances by 50 and divide all capacitances by 50. For a different cutoff frequency, scale each element impedance inversely with frequency. For a 10-Grps cutoff, for example, divide each inductance and capacitance by 10^{10}.

As stated in the main part of the chapter, you may also derive high-pass, bandpass, and bandstop filters from the low-pass prototypes. The relevant transformations are summarized at the very end of this appendix. It's important to emphasize that these transformations preserve the shape of the magnitude response but do not necessarily preserve phase response. This observation is particularly relevant for the Bessel–Thomson filter. One cannot derive a linear-phase bandpass filter directly from the low-pass prototype using the standard transformations, unfortunately. Those requiring a linear phase bandpass filter must derive them using other means, such as direct synthesis.

All of the tables are universal in the sense that one has a choice of starting with either a shunt capacitor or series inductor. The latter choice is indicated in parentheses. You may also freely exchange the input and output ports of any of these filters, with no change in transfer characteristics. This statement applies even to asymmetrical filters (e.g., even-order Butterworths) because this attribute depends not on symmetry but rather on the reciprocal nature of these networks, combined with equal source and load terminations.

In Figure 22.30, the quantities L and C are those for a 1-rps prototype, which explains the initial frequency normalization shown in the first column of the figure. The element values are also normalized to 1 Ω, so perform a final denormalization to the actual impedance level to complete the design:

$$L = RL_n; \tag{157}$$

$$C = (1/R)C_n. \tag{158}$$

Table 22.14. *Normalized element values for constant-k low-pass filters*

$C_1\ (L_1)$	$L_2\ (C_2)$	$C_3\ (L_3)$	$L_4\ (C_4)$	$C_5\ (L_5)$	$L_6\ (C_6)$	$C_7\ (L_7)$	$L_8\ (C_8)$	$C_9\ (L_9)$
1.000	2.000	1.000						
1.000	2.000	2.000	2.000	1.000				
1.000	2.000	2.000	2.000	2.000	2.000	1.000		
1.000	2.000	2.000	2.000	2.000	2.000	2.000	2.000	1.000

Table 22.15. *Normalized element values for Butterworth low-pass filters ($\varepsilon = 1$)*

$C_1\ (L_1)$	$L_2\ (C_2)$	$C_3\ (L_3)$	$L_4\ (C_4)$	$C_5\ (L_5)$	$L_6\ (C_6)$	$C_7\ (L_7)$	$L_8\ (C_8)$	$C_9\ (L_9)$
2.000								
1.414	1.414							
1.000	2.000	1.000						
0.7654	1.848	1.848	0.7654					
0.6180	1.618	2.000	1.618	0.6180				
0.518	1.414	1.932	1.932	1.414	0.518			
0.445	1.247	1.802	2.000	1.802	1.247	0.445		
0.390	1.111	1.663	1.962	1.962	1.663	1.111	0.390	
0.347	1.000	1.532	1.879	2.000	1.879	1.532	1.000	0.347

Table 22.16. *Normalized element values for 0.1-dB–ripple Chebyshev low-pass filter*

$C_1\ (L_1)$	$L_2\ (C_2)$	$C_3\ (L_3)$	$L_4\ (C_4)$	$C_5\ (L_5)$	$L_6\ (C_6)$	$C_7\ (L_7)$	$L_8\ (C_8)$	$C_9\ (L_9)$
1.032	1.147	1.032						
1.147	1.371	1.975	1.371	1.147				
1.181	1.423	2.097	1.573	2.097	1.423	1.181		
1.196	1.443	2.135	1.617	2.205	1.617	2.135	1.443	1.196

Table 22.17. *Normalized element values for 0.5-dB–ripple Chebyshev low-pass filter*

$C_1\ (L_1)$	$L_2\ (C_2)$	$C_3\ (L_3)$	$L_4\ (C_4)$	$C_5\ (L_5)$	$L_6\ (C_6)$	$C_7\ (L_7)$	$L_8\ (C_8)$	$C_9\ (L_9)$
1.596	1.097	1.596						
1.706	1.230	2.541	1.230	1.706				
1.737	1.258	2.638	1.344	2.638	1.258	1.737		
1.750	1.269	2.668	1.367	2.724	1.367	2.668	1.269	1.750

22.8 APPENDIX C: DESIGN TABLES FOR COMMON LOW-PASS FILTERS

Table 22.18. *Normalized element values for 1.0-dB–ripple Chebyshev low-pass filter*

$C_1 (L_1)$	$L_2 (C_2)$	$C_3 (L_3)$	$L_4 (C_4)$	$C_5 (L_5)$	$L_6 (C_6)$	$C_7 (L_7)$	$L_8 (C_8)$	$C_9 (L_9)$
2.024	0.994	2.024						
2.135	1.091	3.000	1.091	2.135				
2.167	1.112	3.094	1.174	3.094	1.112	2.167		
2.180	1.119	3.121	1.190	3.175	1.190	3.121	1.119	2.180

Table 22.19. *Normalized element values for 1-rps Bessel–Thomson low-pass filters*

$C_1 (L_1)$	$L_2 (C_2)$	$C_3 (L_3)$	$L_4 (C_4)$	$C_5 (L_5)$	$L_6 (C_6)$	$C_7 (L_7)$	$L_8 (C_8)$	$C_9 (L_9)$
2.0000								
2.1478	0.5755							
2.2034	0.9705	0.3374						
2.2404	1.0815	0.6725	0.2334					
2.2582	1.1110	0.8040	0.5072	0.1743				
2.2645	1.1126	0.8538	0.6392	0.4002	0.1365			
2.2659	1.1052	0.8690	0.7020	0.5249	0.3259	0.1106		
2.2656	1.0956	0.8695	0.7303	0.5936	0.4409	0.2719	0.0919	
2.2649	1.0863	0.8639	0.7407	0.6306	0.5108	0.3770	0.2313	0.0780

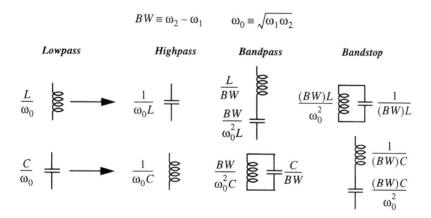

FIGURE 22.30. Summary of transformations from low-pass prototype into other shapes

CHAPTER TWENTY-THREE

MICROSTRIP FILTERS

23.1 BACKGROUND

In this chapter, we consider a collection of methods for designing distributed filters. All derive from lumped prototypes to exploit as much as possible the considerable body of literature on the subject of lumped filter design.[1] In some cases, such translations into distributed form involve straightforward replacement of lumped components by transmission line approximations. In other cases, the translation is based on the recognition that certain principles originally developed in connection with lumped circuits have a broader applicability. Overall, these translations generally provide passband performance that closely matches that of their lumped progenitors. However, distributed filters present a unique challenge: At some point beyond the stopband frequency, their response will pop back up, even if the lumped counterpart's response decays monotonically. The periodic impedance behavior of transmission lines necessarily produces these *re-entrant* modes.[2] Identification and careful management of these modes are some of the preoccupations of the microwave filter engineer. Experience teaches us that, even after a great deal of iterating, we must accept the inevitable: our filters will always behave differently from their lumped prototypes. With our expectations properly lowered, we can now proceed to some derivations.

23.2 DISTRIBUTED FILTERS FROM LUMPED PROTOTYPES

One straightforward method developed for some of the earliest microwave filters still works well for many applications: Simply replace the discrete inductors and capacitors of a lumped prototype with equivalent elements made from transmission line

---

[1] There are direct design methods that do not rely on approximating lumped prototypes. Better filters result, but at the expense of procedures that are difficult to understand or explain intuitively. We focus on indirect methods because they work well enough for all but the most demanding applications and are readily explained with intuitively appealing concepts.

[2] However, it should not be inferred that the overall response is then simply periodic. If the elements are of equal length, then simple periodicity will result.

segments.[3] As discussed in Chapter 7, transmission lines approximate well the behavior of lumped elements if the sections are a suitably small fraction of an electrical wavelength in extent. A short section of open-circuited line functions well as a capacitor, while a short piece of shorted line behaves as an inductor.

However, one important consideration is that, as frequency increases, all lines ultimately cease to be very short relative to a wavelength. The attendant impedance variation consequently alters the filter response. For example, a microstrip low-pass filter may have a response that pops back up again within the nominal stopband. Because such spurious responses are hardly unique to low-pass filters, one must evaluate carefully any proposed realization to assure that all spurious responses are benign in magnitude or location.

To derive one simple method for converting lumped prototypes into distributed filters, first recall that the input impedance of a short piece of open-circuited line is approximately

$$Z \approx \frac{Z_0}{j\omega(l/v)}, \tag{1}$$

so that its equivalent capacitance is

$$C = \frac{l}{vZ_0} = \frac{l\sqrt{\varepsilon_{r,eff}}}{cZ_0}. \tag{2}$$

One can expect about 1.3 pF/cm with 50-Ω lines on FR4, but we are also interested in the maximum practical values we might be able to obtain. With relatively low impedance lines, it might be practical to achieve roughly 4 pF/cm. At 1 GHz, the need to maintain element dimensions well below a wavelength limits us to lengths well below about 15 cm (again, in FR4) or total capacitances below approximately 5–10 pF. This capacitance limit diminishes quadratically as frequency increases because capacitance is proportional to area, which (in turn) is proportional to wavelength squared.

Similarly, for the inductance of a short line terminated in a short circuit, we have

$$L = \frac{lZ_0}{v} = \frac{lZ_0\sqrt{\varepsilon_{r,eff}}}{c}. \tag{3}$$

As we've often cited, a typical value for inductance is roughly of the order of 1 nH/mm for the narrowest (highest-impedance) practical lines in FR4. Again, at 1 GHz, we find a maximum practical inductance value of ~10–20 nH. This approximate inductance limit is inversely proportional to frequency.

A key observation is that these relationships can be reasonable approximations even when the line segments are not terminated in perfect open or short circuits. The

---

[3] See e.g. G. L. Ragan (ed.), *Microwave Transmission Circuits* (MIT Rad. Lab. Ser., v. 9), chap. 10 (by Fano and Lawson), McGraw-Hill, New York, 1948. Also see the chapter by Seymour Cohn in volume 2 of *Very High Frequency Techniques*, Radio Research Laboratory, McGraw-Hill, New York, 1947. These early expositions are necessarily incomplete but historically important.

foregoing equations remain reasonably accurate as long as the segments are terminated in impedances that approximate opens or shorts *relative to the characteristic impedance* of the lines:

$$\frac{Z(z)}{Z_0} = \frac{\frac{Z_L}{Z_0} + j \tan \beta z}{1 + j\frac{Z_L}{Z_0} \tan \beta z}, \tag{4}$$

where we follow the conventions that the coordinate z is a positive value[4] with the load positioned at $z = 0$, and

$$\beta = \omega/v. \tag{5}$$

We therefore conclude that, as long as Z_0 is very different from Z_L, the impedance converges to simple forms. For $Z_L \ll Z_0$, a short line of length l will have a normalized impedance of

$$\frac{Z(z)}{Z_0} \approx j \tan \beta l, \tag{6}$$

which is inductive. For $Z_L \gg Z_0$,

$$\frac{Z(z)}{Z_0} \approx \frac{1}{j \tan \beta l}, \tag{7}$$

which is capacitive. To validate the approximations, we should therefore choose Z_0 as low as possible (or practical) to make a capacitor and choose Z_0 as high as possible to make an inductor.

One cannot specify arbitrarily high characteristic impedances, of course, because there is always a lower bound on the width of lines that may be fabricated reliably. Assuming a typical manufacturing tolerance of 2 mil (50 μm) and supposing that this variation is allowed to represent at most 20% of the total width, one may assume a minimum practical linewidth of about 10 mil (250 μm).[5] Hence, on 1.6-mm FR4, practical line impedances rarely exceed about 200 Ω, with 150 Ω being a commonly encountered maximum value.

There are also practical bounds on the maximum width of the lines because, again, all linear dimensions of a microstrip element must be small compared to a wavelength at all frequencies of interest in order to assure close approximation to lumped element behavior. The associated implicit lower bound on impedance depends on the operational frequency range; but as a general rule, characteristic impedances below approximately 10 Ω are rarely used, with 15 Ω a common value. In realizing microstrip filters, then, it's important to keep in mind that practical impedance levels in FR4 are thus generally within about a factor of 3–4 of 50 Ω.

---

[4] Sorry to switch conventions on you mid-book. Fortunately, it doesn't matter as long as we don't switch conventions mid-derivation.

[5] If cost is not a concern, you may induce some vendors to offer lines as narrow as 1 mil (25 μm) on rigid substrates.

23.2.1 STEPPED-IMPEDANCE FILTERS

Perhaps the simplest method for transforming discrete prototypes into microstrip form uses only the narrowest and widest lines that may be comfortably (or repeatably) fabricated. As in the preceding discussion, the narrow lines implement series inductors and the wide lines implement shunt capacitors. In the *stepped-impedance* filter, lengths are adjusted as necessary to produce the desired component values. There are thus lines of only three widths in such filters (the third is for input/output lines of Z_0 in impedance).

As one might expect, the fundamentally approximate nature of the transformations limits its utility. Stepped-impedance filters are thus best used in applications where one may tolerate relatively large errors relative to the lumped filter prototype's response, or where you don't mind iterating endlessly in an effort to refine the filter. These errors generally increase in significance as one moves above the cutoff frequency, because the true transmission line nature of the segments becomes more apparent as frequency increases. With careful design, the stepped-impedance and lumped-parameter filters might behave similarly below and near the design cutoff frequency. Beyond cutoff, however, the stepped-impedance filter typically fails to roll off as quickly as the prototype and, indeed, the ultimate stopband attenuation may fail to meet specifications. Furthermore, the filter's response may exhibit numerous spurious passbands. Because the individual segments are generally of unequal length, the filter response will not exhibit any simple periodicity.

That said, let's examine how to make a low-pass filter using the stepped-impedance architecture. As a specific example, assume that we desire a cutoff frequency of 1 GHz and that we use a constant-k prototype as the basis for the microstrip filter. If the prototype has two complete T-sections (or three complete π-sections) then the stepped-impedance filter will have seven segments, corresponding to the seven components of the lumped prototype. Assume further that the minimum and maximum realizable line impedances are 15 Ω and 200 Ω.[6] To match the lumped element values of the prototype we require an inductance of 15.915 nH, which we implement with the narrowest available line, whose length is given by

$$l = \frac{vL}{Z_{0,max}} = \frac{f\lambda L}{Z_{0,max}}, \tag{8}$$

which works out to a normalized length for the inductor of about 28.647° at the cutoff frequency.[7] Similarly, the main 6.3662-pF capacitors should have a length

---

[6] Such a high value may be difficult when substrates other than FR4 are used. Most (but thankfully not all) microwave substrates have considerably higher dielectric constants, making it especially challenging to realize high-impedance lines.

[7] Matching the impedances at the cutoff frequency is somewhat arbitrary, but it's a good choice because the behavior in the vicinity of cutoff is often of greatest concern. Also, we provide component values to a ridiculous number of digits to facilitate comparisons you might want to undertake independently.

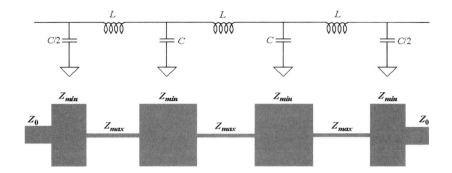

FIGURE 23.1. Stepped impedance filter example (not drawn exactly to scale)

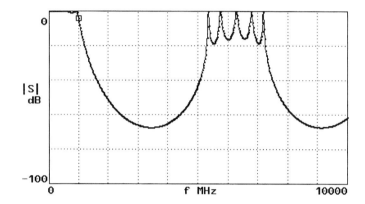

FIGURE 23.2. Stepped-impedance filter behavior over large range of frequency and attenuation (*Puff*)

$$l = vCZ_{0,min} = f\lambda C Z_{0,min}, \tag{9}$$

which corresponds to a normalized length of about 28.742°. The half-section terminating capacitors are exactly half that length.

The layout of the filter appears approximately as sketched in Figure 23.1, where the lumped prototype is also shown for reference. Simulations of this filter with *Puff* show about 3.9 dB of attenuation at the design cutoff frequency of 1 GHz, and a −3-dB bandwidth of 989 MHz. Somewhat different answers would be obtained from a field solver because *Puff* does not take into account the field distortions that accompany step changes in width. The particularly dramatic step changes here cause definitely noticeable effects (such as those arising from an effective shortening of the high-impedance sections, just as in the T-junction shortening we noted in Chapter 7). Also, at frequencies high enough that the capacitive sections are not separated by large distances (say, several dielectric thicknesses), the mutual coupling will alter the response as well. Nevertheless, the basic features we care about here are well captured by the *Puff* simulations.

The simulations (see Figure 23.2) show that the stepped-impedance filter's performance is similar to that of the constant-k prototype, whose attenuation is 10 dB

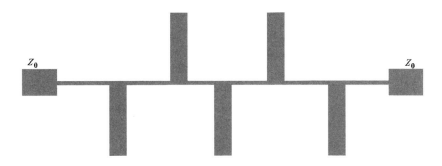

FIGURE 23.3. Stub low-pass filter example (not drawn to scale)

at the cutoff frequency and whose −3-dB bandwidth is 934 MHz. The maximum stopband attenuation exceeds 60 dB but does not exhibit the monotonic increase with frequency of its lumped cousin. Perhaps more important than those differences is the existence of the 2–3-GHz–wide spurious passbands centered around 6.2 GHz for this particular implementation. The lumped constant-k filter, of course, ideally exhibits no such spurious passbands, and this difference in behavior must be taken into account in any practical implementation of distributed filters, stepped-impedance or otherwise.

23.2.2 STUB LOW-PASS FILTER

An alternative implementation continues to use narrow lines as series inductors but realizes the capacitors as open-circuited stubs connected to this inductive backbone. If these stubs can be made narrower than those used in a conventional stepped impedance filter, the resulting structure can correspond more closely to the lumped prototype it's based on – at least at frequencies where the stub lengths remain short relative to a wavelength.

In Figure 23.3, note that the stubs are placed on alternating sides of the main line. This arrangement is not mandatory, but it does reduce the effects of unwanted coupling between adjacent (or even more remote) lines. Such coupling can alter the filter's frequency response in undesired ways. As suggested earlier, line-to-line separations that are at least 4–5 times the dielectric thickness usually suffice to avoid such problems.

The stub values (but not necessarily their shapes) are exactly the same as for the stepped-impedance filter; they're just arranged differently. Simulations of this design reveal a 6.5-dB attenuation at 1 GHz and a 3-dB attenuation at 965 MHz. Thus, the passband characteristics are a little closer to those of the lumped prototype than is the case for the stepped-impedance translation.

Zooming out to see the filter's response over a wide dynamic range of frequency and attenuation (Figure 23.4), we see that the stub filter also has spurious passbands in the same general frequency range as the stepped-impedance implementation. The peaks are smaller and narrower, however, and there is thus less overall transmission by these re-entrant modes. Also, unlike the stepped-impedance filter, the stopband

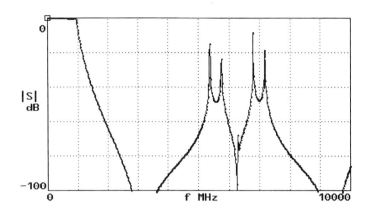

FIGURE 23.4. Stub filter behavior over large range of attenuation and frequency (*Puff*)

attenuation exceeds 100 dB over some frequency interval. Such large attenuations are never observed in practice, of course, for ever-present loss prevents infinitely deep notches. Even if the lines were lossless, there is still essentially no such thing as ten orders of magnitude of isolation in real systems. In any event, devising a clean measurement to verify such an attenuation is itself a significant instrumentation and fixturing challenge.

We've observed in this case that the stub filter seems to be better than the stepped-impedance version. This superiority is observed generally, because the mapping from the lumped prototype is less inexact for the stub implementation than for the stepped-impedance realization. However, this generality does not free you from the obligation to verify it in any case that matters.

If a needed stub is of uncomfortably low impedance then it is best realized as two paralleled stubs, one on each side of the backbone. However, one must worry about the total length of the stubs, because troublesome transverse resonances can occur at higher frequencies where the total length is an odd multiple of a half-wavelength.

A question that is often asked concerns the precise position along the backbone at which a stub is "really" connected. For electrically narrow stubs, it's reasonable to regard the point of attachment as halfway across the stub, widthwise. At higher frequencies or for very low impedance (wide) stubs, it gets progressively more difficult to answer the question satisfactorily. A common solution is to use a *radial* stub, whose narrow point of attachment reduces the uncertainty considerably; see Figure 23.5.

Design degrees of freedom include the radius of the wedge, the angular displacement, and the width (or radius) at the point of attachment (too small a width incurs a penalty in excessive series resistance). Radial stubs may be used in pairs, in a *butterfly* arrangement, to provide stubs of very low impedance (values below 10 Ω are readily achievable). As is apparent from Figure 23.6, it's usually impractical to pack more than a small number of such stubs along a line.

FIGURE 23.5. Filter with radial stub

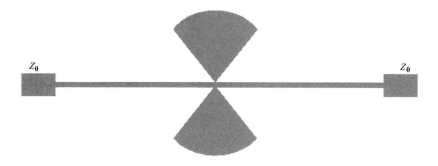

FIGURE 23.6. Line with butterfly stub

The radial stub will behave as a good approximation of a capacitance as long as its radius is very small relative to a wavelength. Its ability to produce a large, high-quality capacitance makes the radial (or butterfly) stub an extremely popular element for power supply bypassing, both in board-level modules and in integrated circuits. In those applications accuracy is rarely important, so a parallel-plate approximation frequently suffices:

$$C \approx \varepsilon \alpha r_L^2 / 2H, \qquad (10)$$

where we have assumed that the outer radius is much greater than the radius at the point of attachment. The dielectric constant in Eqn. 10 is the full (not relative) dielectric constant, and the approximation assumes that fringing is negligible so that the effective and bulk dielectric constants are essentially equal.

For the very fussy, one may consider the radial stub as itself part of a special kind of transmission line, just as an ordinary stub is a part of an ordinary transmission line. There's very little published material that treats the general case, but the following intuitively obvious equations may be found in an HP applications note from days gone by:[8]

$$X = \frac{HZ_0}{\alpha r_i} \frac{\cos(\theta_i - \psi_L)}{\sin(\psi_i - \psi_L)}, \qquad (11)$$

---

[8] Applications Note 976, "Broadband Microstrip Mixer Design – The Butterfly Mixer," 1980. In turn, the note cites J. R. Vinding, "Radial Line Stubs as Elements in Strip Line Circuits," *IEEE NEREM Record,* 1967, pp. 108–9. The equations we present in this chapter are slightly corrected versions of those given in the applications note.

where X is the stub reactance, H is the dielectric thickness, α is the angle (in radians) swept out by the radial stub, r_i is the radius at the point of attachment, and

$$Z_0 = \frac{\eta_0}{\sqrt{\varepsilon_r}} \sqrt{\frac{J_0^2(kr_i) + N_0^2(kr_i)}{J_1^2(kr_i) + N_1^2(kr_i)}}, \tag{12}$$

$$k = \frac{2\pi\sqrt{\varepsilon_r}}{\lambda_0}; \tag{13}$$

$$\theta_i = \tan^{-1}\left[\frac{N_0(kr_i)}{J_0(kr_i)}\right], \tag{14}$$

$$\psi_i = \tan^{-1}\left[-\frac{J_1(kr_i)}{N_1(kr_i)}\right], \tag{15}$$

$$\psi_L = \tan^{-1}\left[-\frac{J_1(kr_L)}{N_1(kr_L)}\right]. \tag{16}$$

Recall that η_0 is the impedance of free space,

$$\eta_0 = \sqrt{\mu_0/\varepsilon_0}. \tag{17}$$

It might also help to know that r_L is the total radius and that J_n and N_n are nth-order Bessel functions of the first and second kind (respectively).

23.2.3 ELLIPTIC, m-DERIVED, AND INVERSE CHEBYSHEV LOW-PASS

As we saw in the previous chapter, lumped filters with nulls (i.e., finite transmission zeros) in their frequency response (such as m-derived, inverse Chebyshev, and elliptic filters) may be implemented either with shunt resonators in the series path or with series resonators connected to ground. Microstrip filters based on the latter configuration tend to be more conveniently realizable. The series resonators are implemented, again, with inductors built out of short segments of narrow line and with capacitors as short segments of wide line; see Figure 23.7.

As with lumped elliptic filters, the most practical microstrip implementations place the lowest-frequency resonator in the center, progressively working toward the input/output ports in alternation with each successive resonator. If you derive the distributed version from a lumped prototype, the proper sequencing will already have been taken care of during the prototype's synthesis.

As an illustrative design, let us translate a simple lumped elliptic filter into distributed form using the approach just outlined. Basing an implementation directly on the example in Section 22.5.4 presents some difficulties because of its use of a parallel-resonant tank as a notch element. Implementing such a network in microstrip form is not trivial. Fortunately, a notch may be produced just as well by a series-resonant branch in shunt with the main filter path, and filter design tools such as *LADDER* synthesize this option as well. The results of that synthesis are shown in Figure 23.8.

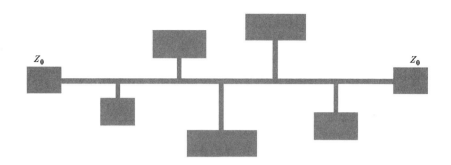

FIGURE 23.7. Example layout for low-pass filters with finite transmission zeros (not drawn to scale)

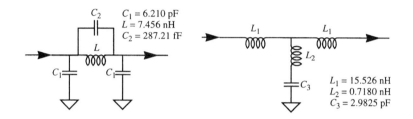

FIGURE 23.8. Lumped low-pass 1-GHz elliptic filter prototypes (shunt trap on left; series trap on right)

In evaluating what follows, bear in mind that the lumped prototype produces a 1-dB passband ripple, a passband edge of 1 GHz, and a stopband that begins at 3 GHz. The minimum attenuation required in the stopband is 30 dB.

As seen in Figure 23.8, the implementation on the right is much more readily translated into microstrip stub form. Following the same procedure as before, the 15.526-nH inductance is implemented by a 200-Ω line whose electrical length is about 27.947° at the cutoff frequency, and the 0.718-nH inductor by a line with a length of 1.292°. Similarly, the 2.9825-pF capacitor is implemented by a 15-Ω line having an electrical length of about 13.465°. Simulations of the filter are shown in the next two figures.

These simulations reveal that the passband error is below 1 dB out to 1.05 GHz, very close to the 1-GHz value of the lumped prototype. The attenuation at the 3-GHz stopband edge is a tiny bit under 40 dB, well in excess of the 30-dB specification. Furthermore, the characteristic stopband notch of an elliptic filter is evident just a little beyond the stopband edge.[9] See Figure 23.9. We do expect the response to pop back up somewhat in the stopband, but this distributed version fails to meet the

---

[9] Again, finite line Q will prevent notches of infinite depth and will also cause insertion loss and other impairments.

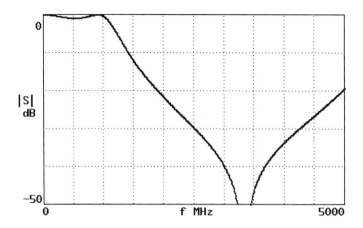

FIGURE 23.9. Detail of passband–stopband transition for stub version of elliptic filter

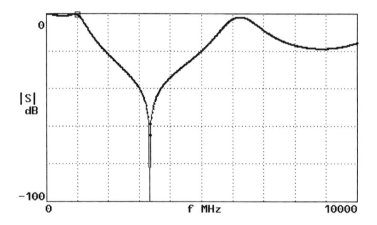

FIGURE 23.10. Stub elliptic filter over large range of frequency and attenuation

30-dB attenuation requirement for frequencies greater than about 4.3 GHz, unlike the lumped prototype.[10]

Examination of the response over a larger range of frequency and attenuation (Figure 23.10) highlights further the differences between a lumped elliptic filter and this stub version. The high-frequency attenuation of this filter is disappointingly small. Again, such differences between the lumped and distributed implementations must be accounted for in any design. For example, if greater ultimate attenuation is required, this elliptic filter could be cascaded with another filter whose passband–stopband transition is lazier but whose stopband performance is superior. By thus

---

[10] In all fairness, however, we should remind you that component parasitics can cause lumped filters to misbehave just as badly, if not worse.

designing two filters, each of which is optimized with a different set of objectives in mind, it may be possible to produce a combination that has all of the desired attributes.

Aside from that fundamental consideration, the layout of such filters can prove troublesome at high frequencies, where the length of the high-impedance inductive backbone shrinks to such an extent that the wide capacitive stubs bump into adjacent structures. Even before explicit collisions occur, coupling between the lines can alter the transfer function in unexpected ways. In such cases, it may be necessary to *lengthen* the backbone by using sections that are wider than the minimum value, at the cost of somewhat worse backbone impedance behavior. Similarly, it may be helpful to implement the capacitive stubs with lines of less than maximum width. Because all of these strategies involve trade-offs of their own, you can expect painful, iterative design of these types of filters, with no guarantee that your design objectives will be met.

23.2.4 COMMENSURATE-LINE FILTERS

From Eqns. 2 and 3, we see that both the line length and characteristic impedance are degrees of freedom; you may vary either or both to produce a desired inductance or capacitance. The stepped-impedance filter arbitrarily uses just two fixed, extreme values of line impedance, varying the length as necessary. A complementary (but still arbitrary) alternative method instead fixes the line length and varies the impedance as necessary. Because all lines are of equal length, the resulting filter is said to use *commensurate lines*.[11] As with the stepped-impedance filter, short segments of shorted line implement inductors, and short pieces of open-circuited line act as capacitors. In Richard's original description of the method, *short* is specifically taken to mean an eighth of a wavelength at the cutoff frequency: Each inductor or capacitor of a lumped prototype is thus replaced by a $\lambda/8$ length of transmission line, whose characteristic impedance is varied to produce the desired component value. Because of the equality of lengths, the resulting filter response is perfectly periodic in frequency (unlike the response of a typical stepped-impedance or stub filter) and may be considered the result of aliasing the lumped prototype's response. After a little thought, you can deduce that the response repeats every $4f_c$, where f_c is the frequency at which the lines are of $\lambda/8$ length. These unavoidable re-entrant modes are usually regarded as undesired, but they are also sometimes exploited (as in the half-wave filter described in Section 23.2.5). It is also sometimes the case that, over some range of frequencies, a *steeper*-than-expected transition between passband and stopband results from the aliasing. The primary virtue of Richard's transformation is the predictability of these

---

[11] P. I. Richard, "Resistor-Transmission Line Circuits," *Proc. IRE*, v. 36, February 1948, pp. 217–20. This paper expands on the methods described in the Cohn reference of footnote 3, and it is the first to offer a coherent theoretical framework for converting lumped filters into distributed ones. Richard has the interesting distinction of having received a Ph.D. from Harvard for this work in two years without first having obtained a bachelor's degree.

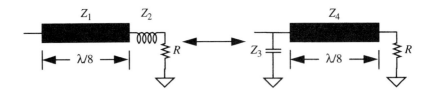

FIGURE 23.11. Circuits for deriving the most useful of Kuroda's identities

re-entrant modes. No fancy simulations are needed to identify where they occur. Beyond that, there is nothing particularly magical about commensurate lines. Indeed, fixing the lengths robs us of an important degree of freedom. Given the constraints on practical widths, losing that degree of freedom may in fact prevent us from ever meeting the design objectives. In such cases, it may be valuable to consider the stub filter as an alternative.

Because of the importance of $\lambda/8$ lines in this method, it's worthwhile examining some of their properties:

$$\frac{Z}{Z_0} = \frac{\frac{Z_L}{Z_0} + j \tan \frac{\pi}{4}}{1 + j \frac{Z_L}{Z_0} \tan \frac{\pi}{4}} = \frac{\frac{Z_L}{Z_0} + j}{1 + j \frac{Z_L}{Z_0}} = \frac{j + \frac{Z_L}{Z_0}}{1 + j \frac{Z_L}{Z_0}}. \tag{18}$$

The wisdom of Richard's choice of $\lambda/8$ is now clear, for the tangent terms become unity and so leave rather simple expressions. When such a line is terminated in a short circuit, the normalized impedance is simply j, or purely inductive. When open-circuited, the impedance is $-j$ (purely capacitive). Thus, implementation of Richard's method involves replacement of capacitors with open-circuited $\lambda/8$ lines and of inductors with shorted ones. Regrettably, there's no practical way to implement the required shorted lines in microstrip. Fortunately, we may exploit transmission line behavior to transform series inductors into shunt capacitors. Commonly known as *Kuroda's identities*, such transformations allow us to avoid having to synthesize grounded elements, making the filter considerably more amenable to fabrication in microstrip form.[12] Formally, there are four Kuroda identities, but half are redundant by reciprocity. Of the remaining two, the one Kuroda identity we will use performs the transformations we seek. The characteristic impedances of the stubs are design degrees of freedom that allow modification of scale factors to produce more practical designs. See Figure 23.11, where the impedances indicated next to the inductance and capacitor are the characteristic impedances of the shorted and open-circuited $\lambda/8$ lines (respectively) that implement those elements.

---

[12] Every standard microwave engineering textbook (and a great many papers) refer to these identities, but almost universally without a specific citation. The first publication with these identities is evidently H. Ozaki and J. Ishii, "Synthesis of a Class of Stripline Filters," *IRE Trans. Circuit Theory*, v. 5, June 1958, pp. 104–9. Ozaki and Ishii state that Kuroda submitted these identities as part of his Ph.D. thesis in 1955. Kuroda himself apparently never published them in any English-language paper.

23.2 DISTRIBUTED FILTERS FROM LUMPED PROTOTYPES

We can derive the relevant identity by determining the input impedances and then setting them equal to each other. For the line loaded by the series LR impedance,

$$Z_{in1} = Z_1 \frac{\frac{R+jZ_2}{Z_1} + j}{1 + j\frac{R+jZ_2}{Z_1}} = Z_1 \frac{R + j(Z_1 + Z_2)}{(Z_1 - Z_2) + jR} = \frac{R + j(Z_1 + Z_2)}{\left(1 - \frac{Z_2}{Z_1}\right) + j\frac{R}{Z_1}}. \quad (19)$$

Similarly, for the other circuit,

$$Z_{in2} = \left[Z_4 \frac{\frac{R}{Z_4} + j}{1 + j\frac{R}{Z_4}} \right] \parallel \frac{Z_3}{j} = \frac{\left[Z_4 \frac{R+jZ_4}{Z_4+jR} \right] Z_3}{j\left[Z_4 \frac{R+jZ_4}{Z_4+jR} \right] + Z_3}$$

$$= \frac{[Z_4(R + jZ_4)]Z_3}{j[Z_4(R + jZ_4)] + Z_3(Z_4 + jR)}, \quad (20)$$

which simplifies to

$$Z_{in2} = \frac{[R + jZ_4]Z_3}{j[R + jZ_4] + Z_3\left(1 + j\frac{R}{Z_4}\right)} = \frac{R + jZ_4}{\left(1 - \frac{Z_4}{Z_3}\right) + j\frac{R}{Z_3}\left[1 + \frac{Z_3}{Z_4}\right]}. \quad (21)$$

Setting the corresponding terms equal in the two input impedance expressions yields:

$$Z_4 = Z_1 + Z_2, \quad (22)$$

$$\frac{Z_4}{Z_3} = \frac{Z_2}{Z_1}, \quad (23)$$

$$\frac{1 + \frac{Z_3}{Z_4}}{Z_3} = \frac{1}{Z_1}. \quad (24)$$

We hope that this seemingly overconstrained equation set (we seek only Z_3 and Z_4, but have three equations) contains no conflicts. Indeed, closer examination reveals that Eqn. 24 contains no information not present in the previous two equations, so there is no problem.

Solving Eqn. 23 for Z_4, setting it equal to Eqn. 22, and then solving for Z_3, we obtain

$$Z_3 = Z_1\left(1 + \frac{Z_1}{Z_2}\right). \quad (25)$$

So, Eqn. 22 and Eqn. 25 describe one of the identities we will use.

We may also solve the foregoing system of equations for Z_1 and Z_2, yielding the same identity in reverse (useful if we start with a shunt capacitance and wish to convert it to a series inductance):

$$Z_1 = Z_3 \parallel Z_4; \quad (26)$$

$$Z_2 = \frac{Z_4^2}{Z_3 + Z_4}. \quad (27)$$

Taken together, Eqn. 22 and Eqns. 25–27 describe the equivalencies shown in Figure 23.12. Again, the indicated impedances are those of the transmission line segments that realize the elements.

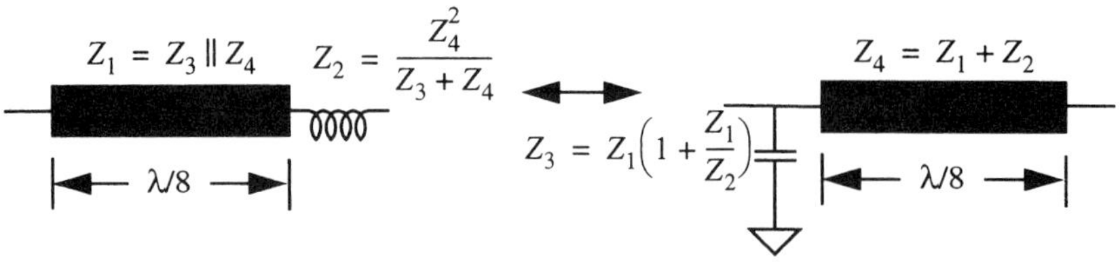

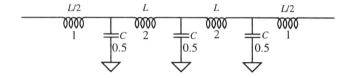

FIGURE 23.12. The relevant Kuroda identities

FIGURE 23.13. Filter prior to Kuroda transformation (all stubs are $\lambda/8$ long at cutoff frequency)

Let's now use this ability to convert between series inductances and shunt capacitances to implement the constant-k filter that we've already translated into various distributed forms. The lumped prototype we've been using is a seventh-order filter with shunt capacitors on the ends. However, it turns out that substantially fewer transformations are required if there are an even number of inductors (there's a complexity associated with a central inductor that requires additional transformations; feel free to try the odd-number-of-inductors case and discover for yourself what happens). Thus, we will use the alternative (but completely equivalent) prototype form with series inductors at the ends. The capacitors are 6.4662 pF and the main inductors are 15.915 nH (half this value for the inductors at the ends).

It is customary to design with normalized element values, deferring denormalization until the very end in order to simplify intermediate calculations. That, in turn, reduces the likelihood of errors, so we'll follow this custom in the sequel.

First let's find the impedance of the line that would implement the main series inductor:

$$Z_2 = \omega L = 100 \, \Omega, \qquad (28)$$

or a normalized value of 2. The inductors on the ends would be implemented with lines whose impedances are half this value. Let's also find the impedance of the lines that would implement the capacitors:

$$Z_C = 1/\omega C = 25 \, \Omega, \qquad (29)$$

for a normalized value of 0.5.

Before transformation by Kuroda's identities, the filter appears as in Figure 23.13. Application of the identities requires that a $\lambda/8$ line segment be connected to each element that is to be transformed, but no ordinary lumped prototype (including that of Figure 23.13) will satisfy this fundamental requirement. The solution is to note

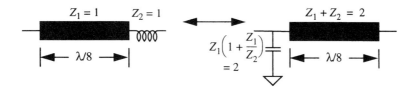

FIGURE 23.14. Application of Kuroda's identities to end inductors of example filter

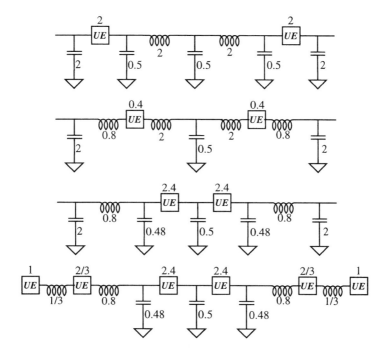

FIGURE 23.15. Sequence of filters generated by repeated application of Kuroda identities

that we may freely add any length of line to the input and output ports of a filter, provided those added lines are of impedance Z_0. The addition of redundant $\lambda/8$-long segments, called *unit elements* in the literature, thus does not alter the filter response at all, but it does enable us to apply the Kuroda relationships to transform the end elements. Furthermore, each application of the Kuroda identity generates a *new* $\lambda/8$ line segment, which can be paired with still another element to be transformed, and so on. Thus we can work our way toward the center from the ends, one transformation at a time, until the elements in the center of the filter have been transformed into shunt form. Then, if necessary, we can work back toward the input and output ports of the filter to (re-)transform any remaining series elements back into shunt stubs.

Thus, in our specific example, the first step is to convert the end inductors into shunt capacitors after adding a unit element (*UE*) to each end, as shown in Figure 23.14. Proceeding in a like manner generates the sequence of equivalent filters displayed in Figure 23.15.

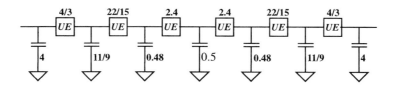

FIGURE 23.16. Final normalized seventh-order low-pass filter after all transformations have been completed

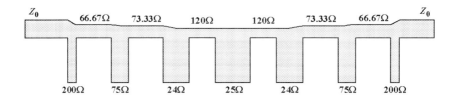

FIGURE 23.17. One possible 1-GHz microstrip stub filter layout (not drawn to scale)

To finish the removal of all series inductances, we need to perform a last transformation to convert the two outer *pairs* of series inductors into shunt capacitive equivalents (these need not be transformed simultaneously, but it's expedient to do so). After taking care of the corresponding transformations, we obtain the final design at last. See Figure 23.16.

The very last steps are to denormalize the element values to obtain actual line impedances, and then to simulate the design to verify that no errors have crept into the process along the way (especially important given the relatively large number of transformations). After computing the denormalized line impedances, a preliminary filter layout follows as shown in Figure 23.17. The center-to-center spacing of the stubs is the same as the stub length (even though the crude manual rendering of the figure may not quite show this).

Notice that the illustrative layout varies only the top boundary to adjust backbone impedances. While not necessarily the electrical optimum, its simplicity has made this choice almost universal.

Simulating this design results in the transmission plotted in Figure 23.18, from which it is seen that the filter performs as expected. The attenuation at 1 GHz is -9.5 dB (compared to the -10-dB value of the constant-k lumped prototype), and the -3-dB point occurs at about 917 MHz (versus the lumped prototype's 934 MHz). Plotting over a larger range of frequency and attenuation (Figure 23.19) clearly shows the expected periodicity of this filter's response.

23.2.5 HALF-WAVE (RE-ENTRANT) "BANDPASS" FILTERS

We've noted that the low-pass filter with commensurate $\lambda/8$ lines has a frequency response that repeats every $4f_c$, where f_c is the low-pass cutoff frequency. It's important

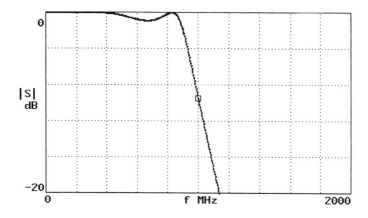

FIGURE 23.18. Simulated performance of final filter after repeated application of Kuroda identities

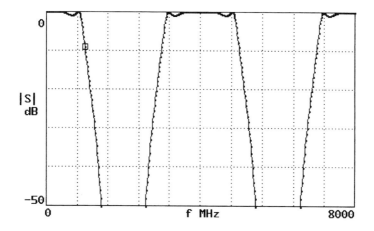

FIGURE 23.19. Filter performance over larger frequency and attenuation range

to emphasize that these periodic re-entrant passbands are not necessarily useless or undesirable. As a specific example to underscore this idea, consider using the first spurious passband of Figure 23.19 to produce a bandpass filter. Its passband shape is approximately that of the constant-k low-pass filter on which it is based. That example is hardly unique: we have noted that *any* commensurate-line low-pass filter using Richard's $\lambda/8$ lines will produce responses with this periodicity, presenting many opportunities for bandpass filtering. For example, at very high frequencies, $\lambda/8$ lines might have impractically small dimensions and so one might use a higher-order passband. This practice of deliberately employing a re-entrant mode is known as *overmoding*. It is especially popular in millimeter- and submillimeter-wave work, where principal-mode structures would be too small for practical fabrication.

In noncritical applications, there is a class of bandpass filters that may be designed easily with pencil and paper. Just as stepped-impedance low-pass filters are trivially

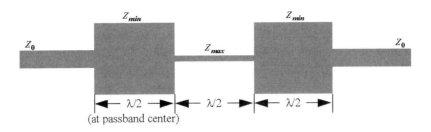

FIGURE 23.20. Half-wave "bandpass" filter
(commensurate stepped-impedance version)

designed, we might imagine basing a bandpass filter on a special case of the stepped-impedance low-pass filter in which the element lengths are commensurate. Note that if lines are $\lambda/8$ in length at a low-pass filter's band edge, then the center of the first bandpass response corresponds to a wavelength of $\lambda/2$. We thus select these commensurate element lengths equal to $\lambda/2$ at the center of the intended passband; see Figure 23.20. When using this structure as a bandpass filter, it's important to keep in mind the fact that it remains a low-pass filter as well and to accommodate this reality in any design.

For an alternative view of how this filter works, recall that a transmission line of length $n\lambda/2$ reproduces at its input the load impedance. For our particular implementation, this condition implies a driving-point impedance that is equal to Z_0, allowing maximum power transfer into the filter and, ultimately, to the load. At frequencies where the electrical length of the filter sections differs significantly from the half-wavelength condition, power is coupled into (and out of) the filter less efficiently because of the gross impedance mismatches at the various interfaces. From this description, we see that the filter indeed behaves as a bandpass filter, with periodically disposed passbands centered around frequencies at which the line lengths are integer multiples of a half-wavelength.

The off-center rejection is maximized by ensuring as great a mismatch as possible between the impedances of the narrow and wide filter sections. This observation implies that better stopband rejection is obtained when each wide half-wavelength section is made as low in impedance (wide) as possible and each narrow section as high in impedance as possible, just as in any stepped-impedance filter. The more extreme the impedance ratio, the deeper the stopband. Furthermore, the stopband rejection increases as the number of sections increases.

The simulation graphed in Figure 23.21 is of the layout in Figure 23.20, using 15-Ω and 200-Ω lines chosen $\lambda/8$ in length at 250 MHz to produce a first re-entrant mode centered about 1 GHz. The responses centered at DC, 1 GHz, and 2 GHz are evident from this simulation. To obtain a close approximation of this response in practical filters, you must make due allowance for the T-junction shortening effect by lengthening the high-impedance lines some small amount (roughly of the order of the dielectric thickness; see Chapter 7 for more refined estimates).

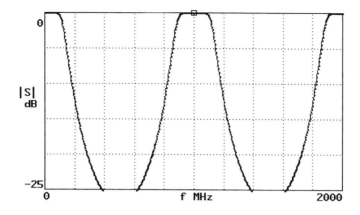

FIGURE 23.21. Response of commensurate-line stepped impedance filter

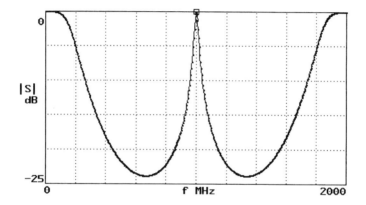

FIGURE 23.22. Yet another simulation of yet another filter

As a final observation we should note that, although we introduced this filter with commensurate lines, it is not an absolute requirement that all lines be of equal length. Just as a crazy example, making the low-impedance sections $\lambda/4$ in length at 1 GHz – and connecting them together with a high-impedance section that is $\lambda/2$ in length – produces extremely narrow passbands centered at odd multiples of 1 GHz and broad passbands centered at even multiples of 1 GHz. See Figure 23.22.

Simulators such as *Puff* are invaluable for rapidly evaluating the effects of varying the length, width, and number of segments. The reader is encouraged to use *Puff* or another appropriate simulation tool to explore how the response shape changes as the lengths are varied and to attempt explaining why the results are what they are.

23.3 COUPLED RESONATOR BANDPASS FILTERS

The half-wave "bandpass" filter exploits a parasitic effect to yield a bandpass response that is more or less an accidental by-product of designing a low-pass filter.

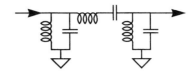

FIGURE 23.23. Classic lumped bandpass filter

We now consider how to design filters that provide the bandpass characteristic directly, rather than incidentally.

23.3.1 LUMPED BANDPASS FILTERS

As we've seen, the classical lumped bandpass filter consists of parallel resonators coupled together with series resonant arms (Figure 23.23). Recall that this structure evolves from a low-pass prototype whose bandwidth and other characteristics are those of the bandpass filter you ultimately want. In this particular example, the low-pass prototype has two shunt capacitors and one series inductor. All three elements are paired with their duals to resonate at the desired center frequency. That is, each of the shunt capacitors is shunted by an appropriate inductance, and a capacitance is placed in series with each series inductance. It is the series resonant section resulting from that last step that unfortunately prevents a straightforward translation of this lumped prototype into microstrip form. Even in lumped implementations, it is considered at least a nuisance to have to focus design attention on two different types of resonators. For example, some parasitic capacitance in shunt with an inductor of a parallel tank may be tolerated, for it may form part of the resonator. However, the inductors in the series arms must have much lower parasitic capacitances, because capacitances there cannot be so readily absorbed.

The first step in solving this problem for both lumped and distributed bandpass filters is to recognize that the series and parallel resonators are duals of each other. That is, if we compute the reciprocal of one network's impedance, we obtain the impedance of the other network (within a scale factor of appropriate dimensions to make the units work out). Specifically, consider the impedance of a series LC resonator,

$$Z_S = sL_s + \frac{1}{sC_s}, \tag{30}$$

and of a parallel LC tank,

$$Z_P = sL_p \parallel \frac{1}{sC_p} = \frac{sL_p}{s^2 L_p C_p + 1}. \tag{31}$$

If we compute the reciprocal of the series network's impedance, say, we obtain

$$Z_{eq} = \frac{K^2}{Z_S} = \frac{K^2}{sL_s + 1/sC_s} = \frac{K^2(sC_s)}{s^2 L_s C_s + 1}, \tag{32}$$

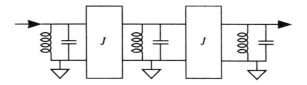

FIGURE 23.24. All-parallel tank bandpass filter using immittance inverters

where we have introduced the factor K (with dimensions of impedance) to fix the units problem. We may also express the relationships in terms of admittances, of course, as follows:

$$Y_{eq} = \frac{J^2}{Y_S}, \tag{33}$$

where once again we have introduced an appropriate scaling factor, J. To keep the units honest, J must have the dimensions of admittance.

Because the forms of Eqn. 31 and Eqn. 32 are the same, we see that we can indeed convert series resonators into parallel ones (and vice versa) provided we have networks that reciprocate impedances. By equating corresponding terms, we can derive the necessary design relationships:

$$K = \sqrt{L_p/C_s} = \sqrt{L_s/C_p}; \tag{34}$$

$$L_s C_s = L_p C_p. \tag{35}$$

Thus, we set the resonant frequencies equal and then choose the transformation factor K (or J) as necessary to satisfy Eqn. 34. As we'll see, an additional degree of freedom remains, facilitating the choice of element values within a practical range. Of course, all of this presupposes that we have impedance inverters at our disposal. Setting aside for the moment the question of how one realizes such impedance or admittance reciprocators (collectively called *immittance inverters*[13]), we see how neatly they solve our practical problem with series resonators. If we replace each series resonator with a parallel resonator sandwiched in between *two* inverters, then circuits to either side of the combination will see a series resonator. We can therefore realize a bandpass filter using only parallel tanks.

For example, we may convert the circuit of Figure 23.23 into an all-parallel resonator equivalent, as shown in Figure 23.24. So, instead of two parallel tanks and one series resonator, we now have three parallel tanks plus two impedance inverters.

In order to enable complete designs, we need to devise ways of realizing these inverters. We've actually already met one form of impedance reciprocator: the $\lambda/4$ transmission line. Of course, it satisfies the $\lambda/4$ condition at only a single frequency,

---

[13] The word *immittance* is a portmanteau that results from combining the words *im*pedance and ad*mittance*. It is frequently misspelled as *immitance*.

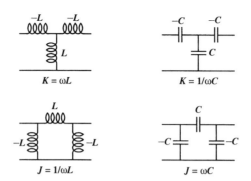

FIGURE 23.25. Commonly used impedance (K) and admittance (J) inverters

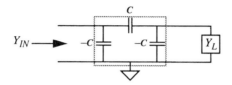

FIGURE 23.26. Capacitive admittance inverter

so such an inverter is not a broadband element. More serious is that the relatively small range of practically realizable characteristic impedances limits the utility of this element as an inverter. That said, it does prove useful on occasion, even though it is often overlooked as an option.

It's possible to overcome some of the limitations of a $\lambda/4$ line by using any one of many lumped networks that approximate its behavior. Examples of both impedance and admittance inverters are shown in Figure 23.25. As can be seen from the examples, J and K are admittance and impedance scaling factors, respectively. Of course, a network that inverts impedances also necessarily inverts admittances, so the distinction is somewhat artificial. The difference is simply the way in which you choose to describe the network's behavior. Thus, for impedance inverters,

$$Z_{eq} = K^2/Z \tag{36}$$

while for admittance inverters we have

$$Y_{eq} = J^2/Y. \tag{37}$$

Selection of one viewpoint over the other is based on convenience.

We'll see that the π-network of capacitors is an especially interesting case for bandpass filters, so let us verify its admittance-inverting properties. From Figure 23.26, it's straightforward to see that the input admittance is

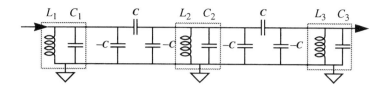

FIGURE 23.27. All-parallel tank bandpass filter using J-inverters, *before* combining capacitors

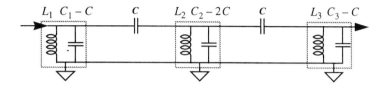

FIGURE 23.28. All-parallel tank bandpass filter using J-inverters, *after* combining capacitors

$$Y_{IN} = \left[\frac{1}{Y_L - sC} + \frac{1}{sC}\right]^{-1} - sC = \frac{(Y_L - sC)sC}{sC + (Y_L - sC)} - sC = -\frac{(sC)^2}{Y_L}. \quad (38)$$

This network is thus indeed an immittance inverter, with a characteristic admittance J equal to the capacitive admittance.

We leave it as an exercise for the reader to verify the other expressions for J and K given in Figure 23.25. It's clear from those expressions that the inversions vary with frequency for these particular networks. Nevertheless, for many narrowband (e.g., 10–20% bandwidth) bandpass filters, the inversions are sufficiently constant over the passband to permit the realization of useful filters. Furthermore, because inductive and capacitive inverters have opposing frequency dependence, alternating them can increase the overall inversion bandwidth by a factor of 2 or more.

Inspection of the networks has undoubtedly left you wondering about the negative inductors and capacitors. Because inverters are never used alone, there is no need to devise a negative element. By choosing the right inverter for a given configuration, the negative components can always be absorbed into the rest of the network, where they simply serve to reduce the value of an existing positive capacitance or inductance. This idea is perhaps best illustrated with a specific example, so let's select a capacitive impedance inverter for the circuit of Figure 23.24 to produce the circuit shown in Figure 23.27.

In this first step, note that the parallel resonators indicated within the highlighted boxes are all tuned to the same frequency as a direct consequence of the low-pass–bandpass transformation. It is worth keeping this fact in mind, for it will prove valuable not only when we consider experimental methods for tuning real filters but also when we explore translating this circuit into various distributed implementations.

The next step is to combine the tank capacitances with the (negative) capacitances of the J-inverters. This is shown in Figure 23.28. After this absorption of the

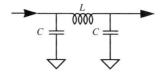

FIGURE 23.29. Low-pass, third-order, 1-dB ripple Chebyshev prototype filter (100 MHz, $Z = 50\ \Omega$)

Table 23.1. *Component values for prototype 100-MHz, 1-dB Chebyshev low-pass filter ($Z = 50\ \Omega$)*

L	C
79.11 nH	64.42 pF

negative capacitances, all network capacitances are positive, as asserted previously. Even though the resonators no longer have equal resonant frequencies, note that short-circuiting the resonators surrounding any given one restores the "lost" capacitance and thus restores the resonant frequency. This constancy suggests that a filter alignment procedure might involve the successive shorting of resonators to permit the tuning of individual resonators. Again, we will expand on this idea when we consider practical methods for tuning filters.

A Detailed Example

To illustrate the design procedure in detail, let's consider how to realize a capacitively coupled bandpass filter. We'll start with a low-pass prototype, apply the low-pass–bandpass transformation, and then use immittance inverters to permit an all-shunt resonator implementation.

Suppose our goal is to design a bandpass filter whose passband extends from 950 MHz to 1.05 GHz. We arbitrarily begin with a third-order 1-dB ripple Chebyshev low-pass prototype featuring a ripple bandwidth of 100 MHz;[14] see Figure 23.29. The element values for this filter are listed in Table 23.1.

Next, we produce a bandpass version by resonating each element at the geometric mean of the upper and lower cutoff frequencies. We would not normally worry about the slight difference between the arithmetic and geometric means because, for

---

[14] We've deliberately chosen a somewhat large ripple in order to make a definite ripple readily visible on magnitude response plots. In practice it is more likely to encounter ripple values of a few tenths of a dB or less, except in noncritical applications.

23.3 COUPLED RESONATOR BANDPASS FILTERS

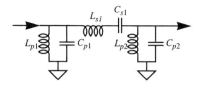

FIGURE 23.30. Prototype bandpass filter (100-MHz bandwidth, 998.75 MHz, $Z = 50\ \Omega$)

Table 23.2. *Component values for 100-MHz–bandwidth, 1-dB–ripple, 1-GHz bandpass filter ($Z = 50\ \Omega$)*

L_{p1}, L_{p2}	C_{p1}, C_{p2}	L_{s1}	C_{s1}
394.242 pH	64.42 pF	79.11 nH	320.993 fF

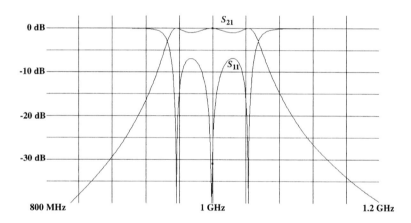

FIGURE 23.31. Magnitude and reflectance of prototype Chebyshev bandpass filter (*RFSim99*)

this 10% bandwidth case, the difference between the two means is small enough to be ignored. However, in the interest of illustrating the procedure precisely, we'll make the distinction and continue to retain more digits than are truly sensible. So here, we produce resonances at about 998.75 MHz; see Figure 23.30. The element values for this bandpass filter are given in Table 23.2.

Just for reference, the response of the prototype bandpass filter is shown in Figure 23.31. The 1-dB passband ripple is evident – as is the 100-MHz bandwidth, which is centered about 0.99875 GHz to an excellent approximation. Close examination of the magnitude response reveals that the lower and upper −1-dB band edges are indeed 950 MHz and 1.05 GHz, as designed.

The next step is using immittance inverters to enable replacement of the series resonant arm with a parallel resonator. Of the many possible choices for the inverter,

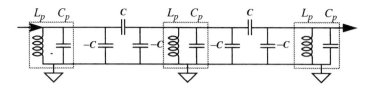

FIGURE 23.32. Bandpass filter *before* combining capacitors

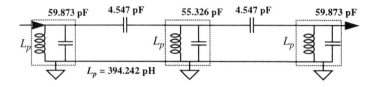

FIGURE 23.33. Bandpass filter *after* combining capacitors

we choose the capacitive π-network, given the amount of time we've spent understanding it and also in anticipation of converting the lumped design into a microstrip version.

In converting the series resonant branch into a parallel equivalent we have an underconstrained problem, for we have the freedom to choose three variables (the parallel LC network element values, as well as the value of J) when creating an equivalent of the original two-element series resonant network. We will arbitrarily make the transformed tank's inductance, L_p, equal to the other inductances (394.242 pH in this case).[15] This choice reduces the number of different components that one must stock (in the case of discrete implementations), design, and characterize. Resonating that capacitance at the center frequency yields the same tank capacitance, C_p, as that of the other two tanks (64.42 pF).

With these element values, the characteristic admittance J of the inverter is then set at

$$ J = \sqrt{\frac{C_{s1}}{L_p}} \approx \sqrt{\frac{320.993 \text{ fF}}{394.242 \text{ pH}}} \approx 28.534 \text{ mS}, \tag{39} $$

corresponding to a characteristic impedance of 35.05 Ω. The required inverter capacitance C is then readily computed as

$$ C = \frac{J}{\omega_0} = \frac{28.534 \text{ mS}}{(2\pi \cdot 0.99875 \times 10^9) \text{ rps}} \approx 4.547 \text{ pF}. \tag{40} $$

See Figure 23.32. After combining elements we obtain the final form of the filter, as shown in Figure 23.33.

---

[15] Because not all low-pass prototypes have equal inductances, not all bandpass filters derived from them will have equal inductances. The particular example we have chosen just happens to be an exception.

23.3 COUPLED RESONATOR BANDPASS FILTERS

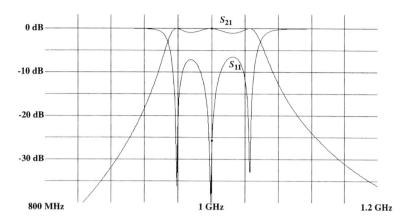

FIGURE 23.34. Magnitude and reflectance of capacitively coupled lumped Chebyshev bandpass filter

As a verification step, consider the simulations (Figure 23.34) of the magnitude response and reflectance of the capacitively coupled filter. As is apparent, the response is quite similar to that of the prototype bandpass filter, but a close examination reveals differences nonetheless. Because the inverter's characteristic admittance increases with frequency, the filter response is less symmetrical, favoring more transmission at higher frequencies. This favoritism is barely noticeable within the passband, although the lower and upper −1-dB band edges do shift upward a tiny bit (here, to 953 MHz and 1.053 GHz) and the passband is no longer equiripple (the ripple is somewhat higher at higher frequencies). For filters of larger fractional bandwidths, the errors may become objectionable. In such cases, one option is to use slightly modified definitions of bandwidth (for the prototype) and center frequency (for the synthesized bandpass filter) in order to reduce the error caused by frequency-dependent inverters. In such cases the low-pass prototype should be designed for a bandwidth

$$(BW) = \frac{f_0^2(f_u - f_l)}{f_l f_u}; \qquad (41)$$

here f_l and f_u are the desired lower and upper band-edge frequencies of the final bandpass filter, whose center frequency f_0 is now given by[16]

$$f_0 = f_l + f_u - \sqrt{(f_u - f_l)^2 + f_l f_u}. \qquad (42)$$

For small fractional bandwidths, these equations converge to the simpler ones we have been using (i.e., geometric-mean center frequency, and equality of low-pass prototype and bandpass bandwidths). For our particular example, these equations tell us that the bandwidth of the low-pass prototype should be set to about 104.3 MHz and that the bandpass filter should be designed for a center frequency of approximately 996.26 MHz.

---

[16] Seymour B. Cohn, "Direct-Coupled-Resonator Filters," *Proc. IRE,* February 1957, pp. 187–96.

Also measurable is a slight increase in the ripple within the passband. Greater differences are apparent further away from the center frequency. Compared with the original bandpass filter, the capacitively-coupled version has better attenuation within the lower stopband, but inferior upper stopband attenuation. These differences do not necessarily represent serious impairments, but their existence must be accounted for. Had we attempted to design a filter with a significantly broader bandwidth, the discrepancies caused by our use of a frequency-dependent immittance inverter would have been considerably more noticeable. Again, these differences must be anticipated and accommodated. If the prototype low-pass filter just barely satisfies requirements, for example, the capacitively coupled version may fail to meet one or more filter specifications.

Summary of Bandpass Filter Design with Immittance Inverters

Because the derivation of the design procedure we've just outlined is spread out over a number of pages, it's helpful for future reference to summarize the results succinctly, in quasirecipe form. Aside from facilitating the design of lumped, capacitively coupled bandpass filters, we'll see that these equations are also directly useful in the design of an important class of fully distributed filters.

The first step is to generate a lumped low-pass prototype whose passband and stopband characteristics (e.g., ripple and bandwidth) are those that the bandpass filter is to possess. For the sake of uniformity in notation, we assume that the prototype filter is normalized to 1-rps bandwidth and 1-Ω impedance, with both inductances and capacitances denoted by the catch-all variable g_k (note: these g are *not* conductances). We will additionally use subscripts p and s to denote elements that are in shunt (parallel) sections and in series branches, respectively.

For a bandpass filter of normalized bandwidth (BW), impedance level Z_0, and center frequency ω_0, the shunt capacitances have a value

$$C_p = \frac{g_{kp}}{\omega_0 Z_0 (BW)} \tag{43}$$

and resonate with added inductors of value

$$L_p = \frac{Z_0 (BW)}{g_{kp} \omega_0}. \tag{44}$$

Similarly, series inductances are given by

$$L_s = \frac{g_{ks} Z_0}{\omega_0 (BW)} \tag{45}$$

and resonate with capacitors of value

$$C_s = \frac{(BW)}{g_{ks} \omega_0 Z_0}. \tag{46}$$

Each series resonator may be replaced by a combination of a shunt resonator and two immittance inverters to permit an all-parallel tank implementation. Arbitrarily choosing to characterize the inverter with an admittance, we have

$$J = \sqrt{C_s/L_p} = \sqrt{C_p/L_s}. \qquad (47)$$

The characteristic impedance of the resonators, Z_{0res}, need not be the same as the external system's impedance. Making them equal does simplify the design by eliminating the need to add impedance matching sections at the input and output, which is probably why most textbooks do not explicitly identify the additional degree of freedom. However, retaining this flexibility is potentially valuable, so for lumped tanks we will write

$$Z_{0res} = \sqrt{L_p/C_p}. \qquad (48)$$

The product of the inverter admittance and the tank impedance then becomes

$$JZ_{0res} = \sqrt{C_s/L_p}\sqrt{L_p/C_p} = \sqrt{C_s/C_p}$$
$$= \sqrt{\frac{(BW)}{g_{ks}\omega_0 Z_{0res}} \bigg/ \frac{g_{kp}}{\omega_0 Z_{0res}(BW)}} = \frac{(BW)}{\sqrt{g_{kp}g_{ks}}}, \qquad (49)$$

which may be readily solved for the inverter admittance given the other parameters. Note that the right-most term in Eqn. 49 is expressed entirely in terms of normalized quantities. Thus, the product JZ_{0res} is readily computed given the normalized bandwidth and low-pass prototype element values. This computation is repeated as necessary to convert each series branch into a shunt one, using the appropriate $g_{kp}g_{ks}$ product at each conversion step.

When the tank impedances are not equal to the system impedance, one may use immittance inverters to provide the required impedance transformations. By analogy with $\lambda/4$ transmission line impedance-matching sections, we see that the desired result is produced by setting the characteristic impedance of the inverter equal to the geometric mean of the impedances that are to be matched. Thus, if the prototype low-pass filter begins with a series inductor of normalized value g_1, then matching to an impedance Z_0 requires

$$J = \frac{1}{\sqrt{Z_0\omega_0 L_1}} = \frac{1}{\sqrt{Z_0 \frac{g_1 Z_{0res}}{(BW)}}} = \frac{1}{Z_0\sqrt{\frac{mg_1}{(BW)}}} = \frac{1}{Z_0}\sqrt{\frac{(BW)}{mg_1}}, \qquad (50)$$

where $m = Z_{0res}/Z_0$. When expressed in the same form as Eqn. 49, this expression becomes

$$JZ_0 = \sqrt{\frac{(BW)}{mg_1}}. \qquad (51)$$

If desired or otherwise appropriate, one may implement these inverters as $\lambda/4$ lines rather than as lumped networks.

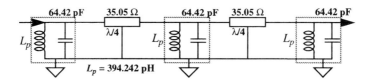

FIGURE 23.35. Bandpass filter with lumped resonators and transmission line inverters

Note that the presence of impedance transformers also allows the source and load impedances to differ from each other, as well as from Z_0. Thus, if g_N is the normalized value of the final prototype filter element, then

$$JZ_0 = \sqrt{\frac{(BW)}{mg_N}}. \qquad (52)$$

Although it is common for the source and load impedances to be equal, this degree of freedom occasionally proves valuable in those instances when they are not.

The next step is to select an inverter topology and determine the inverter-element values needed to produce the required J for each inverter. Then absorb the inverter elements into the rest of the network to eliminate negative element values. As mentioned previously, alternate use of capacitive and inductive inverters extends the bandwidth of the inversion. This option is valuable when designing bandpass filters with large fractional bandwidths (e.g., 30–40%).

Bandpass Filters with Combinations of Lumped and Distributed Elements

With such an emphasis on lumped-element inverters, it's easy to overlook distributed inverters – which, after all, are what got us started on this inversion business in the first place. It is therefore instructive to examine the response of a filter made with three lumped resonators coupled together with a pair of 35.05-Ω $\lambda/4$ lines acting as inverters (see Figure 23.35). Because this characteristic impedance is well within the range of practical realization and also because the resonators are identical in our particular example (instead of *almost* identical), such a filter might be worth considering, especially in view of its reasonable performance.

The magnitude response of this filter is shown in Figure 23.36. As is evident, the passband response indeed corresponds to that of a 1-dB–ripple Chebyshev. Furthermore, the behavior at frequencies well removed from the center is somewhat better than we observe with the lumped-element inverters. However, the 1-dB–ripple passband extends from about 954 MHz to 1045 MHz, so there is some reduction in bandwidth. At the same time, the passband ripple decreases somewhat, to about 0.9 dB. Using Cohn's more sophisticated equations for center frequency and bandwidth would largely correct these slight impairments.

We may produce a completely distributed bandpass filter by replacing the lumped resonators with a transmission line equivalent. To do so is straightforward in principle,

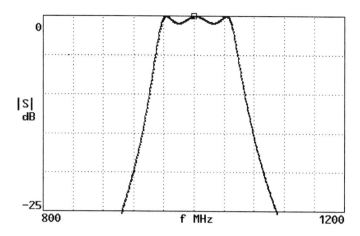

FIGURE 23.36. Response of bandpass filter with transmission line coupled lumped resonators

for we may replace the parallel $L_p C_p$ tank with a shorted $\lambda/4$ line whose characteristic impedance differs only a little from that of the tank:[17]

$$Z_{0stub} = \tfrac{\pi}{4}\sqrt{L_p/C_p}. \tag{53}$$

Just for reference, when replacing a series LC resonator with an open-circuited $\lambda/4$ stub, the corresponding relationship is

$$Z_{0stub} = \tfrac{4}{\pi}\sqrt{L_s/C_s}. \tag{54}$$

These relationships derive from equating the reactance or admittance slope (i.e., dZ/df or dY/df) near the resonant frequency, rather than at values far from resonance (see Section 23.6).

In our example, the lumped tank's characteristic impedance is an already low 2.474 Ω, so the equivalent shorted-line resonator would have an even lower characteristic impedance of 1.943 Ω. Needless to say, implementing such a low line impedance is practically out of the question, but this problem can be solved by simply scaling all impedances upward by a common factor and then performing the necessary downward impedance transformations at the input and output ports. The only limitation is that we must then be mindful not to require *excessive* characteristic impedances of the inverters. Here, we happen to be marginally fortunate, for scaling all impedances upward by a factor of about 5–6 produces resonator and inverter impedances that might just barely lie within the range of practically attainable values for FR4. This case is unusual, for it is all too common for the range of line impedances to exceed by a large factor what can be accommodated. This problem motivates

---

[17] For very crude back-of-the-envelope calculations, you are usually free to neglect the $\pi/4$ factor (it is certainly easier to remember that way) and the $4/\pi$ factor for series equivalents.

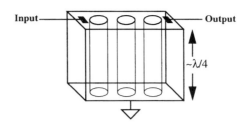

FIGURE 23.37. Typical ceramic dielectric bandpass filter

consideration of alternative distributed resonator filters. One of these retains relatively low-impedance distributed resonators but couples them together with lumped capacitances; the wider range of available values for the latter enables a practical implementation in many cases. In fact, perhaps the most common (in terms of manufacturing volume) microwave bandpass filters use precisely this architecture. The filter is made additionally attractive by employing high-dielectric constant materials to shrink the required volume by large factors. Many cellphone handsets manufactured since the mid-1980s contain at least one filter of this type. Although such filters are not planar, their prominence justifies a brief description.

As seen in Figure 23.37, a typical filter of this type (sometimes called a *monoblock* filter) has a particularly simple structure. A conductive rectangular cavity is filled with a dielectric material, so that only the top surface is not covered with metal. Typical dielectric constants of the materials used in these filters range from about 10 to well beyond 100. While enabling compact filters (e.g., order of 1-cm maximum linear dimension for low-GHz frequencies), most of these dielectrics are also piezoelectric, so one must be mindful of microphonics and even the possibility of generating destructively high voltages when struck.[18] The good news is that the dielectric constants typically exhibit temperature coefficients below 10 ppm/K, thanks to decades of research effort by materials scientists. Additionally, the dielectrics have very low loss, so that resonators possess typical Q–frequency products of several terahertz.

Each shorted $\lambda/4$ resonator is produced by forming a hole in the dielectric material and then plating the walls and bottom of the cylindrical hole with metal. The dielectric constant and the physical dimensions of the resonators together determine the resonator characteristic impedance. The coupling between adjacent resonators is controlled by their mutual proximity, and the capacitive coupling to the input and output ports is provided by placing metallization an appropriate distance away from the end resonators. This results in a structure that may be modeled as shown in Figure 23.38. The high performance of typical units used in handsets is underscored by observing that their insertion loss generally lies between 1 and 2 dB.

---

[18] Dielectric compositions vary, and are varied all the time, but they frequently involve combinations of lead, zirconium, barium, strontium, and titanium.

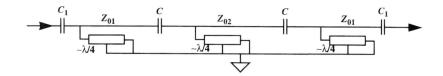

FIGURE 23.38. Approximate model of ceramic dielectric bandpass filter

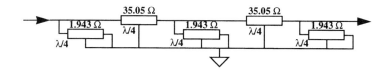

FIGURE 23.39. Impractical all-distributed bandpass filter

The resonators have slightly different characteristic impedances, and they are slightly less than $\lambda/4$ in length because of the action of the negative shunt capacitances of the admittance inverters that couple the resonators together. These effects are readily seen in the all-lumped version of this filter; in fact, the lumped prototype can be translated directly into the form of Figure 23.38 to yield the precise resonator lengths and impedances.

We see that there are many possible variations on the basic coupled resonator theme. Returning now to the relevant subset of those forms that is theoretically amenable to planar implementation, let us simulate the impractical, all-distributed filter (Figure 23.39) without performing any impedance scaling, just for the sake of comparison. Doing so, we discover that both the passband ripple and the 1-dB ripple bandwidth remain smaller than in the all-lumped capacitively coupled implementation. Replacement of the lumped tanks with the shorted lines causes a negligible shift in the passband, as it now extends to roughly 1045 MHz from about 953 MHz (see Figure 23.40). To correct for the ∼8% bandwidth error, one could repeat the entire filter synthesis procedure by starting with a low-pass prototype whose bandwidth is (say) 8–10% larger than the final target. Similarly, any required correction of the center frequency may be effected through an appropriate modification of the center frequency used to compute the initial bandpass prototype from the low-pass prototype.[19]

We see that purely distributed bandpass filters are readily devised in principle, but that highly impractical impedance levels frequently result when following the

---

[19] The Cohn equations reduce, but generally do not eliminate, the relevant errors (typically, the residual error in the final filter's bandwidth is quite a bit larger than the error in the center frequency). Given that using those equations still leaves us with errors and given the availability of modern simulation tools, it is justifiable to proceed as we have, implementing necessary adjustments in subsequent passes.

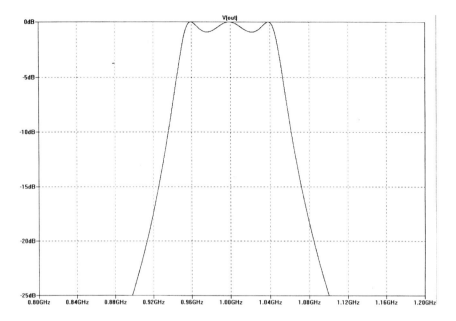

FIGURE 23.40. Magnitude response of impractical all-$\lambda/4$ segment bandpass filter

straightforward synthesis procedures presented. To develop an alternative architecture that gives us the freedom to specify resonator impedances at the outset – and thereby assure that they remain within practical limits – we need to broaden our understanding of coupling's causes and effects. We'll see that a welcome by-product of this broadening is an experimental technique for aligning (tuning) filters so that practical realizations will approximate more closely the theoretical prototypes that inspire them.

23.3.2 ENERGY COUPLING AND MODE SPLITTING

Viewed from a high-level perspective, the goal of lumped filter design is to place the transfer function poles and zeros in whatever configuration provides the desired response shape. This important idea is the basis of all modern lumped filters, bandpass or otherwise. A useful description of how bandpass filters produce a variety of response shapes is that they exploit the *mode splitting* that occurs whenever two or more resonant systems interact.[20] That is, when two identical resonators are connected together in some fashion, the poles of the resulting coupled system do not remain identical. As we'll see momentarily, the stronger the interaction, the wider the induced pole separation. The underlying idea is very similar to the splitting of

---

[20] Actually, this mechanism may be invoked to explain why the poles of a lumped lowpass *RC* line, for example, are similarly not coincident. We will focus on the bandpass case because of its direct relevance to the problem at hand.

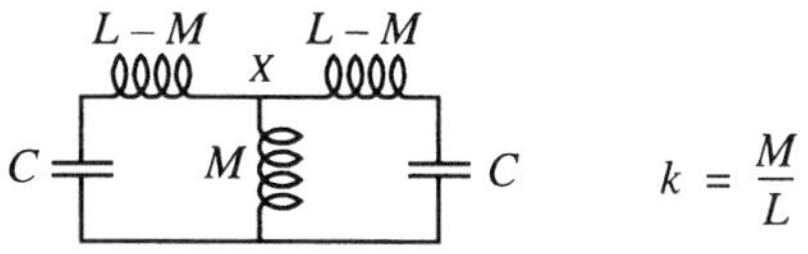

23.3 COUPLED RESONATOR BANDPASS FILTERS

FIGURE 23.41. Magnetically coupled LC resonators

even and odd mode impedances that occurs in coupled lines, an analogy that is worth keeping in mind.

Initially consider the case of two identical, simple LC resonators whose inductors are *magnetically* coupled to each other. We choose the somewhat less familiar (to contemporary students, anyway) magnetic coupling partly for pedagogical reasons (precisely because it's less familiar) and partly to help underscore that the result we will shortly derive is quite general.

We represent the coupled inductors as a transformer and then model the transformer as a T-connection of three inductors; see Figure 23.41. The inductance L is that of each resonator in isolation. The mutual inductance M is some fraction of L and depends on the magnitude (and sign) of the coupling. The magnitude of the coupling coefficient k ranges from zero to unity as the flux linkage of the magnetic fields of the two inductors increases from zero to 100%.[21] In this model, the total series inductance in each of the left- and right-hand subloops is $(L - M) + M = L$ when the other subloop is open-circuited.

To find the resonant frequencies of the resulting fourth-order system[22] one can always employ a brute-force approach: Find the transfer function (first, one needs to define the input and output terminals), then solve for the roots of the denominator polynomial. This method is quite general but also quite cumbersome, particularly for networks of order higher than two or three. Worse, the expenditure of labor is compensated poorly in terms of insights developed. Here, the network happens to be symmetrical, a situation that almost always demands exploitation to simplify analysis and increase the possibility of extracting useful insights.

First recall what poles are. Textbooks tell us that they are the roots of the denominator of the transfer function, but a deeper significance is that they are the *natural frequencies* of a network. What we mean by the term is this: If the system is given

---

[21] We must emphasize that negative values of k violate no laws of nature for, depending on the relative orientation of the inductors, the voltage induced in one coil by currents flowing in the other may be positive or negative. Negative values of M are thus physically realizable, allowing the synthesis of some networks requiring negative inductances. Indeed, generations of Tektronix oscilloscopes depended on just such elements (within capacitively bridged "T-coils") to provide large boosts in vertical amplifier bandwidth, as discussed in Chapter 12.

[22] Despite there being five energy storage elements in the network, the system is nonetheless of the fourth order because not all of the elements are independent. Note, for example, that specifying the initial currents in two of the inductors automatically determines that flowing in the third (by Kirchhoff's current law). Thus, the three inductors actually contribute only two degrees of freedom, diminishing by one the order of the overall network.

FIGURE 23.42. Equivalent network of coupled LC resonators for common-mode initial conditions

some initial energy, then the evolution of the system state *in the absence of any further energy input* takes place with characteristic frequencies whose values are those of the poles. The system state can evolve in an oscillatory fashion, indicating nonzero imaginary parts for at least some poles, or in a manner corresponding to a simple sum of ordinary exponentials, indicating all purely real poles. Cleverly chosen initial conditions might excite only a small subset of all possible modes at a time, thus converting a difficult high-order problem into a collection of more simply solved low-order ones. *Very* clever (or lucky) choices can even result in the excitation of a single mode at a time, making possible the identification of pole frequencies with a minimum of root finding.

We may use this understanding to devise a simple method for finding the poles of our coupled resonator system. First, provide a common-mode (even-mode) excitation by depositing, say, an equal amount of initial charge on the two capacitors. Regardless of what the network does subsequently, we know by symmetry that the capacitor voltages must evolve the same way. Because their voltages are thus always equal, we may short the capacitors together with impunity, resulting in the network shown in Figure 23.42. The common-mode resonant frequency is thus that of a simple LC network:

$$\omega_{cm} = \frac{1}{\sqrt{[(1-k)L/2 + kL]2C}} = \frac{1}{\sqrt{(1+k)LC}}. \quad (55)$$

There are two conjugate imaginary poles of this frequency, so we only need to find the other two poles of this fourth-order network.[23]

Since a common-mode initial condition is so fruitful in discovering two of the poles, it seems reasonable to try a differential initial condition next. Specifically, if one capacitor voltage is initially made equal to some value V and the other to $-V$, then (anti)symmetry allows us to assert that, however the system state evolves from this initial condition, it must do so in a manner that guarantees zero voltage at node X in Figure 23.41. Consequently, the mutual inductance has no current flowing through it, and it may be removed (either by open- *or* short-circuiting it; both actions must, and will, lead to the same answer). Removing that inductance yields the following differential-mode resonant frequency (again, the corresponding poles are conjugate and purely imaginary, with a magnitude of this value):

---

[23] We know that they are purely imaginary because there is no loss. Thus, the energy of the system remains constant for all time.

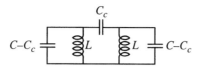

FIGURE 23.43. Capacitively coupled resonators

$$\omega_{dm} = \frac{1}{\sqrt{[2(1-k)L](C/2)}} = \frac{1}{\sqrt{(1-k)LC}}. \quad (56)$$

Now that we've found the pole frequencies, let's see what intuition may be extracted from the exercise. First consider values of k very near zero. In this loosely coupled case the two mode frequencies are nearly the same, because we have two nearly independent and identical tanks. As k increases, however, one resonant frequency decreases while the other increases; mode splitting occurs. The stronger the coupling, the wider the separation in resonant frequencies. This behavior should be familiar, for in coupled transmission lines we've seen that tighter coupling produces wider separation in mode impedances. It is the strong coupling enabled by interdigitation that allows the Lange coupler to exhibit a surprisingly wide bandwidth.

As an illustration that mode splitting is an extremely general consequence of coupling, now consider the dual case of capacitive coupling (Figure 23.43). Here, we choose to express the individual resonator capacitances as a function of the coupling capacitance. One could just as well label each resonator capacitance simply as some initial value C, but the choice shown makes the analogy with the inductive coupling case exact and will also allow us to identify explicitly the presence of a capacitive J inverter. Beyond those considerations, it certainly simplifies the analytical expressions somewhat. In this model, the total capacitance across each tank is $(C - C_c) + C_c = C$ when the other tank is short-circuited (this short-circuit condition is the dual of the open-circuit condition used in analyzing the inductively coupled case). Because of this precise duality, we obtain equations that are isomorphic, as will be seen shortly.

Following an approach analogous to that used to analyze the magnetically coupled case, we find that the two mode frequencies are given by:

$$\omega_{cm} = \frac{1}{\sqrt{(C-C_c)L}} = \frac{1}{\sqrt{(1-k)LC}}; \quad (57)$$

$$\omega_{dm} = \frac{1}{\sqrt{(C+C_c)L}} = \frac{1}{\sqrt{(1+k)LC}}. \quad (58)$$

For these equations, we see that an explicit expression for the coupling coefficient, k, is

$$k = C_c/C. \quad (59)$$

As with the magnetic case, the coupling coefficient cannot exceed unity (if negative element values are disallowed) when expressed in this manner. More important, we see that both magnetic and capacitive coupling give rise to precisely the same splitting of modes. This mechanism is so general that it explains a host of seemingly unrelated phenomena, such as the formation of energy bands in semiconductors (here, the initially identical mode frequencies – energy levels – of free, isolated atoms split as the atoms are brought closer together to form a crystalline solid) and the vibrational modes of coupled spring–mass systems. We may, in fact, generalize and say that mode splitting is due to the coupling of energy.

From Eqns. 57 and 58, it should be clear that one may use a measurement of the two mode frequencies to determine k experimentally. For small values of coupling, the difference in mode frequencies (normalized to their *arithmetic* mean) equals k quite accurately (to the second order, in fact):[24]

$$\frac{\Delta\omega}{\omega_{am}} = 2\left[\frac{\sqrt{1+k} - \sqrt{1-k}}{\sqrt{1+k} + \sqrt{1-k}}\right] = 2\left[\frac{1 - \sqrt{1-k^2}}{k}\right] \approx k. \qquad (60)$$

For filters (as with coupled-line couplers), a coupling coefficient of 0.5 is considered very high. Even for such a large value, however, the actual fractional bandwidth of about 0.536 differs from the coupling coefficient by only 7% or so. For more typical coupling coefficients, the errors will generally be smaller than those due to component tolerances – and frequently smaller than your ability to determine them experimentally to that degree of accuracy. For this reason, many applications notes and textbooks simply assert that the normalized separation in frequencies is equal to the coupling coefficient, even though it's not strictly true.

If you are very fussy, you may use the foregoing equations to derive the following exact expression for the coupling coefficient in terms of the lower and upper mode frequencies ω_1 and ω_2:

$$k = \frac{\omega_2^2 - \omega_1^2}{\omega_2^2 + \omega_1^2}. \qquad (61)$$

One may exploit this relationship to measure coupling directly. One possible experimental procedure is to couple a signal generator very loosely to one resonator (the loose coupling is required to avoid an error-inducing mode splitting of its own) and then observe the response of the coupled resonator with a detector (again, loosely coupled to avoid perturbing the system) while sweeping the signal generator frequency. To a good approximation, the coupling coefficient is the normalized frequency separation between observed response peaks.

---

[24] Many references state that "the" normalized bandwidth is *exactly* equal to k, but the reader is invited to test that assertion by using either the geometric mean, arithmetic mean, or individual (uncoupled) resonator frequency (Dishal's original definition) as the normalizing factor (and there are other possible choices, too). In all cases, the normalized bandwidth differs somewhat from k. Fortunately, however, the error is small enough not to have any serious practical implications. The filter design methods we will consider are accurate only for bandwidths less than about 20% anyway, with corresponding values of k that are similar. For any coupled resonator pair within such filters, the approximation given predicts the mode splitting accurately enough.

23.3 COUPLED RESONATOR BANDPASS FILTERS

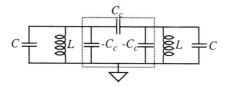

FIGURE 23.44. Redrawn capacitively coupled resonators

To relate all of the foregoing observations about coupling to the capacitively coupled bandpass filter that we've already studied, let's first redraw the two-resonator system as Figure 23.44. This way of drawing the schematic explicitly reveals the trio of capacitors that couple the two resonators to be simply the *J* inverter used to produce the capacitively coupled bandpass filter described in the previous section. Thus, stronger coupling is associated with large inverter admittances. Similarly, inspection of the magnetically coupled case reveals the presence of an inductive *K* inverter, as in the first network of Figure 23.25. In both cases, the tanks connected to each end of the inverter are, once again, resonant at the same frequency. That both capacitive and inductive coupling are equally effective for splitting modes underscores that it is fundamentally the coupling of energy that produces the effect. Although we first observed an identity between immittance inversion and coupling in a lumped context, the generality of energy coupling suggests that this identity is not limited to those lumped examples. Following up on this intuition forms the basis for a class of fully distributed bandpass filters, known as coupled-line filters, in which both magnetic and electric fields simultaneously provide immittance inversion and coupling between transmission line resonators.

23.3.3 MICROSTRIP EDGE-COUPLED BANDPASS FILTERS

Distributed resonators can be coupled together in numerous ways, not only to produce bandpass responses but also to produce bandstop and other response shapes. Research into such filters intensified in the period immediately following the Second World War, and most of the theoretical concepts that form the basis for modern microwave filter design had been worked out by around 1960. Developments during that fertile period are well summarized in a comprehensive volume by Matthaei, Young, and Jones (*MYJ*, or simply "the black book" to microwave cognoscenti), which is a must for anyone who is serious about the subject of microwave filters and the closely related subject of impedance matching.[25] Unfortunately, this tome exists in only one

---

[25] G. L. Matthaei, L. Young, and E. M. T. Jones, *Microwave Filters, Impedance-Matching Networks and Coupling Structures,* McGraw-Hill, New York, 1964 (reprinted in 1980 by Artech House). It should be mentioned that this work also covers important contributions by Seymour Cohn, who did extensive pioneering work on coupled line filters, among others. Cohn led the team at the Stanford Research Institute whose work forms the bulk of the material presented in *MYJ*.

FIGURE 23.45. End-coupled microstrip bandpass filter

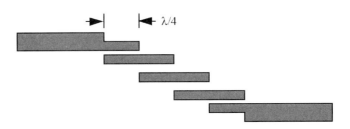

FIGURE 23.46. Classic edge-coupled microstrip bandpass filter

edition, and thus does not cover advances made after about 1960 or so. In particular, key design information relevant to microstrip implementation of filters is absent.

Inspired by a lumped prototype such as that shown in Figure 23.33, one might initially consider coupling resonators together capacitively, perhaps by using the fringing capacitance at the ends to provide the required coupling (either in lieu of, or in addition to, discrete coupling capacitors). Purely end-coupled transmission line filters (see Figure 23.45) are usually difficult to construct, however, as the amount of coupling that one may practically obtain this way is limited by the minimum gaps that may be reliably fabricated.[26]

As it happens, the required gap widths are not merely inconvenient; they're generally impractically small. This problem can be readily solved by using discrete capacitors to supplement end fringing and thereby relax dimensional tolerances. Using discrete capacitors also makes tuning a bit easier to implement. Nevertheless, minimizing the reliance on discrete elements is generally desirable, so it is worthwhile considering methods that eliminate the need for end coupling altogether. We have actually encountered this idea already in connection with coupler design. A powerful solution, then, is to use lateral coupling instead.[27]

A simple (as well as simplified) but practical implementation of this idea is shown in Figure 23.46. As with the end-coupled filter, the resonators remain nominally a half wavelength in extent (with due compensation for end fringing) and overlap each

---

[26] Versions of these filters coupled with discrete capacitors are easily simulated with tools such as *Puff*, however, and thus retain a tutorial value.

[27] Once again, see Seymour Cohn, this time for "Parallel-Coupled Transmission Line Resonator Filters," *IRE Trans. Microwave Theory and Tech.*, v. 6, April 1958, pp. 223–31. Also see H. Ozaki and J. Ishii, "Synthesis of a Class of Stripline Filters," *IRE Trans. Circuit Theory*, v. 5, June 1958, pp. 104–9.

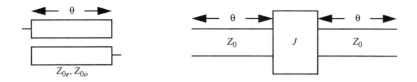

FIGURE 23.47. Symmetrical coupled lines (left) and proposed equivalent model

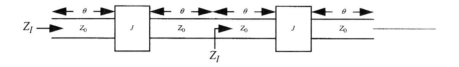

FIGURE 23.48. Infinite cascade of lines and inverters

other by a quarter-wavelength.[28] The detailed response shape is controlled by properly choosing the amount of line-to-line coupling and the characteristic impedances of the lines (for simplicity, all resonator linewidths are shown as equal in the figure).

To devise an explicit design procedure, we exploit directly the notion that coupling and immittance inversion are in fact the same thing, even if the coupling isn't due to a single mechanism. Let us therefore propose that a pair of edge-coupled lines may be modeled as two lines coupled by an immittance inverter, as seen in Figure 23.47.

If the two networks are to be equivalent, one requirement is that they present equal input impedances when terminated in equal load impedances. In studying models for transmission lines (either continuous or artificial), we've seen how the computation of an infinite ladder's input impedance is particularly simple. Thus, although an arbitrary load impedance would suffice in principle, the analysis simplifies immeasurably if we compute the input impedance of an infinite cascade of iterated copies of the network under consideration. We usually call this impedance the "characteristic" impedance of the network, but to avoid confusion with the characteristic impedances of the various transmission lines in the system, we will instead refer to the infinite iterated network's input impedance by another name, the *image* impedance.

Let us first consider the image impedance of the network in Figure 23.48. The image impedance is a function of frequency, but let us evaluate it at the center frequency (where θ is $\pi/2$, corresponding to $\lambda/4$ lines). The image impedance at the center frequency (denoted by the subscript 0) is therefore readily computed in a couple of short steps as

$$Z_{I0} = \frac{Z_0^2}{1/J^2(Z_0^2/Z_I)} = J^2 \frac{Z_0^4}{Z_I} \implies Z_{I0} = JZ_0^2. \tag{62}$$

---

[28] This value of overlap is not a requirement, but as it maximizes the amount of coupling, it maximizes the corresponding required spacing and thus relaxes dimensional tolerances. Hence, it is a near-universal choice.

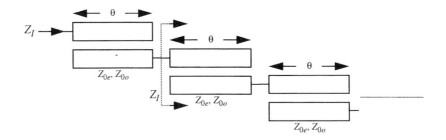

FIGURE 23.49. Infinite cascade of coupled lines

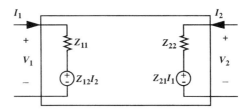

FIGURE 23.50. General impedance two-port

Note that we've deliberately formatted the first equality to facilitate identification of the individual impedance inversions, rather than to maximize compactness.

If we attempt a direct computation of the image impedance of cascaded coupled lines, however, the result is not as readily derived. The reason is that such a structure lacks the symmetry that facilitates the use of even- and odd-mode analysis (see Figure 23.49).

To solve this problem, we deduce the image impedance for this structure *indirectly*, by first deriving a quite general expression for the image impedance in terms of two-port parameters and then figuring out what those parameters are for this coupled-line system. If we choose the right two-port representation, then finding the two-port parameters for our network may involve symmetrical boundary conditions that enable even- and odd-mode decomposition. Specifically, let us consider using an impedance representation,

$$V_1 = I_1 Z_{11} + I_2 Z_{12}, \tag{63}$$

$$V_2 = I_1 Z_{21} + I_2 Z_{22}, \tag{64}$$

where the quantities are as defined in Figure 23.50.

Now consider connecting an impedance of value Z_I, the image impedance, across the output port. For this symmetrical network, the input impedance will also equal the image impedance. To discover the input impedance, we may (for example) apply a test voltage across the input port, compute the current that flows in response, and then take the ratio of voltage to current. Carrying out these steps yields:

23.3 COUPLED RESONATOR BANDPASS FILTERS

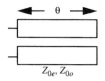

FIGURE 23.51. Coupled lines with open-circuit terminations

$$I_1 = \frac{V_{test} - Z_{12}I_2}{Z_{11}}; \tag{65}$$

$$-I_2 = \frac{Z_{21}I_1}{Z_{22} + Z_I}. \tag{66}$$

Solving these equalities for the ratio V_{test} to I_1, we obtain

$$\frac{V_{test}}{I_1} = Z_{11} - \frac{Z_{12}Z_{21}}{Z_{22} + Z_I} = Z_I, \tag{67}$$

which leads to a quadratic equation in the image impedance,

$$Z_I^2 + Z_I(Z_{22} - Z_{11}) + Z_{12}Z_{21} - Z_{11}Z_{22} = 0. \tag{68}$$

The symmetry of our original network means that the port parameters of its equivalent representation are similarly symmetrical, so that $Z_{11} = Z_{22}$ and $Z_{12} = Z_{21}$. Using these relationships simplifies our task considerably, allowing us to discover readily that

$$Z_I^2 = Z_{11}^2 - Z_{21}^2. \tag{69}$$

Let's now interpret this result. From the defining equations for the two-port model, we see that Z_{11} is the input impedance that we would measure with the output port open-circuited. The *transimpedance* Z_{21} is the ratio of output voltage to input current, also measured with the output port open-circuited. It is fortunate that the two parameters needed to compute the image impedance are both defined under open-circuit conditions, for even- and odd-mode decompositions may then enable rapid discovery of the parameters for our coupled-line case, as we now demonstrate.

First, we find the input impedance of the coupled line pair under the open-circuit condition. As usual, we decompose the calculation into even-mode and odd-mode subcalculations, by first driving both lines with equal currents and then with equal but opposite currents. See Figure 23.51.

Recall that, for an isolated line with an open-circuit termination, the input impedance is

$$Z = -jZ_0 \cot \theta. \tag{70}$$

By analogy, then, we have

$$Z_{11e} = -jZ_{0e} \cot \theta \tag{71}$$

for an even-mode excitation and

$$Z_{11o} = -jZ_{0o}\cot\theta \qquad (72)$$

for an odd. The overall open-circuit input impedance is then the average of the two impedances,

$$Z_{11} = -j\left(\frac{Z_{0e} + Z_{0o}}{2}\right)\cot\theta. \qquad (73)$$

It so happens that Z_{11} is zero at the center frequency, so we won't use this equation now (but we'll need it later).

Now let's consider the transimpedance. Again, we begin with the case of an isolated line. It is straightforward to show that the transimpedance of an open-circuited line is

$$Z_{21} = V_2/I_1 = -jZ_0\csc\theta. \qquad (74)$$

Note that, at the center frequency, the magnitude of the transimpedance is simply the characteristic impedance of the line.

For an even-mode excitation, our coupled lines have a transimpedance

$$Z_{21e} = -jZ_{0e}\csc\theta; \qquad (75)$$

for an odd-mode excitation,

$$Z_{21o} = -jZ_{0o}\csc\theta. \qquad (76)$$

The overall open-circuit transimpedance of our coupled lines is therefore

$$Z_{21} = -j\left(\frac{Z_{0e} - Z_{0o}}{2}\right)\csc\theta. \qquad (77)$$

Evaluating the transimpedance at the center frequency ($\theta = \pi/2$) yields

$$Z_{210} = -j\left(\frac{Z_{0e} - Z_{0o}}{2}\right). \qquad (78)$$

Therefore, from Eqn. 69, we find that the midband image impedance for our coupled lines is simply

$$Z_{I0} = \tfrac{1}{2}(Z_{0e} - Z_{0o}). \qquad (79)$$

Setting this image impedance equal to the corresponding midband value found earlier for the inverter model,

$$Z_{I0} = JZ_0^2, \qquad (80)$$

gives us one equation. We need one more to enable solving for the even- and odd-mode impedances separately. There are several possible choices, but one that makes use of some results we've already derived is to equate Z_{11} for the two networks, again evaluated at (or near) midband.[29]

---

[29] Most treatments are based on the original derivation by Cohn, who matches the image impedance and propagation constant. We've chosen instead to match the input impedances because doing so simplifies the derivation.

23.3 COUPLED RESONATOR BANDPASS FILTERS

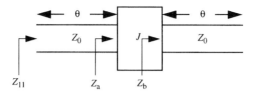

FIGURE 23.52. Equivalent inverter model with open-circuit termination

We've already derived the relevant expression for the coupled line, so all we need to do is derive Z_{11} for the inverter model of Figure 23.52. Working backwards in steps from output to input, we have

$$Z_b = -jZ_0 \cot\theta, \tag{81}$$

$$Z_a = \frac{1}{J^2 Z_b} = \frac{j\tan\theta}{J^2 Z_0}; \tag{82}$$

then,

$$Z_{11} = Z_0 \frac{\frac{Z_a}{Z_0} + j\tan\theta}{1 + j\frac{Z_a}{Z_0}\tan\theta} = Z_0 \frac{\frac{j\tan\theta}{J^2 Z_0^2} + j\tan\theta}{1 + j\left(\frac{j\tan\theta}{J^2 Z_0^2}\right)\tan\theta}$$

$$= jZ_0 \tan\theta \frac{\frac{1}{J^2 Z_0^2} + 1}{1 + j\left(\frac{j\tan\theta}{J^2 Z_0^2}\right)\tan\theta}. \tag{83}$$

Near midband, the tangent factors in the denominator are now very large (infinite at the center frequency), so we may approximate the normalized input impedance there as

$$Z_{11} \approx jZ_0 \tan\theta \frac{\frac{1}{J^2 Z_0^2} + 1}{j\left(\frac{j\tan\theta}{J^2 Z_0^2}\right)\tan\theta} = Z_0 \frac{\frac{1}{J^2 Z_0^2} + 1}{\frac{j\tan\theta}{J^2 Z_0^2}} = Z_0 \frac{1 + J^2 Z_0^2}{j\tan\theta}. \tag{84}$$

Comparing this expression with the corresponding one derived for the coupled line,

$$Z_{11} = -j\left(\frac{Z_{0e} + Z_{0o}}{2}\right)\cot\theta, \tag{85}$$

provides us with the second equation we need: setting them equal to each other yields

$$\frac{1}{2}(Z_{0e} + Z_{0o}) \approx \frac{Z_0 \frac{1 + J^2 Z_0^2}{j\tan\theta}}{-j\cot\theta} = Z_0(1 + J^2 Z_0^2). \tag{86}$$

From equating midband image impedances, we have already deduced that

$$\tfrac{1}{2}(Z_{0e} - Z_{0o}) = JZ_0^2. \tag{87}$$

Addition and subtraction of Eqns. 86 and 87 allow us to solve for the individual mode impedances at last:

$$Z_{0e} \approx Z_0[1 + JZ_0 + (JZ_0)^2]; \tag{88}$$

$$Z_{0o} \approx Z_0[1 - JZ_0 + (JZ_0)^2]. \tag{89}$$

The fact that we are able to derive these equivalences tells us that the inverter-based model is indeed an approximate representation of our original coupled-line system. We only need to be mindful that the approximations hold near midband and become progressively worse as we move away from the center frequency. Thus filters designed with these approximations will exhibit passband performance that conforms well to expectations based on lumped low-pass prototypes, but stopband performance may deviate considerably.

Note that the geometric mean of the mode impedances is not quite equal to Z_0. However, both the geometric and arithmetic means of the mode impedances are equal to each other to the second order.

Here's how to use the equations we've derived. We may compute the J-inverter constants for a lumped prototype bandpass filter. That knowledge, combined with the characteristic impedance of the resonators, allows us to compute the even- and odd-mode impedances for each pair of coupled lines. Finally, we use (say) the Akhtarzad equations to figure out the linewidths and interline spacings to produce the specified even- and odd-mode impedances for microstrip implementations. For stripline, the existence of closed-form analytical expressions for these dimensions simplifies design.

Summary of Design Procedure for Edge-Coupled Bandpass Filters

As with the development of the inverter-coupled lumped bandpass filter, derivation of the design equations for edge-coupled filters is spread out over several pages. To facilitate their actual use, we summarize the design procedure here. Fortunately, using the equations is considerably easier than deriving them.

The design of these filters begins, again, with the parameters of a lumped prototype, and it then proceeds as in designing an ordinary lumped bandpass filter. One minor difference is that the characteristic impedance of a transmission line used as a resonator differs a little bit from that of the equivalent lumped tank, as we have seen for $\lambda/4$ resonators. For an open-circuited $\lambda/2$ line, a $\pi/2$ factor arises from equating the admittance slopes of the line and lumped resonator (see Section 23.6):

$$Z_{0line} = \tfrac{\pi}{2}\sqrt{L_p/C_p} = \tfrac{\pi}{2} Z_{0res}. \tag{90}$$

Accommodating this small modification results in the following sequence of equations. The normalized inverter parameters for the other sections are given as before by

$$J_{j,j+1}Z_{0res} = \frac{(BW)}{\sqrt{g_j g_{j+1}}} \implies J_{j,j+1}Z_{0line} = \frac{\pi}{2}\frac{(BW)}{\sqrt{g_j g_{j+1}}}. \tag{91}$$

The characteristic impedance Z_{0res} of the individual resonators is a free variable, just as in the lumped resonator case. For example, it may be chosen to maximize the resonator Q (to reduce insertion loss and departures from the expected response shape) or the impedance at resonance (to ease the fabrication of the gaps). These two conditions do not necessarily coincide, so a compromise is typically involved. Lossy dielectrics such as FR4 favor somewhat narrower lines than would be the norm for filters built with higher-quality materials in which conductor loss dominates. Nevertheless, since wide (low-impedance) lines require narrow line-to-line spacings in order to produce a given coupling and since narrow (high-impedance) lines can have excessive resistive loss, resonator characteristic impedances are rarely grossly different from about 50–100 Ω in any technology, despite the alleged degree of freedom.

Just as with the lumped bandpass filter, we may use inverters at the input and output ports to couple signals into and out of the filter while simultaneously providing any necessary impedance transformations. The same equations apply (with proper accounting for the difference between resonator and line impedances). Therefore, if we continue to regard the immittance inverters as admittances, then for the input we have

$$J_1 Z_0 = \sqrt{\frac{\pi}{2}\frac{(BW)}{mg_1}}, \tag{92}$$

where g_1 is the normalized value of the first element of the prototype low-pass filter and m is Z_{0line}/Z_0. Similarly, for the output,

$$J_{n+1}Z_0 = \sqrt{\frac{\pi}{2}\frac{(BW)}{mg_n g_{n+1}}}, \tag{93}$$

where g_n is the normalized value of the final prototype element and g_{n+1} is the normalized value of the output termination.

A complication arises, however, if we wish to use coupled lines to implement these impedance transformations, for we have derived an equivalent model only for the case of symmetrical coupled lines (see Figure 23.47). In that model, the immittance inverter is not an isolated element but is instead surrounded by two line segments of equal characteristic impedance. To provide all of the necessary impedance transformations for the most general case (where the resonator impedances differ from the system impedance), the lines surrounding the inverter need to be of differing characteristic impedance, with the system impedance on one side and the resonator impedance on the other. Although one may use an asymmetrical pair of coupled lines for this purpose, there are no simple analytical formulas for their design. We therefore consider some alternatives.

First note that a symmetrical line pair suffices when $m = 1$, so that the design may then proceed without worrying about this complexity. Thus, one simple method for sidestepping the asymmetrical line problem is to choose the line impedances equal to Z_0 throughout. By far, this choice is the most popular one (so much so that it sometimes seems that there is no other option).

Another possibility is to carry out the design with an arbitrary resonator impedance (again chosen, say, to maximize resonator Q) and then provide the necessary matching with something other than an asymmetrical line pair. For example, one may always use a $\lambda/4$ line on each end to provide the necessary match, or suitable lumped immittance inverters.

Simpler than the preceding alternatives is to tap down on the first and last resonators to provide the input and output coupling because the impedance varies continuously from a very high value (near the open end of a $\lambda/2$ line) to almost zero (near the center). Such direct connections obviate the need to design and produce asymmetrical couplers at the input and output ports. At the same time, adjustment of the tapping point to produce the best match is much easier than adjusting the characteristics of asymmetrical coupled lines. The chief disadvantage is that tapping down makes sense only if the resonator impedances exceed Z_0. Fortunately, this requirement is well satisfied in virtually all practical cases (and can certainly be made so).

Next we use the computed inverter constants to deduce the required even- and odd-mode impedances of the coupled resonator sections. From our derivations, we see that these are given approximately by:

$$Z_{0e(j+1)} \approx Z_{0line}[1 + J_{j,j+1}Z_{0line} + (J_{j,j+1}Z_{0line})^2]; \quad (94)$$

$$Z_{0o(j+1)} \approx Z_{0line}[1 - J_{j,j+1}Z_{0line} + (J_{j,j+1}Z_{0line})^2]. \quad (95)$$

Once the mode impedances for each resonator have been found, the method of Akhtarzad (for microstrip) can be used to determine actual layout dimensions of the resonators. Selection and implementation of the input and output coupling structures complete the filter design. A few iterations guided by a field solver will permit any necessary refinements.

A Detailed Design Example

Perhaps more than for the other filters in this book, walking step by step through a complete example is essential for clarifying the design procedure. To facilitate comparisons with the many other examples in this chapter, we continue with the three-resonator 1-dB Chebyshev bandpass filter with a ripple passband extending from 950 MHz to 1050 MHz. Using the geometric mean definition, our center frequency is approximately 998.75 MHz, about which we compute a normalized bandwidth (BW) of about 0.1001. Again, we retain more digits than are practically justified because we wish to illustrate the procedure precisely.

First we find that the normalized element values for the lumped low-pass Chebyshev prototype are $g_1 = 2.013$, $g_2 = 0.989$, and $g_3 = 2.013$. In this particular case, we use the quick filter design tool within *RFSim99* to save us the trouble of computing the values with the equations presented in the previous chapter or looking for a published table with the desired values. In any case, many filter design tables don't have entries for Chebyshev filters with such a large value of ripple.

Next, we select the characteristic impedance of the lines we will use as resonators. Here, assume that 75 Ω satisfactorily balances the desire for low loss with the need for relaxed manufacturing tolerances. Then we may compute the central inverter admittances from

$$J_{j,j+1}Z_{0line} = \frac{\pi}{2}\frac{(BW)}{\sqrt{g_j g_{j+1}}}. \tag{96}$$

For our numbers, we find that $J_{12} = J_{23} = 0.1114/Z_{0line}$, or about 1.4858 mS. The inverter admittances are equal because of the symmetry of the filter, combined with the low filter order we've chosen. Odd-order, constant-k Butterworth and Chebyshev filters with equal terminations are symmetrical.

The even- and odd-mode impedances are readily computed from

$$Z_{0e(j+1)} \approx Z_{0line}[1 + J_{j,j+1}Z_{0line} + (J_{j,j+1}Z_{0line})^2]; \tag{97}$$

$$Z_{0o(j+1)} \approx Z_{0line}[1 - J_{j,j+1}Z_{0line} + (J_{j,j+1}Z_{0line})^2]. \tag{98}$$

Thus, $Z_{0e2} = Z_{0e3} = 84.3$ Ω and $Z_{0o2} = Z_{0o3} = 67.6$ Ω.

Next, we consider the input and output matching inverters. The source and load impedances are Z_0. Then,

$$J_1 Z_0 = J_4 Z_0 = \sqrt{\frac{\pi}{2}\frac{(BW)}{mg_1}} = \sqrt{\frac{\pi}{2}\frac{(BW)}{mg_N}}, \tag{99}$$

which works out to about 0.2282 with $m = 1.5$. These impedance matches are perhaps best provided by direct coupling through tapping, or with a $\lambda/4$ matching section. We will consider both options later.

We now turn to the computation of actual layout dimensions for the resonator sections. The filter core consists of two coupled pairs, but symmetry allows us to cut the work in half. We thus need to find only one linewidth, and one interline spacing, in addition to the physical length of the lines.

To save you the trouble of flipping pages back and forth, we reprise here a subset of the equations of Akhtarzad presented in Chapter 7. Recall in what follows that the ratios W_e/H and W_o/H are those of single isolated microstrip lines whose characteristic impedances are $Z_{0e}/2$ and $Z_{0o}/2$, respectively. In our example, $Z_{0e}/2$ is 42.15 Ω and $Z_{0o}/2$ is 33.8 Ω. Using one of many possible formulas relating microstrip dimensions to characteristic impedance, we find that W_e/H is about 2.38 (for a bulk dielectric constant of 4.6) and W_o/H is approximately 3.31. Better accuracy is possible with more elaborate equations (such as those referenced in Chapter 7), but our aim here is to illustrate the overall design procedure rather than to minimize the difference between theory and practice.

Having completed that step, now comes a somewhat more difficult one. We solve Akhtarzad's equations iteratively for the actual width and spacing for the coupled lines. The equations are

$$\frac{W_e}{H} = \frac{2}{\pi}\cosh^{-1}\left(\frac{2d - g + 1}{g + 1}\right), \tag{100}$$

$$\frac{W_o}{H} = \frac{2}{\pi}\cosh^{-1}\left(\frac{2d - g - 1}{g - 1}\right)$$
$$+ \frac{4}{\pi(1 + \varepsilon_r/2)}\cosh^{-1}\left(1 + 2\frac{W/H}{S/H}\right) \quad \text{if } \varepsilon_r < 6, \tag{101}$$

where

$$g = \cosh\left(\frac{\pi S}{2H}\right), \tag{102}$$

$$d = \cosh\left[\pi\left(\frac{W}{H} + \frac{S}{2H}\right)\right]. \tag{103}$$

A reasonable initial value of S/H to start the iterations is given by

$$\frac{S}{H} \approx \frac{2}{\pi}\cosh^{-1}\left[\frac{\cosh\left(\frac{\pi}{2}\frac{W_o}{H}\right) + \cosh\left(\frac{\pi}{2}\frac{W_e}{H}\right) - 2}{\cosh\left(\frac{\pi}{2}\frac{W_o}{H}\right) - \cosh\left(\frac{\pi}{2}\frac{W_e}{H}\right)}\right]. \tag{104}$$

To deduce a reasonable initial value of W/H, note that the impedance of a line with those dimensions should be roughly twice that of a line of dimensions W_e/H or W_o/H. To the extent that impedances are inversely proportional to the width-to-height ratio to zeroth order, a credible initial value for W/H is, say, $W_e/2H$. Fortunately, convergence does not seem to be overly sensitive to the initial value, so one generally need not obsess over what particular initial value to use.

In iterating with Akhtarzad's equations, it's helpful to note that S/H primarily controls the difference between the computed values of W_e/H and W_o/H and that W/H mainly controls their sum. Thus, one may converge on the correct values more or less orthogonally using this knowledge: compare the correct values of W_e/H and W_o/H with those computed by the equations for given values of W/H and S/H, and then respond accordingly.

After just a minute or two of manually iterating with a spreadsheet using these guiding principles, we find that $W/H = 0.865$ and $S/H = 1.066$.[30] Assuming that the dielectric thickness is 1.6 mm, the lines should be of 1.38-mm width and spaced apart from each other by 1.71 mm. Again, note that we are reporting more digits than are practically significant (it is unlikely that our manufacturing dimensional repeatability is as good as 10 μm, and Akhtarzad's method doesn't account for nonzero conductor thickness in any event).

---

[30] One could also use the solver feature (available as an option in some spreadsheets) to automate the procedure.

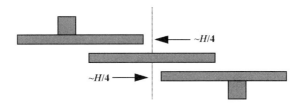

FIGURE 23.53. Layout of resonator strips (tapped input and output couplings shown)

Now that we have the width and spacing, we need to compute the physical length of the lines. We start by computing the effective dielectric constant from[31]

$$\varepsilon_{r,eff} \approx 1 + 0.63 \cdot (\varepsilon_r - 1) \cdot (W/H)^{0.1255} \quad (W/H > 0.6). \quad (105)$$

Continuing to assume a bulk relative dielectric constant of 4.6, we compute an effective dielectric constant of 3.227. At the filter's center frequency of 998.75 MHz, $\lambda/2$ is therefore 83.6 mm. Because of fringing at the ends of the line, the physical layout dimensions must be somewhat less. Our standard rule of thumb is to add $H/2$ per end to the physical length to obtain an estimate of the electrical length. However, in this case, the end fringing is reduced by the flux stolen by adjacent lines. As an ad hoc correction, then, we use a length adjustment of $H/4$ (half the conventional correction) per end that has an adjacent line. Thus, we choose a physical length of 82.8 mm for the center line, and 82.4 mm for the two outer ones. Because we are treating the effect of fringing as equivalent to an extension of length, the layout should reflect this extension consistently as well (see Figure 23.53).

To complete our first-pass design, we need to devise appropriate input and output matching sections. As mentioned before, there is an advantage to directly tapping an appropriate point within the input and output resonators. In lumped implementations, the tapping may be performed either at some point along the inductor or by using a tapped capacitor in the resonator. The impedance is a maximum across the parallel tank, with a value of $Q_L Z_{res0}$, where Q_L is the in-circuit (loaded) Q_L. The minimum, zero, is found at ground. Thus, a continuum of values is available.

A range of impedance levels is similarly found along a distributed resonator. The impedance is a maximum at the open-circuited ends (again with a value $Q_L Z_{res0}$) and is a minimum (~ 0) at the center of the $\lambda/2$ resonant sections. If we assume that currents and voltages vary approximately (co)sinusoidally along the resonant strips, then the impedance should vary in a manner reminiscent of that of a patch antenna's inset feedpoint impedance. Formally, but approximately,[32]

---

[31] As with microstrip characteristic impedance, there are a great many formulas for the effective dielectric constant.

[32] See *Reference Data for Radio Engineers*, 5th ed., International Telephone and Telegraph Corp., 1969. Also see Joseph S. Wong, "Microstrip Tapped-Line Filter Design," *IEEE Trans. Microwave Theory and Tech.*, v. 27, January 1979, pp. 44–50.

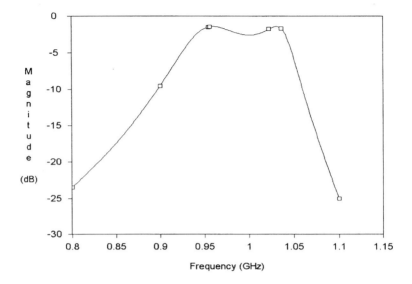

FIGURE 23.54. Simulated filter response (*Sonnet Lite* 9.51); free-space upper boundary, tap at 11 mm

$$(\sin\theta)^2 \approx \frac{\pi}{2}\frac{Z_0}{Q_L Z_{res0}}, \qquad (106)$$

where Q_L is the (doubly) loaded Q of the end resonator,

$$Q_L = \frac{g_1}{2(BW)}. \qquad (107)$$

We ultimately solve for the distance θ (expressed as a phase angle) from the center of the resonator to the tapping point, at which the filter presents an impedance of Z_0 to the external world. For our numbers, Q_L is about 20.1, the resonator impedance is 75 Ω, and the system Z_0 we want to match is 50 Ω. Thus, θ should be about 0.33 radians, corresponding to a calculated tap position of about 8.5 mm (relative to the resonator's center). This position may be used as the starting point for iterations to discover the best tapping point.

In our case, exploration with a field solver reveals that the optimum tapping point is actually about 11 mm from the center of the resonator; see Figure 23.54. Simulating the design with *Sonnet Lite* 9.51 reveals some insertion loss (of the order of 1.4 dB – presumably due to radiation, as the filter is surrounded by walls made of lossless conductors, with only the top open to free space in this simulation). Renormalizing to that loss, the peak-to-peak ripple is about 1.2 dB, and the ripple passband extends from 940 MHz to 1.04 GHz. Thus, the ripple exceeds the design target by a small amount, the center frequency is similarly low by a little bit (about 1%), and the 100-MHz ripple bandwidth is as desired. As expected, the passband conforms closer to expectations than does the transition band or stopband. Despite the deviations, the overall level of performance is satisfactory for a first-pass design, especially in view of the many approximations that we have made to get to this point.

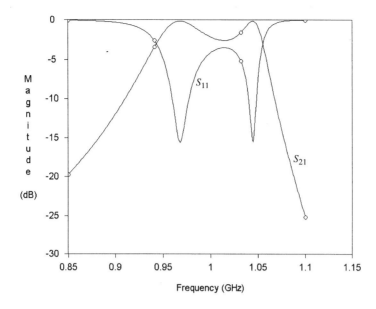

FIGURE 23.55. Simulated response with perfectly conducting top lid; tap at 9 mm

As one final assessment of the design procedure, we re-run the simulation with a slightly different boundary condition. We place a perfectly conducting lid on the box that encloses the filter, with 100 mm of air as the dielectric above the filter conductors. In this case, peak insertion loss indeed drops close to zero, because radiation is now precluded. However, the ripple increases to 2.5 dB. The 2.5-dB ripple passband extends from 950 MHz to about 1054 MHz, while the 1-dB edges occur at 954 MHz and 1051 MHz. So, depending on how you choose to interpret these numbers, the bandwidth is either a little too high or a little too low. Either way, the center frequency appears a little bit on the high side.

In addition to those changes, the best input and output match occurs when the tap positions are moved a little bit closer to the center. In this particular instance, the optimum tap point is approximately 9 mm from center. See Figure 23.55.

The ripple may be reduced by altering the spacing between the resonators. The large value indicates that the coupling is too tight between adjacent resonators, producing excessive mode splitting. Hence, further iterations would explore moving the resonators apart a little bit and then relocating the tap position as necessary to maintain a good match in the process.

Variations on a Theme

The edge-coupled bandpass filter is widely used because it is relatively straightforward to design and it functions quite well. One criticism, however, is that the filter occupies a relatively large area (or an irregularly shaped region), especially if the filter uses many sections. To fill a rectangular space more optimally, it is customary to rotate the layout of the resonator array as shown in Figure 23.56. The input and

FIGURE 23.56. More compact layout of edge-coupled bandpass filter

FIGURE 23.57. Hairpin bandpass filter (input–output couplings not shown)

output coupling lines are frequently bent (as shown) so that the ports are horizontally disposed. The design of such filters proceeds exactly as for the conventional edge-coupled filter. The rotation affects the performance only to the second order, so any necessary adjustments will be quite small.

To shrink the filters even further, one may fold each individual resonator into a hairpin shape;[33] this is shown in Figure 23.57. Note that, as in other bent transmission lines, the resonator sections use mitered bends to minimize discontinuities that would alter the response.

Coupling between the parallel arms within a given hairpin occasionally proves troublesome in these filters, especially if the *slide angle* (slide factor) α is small. The voltages at opposite ends of the hairpin arms are in antiphase, and thus something akin to the classic Miller effect causes the arm-to-arm capacitance to have a seemingly disproportionate effect. The added capacitance lowers the resonant frequency, requiring a shortening of the hairpins to compensate. To minimize this self-coupling problem, one should separate the arms as much as possible by increasing the slide factor. However, because one would then have to reduce the gap between resonators in order to compensate for the smaller lateral overlap, there is a practical limit on the arm separation. In general, arm separations that are about five dielectric thicknesses represent a good compromise between these two conflicting requirements.

There are no simple analytical equations for the design of a hairpin filter, so most engineers design them iteratively, guided by field solvers, or with a CAD tool that automates the procedure. However, it is possible to devise an approximate method that can generate a credible starting point for such iterations. To do so, we first reprise the equation for the coupling between two lines:

---

[33] E. G. Cristal and S. Frankel, "Hairpin Line/Half-Wave Parallel-Coupled-Line Filters," *IEEE Trans. Microwave Theory and Tech.*, v. 20, November 1972, pp. 719–28. For a refined design procedure, see U. H. Geysel, "New Theory and Design for Hairpin-Line Filters," *IEEE Trans. Microwave Theory and Tech.*, v. 22, May 1974, pp. 523–31.

$$\frac{V_{cpld}}{V_{in}} = \frac{j[\tan\theta]\dfrac{Z_{0e}-Z_{0o}}{Z_{0e}+Z_{0o}}}{\dfrac{2Z_0}{Z_{0e}+Z_{0o}}+j\tan\theta} = \frac{[j\tan\theta]C_{F0}}{\sqrt{1-C_{F0}^2}+j\tan\theta}. \tag{108}$$

At midband, the ratio simplifies to

$$\frac{V_{cpld}}{V_{in}} = C_{F0} = \frac{Z_{0e}-Z_{0o}}{Z_{0e}+Z_{0o}}. \tag{109}$$

Even though these equations strictly apply only to terminated coupled lines – rather than to the lines as they are terminated in our filter – we will use them in a way that makes the final result relatively insensitive to this difference. After determining the collection of even- and odd-mode impedances for an ordinary edge-coupled filter, compute the ratio V_{cpld}/V_{in} as given by Eqn. 109 for each coupled line pair. Note that we have denoted with the subscript 0 the coupling factor that applies to the prototype filter.

Next, compute the value of C_F that yields the same magnitude of V_{cpld}/V_{in} when the slide factor α is included:

$$\left|\frac{V_{cpld}}{V_{in}}\right| = \left|\frac{[j\tan(\theta-\alpha)]C_F}{\sqrt{1-C_F^2}+j\tan(\theta-\alpha)}\right|. \tag{110}$$

Setting the magnitude of ratio V_{cpld}/V_{in} equal for the two cases at midband yields

$$C_{F0} = \left|\frac{[j\tan(\frac{\pi}{2}-\alpha)]C_F}{\sqrt{1-C_F^2}+j\tan(\frac{\pi}{2}-\alpha)}\right|. \tag{111}$$

Thus, one first computes C_{F0} as if designing an ordinary edge-coupled filter with zero slide factor. Then one solves Eqn. 111 (e.g., iteratively) for the necessary value of C_F given a nonzero slide angle. That (higher) computed coupling factor is used to deduce the new even- and odd-mode line impedances from

$$\frac{Z_{0en}-Z_{0on}}{Z_{0en}+Z_{0on}} = C_F, \tag{112}$$

where the subscript n denotes *new*.

One additional equation is needed to permit computation of the individual mode impedances. As a first approximation, one may assume that the product of the mode impedances is independent of the slide factor. That additional fact is sufficient to allow a computation of the new mode impedances, after which one may determine the required line and spacing dimensions for the system of hairpin resonators (e.g., again using the method of Akhtarzad).

The computed dimensions are even more approximate than would otherwise be the case because the procedure we've outlined does not account for coupling between the arms of each hairpin. Nevertheless, experience shows that this approach usually generates first-pass designs that are either satisfactory as they are or are readily made so with trivial additional effort.

FIGURE 23.58. Folded hairpin bandpass filter (dimensions not to scale)

As with many other phenomena, what is troublesome in one context may often be exploited in another. Here, the coupling between arms may be regarded as advantageous, for it affords the opportunity for additional size reductions. Once you are philosophically aligned in that direction, your goal shifts toward consideration of ways to enhance, rather than suppress, the effect. One result of that type of thinking is the folded hairpin filter, as seen in Figure 23.58 (input–output couplings not shown).[34]

The inner folded sections within each resonator may be treated as forming a lateral capacitor. They produce an increased capacitive coupling between the arms that lowers the resonant frequency, permitting (actually, requiring) a reduction in the size of the filter. The trade-off is that, as one might expect, there is no simple analytical formula to guide their design. One generally begins with a hairpin prototype and then uses a field solver to evaluate the effect of folding on the resonant frequency. Then, the inter-resonator spacings are continually reduced to maintain the desired coupling. This process is continued until a satisfactory design results or until you tire of trying. Note that one may also use lumped capacitors across the ends of the arms in yet another possible variation on this same theme.

The folded hairpin design is just one of many variations that have evolved over the decades. To underscore that creativity has hardly been exhausted, we present two more configurations, without analysis.

One is the interdigital filter (which predates the folded hairpin design by decades). One may think of this filter as resulting from folding on itself each $\lambda/2$ resonator of a conventional edge-coupled design. This folding produces a series of coupled $\lambda/4$ resonators, as seen in Figure 23.59. The design equations are essentially the same as those for the $\lambda/2$ edge-coupled filter. There is a small modification necessitated by the factor-of-2 difference in reactance slopes, but other than that minor detail, the design proceeds as before.

It is inconvenient that the interdigital filter requires ground connections, and on alternate ends of the resonators at that, so this filter is not widely used in microstrip form. Nonetheless, once you have broadened your thinking to consider grounded

---

[34] M. Sagawa, K. Takahashi, and M. Makimoto, "Miniaturized Hairpin Resonator Filters and Their Applications to Receiver Front-end MIC's," *IEEE Trans. Microwave Theory and Tech.*, v. 37, no. 12, December 1989, pp. 667–70.

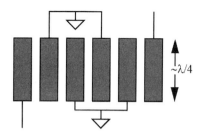

FIGURE 23.59. Interdigital filter

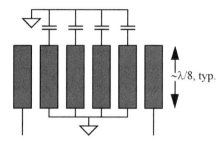

FIGURE 23.60. Combline filter

connections, it is worthwhile considering an alternative arrangement of the resonators in which all of the ground connections are on one side; see Figure 23.60.

The resonators of a *combline* filter are usually implemented as shown in the figure, with line sections that are shorter than $\lambda/4$ (typically $\lambda/8$, in fact) acting in concert with discrete capacitors. The use of such electrically short lines defers the appearance of parasitic passbands while simultaneously shrinking the overall filter. In addition, the use of discrete capacitors facilitates tuning. Finally, for lossy dielectrics (such as FR4), the overall tank Q is increased by using good discrete capacitors as part of the resonator, thus decreasing insertion loss (and otherwise reducing the difference between actual performance and theoretical expectations). Performance normally associated with substantially more exotic (and expensive) board dielectrics are possible with this architecture. Such advantages are sometimes sufficiently compelling to make combline filters attractive, despite the inconvenient requirement for grounded connections.

23.4 PRACTICAL CONSIDERATIONS

There are several serious practical difficulties that every filter designer encounters when trying to build a real filter. One is the accommodation of manufacturing tolerances. The more critical a filter's performance characteristics, the more care must be exercised in designing, building, and tuning the circuit – as is evident from examining the many Monte Carlo simulations presented in the previous chapter. For critical applications, the relatively simple procedures we have presented will usually produce designs that do not meet all specifications. Refinement of such designs is best guided by intuition and subsequently verified by rigorous tools, such as field solvers.[35]

Another challenge is developing a sensible procedure for tuning a filter once you've built it. For low-order filters, it costs little to add enough margin to the design to accommodate manufacturing tolerances and thus obviate the need for tuning. For demanding applications, however, this option may be unacceptable for any number of reasons, and you may still be obligated to devise and implement a tuning procedure.

A modified version of a clever alignment method described by Dishal in 1949 works (in principle) for any coupled-resonator filter, whether implemented in lumped or distributed form, and it also exploits a couple of observations we made earlier.[36] One is that the resonant frequencies of the tanks in isolation are all equal. Another is that the coupling networks (J- and K- immittance inverters) act approximately as $\lambda/4$ lines. To see how far one can go with just these two facts, consider observing the input impedance while adjusting the tank tuning. An inexpensive way to do so is to set a signal generator to the desired center frequency and then place a slotted line between it and the filter input. The slotted line does not have to be calibrated, so a homemade version such as described in Chapter 8 is suitable. Then, *grossly detune* all filter tanks (e.g., by approximating, as well as you can, a short across them). Slide the probe of the slotted line until you find a minimum and lock it there for the remainder of the tuning procedure. Tune the first resonator (i.e., the one closest to the input) to provide a peak probe indication (caused by a corresponding peak in the impedance). Then adjust the next resonator's tuning to provide a *minimum* response. A minimum results because, when the second resonator is tuned correctly, its impedance is a maximum. When transformed through the $\lambda/4$ line–like action of the immittance inverter, the input impedance becomes a minimum. A little thought allows us to conclude that we will have achieved our tuning goals if we adjust for a maximum probe indication when tuning odd-numbered tanks and for a minimum when tuning even-numbered tanks. This method is appealing because it does not require a two-port characterization of the filter and demands only n adjustments for

---

[35] Such tools were once priced well beyond the budgets of students and hobbyists. Fortunately, that situation has changed dramatically in recent years, with several vendors offering free or inexpensive student and demo versions of their field solvers. Representative is *Sonnet Lite* (included in the CD-ROM collection accompanying this book), which is a free electromagnetic field solver capable of quite advanced analysis. With this tool, it is straightforward to accommodate fringing, coupling, and other effects quite accurately, making the design of microstrip filters a much less painful affair than it used to be. The availability of these programs is revolutionizing the engineering practice of student and hobbyist alike. Even the weekend experimenter can now rapidly produce designs of a sophistication that would have been almost unthinkable only a few years ago.

[36] Milton Dishal, "Alignment and Adjustment of Synchronously Tuned Multiple-Resonant-Circuit Filters," *Proc. IRE,* November 1951, pp. 1448–55, and "Alignment and Adjustment of Synchronously Tuned Multiple-Resonant-Circuit Filters," *Electrical Commun.*, June 1952, pp. 154–64.

an n-resonator filter. The only calibrated equipment required is a (fixed-frequency) signal generator.

The foregoing procedure assures that the resonators are tuned properly, but we aren't quite done aligning the filter because we must still adjust the inter-resonator coupling coefficients. The procedure outlined in the earlier discussion on mode splitting may be used as the basis for such adjustments. After the couplings have been set properly, it's a good idea to recheck the resonator tunings. Because inter-resonator couplings are generally small, their adjustment usually perturbs the tuning by negligible amounts. However, it's always best to verify this expectation, especially if gross adjustments in coupling were required in the previous step or if the filter specifications are tight. Even if some resonator retuning is necessary, the coupling factors rarely need a second adjustment iteration. In any event, convergence on the proper alignment is extremely rapid with this technique, so very few iterations are required in practice to complete the procedure.

Another important observation is that all of our design equations and procedures have fancifully assumed infinite unloaded Q for all elements. For filters whose poles have Q-values well below the unloaded Q of the elements, the discrepancies will not be serious. As the order and sophistication of a filter increase, however, the need for high-Q poles (and zeros) also increases, making the issue of parasitic dissipation progressively more serious. Insertion loss will increase, passband accuracy will degrade, and what should have been transmission nulls will fail to attenuate by infinite amounts. With more modern synthesis methods, it is possible to accommodate some lossiness during the design process (e.g., by altering L/C ratios so that the combination of circuit and self-loading ultimately results in the same in-circuit loaded Q as in a design with lossless elements), but at the expense of increased insertion loss.[37] Nothing can be done about the transmission nulls, regrettably. In any case, there exists for every filter some critical Q below which acceptable filter behavior is impossible to attain. In such instances, the only alternatives are to implement the filter in a better technology. In that context, one should not overlook the possibility of using a combination of lumped and distributed elements, such as the combline architectures.

23.5 SUMMARY

We've seen that insights developed in the lumped domain may be carried over to the distributed domain, allowing us to comprehend the operational principles underlying

---

[37] This method is known as *predistortion* in the literature. The extensive tabulated designs found in Zverev (*Handbook of Filter Synthesis,* Wiley, New York, 1967) include filters that accommodate finite-Q elements through predistortion. Once again, we have Dishal to thank for providing the first explicit equations for predistortion, in "Design of Dissipative Band-pass Filters Producing Desired Exact Amplitude–Frequency Characteristics," *Proc. IRE,* September 1949, pp. 1050–69.

many important microstrip filters. Often, we may even derive analytical design formulas from those that apply to lumped prototypes.

We've also seen that the relatively high electric fields produced within the resonant sections of many coupled line filters implies particular sensitivity to the moderately high dielectric loss of FR4. Replacing some of the lossy line with a low-loss discrete capacitor improves performance, as in the capacitively tuned combline filter. For this reason, critical filters (and resonators) are perhaps best realized with this sort of hybrid microstrip–discrete combination if a relatively lossy material such as FR4 is to be used.

23.6 APPENDIX: LUMPED EQUIVALENTS OF DISTRIBUTED RESONATORS

In many of the design approaches, we match the characteristics of a resonant transmission line with those of an equivalent lumped resonator. To do so, we match the behavior of the two networks near resonance. Specifically, we match the *slope* of the reactance (or admittance) at the resonant frequency. We start, as usual, with the expression for the input impedance of a loaded transmission line:

$$\frac{Z(z)}{Z_0} = \frac{\frac{Z_L}{Z_0} + j \tan \beta z}{1 + j \frac{Z_L}{Z_0} \tan \beta z}. \tag{113}$$

For an open-circuited line, we have

$$\frac{Z(z)}{Z_0} = \frac{1}{j \tan \beta z}. \tag{114}$$

To facilitate comparisons with lumped tanks, we will actually consider the admittance,

$$\frac{Y(z)}{Y_0} = j \tan \beta z. \tag{115}$$

Now,

$$\beta = \omega/v \tag{116}$$

and

$$\lambda = 2\pi v/\omega, \tag{117}$$

so a line that is $\lambda/2$ long at the center frequency has a length expressed as

$$\lambda_0/2 = l = \pi v/\omega_0. \tag{118}$$

Thus,

$$Y(l) = jY_0 \tan \frac{\omega \pi}{\omega_0}. \tag{119}$$

Now let us evaluate the derivative of the admittance at the center frequency:

$$\left. \frac{dY}{d\omega} \right|_{\omega=\omega_0} = jY_0 \frac{\pi}{\omega_0}. \tag{120}$$

23.6 APPENDIX: LUMPED EQUIVALENTS OF DISTRIBUTED RESONATORS

Performing the same calculation for a parallel LC tank, we find that

$$\left.\frac{dY}{d\omega}\right|_{\omega=\omega_0} = j2C. \tag{121}$$

Equating the two admittance slopes yields the relationship we have used repeatedly:

$$Y_0(\pi/\omega_0) = 2C \implies Z_0 = \tfrac{\pi}{2}\sqrt{L/C} = \tfrac{\pi}{2}Z_{0res}. \tag{122}$$

That is, a $\lambda/2$ transmission line of characteristic impedance Z_0 acts as a parallel LC tank whose impedance is $(2/\pi)Z_0$.

INDEX

abbreviations, and prefixes, 70
absolute temperature, 70
accumulator (ACC), and direct digital synthesis, 559, 560
"Accurate and Automatic Noise Figure Measurements" (Hewlett-Packard, 1980), 492
acoustic horn, 99n14
acoustics, and history of microwave engineering, 6n15
active bias, and amplifiers, 379–81
active double-balanced mixer, 321–6
adaptive bias, and power amplifiers, 666–7
adhesion promoters, 158n2
adjacent channel power ratio (ACPR), 674–6
admittance, and Smith chart, 64, 65–6
Advanced Mobile Phone Service (AMPS), 25
Agilent, 310n11, 467, 623; see also Hewlett-Packard
air-core solenoids, 140–4
air-dielectric transmission line, and slotted-line system, 249–50
Airy, George B., 591
Akhtarzad, Sina, 216, 218, 225, 833–4
Alexanderson, Ernst F. W., 10
Alexanderson alternator, 11
Alexopoulos, N. G., 710n23
alumina, and PC boards, 161–2, *237*
aluminum, and skin depth, 126
American Radio Relay League, 689n3
America's Cup yacht race, 10
Ampère's law, 42, 125n2
Amphenol Corporation, 109, 112
amplifiers
    bandwidth extension techniques, 381–94
    basic configurations, 369
    bridged T-coil transfer function, 427–39
    combiners and splitters, 183
    couplers, 225
    Gunn diodes, 294
    impedance and stability, 420–7
    low-noise amplifiers (LNAs), 440–71
    microwave biasing, 370–81
    neutralization and unilateralization, 417–20
    on-board transformers, 152
    parametric amplification, 282–4
    regenerative amplifier/detector, 15–18
    shunt-series amplifier, 395–413
    superregenerative amplifier, 59n12
    tuned amplifiers, 413–17
    tunnel diodes, 287
    see also power amplifiers
Ampliphase system, 660, 661, 662
amplitude-to-phase (AM–PM) conversion, 576
amplitude response, and phase noise, 597–9
AM radio
    Christmas Eve broadcast in 1906, 11
    coil design, 145
    crystal radios, 299
    dipole antenna, 697
    frequency bands, 39
    loop antennas, 706
    neutralization, 418n19
    outphasing modulation, 660
    single-diode demodulator, 331
    see also radio
Amstutz, Pierre, 760n34
analog multiplier, as phase detector, 537–8, 604–5
analog phase interpolation (API), 557
analog scopes, 614–17
ANSI (American National Standards Institute), 70
antenna gain, 696
antennas
    biconical antenna, 720–1
    characteristics of, 695–7
    circuit theory, 688–90
    dipole antenna, 697–707
    length of, 40, 693, 698–9, 705, 720
    microstrip patch antenna, 707–20
    Poynting's theorem, 690–1
    radiation, 691–5
Antennas (Kraus, 1988), 689n3
Antenna Theory (Balanis, 1996), 689n3
Antenna Theory and Design (Stutzman & Thiele, 1998), 689n3
APC connectors, 112–13
APLAC (simulator), 215n43
AppCAD (simulator), 310n11, *371,* 372, 377–8, 467
arbitrary noise signal, 585
arbitrary termination, of transmission line, 52–3
Archimedean spirals, 719

arc technology, for industrial illumination, 10n21
area-coupled lines, 222
Armstrong, Edwin Howard, 15–18, 305
Army Signal Corps, 16, 23
arrays, of antennas, 721
ARRL Antenna Handbook, 689n3
artificial lines, 54–8
The Art of Electronics (Horowitz & Hill), 333
"The Art of Phase Noise Measurement" (Hewlett-Packard, 1984), 612
Atalla, Martin, 343
Atlantic Cable Project, 724
AT&T, 14, 18, 25n61
attenuation
  coaxial cable, 115, 120–1
  connectors, 108–14
  standardized impedances, 72–3
  waveguide, 120
audions, 13–14, 15–16
automatic gain control (AGC), and PIN diodes, 288
automobile anticollision radar systems, 40
autotransformer, 418
available power gain, 71
avalanche breakdown, and pulse generator, 272–4
avalanching
  diodes, 290, 295–6
  sampling scopes, 618–19
Avantek, 457
average power, and lumped circuit theory, 691
AWG (American wire gauge), 144

Babinet, Jacques, 717n31
Babinet's principle, 717–18
backward diode, 287
backward-wave coupler, 215, 225, 226–7
Bahl, I., 169n15
Ball, J. A., 779n50
ballasting, and thermal runaway, 681
balun
  broadband transmission line transformers, 152–3, 154, *155*
  combiners, splitters, and couplers, 191–201
  narrowband transmission line transformers, 156
band edges, and broadband impedance matching, 104–5, 107
bandpass constant-k filter, 736–8
bandpass filters
  coupled resonator bandpass filters, 803–41
  half-wave (re-entrant) bandpass filters, 800–3
  low-pass prototypes, 781
  single-sided bandwidth, 512n11
bandstop filter, 738, 781
bandwidth
  amplifiers and extension techniques, 381–94
  antenna length, 705
  constant-k filter, 736–8
  shunt-peaked amplifiers and enhancement, 385, 388–94
  see also ripple values; ultrawideband systems
Bardeen, John, 341
Barrett, Robert M., 162
Barrow, Wilmer L., 18, 118
basewidth modulation, of bipolar transistors, 354
beamwidth parameter, and antenna performance, 696

Becker, Joseph, 300
Bell, Alexander Graham, 23–4n56, 30–1, 36, 70n8
Bell Laboratories, 22, 24–5, 60–1, 276, 296, 724n5, 757
bends, and line-to-line discontinuities in transmission lines for PC boards, 172–3
beryllia, and PC boards, 162, *237*
Bessel-Thomson filters, 761–6, 767, 781, *782*
BFL inductor, 632–3
Bhartia, P., 169n15
bias and biasing
  amplifiers, 370–81, 456–7
  Colpitts oscillator, 506
bias voltage, and Gunn diode, 295
biconical antenna, 720–1
bi-fin antenna, 721
binary arrays, and Wilkinson combiners, 188–9
binocular core, of broadband transformer coupler, 206
binomial transformers, 99n13
bipolar cascomp, 325
bipolar transistors
  biasing, 370–6
  Class E amplifier, 645–6
  distributed amplifiers, 411
  FET models, 361–8
  history of, 341–51
  low-noise amplifier, 445–51, 468
  power amplifiers, 680–1
  small-signal models, 352–61
Black, Harold, 655
BNC connector
  microstrip slotted-line system, 268, 270
  moding and attenuation, 111, 112
  PC boards, 177–8
Bode, Hendrik W., 93
bolometer, 6, 7
Boot, Henry A. H., 21
bornite, and crystal radios, 298
Bose, Jagadish Chandra, 5–6, 279–80
bow-tie antenna, 721
Bragg cutoff, 408
branchline coupler, 196–9, 228–9
Branly, Edouard, 3–4
Brattain, Walter, 300, 341
Braun, Ferdinand, 6n12, 279, 297, 614–15
breakdown phenomena, and power amplifier, 680–1
breakdown voltages, and snap diodes, 292
bridged T-coil amplifiers
  common-mode response, 429–30
  complete transfer function, 431–2
  differential-mode response, 428–9
  maximally flat delay, 436–7
  maximally flat magnitude response, 432–6
  maximum bandwidth, 437–8
  two-port bandwidth enhancement, 392–4, *395*
broadband amplifiers, 411–13
broadband directional coupler, 203–4
broadband impedance-matching techniques, 93–107
"Broadband Microstrip Mixer Design – The Butterfly Mixer" (Hewlett Packard, 1980), 791
broadband mixer, 333, 336
broadband patch antennas, 716–20
broadband transmission line transformers, 152–5

broadside-coupled lines, 222n50
Brune, Otto, 766
buffer space, of N connector, 110
butterfly arrangement, of radial stubs, 790–1
Butterworth condition, 99
Butterworth filters
    approximations, 740–4, 746–7
    Chebyshev filter, 752
    delay-magnitude counterparts, 765
    design tables, 781, *782*
    lossless ladder network synthesis, 767, 769–70, 773

cable television, 118, 412; *see also* television
CAD tools, and loop filters for PLLs, 549–50
Cady, Walter G., 516, 522n22
cameras, and xenon flash circuitry, 272
Campbell, George Ashley, 724, 725, 726, 727n11
capacitance, of resistors, 129–33
capacitive coupling, and mode splitting, 821–3
capacitive degeneration, and mixers, 320n24
capacitive loading, of dipole antennas, 702–4
capacitors, and passive components, 133–8
capacity hats, 703, *704*, 722
carborundum (SiC), and crystal detectors, 7, 10n19, 298
Cartesian feedback, and power amplifiers, 664–6, 669
cascading systems
    Friis's formula for noise figure, 479–80
    image impedance of microstrip filters, 826
    low-noise amplifier (LNA), 464–7
    power boost for amplifiers, 670
cascode amplifier, 411n14, 417, 454
cascoding, and PLL charge pump, 547
"cascomp" circuit, and mixers, 325
CAT-5 computer networking cable, 115
catwhisker
    crystal radios, 299–300, 301, 302, 303
    development of semiconductor detectors, 7, 9, 18
    Schottky diode, 281
Cauer, Wilhelm, 756–7, 766, 767
Cauer filters, 755–60
cavity magnetron, 20–2
cavity resonators, 183
C connector, 111n8
cell phones
    bandpass filters, 816
    connectors, 113
    development of, 24–5, 26
    frequency bands, 39
    heterojunction bipolar transistor (HBT), 347–9
center pins, of connectors, 112, 114
ceramic capacitors, 134–6
Chain Home Low, 18
chalcopyrites, and crystal radios, 298
channel spacing counter, and integer-N synthesizer, 555
Chaplin, G. B. B., 618n6
characteristic impedance
    distributed filters, 786
    image impedance, 825–8
    infinite transmission line, 45
    lossy transmission line, 47

lumped bandpass filters, 815–16
lumped impedance matching, 78
charge pumps, and phase-locked loops, 544–51
Chebyshev, Pafnuti L'vovich, 744–5n21
Chebyshev approximation, 99, 106
Chebyshev filters
    approximations, 744–53
    delay counterpart, 765
    design tables, 781, *782*
    lossless ladder network synthesis, 767
    ripple value, 749, 832–3
Chireix, Henri, 659
choke, and bipolar transistor biasing, 372–3
circles, and noise figures, 443, *444*
circuits and circuit theory, 688–90; *see also* RF circuits
Clapp, James K., 520n20
Clapp oscillator
    Colpitts oscillator, 520
    phase noise, 593–4
    tapped resonators, 526, 527, 528
Clarke, Kenneth K., 510n9
Class A amplifiers
    adaptive bias, 666
    cascading, 670
    defining characteristics, 632–5, 687
    load-pull characterization, 683, *686*
    modulation of, 650–1
Class AB amplifiers, 640, 648, 651, 670
Class B amplifiers, 687
    defining characteristics, 635–7, 687
    Doherty amplifier, 668–9
    modulation of, 650, 653
Class C amplifiers, 637–40, 651–2, 670, 687
Class D amplifiers, 640–2, 687
Class E amplifiers, 642–6, 687
Class F amplifiers, 646–50, 687
Class S amplifiers, 669
clocks, and escapement, 591; *see also* digital watches
close-in phase noise, 591–2
CMOS circuit, and PLLs, 547, 565–8
coaxial cables
    attenuation, 120–1
    power-handling capability, 71–2
    properties of, 121–2
    reasons for use of, 115–17
    skin effect, 127
    types of, 117–18
    waveguides and moding, 118–20
coaxial feed, for patch antenna, 715
coaxial slotted line, *250*
codirectional coupler, 220–2
C0G ceramic capacitor, 135
coherer, 3–5, 11
Cohn, Seymour B., 163n9, 823n25, 828n29
Cohn equations, 817n19
coil design, and formulas for inductance, 145
coinage, and crystal detectors, 9n17, 303
collector–base capacitance, and bipolar noise model, 447
collector load, and low-noise amplifier, 455–6
collector voltage, and power amplifiers, 633, 636, 637–9, 642, 644, 647, *649*
color television, 530–1

Colpitts, Edwin Henry, 501n5
Colpitts oscillator
   describing functions, 501–15
   Gunn diodes, 293
   LC feedback oscillators, 520, 522, 528
   phase noise, 576, 589–90, 592–4
combination synthesizers, 558–9
combiners, and PC boards, 183–230
combline filter, 841
commensurate-line filters, 795–800
common-emitter amplifier, 413–14, *419*
common-mode choke, 153
common-mode response, and bridged T-coil transfer function, 429–30
commutating multiplier, 539–40
compensation, and time-domain reflectometer, 245–6
complementary modulus, and lumped filters, 777
complete elliptic function, and impedance of stripline, 164
complete transfer function, and bridged T-coil transfer function, 431–2
complex mixers, 338
composite constant-k and m-derived filter, 735
composite second-order (CSO) distortion, 412
composite triple beat (CTB) distortion, 412–13
compression point, of RF signal in mixers, 308–9
computers, and CRTs, 614
   see also Macintosh computers; simulation and simulators
Concelman, Carl, 111
conductance balance and equalization, and broadband impedance matching, 106–7
conductor, coplanar, 168–70
conductor loss, 115–16, 137
conformal mapping, CPW lines and CPS conductors, 169–70
connectors
   moding and attenuation, 108–14
   nonlinear effects, 114–15
   transmission lines for PC boards, 176–8
connector savers, 113
constant current modulator, and power amplifiers, 652
constant-envelope amplifiers, 648
constant-k filters
   design tables, 781, *782*
   lumped prototype, 798
   stepped-impedance filters, 787–9
   transmission lines, 726–38
consumer products, and history of microwave technology, 24–7
   see also cell phones; radio; television; toys
continued fraction expansion, and lumped filters, 768
control-line ripple, loop filters and charge pumps for PLLs, 546–51
convergence, and broadband impedance matching, 106–7
conversion gain, and mixers, 307
coplanar waveguide (CPW) and coplanar strip (CPS), 168–70
copper
   air-core solenoids, 143, 144
   crystal detectors, 9n17
   cuprous oxide as semiconductor, 300–1, 302
   green patina, 114n9

PC boards, 158
   skin depth, 126
copper foil tape, and microstrip, 165, 268
corporate arrays, and Wilkinson combiners, 188–9
corporate combiner, 671–2
correction factors, and fringing capacitance, 233–7
correlation admittance, and two-port noise theory, 442
corrosion, and nonlinearity in connectors, 114
coupled resonator bandpass filters, 803–41, 843
coupled wires
   interline spacing, 222–3
   magnetically coupled conductors, 151–2
couplers, and PC boards, 183–230
coupling coefficient, for microstrip filters, 821–2
coupling factor, and directional coupler, 201–7, 215, 218
Cripps, S. L., 683
crossed-field device, and cavity magnetron, 20
crossover frequency, and loop filters for PLLs, 550
CRTs, and analog scope, 614–17
cryogenics, and pulse generators, 239n1
crystal detectors, 5–10, 275
crystal oscillator, 521–3
crystal radios, 9, 297–304, 331
crystal rectifier, 298n27, 522n22
Curie, Jacques and Pierre, 516n14
current-mode mixers, 328n29
cutoff frequency, of lumped lines, 55–6
CW oscillator, 294
cyclostationarity, and phase noise, 575–6, *590,* 593, 595
cyclotron, 10n21
cylindrical conductors, *124,* 127–8, 132–3
cylindrical shields, and inductors, 146

damping ratio, and Chebyshev filter, 745–6
Darlington, Sidney, 725, 766, 768, 770
data sheets
   model parameters for bipolar transistors, 360–1
   stability of amplifiers, 425
DC model, for bipolar transistors, 352–6
DC resistivity, of copper, 158
de Bellescize, H., 529, 530
De Forest, Lee, 10, 12–14, 15–16
delay, and lumped versus distributed circuits, 44
delay line discriminator, and phase noise measurement, 607–9, 612
De Loach, W. H., 297
delta-sigma techniques, for nonuniform distribution of noise, 558
depletion-mode FET, and biasing, 376–9
The Design of CMOS Radio-Frequency Integrated Circuits (Lee, 2004), 129n3, 133n6, 396n8, 444n3, 494n1, 530n2
The Design of Low-Noise Oscillators (Hajimiri & Lee, 1999), 584n5, 596n11, 602n1
design tables, for low-pass filters, 781–3
dielectric capacitors, 133–4
dielectric constants, of materials used in microwave circuits, 237
dielectric loss
   capacitors, 137
   coaxial cable, 116, 121
   PC boards, 159

INDEX

differential-mode response, and bridged T-coil transfer function, 428–9
digital-to-analog converter (DAC), and direct digital synthesis, 559, 560
digital oscillator, 523
digital watches, and resonance of crystals, 516n15
diode cutoff frequency, 277–8
diode detectors, and noise figure measurement, 490
diodes
 backward diode, 287
 Gunn diodes, 293–5, 297
 IMPATT diodes, 295–7
 impedance matching, 101
 junction diodes, 276–9
 MIM diodes, 295
 noise diodes, 289–90
 nonlinear junction capacitance, 331
 "penny" diodes and crystal radios, 297–304
 PIN diodes, 287–9
 Schottky diodes, 279–81
 snap diodes, 290–2
 tunnel diodes, 284–7
 use of term, 275
 varactors, 281–4
diplexer, 335n37, 336
dipole antenna, 132–3, 689–90, 697–707
direct-conversion receivers, 465, 529–30
direct digital synthesis (DDS), 559–60
directional couplers, 201–25, 255, 259–60
directivity, of antennas, 695
discontinuities, and time-domain reflectometer, 238–43
Dishal, Milton, 842, 843n37
disk terminator, and codirectional coupler, 221–2
distributed active transformer (DAT), 673, *674*
distributed amplifier, 406–11
distributed bandpass filters, 804–5
distributed circuits, 41–4
distributed combiners, 184–91
distributed filters, 784–803, 814–18; *see also* microstrip filters
distributed resonators, 844–5
dithering moduli, and synthesizers, 556–8
divider "delay," and PLL synthesizers, 551–3
Doherty amplifier, 667–9
Dolbear, Amos E., 31
doping
 bipolar transistors, 346, 347
 profiles of diodes, 281–2, 286
double-balanced diode mixer, 334–5
double-sideband (DSB) noise, 308, 491–2
double-stub tuner, 92, *93*
Douglas, Alan, 280n7
downward impedance transformer, *81*
drain modulation, and power amplifiers, 653–4, 667, 670
drive amplitude, and oscillators, 496, 500
driving-point impedance, and RF circuits, 44–6
dry ice, and ethanol, 484
dual-gate FET mixer, 317
Duddell, William, 10n21
Dunwoody, Henry Harrison Chase, 7, 298
duplexer, and diplexer, 335n37
dynamic elements, of FET circuits, 364–8

Early, James, 354
Early effect, and bipolar transistors, 354, 446n5
earphone, for crystal radio, 303–4
"E-band" spectrum, 26, 41
eddy current loss, and transformers, 150–1
edge-coupled bandpass filters, 823–41
Edison, Thomas, 11–12
"Edison effect," 12
effective gain, and drive amplitude of oscillator, 496
effective series resistance (ESR)
 dielectric loss in capacitors, 137
 solenoids, 143
effective width, of microstrip, 166
efficiency-boosting techniques, for power amplifiers, 666–9
E-field, and waveguides, 119
Einstein, Albert, 3n4
EIRP (effective isotropically radiated power), 28
electrical reference plane, and slotted line, 249
electric wave filters, 726
electromagnetic energy, and nature of radiation, 691–5
electromagnetic interference (EMI), and ferrite beads, 139
electromagnetic waveguides, 6n15
electron, and history of vacuum tube, 11
Electronics Industry Association, 134
Electronics magazine, 61
electrostatically deflected CRTs, 614, 615
ELI, as mnemonic, 691n8
elliptical pole distribution, and Chebyshev filters, 752
elliptic filters, 755–60, 764–5, 774–81, 792–5
elliptic functions, CPW lines and CPS conductors, 169–70
Elwell, Cyril, 10
Emde, F., 775
emitter-coupled amplifier, 417–18
emitter degeneration inductance, and low-noise amplifier, 455
energy
 parametric amplifier, 283
 transmission lines, 46
 wires for antennas, 690–1
 see also electromagnetic energy
energy coupling, and microstrip filters, 818–23
Engelmann, H. F., 163n8
Engen, G., 261n6
enhancement-mode MESFETs, 346
entertainment, and commercial potential of radio, 17
envelope detector, and crystal radio, 331
envelope elimination and restoration (EER), and power amplifiers, 658–9, 663, 667
envelope feedback, and linearization of power amplifiers, 654–5
envelope impedance, and Colpitts oscillator, 512
envelope loop transmission, 509–10
equiangular spiral antenna, 718
equipartition theorem, of thermodynamics, 580
equivalent isotropic power density (EIPD), 695
error sources
 impedance measurement, 251–4, 256–63
 noise figure measurement, 487–90
 phase noise measurement, 611–12
error vector magnitude (EVM), 677–8
Esaki, Leo, 284–5, 349

escapement, in mechanical clocks, 591
ethanol, and dry ice, 484
Europe, and color television, 531
even-mode excitations, for combiners and couplers, 186, 210–11, 212–15
Ewing, Gerald D., 643n12
excess amplitude, and phase noise, 597–8
excess noise ratio (ENR), and noise figure measurement, 475–6, 489
exclusive-or gate, and phase-locked loops, 540–2
expansion ratio, and equiangular spiral antenna, 718
extended-range phase detector, 543–4
external noise, and noise figure measurement, 487–8
eyeball techniques, for noise figure measurement, 492–3
EZNEC (antenna analysis program), 704

Fairchild Semiconductor, 343
Fano, Robert M., 93
Faraday's law, 42, 123
F connectors, 114
Federal Communications Commission (FCC)
    allocation of spectrum, 26, 27–8
    CTB ratio for cable TV equipment, 412
    phased-locked loops (PLLs) and television, 530
Federal Telecommunications Laboratory (ITT), 162
Federal Telegraph (California), 10, 14
feedback
    mixers, 324–5
    oscillator design, 494–5
    see also Cartesian feedback; envelope feedback; polar feedback
feedforward, and linearity, 325, 655–6
ferrite beads, and inductors, 138–9
Fessenden, Reginald, 10, 16, 17, 305n1
FET models, for bipolar transistors, 361–8, 468–70
Feynman, Richard, 1n2, 233, 234, 693n10
field-effect transistors, 342
"50 Years of RF and Microwave Sampling" (*IEEE Trans. Microwave Theory and Tech.*, 2003), 629
figures of merit, and high-frequency transistors, 359–60
fill factor, for planar spirals, 232
filters
    impedance matching, 75
    spectrum analyzers, 626–7
    transmission lines, 726–38
    see also loop filters; low-pass filters; lumped filters; microstrip filters
finite-length transmission lines, 51–3
first-order phased-locked loop, 533–4
Fitzgerald, George Francis, 2
fixed-source impedance, 75
fixturing loss, and noise figure measurement, 488
flare angle, and equiangular spiral antenna, 718
flat sheets, and formulas for inductance, 140, 230–1
Fleming, John Ambrose, 12
Fleming valve, 12–14
FM radio
    frequency bands, 39
    mixers, 311–12
    PLL-like circuits, 532
    tuned amplifiers, 414
    see also radio

forward recovery, of junction diode, 277–8
forward-wave coupler, 220
Fourier series, 98–9, 585
four-resonator system, for radio, 34–5
4046 CMOS PLL, 565–8
fractal antennas, 720
fractional-N synthesizers, 557, 565
frame counter, and integer-N synthesizer, 555
France, and color television, 531
Frederick, Raymond E., 203
"free-energy" radio, 9
free-running pulse generator, 271–4
frequency aperture, of vector network analyzer, 257
frequency bands, for RF and microwave circuits, 38–41
frequency detectors, and phase-locked loops, 544
frequency-division duplexing (FDD), 23, 25
frequency-division multiple access (FDMA), 23, 25
frequency domain, of oscillators, 495
frequency-hopped spread-spectrum systems, 559
frequency-independent delay, and transmission lines, 50
frequency multiplier, and snap diode, 292
frequency pulling, and Colpitts oscillator, 514
frequency range, and shunt-series amplifier, 405–6
frequency response, and vector network analyzer, 260
frequency sensitivity, and impedance matching, 97
frequency synthesizers, 529, 551–60
Fresnel lenses, 38
FR4 (flame-retardant formulation number 4)
    patch antenna, 712–15
    PC boards, 159–60, 161, 165, 166, 179, 183, 237, 268, 270
Friis, Harald T., 441n1, 472
Friis formula, 27, 479–80, 485
fringing capacitance, and correction factors, 233–7
Fuller, Leonard, 10n21
"Fundamentals of RF and Microwave Noise Figure Measurements" (Hewlett-Packard, 1983), 492

gain boost, by injection locking for power amplifiers, 671
    see also maximum available gain; power gain
galena bolometer, 5–6, 7
galena crystal detectors, 9, 298, 299, 302–3
gallium arsenide (GaAs)
    Gunn diodes, 293, 294
    transistors, 24, 346, 348
gallium nitride (GaN), and transistors, 351
gamma (γ): see propagation constant
gated impedance, 265
Gauss's law, 42
Geiger, P. H., 300
General-Admiral Apraksin (ship), 32n73
germanium, and transistors, 343, 346, 348
GGB Industries, 623
Ghinoe, G., 169n15
Gibbs, Willard, 1n1
"giga-," pronunciation of, 70
Gilbert mixers, 318n22, 322–3
Ginzton, E. L., 406
Global Positioning System (GPS), 39, 113, 677
Goddard, Robert, 14

golden ratio, 45
grading layers, and heterojunction bipolar transistors, 348
The Great Eastern (ship), 724n3
grid-dip oscillator (GDO), 266–7
Grieg, D. D., 163n8
Grondahl, L. O., 300
ground leads, for probes, 621
G10, and PC boards, 159
Guanella, G., 152–3, 155
Guillemin, E. A., 751n25
Gunn, J. B., 293
Gunn diodes, 293–5, 297
gyration, of base-emitter capacitance in amplifiers, 423

hairpin filter, 838–40
Hajimiri, A., 583n4, 597n12
half-section, and termination of lumped delay lines, 56–8
half-wave "bandpass" filters, 800–3
half-wave patch antenna, 711–13
half-wave transmission lines, 88–9
Hammerstad, E. O., 171
Hamming, R. W., 748n23
hams (radio amateurs), 17
Handbook of Filter Synthesis (Zverev, 1967), 843n37
Hansen, William, 19
hard substrates, and PC boards, 161–2
harmonic energy, and snap diode, 292
harmonic telegraph, 23–4n56
Hartley, Ralph Vinton Lyon, 501n5
Hartley oscillator, 519, 528
Hawks, Ellison, 30n67
headphones, and crystal radio, 303–4
Heaviside, Oliver, 1n1, 18n38, 50–1, 690n5, 724
Hegazi, E., 596n11
Heising modulation, 653n19
Herrold, Charles, 17
Hertz, Heinrich, 2–3, 31, 36, 723–4
Hess, Donald T., 510n9
heterodyne principle, 16, 305n2
heterojunction bipolar transistor (HBT), 347–9
heterostructure FET (HFET), 349
Hewlett, W. R., 406
Hewlett-Packard, 101, 112, 291, 492, 612, 791n8
Hewlett-Packard Journal, 246
high electron mobility transistor (HEMT), 349
higher-order poles, loop filters and charge pumps for phase-locked loops, 546–51
high-frequency models, of bipolar transistors, 356–60
high-K (high dielectric constant) ceramics, and capacitors, 135–6
high-pass constant-k filter, 735–6, 781
high-power systems, and impedance matching, 75
Hilbert, David, 756n29
Hilsum, C., 293
history, of microwave technology
 consumer microwave products, 24–7
 debate on invention of radio, 29–36
 early years of, 1–18
 lumped filters, 723–6
 modern microwave diode, 275

 phased-locked loops (PLLs), 529–32
 Smith chart, 60–1
 transistors, 341–51
 World War II, 18–23
Hoer, C., 261n6
Hollmann, Hans E., 21
homodyne receiver, 529–30
homojunctions, and transistors, 347
Houck, Harry, 16
HP8970 noise figure meter, 486, 492
Hughes, David Edward, 30
Hull, Albert W., 20
Hülsmeyer, Christian, 18n40
humans
 brains as coherers, 5
 crystal radios and auditory system, 299
 injection locking and circadian rhythms, 530n3
Hurwitz (minimum phase) polynomial, 770
hybrid combiners, splitters, and couplers, 191–201
hybrid parameters, 67
hyperabrupt junctions, 277, 357
hysteresis loss, and transformers, 150

ICE, as mnemonic, 691n8
ideality factor, and diodes, 276, 281
IEEE: *see* Institute of Electrical and Electronics Engineers
image frequency, and mixers, 307–8
image impedance, and microstrip filters, 825–8
image parameter filters, 726–38
image termination, and mixers, 335–8
immittance inverters, and coupled resonator bandpass filters, 805–7, 809–10, 812–14
IMPATT diodes, 295–7
impedance
 amplifiers and strange impedance behaviors (SIBs), 420–7
 grid-dip oscillator, 266–7
 iterated structures and driving-point, 44–6
 line-to-line discontinuities, 170–6
 microstrip "slotted"-line project, 268–71
 microstrip and transmission lines on PC boards, 165–6
 reactive components of antenna, 701–2
 sampling scopes and measurement of, 623
 slotted line, 246–54
 Smith chart, 60–6
 standardized for RF instruments and coaxial cables, 71–3
 stripline and PC boards, 163–4
 sub-nanosecond pulse generators, 271–4
 summary of calibration methods, 264–5
 SWR meter, 265–6
 time-domain reflectometer (TDR), 238–46
 vector network analyzer (VNA), 254–64
 see also characteristic impedance; impedance matching
impedance matching
 importance of in RF engineering, 74–5
 large-signal power amplifiers, 682–3
 maximum power transfer theorem, 75–7
 methods of, 77–107
impulse sensitivity function (ISF), and phase noise, 584–8, 591, 595, 600

indium phosphide (InP)
  Gunn diodes, 293
  heterojunction bipolar transistors, 348
inductance
  formulas for, 139–47, 230–3
  low-noise amplifier and emitter degeneration, 455
  resistors, 129–33
inductive degeneration, and single-balanced mixer, 320
inductive loading, of dipole antennas, 704–6
inductors, and passive components, 138–47
infinite ladder network
  constant-k filter, 728
  transmission lines, 45–6
injection locking, and sawtooth generators, 530
input–output impedances, and shunt-series amplifier, 399–401
input–output resistances, and shunt-series amplifier, 397–9
input-referred third-order intercept (IIP3), and low-noise amplifier, 459–60, 462, 463–4
insertion loss, of connectors, 112
insertion power gain, 71
instability, and power amplifiers, 679
Institute of Electrical and Electronics Engineers (IEEE), 70
Institute of Radio Engineers (IRE), 343n5, 473; *see also* Institute of Electrical and Electronics Engineers
insulators, and FR4, 161
integer-N synthesizer, 555
interconnect, and lumped versus distributed circuits, 44
interdigital filter, 840–1
intermediate frequency (IF)
  amplifiers, 484
  mixers, 305–6
intermediate resistance, and π-match circuit, 84
intermodulation distortion, in broadband amplifiers, 411–13
intermodulation (IM) products, 313
inverse Chebyshev filters, 745, 753–5, 792
inverse Class F amplifier, 648–9
IP3: *see* third-order intercept
IRE National Conference on Airborne Electronics (1954), 343n5
iron pyrites, and crystal radios, 298
Ishii, J., 796n12
ISM (industrial-scientific-medical) band, 27–8
isolation factor (IF)
  directional coupler, 202
  filters for spectrum analyzers, 626–7
  mixers, 308–10
isotropic antenna, 695
iterated structures, and driving-point impedance, 44–6

Jackson, D. R., 710n29
Jahnke, E., 775
Jamieson, H. W., 162n7
Jansen, R. H., 215n43
Jasberg, J. H., 406
Jell-O™ transistors, 414
JFET-based "grid"-dip oscillator, 267

J-inverters, 807–8, 810
Jones, E. M. T., 227n57, 823–4
junction bipolar transistor, 342, *343*
junction diodes, 276–9
junction FETs, 342, *343*, 468

Kahn, Leonard, 658
Kahng, Dawon, 343
Kahrs, Mark, 629
Kelvin, Lord (William Thomson), 724
kelvins, and absolute temperature, 70
Kennedy, John F., 36, 305n2
Kirchhoff's laws, 41–3, 124, 688
Kirschning, M., 215n43
Klopfenstein, R. W., 100
klystron, 19–20
Kroemer, Herbert, 347n9
Kuroda's identities, 796, 799

LADDER (program), 755, 760, 764, 774, 792–3
Laguerre, E. N., 770n44
Landon, V. D., 742n19
Lange, Julius, 223
Lange coupler, 223–5, 226, 228
Langmuir, Irving, 14–15
L-antenna, 703–4
Laplace transforms, 50n8
large-signal impedance matching, and power amplifiers, 682–3
large-signal operating regimes, for power amplifiers, 630–1
large-signal performance, and linearity of low-noise amplifier, 457–62
large-signal transconductor, *498*
lateral coupling, and microstrip filters, 823–4
Lawrence, Ernest O., 10n21
LC oscillators, 519–21, 528, *585*, 592–7
L/C ratio, and artificial lines, 56, 57–8
LDMOS (laterally diffused MOS), 344–5
leakage inductance, 150
Lectures on Physics (Feynman), 233
Leeson, D. B., 581
Leeson model, of phase noise, 581–2
Legendre, A.-M., 777
letter designations, for frequency bands, 39
Lévy, Lucien, 305n2
light bulb, 11, 301
limiting amplifier, 658
LINC (linear amplification with nonlinear components), 659–62
linear, time-varying (LTV) system, and oscillators, 576, 584, 594, 597, 600
linear, time-invariant (LTI) theories, of phase noise, 575–6, 582–92
linearity
  diode detectors and noise figure measurement, 490
  low-noise amplifier (LNA), 457–62, 465–7
  mixers, 305, 308–10, 312–17, 323–6
  oscillators, 494–5
  phased-locked loops (PLLs), 532–6
  time variation in phase noise, 582–92
  see also nonlinearity
linearization, and modulation of power amplifiers, 654–6

line-to-line discontinuities, and transmission lines for PC boards, 170–6
Linvill stability factor, 425
Lippman, Gabriel, 516n14
Lissajous figures, 749
lithium niobate, and surface acoustic wave devices, 518–19
L-match
  Class E power amplifier, 645
  lumped impedance matching, 80–4
  shunt-series amplifiers, 400
LNA: *see* low-noise amplifier
  load inductor design, and Colpitts oscillator, 505–6
  load-pull characterization, of power amplifiers, 683–7
local oscillator (LO), 16, 305, 529–30
Lodge, Oliver, 2, 3, 4n8, 31–2, 36, 720–1, 724
lodgian waves, 2
logarithmic spiral antenna, 718
logarithmic warping, of IF filter, 627
log-periodic dipole array, 720
Logwood, Charles, 14
Loomis, Mahlon, 29
loop antenna, 706–7
loop filters, and phase-locked loops, 544–51, 568–73
LO phase noise, 574–5
Lorentzian spectrum, 592
lossless ladder network synthesis, and lumped filters, 766–74
lossless LC tank, and impulse responses, 583
loss tangent, and capacitors, 137
lossy transmission line, 46, 47
low-frequency gain, and shunt-series amplifier, 397–9
low-noise amplifier (LNA)
  biasing, 456–7
  bipolar noise model, 445–51, 468
  collector load, 455–6
  emitter degeneration inductance, 455
  FET noise parameters, 468–70
  linearity and large-signal performance, 457–62
  minimum noise figure (NF), 440
  model parameters for practice, 470–1
  narrowband LNA, 451–4
  noise temperatures, 478
  phase noise measurement, 608
  spurious-free dynamic range (SFDR), 462–4
  two-port noise theory, 440–5
low-pass filters
  Colpitts oscillator, 512n11
  design tables for, 781–3
  stepped-impedance architecture, 787
low-temperature co-fired ceramic (LTCC), 162
LRM (line-reflect-match) method, of impedance measurement, 262–3, 264–5
LTSpice (simulation), *389, 395, 403, 404, 438*
lumped bandpass filters, 804–18
lumped circuits, 41–4
lumped coupling capacitors, 209
lumped delay lines, 55–8
lumped filters
  bandpass filters with combinations of lumped and distributed elements, 814–18
  classification and specifications, 738–40
  common approximations, 740–66

  distributed filters, 784–803, 814–18, 844–5
  elliptic integrals, functions, and filters, 774–81
  history of, 723–6
  network synthesis, 766–74
  transmission lines, 726–38
lumped impedance matching methods, 78–87
lumped model, for lossy transmission line, 46
lumped narrowband directional coupler, 207
lumped two-way Wilkinson splitter/combiner, *191*

Maas, Stephen, 334n36
Macintosh computers, and shunt peaking, 382
magnetically coupled conductors, 147–56
magnetic cores, coils wound on, 145–6
magnetic loop antenna, 706–7, 722
magnetic materials, and nonlinearity in connectors, 114–15
magnetic resonance imaging, 27
magnetizing inductance, 150
magnetrons, 20–3
Maliarov, A., 21n46
Manhattan Project, 22
Marchand, Nathan, 156
Marchand balun, 156, 334, 336
Marconi, Guglielmo, 3–5, 32–3, 34–6, 699, 724
Marconi Wireless Telegraph Corporation of America, 34
Margarit, M. A., 594, *596*
Marks, R. B., 262n7
Massachusetts Institute of Technology (MIT): *see* Rad Lab
matched termination, and transmission line, 51
Matthaei, G. L., 227n57, 823–4
maximally flat delay, and bridged T-coil transfer function, 436–7
maximally flat magnitude response, and bridged T-coil transfer function, 432–6
maximum available gain (MAG), and amplifiers, 426
maximum bandwidth, and bridged T-coil transfer function, 437–8
maximum collector efficiency, for Class B amplifier, 637
maximum power transfer theorem, and impedance matching, 75–7
Maxwell, James Clerk, 1–4, 43n4
Maxwell's equations, 37, 41–3, 50n8
McCandless, H. W., 13
McKinsey and Company, 25n61
MC12181 frequency synthesizer, 561–3
MCX connector, 113
m-derived filters, 730–5, 738, 792
m-derived half-section, for line termination, 57–8
Mead, Carver, 24, 346
mechanical clocks, and escapement, 591
Medhurst, R. G., 142n14–15, 143
megalodges, 2
MESFET (metal-semiconductor FET)
  biasing, 378–9
  bipolar transistors, 346–7, 348, 350, 363–4, 368
  cell-phone amplifiers, 24
  noise parameters, 468
  traveling-wave amplifiers, 408–9
metamorphic HEMTs (mHEMTs), 349, 350
metric ruler, and microstrip "slotted"-line project, 269

mica, and capacitors, 133, *237*
microchannel plate CRTs, 616, 617
microhenries, and inductance, 141
microstrip
  bow-tie antenna, 721
  branchline coupler, 196–9
  Colpitts oscillator, 504
  coupled resonator bandpass filters, 803–41
  directional couplers, *209*, 218–20
  impedance measurement, 263–4, 268–71
  lumped analogue of, 200
  PC boards, *163*, 164–8
  vector network analyzer, 263–4
microstrip filters
  coupled resonator bandpass filters, 803–41
  distributed filters from lumped prototypes, 784–803
  practical considerations in design, 841–3
microstrip patch antenna, 707–20
microwave circuits: *see* circuits and circuit theory; RF circuits
Microwave Filters, Impedance Matching Networks, and Coupling Structures (Matthaei, Young, & Jones, 1980), 227n57, 823–4
microwave ovens, 24, 39
microwave spectroscopy, 23n50
Miller, F. H., 775n47
Miller effect, 399–400, 415, 417, 423
MIM diodes, 295
minimum noise figure (NF), and low-noise amplifer, 440
minority-carrier device, junction diode as, 278
mitered bend
  time-domain reflectometer, 245
  transmission lines, 172–3
mixers
  conversion gain, 307
  linearity and isolation, 308–10
  multiplier-based mixers, 317–40
  noise figure, 307–8
  noise figure measurement, 491–2
  nonlinearity and time variation, 312–17
  spurs, 310–12, 339–40
  superheterodyne receiver, 305–6
MMCX connector, 113
mobile communications, and World War II, 23
Mobile Telephone Service (MTS), 23
modeling, of transistors, 351–68
mode splitting, and microstrip filters, 818–23
moding
  coaxial cable, 115, 118–20
  connectors, 108–14
  microstrip, 166–8
  see also overmoding
modular angle, and lumped filters, 777
modulation, of power amplifiers, 650–78
modulation-doped FET (MODFET), 349
monoblock filter, 816
monopole antenna, 699–700, 705–6
monotonicity condition, and impedance matching, 98–9
Monte Carlo analyses, 564, 744, *754, 761,* 764–5
Moore's law, 344
MOS devices, and breakdown phenomena, 680

MOSFETs
  bipolar transistors, 343–7, 348, 363–4, 368
  noise parameters, 468
  stability problem, 679
Motorola, 23, 24, 561
Mstrip40 (program), 709, 711
multiple-stub tuners, 91–2
multiplier-based mixers, 317–40
multiplying DAC (MDAC), 560
multisection coupler, *226*
multisection stepped-impedance transformer, 94–5, 97
mutual inductance, 148–9

Nagaoka, S., 141–2
Nahin, Paul J., 724n5
Naldi, C., 169n15
narrowband LNA, 451–4
narrowband transmission line transformers, 155–6
National Bureau of Standards (NBS), 476
National Institute for Standards and Technology (NIST), 476
National Semiconductor, 563
National Television Systems Committee (NTSC), 531
N connector, 110–11, 112
negative feedback, and mixers, 324–5
negative impedance converter (NIC), 524
negative resistance
  amplifiers, 422, 424, 453
  oscillators, 524–8
negative voltages, and Colpitts oscillator, 504
Neill, Paul, 110, 111
neper (Np), and amplitude change, 182n25
network analyzer, 178, 246; *see also* vector network analyzer
network synthesis, and lumped filters, 766–74
neutralization, and amplifiers, 417–20
9913 coaxial cable, 121–2
NIST standards, for noise diodes, 290
Noe, J. D., 406
noise
  delta-sigma techniques for nonuniform distribution, 558
  FET parameters, 468–70
  model of for bipolar transistor, 445–51, 468
  phase-locked loops (PLLs), 536–7
  see also noise factor; noise figure; phase noise; thermal noise
noise diodes, 289–90, 476
noise factor
  low-noise amplifier (LNA), 464–5
  two-port noise theory, 441–3
noise figure (NF)
  measurement of, 472–93
  mixers, 307–8, 322–3
  noise temperature, 444–5
noise measure, 480–1
noise modulation function (NMF), 589
nominal detectable signal (NDS), 490
nonlinearity
  connectors, 114–15
  mixers, 308–17, 339
  oscillators, 495, 582
  see also linearity

nonplanar transformer, 152
nonquasistatic (NQS) effects, and FET circuits, 367–8
Nordic Mobile Telephone System (NMT-450), 25
normalized power efficiency, for power amplifiers, 641, 648
notched transmission line, 174, *175*
NP0 (negative-positive-zero) designation, for temperature coefficient, 135
nuclear magnetic resonance, 27
numerical evaluation, of elliptic functions, 778–80
N-way Wilkinson splitter/combiner, *189, 190*

Occam's razor, 43n4
odd-mode excitations, for combiners and couplers, 186–7, 210, 212–15
offset synthesizer, 558–9
Ohl, Russell, 18, 275, 276
Ohm's law, 123–4
omnidirectional antenna, 696
1-dB compression point, 457
1N34A germanium diode, 300, 302
$1:n$ transformer, 148
op-amp, and active bias, 380
optimum source admittance, and two-port noise theory, 443–4
optimum source resistance, and bipolar noise model, 449–51
Orchard, H. J., 760, 780
orthogonality, and Cartesian feedback, 665
oscillators
    description of functions, 495–515
    Gunn diodes, 293, 294
    IMPATT diodes, 297
    impedance measurement and grid-dip, 266–7
    linear, time-varying (LTV) systems, 576, 584, 594, 597, 600
    linearity, 494–5
    negative resistance oscillators, 524–8
    nonlinearity, 582
    phase noise, 574–600
    PLLs and frequency synthesis, 551
    resonators, 515–19
    tuned oscillators, 501, 509, 519–24
    tunnel diode, 286
    voltage-controlled oscillators, 281
    zero temperature coefficient, 134
    see also voltage-controlled oscillator
oscilloscopes
    analog scopes, 614–17
    common-mode chokes, 153
    distributed amplifiers, 411
    noise figure measurement, 493
    sampling scopes, 617–25
    snap diodes, 291
    time-domain reflectometer, 238, 239
    two-port bandwidth enhancement, 392, 393n6
Osmani, R. M., 225
Ou, W. P., 224–5
outphasing modulation, and power amplifiers, 659–62
overmoding
    half-wave "bandpass" filters, 801
    waveguides, 119–20
Ozaki, H., 796n12

Padé approximants, 763n3
parameter extraction, and time-domain reflectometer, 243–5
parametric amplifiers, 282–4, 331n34
parametric converters, 307n6
Parseval's theorem, 587–8
passive components
    capacitors, 133–8
    combiners, splitters, and couplers, 183–230
    inductors, 138–47
    magnetically coupled conductors, 147–56
    resistors, 129–33
    resonators, 181–3
    skin effect, 123–9
    transmission line segments, 178–81
passive double-balanced mixer, 326–30
passive filters, 723
passive probe, for sampling oscilloscope, 623–4
patch antennas, 707, 722
PC (printed circuit) boards
    capacitors, 136
    combiners, splitters, and couplers, 183–230
    general characteristics of, 158–62
    resonators, 181–3
    transmission lines, 162–81
pencil lead, and crystal radio, 301, 302
"penny" diodes, 297–304
Penzias, Arno, 24
Percival, William S., 406
performance metrics, and power amplifiers, 673–8
phase-alternating line (PAL) system, 531
phase detector
    phased-locked loops (PLLs), 532–3, 537–44, 565–7
    phase noise measurement, 604–7
    vector network analyzer, 257
phase distortion, and shunt-peaked amplifiers, 385
phase equalizer, 765
phase-frequency detectors, 544
phase-locked loop (PLL)
    design example, 561–4
    frequency synthesis, 551–60
    history of, 529–32
    inexpensive design as lab tutorial, 565–73
    linearized model, 532–6
    loop filters and charge pumps, 544–51
    phase detectors, 537–44
    phase noise measurement, 604–7
    rejection of noise on input, 536–7
phase margin, and loop filters for PLLs, 550
phase noise
    amplitude response, 597–9
    detailed analysis of, 579–82
    LC oscillators, 592–7
    linear, time-invariant (LTI) theories, 575–6, 582–92
    measurement of, 601–12
    reciprocal mixing, 574–5
    simulation, 600
    tank resistance, 576–9
"Phase Noise Characterization of Microwave Oscillators" (Hewlett-Packard, 1983), 612
phenolic, and PC boards, 160
phosphor-bronze wire, and crystal radio, 301, 303
photophone, 30–1, 36
Pickard, Greenleaf Whittier, 7, 9, 280, 298, 300

piecewise approximation, and mixers, 326, *327*
Pierce, George Washington, 298n27, 522n22
Pierce, John R., 20
Pierce oscillator, 522–3, 528
piezoelectricity, 516n14
π-match circuit, 84–5
pinchoff voltage, and FETs, 362–3
PIN diodes, 287–9
planar spirals, and formulas for inductance, 231–3
PLL: *see* phase-locked loop
PLL_LpFltr program, 563
PNP transistor, 380
point-contact bipolar transistor, 341
point-contact crystal detectors, 5–6
point-contact diodes, 276
polar feedback, and power amplifier, 657, 663–4
pole zero doublet, 389
police radar, and frequency bands, 39–40
polycrystalline wafers, and diodes, 280–1
polynomials, and lumped filters, 750–1, 757, 762–3, 770
polyphenylene oxide (PPO), and PC boards, 160–1
polystyrene capacitors, 134
Popov, Alexander, 32, 36
port assignments, and broadband directional couplers, 204
postdistorter, and power amplifier, 656
potentiometer, and PIN diode, 289
Poulsen, Valdemar, 10
power-added efficiency (PAE), and power amplifiers, 679
power amplifiers
  breakdown phenomena, 680–1
  classical topologies, 631–50
  instability, 679
  large-signal impedance matching, 682–3
  load-pull characterization, 683–7
  modulation, 650–78
  power-added efficiency (PAE), 679
  small- versus large-signal operating regimes, 630–1
  thermal runaway, 681–2
power backoff, and modulation of power amplifiers, 651, 654
power factor, and capacitors, 137
power gain, 70–1, 74–5; *see also* gain boost
power spectral density (PSD), 608
power supply terminal, and modulation of power amplifiers, 652
power transfer function, and elliptic filters, 758
Poynting, John Henry, 690n5
Poynting's theorem, 690–1
predistorter
  microstrip filters, 843n37
  mixers, 324
  power amplifiers, 656–8
prefixes, and abbreviations, 70
Principles of Wireless Telegraphy (Pierce, 1909), 522n22
printed circuit boards: *see* PC boards
probability density function (PDF), 668
probes
  microstrip "slotted"-line project, 269
  sampling oscilloscopes, 620–5
propagation constant, for transmission lines, 46–51

proximity effect, in conductors, 128
pseudomorphic HEMTs, 349–50
PTFE (polytetrafluoroethylene)
  capacitors, 134
  dielectric constant of, *237*
  PC boards, 160
Puff (simulator), 173, 216, 218, 219, 730, 788
pulse generators, and impedance measurement, 238–40, 271–4
pulse impedance, of lossy transmission line, 47
pulse response, of shunt-peaked amplifiers, 386–7
pulsewidth modulation (PWM), 669–70
Pupin, Michael Idvorsky, 51, 724
Pupin coils, 51

Quackenbush, E. Clark, 109
quadrature coupler, 201, 338
quadrature oscillators, 523–4
quadrature phase detector, 538
quarter-wave resonators, 515–16
quarter-wave transmission line transformer, 87–9
quartz, and PC boards, 162
quartz resonators, 9n17, 516–18
quasi-TEM propagation, and microstrip, 164
Quinn, Pat, 325
Q-value, of resonators, 181–3

radar
  frequency classification system, 39–40
  history of microwave technology and, 18–27, 275
  single-diode mixers, 330
  video detector, 627n13
radial stub, and stub low-pass filter, 790–2
radiation, and antennas, 691–5, 707–8
radiation resistance, of dipole antenna, 698–701
radio
  America's Cup yacht race in 1901, 10
  broadcast, rapid rise of, 17
  invention of, debate on, 29–36
  UHF connector, 109–10
  World War II and development of, 23
  see also AM radio; crystal radio; FM radio
radioastronomy, 24, 308n8, 331n34, 339
Radio Day (Russia), 32
radio-frequency interference (RFI), and ferrite beads, 139
Rad Lab (Massachusetts Institute of Technology), 22, 23, 61, 275, 282
Ramanujan, S., 780–1
Randall, John T., 21
rat-race hybrid, 193–4, *332*, 333
Rayleigh, Lord (John William Strutt), 6, 18n39, 99n14, 118n15
razor blades, and crystal radio, 300
RCA, 16, 17, 660
Read, W. T., 295–6
read-only memory (ROM) lookup table, and direct digital synthesis, 559, 560
reciprocal mixing, and phase noise, 574–5
rectangular spiral, and broadband antenna, 720
Reddick, H. W., 775n47
re-entrant modes, and distributed filters, 784
reflectance, and impedance matching, 95–6
reflection, and vector network analyzer, 258

reflection coefficient, and termination impedance, 240
regenerative amplifier/detector, 15–18
relaxation oscillators, 501
repeller, 20
resistance
    bipolar noise model and optimum source, 449–51
    skin depth, 126–7
    see also negative resistance
resistive combiners, 184
resistors
    inductance and capacitance, 129–33
    PIN diodes, 289
    thermal noise, 446
resolution bandwidth, of spectrum analyzer, 626–7
resonant circuits, and lumped impedance matching, 78–80
resonator frequency discriminator, and phase noise measurement, 609–11
resonators
    coupled pairs in microstrip filters, 818–23
    coupled resonator bandpass filters, 803–41
    distributed resonators, 844–5
    oscillators, 515–19
    PC boards and passive elements, 181–3
    tuning of, 843
response calibration, and impedance, 264
return loss (RL), 66, 76
reverse recovery
    junction diode, 277–8
    snap diode, 291–2
RF choke (RFC), 476, 643, 652
RF circuits
    artificial lines, 54–8
    definition of, 37–8
    driving-point impedance of iterated structures, 44–6
    frequency bands, 38–41
    impedance transformations, 424
    lumped versus distributed circuits, 41–4
    transmission lines, 46–54
    see also circuits and circuit theory
RF signal, and mixers, 307
RFSim99 (simulator), 217n45, 218, 310n11, 744
RG8 coaxial cable, 122
RG-*n*/U coaxial cable, 117
RG174 coaxial cable, 121
Richard, P. I., 795–6
Righi, Augusto, 3
rigid coaxial cables, 117, 122
ring hybrid, 193–4, 199, 201, 332–3
ring oscillator, *585*
ripple values
    Chebyshev filters, 749, 832–3
    lumped bandpass filters, 808–9, 817
Rodwell, Mark, 619n9
RO4003, and PC boards, 160, 161, 270
Rogers Corporation, 160
rolloff behavior, and lumped filters, 727
root locus
    oscillator, 494–5, 513
    second-order PLL, *536*
rotation, and broadband impedance matching, 106
Round, Henry J., 16
round wires, and formulas for inductance, 230–1
Rumsey, V. H., 162n7
Russia, and development of radio, 32
Ruthroff, C. L., 152, 154–5
Rutledge, David B., 752n26

safety
    beryllia, 162
    ethanol and dry ice, 484
safety pin, and crystal radio, 301, 302
St. Louis (Missouri), and mobile telephone service, 23
sample-and-hold (S/H) operation, and phase-locked loops, 552
sampling scopes, 617–25
Samuel, Arthur L., 20
sapphire, and PC boards, 162
Sarnoff, David, 16
satellite communications systems, 477, 532
sawtooth generators, 530
scattering parameters: *see* S-parameters
Schiffman "alligator clip" broadband coupler, *229*
Schmitt trigger, 497
Schottky, Walter, 305n2
Schottky diodes
    crystal detectors, 8
    history and characteristics of, 279–81
    MESFETs, 363
    microstrip "slotted"-line project, 269–70
scientific instrumentation, and PLLs, 532
SCR-584 gun-laying radar, 22
second-order PLL, 535–6, 568–73
second stage contribution, and noise figure measurement, 488–9
selenium, and television, 30–1n69
self-bias, and depletion-mode FETs, 376
semiconductor detector, 7–9
semiconductors, and surface phenomena, 341
semi-infinite conductive block, *125*
semirigid coaxial cables, 117, 122
sequential phase detectors, 542–4
series inductor, 781
series-peaking amplifiers, 381–8, 391–3
75-Ω coaxial cables, 118
75-Ω connectors, 114
shape factor, and lumped filters, 738–9
sheet resistivity, of copper, 158
Shockley, William, 341–2, 343n5, 347
shockwave transmission line, 619–20
short/open/load/thru (SOLT) technique, for impedance measurement, 261, 263, 264
shortwave radio, and hams, 17
shot noise, and bipolar transistor, 446
shunt capacitor, and design tables for low-pass filters, 781
shunt impedances, and Smith chart, 64–5
shunt inductance, and broadband impedance matching, 103–4
shunt-peaked amplifiers, 381–90
shunt resonator, and band edges, 107
shunt-series amplifier, 395–413
sidebands, and mixers, 307
sidelobe radiation, of antenna, 696
sidetone, and telephone, 192n31

signal generator, and noise figure measurement, 482–3
signal-to-noise ratio (SNR)
  mixers, 307
  noise figure measurement, 472
silica, and PC boards, 162, *237*
silicon
  crystal radios, 298
  semiconductor detectors, 7, *8*
  transistors, 343–4, 346, 348, 350, 351
silver, tarnish on, 114n9
simulation and simulators, and phase noise, 600
  see also APLAC; AppCAD; LADDER; LTSpice; Puff; RFSim99; Sonnet Lite; Spectre; Spice
single-balanced mixer, 318–20
single-diode mixer, 330–1
single-layer solenoid, 140, *141*
single loop, and formulas for inductance, 146–7, 231
single-pole low-pass filter, 739
single-sideband noise figure (SSB NF), 308, 491–2
single-stage quarter-wave transformer, 95
single-stub tuner, and impedance matching, 89–91, *103*
size reduction, for couplers, 226–30
skin effect, and passive components, 123–9, 158n2
skin loss, and coaxial cables, 120, 121
slide angle, and microstrip filters, 838
slider assembly, and microstrip "slotted"-line project, 269–70
slotted line, and impedance measurement, 246–54, 268–71
SMA (sub-miniature, type A) connector, 111–12, 113, 177, 270
small-reflection approximation, *97, 98*
small-signal models, for bipolar transistors, 352–61, 446
small-signal operating regimes, for power amplifiers, 630–1
SMC connector, 112
Smith, Phillip Hagar, 60–1, 66n5, 72n9
Smith chart
  double-stub match, *93*
  introduction to, 60–6
  L-match, 82–4
  load-pull contours of power amplifiers, 685
  multiple-stub tuners, 92
  shunt-series amplifier, 405
  single-stub match, 90
  vector network analyzer, 258
snap diodes, 290–2
Sokal, A. D. and N. O., 643n12
Sonnet Lite (simulation), 217n45, 225, 836, 842n35
Sontheimer, Carl G., 203
Sony Corporation, 284
source mismatch, and vector network analyzer, 260
Southworth, George C., 18, 118, 275, 280
space communications systems, and noise temperature, 477
S-parameters, 66–9
spark signal, ultraband spectrum of, 10
specified load capacitance, of Pierce oscillator, 523
spectral regrowth, and adjacent channel power ratio, 676
Spectre (simulator), 596

spectrum allocation, RF circuits and upper limits of, 40; *see also* Federal Communications Commission
"Spectrum Analysis … Spectrum Analyzer Basics" (Hewlett-Packard, 1974), 629
"Spectrum Analyzer Measurement Seminar" (Hewlett-Packard, 1988), 629
spectrum analyzers
  architectures for building, 625–9
  noise figure measurement, 485, 492
  performance limits, 628–9
  phase noise measurement, 603, 604, 611–12
Spencer, Percy, 24
Spice simulations, 292, 460, 461, 564
splitters, and PC boards, 183–230
spot noise figure, 479
spurious-free dynamic range (SFDR), and low-noise amplifier, 462–4
spurs, and mixers, 310–12, 339–40
Sputnik, 24
square-law mixer, 313–17
square spiral, and broadband antenna, 720
squegging, and oscillator design, 509
stability, and impedance transformations in amplifiers, 420–7
standing-wave ratio (SWR)
  definition of, 76–7
  dipole antenna, 699
  slotted line, 248
  Smith chart, 66
Stanford University, 159, 178
start-up, of Colpitts oscillator, 507–8
static moduli, and phase-locked loops (PLLs), 553–6
steady-state error, and first-order PLL, 534
step discontinuity, and transmission lines, 174
stepped-impedance filters, 787–9
step recovery diodes, 290
Stern stability factor, 425
Stoney, George Johnstone, 11
stopband attenuation, and m-derived filters, 733
Storch, L., 761, 762
strange impedance behaviors (SIBs), and amplifiers, 420–7
stripline, and transmission lines for PC boards, 163–4, 166–8
stub low-pass filter, 789–92
Stutzman, W. L., 689n3, 700n17
subcarrier frequency, and color television, 531
subharmonic mixers, 338–40
sub-nanosecond pulse generators, 271–4
superconductors, and pulse generators, 239n1
superheterodyne receiver, 16, 305–6, 530
superregenerative amplifier, 59n12
supply pushing, and Colpitts oscillator, 514–15
Supreme Court, 1943 decision on invention of radio, 34, 724n2
surface acoustic wave (SAW) devices, 518–19
surface-mount inductors, 138–9
surface-mount resistors, 130–3
surface roughness, and PC boards, 159, 162
"surface state" hypothesis, 341
surge impedance, of loss transmission line, 47
sweep generator, and CRT, 614
sweep oscillators, and television, 530

INDEX

switched-mode power converters, 633n2, 642
switching-type amplifiers, and instability, 679
SWR meter, and impedance measurement, 265–6
symmetry problems, in couplers, 223
synthesizers
    combination synthesizers, 558–9
    dithering moduli, 556–8
    static moduli, 553–6
    see also phase-locked loops
syntony, and tuned circuits, 724

tail noise, and phase noise, 595–6
tangential signal sensitivity (TSS), 490
tank inductor, and Colpitts oscillator, 506–7
tank resistance, and phase noise, 576–9
tapered acoustic horn, 99
tapered transmission line, 94–5, 174, *175*
tapped capacitors, and impedance matching, 86–7
tapped inductor, 87
tapped resonators, 526, 528, 594–5
tapped transformers, and neutralization of amplifiers, 419
T-coil network, and bandwidth enhancement of amplifiers, 392–4, 427–39
T-combiner, 184–5
Teal, Gordon, 343
Teflon: *see* PTFE
Teflon dielectric semirigid coaxial cable, 117
Tektronix, 392, 393n6, 411, 616, 819n21
telegraph, 10, 14, 31
telephone
    hybrid couplers, 191–2
    lumped inductances along transmission lines, 51
    port isolation, 192n31
    see also cell phones; Nordic Mobile Telephone System
television
    bow-tie antenna, 721
    CRTs, 614
    dipole antenna, 697
    first proposals for broadcast, 18
    frequency bands, 39
    phased-locked loops (PLLs), 530–1
    selenium and patents for, 30–1n69
    series- and shunt-peaking amplifiers, 381, 382
    see also cable television
temperature
    Gunn diode, 295
    noise factor, 441, 444–5, 477–9
    noise figure measurement, 472–3, 475, 490
    thermal runaway of power amplifiers, 681–2
    see also absolute temperature; thermal noise
temperature coefficient (TC), and capacitors, 134–5
temperature-limited diodes, 483
10-dB coupler, 260
tennantite/tetrahedrite, and crystal radios, 298
Terman–Woodyard composite amplifiers, 667–9
Tesla, Nikola, 34, 36, 724
Tesla coil, 34
test voltage, of tuned amplifier, 415–16
Texas Instruments, 343
thermal noise
    bipolar transistors, 446
    FET circuits, 367

thermal runaway, and power amplifiers, 681–2
thermodynamics, and equipartition theorem, 580
Thévenin resistance, 728
Thiele, G. A., 689n3, 700n17
third-order intercept (IP3), 323–4, 457–62
Thompson, Elihu, 10n21
Thomson, J. J., 11–12
Thomson, W. E., 762
3-dB coupler, 223
3-D radiation pattern, 697
thru-reflect-line (TRL) method, of impedance measurement, 261–2, 263, 264
time-dependent dielectric breakdown (TDDB), 680
time-division multiple access (TDMA), 31
time-domain reflectometer (TDR), and impedance measurement, 178, 238–46, 265
time-domain simulators, and mixers, 323–4
time variation
    mixers, 312–17
    phase noise, 575, 582–92
    see also linear, time-varying (LTV) system; linear, time-invariant (LTI) theories
time-varying phase noise model, 575
TITO (tuned input–tuned output) oscillator, 520–1, 528
Tizard mission, 21–2
T-junction shortening, of transmission lines, 175–6
T-match, and impedance matching, 85–6
TNC connector, 112
toroidal inductor, 146
toys
    coherer, 11
    phenolic PC boards, 160
tracking generator, 628
transatlantic communication, 5n9, 724
transconductance, and bipolar transistors, 354
transconductors, 319, 355
transducer power gain, and S-parameters, 68–9, 71
transformers, and magnetically coupled conductors, 147–51
transient impedance, and lossy transmission line, 47
transimpedance, and microstrip filters, 827–8
transistors
    avalanche mode operation, 273
    gallium arsenide (GaAs), 24
    history of microwave engineering, 341–51
    modeling, 351–68
    universal describing function for, 498–500
transit time effects, and FET circuits, 367–8
transmission lines
    broadband impedance matching and tapered, 94–5
    Class F amplifier, 650
    distributed filters, 785
    finite-length, behavior of, 51–3
    impedance transformers, 87–92
    infinite ladder network, 45–6
    lumped coupling capacitors, 209
    lumped filters, 726–38, 740, 785
    microwave circuits and behavior of, 38
    passives made from segments of, 178–81
    PC boards, 162–78
    propagation constant, 46–51
    sampling scopes, 619–20
    summary of equations for, 53–4
    transformers, 152–5